HANDBOOK OF ALGAL BIOFUELS

HANDBOOK OF ALGAL BIOFUELS

Aspects of Cultivation, Conversion, and Biorefinery

Edited by

MOSTAFA EL-SHEEKH
Botany and Microbiology Department, Faculty of Science, Tanta University, Tanta, Egypt

ABD EL-FATAH ABOMOHRA
New Energy and Environmental Laboratory (NEEL), Department of Environmental Engineering, School of Architecture and Civil Engineering, Chengdu University, Chengdu, P.R. China

ELSEVIER

Elsevier
Radarweg 29, PO Box 211, 1000 AE Amsterdam, Netherlands
The Boulevard, Langford Lane, Kidlington, Oxford OX5 1GB, United Kingdom
50 Hampshire Street, 5th Floor, Cambridge, MA 02139, United States

Copyright © 2022 Elsevier Inc. All rights reserved.

No part of this publication may be reproduced or transmitted in any form or by any means, electronic or mechanical, including photocopying, recording, or any information storage and retrieval system, without permission in writing from the publisher. Details on how to seek permission, further information about the Publisher's permissions policies and our arrangements with organizations such as the Copyright Clearance Center and the Copyright Licensing Agency, can be found at our website: www.elsevier.com/permissions.

This book and the individual contributions contained in it are protected under copyright by the Publisher (other than as may be noted herein).

Notices

Knowledge and best practice in this field are constantly changing. As new research and experience broaden our understanding, changes in research methods, professional practices, or medical treatment may become necessary.

Practitioners and researchers must always rely on their own experience and knowledge in evaluating and using any information, methods, compounds, or experiments described herein. In using such information or methods they should be mindful of their own safety and the safety of others, including parties for whom they have a professional responsibility.

To the fullest extent of the law, neither the Publisher nor the authors, contributors, or editors, assume any liability for any injury and/or damage to persons or property as a matter of products liability, negligence or otherwise, or from any use or operation of any methods, products, instructions, or ideas contained in the material herein.

British Library Cataloguing-in-Publication Data
A catalogue record for this book is available from the British Library

Library of Congress Cataloging-in-Publication Data
A catalog record for this book is available from the Library of Congress

ISBN: 978-0-12-823764-9

For Information on all Elsevier publications
visit our website at https://www.elsevier.com/books-and-journals

Publisher: Candice Janco
Acquisitions Editor: Peter Adamson
Editorial Project Manager: Andrea R. Dulberger
Production Project Manager: Prasanna Kalyanaraman
Cover Designer: Victoria Pearson

Typeset by MPS Limited, Chennai, India

Contents

List of contributors xi
About the editors xv

1. Cyanoprokaryotes and algae: classification and habitats 1

Abdullah A. Saber, Ahmed A. El-Refaey, Hani Saber,
Prashant Singh, Sanet Janse van Vuuren and Marco Cantonati

1.1 Introduction 1
1.2 Key taxonomic characteristics of cyanoprokaryotes and algae 2
1.3 Cyanoprokaryotes: taxonomic classification history, modern age, and perspectives 2
1.4 History and present-day algal taxonomy 11
1.5 Global distribution and habitats of cyanoprokaryotes and algae 14
1.6 Conclusions and perspectives 31
References 32

2. Global seaweeds diversity 39

Mohamed S.M. Abdel-Kareem and Alaa A.F. ElSaied

2.1 Introduction 39
2.2 Basis of seaweeds classification 40
2.3 The diverse groups of seaweeds 41
2.4 Seaweeds composition based on classification 42
2.5 Global distribution of seaweeds 46
2.6 Symbiotic relation of seaweeds with other marine organisms 49
2.7 Conclusion 50
References 50

3. Biochemical compounds of algae: sustainable energy sources for biofuel production 57

İlknur Ak, Edis Koru, Gülen Türker, Ekrem Cem Çankırılıgil and
Macide Güneş Dereli

3.1 Introduction 57

3.2 Lipids and fatty acids in algae 58
3.3 Carbohydrates 66
3.4 Proteins 73
3.5 Conclusion 73
References 74

4. Algal physiology and cultivation 79

Kushi Yadav, Reetu, Shrasti Vasistha and Monika Prakash Rai

4.1 Introduction 79
4.2 Photosynthetic efficiency of algae 80
4.3 Factors influencing growth and biochemical composition 81
4.4 Microalgae cultivation system 83
4.5 Artificial growth of seaweeds 86
4.6 Cost analysis of algae cultivation 87
4.7 Integrated cultivation system 88
4.8 Conclusion 90
Acknowledgments 91
References 91

5. Genetic manipulation of microalgae for enhanced biotechnological applications 97

Ashutosh Pandey, Gaurav Kant, Shadma Afzal,
Manish Pratap Singh, Nand Kumar Singh, Sanjay Kumar and
Sameer Srivastava

5.1 Introduction 97
5.2 Genetic modification of algae for the generation of energy and value-added metabolites 98
5.3 Advance methods used in genome editing 103
5.4 Future perspectives of genetic engineering in microalgae 110
5.5 Concluding remarks 114
Acknowledgments 114
References 114

6. The current status of various algal industries 123

Ahmed E. AlProl and Marwa R. Elkatory

6.1 Introduction 123
6.2 The algae industry 124
6.3 Main chemical compounds and bioactive compounds in algae 125
6.4 Energy production 126
6.5 Algae-based nonenergy field 131
6.6 Cosmetics 135
6.7 Food ingredients and polymers 136
6.8 Algae industrial companies 137
6.9 Wastewater treatment by marine algae 138
6.10 Conclusion 141
References 142

7. Algal biomass as a promising tool for CO_2 sequestration and wastewater bioremediation: an integration of green technology for different aspects 149

Reda M. Moghazy, Sayeda M. Abdo and Rehab H. Mahmoud

7.1 Introduction 149
7.2 Strategies of carbon dioxide sequestration 150
7.3 Carbon dioxide biosequestration using microalgae 151
7.4 Bioremediation: an ecofriendly approach for wastewater treatment 152
7.5 Algae-based wastewater treatment plants 153
7.6 Algal bacterial symbiosis system for wastewater treatment: role and effect of carbon dioxide 154
7.7 Biosorption and bioaccumulation 157
7.8 Conclusion 162
References 162

8. Application of halophilic algae for water desalination 167

Shristy Gautam and Dhriti Kapoor

8.1 Introduction 167
8.2 Marine environment 169
8.3 Isolation of halophilic microalgae 171
8.4 Efficiency of microalgae and seaweeds for water desalination 173

8.5 Economic feasibility 174
8.6 Role of algal biofuel in desalination process 175
8.7 Conclusion 176
References 176

9. Biofuel versus fossil fuel 181

Ahmed I. EL-Seesy, Mostafa E. Elshobary and Zhixia He

9.1 Introduction 181
9.2 Mechanisms of fossil fuel and biofuel production 183
9.3 Algal biomass 183
9.4 Algal biodiesel and other types of physicochemical properties 186
9.5 Combustion and emission parameters 186
9.6 Conclusions 187
Acknowledgments 190
References 190

10. Algae for biodiesel production 195

Mohammadhosein Rahimi, Fateme Saadatinavaz and Mohammadhadi Jazini

10.1 Introduction 195
10.2 Lipids in algal biomass 197
10.3 Different methods of transesterification 208
10.4 Biodiesel characteristics 211
10.5 Economic feasibility 217
10.6 Conclusions and perspectives 217
References 217

11. Eco-friendly biogas production from algal biomass 225

Mohamed A. Hassaan, Marwa R. Elkatory, Ahmed El Nemr and Antonio Pantaleo

11.1 Introduction 225
11.2 Structural and chemical composition of seaweeds 227
11.3 Anaerobic digestion 228
11.4 Pretreatments 233
11.5 The challenges of biogas production from algae 243
11.6 Conclusion 244
References 244

Contents

12. Algal biomass for bioethanol and biobutanol production 251

Marwa R. Elkatory, Mohamed A. Hassaan and Ahmed El Nemr

12.1 Introduction 251
12.2 Current biofuels status 253
12.3 Bioalcohols 254
12.4 Microalgae 256
12.5 Biofuel production processes 259
12.6 Bioethanol from microalgae 265
12.7 Biobutanol from microalgae 265
12.8 Macroalgae 267
12.9 Aquaculture seaweed cultivation 269
12.10 Bioethanol from macroalgae 270
12.11 Biobutanol from macroalgae 271
12.12 Conclusion 273
References 273

13. Thermochemical conversion of algal biomass 281

Sabariswaran Kandasamy, Narayanamoorthy Bhuvanendran, Mathiyazhagan Narayanan and Zhixia He

13.1 Introduction 281
13.2 Thermochemical conversion 282
13.3 Pyrolysis 283
13.4 Hydrothermal liquefaction 287
13.5 Gasification 291
13.6 Torrefaction 292
13.7 Direct combustion 293
13.8 Economic feasibility 293
13.9 Conclusion 294
Acknowledgment 295
References 295

14. Direct biohydrogen production from algae 303

Eithar El-Mohsnawy, Ali Samy Abdelaal and Mostafa El-Sheekh

14.1 Introduction 303
14.2 Biohydrogen as efficient future fuels 305
14.3 Direct cellular biohydrogen production 306
14.4 Photosynthetic hydrogen production— cyanobacteria 314
14.5 Enhancing hydrogen production in microalgae by gene technology 316

14.6 Biosystem and semiartificial system for photocurrent and biohydrogen productions 320
14.7 Conclusion 326
References 326

15. Biojet fuels production from algae: conversion technologies, characteristics, performance, and process simulation 331

Medhat Elkelawy, Hagar Alm-Eldin Bastawissi, Ahmed Mohamed Radwan, Mohamed Taha Ismail and Mostafa El-Sheekh

15.1 Introduction 332
15.2 Biomass jet fuel conversion pathways 333
15.3 Algae biojet fuel 345
15.4 Biojet fuel performance characteristics 346
15.5 Fuel compatibility with current fueling system of aircraft 350
15.6 Process simulation 355
15.7 Conclusions 356
References 357

16. Photosynthetic microalgal microbial fuel cells and its future upscaling aspects 363

Mohd Jahir Khan, Vishal Janardan Suryavanshi, Khashti Ballabh Joshi, Praveena Gangadharan and Vandana Vinayak

16.1 Introduction 363
16.2 What are photosynthetic microalgal microbial fuel cell 366
16.3 Effect of light on the performance of photosynthetic microalgal microbial fuel cells 370
16.4 DNA sequencing of microbial genomes 373
16.5 Integrated approaches of photosynthetic microalgal microbial fuel cells 376
16.6 Future of photosynthetic microalgal microbial fuel cells using diatoms 377
16.7 Conclusions 379
Acknowledgments 380
Conflict of Interest 380
References 380

17. Sequential algal biofuel production through whole biomass conversion 385

Mahdy Elsayed and Abd El-Fatah Abomohra

17.1 Introduction 385
17.2 Different processes of algal biofuel production 387
17.3 Recent trends in sequential algal biofuel production 396
17.4 Conclusion 400
References 400

18. By-products recycling of algal biofuel toward bioeconomy 405

Hanan M. Khairy, Heba S. El-Sayed, Gihan M. El-Khodary and Salwa A. El-Saidy

18.1 Introduction 405
18.2 Applications of algae by-products 409
18.3 Microalgal by-products of biomasses conversion processes 411
18.4 By-products from ethanol production 411
18.5 Glycerol by-products of biodiesel productions 412
18.6 By-products from bio-oil fuel production 412
18.7 Microalgal-based protein by-products 413
18.8 Environmental impact of biodiesel and by-products 413
18.9 Economic feasibility of microalgae biodiesel 414
18.10 Future research focus and perspectives 415
18.11 Conclusion 416
References 416

19. Harnessing solar radiation for potential algal biomass production 421

Imran Ahmad, Norhayati Abdullah, Mohd Danish Ahmad, Iwamoto Koji and Ali Yuzir

19.1 Introduction 421
19.2 Solar cells 423
19.3 Solar panel 426
19.4 Different applications of solar radiation 426
19.5 Conversion of solar radiation to algal biomass 430
19.6 Solar tracking system 433
19.7 Solar to heat for thermochemical conversion of algal biomass 438
19.8 Technoeconomic considerations for different routes of conversion of algae 441
19.9 Conclusion 443
Acknowledgment 444
Conflict of interest 444
References 444
Further reading 449

20. Physical stress for enhanced biofuel production from microalgae 451

Sivakumar Esakkimuthu, Shuang Wang and Abd El-Fatah Abomohra

20.1 Introduction 451
20.2 Nutrient stress 452
20.3 Physical stress 454
20.4 Challenges and future directions of physical stress 467
20.5 Conclusion 469
References 469

21. Microalgal—bacterial consortia for biomass production and wastewater treatment 477

Muhammad Usman Khan, Nalok Dutta, Abid Sarwar, Muhammad Ahmad, Maryam Yousaf, Yassine Kadmi and Mohammad Ali Shariati

21.1 Introduction 477
21.2 Interchange of substrates, intercellular communication, and horizontal gene transfer in microalgal—bacterial consortia 479
21.3 Distribution and role of microalgal—bacterial consortia in the wastewater treatment 481
21.4 Biofuel and bioproducts generation by microalgal—bacterial consortia 486
21.5 Reduction in CO_2 emission and electricity generation 487
21.6 Role of lipase in wastewater treatment 488
21.7 Conclusions 493
Acknowledgments 494
References 494

22. Process intensification for sustainable algal fuels production 503

Hector De la Hoz Siegler

22.1 Introduction 503
22.2 Intensification of photobioreactors 506

22.3 Harvesting 512
22.4 Biomass conversion to biofuel 512
22.5 Conclusion 517
References 517

23. Life cycle assessment for microalgae-derived biofuels 523

Elham Mahmoud Ali

23.1 Introduction 523
23.2 Pros and cons of algal biofuel production 524
23.3 Life cycle assessment approach 526
23.4 Benefits on application of life cycle assessment for microalgal biofuel commercial production 536
23.5 Current scenario on production and application of biofuels 538
23.6 Future prospective 539
23.7 Conclusions 540
Acknowledgment 541
References 541

24. An overview of the algal biofuel technology: key challenges and future directions 547

Kushi Yadav, Reetu and Monika Prakash Rai

24.1 Introduction 547
24.2 Challenges 548
24.3 Lipid extraction 555
24.4 Future perspectives 559
24.5 Conclusions 560
Acknowledgment 560
References 560

25. History and recent advances of algal biofuel commercialization 567

Ali Noor and Fouzia Naseer

25.1 Introduction and history of biofuel production 567
25.2 Recent advancement in large-scale biofuel production 568
25.3 Pilot-scale and large-scale trials of algal biofuel production 569
25.4 Top companies of biofuel production from different feedstocks 570

25.5 Top companies of algal products commercialization 573
25.6 Top companies of biofuel production from algae 575
25.7 Biofuel production and its impact on environment 578
25.8 Challenges of biofuel commercialization from algae 581
25.9 Future prospective of biofuel 581
References 583

26. Biointelligent quotient house as an algae-based green building 587

Anas Tallou, Khalid Aziz, Mounir El Achaby, Sbihi Karim and Faissal Aziz

26.1 Introduction 587
26.2 Green buildings 589
26.3 Renewable energy applications in green buildings 590
26.4 Biointelligent quotient in Hamburg, Germany 593
26.5 University of technology Sydney green building case study 594
26.6 Conclusion 595
References 597

27. National Renewable Energy Laboratory 599

Sbihi Karim, Aziz Faissal and El Baraka Noureddine

27.1 Introduction 599
27.2 National Renewable Energy Laboratory 600
27.3 History of National Renewable Energy Laboratory algal biofuels projects 603
27.4 Principal project 605
27.5 Microalgae isolation and characteristics during the project 606
27.6 Limitation of industrial application 610
27.7 Conclusion of the project 610
References 612

28. Aquatic species program 615

Faissal Aziz, Anas Tallou, Karim Sbihi, Khalid Aziz and Nawal Hichami

28.1 Introduction 615
28.2 Introduction to US department of energy 616
28.3 History of the algae species program 619

28.4 Microalgal isolation and characteristics 623
28.5 Relationship between National Renewable Energy Laboratory and algae species program 627
28.6 Limitations of industrial applications 628
28.7 Conclusions of the project 630
Acknowledgments 631
References 632

29. Algal fuel production by industry: process simulation and economic assessment 635

Sayeda M. Abdo, Entesar Ahmed, Sanaa Abo El-Enin, Guzine El Diwan, K.M. El-Khatib, Gamila H. Ali and Rawheya A. Salah El Din

29.1 Introduction 635

29.2 Life cycle assessment toward microalgae industrialization 636
29.3 Operating conditions 639
29.4 Algal biodiesel 641
29.5 Process simulation 644
29.6 Process description 644
29.7 Economic assessment 646
References 650

Index 653

List of contributors

Ali Samy Abdelaal Department of Genetics, Faculty of Agriculture, Damietta University, Damietta, Egypt

Mohamed S.M. Abdel-Kareem Botany and Microbiology Department, Faculty of Science, Alexandria University, Alexandria, Egypt

Sayeda M. Abdo Water Pollution Research Department, National Research Centre, Cairo, Egypt

Norhayati Abdullah UTM International, Level 8, Menara Razak, Universiti Teknologi Malaysia, Jalan Sultan Yahya Petra, Kuala Lumpur, Malaysia

Abd El-Fatah Abomohra Department of Environmental Engineering, School of Architecture and Civil Engineering, Chengdu University, Chengdu, P.R. China; Botany and Microbiology Department, Faculty of Science, Tanta University, Tanta, Egypt

Shadma Afzal Department of Biotechnology, Motilal Nehru National Institute of Technology Allahabad, Prayagra, India

Imran Ahmad Algae and Biomass, Research Laboratory, Malaysia-Japan International Institute of Technology (MJIIT), Universiti Teknologi Malaysia, Jalan Sultan Yahya Petra, Kuala Lumpur, Malaysia

Mohd Danish Ahmad Department of Post-Harvest Engineering and Technology, Aligarh Muslim University, Aligarh, India

Muhammad Ahmad School of Chemistry and Chemical Engineering, Beijing Institute of Technology, Beijing, P.R. China

Entesar Ahmed Botany and Microbiology Department, Faculty of Science, Al-Azhar University (Girls Branch), Cairo, Egypt

İlknur Ak Faculty of Marine Science and Technology, Department of Aquaculture, Çanakkale Onsekiz Mart University, Çanakkale, Turkey

Elham Mahmoud Ali Department of Aquatic Environment, Suez University (SU), Suez, Egypt; Department of Environmental Studies, The National Authority for Remote Sensing and Space Sciences (NARSS), Cairo, Egypt

Gamila H. Ali Water Pollution Research Department, National Research Centre, Giza, Egypt

Ahmed E. AlProl National Institute of Oceanography and Fisheries, Hurghada, Egypt

Faissal Aziz Laboratory of Water, Biodiversity, and Climate Change, Faculty of Sciences Semlalia, Cadi Ayyad University, Marrakech, Morocco; National Center for Research and Studies on Water and Energy (CNEREE), Cadi Ayyad University, Marrakech, Morocco

Khalid Aziz Materials, Catalysis and Valorization of Natural Resources, Faculty of Sciences, University Ibn Zohr, Agadir, Morocco

Hagar Alm-Eldin Bastawissi Mechanical Power Engineering Department, Faculty of Engineering, Tanta University, Tanta, Egypt

Narayanamoorthy Bhuvanendran Institute for Energy Research, Jiangsu University, Zhenjiang, P.R. China

Ekrem Cem Çankırılıgil Sheep Breeding Research Institute, Bandırma, Balıkesir, Turkey

Marco Cantonati MUSE–Museo delle Scienze, Limnology & Phycology Section, Corso del Lavoro e della Scienza 3, Trento, Italy

Hector De la Hoz Siegler Department of Chemical and Petroleum Engineering, University of Calgary, Calgary, AB, Canada

Macide Güneş Dereli Faculty of Pharmacy, Department of Pharmacology, Istanbul University, Istanbul, Turkey; Graduate School of Health Sciences, Istanbul University, Istanbul, Turkey

Nalok Dutta Bioproducts Science and Engineering Laboratory, Washington State University, Richland, WA, United States

Mounir El Achaby Materials Science and Nano-Engineering Department, Mohammed VI Polytechnic University, Benguerir, Morocco

Guzine El Diwan Chemical Engineering and Pilot Plant Department, National Research Centre, Cairo, Egypt

Ahmed El Nemr National Institute of Oceanography and Fisheries, Cairo, Egypt

Sanaa Abo El-Enin Chemical Engineering and Pilot Plant Department, National Research Centre, Cairo, Egypt

Marwa R. Elkatory Advanced Technology and New Materials Research Institute (ATNMRI), City for Scientific Research and Technological Applications (SRTA-City), Alexandria, Egypt

Medhat Elkelawy Mechanical Power Engineering Department, Faculty of Engineering, Tanta University, Tanta, Egypt

K.M. El-Khatib Chemical Engineering and Pilot Plant Department, National Research Centre, Cairo, Egypt

Gihan M. El-Khodary Department of Zoology, Faculty of Science, Damanhour University, Damanhour, Egypt

Eithar El-Mohsnawy Botany and Microbiology Department, Faulty of Science, Kafrelsheikh University, Kafr Al Sheikh, Egypt

Ahmed A. El-Refaey Algae Lab, Botany and Microbiology Department, Faculty of Science, Al-Azhar University, Cairo, Egypt

Salwa A. El-Saidy Department of Zoology, Faculty of Science, Damanhour University, Damanhour, Egypt

Alaa A.F. ElSaied Botany and Microbiology Department, Faculty of Science, Alexandria University, Alexandria, Egypt; Biology Department, Unaizah College of Sciences and Arts, Qassim University, Unaizah, Kingdom of Saudi Arabia

Heba S. El-Sayed National Institute of Oceanography and Fisheries (NIOF), Cairo, Egypt

Mahdy Elsayed Department of Agricultural Engineering, Faculty of Agriculture, Cairo University, Giza, Egypt

Ahmed I. EL-Seesy Institute for Energy Research, Jiangsu University, Zhenjiang, P.R. China; Department of Mechanical Engineering, Benha Faculty of Engineering, Benha University, Benha, Egypt

Mostafa El-Sheekh Department of Botany, Faculty of Science, Tanta University, Tanta, Egypt

Mostafa E. Elshobary School of Food and Biological Engineering, Jiangsu University, Zhenjiang, P.R. China; Department of Botany and Microbiology, Faculty of Science, Tanta University, Tanta, Egypt

Sivakumar Esakkimuthu New Energy Department, School of Energy and Power Engineering, Jiangsu University, Jiangsu, P. R. China

Aziz Faissal National Center for Research and Study on Water and Energy (CNEREE), University Cadi Ayyad, Marrakech, Morocco; Laboratory of Water, Biodiversity and Climate Changes, Semlalia Faculty of Sciences, Marrakech, Morocco

Praveena Gangadharan Department of Civil Engineering, Indian Institute of Technology, Palakkad, India

Shristy Gautam Department of Molecular Biology and Genetic Engineering, School of Bioengineering and Biosciences, Lovely Professional University, Phagwara, India

Mohamed A. Hassaan National Institute of Oceanography and Fisheries, Cairo, Egypt

Zhixia He Institute for Energy Research, Jiangsu University, Zhenjiang, P.R. China; School of Energy and Power Engineering, Jiangsu University, Zhenjiang, P.R. China

Nawal Hichami Laboratory of Biotechnology and Sustainable Development of Natural Resources, Sultan Moulay Slimane University, Beni Mellal, Morocco

Mohamed Taha Ismail Mechanical Power Engineering Department, Faculty of Engineering, Tanta University, Tanta, Egypt

Mohammadhadi Jazini Department of Chemical Engineering, Isfahan University of Technology, Isfahan, Iran

Khashti Ballabh Joshi School of Chemical Science and Technology, Department of Chemistry, Dr. Harisingh Gour Central University Sagar, Sagar, India

Yassine Kadmi LASIRE CNRS UMR 8516, Sciences and Technologies, University of Lille, Lille, France

Sabariswaran Kandasamy Institute for Energy Research, Jiangsu University, Zhenjiang, P.R. China

Gaurav Kant Department of Biotechnology, Motilal Nehru National Institute of Technology Allahabad, Prayagra, India

Dhriti Kapoor Department of Botany, School of Bioengineering and Biosciences, Lovely Professional University, Phagwara, India

Sbihi Karim Laboratory of Biotechnology, Materials and Environment, Natural Substances and Environment Unit, Faculty Polydisciplinary of Taroudant, University Ibn Zohr, Taroudant, Morocco; National Center for Research and Study on Water and Energy (CNEREE), University Cadi Ayyad, Marrakech, Morocco

Hanan M. Khairy National Institute of Oceanography and Fisheries (NIOF), Cairo, Egypt

Mohd Jahir Khan Diatom Nanoengineering and Metabolism Laboratory (DNM), School of Applied Sciences, Dr. Harisingh Gour Central University Sagar, Sagar, India

Muhammad Usman Khan Bioproducts Science and Engineering Laboratory, Washington State University, Richland, WA, United States; Department of Energy Systems Engineering, Faculty of Agricultural Engineering and Technology, University of Agriculture, Faisalabad, Pakistan

Iwamoto Koji Algae and Biomass, Research Laboratory, Malaysia-Japan International Institute of Technology (MJIIT), Universiti Teknologi Malaysia, Jalan Sultan Yahya Petra, Kuala Lumpur, Malaysia

Edis Koru Faculty of Fisheries, Department of Aquaculture, Ege University, İzmir, Turkey

Sanjay Kumar School of Biochemical Engineering, Indian Institute of Technology (BHU) Varanasi, Varanasi, India

Rehab H. Mahmoud Water Pollution Research Department, National Research Centre, Cairo, Egypt

Reda M. Moghazy Water Pollution Research Department, National Research Centre, Cairo, Egypt

Mathiyazhagan Narayanan PG and Research Centre in Biotechnology, MGR College, Adhiyamaan Educational Research Institute, Krishnagiri, India

Fouzia Naseer Department of Botany, University of Karachi, Karachi, Pakistan

Ali Noor Department of Biological Sciences, Karakoram International University, Gilgit-Baltistan, Pakistan

El Baraka Noureddine Laboratory of Biotechnology, Materials and Environment, Physicochemistry of Natural Environments, Materials and Environment Team, Polydisciplinaire Faculty of Taroudant, Taroudant, Morocco

Ashutosh Pandey Department of Biotechnology, Motilal Nehru National Institute of Technology Allahabad, Prayagra, India; Department of Biotechnology, IMS Engineering College (Affilliated to Dr. APJ Abdul Kalam Technical University, Lucknow) Ghaziabad, India

Antonio Pantaleo Agriculture and Environmental Sciences Department, Bari University, Bari, Italy

Ahmed Mohamed Radwan Mechanical Power Engineering Department, Faculty of Engineering, Tanta University, Tanta, Egypt

Mohammadhosein Rahimi Department of Chemical Engineering, Isfahan University of Technology, Isfahan, Iran

Monika Prakash Rai Amity Institute of Biotechnology, Amity University, Noida, India

Reetu Amity Institute of Biotechnology, Amity University, Noida, India

Fateme Saadatinavaz Department of Chemical Engineering, Isfahan University of Technology, Isfahan, Iran

Abdullah A. Saber Botany Department, Faculty of Science, Ain Shams University, Cairo, Egypt

Hani Saber Department of Botany and Microbiology, Faculty of Science, South Valley University, Qena, Egypt

Rawheya A. Salah El Din Botany and Microbiology Department, Faculty of Science, Al-Azhar University (Girls Branch), Cairo, Egypt

Abid Sarwar Department of Irrigation and Drainage, Faculty of Agricultural Engineering and Technology, University of Agriculture, Faisalabad, Pakistan

Karim Sbihi Laboratory of Biotechnology, Materials and Environment, Natural Substances and Environment Unit, Faculty Polydisciplinary of Taroudant, University Ibn Zohr, Taroudant, Morocco

Mohammad Ali Shariati Kazakh Research Institute of Processing and Food Industry (Semey Branch), Semey, Kazakhstan

Manish Pratap Singh Department of Biotechnology, Motilal Nehru National Institute of Technology Allahabad, Prayagra, India

Nand Kumar Singh Department of Biotechnology, Motilal Nehru National Institute of Technology Allahabad, Prayagra, India

Prashant Singh Laboratory of Cyanobacterial Systematics and Stress Biology, Department of Botany, Institute of Science, Banaras Hindu University, Varanasi, India

Sameer Srivastava Department of Biotechnology, Motilal Nehru National Institute of Technology Allahabad, Prayagra, India

Vishal Janardan Suryavanshi School of Chemical Science and Technology, Department of Chemistry, Dr. Harisingh Gour Central University Sagar, Sagar, India

Anas Tallou Polydisciplinary Laboratory of Research and Development, Faculty of Sciences and Techniques, Sultan Moulay Slimane University of Beni Mellal, Beni-Mellal, Morocco

Gülen Türker Department of Food Technology, Çanakkale School of Applied Sciences, Çanakkale Onsekiz Mart University, Çanakkale, Turkey

Sanet Janse van Vuuren Unit for Environmental Sciences and Management, North-West University, Potchefstroom, South Africa

Shrasti Vasistha Amity Institute of Biotechnology, Amity University, Noida, India

Vandana Vinayak Diatom Nanoengineering and Metabolism Laboratory (DNM), School of Applied Sciences, Dr. Harisingh Gour Central University Sagar, Sagar, India

Shuang Wang New Energy Department, School of Energy and Power Engineering, Jiangsu University, Jiangsu, P. R. China

Kushi Yadav Amity Institute of Biotechnology, Amity University, Noida, India

Maryam Yousaf School of Chemistry and Chemical Engineering, Beijing Institute of Technology, Beijing, P.R. China; Department of Chemistry, University of Agriculture Faisalabad, Faisalabad, Pakistan

Ali Yuzir Department of Chemical and Environmental Engineering (ChEE), Malaysia-Japan International Institute of Technology (MJIIT), Universiti Teknologi Malaysia, Jalan Sultan Yahya Petra, Kuala Lumpur, Malaysia

About the editors

Mostafa El-Sheekh is a professor of Phycology at Tanta University, Egypt. He has nearly 38 years of experience of research and teaching in the field of algae, microbiology, and its applications. Dr. El-Sheekh served as the Vice Dean and Dean Faculty of Science, Tanta University, Egypt, and as a Cultural counselor in the Egyptian Embassy in Yemen and Uzbekistan. He also served as Vice President of Tanta University for postgraduate studies and research. He holds a PhD in Phycology from Tanta University and Göttingen University, Germany, as a Chanel System fellowship. He is the recipient of fellowships in Germany, Hungary, India, Sweden, the Czech Republic, and Japan. He is the author of more than 210 papers, 13 book chapters, and four books. He also serves as an editorial member and reviewer for more than 92 peer-reviewed journals. He has published several papers on the production of biodiesel, ethanol, and hydrogen from algae. He was included in the World's Top 2% of Scientists List published by Stanford University in 2020 and 2021.

Prof. Abd El-Fatah Abomohra is a professor of Environmental Engineering and has been the Head of New Energy and Environmental Laboratory (NEEL) at Chengdu University, Chengdu, China, since 2019. He received his PhD in 2012 through a cooperation program between Tanta University (Egypt) and Hamburg University (Germany) funded by Deutscher Akademischer Austauschdienst (DAAD). He has 20 years' work experience at different universities as a research assistant, lecturer, associate professor, and professor. He has authored seven books and more than 100 contributions published in SCI index journals. He was included in the World's Top 2% of Scientists List published by Stanford University in 2020 and 2021. His research group primarily works on bioenergy production from different biomass feedstocks and waste.

CHAPTER 1

Cyanoprokaryotes and algae: classification and habitats

Abdullah A. Saber[1], Ahmed A. El-Refaey[2], Hani Saber[3], Prashant Singh[4], Sanet Janse van Vuuren[5] and Marco Cantonati[6]

[1]Botany Department, Faculty of Science, Ain Shams University, Cairo, Egypt [2]Algae Lab, Botany and Microbiology Department, Faculty of Science, Al-Azhar University, Cairo, Egypt [3]Department of Botany and Microbiology, Faculty of Science, South Valley University, Qena, Egypt [4]Laboratory of Cyanobacterial Systematics and Stress Biology, Department of Botany, Institute of Science, Banaras Hindu University, Varanasi, India [5]Unit for Environmental Sciences and Management, North-West University, Potchefstroom, South Africa [6]MUSE—Museo delle Scienze, Limnology & Phycology Section, Corso del Lavoro e della Scienza 3, Trento, Italy

1.1 Introduction

Algae, including the evolutionarily primitive photoautotrophic cyanoprokaryotes, colonize all abiotic components of planet Earth, namely, air, land, and water. This highly heterogeneous group of organisms is indeed of universal distribution. Most species are morphologically and physiologically adapted to the biomes in which they live. Although many species can live in aerial and terrestrial environments, they are by far the most abundant in aquatic habitats [1]. Starting from the late nineteenth century onwards, the taxonomy of cyanobacteria and algae has been punctuated by a series of major breakthroughs. Great progress during the last century has been fostered by the advent of the electron microscope technique, which led to many new discoveries such as detailed structures of plastids and flagella [2,3]. The development of molecular and phylogenetic tools throughout recent decades not only refined our understanding of the immense diversity of cyanobacteria and algae but also enabled a revolution of ideas with regard to their taxonomy

Handbook of Algal Biofuels
DOI: https://doi.org/10.1016/B978-0-12-823764-9.00024-8

© 2022 Elsevier Inc. All rights reserved.

and classification [4–7]. In other words, the discrimination of morphologically similar species and genera has nowadays become much easier in the context of applying the integrative polyphasic approach.

Throughout this chapter, we present an overview of the key taxonomic characteristics of cyanoprokaryotes and algae. We also present exhaustive insights into the history and modern classification systems, particularly for the blue–green algae. The diverse habitats and global distribution of these intriguing microorganisms have also been our interest and we outline and briefly discuss them across all biomes and also provide some details on the species inhabiting the hot and cool spring habitats.

1.2 Key taxonomic characteristics of cyanoprokaryotes and algae

Algae are a highly heterogeneous group comprising cyanoprokaryotes, characterized by cells mainly lacking membrane-bound organelles, and eukaryotic algae, with cells containing true-shaped organelles [8]. They have representatives in all inland waters and are distributed between eight and 12 major evolutionary lineages [9]. They are classified based on different taxonomic criteria, namely, cell structures (such as plastids, nucleus, composition of cell walls, etc.), physiological characteristics (such as types of pigments, storage products, etc.), or morphological features such as number and position of flagella. The nature of photosynthetic pigments was the primary base for algal classification into different divisions. However, more ultrastructural cellular details, with keystone taxonomic value, have been unveiled with the advent of the electron microscope [8]. For a description of the taxonomic features of the major algal groups treated in this section, we follow and recommend recent phycology texts [9,10]. In general, there is little consensus among taxonomists about the exact number of algal divisions, and the rapidly growing polyphasic approaches are most likely to change our understanding with regard to the classification systems of algae.

The key taxonomic features of the main algal groups in this chapter are summarized in Table 1.1.

1.3 Cyanoprokaryotes: taxonomic classification history, modern age, and perspectives

1.3.1 History of cyanobacterial taxonomy

Cyanobacterial taxonomy has been a challenging field, with numerous criteria being developed and revised in order to attain a system that is stable and consistent. The latter decades of the 19th century were characterized by the documentation of two tribes of blue–green algae, the Coccogoneae (having unicellular reproductive bodies) and Hormogoneae (having few short fragments, like hormogones) [12]. The taxonomy of the Coccogoneae was further elaborated in other works [13]. Interestingly, even before all these developments, Nägeli had documented the unicellular Chroococcaceae already in 1849 [14]. The Hormogoneae tribe was

1.3 Cyanoprokaryotes: taxonomic classification history, modern age, and perspectives

TABLE 1.1 Major taxonomic characteristics of cyanoprokaryotes and algae presented in this chapter [10], [11].

Algal groups	Photosynthetic pigments	Plastid outer membranes	Storage product	Cell wall structure	Flagella (number and insertion)
Cyanoprokaryotes	Chl *a*, c-phycocyanin, c-phycoerythrin, myxoxanthin, myxoxanthophyll, carotenes, flavacene	0	Cyanophycean starch	Peptidoglycan matrices	Absent
Red algae	Chl *a*, r-phycoerythrin, r-phycocyanin, α/β-carotenes, lutein, zeaxanthin, violaxanthin	2	Floridean starch in cytoplasm outside the chromatophores	Galactose polymer matrix, in addition to cellulose and pectin	Absent
Diatoms	Chl *a*, *c*, fucoxanthin, diatoxanthin, diadinoxanthin	4	Chrysolaminarin and oils	Siliceous frustules	One (only rarely), anterior
Xanthophytes and Raphidophytes	Chl *a*, *c*, heteroxanthin, diadinoxanthin, vaucheriaxanthin	4	Chrysolaminarin	Mostly cellulosic walls, some naked	Two unequal, tinsel and whiplash, apically inserted
Eustigmatophytes	Chl *a*, vaucheriaxanthin, violaxanthin	4	Chrysolaminarin	Mostly cellulosic walls	One to two unequal, long forward-directed tinsel flagellum and a shorter backward-directed smooth flagellum
Chrysophytes	Chl *a*, *c*, 3-carotenes, fucoxanthin, lutein and neofucoxanthin	4	Chrysolaminarin	None, scales, lorica	Two unequal, one short smooth and directed laterally or posteriorly, and the other long, directed anteriorly, and pleuronematic (tinsel, hairy, or flimmer flagellum) bearing two rows of tripartite hairs attached through the flagellar membrane to specific outer doublets

(Continued)

Handbook of Algal Biofuels

4 1. Cyanoprokaryotes and algae: classification and habitats

TABLE 1.1 (Continued)

Algal groups	Photosynthetic pigments	Plastid outer membranes	Storage product	Cell wall structure	Flagella (number and insertion)
Haptophytes	Chl *a*, *c*, fucoxanthin, diadinoxanthin	4	Chrysolaminarin	Nonsiliceous scales	Two smooth flagella of equal or subequal length equal + haptonema
Synurophytes	Chl *a*, *c*, fucoxanthin	4	Chrysolaminarin	Siliceous scales	Two unequal, apically inserted
Dinoflagellates	Chl *a*, *c*, peridinin	3	Oil and starch	Cellulosic theca	Two unequal (heterokont), lateroventral
Brown algae	Chl *a*, *c*, fucoxanthin	4	Laminarin	Alginate matrices	Two unequal, whiplash and tinsel, transverse + longitudinal
Euglenoid algae	Chl *a*, *b*, 3-carotenes, lutein, zeaxanthin, neoxanthin, astaxanthin	3	Paramylon granules	Elastic or firm pellicle and lorica	One to two tinsel, apically inserted
Green algae	Chl *a*, *b*, α, 3 and γ-carotenes, lutein, zeaxanthin, neoxanthin, violaxanthin	2	Starch grains within the chloroplast envelopes	Cellulose, hemicellulose, and pectin	Zero to many, whiplash, anterior

Rewritten and edited after R.G. Sheath, J.D. Wehr, Introduction to the freshwater algae, in: J.D. Wehr, R.G. Sheath, J.P. Kociolek, (Eds.), Freshwater Algae of North America: Ecology and Classification, Academic Press, San Diego, California, 2015, pp. 1—11. D. Sahoo, P. Baweja, General characteristics of algae, in: D. Sahoo, J. Seckbach, (Eds.), The Algae World, Springer, Dordrecht, 2015, pp. 3—29.

further subjected to more attention in the works of Thuret [12], and Bornet and Flahault [15—20]

Understanding of taxonomy in these early times focused on exploring certain prominent and visible morphological characters. Branching was, amongst others, an important morphological trait that was considered to differentiate filamentous forms of blue—green algae. Broadly speaking, filamentous forms included both the unbranched and branched forms, with the branched forms being further divided into either true branched or false branched forms [21]. The orders Nostocales and Stigonematales were thus recognized in many different systematic studies [22—27]. Another important criterion used for classifying the blue—green algae was the absence or presence of heterocytes, which led to the formation of the families Homocysteae and Heterocysteae in the order Hormogonales. While this system, in the modern times looks satisfactory, it was not accepted by many phycologists in the early years of the 20th century. Prominent proponents of this system were Bornet and Flahault [16—20] and Setchell and Gardner [28]. Elenkin [29] supported this system too, and in fact recognized

Heterocysteae and Aheterocysteae, with the latter group having 14 distinct families under the order Oscillatoriales. Thus it is evident that from very early times the presence or absence of different morphological characters intrigued cyanobacterial taxonomists, encouraging them to dive deeper into solving the problems of cyanobacterial classification. During the course of these studies, many new schools of thought emerged, and some of them in the early ages were mentioned above. Moving ahead of the findings of Bornet and Flahault [16—20] and Borzì [30—33], it was Geitler [22—24] who developed a new system of classification in a series of studies and revisions which ultimately led him to recognize four orders: Chroococcales, Dermocarpales, Pleurocapsales, and Hormogonales. At the same time, Frémy [34,35] recognized three orders: Chroococcales, Chamaesiphonales, and Hormogonales. He subdivided the Hormogonales into Homocysteae and Anhomocysteae. This system was also adopted, with some modifications, in the work of Elenkin [29]. Later on Fritsch [25—27], with an enhanced focus on the heterocytous forms, determined the five orders Chroococcales, Chamaesiphonales, Pleurocapsales, Nostocales, and Stigonematales. In 1959 Desikachary [21] broadly accepted Fritsch's classification, especially in the context of the recognition of the Stigonematales as the most advanced group of cyanobacteria, along with supporting the separation of the Nostocales and Stigonematales. However, differences in the classification of the genera *Mastigocladus* and *Brachytrichia* were deviations from the work of Fritsch. Thus even after having differences at the family level of classification of some groups, Desikachary too recognized five orders: Chroococcales, Chamaesiphonales, Pleurocapsales, Nostocales, and Stigonematales. Desikachary's work also pointed to the importance of physiological, ecological, and cytological features of blue—green algae, which in the coming modern times eventually became part of the accurate taxonomic identification of this prokaryotic algal group. Overall, starting from the late 19th century onwards, the first half of the 20th century witnessed rigorous taxonomic studies from many phycologists all around the globe, which ultimately led to an enhanced basic understanding of the taxonomy and identity of cyanobacteria with a common consensus being centered around unicellular and colonial forms, filamentous nonheterocytous forms, unbranched filamentous heterocytous forms, and branched filamentous heterocytous forms. With the introduction of better planned studies and more clarity on gene markers, in particular during the last four decades, modern taxonomic assessments started to emerge.

1.3.2 The modern age of cyanobacterial taxonomy

The modern age of cyanobacterial taxonomy was characterized by the introduction of more clear and concise descriptions of various taxonomic entities. One of the initial contributions was the work of Prescott [36], who studied all algae, with cyanobacteria constituting a part of his studies. He recognized three main cyanobacterial orders, namely, the Chroococcales, Chamaesiphonales, and Hormogonales, which was basically in agreement with the classification of Frémy [34]. The Hormogonales were subdivided into six families: Oscillatoriaceae, Nostochopsaceae, Stigonemataceae, Rivulariaceae, Scytonemataceae, and Nostocaceae. Another important work of Bourrelly [2] was in congruence with the five order system of Desikachary [21]. Eventually, one of the most well-received and followed contributions came from the work of Rippka et al. [37] in which the cyanobacteria were divided into five sections.

In a study based on 178 strains of cyanobacteria, Rippka et al. [37] aimed to provide stable and consistent generic identities. In total, five sections were proposed, giving recognition to 22 genera. Importantly, this system attempted to identify cyanobacteria in laboratory cultures and took into consideration the morphological changes which were encountered when comparing natural samples with laboratory-grown cultures. Notably, this was also one of the first planned studies attempting to recognize strains on the basis of culture characteristics, rather than the appearance of the samples in nature. Thus the morphological plasticity with change in environmental and culture conditions was addressed in this study. Section I comprised unicellular cells that reproduced by binary fission or budding. Section II also consisted of unicellular cells that reproduced by multiple fissions, leading to the formation of baeocytes. Sections III, IV, and V all contained filamentous cyanobacteria. Section III consisted of taxa not forming heterocytes, akinetes, or hormogonia. Sections IV and V were heterocytous, could form akinetes, and also reproduced by hormogonia formation. Filaments in Section IV divided in only one plane (unbranched), while those in Section V showed division in more than one plane (branched). Interestingly, this system was also an attempt to discuss the bacterial treatment of cyanobacteria rather than following the botanical code, as was done previously by most taxonomists [2,21,22].

After the system of Rippka et al. [37], further progress was made with phylogenetic perspectives being incorporated in the work of Castenholz [38]. This work supported the hypothesis of recognizing small genera [39] as compared to having fewer genera consisting of many species. In this system, emphasis was placed on describing and identifying stable culture characteristics of taxa. Also, in a well-envisioned outlook, the incorporation of more extensive DNA and phylogenetic data was anticipated in future studies. The five orders established in the previous works were referred to as subsections I-V in this system.

1.3.3 Development of the polyphasic approach

The era from 1980 onwards generally saw an increased focus on understanding the genetic characters of cyanobacteria, and taxonomy was not left untouched by these new methods. The better understanding of morphological features and usage of molecular tools (which later extended to phylogenetic methods) resulted in a refreshed focus on issues of cyanobacterial identification and taxonomy. The main methods contributing to these developments were the introduction of electron microscopy and 16 S rRNA gene-based phylogeny. The usage of these strategies helped in the separation of the orders Synechococcales and Chroococcales, the establishment of the separate families of Pseudanabaenaceae and Leptolyngbyaceae, the establishment and better clarity of the entire heterocytous cluster, and the description of numerous new genera and species [40].

Looking closely at the current developments, there are usually two major ways thorough which cyanobacterial taxonomic studies are still progressing. The first school of thought comprises mainly field specialists and ecologists emphasizing the study of natural populations, and hence assigning names usually on the basis of morphological criteria. Genetic studies and phylogenetic interpretations are usually sporadic and do not follow a particular pattern, with no attention being paid to the identification of monophyletic

Handbook of Algal Biofuels

1.3 Cyanoprokaryotes: taxonomic classification history, modern age, and perspectives

genetic clusters. This approach is clearly informative in terms of discussing a lot of information about the naturally occurring populations, in terms of their morphology and ecology, however, the morphological plasticity and changes that occur in long-term laboratory subculturing are neglected or minimized in most of these studies. Hence, it may eventually result in assigning erroneous, ambiguous, or confusing traits to the taxa being studied. The second pattern of assessment that has come into relevance is the extensive evaluation of isolated laboratory-grown strains that are eventually sequenced and of which the phylogeny is determined. Ultimately, a strain is assigned to a particular cluster, based on its position in the phylogeny. While this second approach clearly defines the traits of laboratory-grown cultures, somehow the naturally occurring samples are not paid equal attention, again resulting in ambiguity in taxonomy [40]. We need to understand that a careful balance is required between both these approaches, and this is where the proper understanding of the polyphasic approach is essential. While it is commonly agreed that this type of balance is necessary, achieving this balance for correct identification is still not universally accepted or understood. Discrete plans of work and methods to establish the identity of particular taxa, in coherence with a polyphasic approach, are hence lacking. But, this level of deviation must be accepted considering the ecological and morphological diversity of cyanobacteria [40]. Also, the use of 16 S rRNA genes as molecular markers is also essential, at least in initial studies, as this gene evolves relatively slowly [40].

Connected to the lack of a consensus scheme in the polyphasic approach, it is equally important to realize the uneven weighting of similar characters in different taxonomic groups. Morphological and ecological characters that may be important in one group of cyanobacteria may not hold much relevance in another group (for example, false branching is important in *Scytonema*, life cycle events are useful in *Nostoc*). It must be understood that the differential evolution of these characters may also reflect in phylogenies with higher and better taxon sampling. Thus a universal "fit for all sizes" mode of understanding cyanobacterial taxonomic issues may never come into existence as these organisms are simply too diverse and complex. Examples of some of these facets include the presence of complicated life cycles in the genus *Nostoc*, but their absence in the morphologically closely related *Nostoc*-like genera, such as *Aliinostoc* or *Desikacharya*. The presence/absence of gas vesicles is another characteristic that may, or may not, have taxonomic value in different cyanobacterial groups. Due to all these complications, it is desirable to use polyphasic approaches in modern taxonomic schemes, which promotes the combined usage of morphological, ecological, molecular, and phylogenetic methods. Though not fully tested, a broad consensus has been reached in the usage of the polyphasic approach with the genetic criterion being the primary method, and morphological, ecological, and ecophysiological methods being secondary criteria for identifying cyanobacterial taxa. Also, it has been reasonably established that these secondary methods could vary in different groups and lineages [5].

1.3.4 The modern classification of cyanobacteria

With the introduction of modern methods and better understanding of the polyphasic approach, Komárek et al. [4] put forward a modern classification system of cyanobacteria. This was the result of decades of improvements, revisions, and the hard work of many

Handbook of Algal Biofuels

cyanobacterial taxonomists, with Prof. Jiří Komárek leading the way. This was also a result of the completion of the monographic series on cyanobacteria in the *Süßwasserflora von Mitteleuropa* in which the unicellular, nonheterocytous, and heterocytous cyanobacterial groups were described in detail in three parts [41—43]

The new classification system is based on a few points which need elaboration before discussing the insights of the modern system. The new system basically supports the idea of attaining monophyly at different levels, which could also reflect evolutionary history. Instead of creating larger clusters having a large number of different or unrelated taxa, it is better to have smaller monophyletic entities consisting of related species. The new system keeps the scope open for further revisionary works to attain stability along with well-supported monophyly. It is thus evident that in comparison to the older classification schemes, the new system focuses on phylogenies and the usage of the polyphasic approach. Clarity and consensus are still issues, but the enhanced scope and flexibility of revisions makes the new system of classification much better and comprehensive. It is also important to mention the recommendations of Hoffmann et al. [44,45] that led to the first modern systematic scheme of cyanobacteria in which the class Cyanophyceae was divided into four subclasses: Gloeobacteriophycidae, Synechococcophycidae, Oscillatoriophycidae, and Nostochophycidae. This system differed from all the earlier systems in being based more on phylogeny and modern methods.

The new system of classification also discussed the problematic areas in a more realistic and practical way by reflecting on five broad categories according to the taxonomic clarity/ambiguities present. Accordingly, category I includes all genera which were strongly supported by 16 S rRNA gene phylogeny, and the type species were sequenced. Taxa present in this category are usually strongly supported by polyphasic evaluation and the taxonomy of such genera, at least at present, is more or less stable. Some representatives of this category include *Brasilonema*, *Mojavia*, *Cyanothece*, *Cylindrospermum*, *Microcystis*, and *Halotia*. Category II includes genera whose type species were not studied using molecular methods. Modern methods have been used for evaluating these genera but the ambiguity over the *sensu stricto* cluster and the positioning of the type species make these genera complicated and challenging. Examples of category II include *Aulosira*, *Hyella*, *Myxosarcina*, *Petalonema*, and *Schizothrix*. Category III comprises interesting and complicated larger taxonomic entities like *Calothrix*, *Nostoc*, *Leptolyngbya*, *Anabaena*, *Oscillatoria*, *Pseudanabaena*, *Trichormus*, and *Synechococcus*. All these genera are cosmopolitan, traditionally defined, and not monophyletic. These genera usually have a large number of morphotaxa, are often incorrectly identified and have the typical problem of very closely related morphogenera. These genera must be studied further using the polyphasic approach and many revisionary attempts are anticipated in this category. Category IV comprises taxa that lack significant molecular assessment, although they have been described many years ago. The reasons for the lack of molecular data include difficulties in isolation, purification, and cultivation, or sometimes the rare occurrence of many representatives. Some members of this group are doubtful or improperly diagnosed genera. Prominent members of this group are *Asterocapsa*, *Cyanosarcina*, *Desmosiphon*, *Lithomyxa*, *Placoma*, and *Loriella*. Finally, category V is comprised of members which are taxonomically invalid and currently have no nomenclatural standing. Examples include *Exococcus*, *Gervasia*, *Haplonema*, and *Lagerheimiella*.

The new system of classification mentions the orders Gloeobacterales, Synechococcales, Chroococcales, Spirulinales, Pleurocapsales, Chroococcidiopsidales, Oscillatoriales, and Nostocales (Fig. 1.1). It is notable that the basis of separation at this level of organization was the ultrastructural pattern of thylakoids and preliminary phylogeny. Thus modern taxonomy is clearly a reflection of the evolutionary tendencies of cyanobacteria and supports the idea of genera being monophyletic. It also gives equal weight to understanding the morphological intricacies along with documentation of the habitats and ecological niches. The present system also addresses the issue of cryptogenera (morphologically indistinct, phylogenetically distinct) and the challenges encountered in their description, especially by using the general methods of taxonomic description. Also, morphogenera (differing in morphology, but molecular data insufficient) represent another case study and a challenging taxonomic issue.

Thus the modern taxonomy is indeed a consideration of all the taxonomic reflections that started in the 19th century and are still continuing in the 21st century. In simple terms, this modern system of classification introduces the following aspects that must be tested and also scrutinized by contemporary taxonomists:

1. Usage of the polyphasic approach is essential and it must be applied judiciously as per the taxa under investigation.
2. Molecular and phylogenetic evidences are the primary criteria for establishing taxonomic entities, but they must be accompanied by thorough secondary evidence from morphological and ecological attributes.
3. Cyanobacterial taxonomy will undergo further revisions in the coming decades and there is ample scope for the discovery and creation of more monophyletic generic units.

1.3.5 Present status and the future of cyanobacterial taxonomy

The period after the new classification system implemented was characterized by more clarity and thus enhanced efforts to solve confusing taxonomic issues. This led to the establishment of many new families, genera, and species. The major family level changes included the establishment of the families Oculatellaceae and Trichocoleaceae [46] in the order Synechococcales, the family Aliterellaceae [47] in the order Chroococcidiopsidales, the family Desertifilaceae [48] in the order Oscillatoriales, and the families Calotrichaceae [49], Cyanomargaritaceae [50], Dapisostemonaceae [51], Fortieaceae [4], Geitleriaceae [52], and Heteroscytonemataceae [53] in the order Nostocales. Apart from this, an astonishing number of new genera have also been described (more than 80) since the publication of Komárek et al. [4]. Thus it is evident that the new classification system has indeed shown a way for all modern-day cyanobacterial taxonomists to discover and describe cyanobacterial diversity with continued efforts in an efficient and better directed manner. The near future may entail the introduction of more studies, for instance characterization of the secondary metabolites and unique bioorganic compounds, which could enhance our understanding of cyanobacterial taxonomy. With increased efforts of using genomics-based approaches it is also essential to maintain the nomenclatural rules and most importantly describe the taxa with typification in as much detail as possible [6]. It is advisable to adopt caution while using these modern methods as these methods can sometimes create more

10   1. Cyanoprokaryotes and algae: classification and habitats

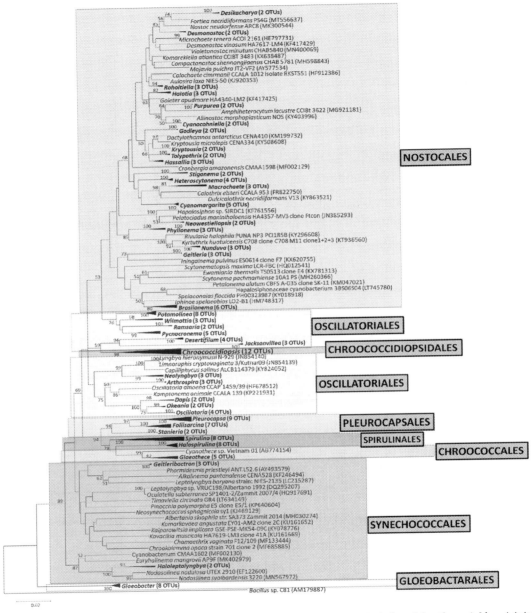

FIGURE 1.1 Phylogenetic positioning of different orders of cyanobacteria inferred by the neighbor joining method based on the 16 S rRNA gene. The evolutionary distances were computed using the Kimura 2-parameter method and are in the units of the number of base substitutions per site. The rate variation among sites was modeled with a gamma distribution (shape parameter = 5). The analysis involved 212 nucleotide sequences. Bar, 0.02 changes per nucleotide position. No bootstrap values <50 are shown. Evolutionary analyses were conducted in MEGA5.

confusion due to the absence of proper descriptions and types. This was evident in the study of Walter et al. [54] where genomic analysis resulted into the invalid description of 33 new genera and 28 new species. At present, the careful usage of the polyphasic approach, observance of the nomenclatural rules, and comparative evaluation with all closely related taxa are the way through which cyanobacterial taxonomy can move ahead in a positive direction.

Similar to the cyanophycean taxonomy, the issues of morphological plasticity, the huge amount of heterogeneity, extensive reproductive mechanisms, and complicated life cycles are common characters affecting the taxonomy and systematics of eukaryotic algae. While understanding these features is absolutely essential, synchronizing the evolution of these characters with the modern concepts of polyphasic taxonomy and phylogenetic tools is another aspect that makes the present-day taxonomy of eukaryotic algae a challenging field of study. At present, the modern understanding of green algal systematics divides the Chlorophyceae into two lineages: the Cholorophyta and the Streptophyta. The brown algal systematics has also very recently seen major developments [55] based on 12 molecular markers. The taxonomy and phylogeny of the Phaeophyceae is also complicated, taking into account the presence of life cycle variations in different groups and the complicated process of sexual reproduction. Also, the extensive influence of speciation based on biogeographic patterns is another aspect that makes the taxonomy of Phaeophyceae interesting. Similarly, the current understanding of the taxonomy of red algae is again very complicated as a result of the occurrence of pit connections, pit plugs, and a triphasic life cycle with postfertilization changes. At present the phylum Rhodophyta is divided into two subphylums Rhodophytina and Cyanidiophytina using combined plastid protein sequences [56]. Nonetheless, in spite of the enhanced usage of modern genomic tools and phylogenetic methods, the taxonomy of all eukaryotic algae is still undergoing many revisions and hence just like the cyanophyceae, usage of a polyphasic approach is always recommended for all the eukaryotic algae. The upcoming section on eukaryotic algal taxonomy hence aims to capture the essence of the history of development of well established eukaryotic taxonomic classifications in a nutshell. It must be noted that taxonomy as a branch of science is ever changing and ever challenging, hence it is impossible to discuss in detail the central tenets of eukaryotic algal taxonomy in a single communication.

1.4 History and present-day algal taxonomy

The foundation of modern biological systematics and nomenclature was laid by the Swedish botanist Carl Linnaeus who published the book "Species Plantarum" in 1753. Linnaeus [57] classified the plant kingdom into 25 classes based on sexual reproduction. He used the term "algae" for the first time as one of the orders in the class Cryptogamia (plants with hidden reproduction) which included the flowerless plants, but also the genera *Conferva*, *Ulva*, *Fucus*, and *Chara*, which are now considered algae. Augmentation of our better understanding of the taxonomic classifications of algae, led to the classification of algae into eight and 12 major evolutionary lineages [9].

Classification of algae varies continuously due to the improvement in modern classification techniques resulting in the reclassification and the identification of new species. After

12   1. Cyanoprokaryotes and algae: classification and habitats

the huge efforts exerted by algal taxonomists, starting from the 19th century to date, we simply confirm that the accurate classification of algae has to be conducted based on integrative multifaceted approaches, that is, morphotaxonomy, autecology, molecular studies and DNA fingerprinting, phylogeny, ultrastructure of cell organelles, following life cycle stages, physiological characteristics, and biochemical constituents.

The most acceptable algal classification systems proposed by the algologists are discussed below:

Harvey's classification [58]: Algae were classified for the first time into four main groups: Chlorospermae (green algae), Melanospermae (brown algae), Rhodospermae (red algae), and Diatomacea (diatoms), based on their pigmentation. In 1843 Kützing published his treatise "Phycologia generalis oder Anatomie, Physiologie und Systemkunde der Tange" in which 96 families were recognized, 62 families of these were new. This work is of major significance in the history of algal classification.

Eichler's classification [59]: The author created the new division "Thallophyta" and classified algae and fungi together in this division. The five-group system of algal classification was proposed on the basis of their color and algae were divided into Cyanophyceae (blue−green algae), Diatomeae (diatoms), Chlorophyceae (green algae), Phaeophyceae (brown algae), and Rhodophyceae (red algae).

Engler and Prantl's classification [60]: Algae and fungi were grouped together under Euthallophyta and different algal groups were identified, including Schizophyta (blue−green algae), Flagellatae, Dinoflagellata (flagellate protists), diatoms, Chlorophyta (Conjugatae and Chlorophyceae), Charophyta, Phaeophyta, and Rhodophyta.

West's classification [61]: Algae were divided into four groups on the basis of reproductive structures and presence or absence of flagella: Akontae (flagella absent), Isokontae (flagella of equal size), Heterokontae (flagella of unequal size), and Stephanokontae (flagella crowned).

Pascher's classification [62]: The first evolutionary scheme of algal classification was proposed based on the phylogeny and relationships among the different algal groups, and algae were classified into eight divisions which were, in turn, subdivided into classes: Cyanophyta, Rhodophyta, Phaeophyta, Chrysophyta, Pyrrophyta, Euglenophyta, Charophyta, and Chlorophyta.

Tilden's classification [63]: Algae were classified into five groups on the basis of reserve foods, pigmentation, and flagellar structure, number, insertion, and arrangement, namely, Myxophyceae, Rhodophyceae, Phaeophyceae, Chrysophyceae, and Chlorophyceae. According to this system, pigments are of fundamental significance in the development and advancement of algal members, and thus this system supported the retention of algae names based on color.

Fritsch's classification [64]: The most acceptable and comprehensive algal classification scheme was proposed in this work. Eleven classes were proposed, based on a combination of different characteristics, including the presence or absence of an organized nucleus, photosynthetic pigments, storage products, thallus organization, flagellar arrangement (number and insertion), and reproduction. These classes are Myxophyceae (Cyanophyceae or blue−green algae), Rhodophyceae (red algae), Phaeophyceae (brown algae), Euglenophyceae (euglenoids), Chloromonadineae, Dinophyceae (dinoflagellates),

Handbook of Algal Biofuels

1.4 History and present-day algal taxonomy

Cryptophyceae, Bacillariophyceae (diatoms), Chrysophyceae (golden-brown or golden algae), Xanthophyceae (yellow-green algae), and Chlorophyceae (green algae).

Smith's classification [65]: Pascher's classification system was followed with some modifications. Algae were classified into seven major divisions, each with one or more classes. The seven algal divisions proposed were: Chlorophyta, Euglenophyta, Pyrrophyta, Chrysophyta, Phaeophyta, Cyanophyta, and Rhodophyta.

Papenfuss's classification [66]: Pascher's and Fritch's classification systems were criticized and algae were classified based on their evolutionary relationships. Seven divisions were proposed, namely, Chlorophycophyta, Charophycophyta, Euglenophycophyta, Chrysophycophyta, Pyrrophycophyta, Phaeophycophyta, and Rhodophycophyta, each with one or more classes. He also classified the blue—green algae (Myxophyceae) in a separate phylum, Schizophyta, together with eubacteria.

Chapman's classification [67]: The four major algal divisions scheme was proposed based on pigmentation, morphological differences, and phylogenetic relationships. Each division was further divided into one or more classes as follows: Myxophycophyta (Myxophyceae), Euphycophyta (Chlorophyceae, Phaeophyceae, and Rhodophyceae), Chrysophycophyta (Chrysophyceae, Xanthophyceae, and Bacillariophyceae) and Pyrrophycophyta (Cryptophyceae and Dinophyceae).

Prescott's classification [68]: Algae were divided into nine major phyla with different classes based on a combination of taxonomic characters, such as the presence or absence of a true nucleus, photosynthetic pigments, biochemical nature of the cell wall, reserve foods, life history, and reproduction. These phyla were Chlorophyta (Chlorophyceae and Charophyceae), Euglenophyta, Chrysophyta (Bacillariophyceae, Chrysophyceae, and Xanthophyceae), Pyrrophyta (Desmophyceae and Dinophyceae), Phaeophyta (Isogeneratae, Heterogeneratae, and Cyclosporae), Rhodophyta (Bangioideae and Florideae), Cyanophyta (Coccogoneae and Hormogoneae), Cryptophyta, and Chloromonadophyta.

Round's classification [69]: Algae were divided into two major groups: phylum Cyanophyta in Prokaryota and all other algae in Eukaryota. He classified eukaryotic algae into 11 phyla: Chlorophyta, Charophyta, Euglenophyta, Prasinophyta, Xanthophyta, Haptophyta, Dinophyta, Bacillariophyta, Chrysophyta, Phaeophyta, and Rhodophyta.

Bold and Wynne's classification [70]: Papenfuss's classification system was followed and the term "phyco-" was used before "phyta" in naming algal divisions. Nine divisions of algae were proposed: Cyanochloronta, Chlorophycophyta, Charophycophyta, Euglenophycophyta, Phaeophycophyta, Chrysophycophyta, Pyrrhophycophyta, Cryptophycophyta, and Rhodophycophyta.s

Parker's classification [71]: Algae were mainly divided into Prokaryota and Eukaryota on the basis of the presence or absence of membrane-bounded organelles. Prokaryota was further divided into two divisions Cyanophycota (Cyanophyceae) and Prochlorophycota (Prochlorophyceae). Eukaryota included Rhodophycota (Rhodophyceae), Chromophycota (with the classes Chrysophyceae, Prymnesiophyceae, Xanthophyceae, Eustigmatophyceae, Bacillariophyceae, Dinophyceae, Phaeophyceae, Raphidophyceae, and Cryptophyceae), Euglenophycota (Euglenophyceae), and Chlorophyta with three classes, namely, Chlorophyceae, Charophyceae, and Prasinophyceae.

Lee's classification [8]: This classification system is currently widely accepted. Lee divided algae into two major groups, namely, Prokaryota and Eukaryota, which were

14   1. Cyanoprokaryotes and algae: classification and habitats

further divided into several divisions and classes. Prokaryota only comprised one division, named Cyanophyta with the single class Cyanophyceae. Eukaryota were divided into three main groups, based on structure of the chloroplast envelope membranes:

Group I—algal divisions with chloroplasts surrounded by a double-membraned chloroplast envelope, including Glaucophyta, Rhodophyta, and Chlorophyta.

Group II—algal divisions distinguished by having double-membraned chloroplasts surrounded by one membrane of the chloroplast endoplasmic reticulum. These include Euglenophyta (euglenoids) and Dinophyta (dinoflagellates).

Group III—algal divisions having double-membraned chloroplasts surrounded by two membranes of the chloroplast endoplasmic reticulum envelope. These include Cryptophyta (cryptophytes), Prymnesiophyta (haptophytes), and Heterokontophyta (heterokonts). The latter division was further divided into several classes, including Chrysophyceae, Synurophyceae, Bacillariophyceae, Phaeophyceae, Eustigmatophyceae, Xanthophyceae, and Raphidophyceae. The key taxonomic characteristics for each algal class have been discussed in Section 1.2.

1.5 Global distribution and habitats of cyanoprokaryotes and algae

Algae, including the evolutionarily primitive photoautotrophic cyanoprokaryotes, colonize all three abiotic components of planet Earth, namely, air, land, and water. These organisms are abundant, widely distributed, several are cosmopolitan, and, as long as sunlight and water cooccur, they can occupy all biomes around the globe. Most species are morphologically and physiologically adapted to the specific biotopes in which they live. Although many algae can live in aerial and terrestrial environments, they are by far the most abundant in aquatic habitats.

1.5.1 Aerial (subaerial or aerophytic) algae

Aerial algae inhabit any object in the air, above the soil or may be found on the surface of water. Different types of aerial habitats include exposed bedrock, the soil's surface, terrestrial bryophytes, tree bark, and anthropogenic structures. Moisture, needed to sustain life, can originate from groundwater seep, precipitation, humidity, or waterfall spray and can be highly variable, ranging from perennially moist to mostly dry [72].

Light, rain, and air humidity are considered to be the most influential factors affecting aerial algae [73], and aerial algal communities are usually adapted to these conditions. Adaptations include mechanisms to retain moisture or limit moisture loss. Some adaptations include a reduction in cell size, morphological changes (especially common in diatoms), as well as mucilage production that aids water retention [74]. As solar radiation is much more intense in aerial habitats, some of these algae produce internal or external protective pigments to protect them against harmful UV-radiation [72].

The terms "euaerial" and "pseudoaerial" are often used to distinguish between different types of aerial algae [75]. Algae inhabiting elevated objects, receiving only atmospheric moisture, are termed euaerial, while algae inhabiting areas that receive a fairly constant

Handbook of Algal Biofuels

supply of moisture from seeping groundwater, surface runoff, or waterfall spray are termed pseudoaerial.

On the basis of the habitats and substrates that free-living aerial algae occupy, they can be subdivided into the following types:

- *Airborne algae*—the presence of airborne algae has long been recognized, since the publication by Ehrenberg [76] in which 18 diatoms in air dust samples, collected by Darwin, were described. Airborne cyanobacteria and algae can be found in sea spray and as minute particles blown around by wind. The most common airborne genera include *Gloeocapsa*, *Chlorella*, *Chlorococcum*, *Scenedesmus*, *Acutodesmus*, and *Desmodesmus*. An extensive review of airborne cyanobacteria and algae, and their health effects, was published by Genitsaris et al. [77]. This review also presented a table in which all airborne cyanobacteria and algal taxa reported by investigators of aerial algae are summarized. A recent study by Saber [78] on the airborne algae of the El-Farafra Oasis (Western Desert of Egypt) revealed five taxa, namely, *Myxosarcina chroococcoides* (nowadays known as *Cyanosarcina chroococcoides*), *Oscillatoria acuminata* (currently *Oxynema acuminatum*), *Lyngbya limnetica* (currently regarded as *Planktolyngbya limnetica*), *Schizothrix braunii*, and *Westiellopsis prolifica*. He concluded that the airborne algal propagules were exfoliated from soil surfaces and other aquatic ecosystems into the air.
- *Lithophytic (epilythic and endolythic) algae*—grow on rocks, stones, bricks, or cement and are commonly found on stone walls, monuments, and walkways. *Vaucheria*, *Nostoc*, *Gloeocapsa*, and many other algae are common genera found on wet rocks. Black spots on stone walls in rainy seasons are often caused by growths of *Scytonema*. Caves are habitats generally characterized by low light intensities. Pseudoaerial cyanobacteria and algae growing against cave walls and shaded overhangs (Fig. 1.2A−B), receiving moisture from dripping rocks and spray from waterfalls, can also be classified as lithophytic algae, as they represent forms living in the transition zone between aerial and terrestrial (rock) habitats.
- *Epixylous algae*—grow on dead wood such as poles, posts, and doors.
- *Epimetallous algae*—grow attached to metal surfaces.
- *Epiphytic algae*—grow on living plants and are often found in trees. Epiphyllous algae, such as *Trentepohlia*, grow on the leaves of plants, while epiphloeophytic algae (also known as epiphellous or corticolous algae) are commonly found on bark, stems, or trunks of trees. Algae growing on the bark of trees are often mixed with mosses and liverworts and include several species of cyanobacteria (*Aphanothece*, *Chroococcus*, *Lyngbya*, *Phormidium*, *Porphyrosiphon*, *Hapalosiphon*, *Hassalia*, *Nostoc*, *Scytonema*, and *Stigonema*), as well as eukaryotic green algae, such as *Trentepohlia* and *Klebsormidium* [11]. Epiphytic algae use their hosts for aerial support, but obtain their resources from the atmosphere and because they are excellent examples of commensalism, some aerial epiphytes will be further discussed in Section 1.5.4.
- *Epizoic algae*—grow on land animals. The most well-known examples are the red alga *Rufusia pilicola* [80], the cyanobacterium *Oscillatoria pilicola* [81], and the green alga *Trichophilus welckeri* [79] found growing in the fur of certain sloth species. There are grooves on the surface of the hair that retain water, allowing the growth of these algae that turn the hair green (Fig. 1.2C−E).

FIGURE 1.2 Representative habitats of cyanoprokaryotes and algae. (A–B) Aerial (lithophytic) algae, including a population of the heterocytous cyanoprokaryote *Petalonema alatum* (B), growing against an overhang that receives water from a seepage and dripping rocks, the Drakensberg area, South Africa. (C–E) Hairs of the sloth *Choloepus hoffmanni* covered with the aerial (epizoic) green *Trichophilus*-like alga. (D–E) Close-up views showing details of the *Trichophilus*-like alga [79]. (F) Terrestrial (cryptophytic) cyanobacteria growing just below the surface of the soil. (G) The pennate diatom *Surirella* sp. acting as a substrate for the epiphytic green alga *Characium* sp. (H–I) Pool in the Kruger National Park, South Africa, discolored red due to *Euglena sanguinea* blooms. *E. sanguinea* single cell (Fig. I) with red astaxanthin pigments and white paramylon granules. (J–K) Floating masses of *Spirogyra* sp. and *Zygnema* sp. in an urban lentic pond in Potchefstroom, South Africa. (L–M) Bloom of the saxitoxins-producing cyanobacterium *Dolichospermum circinale* in the Fitzroy River Barrage, Central Queensland, Australia. (M) Close-up view of *D. circinale* trichomes depicting akinetes and heterocytes. Scale bars = 20 μm. *Source: (A–B) Anatoliy Levanets. (D–E) From M. Suutari, M. Majaneva, D.P. Fewer, B. Voirin, A. Aiello, T. Friedl, et al., Molecular evidence for a diverse green algal community growing in the hair of sloths and a specific association with Trichophilus welckeri (Chlorophyta, Ulvophyceae), BMC Evol. Biol. 10 (1) (2010) 86–97. (F) S. Janse van Vuuren. (G) Jonathan Taylor. (H–I) J. Taylor and Marno Laubscher. (J–K) S. Janse van Vuuren. (L–M) Glenn McGregor.*

1.5.2 Terrestrial algae

Cyanobacteria and eukaryotic algae occur in every terrestrial habitat on our planet. Terrestrial algae can occur on the surface of the soil or at depths upto several centimeters in the soil or in/on soil crusts (Fig. 1.2F). Similar to aerial algae, it is well-established that solar radiation, water, and temperature are the most important abiotic factors governing the distribution, metabolism, and life history strategies of terrestrial algae [82].

The algal flora of the soil includes members of the Cyanophyta, Chlorophyta, Euglenophyta, Chrysophyta, and Rhodophyta [82]. Cyanobacteria and eukaryotic green algae are usually the most common taxa in terrestrial environments. Green algal genera commonly encountered in soils are *Chlamydomonas*, *Chlorella*, *Chlorococcum*, and *Tetracystis* [1], whilst common cyanobacteria include *Anabaena*, *Gloeocapsa*, *Microcoleus*, *Nostoc*, *Phormidium*, *Westiellopsis*, and *Scytonema* [83,84]. A study by Saber et al. [85] on Egyptian hyperarid desert habitats unveiled four interesting green algal isolates, based on morphological and molecular evidence, and one of them was a genus and species new to science described as *Pharao desertorum*. In general, the biodiversity of soil algae is still grossly understudied, and there are likely still many interesting species to be described using combined polyphasic approaches.

Different authors have different viewpoints about the classification of terrestrial algae. Petersen [75] defined three major categories based on their habitats, namely:

- *Aeroterrestrial algae*—occurring on substrates that are elevated above the ground.
- *Hydroterrestrial algae*—growing on permanently wet soil.
- *Euterrestrial algae*—including both epiterranean and subterranean forms. Tiffany [86] used the term **edaphophytic** (soil) **algae** for these algae and subdivided them into:
  - **Saphophytic algae**, representing surface forms. Examples include many species of cyanobacteria found upon the surface of the soil. Besides these, *Botrydium*, *Fritschiella*, *Mesotaenium*, *Oedocladium*, *Protosiphon*, *Vaucheria*, and many other algae, grow on the surface of wet soils.
  - **Cryptophytic algae**, representing subsurface forms. *Anabaena*, *Calothrix*, *Cylindrospermum*, *Nodularia*, *Scytonema*, *Stigonema*, and *Trichormus* have been reported from rice paddy fields, where heterocytous forms fix atmospheric nitrogen in the soil.

Some terrestrial habitats are extremely hostile environments, ranging from very arid areas, rocks in hot and cold deserts, Antarctic soils, and highly acidic postmining sites. Extreme fluctuations in environmental conditions, such as aridity and/or low or high levels of temperature or light intensity, can result in stress, leading to morphological and physiological adaptations. Cyanobacteria produce mucilage, thick sheaths, or protective pigments as adaptations [83]. The green alga *Zygogonium* forms thick mats with extremely high water-holding capacity [87], and the species *Zygogonium ericetorum* produces reduced cytoplasm, thicker walls, and solutes with UV-absorbing capacities [88]. Several green algal taxa can survive extended periods without moisture, after which they are able to recover upon receiving moisture. Friedmann et al. [89] classified the desert soil algae into five categories:

- *Endedaphic algae*—living in the desert soils.
- *Epidaphic algae*—living on the surface of desert soils.
- *Hypolithic algae*—living on the lower surface of stones on desert soils.

18   1. Cyanoprokaryotes and algae: classification and habitats

- *Chasmolithic algae*—living in the rock fissures in desert soils.
- *Endolithic algae*—penetrate rocks as they grow.

A variety of algae can also live in symbiosis with terrestrial plants. These symbiotic terrestrial algae are discussed, together with aerial and aquatic symbiotic algae, in Section 1.5.4.

1.5.3 Aquatic algae

The majority of algae are aquatic, yet the word "aquatic" is almost limited and restricted in its ability to encompass the diversity and complexity of these habitats. A multitude of habitats can be distinguished, depending on the classification system used.

Classification of habitats can be based on the position where the cyanobacteria or algae live in the water body (e.g., zone in the sea, lake, dam, river, or pond), the way that they live (epiphytic, epizoic, symbiotic, parasitic, etc.), or how they maintain their position in the water column (e.g., planktonic vs. benthic algae).

In both marine and freshwater habitats **epiphytic algae** are found living upon other species of algae (Fig. 1.2G). Larger algal forms, such as *Cladophora*, *Chara*, or *Nitella*, may serve as substrata for epiphytic diatoms and other algae, such as *Coleochaete nitellarum* [1]. *Chaetonema* is found to be epiphytic on the mucilaginous masses of *Tetraspora* and *Batrachospermum* [11]. In the marine environment more than 50 epiphytic algal species have been reported growing on the stipes and holdfasts of *Ecklonia maxima* (sea bamboo—a species of kelp [90]). Species of *Carpoblepharis*, *Polysiphonia virgate*, and *Suhria vittata* have been identified as the most important kelp epiphytes [91]. Several algae, of which some are mentioned in Section 1.5.4, can also live as epiphytes on aquatic plants. Species of *Coleochaete* are, for instance, epiphytic on *Ipomoea*, *Typha*, *Vallisneria*, and several other aquatic plants. *Rivularia* species are commonly found as epiphytes on *Potamogeton pectinatus* [92].

Epizoic algae include a large amount of species living on aquatic animals such as turtles, snails, shellfish, and fish, both in marine and freshwater environments. *Basicladia*, *Dermatophyton*, *Oscillatoria*, and *Protoderma* are attached to the backs of turtles [11]. Kanjer et al. [93] studied the diatom community structure on the skin of loggerhead sea turtle heads and 113 highly specialized diatom taxa were recorded. Out of these taxa, the probably obligate epizoic diatoms *Achnanthes elongata*, *Chelonicola* sp., and *Poulinea lepidochelicola* contributed upto 97.1% of the total diatom abundance. Several members of the Ulotrichales (*Microspora*, *Ulothrix*, *Uronema*), Cladophorales (*Cladophora* and *Rhizoclonium*), Chaetophorales (*Chaetophora*, *Coleochaeta*, and *Stigeoclonium*), and Oedogoniales (*Oedogonium* spp.) grow upon mollusk shells [11]. *Rhopalosolen saccatus* was observed growing on at least two cladocerans (*Daphnia similis* and *Simocephalus vetulus*) by Holland and Hergenrader [94] while *Characium* and *Characiopsis* occur on the legs of *Branchipus* (fairy shrimp [95]) as well as on *Anopheles* larvae [96]. *Stigeoclonium* is epizoic on the gills of fish [95].

Endophytic and **endozoic** algae are symbiotic and live in close association with other algae, plants and animals (Section 1.5.4), while **parasitic** algae (discussed in Section 1.5.5) may exploit their hosts to survive.

Handbook of Algal Biofuels

The majority of aquatic algae are, however, free-living and they can live in a wide range of salinities, ranging from freshwater, brackish water, marine water, to halophilic environments, characterized by extremely high salt concentrations. In this section two major habitats of free-living aquatic algae will be discussed, namely, marine and freshwater habitats, while halophilic environments and halophilic algae will be discussed in Section 1.5.6, as part of extreme environments.

- *Marine and salt water habitats*—they live in many different habitats within the sediment of intertidal areas, as well as in the open water of oceans. These habitats include:
  - **Estuaries**—besides being planktonic, algae also inhabit the top millimeters of sediment where they live interstitially between the sediment grains enabling them to conduct photosynthesis.
  - **Sand flats and saltmarshes**—in these areas algae can be benthic (living in or on the bottom, often attached to a substrate) or planktonic (free-floating), and in both cases the assemblage is often dominated by diatoms.
  - **Muddy shores**—large quantities of various cyanoprokaryotes and algae such as diatoms, dinoflagellates, and filamentous green and brown algae live interstitially within sediment particles.
  - **Bare soft substrates**—soft substrates are defined as all areas of nonvegetated fine-sediment bottom occurring within estuarine and marine waters below low tide level. Examples of soft substrates include mud, ooze, silt, sand, shell grit, and finer gravels.
  - **Oceans**—in the ocean most marine algae live in the subtidal zone while other species can be found in the splash zone. The open water of oceans is mostly dominated by planktonic algae.
  - As a rule, red algae (Rhodophyta), dinoflagellates (Dinophyta), and brown algae (Phaeophyceae) are generally more diverse in marine than in fresh waters [1].
- *Freshwater habitats*—these habitats are diverse and differ considerably between different types of waterbodies. **Epilithic habitats** in the freshwater environment include submersed stones, boulders, and bedrock to which a variety of algae can attach. Algal taxa such as *Chamaesiphon* spp., *Gongrosira incrustans*, *Heribaudiella fluviatilis*, *Rivularia* spp., and *Tolypothrix distorta*, occurring in stony streams, are commonly reported from epilithic habitats [97]. **Epipelic habitats** are home to a variety of benthic algae colonizing sediments and mud, whilst **epipsammic habitats** consist of a layer of sand to which benthic algae attach. The abovementioned habitats can be found in both standing (lentic) and flowing (lotic) waters.
  - **Lentic water bodies**—lentic water bodies consist of stagnant (or extremely slow flowing) water and include ditches, seeps, lakes, reservoirs (dams), ponds, pools, wetlands, swamps, and marshes. The size of lentic waterbodies may range extensively, from extremely small rainwater pools to larger lakes, reservoirs, or dams (man-made structures), to extremely large natural lakes such as Lake Baikal or Tanganyika. Small pools and ponds may have distinctive algal assemblages, as the surrounding land use may influence the composition of the algal assemblage [1]. Bright red blooms by some *Euglena* species, for example, *E. sanguinea*, a toxic species, are often observed in ponds from which animals drink, wetlands, waterholes on golf

courses, and sewage lagoons [98]. The red discoloration of the water is the result of a mixture of carotenoid pigments (astaxanthin being the most abundant) inside the algae. *E. sanguinea* has recently formed extensive blooms in small ponds from which animals drink in the Kruger National Park, South Africa [99] (Fig. 1.2H—I). Other algal genera commonly encountered in small stagnant pools or ponds, include *Chlamydomonas*, *Desmodesmus*, *Hydrodictyon*, *Pediastrum*, *Scenedesmus*, *Spirogyra*, *Volvox*, and *Zygnema* (Fig. 1.2J—K).

In larger, deep lentic water bodies, such as lakes, dams, and reservoirs, at least four major zones have been identified, in which different algal assemblages may occur:

1. **Littoral zone**—this is near the shore of the lake, it is usually shallow, so that sunlight can penetrate all the way to the sediments, allowing an abundance of aquatic plants (macrophytes) and algal growth. Cyanobacteria and algae are often intermingled with macrophytes that provide a sheltered environment. Many vertebrates (waterbirds, mammals), invertebrates (e.g., insects, zooplankton), snails and fish also live in the littoral zone, feeding on algae living in the stable environment between the macrophytes. The littoral zone is home to a variety of algae, including amongst others a mixture of (pseudo) planktonic (free floating) forms, benthic (attached) species, symbiotic and parasitic species.
2. **Pelagic (limnetic) zone**—this is the open water area of the lake/reservoir where light does not penetrate to the bottom. The pelagic zone can be subdivided into the euphotic zone or epilimnion (zone in which light penetrates, down to the depth where photosynthetic active radiation reaches 1% of radiation incident on the water surface) and the profundal zone or hypolimnion (no light penetration). Rooted plants cannot grow in the pelagic zone as it is generally too deep for roots to reach the bottom. Microalgae in the euphotic zone, are mostly dominated by planktonic forms (phytoplankton), exhibiting different adaptations to avoid sinking. Numerous algae (neuston) can also attach to the surface of the air/water interface where they keep their position by means of the surface tension.
3. **Profundal zone**—this zone represents the deepwater layers of the hypolimnion below the euphotic zone. The profundal zone is usually dark in deep lakes as sunlight does not penetrate to such a depth. Photosynthetic algae are typically absent (though some algae can switch to heterotrophic metabolism to temporarily survive in dark conditions) and heterotrophic forms dominate this layer.

   Several authors identified depth-distribution zones while analyzing the spatial structure of algal assemblages along littoral depth gradients (e.g., Cantonati et al. [100] for diatoms; Cantonati et al. [101] for soft algae and cyanoprokaryotes). Most frequently three depth distribution zones were identified and named: shallow, mid-depth, and deep. Generally, some disturbance factor is found to be most determinant in the shallow zone, whilst the mid-depth and deep zones are stable, with the latter being affected by extreme light reduction. In a classic paper, Kann and Sauer [102] described a "remarkable periphyton of different colorings composed of pigmented algae typically found in the deeper part of lakes" ("*Rotbunte Tiefenbiocönose*").

   — **Lotic water bodies**: In contrast to lentic water bodies, lotic water bodies consist of flowing water, such as those in creeks, brooks, springs, streams, and rivers, and it may harbor different types of algal assemblages than those found in standing waters. In general, lotic bodies may include a combination of different types of habitats, depending on the flow rate of the water. If the water is slow-flowing in particular areas of, for examle, small streams or large rivers, it can mimic lentic habitats and similar algal assemblages may be found.

Groundwater-dependent habitats, such as fountains, springs, and drilled wells, are important freshwater sources and some may represent extreme habitats (e.g., thermal-to-hot and acidic springs; see Section 1.5.6), sustaining unique algal assemblages.

1. **Spring habitats**—springs are keystone ecosystems that are rapidly getting impaired and disappearing because of unchecked appropriation of groundwater and site-specific habitat destruction [103]. Since springs are ecotones (i.e., transitional environments) linking groundwater and surface water systems [104], their study requires the integration of multiple disciplines, particularly ecology and hydrogeology [105]. Springs harbor a disproportionately high gamma-biodiversity [106] due to their intrasite substrate heterogeneity and intersite diversity of habitat conditions, as related to extensive variation in their geological age and aquifer geochemistry, and to their widespread distribution across many climatic zones, geological provinces, and biogeographic regions [104] Although some (e.g., geothermal) springs are inhospitable and colonized only by specialized, extremophilic microbial species (these will be discussed in Section 1.5.6), other springs contain multiple microhabitats and consequently diverse assemblages of microbial, plant, and animal life [103,104]. According to Cantonati et al. [107,108], the most common and characteristic algae and cyanoprokaryotes of major spring types are as follows. Iron springs have low pH, high iron and sulfates, and host a few species of filamentous chlorophytes and xanthophytes (*Tribonema* spp.), plus abundant growth of iron bacteria; diatoms are typically represented by species-poor assemblages with low numbers of individuals. Seepages and pool springs have low current flow, fine-particle-size substrata, and mostly slightly acidic pH; cyanobacteria are often lacking, and filamentous chlorophytes (*Spirogyra* spp., *Mougeotia* spp., *Microspora* spp.) dominate; diatom communities are characterized by the mire-species, such as *Frustulia crassinervia*, acidophilous taxa, such as *Tabellaria flocculosa*, very low-alkalinity indicators, such as *Psammothidium acidoclinatum*, and many *Eunotia* species (e.g., *E. borealpina*, *E. exigua*, *E. tenella*). Flowing springs on siliceous substratum are often located at higher elevations, with low to very low conductivities, and frequently with higher discharge; the benthic algae are dominated by rheophilic cyanobacteria, and by the rheobiontic chrysophyte *Hydrurus foetidus*; the cyanobacterium *Tapinothrix janthina* is also common whilst in the carbonate flowing springs it was replaced by *Tapinothrix varians*; common diatoms are *Odontidium mesodon*, *O. hyemale*, *Eunotia minor*, *Navicula exilis*, and *Planothidium lanceolatum*. Mid- to high-altitude, oligotrophic, carbonate flowing springs with medium conductivities and/or affected by seasonal desiccation host benthic-algae assemblages dominated by cyanobacteria (Chroococcales and Oscillatoriales including the rheophilic *T. varians*); common diatoms are *Achnanthidium pyrenaicum*, *A. lineare*, *A. pfisteri*, *Gomphonema elegantissimum*, *Nitzschia fonticola*, *Humidophila perpusilla*, *H. contenta*, *Planothidium frequentissimum*, *Meridion circulare*, and *Achnanthidium dolomiticum*. Low-altitude, mostly shaded and nitrate-enriched carbonate rheocrenes, with medium—high conductivities

22   1. Cyanoprokaryotes and algae: classification and habitats

include mainly cyanobacteria (*Pleurocapsa minor*, the eutraphentic *Phormidium retzii*), and red algae (*Audouinella* spp., *Hildenbrandia rivularis*); diatoms include *Cocconeis euglypta*, *C. lineata*, *C. pseudolineata*, *Amphora pediculus*, *A. inariensis*, *Caloneis fontinalis*, *Reimeria* spp., and *Eunotia arcubus*. Hygropetric rheocrenes, with mid-high conductivities and low to moderate current flow, are characterized mainly by cyanobacteria (*Rivularia* spp. and *Plectonema tomasinianum* but also *Ammatoidea normanni*, and the pseudaerial *Calothrix parietina*); diatoms include *Encyonopsis microcephala*, *E. cesatii*, *Gomphonema lateripunctatum*, *Delicatophycus delicatulus*, *D. minutus*, *Denticula tenuis*, and *Cymbopleura austriaca*. The most widespread benthic algae in limestone-precipitating springs [109], located at low altitudes and with high conductivities, are the desmid *Oocardium stratum*, the cyanoprokaryotes *Phormidium incrustatum* and *Tapinothrix crustacea*; other characteristic cyanobacteria belong to the genera *Scytonema*, *Dichothrix*, *Schizothrix*, *Gloeocapsopsis*, and *Gloeothece*; diatoms include *Achnanthidium trinode*, *Brachysira calcicola*, *Denticula elegans*, and *Cymbella diminuta*.

2. **Stream and river habitats**—planktonic algae are common in large rivers (in streams there may be algal drift) as they are kept in suspension by the continuous mixing of the flowing water (Fig. 1.2L—M). However, streams and rivers represent complex habitats harboring a variety of other microhabitats, such as those sheltered by large boulders and stones. Cyanobacteria and algae living in these sheltered microhabitats may resemble those found in lentic systems. Sheltered microhabitats inside slow-flowing water may also be the home of various benthic species, for example, *Chara*, *Chamaesiphon*, and *Cladophora*, growing attached to substrates in the water. Small microscopic algae, for instance the diatom *Gomphonema*, may also attach to substrates (stones or boulders in the riverbed) by mucilage stalks (Fig. 1.3A—B). In sheltered patches, filamentous algae may also occur where they act as substrates for a variety of other epiphytic algal species.

Streams and rivers may also consist of stretches of slow-flowing water, alternating with areas marked by rapid and fast-flowing patches of water. Large filamentous algae, such as *Enteromorpha* (currently moved taxonomically to *Ulva*) and *Vaucheria*, are generally found in slow-flowing water, while fast-flowing water is often characterized by fluviatile algae, such as *Ulothrix*, which is common in mountain falls [110]. Genera such as *Stigeoclonium*, *Batrachospermum* and *Lemanea* were also reported from several swift running streams [111]. *Cladophora glomerata* is very common in flowing water, for example, the Nile River, where it is often responsible for nuisance conditions. Filaments of *Cladophora glomerata* are particularly abundant in concrete-lined canal systems, for example, irrigation canals (Fig. 1.3C—D), or water features such as fountains (Fig. 1.3E—F). At a young stage the filaments are anchored to the concrete by means of rhizoids, but as they mature they will detach, becoming free-floating.

1.5.4 Symbiotic algae

Symbiotic algae are widespread and common and may be found in all three of the abovementioned habitats (aerial, terrestrial, and aquatic), where they live in symbiotic relationships with a variety of organisms, including plants, animals such as ciliates,

FIGURE 1.3 Representative habitats of cyanoprokaryotes and algae. (A) Stones covered with growths of freshwater, benthic diatoms in a stream bed. (B) *Gomphonema truncatum* var. *turgidum*, a pennate diatom, attached to the substrate by mucilage stalks. (C–D) *Cladophora glomerata* filaments causing nuisance conditions in irrigation channels. (E–F) *C. glomerata* filaments growing against the concrete lining of a water feature in front of the main building of the North-West University, Potchefstroom, South Africa. (G–H) Stones covered with crustose and foliose lichens (symbiosis between algae or cyanobacteria and fungi) in the botanical garden of the North-West University, Potchefstroom. (I) Stem of a tree where the bark is covered with various lichens. (J) Mass development of *Azolla filliculoides* (harboring the symbiotic nitrogen-fixing *Trichormus azollae*, Fig. K) in a roadside pond between Frankfort and Heilbron, South Africa. Scale bars = 20 μm. *Source: (A) S. Janse van Vuuren. (B) J. Taylor. (C–D) K. du Plessis. (E–H) S. Janse van Vuuren. (I) A.A. Saber. (J) S. Janse van Vuuren. (K) S. Janse van Vuuren and J. Taylor.*

sponges, and mollusks, as well as fungi. In these symbiotic relationships, the cyanobacteria or algae supply organic substances, derived from photosynthesis to the host organism which, in turn, provides protection to the algal cells.

The most striking example of symbiosis in **aerial environments** is in the form of lichens (Fig. 1.3G–I), where cyanobacteria or algae live in symbiotic associations with fungi. Various cyanobacteria genera, for example, *Chroococcus, Gloeocapsa, Microcystis, Nostoc,* and *Scytonema,* have been isolated from lichens [110]. *Chlorella, Coccomyxa, Palmella,* and *Protococcus* are green algal symbionts also found in lichens [110]. The most common alga in terrestrial lichens is a unicellular green alga belonging to the genus *Trebouxia* [112].

In **terrestrial environments** there are numerous examples of cyanobacteria and algae that live in symbiosis with plants or animals. A few examples are listed below:

- *Endophytic algae*—cyanobacteria or algae living inside the tissue of plants.
  - *Nostoc* lives in symbiosis with thalloid liverworts (*Blasia pusilla* and *Cavicularia densa*) and all hornworts (e.g., *Anthoceros, Notothylas,* and *Phaeoceros* spp.), where the *Nostoc* colonies fix nitrogen for their hosts. *Nostoc* can also be found in the leaves of a variety of terrestrial bryophytes (feather mosses and *Sphagnum*), ferns, cycads, and wetland plants such as *Gunnera.*
  - *Nostoc cycadae* is found in the coralloid roots of the cycad, *Cycas.*
  - Filamentous cyanobacteria were also found inside the land plant, *Aglaophyton major.*
- *Endozoic algae*—cyanobacteria or algae that live inside the body of animals.
  - There are about 14 species in the family Oscillatoriaceae found in the digestive and respiratory tracts of various vertebrates [110,113]. However, it seems as if endozoic algae are much less common in terrestrial environments compared to aquatic environments.
  - *Anabaeniolum* lives in unicellular animals and it was also found living in the digestive tract of mammals, including man [113].

Aquatic environments (both marine and freshwater) are characterized by many symbiotic relationships between cyanobacteria or algae on the one hand, and aquatic plants or animals on the other hand. Examples are listed below:

- *Lichens*—although *Trebouxia* is the most common algal genus in terrestrial lichens, it is rarely a phytobiont in aquatic lichens [114]. The cyanobacteria *Calothrix* (in *Lichina*), *Nostoc* (in *Pyrenocollema*), *Stigonema* (in *Ephebe*), and the green algae *Stichococcus* (in *Staurothele*), and *Dilabifilum* and *Coccobotrys* (in *Verrucaria*) are common in aquatic lichens [114]. *Heterococcus* is a yellow–green alga, also acting as the photobiont in the lichen *Verrucaria* [114].
- *Endophytic algae*—living inside plants or other algae in aquatic environments, include the following:
  - *Cyanocyta korschikoffiana* is a cyanobacterial symbiont living inside *Cyanophora paradoxa.*
  - Several algae live in a close association with other algae, for example, the green algae *Blastophysa rhizopus, Bolbocoleon piliferum* and *Ulvella leptochaete* were found living inside macroscopic seaweed *Grateloupia lanceolata* in the Indian Ocean and Jeju Island [115], where they are believed to confer ecological advantages to the host, such as disease resistance.

- *Trichormus azollae* occurs inside the leaves of the red water fern *Azolla* (Fig. 1.3J—K), while *Chlorochytrium* lives endophytically inside the water fern *Lemna* and the aquatic angiosperm *Ceratophyllum* [110].
- **Endozoic algae**—living inside animals in the aquatic environments, include the following:
  - Zoochlorellae—these are small, nonmotile (often *Chlorella*) or sometimes, flagellated species (e.g., *Tetraselmis*, *Carteria*) that live within the bodies of various freshwater protozoans and invertebrates. An example is the green hydra (*Hydra viridissima*, also known as *H. viridis* or *Chlorohydra viridissima*) in which *Chlorella vulgaris* live. *Chlorella* can also live in symbiosis with ciliates such as *Ophrydium* and *Stentor* species [116]. *Chlorella* supplies the hosts with the products of photosynthesis. In marine environments zoochlorellae can co-occur with zooxanthellae inside sea anemones.
  - Zooxanthellae—these are dinoflagellates that live in symbiosis with freshwater or marine invertebrates, such as fresh and seawater sponges, corals, jellyfish, sea anemones, and nudibranchs (mollusks).
  - Green algae were found living as symbionts inside the cells of salamander embryos [117].
  - *Vaucheria litorea* lives in symbiosis with a sacoglossan marine mollusk, *Elysia chlorotica* [118]. *Vaucheria* acts as food for the mollusk, but only the chloroplasts of *Vaucheria* are seized by the host. The term "kleptoplasty" (retaining of "stolen plastids") is used to define the plastid symbiosis.

1.5.5 Parasitic algae

Although parasitic algae are not common and diverse, there are a few algae parasitizing upon other organisms. Of all the parasitic algae encountered, filamentous green algae of the Order Trentepohliales are probably the most common and well-known [119]. These parasitic algae are usual aerial and often have a bright orange color due to the presence of carotenoid pigments. The genus *Cephaleuros* is well-known and well-researched as it is often responsible for severe plant damage [120]. It is found on the leaves and stems of tea, coffee, mango, guava, avocado, and citrus fruits where it causes bright orange spots. These spots reduce the esthetic appearance and marketability of the host plants. As the spots look similar to rust fungi, the spots are commonly referred to as "red rust". C. biolophus, C. minimus, C. parasiticus and C. pilosa are regarded as the most damaging species [121], growing intercellularly in the host tissues. The remaining nine species of the genus *Cephaleuros* grow between the epidermis and the cuticle, and they are responsible for limited to no damage to the host. *Stomatochroon*, another genus in the same order, was described as an obligate endophyte (parasite) by Chapman and Good [122]. *Stomatochroon* lives in the substomatal chambers of its host, where it may be responsible for incidental water soaking of the host tissues [119]. The genus *Phyllosiphon* thrives as an endophytic parasite in the leaves of various representatives of the Araceae, such as *Arisarum vulgarum* [123] and *Arum italicum* [124].

Red algae are sometimes found as parasites on other red algae [125]. Parasitic red algae mainly belong to the orders Ceramiales, Corallinales, Gigartinales, Gracilariales, Halymeniales,

Palmariales, Plocamiales, and Rhodymeniales [126]. Goff [125] designated red algal parasites, taxonomically closely related to their hosts, as "adelphoparasites," while the minority, which are more distantly related to their hosts, were designated as "alloparasites." Red algal parasites have a unique ability to transfer organelles (e.g., mitochondria, plastids, and nuclei) into their host cells—in this way they can control the host cells for their own benefits [127]. *Polysiphonia lanosa* is a small red alga, that grows as an epiphyte/parasite on *Ascophyllum* (a common cold-water seaweed) and occasionally also on *Fucus* [128]. It is considered by some to be parasitic because its rhizoids penetrate the host, but others are of the opinion that it only receives structural support and acts as an epiphyte [128]. An interesting fact is that some species of *Polysiphonia* act as hosts for other red algal parasites, for example, *P. caespitosa* serves as a host for *Aiolocolax pulchella*, a smaller red microalgal parasite [129]. Preuss et al. [129] also compiled a list of 120 parasitic red algal species and host species, on which phylogenetic analyses have been conducted.

Planktonic, as well as benthic forms of parasitic diatoms, can live inside Antarctic sponges. Diatoms enter the body of the sponges through pinacocytes [130], flat epithelial-like cells, forming the outermost layer of sponges. Drebes [131] estimated that approximately 7% of dinoflagellates have parasitic life strategies. According to Coats [132] they can infect other protists, cnidarians, crustaceans, and fish. Marine crustaceans are parasitized by two orders of dinoflagellates, namely, Blastodinida and Syndinida. *Blastodinium* species live as parasites in the intestines of marine, planktonic copepods [133]. *Amoebophrya* is a parasitic dinoflagellate that infects free-living dinoflagellates, amongst others species responsible for harmful algal blooms, such as *Akashiwo sanguinea*, which is responsible for red tides in the ocean [134]. Further studies may determine if the *Amoebophrya* can be used as a control against harmful algal blooms.

Zooparasitic algae (algae parasitic on animals) can be extremely problematic. Of particular concern are those that infect coral reefs worldwide. The cyanobacterium *Phormidium corallyticum* is a parasitic pathogen responsible for the so-called "black band disease", resulting in the death of coral reefs [135]. Protothecosis, a rare infection in humans, is caused by the heterotrophic, nonphotosynthetic green algae *Prototheca wickerhamii* and *P. zopfii* [136].

1.5.6 Algae living in extreme environments

Algae living in these environments are biologically adapted to extreme conditions and despite difficulties, they often flourish in these unique environments. Among these habitats are several places which are, from a human perspective, inhospitable such as extremely hot and cold environments (hot springs, deserts, alpine areas, polar regions, permafrost zones, ice, and snow), environments with extremely high salinities, and acidic or alkaline habitats. Other extreme environments that will not be discussed here include anoxic conditions, high and low light intensity environments, and habitats with extreme metal (e.g., iron) concentrations. Cyanobacteria and algae occurring in extreme environments are termed "**extremophiles**".

In the following paragraphs different types of extremophiles will be discussed, with special emphasis on hot spring habitats.

- ***Thermophilic and cryophilic (or psychrophilic) algae***—thermophiles are organisms preferring elevated temperatures and they often grow in hot environments, particularly hot springs that can emit water near to boiling point (90°C or higher [137]). Thermophilic cyanobacteria can flourish in water temperatures close to 75°C [113,137] whilst the highest temperature recorded for algae is in the upper 50°C [138]. Of the cyanobacteria, *Synechococcus* is probably the most thermophilic [137]. Other thermophilic cyanobacteria include *Oscillatoria terebriformis* (currently regarded as a *Phormidium terebriforme*), *Heterohormogonium schizodichotomum* (nowadays known as *Johannesbaptistia schizodichotoma*), *Synechococcus elongatus*, *Mastigocladus laminosus*, *Phormidium tenue* (the taxonomic name currently accepted *Leptolyngbya tenuis*), and *Schizothrix calcicola*. In Egypt, one of the most famous mineral hot springs, with a constant water temperature of c.70°C, is Hammam Faraun. It is located on the shore of the eastern side of the Gulf of Suez in the Red Sea. The springhead is located inside a limestone cave in a mountain close to the shore, and its hot water flows slowly through shallow rivulets towards the sea (Fig. 1.4A). Thick bluish—green mass growths of cyanoprokaryote assemblages (Fig. 1.4B) are common and physiologically adapted to this thermal environment, and *Geitlerinema sulphureum* (previously identified as *Oscillatoria geminata* f. *sulfurea*) is the dominant species [140]. Thermophilic algae belong to Chlorophyceae (Zygnematales, such as *Mougeotia*) and Bacillariophyceae. Recent techniques and approaches in microbiology (e.g., shotgun metagenomics, high-throughput culturing, comparative genomics) have allowed comprehensive and in-depth studies of the spring microbiome that were previously impossible. Hot (geothermal) springs, which represent discrete, relatively homogeneous, geographically widely distributed habitats covering ample geochemical gradients, are model ecosystems to study microbial biogeography. Hot springs are also highly promising sites for the discovery of organisms and compounds with useful biotechnological applications, for example, thermostable enzymes, that is, thermozymes [141], and extremophiles that can be used for bioremediation [142], etc. Hot springs were also found to be organic chemodiversity hotspots, that is, sites with a unique chemodiversity of dissolved organic matter [143].

At the other end of the thermometer are the **cryophiles** (also known as **psychrophiles**), growing in cold habitats, such as the Arctic, Antarctica, and the permafrost areas of Alaska, Russia, and Canada. They can tolerate extremely low temperatures, as long as the cytoplasm does not freeze inside the cells. In some cases, the cells produce compounds to provide an "antifreeze" effect. Most snow algae belong to the green algal genera *Chlamydomonas* and *Chloromonas* [144]. Depending on the type of algae and the dominant pigments inside the cells, the algae may be responsible for a discoloration of the snow or ice ranging from the very common orange/pink/red colors, to green, yellow, purple, and brownish to black colors. **Red snow** (termed "watermelon snow" or "blood snow") is the most common form of colored snow and it is most often caused by blooms of *Chlamydomonas nivalis* [145], but many other species, such as *C. sanguinea*, *Chloromonas nivalis*, *C. nivalis* subsp. *tatrae* (Fig. 1.4C—G; after Procházková et al. [139]), *C. brevispina*, *C. polyptera*, *C. rubroleosa*, *Chlorosarcina antarctica*, *Smithsonimonas abbotii*, *Chlainomonas kolii*, *C. rubra*, *Sanguina nivaloides*, and *S. aurantia* may also discolor snow in various shades of red [146]. **Green snow** is also

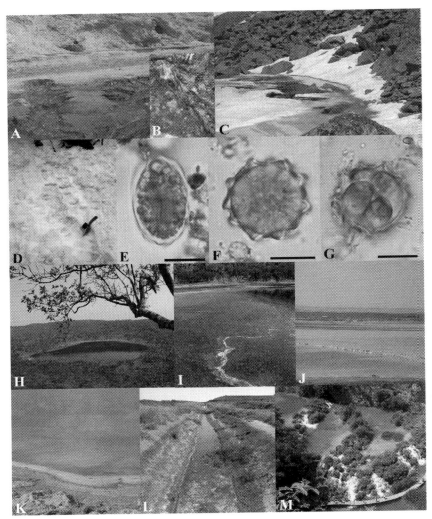

FIGURE 1.4 Habitats of cyanoprokaryotes and algae in extreme environments. (A) The helocrenic hot spring Hammam Faraun, Egypt. (B) Thick bluish-green scums of cyanoprokaryotes growing in the spring streams mainly dominated by *Geitlerinema sulphureum*. (C–G) Habitats and light micrographs of the red snow-coloring green alga *Chloromonas nivalis* subsp. *tatrae* [139]. The vegetative cell with reddish astaxanthin pigments and central greenish chloroplasts (E), aplanozygote in upright position (F), and four daughter cells with smooth walls in apical view surrounded by the mother cell wall (G). (H–I) Tswaing crater, a halophilic impoundment caused by the impact of a meteorite, located near Pretoria (South Africa), dominated by *Limnospira fusiformis*. (J–K) El Sabkha Lake, one of Wadi El Natrun soda lakes (Egypt), dominated by *Dunaliella salina*. (L) Acidophilic algae inhabiting walls of a canal transporting gold mine effluents, responsible for acid mine drainage, in Gauteng Province, South Africa. (M) Plitvice Lake (Croatia), a typical example of marl lakes rich in calcium carbonate. *Source: (A,B, J–K) A.A. Saber. (C–G) From L. Procházková, D. Remias, T. Řezanka, L. Nedbalová, Chloromonas nivalis subsp. tatrae, subsp. nov. (Chlamydomonadales, Chlorophyta): re–examination of a snow alga from the High Tatra Mountains (Slovakia), Fottea 18 (1) (2018) 1–18. (H–I) Brendan Hohls. (L) J. Taylor. (M) S. Janse van Vuuren.*

1.5 Global distribution and habitats of cyanoprokaryotes and algae 29

fairly common, especially in the arctic and antarctic regions and it is caused by *Chlamydomonas yellowstonensis*, as well as species of the genera *Koliella*, *Raphidonema* and *Carteria* [147]. **Yellow snow** was described by Remias et al. [148] in polar summer snowfields and it was caused by *Hydrurus*-related golden algae (Chrysophyta). Chrysophyta are, however, not common snow algae, although *Chromulina* was described from snow by Stein [149] and *Ochromonas smithii* and *O. itoi* by Tanabe et al. [150]. Kawecka [147] mentioned that yellow snow can also be caused as a result of blooms of *Cystococcus nivicola* and yellowish–green snow by *Chloromonas rostafinski*. **Blue snow** is often caused by mass developments of *Dactylococcopsis hungarica* (currently regarded as *Rhabdogloea hungarica*) [147]. **Purple–brown** snow is caused by *Ancylonema nordenskioeldii* [147]. Although the composition of **black snow** is not precisely known, it was proposed that it can be caused by *Scotiella nivalis* (an old synonym of *Chloromonas nivalis*) and *Raphidonema brevirostre* [67]. *Nostoc* and the desmid *Mesotaenium* were found to dominate on black ice by Kol and Peterson [151], and *Mesotaenium berggrenii* to cause greyish discolorations of glaciers [152].

- *Halophilic algae*—halophilic organisms can tolerate salt ranges that relatively few organisms have been able to adapt to, and can survive. The salt concentrations in these water bodies may exceed that of the ocean by five or more times. Halophilic cyanobacteria and other algae were identified from many coastal and other saline lakes, of which the most famous include the Dead Sea (Jordan, Israel), the Great Salt Lake (USA), the alkaline salt marshes and lakes in Wadi El Natrun (Egypt) (Fig. 1.4H–I), Hutt Lagoon (Australia), and the inland saltern of La Mala (Spain) [153,154].

Oren [155] presented a review on the occurrence and properties of halophilic cyanobacteria. *Aphanothece halophytica* is widespread and it is probably the best known amongst the unicellular, halophilic cyanobacteria. Common filamentous cyanobacteria include *Microcoleus chthonoplastes* (currently identified as *Coleofasciculus chthonoplastes*) that can survive in salinities of 200 g/L [156]. *Halospirulina tapeticola* is also a common cyanobacterial inhabitant of halophilic environments. Of the eukaroytes, *Dunaliella* (green microalga) has been described by Javor [156] as the best known, and possibly the most ubiquitous, eukaryotic microorganism in hypersaline environments. The most noticeably species of *Dunaliella* in halophilic habitats are *Dunaliella salina* (salt optimum at about 120 g/L) and *D. viridis* (optimum of 60 g/L NaCl, although it can tolerate salt concentrations upto about 30 g/L [157]). Litchfield [158] determined that *Dunaliella* grows under a broad salt range, ranging from 0%–30% NaCl. *Dunaliella parva* is also an important inhabitant of the Dead Sea [159], a halophilic environment so harsh that relatively few algae can survive here. *Dunaliella* cells produce orange–red β-carotenoid compounds, often leading to an orange to orange–brown to pink discoloration of waterbodies. The pink, lilac, or sometimes red discoloration of Hutt Lagoon by *D. salina* makes it a popular Australian tourist attraction.

Balakrishnan [160] observed that species diversity decreases sharply if the salinity of the water increases. According to Wehr and Sheath [1] mildly saline lakes (TDS 500–2000 mg/L) are inhabited by many algae, and a variety of diatoms (e.g., *Amphora*, *Campylodiscus*, *Cyclotella*, *Fragilaria*, and *Rhopalodia*) and green algae (*Crucigenia*, *Pediastrum*, *Oocystis*, and *Sphaerocystis*) can be found. With elevated nutrient concentrations cyanobacteria (*Anabaena*, *Aphanizomenon*,

Handbook of Algal Biofuels

30   1. Cyanoprokaryotes and algae: classification and habitats

Lyngbya, Microcystis, and *Oscillatoria*) are also common (Fig. 1.4K−L). Under strong saline conditions (2−20 g/L), many of these species are eliminated and may disappear. In hypersaline conditions (> 20−600 g/L) algal diversity is usually low, commonly favoring diatoms, such as *Anomoeoneis sphaerophora, Navicula subinflatoides,* and *Nitzschia frustulum,* cyanobacteria such as *Nodularia spumigena* and *Aphanothece halophytica,* and green algae such as *Ctenocladus circinatus* [161], and *Dunaliella* species (e.g., *D. salina, D. viridis, D. parva*).

- *Acidophilic and alkaliphilic algae*—many algae are able to grow at pH values as low as 4, but the majority of these can also grow at neutral or slightly alkaline pH levels—these organisms are regarded as **acidotolerant**. On the other hand, there are the **acidophilic** organisms, which are those growing at pH values below 3, and they are unable to live at neutral pH levels [162]. Some of these acidophilic algae are adapted to live at pH values as low as 0.05. Natural niches where acidophiles can be found are volcanic areas, hydrothermal sources, deep-sea vents, metal mining activities (Fig. 1.4L), acidic hot springs, abandoned coal mines affected by acid mine drainage, or in the stomachs of animals. According to Johnson [163] acidophilic algae include eukaryotic Chlorophyta (*Chlamydomonas acidophila* and *Dunaliella acidophila*), Chrysophyta (*Ochromonas* sp.), Rhodophyta (*Cyanidium caldarium* and *Galdieria sulphuraria*), Euglenophyta (*Euglena mutabilis*), and certain diatoms (e.g., *Eunotia* spp.). A filamentous alga, *Zygnema circumcarinatum,* has been found in abundance on surface streamer growths in an extremely acidic (pH ∼2.7) metal-rich stream draining a mine in southwest Spain [163]. More mildly acidic habitats (about 4 < pH < 6), such as mires, dystrophic lakes, and seepages, can host rich algal assemblages, that include, in particular, diverse diatom communities that can be extremely rich in Red List species belonging to threatened categories [164]. The diatom genera *Eunotia* and *Microfissurata* (a bryophilous taxon described from acidic pool- and seepage springs [165]) typically occur in these very low-alkalinity environments.

- *Alkaline habitats*—these are typical extreme environments which include naturally occurring soda lakes, marl lakes, desert soils, and artificially occurring industrial-derived waters. Microalgae growing optimally in excess of pH 8 are defined as **alkaliphiles**.
  - **Soda lakes** are alkaline lakes, typically with a pH above 9. They are usually dominated by carbonate salts (sodium carbonate and related salt complexes), giving rise to high levels of alkalinity (Fig. 1.4J−K). In addition, many soda lakes also contain high sodium chloride concentration together with dissolved salts, making them saline or hypersaline lakes as well. Due to their hypersalinity and high alkalinity soda lakes are considered some of the most extreme aquatic environments on Earth. The most common cyanobacterium in soda lakes is *Arthrospira fusiformis,* formerly known as *Spirulina platensis* (currently taxonomically accepted name: *Limnospira fusiformis*). This species is adapted to a wide range of habitats, ranging from freshwater−alkaline to saline−alkaline and even hypersaline environments [166]. Other common cyanobacteria in soda lakes include species of the genera *Anabaenopsis, Chroococcus, Cyanospira, Haloleptolyngbya, Synechococcus,* and *Synechocystis* [167]. Eukaryotic algae found by Schagerl et al. [167] in Kenyan soda lakes, and Afonina and Tashlykova [168] in soda lakes from Russia, include

Handbook of Algal Biofuels

representatives of Cryptophyta (*Cryptomonas*), Chlorophyta (*Ankyra ancora, Lemmermannia komarekii, Monoraphidium arcuatum, M. contortum, M. minutum, Oocystis borgei, O. submarina, Pseudoschroederia robusta*, and *Schroederia setigera*), Bacillariophyta (*Cyclotella, Cocconeis placentula, Nitzschia*, and *Navicula*) and Euglenophyta (*Euglena* spp. and *Phacus limnophilus*).

— **Marl lakes** (e.g., Plitvice Lake in Croatia; Fig. 1.4M) are clear, alkaline systems with deposits of $CaCO_3$ on sediments and plants, often rendering the sediments at the bottom white in color. Marl lakes have low or moderate phytoplankton productivity, due to the coprecipitation of phosphorus with carbonates. Littoral *Chara* growths, encrusted with marl, are extremely common in these lakes [169].

Besides the aforementioned ecosystems, a new cyanobacterium genus, *Chakia*, was recently described from an **alkaline marsh** [170]. Sometimes environments may be characterized by more than one extreme factor. Cyanobacteria and algae dwelling in those habitats are referred to as **polyextremophiles** [171]. Higher temperature may, for example, be accompanied by acidity—such habitats will be occupied by thermoacidophilic algae. Extremely hot or cold environments, such as deserts near polar regions, are often exposed to UV irradiation (home to xerophilic cyanobacteria and algae). Apart from the habitats discussed above there are ample other extreme environments (and combinations of extreme environments) in which cyanobacteria and algae may thrive. They can survive in hostile habitats under conditions seeming to be impossible for sustaining life. As long as water, sunlight, and nutrients are present, cyanobacteria and algae can adapt and make any place on Earth their home.

1.6 Conclusions and perspectives

Cyanoprokaryotes and eukaryotic algae have been undergoing extensive taxonomic revisions during the last two centuries. The advent of polyphasic taxonomy, resulting primarily from an integrative combination of molecular, morphological, and ecological traits, largely refined our understanding of the classification of cyanobacteria and algae, and also resolved taxonomic problems related to many cryptic entities. This development led to the discovery and establishment of many new species, genera, families, and orders. However, we expect further in-depth taxonomic revisions to be conducted in the coming decades, including the use of distinctly variable gene markers and more weighted taxonomic attributes, such as secondary metabolites, to greatly reduce confusing taxonomic issues. With respect to habitats and geographical distribution of cyanobacteria and algae, several extensive taxonomic studies have been conducted, primarily in developed countries, giving a good picture of the diversification of these microorganisms and their distributional patterns in different biotopes. However, many hidden species still await discovery by the curious and contemporary taxonomists, particularly in Africa, South America, and some Asian countries. Broadly speaking, discovering interesting and novel species, particularly in the aforementioned continents, will not only update our knowledge on the taxonomy of cyanobacteria and algae, but also enable us to ascertain ecological niches of geographically limited species.

References

[1] J.D. Wehr, R.G. Sheath, Habitats of freshwater algae, in: J.D. Wehr, R.G. Sheath, J.P. Kociolek (Eds.), Freshwater Algae of North America: Ecology and Classification, Academic Press, San Diego, California, 2015, pp. 13–74.

[2] P. Bourrelly, Les algues d'eau douce III, N. Boubée & Cie, Paris, 1970.

[3] J. Komárek, J. Čáslavská, Thylakoidal patterns in oscillatorialean genera, Arch. Hydrobiol. 64 (1991) 267–270.

[4] J. Komárek, J. Kaštovský, J. Mareš, J.R. Johansen, Taxonomic classification of cyanoprokaryotes (cyanobacterial genera) 2014, using a polyphasic approach, Preslia 86 (4) (2014) 295–335.

[5] J. Komárek, A polyphasic approach for the taxonomy of cyanobacteria: principles and applications, Eur. J. Phycol. 51 (2016) 346–353.

[6] J. Komárek, Quo vadis, taxonomy of cyanobacteria, Fottea 20 (1) (2019) 104–110. 2020.

[7] K. Fučíková, P.O. Lewis, S. Neupane, K.G. Karol, L.A. Lewis, Order, please! Uncertainty in the ordinal-level classification of Chlorophyceae, PeerJ 7 (2019) e6899. Available from: https://doi.org/10.7717/peerj.6899.

[8] R.E. Lee, Phycology, 4th (ed.), Cambridge University Press, New York, USA, 2008.

[9] L.E. Graham, J.M. Graham, L.W. Wilcox, Algae, second (ed.), Benjamin Cummings, San Francisco, 2008.

[10] R.G. Sheath, J.D. Wehr, Introduction to the freshwater algae, in: J.D. Wehr, R.G. Sheath, J.P. Kociolek (Eds.), Freshwater Algae of North America: Ecology and Classification, Academic Press, San Diego, California, 2015, pp. 1–11.

[11] D. Sahoo, P. Baweja, General characteristics of algae, in: D. Sahoo, J. Seckbach (Eds.), The Algae World, Springer, Dordrecht, 2015, pp. 3–29.

[12] G. Thuret, Essai de classification des Nostochinées, Ann. Sci. Nat. Bot. Sér. 6 (1) (1875) 372–382.

[13] A. Hansgirg, P. der Algenflora von Böhmen, Zweiter Theil welcher die blaugrünen Algae (Myxophyceen, Cyanophyceen), nebst Nachträgen zum ersten Theile und einer systmatischen Bearbeitung der in Böhmen verbreiten saprophytischen Bacterien und Euglenen enhält. Mit den opiz-preise Gekrönte Arbeit auf Kosten des Opizfondes, Arch. für die naturwissenschaftliche Landesdurchforsch. von. Böhmen 8 (4) (1892) 1–268.

[14] C. Nägeli, Gattungen einzelliger Algen, physiologisch und systematisch bearbeitet., Friedrich Schulthess, 1849.

[15] É. Bornet, C. Flahault, Tableau synoptique des Nostochacées filamenteuses hétérocystées, M. Soc. Natl. Sci. Nat. Math. Cherb. 25 (1885) 195–223.

[16] É. Bornet, C. Flahault, Revision des Nostocacées hétérocystées contenues dans les principaux herbiers de France, Ann. Sci. Nat. Bot. Septième Série 3 (1886) 323–381.

[17] É. Bornet, C. Flahault, Revision des Nostocacées hétérocystées contenues dans les principaux herbiers de France (deuxième fragment), Ann. Sci. Nat. Bot. Septième Série 4 (1886) 343–373.

[18] É. Bornet, C. Flahault, Revision des Nostocacées hétérocystées contenues dans les principaux herbiers de France (Troisième fragment), Ann. Sci. Nat. Bot. Septième Série 5 (1886) 51–129.

[19] É. Bornet, C. Flahault, Revision des Nostocacées hétérocystées contenues dans les principaux herbiers de France (quatrième et dernier fragment), Ann. Sci. Nat. Bot. Septième Série 7 (1886) 177–262.

[20] É. Bornet, C. Flahault, Note sur deux nouveaux genres d'algues perforantes, J. Bot. (Morot) 2 (1888) 161–165.

[21] T.V. Desikachary, Cyanophyta. I.C.A.R. Monographs on Algae, New Delhi, 1959.

[22] L. Geitler, Cyanophyceae, in: A. Pascher (Ed.), Süswasserflora 12, Gustav Fischer, Verl, Jena, 1925, p. 481.

[23] L. Geitler, Cyanophyceae, Rabenhorst's Kryptogamen Flora von Deutschland, Vol. 14, Akad. Verlagsges, Leipzig, 1932, p. 1196.

[24] L. Geitler, S. Klasse Schizophyceae, in: A. Engler, K. Prantl (Eds.), Natürliche Pflanzenfamilien 1b, Duncker & Humblot, Berlin, 1942, pp. 1–232.

[25] F.E. Fritsch, The interrelations and classification of the Myxophyceae (Cyanophyceae), N. Phytol. 41 (2) (1942) 134–148.

[26] F.E. Fritsch, Present-day classification of algae, Bot. Rev. 10 (1944) 233–277.

[27] F.E. Fritsch, The Structure and Reproduction of the Algae. Volume II. Foreword, Phaeophyceae, Rhodophyceae, Myxophyceae, Cambridge University Press, Cambridge, 1945, pp. 1–939.

[28] W.A. Setchell, N.L. Gardner, The marine algae of the Pacific coast of North America. Part I. Myxophyceae, Univ. Calif. Publ. Bot. 8 (1) (1919) 1–138.

[29] A.A. Elenkin, Monografia algarum cyanophycearum aquidulcium at terrestrium in finibus URSS inventarum [Blue-green algae of the USSR]. Izdat. Akad. Nauk SSSR, 1936–1949.

References

[30] A. Borzí, Studi algologici: saggio di richerche sulla biologia delle alghe. Fasc. II, Alberto Reber Libreria Carlo Clausen, Palermo, 1895, pp. 121–378.

[31] A. Borzì, Studi sulle Mixoficee. I. Cenni generali - Systema Myxophycearum, Nuovo G. Botanico Italiano Ser. 2 (21) (1914) 307–360.

[32] A. Borzì, Studi sulle Mixoficee. II. Stigonemaceae, Nuovo Giorn. Bot. Ital. Ser. 2 (23) (1916) 559–588.

[33] A. Borzì, Studi sulle Mixoficee. II (continuaz.), Nuovo Giorn. Bot. Ital. Nuova Ser. 24 (1917) 17–25.

[34] P. Frémy, Les Nostocacées de la Normandie, Not. Mem. Doc. Soc. Agric. Archéol. Hist. nat. Manche 41 (1929) 197–228.

[35] P. Frémy, Cyanophycées des côtes d'Europe, M. Soc. Nat. Sci. Nat. Math. Cherbg. 41 (1934) 1–235.

[36] G.W. Prescott, Algae of the Western Great Lakes Area, Wm. C. Brown, Dubuque, Iowa, 1962.

[37] R. Rippka, J. Deruelles, J.B. Waterbury, M.R. Herdman, R.Y. Stanier, Generic assignments, strain histories and properties of pure cultures of Cyanobacteria, J. Gen. Microbiol. 111 (1979) 1–61.

[38] R.W. Castenholz, Phylum BX. Cyanobacteria, in: D.R. Boone, R.W. Castenholz (Eds.), Bergey's Manual of Systematic Bacteriology, Springer, New York, 2001, pp. 473–599.

[39] K. Anagnostidis, J. Komárek, Modern approach to the classification system of the cyanophytes 1: introduction, Algol. Stud. 38/39 (1985) 291–302.

[40] J. Komárek, Several problems of the polyphasic approach in the modern cyanobacterial system, Hydrobiologia 811 (2018) 7–17.

[41] J. Komárek, K. Anagnostidis, Cyanoprokaryota 1. Chroococcales, in: H. Ettl, G. Gärtner, H. Heynig, D. Mollenhauer (Eds.), Süsswasserflora von Mitteleuropa 19/1, Gustav Fischer, Jena-Stuttgart-Lübeck-Ulm, 1998, p. 548.

[42] J. Komárek, K. Anagnostidis, Cyanoprokaryota. 2. Oscillatoriales, in: B. Büdel, L. Krienitz, G. Gärtner, M. Schagerl (Eds.), Süsswasserflora von Mitteleuropa 19/2, Elsevier/Spektrum, Heidelberg, 2005, p. 759.

[43] J. Komárek, Cyanoprokaryota. 3. Heterocytous genera, in: B. Büdel, G. Gärtner, L. Krienitz, M. Schagerl (Eds.), Süswasserflora von Mitteleuropa/Freshwater flora of Central Europe, Springer Spektrum Berlin, Heidelberg, 2013, p. 1130.

[44] L. Hoffmann, J. Komárek, J. Kaštovský, System of cyanoprokaryotes (cyanobacteria) - state in 2004, Algol. Stud. 117 (2005) 95–115.

[45] L. Hoffmann, J. Komárek, J. Kaštovský, Proposal of cyanobacterial system-2004, in: B. Büdel, L. Krienitz, G. Gärtner, M. Schagerl (Eds.), Süsswasserflora von Mitteleuropa 19/2, Elsevier/Spektrum, Heidelberg, 2005, pp. 657–660.

[46] T. Mai, J.R. Johansen, N. Pietrasiak, M. Bohunická, M.P. Martin, Revision of the Synechococcales (Cyanobacteria) through recognition of four families including Oculatellaceae fam. nov. and Trichocoleaceae fam. nov. and six new genera containing 14 species, Phytotaxa 365 (1) (2018) 1–59.

[47] J. Rigonato, W.A. Gama, D.A. Alvarenga, L.H.Z. Branco, F.P. Brandini, D.B. Genuário, et al., *Aliterella atlantica* gen. nov., sp. nov., and *Aliterella antarctica* sp. nov., novel members of coccoid Cyanobacteria, Int. J. Syst. Evol. Microbiol. 66 (2016) 2853–2861.

[48] P. Hašler, D. Casamatta, P. Dvořák, A. Poulíčková, *Jacksonvillea apiculata* (Oscillatoriales, Cyanobacteria) gen. & sp. nov.: a new genus of filamentous, epipsamic cyanobacteria from North Florida, Phycologia 56 (3) (2017) 284–295.

[49] A. Saraf, A. Suradkar, H.G. Dawda, L.A. Gaysina, Y. Gabidullin, A. Kumat, et al., Phylogenetic complexities of the members of Rivulariaceae with the re-creation of the family Calotrichaceae and description of *Dulcicalothrix necridiiformans* gen nov, sp nov., and reclassification of *Calothrix desertica*, FEMS Microbiol. Lett. 366 (19) (2019) fnz219.

[50] S. Shalygin, R. Shalygina, J.R. Johansen, N. Pietrasiak, E. Berrendero Gómez, M. Bohunická, et al., *Cyanomargarita* gen. nov. (Nostocales, Cyanobacteria): convergent evolution resulting in a cryptic genus, J. Phycol. 53 (2017) 762–777.

[51] G.S. Hentschke, J.R. Johansen, N. Pietrasiak, M.F. Fiore, J. Rigonato, C.L. Sant'Anna, et al., Phylogenetic placement of *Dapisostemon* gen. nov. and *Streptostemon*, two tropical heterocytous genera (Cyanobacteria), Phytotaxa 245 (2) (2016) 129–143.

[52] C.H. Kilgore, J.R. Johansen, T. Mai, T. Hauer, D.A. Casamatta, C.A. Sheil, Molecular characterization of *Geitleria appalachiana* sp. nov. (Nostocales, Cyanobacteria) and formation of Geitleriaceae fam. nov, Fottea 18 (2) (2018) 150–163.

[53] B.C. Sendall, G.B. McGregor, Cryptic diversity within the Scytonema complex: characterization of the paralytic shellfish toxin producer *Heterosyctonema crispum*, and the establishment of the family Heteroscytonemataceae (Cyanobacteria/Nostocales), Harmful Algae 80 (2018) 158−170.

[54] J.M. Walter, F.H. Coutinho, B.E. Dutilh, J. Swings, F.L. Thompson, C.C. Thompson, Ecogenomics and taxonomy of cyanobacteria phylum, Front. Microbiol. 8 (2017) 2132. Available from: https://doi.org/10.3389/fmicb.2017.02132.

[55] T.T. Bringloe, S. Starko, R.M. Wade, C. Vieira, H. Kawai, O. De Clerck, et al., Phylogeny and evolution of the brown algae, Crit. Rev. Plant. Sci. 39 (4) (2020) 281−321.

[56] H.S. Yoon, G.C. Zuccarello, D. Bhattacharya, Evolutionary history and taxonomy of red algae, in: D.J. Seckbach, Chapman (Eds.), Cellular Origin, Life in Extreme Habitats and Astrobiology, Springer, New York, 2014, pp. 25−42.

[57] C. Linnaeus, Species Plantarum, exhibentes Plantas rite Cognitas, ad Genera Relatas, cum Differentiis Specificis, Nominibus Trivialibus, Synonymis Selectis, Locis Natalibus, secundum Systema Sexuale digestas. Salvius, Stockholm, 1753.

[58] W.H. Harvey, Algae, in: J.T. Mackay (Ed.), Flora Hibenica, Comprising the Flowering Plants, Ferns, Characeae, Musci, Hepaticae, Lichenes, and Algae of Ireland Arranged according to the Natural System with a Synopsis of the Genera according to the Linnaean System, Volume 2, Curray, Dublin, 1836, pp. 157−256.

[59] A.W. Eichler, Syllabus der Vorlesungen über Phanerogamenkunde, Harvard University, Library of the Gray Herbarium, 1883, p. 36.

[60] A. Engler, K. Prantl, Die Naturlichen Pfl anzenfamilien, 23 Vol, W. Engelmann, Leipzig, 1887−1915.

[61] G.S. West, Algae, Vol I, in Cambridge Botanical handbooks, Cambridge University Press, London, 1916, p. 475.

[62] A. Pascher, Eine neue braune Fadenalge des Süßwassers Chrysophycee, 1931.

[63] J.E. Tilden, A classification of the algae based on evolutionary development, with special reference to pigmentation, Bot. Gaz. 95 (1) (1933) 59−77.

[64] F.E. Fritsch, The structure and reproduction of algae, Vol I, Cambridge University Press, Cambridge, UK, 1935.

[65] G.M. Smith, Cryptogamic Botany: Algae and Fungi, McGraw-Hill Company, 1955.

[66] G. Papenfuss, Classification of the Algae: A Century of Progress in the Natural Sciences, 1853−1953, California Academy of Sciences, San Francisco, USA, 1955, pp. 115−224.

[67] V.J. Chapman, The Algae, Macmillan Publishers, London, 1962, p. 492.

[68] G.W. Prescott, The Algae: A Review, Thomas Nelson and Sons, London, 1969.

[69] F.E. Round, The Biology of the Algae, Edward Arnold Publishers, London, 1973, p. 585.

[70] H.C. Bold, M.J. Wynne, Introduction to Algae: Structure and Reproduction, Prentice Hall of India, New Delhi, 1978, p. 720.

[71] S.P. Parker, Synapsis and Classification of Living Organisms, Vol 1 & 2, McGraw Hill, New York, 1982, p. 2424.

[72] J.A. Ress, The ecology of aerial algae, Ph.D. dissertation, Biological Sciences, Bowling Green State University, USA, 2012, pp. 163.

[73] M.V.N. Panikkar, A. Jackson, Biodiversity of subaerial algae, in: Proceedings of the National Seminar on 'The Taxonomy and Ecology of Freshwater Algae', St. Albert's College, Ernakulam, India, February 2014.

[74] J.F. Gerrath, Conjugating green algae and desmids, in: J.D. Wehr, R.G. Sheath (Eds.), Freshwater Algae of North America: Ecology and Classification, Academic Press, San Diego, California, 2003, pp. 353−382.

[75] J.B. Petersen, Studies on the biology and taxonomy of soil algae, Dansk. Bot. Arkiv. 8 (1935) 1−180.

[76] G.G. Ehrenberg, Bericht üeber die zur Bekanntmachung geeigneten Verhandunger der Konigl Preuss, Akad. Wiss. Berl. 9 (1844) 194−207.

[77] S. Genitsaris, K. Kormas, M. Moustaka-Gouni, Airborne algae and cyanobacteria: occurrence and related health effects, Front. Biosci. (Elite Ed.) 3 (2) (2011) 772−787.

[78] A.A. Saber, Algae of El-Farafra Oasis, Ph.D. Thesis, Faculty of Science, Ain Shams University, Cairo, Egypt, 2016, p. 369.

[79] M. Suutari, M. Majaneva, D.P. Fewer, B. Voirin, A. Aiello, T. Friedl, et al., Molecular evidence for a diverse green algal community growing in the hair of sloths and a specific association with *Trichophilus welckeri* (Chlorophyta, Ulvophyceae), BMC Evol. Biol. 10 (1) (2010) 86−97.

References

[80] D.E. Wujek, P. Timpano, *Rufusia* (Porphyridiales, Phragmonemataceae), a new red alga from sloth hair, Brenesia 25/26 (1986) 163–168.

[81] D.E. Wujek, T.A. Lincoln, Ultrastructure and taxonomy of *Oscillatoria pilicola*, a new species of bluegreen alga from sloth hair, Brenesia 29 (1988) 1–6.

[82] B. Metting, The systematics and ecology of soil algae, Bot. Rev. 47 (2) (1981) 195–312.

[83] A. Pentecost, B.A. Whitton, Subaerial cyanobacteria, in: B.A. Whitton (Ed.), Ecology of Cyanobacteria II. Their Diversity in Time and Space, Springer Science + Business Media B.V., New York, 2012, pp. 291–316.

[84] A.A. Saber, M. Cantonati, J. Mareš, A. Anesi, G. Guella, Polyphasic characterization of *Westiellopsis prolifica* (Cyanobacteria) from the El-Farafra Oasis (Western Desert, Egypt), Phycologia 56 (6) (2017) 697–709.

[85] A.A. Saber, K. Fučíková, H.A. McManus, G. Guella, M. Cantonati, Novel green algal isolates from the Egyptian hyper-arid desert oases: a polyphasic approach with a description of *Pharao desertorum* gen. et sp. nov. (Chlorophyceae, Chlorophyta), J. Phycol. 54 (3) (2018) 342–357.

[86] L.H. Tiffany, Ecology of freshwater algae, in: G.M. Smith (Ed.), Manual of Phycology, Chronica Botanica Company, Waltham, Massachusetts, 1951, pp. 293–311.

[87] R. Lynn, T.D. Brock, Notes on the ecology of a species of *Zygogonium* (Kutz.) in Yellowstone National Park, J. Phycol. 5 (1969) 181–185.

[88] A. Holzinger, A. Tschaikner, D. Remias, Cytoarchitecture of the desiccation-tolerant green alga *Zygogonium ericetorum*, Protoplasma 243 (2010) 15–24.

[89] E.I. Friedmann, Y. Lipkin, R. Ocampo-Paus, Desert algae of the Negev (Israel), Phycologia 7 (1967) 185–200.

[90] H. Drummond, Distribution and biomass of epiphytic seaweeds on the kelp *Ecklonia maxima* (Osbeck) Papenfuss, and the potential effects of two kelp-harvesting methods in the Western Cape, Honors dissertation, University of Cape Town, South Africa, 2001, pp. 30.

[91] H. Stegenga, J.J. Bolton, R.J. Anderson, Seaweeds of the South African West Coast, Contri. Bolus Herb. 18 (1997) 640.

[92] F.M. Hassan, M.M. Salah, J.M. Salman, Quantitative and qualitative variability of epiphytic algae on three aquatic plants in Euphrates River, Iraq. Iraq. J. Aqua. 1 (2007) 1–16.

[93] L. Kanjer, R. Majewska, B. Van de Vijver, R. Gračan, B. Lazar, S. Bosak, Diatom diversity on the skin of frozen historic loggerhead sea turtle specimens, Diversity 12 (2020) 383. Available from: https://doi.org/10.3390/d12100383.

[94] R.S. Holland, G.L. Hergenrader, *Rhopalosolen saccatus* Fott: an epizoophyte from an alkaline lake in Nebraska, Am. Midl. Nat. 106 (2) (1981) 403–405.

[95] A.K. Awasthi, Textbook of Algae, Vikas Publishing House Pvt Ltd, New Delhi, India, 2015, p. 407.

[96] M.O.P. Iyengar, M.O.T. Iyengar, On a *Characium* growing on *Anopheles* larvae, N. Phytol. 31 (1) (1932) 66–69.

[97] M. Kahlert, A.T. Hasselrot, H. Hillebrand, K. Pettersson, Spatial and temporal variation in the biomass and nutrient status of epilithic algae in Lake Erken, Sweden, Freshw. Biol. 47 (2002) 1191–1215.

[98] M.D. Oliveira, D.F. Calheiros, Flood pulse influence on phytoplankton communities of the south Pantanal floodplain Brazil, Hydrobiologia 427 (2000) 101–112.

[99] S. Janse van Vuuren, A. Levanets, Mass developments of *Euglena sanguinea* Ehrenberg in South Africa, Afr, J. Aquat. Sci. 46 (1) (2021). Available from: https://doi.org/10.2989/16085914.2020.1799743.

[100] M. Cantonati, S. Scola, N. Angeli, G. Guella, R. Frassanito, Environmental controls of epilithic diatom depth-distribution in an oligotrophic lake characterised by marked water-level fluctuations, Eur. J. Phycol. 44 (2009) 15–29.

[101] M. Cantonati, G. Guella, J. Komárek, D. Spitale, Depth-distribution of epilithic cyanobacteria and pigments in a mountain lake characterized by marked water-level fluctuations, Freshw. Sci. 33 (2014) 537–547.

[102] E. Kann, F. Sauer, Die "Rotbunte Tiefenbiocönose" (Neue Beobachtungen in österreichischen Seen und eine zusammenfassende Darstellung), Arch. Hydrobiol. 95 (1982) 181–195.

[103] M. Cantonati, R.J. Fensham, L.E. Stevens, R. Gerecke, D.S. Glazier, N. Goldscheider, et al., Urgent plea for global protection of springs, Biol. Conserv. (2020). Available from: https://doi.org/10.1111/cobi.13576.

[104] D.S. Glazier, Springs, in: S.A. Elias (Ed.), Reference Module in Earth Systems and Environmental Sciences, Elsevier, Waltham MA, USA, 2014, pp. 1–78.

[105] M. Cantonati, L.E. Stevens, S. Segadelli, A.E. Springer, N. Goldscheider, F. Celico, et al., Ecohydrogeology: the interdisciplinary convergence needed to improve the study and stewardship of springs and other groundwater-dependent habitats, biota, and ecosystems, Ecol. Indic. 110 (2020). Available from: https://doi.org/10.1016/j.ecolind.2019.105803.

[106] M. Cantonati, S. Segadelli, S. Spitale, J. Gabrieli, R. Gerecke, N. Angeli, et al., Geological and hydrochemical prerequisites of unexpectedly high biodiversity in spring ecosystems at the landscape level, Sci. Total. Environ. 740 (2020) 140−157.

[107] M. Cantonati, E. Rott, D. Spitale, N. Angeli, J. Komárek, Are benthic algae related to spring types? Freshw. Sci. 31 (2012) 481−498.

[108] M. Cantonati, N. Angeli, E. Bertuzzi, D. Spitale, H. Lange-Bertalot, Diatoms in springs of the Alps: spring types, environmental determinants, and substratum, Freshw. Sci. 31 (2012) 499−524.

[109] M. Cantonati, S. Segadelli, K. Ogata, H. Tran, D. Sanders, R. Gerecke, et al., A global review on ambient limestone-precipitating springs (LPS): hydrogeological setting, ecology, and conservation, Sci. Total. Environ. 568 (2016) 624−637.

[110] B.P. Pandey, Botany for Degree Students, Biodiversity, S Chand and Company Limited, Ram Nagar, New Delhi, 2017, p. 673.

[111] A.A. Saber, M. Cantonati, M.L. Vis, A. Anesi, G. Guella, Multifaceted characterization of a *Lemanea fluviatilis* population (Batrachospermales, Rhodophyta) from a glacial stream in the south−eastern Alps, Fottea 16 (2) (2016) 234−243.

[112] P. Bubrick, M. Galun, A. Frensdorff, Observations on free-living *Trebouxia* De Puymaly and *Pseudotrebouxia* Archibald, and evidence that both symbionts from *Xanthoria parietina* (L.) Th. Fr. can be found free-living in nature, N. Phytol. 97 (3) (2006) 455−462.

[113] P.C. Trivedi, S. Pandey, S. Bhadauria, Text Book of Microbiology, Aavishkar Publishers, Jaipur, India, 2010, p. 446.

[114] D.L. Hawksworth, Freshwater and marine lichen-forming fungi, in: K.D. Hyde, W.H. Ho, S.B. Pointing (Eds.), Aquatic Mycology across the millenium, Fungal Divers. 5 (2000) 1−7.

[115] C. Kim, Y.S. Kim, Effects of temperature and irradiance on the growth and infection of three endophytic green algae, Korean J. Fish. Aquat. Sci. 48 (1) (2015) 88−95.

[116] S. Woelfl, P. Garcia, C. Duarte, *Chlorella*-bearing ciliates (*Stentor*, *Ophrydium*) dominate in an oligotrophic, deep North Patagonian lake (Lake Caburgua, Chile), Limnologic.40 (2010) 134−139.

[117] R. Kerney, Symbiosis between salamander embryos and green algae, Symbiosis 54 (3) (2011) 107−119.

[118] B.J. Green, W.Y. Li, J.R. Manhart, T.C. Fox, E.J. Summer, R.A. Kennedy, et al., Mollusc-algal chloroplast endosymbiosis. Photosynthesis, thylakoid protein maintenance, and chloroplast gene expression continue for many months in the absence of the algal nucleus, Plant. Physiol. 124 (2000) 331−342.

[119] F. Brooks, Plant-parasitic algae (Chlorophyta: Trentepohliales) in American Samoa, Pac. Sci. 58 (3) (2004) 419−428.

[120] R.B. Marlatt, S.A. Alfieri, Hosts of a parasitic alga, *Cephaleuros Kunze*, in Florida, Plant. Dis. 65 (1981) 520−522.

[121] R.H. Thompson, D.E. Wujek, Trentepohliales: *Cephaleuros*, *Phycopeltis*, and *Stomatochroon*: Morphology, Taxonomy, and Ecology, Science Publishers Inc, Enfield, New Hampshire, USA, 1997, p. 149.

[122] R.L. Chapman, B.H. Good, Subaerial symbiotic green algae: interactions with vascular plant hosts, in: L.J. Goff (Ed.), Algal Symbiosis: A Continuum of Interaction Strategies, Cambridge University Press, Cambridge, 1983, pp. 173−203.

[123] M. Aboal, O. Werner, Morphology, fine structure, life cycle and phylogenetic analysis of *Phyllosiphon arisari*, a siphonous parasitic green alga, Eur. J. Phycol. 46 (3) (2011) 181−192.

[124] K. Procházková, Y. Nemcova, J. Neustupa, *Phyllosiphonari* sp. nov. (Watanabea clade, Trebouxiophyceae), a new parasitic species isolated from leaves of *Arum italicum* (Araceae), Phytotaxa 283 (2) (2016) 143−154.

[125] L.J. Goff, The biology of Parasitic red algae, in: F.E. Round, D.J. Chapman (Eds.), Progress in Phycological Research, Elsevier Biomedical Press, Amsterdam, 1982, pp. 289−369.

[126] E.D. Salomaki, C.E. Lane, Are all red algal parasites cut from the same cloth? Acta Soc. Bot. Pol. 83 (2014) 369−375.

[127] E.D. Salomaki, K.R. Nickles, C.E. Lane, The ghost plastid of *Choreocolax polysiphoniae*, J. Phycol. 51 (2015) 217−221.

[128] D.J. Rawlence, A.R.A. Taylor, The rhizoids of *Polysiphonia lanosa*, Can. J. Bot. 48 (3) (2011) 607−611.

[129] M. Preuss, W.A. Nelson, G.C. Zuccarello, Red algal parasites: a synopsis of described species, their hosts, distinguishing characters and areas for continued research, Bot. Mar. 60 (1) (2017) 13−25.

[130] E. Gaino, G. Bavestrello, R. Cattaneo-Vietti, M. Sara, Scanning electron microscope evidence for diatom uptake by two Antarctic sponges, Polar Biol. 14 (1994) 1−4. Available from: https://doi.org/10.1007/BF00240273.

References

[131] G. Drebes, Life cycle and host specificity of marine parasitic dinophytes, Helgol. Meeresunters. 37 (1984) 603–622.

[132] D.W. Coats, Parasitic life styles of marine dinoflagellates, J. Eukaryot. Microbiol. 46 (4) (1999) 402–409.

[133] A. Skovgaard, S.A. Karpov, L. Guillou, The parasitic dinoflagellates *Blastodinium* spp. inhabiting the gut of marine, planktonic copepods: morphology, ecology, and unrecognized species diversity, Front. Microbiol. 3 (2012) 305. Available from: https://doi.org/10.3389/fmicb.2012.00305.

[134] M. Johansson, D.W. Coats, Ciliate grazing on the parasite *Amoebophrya* sp. decreases infection of the red-tide dinoflagellate *Akashiwo sanguinea*, Aquat. Microb. Ecol. 28 (1) (2002) 69–78.

[135] L.L. Richardson, K.G. Kuta, Ecological physiology of the black band disease cyanobacterium *Phormidium corallyticum*, FEMS Microbiol. Ecol. 43 (2003) 287–298.

[136] C. Lass-FlörI, A. Mayr, Human Protothecosis, Clin. Microbiol. Rev. 20 (2007) 230–242.

[137] J. Seckbach, Algae and Cyanobacteria in Extreme Environments, Springer, Dordrecht, The Netherlands, 2007, p. 786.

[138] T.D. Brock, Life at high temperatures, Science 58 (3804) (1967) 1012–1019.

[139] L. Procházková, D. Remias, T. Řezanka, L. Nedbalová, *Chloromonas nivalis* subsp. *tatrae*, subsp. nov. (Chlamydomonadales, Chlorophyta): re–examination of a snow alga from the High Tatra Mountains (Slovakia), Fottea 18 (1) (2018) 1–18.

[140] A.F. Hamed, Biodiversity and distribution of blue-green algae/cyanobacteria and diatoms in some of the Egyptian water habitats in relation to conductivity, Aust. J. Basic. Appl. Sci. 2 (1) (2008) 1–21.

[141] G. Kaushal, J. Kumar, R.S. Sangwan, S.P. Singh, Metagenomic analysis of geothermal water reservoir sites exploring carbohydrate-related thermozymes, Int. J. Biol. Macromol. 119 (2018) 882–895.

[142] R. Orellana, C. Macaya, G. Bravo, F. Dorochesi, A. Cumsille, R. Valencia, et al., Living at the frontiers of life: extremophiles in chile and their potential for bioremediation, Front. Microbiol. 9 (2018) 2309. Available from: https://doi.org/10.3389/fmicb.2018.02309.

[143] M. Gonsior, N. Hertkorn, N. Hinman, S.E.M. Dvorski, M. Harir, W.J. Cooper, et al., Yellowstone hot springs are organic chemodiversity hot spots, Sci. Rep. 8 (2018) 14155. Available from: https://doi.org/10.1038/s41598-018-32593-x.

[144] D. Remias, U. Lutz-Meindl, C. Lutz, Photosynthesis, pigments and ultrastructure of the alpine snow alga *Chlamydomonas nivalis*, Eur. J. Phycol. 40 (3) (2005) 259–268.

[145] R. Gradinger, D. Nürnberg, Snow algal communities on arctic pack ice floes dominated by *Chlamydomonas nivalis* (Bauer) Wille. Proc. NIPR Symp, Polar Biol. 9 (1996) 35–43.

[146] K. Procházková, T. Leya, H. Křížková, L. Nedbalová, *Sanguina nivaloides* and *Sanguina aurantiagen* et spp. nov. (Chlorophyta): The taxonomy, phylogeny, biogeography and ecology of two newly recognized algae causing red and orange snow, FEMS Microbiol. Ecol. 95 (2019) fiz064. Available from: https://doi.org/10.1093/femsec/fiz064.

[147] B. Kawecka, Ecology of snow algae, Pol. Polar Res. 7 (4) (1986) 407–415.

[148] D. Remias, S. Jost, J. Boenigk, J. Wastian, C. Lütz, Hydrurus-related golden algae (chrysophyceae) cause yellow snow in polar summer snowfields, Phycol. Res. 61 (2013) 277–285.

[149] J.R. Stein, A *Chromulina* (Chrysophyceae) from snow, Can. J. Bot. 41 (1963) 1367–1370.

[150] Y. Tanabe, T. Shitara, Y. Kashino, Y. Hara, S. Kudoh, Utilizing the effective xanthophyll cycle for blooming of *Ochromonas smithii* and *O. itoi* (Chrysophyceae) on the snow surface, PLoS One 6 (2) (2011) e14690. Available from: https://doi.org/10.1371/journal.pone.0014690.

[151] E. Kol, J.A. Peterson, Cryobiology, in: G.S. Hope, J.A. Peterson, I. Allison, U. Radok (Eds.), The Equatorial Glaciers of New Guinea, A.A. Balkema, Rotterdam, The Netherlands, 1976, pp. 81–91.

[152] D. Remias, A. Holzinger, C. Lütz, Physiology, ultrastructure and habitat of the ice alga *Mesotaenium berggrenii* (Zygnemaphyceae, Chlorophyta) from glaciers in the European Alps, Phycologia 48 (4) (2009) 302–312.

[153] Z. Mianping, Discovery and study of halobacteria and halophilic algae in the ZabuyeCaka, in: Z. Mianping (Ed.), An Introduction to Saline Lakes on the Qinghai-Tibet Plateau, Monographiae Biologicae, Kluwer Academic Publishers, Dordrecht, The Netherlands, 1997, pp. 181–199.

[154] I. Ali, S. Prasongsuk, A. Akbar, M. Aslam, P. Lotrakul, H. Punnapayak, et al., Hypersaline habitats and halophilic microorganisms, Maejo. Int. J. Sci. Technol. 10 (3) (2016) 330–345.

[155] A. Oren, Salts and brines, in: B.A. Whitton, M. Potts (Eds.), Ecology of Cyanobacteria: Their Diversity in Time and Space, Kluwer Academic Press, Dordrecht, The Netherlands, 2000, pp. 283–306.

[156] B. Javor, *Dunaliella* and other halophilic, eucaryotic algae, in: B. Javor (Ed.), Hypersaline Environments, Brock/Springer Series in Contemporary Bioscience, Springer, Berlin, Heidelberg, 1989, pp. 147—158.

[157] L.J. Borowitzka, The microflora, Adapt. life extremely saline lakes, Hydrobiologia 81 (1981) 33—46.

[158] C.D. Litchfield, Survival strategies for microorganisms in hypersaline environments and their relevance to life on early Mars, Meteorit. Planet. Sci. 33 (4) (1998) 813—819.

[159] A. Oren, D. Ionescu, M. Hindiyey, H. Malkawi, Microalgae and cyanobacteria of the Dead Sea and its surrounding springs, Isr. J. Plant. Sci. 56 (2008) 1—13.

[160] S. Balakrishnan, P. Santhanam, S. Jeyanthi, M. Divya, M. Srinivasan, Preliminary screening of halophilic microalgae collected from different salt pans of Tuticorin, southeast coast of India, Proc. Zool. Soc. 72 (1) (2019) 90—96.

[161] J.E. Reuter, C.L. Rhodes, M.E. Lebo, M. Klotzman, C.R. Goldman, The importance of nitrogen in Pyramid Lake (Nevada, USA), a saline, desert lake, Hydrobiologia 267 (1993) 179—189.

[162] H. Gimmler, Acidophilic and acidotolerant algae, in: L.C. Rai, J.P. Gaur (Eds.), Algal Adaptation to Environmental Stresses, Springer, Berlin, Heidelberg, 2001, pp. 259—290.

[163] D.B. Johnson, Extremophiles: acidic environments, in: M. Schaechter (Ed.), Encyclopedia of Microbiology (third ed.), Elsevier, Oxford, United Kingdom, 2009, pp. 107—126.

[164] M. Cantonati, H. Lange-Bertalot, F. Decet, J. Gabrieli, Diatoms in very-shallow pools of the site of community importance Danta di Cadore Mires (south-eastern Alps), and the potential contribution of these habitats to diatom biodiversity conservation, Nova Hedwig. 93 (2011) 475—507.

[165] M. Cantonati, B. Van de Vijver, H. Lange-Bertalot, *Microfissurata* gen. nov. (Bacillariophyta), a new diatom genus from dystrophic and intermittently-wet terrestrial habitats, J. Phycol. 45 (2009) 732—741.

[166] P. Dadheech, A. Ballot, P. Casper, K. Kotut, E. Novelo, B. Lemma, et al., Phylogenetic relationship and divergence among planktonic strains of *Arthrospira* (Oscillatoriales, Cyanobacteria) of African, Asian and American origin, deduced by 16S—23S ITS and phycocyanin operon sequences, Phycologia 49 (2010) 361—372.

[167] M. Schagerl, A. Burian, M. Gruber-Dorninger, S. Oduor, M.N. Kaggwa, Algal communities of Kenyan soda lakes with a special focus on *Arthrospira fusiformis*, Fottea 15 (2) (2015) 245—257.

[168] E.Y. Afonina, N.A. Tashlykova, Fluctuations in plankton community structure of endorheic soda lakes of southeastern Transbaikalia (Russia), Hydrobiologia 847 (2020) 1383—1398.

[169] A. Pentecost, J.E. Andrews, P.F. Dennis, A. Marca-Bell, S. Dennis, Charophyte growth in small temperate water bodies: Extreme isotopic disequilibrium and implications for the palaeoecology of shallow marl lakes, Palaeogeogr. Palaeoclimatol, Palaeoecol. 240 (2006) 389—404.

[170] J. Komárková, E. Zapomělová, J. Komárek, *Chakia* (cyanobacteria), a new heterocytous genus from Belizean marshes identified on the basis of 16S rRNA gene, Fottea 13 (2013) 227—233.

[171] J. Seckbach, P.H. Rampelotto, Polyextremophiles, in: C. Bakermans (Ed.), Microbial Evolution under Extreme Conditions, De Gruyter, Berlin, 2015, pp. 153—170.

CHAPTER 2

Global seaweeds diversity

Mohamed S.M. Abdel-Kareem[1] and Alaa A.F. ElSaied[1,2]

[1]Botany and Microbiology Department, Faculty of Science, Alexandria University, Alexandria, Egypt [2]Biology Department, Unaizah College of Sciences and Arts, Qassim University, Unaizah, Kingdom of Saudi Arabia

2.1 Introduction

Seaweeds are collectively named thallophytes because they lack true roots, stems, or leaves. All seaweed species contain chlorophyll "a" as the primary photosynthetic pigment. All cells of their reproductive structures are fertile, except for some species, such as *Chara*, which have a sterile wall around their reproductive structures. Also they do not form embryos. While structural sophistication is usually lacking among seaweeds, still some exhibit noticeable specialization and differentiation in their vegetative cells. Parenchymatous growth is recorded only in some species, and both monogenetic and digenetic life cycles are observed [1].

Despite the rapid progress in all aspects of scientific research tools, many aquatic seaweeds species have not yet been fully discovered, presenting great mystery to marine biologists [2]. As it is believed by many scientists that the first form of life appeared in oceans, the number of aquatic species has been found to represent three quarters of all known terrestrial species. Seaweeds show great diversity in terms of their morphology, anatomy, life span, distribution, and chemical composition, which is a major taxonomic challenge for marine biologists. Seaweeds are responsible not only for the primary production but also for providing the habitat needed by near-shore benthic populations [3].

A considerable number of compounds that have been extracted from seaweed have been found to have stabilizing and stiffening properties, thus they are used in the food industry and as precursors for cosmetics [4]. Some compounds of seaweeds have led to new opportunities in medical advances, as bioactive compounds [5], as a CO_2 sink [6], and as biofuel [7]. In addition, they have been widely used as a model organism to study biogeographic patterns and for testing various ecological theories [8].

This chapter is a summary of the current status of the global biodiversity data of seaweeds, with special reference to the basis of their classification, their diverse groups, their composition based on classification, global distribution, and their symbiotic relation with other marine organisms.

2.2 Basis of seaweeds classification

Taxonomy is always claimed to be retrogressive and outmoded; this is because it is unable to provide solutions to continuous, unexpected changes in the classification scheme of algae due to biodiversity catastrophes. The discovery of new species or genera in different localities around the world is a continuous process, but determining their taxonomy and introducing them into the general scheme may consume a lot of effort and time [9]. While this may be true for the lower taxonomic ranks (species and genera), higher taxa (family, order, and classes) exhibit no or a negligible updates. In addition, the polyphyletic structure of the algal group is somewhat contradictory with the conventional taxonomic groupings, but they are still useful for identifying the general character and degree of organization. However, taxonomic opinion can shift as knowledge accumulates [1].

The classification of seaweeds is primarily based on their morphological and anatomical structures. The morphology of the vegetative phase is not a reliable taxonomic criterion since it is not stable, since it varies greatly with the fluctuation of surrounding ecological factors. Taxonomic problems are not recent, as they have been frequently reported in the old literature. In his book published in 1935 Fritsch proposed a classification system that would divide algae into 11 classes based on pigmentation, flagella type, food reserve materials, thallus structure, and reproduction [10].

A review of previously published taxonomic studies of seaweeds showed that they could show a range of shape forms such as corticated, filamentous, foliose, siphonous, leathery, and crustose thallus. For example, the sporophyte of *Undaria* (brown algae) is a large and leathery thallus, whereas the gametophytic phase is microscopic and filamentous. A habitat with low disturbance and high production capacity is often dominated by corticated or leathery red macrophytes, such as *Gracilaria* and *Hypnea*, and brown species, such as *Fucus* [11]. In contrast, areas with higher disturbance potential are dominated by filamentous species including many red algae (such as *Polysiphonia*), Ochrophytes (such as *Ectocarpus*), and green algae (such as *Cladophora* and *Chaetomorpha*). However, areas with production capacity and a little disturbance are dominated by corticated foliose phyaephyceae species, such as *Padina* and *Dictyota*, chlorophyceae species, such as *Ulva*, and a few rhodophyceae species, such as *Porphyra*.

The observation and discovery of new algal communities, as well as using new modern molecular technologies, may lead to a reasonable doubt about the taxonomic status of known algae species. This is why almost all phycologists considered the traditional classification methods unreliable, shifting their methods, step by step toward more recent and sophisticated technologies to obtain more reliable results. Consequently, there is an urgent need for an international agreement by certain scientific and/or academic institutes, as an international taxonomic reference, to handle and publish the new taxonomic data on various respected international taxonomic websites, so that it can be freely available to the phycologist community [9].

Although, there are a large number of published identification and taxonomic keys, there are still several classification problems at all taxonomic levels [2]. This is because taxonomic schemes are almost all international but the identification keys are regional, which may lead to the existence of more than one name for the same algal species, or even to species misidentification. The rapid progress in the technology of DNA analysis has resulted in bigger classification problems, especially since algae do not have a single taxonomic character in common except their autotrophy [12].

Since the classification of algae according to their morphological and anatomical characteristics is a classical method and is not sufficient to provide reliable taxonomic status, the use of DNA-based molecular techniques has been considered more reliable and accepted in taxonomic identification processes [13]. Progress within DNA modern techniques, such as barcoding systematics, has resulted in various approaches in algal taxonomy [14], and the vast progress in molecular DNA techniques and their application for the classification of algae have resulted in a very rapid change in the scientific names of many algal species. This is true at different levels of the taxonomic ranks, and therefore nomenclature and classification of algae species are expected to face serious challenges in the next few years [9,15]. There is no clear definable general classification scheme appropriate to meet the classification criteria of all known algal species, this is because taxonomy as a science is always subject to continuous and rapid revision at all stages, with new genetic and ultra-structural facts being pursued every day [1].

2.3 The diverse groups of seaweeds

Although the algal taxonomy has been subjected to a drastic change over the last 30 years, there is no international classification system that has been accepted by all scientists working in the field of the taxonomy of algae [16]. A number of systematic schemes have been appeared in previously published studies, for example, Bold and Wynne [17], South and Whittick [18], Dawes [19], Margulis et al. [20], Hoek et al. [21], Johri et al. [22], Barsanti and Gualtieri [1], Lee [23], and Graham et al. [24]. Pereira and Neto [16] suggested an adopted classification scheme, in which the three groups of seaweeds are located within Chlorophyta (green algae), Ochrophyta (brown algae, formerly known as Phaeophyta), and Rhodophyta (red algae). However, it is not an integrated unified system; instead, there are individual schemes (from various sources), each for a certain group of seaweeds. This adopted classification scheme has been compiled together and follows the citations reported in the work of Yoon et al. [25] for red algae, Leliaert et al. [26] for green algae, and those of Riisberg et al. [27] and Yoon et al. [28] for Ochrophyta, as well as data published on the website AlgaeBase [29].

Since the majority of recognized algal species have been identified and classified on a morphological basis, without making use of the available molecular information, phycologists face serious taxonomic issues. It has been stated that there is "a widespread perception of taxonomy as a fundamentally retrogressive and outmoded science unable to cope with the current biodiversity crisis" [9]. The current situation reflects a serious taxonomic problem, in that a lot of the algal DNA sequences in the GenBank are anonymous, that is, they do not have a scientific name to refer to. For example, Ochrophyta has a complex

42   2. Global seaweeds diversity

systematic history as reported in the literature [30–34]. Most of these publications depended on a few morphoanatomical and reproductive features [35]. Thus classification systems of brown algae have been subject to several revisions due to the rapid progress in the new molecular technologies. Despite the application of the new molecular technologies in phaeophycean algae classification, phycologists still have blurred vision with regard to their classification. A lot of contradictory and unexpected data on the higher taxonomic ranks of phaeophyceae species have emerged over time, and hence it has become more difficult to provide an accurate taxonomic ranking for a certain genus [36]. These classification difficulties have shown the deep need to revise the generic and suprageneric classification of the phaeophyceae species [36,37].

Overall, there have been many respected efforts published recently concerning the taxonomy of algae, linking their morphological and molecular data together to obtain a reliable accepted classification system. It is necessary to point out the importance of taxonomic websites for keeping track of the new taxonomic statuses of the already known as well as the newly discovered species. For example, "AlgaeBase" [29] represents one of the most valuable scientific initiatives in this regard, providing phycologists with the recent updated nomenclature of a taxon. Such scientific websites should be constantly updated and supported by governments and scientists, particularly those in the field of the taxonomy of algae. [15].

2.4 Seaweeds composition based on classification

Seaweeds are a rich natural source of precious nutrients such as proteins, fiber, carbohydrates, lipids, minerals, and vitamins, and some have been shown to have bioactive potential [38–40]. Some of these compounds have been found to be more significant than others according to their pharmaceutical, medicinal, and commercial applications [41], for example, agar, carrageenan (extracted from red seaweeds), and alginates (extracted from brown seaweeds). Many economically chemical compounds are therefore considered to be a significant source of renewable natural resources [42], and different algal species showed a great variation in their content of such chemical compounds [43]. In fact, each algae group is characterized by specific chemical constituents [44], this individuality is the foundation of the mechanism of classification. These chemical compounds are mainly used for human consumption, animal feed, manure (in many countries), and are widely used in the food, textiles, dairy, paper, and confectionery industries. Seaweed extracts are also a very important component of the mast biostimulant commodity currently on the market [45]. Seaweed compounds have been recognized for decades for their bioactivity, as reported in several published works. These include bioactive compounds, such as polysaccharides, lipids, and polyphenols with antibacterial, antifungal, and antiviral properties [46].

Chemical compounds derived from algae may take one of the following forms: structural, such as cell walls, pigments, photosynthetic storage materials, or metabolic by-products. Variation in cell wall chemical composition, pigment ratios, photosynthetic storage products, and metabolic by-products enables phycologists to distinguish and compare various algal species in order to create a solid taxonomic scheme [29,47]. Thus almost all phycologists view these unique chemical compounds as taxonomic markers [40], classifying seaweeds

Handbook of Algal Biofuels

into three main categories: Rhodophyta, Ocrophyta (formerly Phaeophyta), and Chlorophyta, based on their pigment profiles [42].

2.4.1 Cell wall

Cellulose is the main cell wall component of most rhodophyceae species, while in the haploid phase of some Bangiales species, such as *Porphyra*, a-1,3-linked xylan is the main component [48]. In most red algae, amorphous mucilage may be one of two types: agar or carrageenan, which may account for up to two-thirds of the cell wall dry weight. Another characteristic feature of most Rhodophyceae species is the presence of an external cuticle layer made of protein. [49].

The Phaeophycean algae cell has a multilayer cellulose wall [50]. The sugar residues of alginic acid and fucoidan form an amorphous part of the cell wall, while the mucilaginous part and the cuticle are formed mainly of alginic acid [51]. The ratio of alginic acid to fucoidan has been found to differ depending on the species and environmental factors. The distinctive morphological concentric bands of needle-shaped aragonite crystals on the surface of some phaeophyceae species, such as *Padina*, are due to the calcification of their cell wall with calcium carbonate deposition [52]. Cellulose in the cell walls of Chlorophycean seaweed species is usually the main polysaccharide component, meanwhile in the majority of species belonging to Caulerpales cellulose is replaced by sugar residues of xylans or mannans [53].

2.4.2 Pigments

Chlorophyll "a" is a common pigment in all algae, as discussed in Chapter 1, Cyanoprokaryotes and Algae: Classification and Habitats. The pigmentation of the three major phyla of seaweeds [16] is shown in Table 2.1. The pigment profiles recorded for Chlorophycean species are identical to those in higher plants; both have chlorophyll "a" and "b." Siphonaceous algae, as well as unicellular ones, are characterized by their content of siphonoxanthin and its ester siphonin [54]. Each pigment may characterize a certain group of algae, chlorophyll "c" for the brown algae, phycobilin for the red algae, and finally chlorophyll "b" for the green algae. However, brown and green algae show a little similarity with regard to pigments, both containing carotenoids in their cells.

TABLE 2.1 The pigmentation of the three major phyla of seaweeds.

Phylum	Chlorophylls	Phycobilins	Carotenoids	Xanthophylls
Rhodophyta	a, d	Allophycocyanin Phycoerythrins Phycocyanins	α-, β-Carotene	Lutein
Ochrophyta	a, c_1, c_2, c_3	Absent	α-, β-, ε-Carotene	Fucoxanthin, violaxanthin, diadinoxanthin, heteroxanthin
Chlorophyta	a, b	Absent	α-, β-, γ-Carotene	Lutein, prasinoxanthin

Phycobiliproteins, such as R-phycocyanin, allophycocyanin, and phycoerythrins, give the red algae its unique pinkish color. In a combination of different pigment ratios of four individuals namely, *Porphyridium*, *Rhodosorus*, *Rhodochorton*, and *Smithora*, B-phycoerythrin is formed, which is present in the more primitive red species, while R-phycoerythrin characterizes the higher ones. Some Rhodophyceae species may have a special type of pigment. *Porphyridium*, *Porphyra*, and *Polysiphonia* contain C-phycoerythrin [55].

2.4.3 Carbohydrates

Generally, the percentage of carbohydrates content in seaweed tissues varies greatly between species. The Rhodophyceae species contain a variety of carbohydrates, such as floridean starch, cellulose, xylan, and mannan. The content of agar and carrageenan could affect fiber solubility [56]. Ochrophytes species have been recognized by their content of some unique sulfated carbohydrates, such as fucoidan, laminarin, and alginate. Fibers extracted from brown algae have been found to be composed primarily of cellulose and various insoluble alginate salts, while their amorphous fraction is made up of water-soluble alginates and/or fucoidan. Their mineral content (Na, K, Ca, Mg, Fe, Zn, etc.) shows high variability according to species, locality, seasons, environmental, and physiological factors [57].

2.4.3.1 *Ulvan*

Ulvan may constitute up to 29% of the total dry weight of some Chlorophycean [58]. It is composed primarily of repeated sulfated disaccharides, namely aldobiuronic acids, which comprise two types: A (ulvanobiuronic acid 3-sulfate) and B (ulvanobiuronic acid 3-sulfate) [59].

2.4.3.2 *Laminarin*

Laminarins are a storage glucan extracted from tissues of brown algae [60]. The degree of branching of Laminarin is the key factor determining its solubility. Laminarins have shown interesting biological activities [61].

2.4.3.3 *Fucans*

Fucans represent one of the most significant sulfated polysaccharides groups, which were first isolated by Kylin in 1913 [62] and have been found to consist of a repeated unit of fucose. Fucans are a good example of fucoidan, which are extracted from phaophyceae species such as *Fucus vesiculosus*, and *Ascophyllum nodosum*. Fucan consists primarily of $(1 \rightarrow 2)$-linked 4-O-sulfated fucopyranose residues [63].

2.4.3.4 *Alginates*

Alginic acid is an organic compound, made up of two monomers, namely D-mannuronic acid and L-guluronic acid [64]. It is derived from brown algae in the form of a mixture of different mineral salts. This is achieved by the conversion of cell wall insoluble alginates into soluble ones.

2.4.3.5 Agar

Agar is a colloid derived from several types of Rhodophyta. It is used as a gelling agent, and also as a substrate for microbial cultivation. It is a polysaccharide, composed mostly of D- and L-galactose groups. Agarophytes is a term denoting the rhodophyceae genera used to extract agar. Two fractions have been obtained from agar by a fractionation process. The first fraction is agarose which is composed of repeating units of alternating 1,3-linked β-D-galactopyranose and 1,4-linked 3,6-anhydro-α-L-galactopyranose. Agaropectin is the second fraction, which has a more complex chemical composition, made up of various residues of sulfuric, pyruvic, and uronic acids, in addition to D-galactose and 3,6-anhydro-L-galactose [65].

2.4.3.6 Carrageenan

Carrageenan is another essential important sulfated polysaccharide from rhodophycean algae. There are about 15 chemical structures representing different types of carrageenan, varying in the number and position of sulfate groups, the presence of 3,6-anhydro-D-galactose, and the conformation of the pyranosidic ring [66]. Out of the 15 types of carrageenan only three are used commercially, and they are denoted as kappa, iota [67], and lambda [67,68]. The gelling ability of carrageenan depends on its helical structure conformation. Lambda-carrageenan is well-known as a thickening agent, because the of the absence of the 3,6-anhydro bridges, which disturb its helical conformation [68].

2.4.4 Lipids

The content of lipids in seaweeds is relatively low but it is still considered important to human health as a food source [69]. Some studies [38,47,71−74] and many others have highlighted lipid content between <1% and 20%. Polyunsaturated fatty acids account for almost 50% of the total lipids produced by seaweed. Eicosapentaenoic acid and arachidonic acid are very significant examples of bioactives with the capacity to regulate blood pressure and blood clotting and prevent the risk of cardiovascular disease, osteoporosis, and diabetes [70]. Furthermore, red and brown seaweeds are rich in EPA and AA, and green seaweeds like *Ulva pertusa* predominantly contain hexadecatetraenoic, oleic, and palmitic acids [38,71].

Various algal species have demonstrated variance in their lipid and fatty acid content as they are greatly influenced by variations in environmental factors such as light intensity, seawater salinity, and temperature. For example, a high content of total fatty acids of the green species *Ulva pertusa* was obtained as a result of exposure to high salinity, whereas low salinity and high light intensity resulted in a low total fatty acid content of the same species [72]. Different algal species respond differently to the same environmental element. For example, high light intensity elevates the levels of the majority of saturated fatty acids in *U. Pertusa*, but has the reverse effect on the production of nearly all the fatty acids in *Sargassum piluliferum*. In addition, temperature fluctuation exerts a significant influence on the fatty acid levels in different algal species [38]. A seasonal variation of the total lipid content of brown seaweeds was also reported [73]. In all species, the growth period between winter and spring displayed the maximum lipid content and the overall lipid content of many brown algae species was close to 10% by their dry weight.

2.5 Global distribution of seaweeds

The distribution of seaweeds is limited from the lower intertidal to the shallow subtidal zones of the marine environment [74]. Many of Rhodophycean species have been found to grow in subtidal waters, such as *Gracilaria, Gelidiella, Pterocladia, Hypnea*, and *Porphyra*, while Phaeophycean species were found to inhabit the tidal or upper subtidal zone, such as *Sargassum, Ascophyllum, Laminaria, Turbinaria*, and *Dictyota*. Many green seaweeds are most commonly found in the intertidal zone, such as *Ulva, Chaetomorpha, Codium*, and *Caulerpa*. Some marine algae species have been widely spread (e.g., *Digenea simlex* in all warmer seas), other species are cosmopolitan (*Ulva lactuca, Colpomenia sinusa, Ectocarpus siliculosus, Gracilaria confervoides, Hypnea musciformis*), while most species are present only in limited geographical areas [10]. Littler et al. [75] clarified the superiority of leathery algae under conditions of high productivity capacity because they contain more nonphotosynthetic tissue than other classes of algae. Brown species, such as *Undaria, Sargassum*, and *Fucus*, flourish in nutrient-rich regions with moderate temperature. Meanwhile, areas characterized by high physical disturbance and low productivity have been dominated by Crustose algae. *Lithophyllum yessoense* is the only crustose species that has succeeded in propagating in the Mediterranean Sea; articulated calcareous species (green species *Halimeda* and the red species *Corallina*) failed to do so, due to their calcification which affects their physiological tolerance [11].

Geographical communities should be defined simply enough to be understood and detailed enough to provide realistic data about their structure and functional components [11]. Chlorophyceae spices have been found to flourish in freshwater rather than in the oceans. *Ulva* is the most common green algae in a wide range of habitats. Ochrophyta species (brown seaweeds) inhabit the cold coastal waters and their stipe reaches up to 35 m in length, so they have been known as "kelps." In contrast, Rhodophyta prefer the dark ocean bed, maybe because they have an efficient photosynthetic apparatus [76].

Seaweed biogeographers typically divide the world's oceans into seven large regions: Arctic and Antarctic polar regions, cold and warm temperate regions of both hemispheres, and the tropical regions of the Atlantic and Indo-Pacific [77]. There are no sharp lines between these regions since they are always subject to changes in species structure, temperature adaptation [78], natural barriers [79], and species spreading limits [80].

2.5.1 Temperature

It is well-known that the coasts located in various environments, and those found in regions with similar climatic conditions but located in different oceans or hemispheres are populated by dissimilar seaweed flora [81]. Seaweeds occupy primarily the photic zone, where temperature variations affect the spreading of species similar to those patterns observed in oceans [82]. The spreading limits of each species strictly follow the main marine isotherms [78], resulting in a firm correlation with the temperature signatures of major ocean currents [83]. Such trends are defined by the two main temperature limits: (1) the lethal boundary, the limiting factor which reflects the ability of a given seaweed to stand and adapt to unfavorable conditions; and (2) the growth and reproduction limits, which reflect the capacity of the species to survive and reproduce under favorable conditions [78]

In the case of continuous open oceans, the distribution of most seaweed species would be cosmopolitan according to their thermal limits [84,85]. Both climate and temperature have a considerable influence on the global distribution of seaweeds and, as a consequence, the greenhouse effect has resulted in a significant shift in the distribution of algae [86]. Overall, climate change has resulted in the shifting of local diversity and the distribution of seaweed species to become global [87].

2.5.2 Barriers

Spreading of algal species has been controlled by two types of barriers: hard (physical) or soft (hydrographic features) [84], according to their underlying mechanism [79]. Both hard and soft barriers are not permanent, particularly over geological timescales. For example, the disturbance of ice cover and sea level has led to significant changes in marine algae biogeography [88]. Extensions and contractions of algae populations are significantly affected by physical ecological factors, which may reduce the rate of distribution of species relative to the rate of introduction [89].

Seaweeds have been exploited for a long time for two main purposes: taxonomic studies and commercial applications. A variety of interlinked ecological factors have had a significant impact on the global distribution of seaweed across diverse ecosystems. Nevertheless, as a primary producer [3], seaweed is an important biotic element in aquatic ecosystems. As a result, there was a great need to study how the modification of the abiotic elements of the ecological influences the distribution of the marine algae population. The use of seaweeds as a sustainable source for many natural products has given rise to the desire to maximize their production in relation to the local environment [90]. Effects of biotic and abiotic ecological elements on the distribution of algae have captured the attention of scientists, recognizing that these ecological connections are often dynamic and interconnected [91]. For example, although life cycles of seaweed species are quite simple, most of its phases are microscopic, which represents a considerable obstacle for their ecological study. In addition, introduced algae (invaders) can represent another significant biological factor that interferes with abiotic elements when considering the distribution of algae [92]. Taxonomic studies data may be used as an indicator to classify potential invasive species and to support or reject the presumption that a particular algal family is more or less likely to be invasive [93].

Community patterns of seaweed by groupings, based on their response to environmental impacts, will provide significant insight into the distribution of community seaweeds [11]. In general, the number of species in marine aquatic systems increases toward the tropics [94]. In addition, low and high species richness has been recorded across temperate and tropical areas [95]. Although the trend of latitude gradients of marine algae biodiversity appears to be general, it could not be described as "global" [96]. In contrast, no evidence of such a latitudinal trend has been shown in other studies of seaweeds [97].

Coastal rocky shores of the intertidal zone are the main habitat of seaweeds, which interact with the surrounding marine biota. Hence, seaweeds represent the essential biotic element of the ecosystem affecting the overall coastal biodiversity [98]. Seaweeds have been found to be subjected to threat in various places of the world, where different human destructive activities increased. Although much significant published data discuss the distribution of seaweeds according to the disturbance in their ecosystem, there is no comprehensive strong single ecological scheme to explain the mechanism of seaweeds global distribution. This is why all published and forthcoming data must be carried out as collaborative research in the context of a single global project. Of course, such an enormous study needs to be funded and coordinated by international organizations and requires the cooperation of a large team, including marine biologists and oceanographers, to cover various marine algae habitats.

There are many environmental problems which could affect the distribution of seaweeds, the most important are overfishing and pollution. Overfishing of herbivorous fishes, as well as excess pollutants in aquatic marine ecosystems, is resulting in the destruction of the reef, followed by a gradual replacement of dominant corals by macroalgae. The growth of macroalgae in coral reefs is often detrimental by increasing the process of reef degradation [91]. In reality, macroalgae, besides being destructive to the reef, show the opposite trend in some cases. For example, Coralline macroalgae, in particular, crustose calcareous (Order Corallinales) could share in rebuilding the destructive reef through the deposition of calcium carbonate [99]. Noncoralline algae also have a significant role, especially in deep areas, forming an elevated conformation of algal masses. Some calcareous algae increase marine sediments, filling interspaces of the reef [100]. Carbonate salts of calcium, such as aragonite, calcite, and magnesium calcite, are formed in algal tissues representing a defensive mechanism against grazing, wave shock, and providing mechanical support [91]. Investigation of the ecological distribution of seaweed on a taxonomic basis is quite confusing, where even species belonging to the same higher taxa react differently to changes of the biotic or abiotic elements of the ecosystem. In fact, using "functional groupings" is much more practical, since it is based on seaweeds' ecological features, such as form, size, toughness, photosynthetic efficiency, and grazing resistance. As a matter of fact, functional groups provide a deeper understanding of how different algal populations distribute and react to ecological stresses. This is probably because algal species share the same ecological features that show similar responses to ecological pressures. This functional approach provides a solution to the difficulty of identifying algae at the field level, which makes this approach more accepted to characterize algae communities in ecological studies [91].

Offshore reefs tend to support the growth of fleshy and crusty calcareous brown and red algae compared to inshore reefs [101]. Reef zones have been found to be dominated by many algal turfs [102] which have been influenced by grazing and water quality. Populations of algal turfs have demonstrated marked differences across the continental shelf. Crustose calcareous algae have dominated the other species growing offshore [103]. Inshore reefs typically have a rich abundance of fleshy brown macroalgae, especially *Sargassum*, *Turbinaria*, and *Cystoseira* [104]. Occasionally, the absence of grazing has a greater impact on the diversity and productivity of *Sargassum* in inshore reefs than the presence of a high concentration of nutrients in coastal marine environments, which may be mistaken as an indication of eutrophication [105]. However, lacking enough historical evidence, it difficult to decide if existing abundances of algae are normal behavior or resulting from human activities. The turf algae growing on inshore reefs are totally different from those growing on offshore reefs. Depth, wave exposure, and grazing are the key factors controlling the composition and abundance of benthic macroalgae within the reef [106].

2.5.3 Seaweeds production

Seaweeds production provides important opportunities for developing countries, and is accordingly growing rapidly in Africa, South America, and Southeast Asia [107]. However, the steep growth in the use of marine biological resources represents a fundamental change in the way humans derive benefits from the oceans [108]. Seaweed aquaculture, growing at 7.5% per year [109], is becoming an important component of marine aquaculture, driven by a divergence of the demand for seaweed products from traditional uses to bioenergy [110], cosmetics [111], and biomedicine applications [112]. According to Mazarrasa et al. [113], the cultivation of seaweed is more widespread than innovation, as seaweed is grown in twice as many nations as those filing patent claims. At the same time, seaweed production is traditionally important in Southeast Asian countries, especially China and Japan.

2.6 Symbiotic relation of seaweeds with other marine organisms

Considering the relationships and interactions between different species living in the same ecosystem, each organism has its own unique relationship with the surrounding biota, creating its own microenvironment, which depends on the morphology of the thallus, the surface texture, and the internal and external metabolic activity of the algae. The seaweed surface affords an appropriate substratum for the growth of microorganisms and also secretes various organic nutritious materials that increase the growth of bacterial colonies, forming microbial biofilms [114]. The microbiota inhabiting the seaweed surface is extremely diverse, competitive, and consists of a group of microorganisms comprising bacteria, fungi, diatoms, protozoa, spores, and larvae of marine invertebrates [115].

Seaweed-associated bacteria play a key role in the morphogenesis and development of seaweed in direct and/or indirect ways. The most common bacterial species are members of Proteobacteria and Firmicutes. They contain plant growth-promoting agents, quorum-sensing signaling molecules, bioactive compounds, and other important molecules that are responsible for the natural morphology, production, and growth of algae [116].

Seaweeds may develop different relations with another biota living in the same habitat, which may be a mutualism relationship with various microorganisms (bacteria, fungi, protozoa), or even a commensalism relationship. Seaweed–bacterial relations show a wide spectrum, ranging from passive and random epibiosis to highly specific and obligate symbiosis [117]. In the case of a mutualistic relationship with a bacterial community that protects seaweed from biofouling, seaweed (as a host) at the time is not only physical shelter but also provides the microorganism with nutrient needs. These relationships tend to be dominated by a variety of secondary chemical metabolites which are secreted only by seaweed, or even by both species [118].

Seaweed is often in direct competition with other benthic space species, therefore it is often susceptible to epibiosis [119]. Algae are, therefore, continually subjected to invasion by the surrounding symbiotic and herbivorous organisms [120]. Seawater is considered to be a poor environment due to its low concentration of nutrients, whereas seaweed surfaces are often covered by a large amount of organic exudate which represents a rich nutritious source for other organisms [121]. Biofilm formed by microorganisms on the surface of the seaweed acts as a shield against different environmental stresses [115]. Despite the protective benefit of the epibiosis process, biofilm represents a serious threat to marine algae [122], where it has increased hydrodynamic drag on the basibionts (the alga) resulting in a reduction in its buoyancy and tissue elasticity, and increased grazing [118]. The rate of gaseous exchange and light penetration is negatively affected by biofilm formation, which severely reduces photosynthetic assimilation [122]. Biofilm, on the other hand, represents a fertile substrate for other fouling invaders, such as diatoms, and invertebrate larvae [123], ultimately leading to the destruction of seaweed due to the secretion of a variety of toxins, digestive enzymes, and inhibitors as part of their defense mechanism [124]. That is why many biologists describe seaweed ecosystems as "complex" and "dynamic," because they consist of a complex interconnected multirelationship between a variety of organisms [125].

In coastal marine ecosystems, grazing represents a vital threat to seaweeds [126]. Seaweeds have a range of effective chemicals against herbivores [127]. It has been reported that the secretion of protective exudates in response to grazing triggers happened within a short time period [128]. An interesting finding revealed that some seaweed species have the ability to detect herbivory induction compounds dissolved in the surrounding aquatic ecosystem [129]. However, only one species, namely *Ascophyllum nodosum* (Ocrophyta), has been shown to be able to detect herbivore-induced waterborne signals [130].

2.7 Conclusion

Ecological studies on seaweeds in a specific ecosystem must be designed cautiously, taking into consideration all possible factors. The exploitation of natural resources for human benefit should be done wisely, considering the sustainability of the resource itself. In fact, marine ecosystems are naturally very organized and intact, but the accumulation of small disturbances may eventually lead to the elimination of humans themselves from the planet. This is because the microecosystem of seaweed is fragile and holding it in balance is a very critical process. Responsible corporations, the popularization of seaweed products, and scientific awareness of how to ensure the wise use of available natural resources will undoubtedly help to create a sustainable future for humans.

References

[1] L. Barsanti, P. Gualtieri, Algae, Anatomy, Biochemistry and Biotechnology, CRC, Taylor & Francis, Boca Raton, FL, 2006.

References

[2] T. Silberfeld, F. Rousseau, B. de Reviers, An updated classification of brown algae (Ochrophyta, Phaeophyceae), Cryptogamie, Algologie 35 (2) (2014) 117–156.

[3] K.H. Mann, Seaweeds: their productivity and strategy for growth, Science 182 (4116) (1973) 975–981.

[4] C. Wiencke, C.D. Amsler, Seaweeds and their communities in Polar Regions, Seaweed Biology, Springer, Berlin, Heidelberg, 2012, pp. 265–291.

[5] T.H. Ranahewa, A.D. Premarathna, L.J. Jayasooriya, R.R. Wijesundara, R.P. Rajapakse, 2016. In-vitro anti-cancer and cytotoxic properties of aqueous seaweed extracts on BHK and HeLa cell lines, in: Proceedings of the International Research Symposium on Pure and Applied Sciences, Faculty of Science, University of Kelaniya, Sri Lanka, pp. 07.

[6] D. Muraoka, Seaweed resources as a source of carbon fixation, Bull. Fish. Res. Agen. (Supplement 1)(2004) 59–63.

[7] S. Bastianoni, F. Coppola, E. Tiezzi, A. Colacevich, F. Borghini, S. Focardi, Biofuel potential production from the Orbetello lagoon macroalgae: a comparison with sunflower feedstock, Biomass Bioenergy 32 (7) (2008) 619–628.

[8] A.D. Premarathna, A.M. Kumara, A.P. Jayasooriya, D.E. Jayanetti, R.B. Adhikari, L. Sarvananda, et al., Distribution and diversity of seaweed species in south coastal waters in Sri Lanka, J. Oceanogr. Mar. Res. 7 (2020) 196–202.

[9] O. De Clerck, M.D. Guiry, F. Leliaert, Y. Samyn, H. Verbruggen, Algal taxonomy: a road to nowhere? J. Phycol. 49 (2) (2013) 215–225.

[10] F.E. Fritsch, The Structure and Reproduction of the Algae, Vols. I & II, Cambridge University Press, Cambridge, London, 1935.

[11] R.S. Steneck, M.N. Dethier, A functional group approach to the structure of algal-dominated communities, Oikos. 69 (3) (1994) 476–498.

[12] H.D. Kumar, H.N. Singh, A Textbook on Algae, 1st (ed.), The Macmillan Press Litd, London and Basingstoke, 1979.

[13] S.Y. Kim, A. Manghisi, M. Morabito, E.C. Yang, H.S. Yoon, K.A. Miller, et al., Genetic diversity and haplotype distribution of *Pachymeniopsis gargiuli* sp. *nov.* and *P. lanceolata* (Halymeniales, Rhodophyta) in Korea, with notes on their non-native distributions, J. Phycol. 50 (5) (2014) 885–896.

[14] M.A. Alshehri, A.T. Aziz, O. Alzahrani, A. Alasmari, S. Ibrahim, G. Osman, et al., DNA-barcoding and species identification for some Saudi Arabia seaweeds using rbcL gene, J. Pure Appl. Microbiol. 13 (4) (2019) 2035–2044.

[15] J.J. Bolton, The problem of naming commercial seaweeds, J. Appl. Phycol. 32 (2020) 751–758.

[16] L. Pereira, J.M. Neto, Marine Algae: Biodiversity, Taxonomy, Environmental Assessment, and Biotechnology, CRC Press, 2015, p. 398.

[17] H.C. Bold, M.J. Wynne, Introduction to the Algae, Prentice Hall, Englewood Cliffs, NJ, 1978.

[18] G.R. South, A. Whittick, Introduction to Phycology, Blackwell Scientific Publ, Oxford, 1987.

[19] C.J. Dawes, Marine Botany, Hohn Willey & Sons, New York, 1998.

[20] L. Margulis, J.O. Corliss, M. Melkonian, D.J. Chapman, Handbook of Protoctista, Jones and Barleit Publ., Boston, MA, 1989.

[21] V.C. Hoek, D.G. Mann, H.M. Jahns, Algae, An Introduction to the Phycology, Cambridge University Press, London, 1995.

[22] R.M. Johri, S. Lata, S. Sharma, A Textbook of Algae, Dominant Cop., New Delhi, 2004.

[23] R.E. Lee, Phycology, 4th (ed.), Cambridge University Press, Cambridge, 2008. 3(2).

[24] L.E. Graham, J.M. Graham, L.W. Wilcox, Algae, Second edition, Benjamim Cumming, Pearson, San Francisco, CA, 2009.

[25] H.S. Yoon, K. Muller, R. Sheath, F. Ott, C. Bhattacharya, Defining the major lineages of red algae (Rhodophyta), J. Phycol. 42 (2) (2006) 482–492.

[26] F. Leliaert, D.R. Smith, H. Moreau, M.D. Herron, H. Verbruggen, C.F. Delwiche, et al., Phylogeny and molecular evolution of the green algae, Crit. Rev. Plant. Sci. 31 (2012) 1–46.

[27] I. Riisberg, R.J.S. Orr, R. Kluge, K. Shalchian-Tabrizi, H.A. Bowers, V. Patil, et al., Seven gene phylogeny of heterokonts, Protist 160 (2009) 191–204.

[28] H.S. Yoon, R.A. Andersen, S.M. Boo, D. Bhattacharya, Stramenopiles. 721–731, In: M. Schaechter (Ed.), Encyclopedic of Microbiology, Vol. 5, Elsevier, Oxford, 2009.

2. Global seaweeds diversity

[29] M.D. Guiry, G.M. Guiry, AlgaeBase. World-Wide Electronic Publication, National University of Ireland, Galway, 2020. <https://www.algaebase.org>.

[30] G.F. Papenfuss, Phaeophyta, In: G.M. Smith (Ed.), Manual of Phycology—An Introduction to the Algae and Their Biology, A New Series of Plant Science Books, Vol. 27, The Chronica Botanica Co., Waltham, MA, 1951a, pp. 119–158.

[31] G.F. Papenfuss, Problems in the classification of the marine algae, Svensk. Bot. Tidsk. 45 (1951) 4–11.

[32] G.F. Papenfuss, Problems in the classification of the marine algae of the tropical and southern pacific, in: Proceedings of the Seventh International Botanical Congress, Stockholm, 1950, 1953, pp. 822–823.

[33] R.F. Scagel, The Phaeophyceae in perspective, Oceanogr. Mar. Biol. Annu. Rev. 4 (1966) 123–194.

[34] M.J. Wynne, S. Loiseaux, Recent advances in life history studies of the Phaeophyta, Phycologia 15 (1976) 435–452.

[35] B. de Reviers, F. Rousseau, Towards a new classification of the brown algae, Prog. Phycol. Res. 13 (1999) 107–201.

[36] T. Silberfeld, J.W. Leigh, H. Verbruggen, C. Cruaud, B. de Reviers, F. Rousseau, A multi-locus time-calibrated phylogeny of the brown algae (Heterokonta, Ochrophyta, Phaeophyceae): investigating the evolutionary nature of the "brown algal crown radiation", Mol. Phylogenet. Evol. 56 (2010) 659–674.

[37] T. Silberfeld, M.F.L.P. Racault, R.L. Fletcher, A. Couloux, F. Rousseau, B. de Reviers, Systematics and evolutionary history of pyrenoid-bearing taxa in brown algae, Eur. J. Phycol. 46 (2011). 362–278.

[38] J. Ortiz, N. Romero, P. Robert, J. Araya, J. Lopez-Hernández, C.E. Bozzo, et al., Dietary fiber, amino acid, fatty acid, and tocopherol contents of the edible seaweeds *Ulva lactuca* and *Durvillaea antarctica*, Food Chem. 99 (2006) 98–104.

[39] H. Yaich, H. Garna, B. Bchi'r, S. Besbes, M. Paquot, A. Richel, et al., Chemical composition and functional properties of dietary fiber extracted by Englyst and Prosky methods from the alga *Ulva lactuca* collected in Tunisia, Algal Res. 9 (2015) 65–73.

[40] H.M. Mwalugha, J.G. Wakibia, G.M. Kenji, M.A. Mwasaru, Chemical composition of common seaweeds from the Kenya coast, J. Food Res. 4 (6) (2015) 28.

[41] I.A. Abbott, Food and food products from seaweeds, In: C.A. Lembi, J.R. Waaland (Eds.), Algae and Human Affairs, Cambridge University Press, Phycological Society of America, 1989, p. 141.

[42] A. Diharmi, D. Fardiaz, N. Andarwulan, Chemical and minerals composition of dried seaweed *Eucheuma spinosum* collected from Indonesia coastal sea regions, Int. J. Ocean. Oceanogr 13 (1) (2019) 65–71.

[43] E. Marinho-Soriano, P.C. Fonseca, M.A.A. Carneiro, W.S.C. Moreira, Seasonal variation in the chemical composition of two tropical seaweeds, Bioresour. Technol. 97 (18) (2006) 2402–2406.

[44] P. Burtin, Nutritional value of seaweeds, Elec J. Env. Agric. Food Chem. 2 (4) (2003) 498–503.

[45] S. Holdt, S. Kraan, Bioactive compounds in seaweed: functional food applications and legislation, J. Appl. Phycol. 23 (2011) 543–597.

[46] C.S. Kumar, P. Ganesan, N. Bhaskar, In vitro antioxidant activities of three selected brown seaweeds of India, Food Chem. 107 (2008) 707–713.

[47] C. Dawczynski, R. Schubert, G. Jahreis, Amino acids, fatty acids, and dietary fiber in edible seaweed products, Food Chem. 103 (3) (2007) 891–899.

[48] E. Frei, R.D. Preston, Non-cellulosic structural polysaccharides in algal cell walls. II. Association of xylan and mannan in *Porphyra umbilicalis*, Proc. R. Soc. Lond. [B] 160 (1964) 314–327.

[49] J.S. Craigie, J.A. Correa, M.E. Gordon, Cuticles from *Chondrus crispus*, J. Phycol. 28 (1992) 777–786.

[50] B. Kloareg, R. Quatrano, Structure of the cell walls of marine algae and ecophysiological function of the matrix polysaccharides, Ann. Rev. Ocean. Mar. Biol. 26 (1988) 259–315.

[51] L.V. Evans, M.S. Holligan, Correlated light and electron microscope studies on brown algae. I. Localization of alginic acid and sulphated polysaccharides in *Dictyota*, N. Phytol. 71 (1972) 1161–1172.

[52] M. Benita, Z. Dubinsky, D. Iluz, Padina pavonica: morphology and calcification functions and mechanism, Am. J. Plant. Sci. 9 (6) (2018) 1156–1168.

[53] H.J. Huizing, H. Rietema, J.H. Sietsma, Cell wall constituents of several siphonous green algae in relation to morphology and taxonomy, Br. Phycol. J. 14 (1979) 25–32.

[54] Y. Yoshii, S. Takaichi, T. Maoka, I. Inouye, Photosynthetic pigment composition in the primitive green alga Mesostigma viride (Prasinophyceae): phylogenetic and evolutionary implications, J. Phycol. 39 (2003) 570–576.

[55] E. Gantt, S.F. Conti, Ultrastructure of blue green algae, J. Bacteriol. 97 (1969) 1486–1493.

References

[56] A. Jimenez-Escrig, F.J. Sanchez-Muniz, Dietary fiber from edible seaweeds: chemical structure, physicochemical properties and effects on cholesterol metabolism, Nut. Res. 20 (2000) 585−598.

[57] A.R. Circuncisão, M.D. Catarino, S.M. Cardoso, A. Silva, Minerals from macroalgae origin: health benefits and risks for consumers, Mar. Drugs 16 (11) (2018) 400−429.

[58] A. Robic, D. Bertrand, J.F. Sassi, Y. Lerat, M. Lahaye, Determination of the chemical composition of ulvan, a cell wall polysaccharide from *Ulva* spp. (Ulvales, Chlorophyta) by FT-IR and chemometrics, J. Appl. Phyc. 21 (2009) 451−456.

[59] G.L.,Y.U. Jiao, G.L. Zhang, J.Z. Ewart, H. S., Chemical structures and bioactivities of sulfated polysaccharides from marine algae, Mar. Drugs 9 (2) (2011) 196−223.

[60] B.A. Stone, A.E. Clarke, Chemistry and Biology of (1−3)-Glucans, La Trobe University Press, Melbourne, 1992, pp. 1−803.

[61] T.N. Zvyagintseva, L.A. Elyakova, V.V. Isakov, Enzyme transformation of Laminarans into $1 \rightarrow 3$, $1 \rightarrow 6$-3-D-glucans, having immunostimulating action, Bioorg. Khim. 21 (1995) 218−225.

[62] H. Kylin, Biochemistry of sea algae, Z. Phys. Chem. 83 (1913) 171−197.

[63] L. Chevolot, B.J. Mulloy, A. Ratiskol, Foucault, S. Colliec-Jouault, A disaccharide repeat unit is the major structure in fucoidans from two species of brown algae, Carb. Res. 330 (4) (2011) 529−535.

[64] L. Pereira, Estudos em macroalgas carragenófitas (Gigartinales, Rhodophyceae) da costa portuguesa—aspectos ecológicos, bioquímicos e citológicos (Ph.D. thesis), Departamento de Botânica—FCTUC, Universidade de Coimbra, Coimbra, 2004, p. 293.

[65] H.J. Bixler, H. Porse, A decade of change in the seaweed hydrocolloids industry, J. Appl. Phycol. 23 (2011) 321−335.

[66] T. Chopin, B.F. Kerin, R. Mazerolle, Phycocolloid chemistry as taxonomic indicator of phylogeny in the Gigartinales, Rhodophyceae: a review and current developments using Fourier transform infrared diffuse reflectance spectroscopy, Phycol. Res. 47 (1999) 167−188.

[67] L. Pereira, Population studies and carrageenan properties in eight gigartinales (Rhodophyta) from western coast of Portugal, Sci. World J. 2013 (2013) 11. Article ID 939830.

[68] V.F. de Velde, G.A. de Ruiter, Carrageenan, In: A. Steinbüchel, S. De Baets, E.J. VanDamme (Eds.), Biopolymers. Vol. 6: Polysaccharides. II: Polysaccharides. Eukaryotes, Wiley-VCH, Weinheim, Germany, 2002, pp. 245−274.

[69] Y. Peng, J. Hu, B. Yang, X. Lin, X. Zhou, X. Yang, et al., Chemical composition of seaweeds, Seaweed Sustainability, Elsevier Inc, 2015. Available from: http://doi.org/10.1016/B978-0-12-418697-2.00005-2.

[70] H. Maeda, T. Tsukui, T. Sashima, M. Hosokawa, K. Miyashita, Seaweed carotenoid, fucoxanthin, as a multifunctional nutrient, Asia Pac. J. Clin. Nutr. 17 (1) (2008) 196−199.

[71] M.H. Norziah, C.Y. Ching, Nutritional composition of edible seaweed Gracilaria changgi, Food Chem. 68 (2000) 69−76.

[72] E.A.T. Floreto, S. Teshima, The fatty acid composition of seaweeds exposed to different levels of light intensity and salinity, Bot. Mar. 41 (1998) 467−481.

[73] M. Terasaki, A. Hirose, B. Narayan, Y. Baba, C. Kawagoe, H. Yasui, et al., Evaluation of recoverable functional lipid components of several brown seaweeds of Japan with special reference to fucoxanthin and fucosterol contents, J. Phycol. 45 (4) (2009) 974−980.

[74] A.D. Premarathna, T.H. Ranahewa, S.K. Wijesekera, R.R. Wijesundara, A.P. Jayasooriya, V. Wijewardana, et al., Wound healing properties of aqueous extracts of *Sargassum illicifolium*: an in vitro assay, Wound Med. 24 (1) (2019) 1−7.

[75] M.M. Littler, D.S. Littler, E.A. Titlyanov, Producers of organic matter on tropical reefs and their relative dominance, Mar. Biol. 6 (1991) 3−14.

[76] S.K. Kim, Handbook of Marine Macroalgae: Biotechnology and Applied Phycology, Wiley-Blackwell, 2012.

[77] I. Bartsch, C. Wiencke, T. Laepple, Global seaweed biogeography under a changing climate: the prospected effects of temperature, In: C. Wiencke, K. Bischof (Eds.), Seaweed Biology. Berlin Heidelberg, Springer, 2012, pp. 383−406.

[78] C.V. Hoek, The distribution of benthic marine algae in relation to the temperature regulation of their life histories, Biol. J. Linn. Soc. 18 (1982) 81−144.

[79] P.F. Cowman, D.R. Bellwood, Vicariance across major marine biogeographic barriers: temporal concordance and the relative intensity of hard versus soft barriers, Proc. R. Soc. B. 280 (2013) 1541.

[80] J.J. Wiens, The niche, biogeography, and species interactions, Phil. Trans. R. Soc. B 366 (2011) 2336−2350.

[81] C.V. Hoek, World-wide latitudinal and longitudinal seaweed distribution patterns and their possible causes, as illustrated by the distribution of Rhodophytan genera, Helgolander Meeresunters. 38 (1984) 227–257.

[82] W.H. Adey, R.S. Steneck, Thermogeography over time creates biogeographic regions: a temperature/space/time-integrated model and an abundance-weighted test for benthic marine algae, J. Phycol. 37 (2001) 677–698.

[83] T. Wernberg, M.S. Thomsen, S.D. Connell, B.D. Russell, J.M. Waters, G.C. Zuccarello, et al., The footprint of continental-scale ocean currents on the biogeography of seaweeds, PLoS One 8 (11) (2013) e80186.

[84] A.A. Myers, Biogeographic barriers and the development of marine biodiversity, Estu Coast. Shelf Sci. 44 (1997) 241–248.

[85] B. Gaylord, S.D. Gaines, Temperature or transport? Range limits in marine species mediated solely by flow, Am. Nat. 155 (2000) 769–789.

[86] C.D.G. Harley, K.M. Anderson, K.W. Demes, J.P. Jorve, R.L. Kordas, T.A. Coyle, et al., Effects of climate change on global seaweed communities, J. Phycol. 48 (2012) 1064–1078.

[87] L. Duarte, R.M. Viejo, B. Martínez, M. de Castro, M. Gomez-Gesteira, T. Gallardo, Recent and historical range shifts of two canopy-forming seaweeds in North Spain and the link with trends in sea surface temperature, Acta Oecol. 51 (2013) 1–10.

[88] S.C. Straub, M.S. Thomsen, T. Wernberg, The dynamic biogeography of the anthropocene: the speed of recent range shifts in seaweeds, Seaweed Phylogeography: Adaptation and Evolution of Seaweeds Under Environmental Change, Springer, The Netherlands, 2016. 63–93.

[89] C.J.B. Sorte, S.L. Williams, J.T. Carlton, Marine range shifts and species introductions: comparative spread rates and community impacts, Glob. Ecol. Biogeogr. 19 (2010) 303–316.

[90] M.L. Wells, P. Potin, J.S. Craigie, J.A. Raven, S.S. Merchant, K.E. Helliwell, et al., Algae as nutritional and functional food sources: revisiting our understanding, J. Appl. Phycol. 29 (2017) 949–982.

[91] G. Diaz-Pulido, L. McCook, Macroalgae (Seaweeds), In: A. Chin (Ed.), The State of the Great Barrier Reef On-line, Great Barrier Reef Marine Park Authority, Townsville, 2008.

[92] S.L. Williams, J.E. Smith, A global review of the distribution, taxonomy, and impacts of introduced seaweeds, Ann. Rev. Ecol. Evol. Syst. 38 (1) (2007) 327–359.

[93] J.L. Lockwood, Using taxonomy to predict success among introduced avifauna: relative importance of transport and establishment, Conserv. Biol. 13 (1999) 560–567.

[94] F.G. Stehli, A.L. McAlester, C.E. Helsey, Taxonomic diversity of recent bivalves and some implications for geology, Geol. Soc. Am. Bull. 78 (1967) 466.

[95] B. Konar, K. Iken, J.J. Cruz-Motta, L. Benedetti-Cecchi, A. Knowlton, G. Pohle, et al., Current patterns of macroalgal diversity and biomass in northern hemisphere rocky shores, PLoS One 5 (10) (2010) 1–8.

[96] B. Santelices, J.J. Bolton, I. Meneses, Marine algal communities, In: J.D. Witman, R. Kaustuv (Eds.), Marine Macroecology, University of Chicago Press, Chicago, IL, 2009, pp. 153–192.

[97] J.J. Bolton, Global seaweed diversity: patterns and anomalies, Bot. Mar. 37 (1994) 241–245.

[98] S. Satheesh, S.G. Wesley, Diversity and distribution of seaweeds in the Muttom coastal waters, south-west coast of India, Biodiv. J. 4 (1) (2013) 105–110.

[99] M.M. Littler, D.S. Littler, Impact of CLOD pathogen on Pacific coral reefs, Science 267 (1995) 1356–1360.

[100] J.F. Marshall, P.J. Davies, *Halimeda* bioherms of the northern Great Barrier Reef, Coral Reefs 6 (1988) 139–148.

[101] L.J. McCook, G. De'ath, I.R. Price, G. Diaz-Pulido, J. Jompa, Macroalgal resources of the Great Barrier Reef: taxonomy, distributions and abundances on coral reefs, Report to the Great Barrier Reef Marine Park Authority, Townsville, 2000.

[102] D.W. Klumpp, A.D. McKinnon, Community structure, biomass, and productivity of epilithic algal communities on the Great Barrier Reef: dynamics at different spatial scales, Mar. Ecol. Prog. Ser. 86 (1992) 77–89.

[103] G. Diaz-Pulido, L.J. McCook, The fate of bleached corals: patterns and dynamics of algal recruitment, Mar. Ecol. Prog. Ser. 232 (2002) 115–128.

[104] B. Schaffelke, D.W. Klumpp, Biomass and productivity of tropical macroalgae on three nearshore fringing reefs in the central Great Barrier Reef, Australia, Bot. Mar. 40 (1997) 373–383.

[105] L.J. McCook, Effects of herbivores and water quality on *Sargassum* distribution on the central Great Barrier Reef: cross-shelf transplants, Mari. Ecol. Prog. Ser. 139 (1996) 179–192.

[106] L.J. McCook, Effects of herbivory on zonation of *Sargassum* spp. within fringing reefs of the central Great Barrier Reef, Mar. Biol. 129 (1997) 713–722.

[107] G.H. Wikfors, M. Ohno, Impact of algal research in aquaculture, J. Phycol. 37 (2002) 968–974.

References

[108] C. Duarte, N. Marba, M. Holmer, Rapid domestication of marine species, Science 316 (2007) 382–383.

[109] C. Duarte, et al., Will the oceans help feed humanity? Bioscience 59 (2009) 967–976.

[110] S. Kraan, Mass-cultivation of carbohydrate rich macroalgae, a possible solution for sustainable biofuel production, Mitig. Adapt. Strateg. Glob. Change 18 (2013) 27–46.

[111] A. Kijjoa, P. Sawangwong, Drugs and cosmetics from the sea, Mar. Drugs 2 (2004) 73–82.

[112] A.J. Smit, Medicinal and pharmaceutical uses of seaweed natural products: a review, J. Appl. Phycol. 16 (2004) 245–262.

[113] I. Mazarrasa, Y. Olsen, E. Mayol, N. Marba, Rapid growth of seaweed biotechnology provides opportunities for developing nations, Nat. biotechnol. 31 (7) (2013) 591–592.

[114] R.P. Singh, Studies on Certain Seaweed–Bacterial Interaction From Saurashtra Coast (Ph.D. thesis), 2013, 161 p. <http://ir.inflibnet.ac.in:8080/jspui/handle/10603/9191>.

[115] C. Burke, T. Thomas, M. Lewis, P. Steinberg, S. Kjelleberg, Composition, uniqueness and variability of the epiphytic bacterial community of the green alga *Ulva australis*, ISME J. 5 (2011) 590–600.

[116] R.P. Singh, C.R.K. Reddy, Seaweed–microbial interactions: key functions of seaweed-associated bacteria, FEMS Microbiol. Ecol. 88 (2014) 213–230.

[117] M.M. Bengtsson, Bacterial biofilms on the kelp *Laminaria hyperborea*. Dissertation for the degree philosophiae doctor (PhD) at the University of Bergen, Norway (2011) 140.

[118] F.G. Saavedra, Associations between microbes and macroalgae: host, epiphyte and environmental effects, zur Erlangung des Doktorgrades der Mathematisch Naturwissenschaftlichen Fakultät der Christian-Albrechts-Universität zu Kiel, Germany, 2011.

[119] C. Lam, A. Stang, T. Harder, Planktonic bacteria and fungi are selectively eliminated by exposure to marine macroalgae in close proximity, FEMS Microbiol. Ecol. 63 (2008) 283–291.

[120] K. Bouarab, P. Potin, F. Weinberger, J. Correa, B. Kloareg, The *Chondrus crispus-Acrochaete operculata* host-pathogen association, a novel model in glycobiology and applied phycopathology, J. Appl. Phycol. 13 (2001) 185–193.

[121] A.L. Lane, J. Kubanek, Secondary metabolite defenses against pathogens and biofoulers, 229–243, In: C.D. Amsler (Ed.), Algal Chemical Ecology, Springer-Verlag, Berlin, 2008, p. 313.

[122] M. Wahl, Ecological lever and interface ecology: epibiosis modulates the interactions between host and environment, Biofouling 24 (2008) 427–438.

[123] T. Harder, Marine epibiosis: concepts, ecological consequences, and host defence, In: H.C. Flemming, P.S. Murthy, R. Venkatesan, K. Cooksey (Eds.), Marine and Industrial Biofouling, Springer-Berlin Heidelberg, 2009. 219–231.

[124] E.P. Ivanova, I.Y. Bakunina, T. Sawabe, K. Hayashi, Y.V. Alexeeva, N.V. Zhukova, et al., Two species of culturable bacteria associated with degradation of brown algae *Fucus evanescens*, Microb. Ecol. 43 (2002) 242–249.

[125] C. Holmström, S. Egan, A. Franks, S. McCloy, S. Kjelleberg, Antifouling activities expressed by marine surface associated *Pseudoalteromonas* species, FEMS Microbiol. Ecol. 41 (2002) 47–58.

[126] M.E. Hay, Fish-seaweed interactions on coral reefs: effects of herbivorous fishes and adaptations of their prey, In: P.F. Sale (Ed.), The Ecology of Fishes on Coral Reefs, Academic Press, San Diego, 1992, pp. 96–119.

[127] V.J. Paul, E. Cruz-Rivera, R.W. Thacker, Chemical mediation of macroalgal-herbivore interactions: ecological and evolutionary perspectives, In: J.B. McClintock, B.J. Baker (Eds.), Marine Chemical Ecology, CRC Press, Boca Raton, FL, 2001, pp. 227–265.

[128] K. Hammerstrom, M.N. Dethier, D.O. Duggins, Rapid phlorotannin induction and relaxation in five Washington kelps, Mar. Ecol. Prog. Ser. 165 (1998) 293–305.

[129] G.A. Pearson, E.A. Serrao, S.H. Brawley, Control of gamete release in fucoid algae: sensing hydrodynamic conditions via carbon acquisition, Ecology 79 (1998) 1725–1739.

[130] G.B. Toth, H. Pavia, Water-borne cues induce chemical defense in a marine alga (*Ascophyllum nodosum*)., Proc. Natl. Acad. Sci. United States A. 97 (2000) 14418–14420.

CHAPTER 3

Biochemical compounds of algae: sustainable energy sources for biofuel production

İlknur Ak[1], Edis Koru[2], Gülen Türker[3], Ekrem Cem Çankırılıgil[4] and Macide Güneş Dereli[5,6]

[1]Faculty of Marine Science and Technology, Department of Aquaculture, Çanakkale Onsekiz Mart University, Çanakkale, Turkey [2]Faculty of Fisheries, Department of Aquaculture, Ege University, İzmir, Turkey [3]Department of Food Technology, Çanakkale School of Applied Sciences, Çanakkale Onsekiz Mart University, Çanakkale, Turkey [4]Sheep Breeding Research Institute, Bandırma, Balıkesir, Turkey [5]Faculty of Pharmacy, Department of Pharmacology, Istanbul University, Istanbul, Turkey [6]Graduate School of Health Sciences, Istanbul University, Istanbul, Turkey

3.1 Introduction

The limited fossil fuel reserves have declined rapidly due to the increasing energy demand [1]. Moreover, these sources are not renewable and they cause environmental problems such as greenhouse gases [2]. As a consequence of these general concerns, scientists have focused on using algae as a renewable energy source. Biofuel, which is produced from biomass such as algae and plants, is an alternative solution for the energy crisis and climate change problems. Biofuel is divided into three generations based on their features. The first-generation fuels are known to compete with food production since they need agricultural lands and freshwater. The second-generation biofuels, which are converted to energy by biological and thermochemical methods, also cause the same dilemma, and they are more restricted because of the higher costs of lignin removal [3]. However, the third-generation feedstocks, such as microalgae and macroalgae, are becoming more attractive currently due to their advantages, such as no competition for arable land, reduction of global warming, and cultivation capacity throughout the year [4–6]. In the last few years,

Handbook of Algal Biofuels
DOI: https://doi.org/10.1016/B978-0-12-823764-9.00026-1

© 2022 Elsevier Inc. All rights reserved.

58   3. Biochemical compounds of algae: sustainable energy sources for biofuel production

both micro- and macroalgae have been considered as third-generation fuel sources due to their high biomass productivity and high lipid content [4,7].

In the history of life on Earth, photosynthetic bacteria (cyanobacteria or blue-green algae) developed. These organisms were able to grow in abundance in extreme habitats, including hot water geysers, hypersaline waters, and acidic or alkali waters [8]. They constitute the most important actors of the first change in climate that occurred billions of years ago. They possess an excellent adaptation capacity to many environments by changing their metabolism and chemical composition [9]. This feature renders algae unique. Due to that adaptation capacity, they are rich in valuable compounds like hydrocolloids, fatty acids (FAs), amino acids, pigments, and polyphenols that are used as animal nutrition, jelling agents, nutritional supplements, pharmaceuticals, and biofuels. For instance, *Botryococcus braunii* is regarded as a renewable biodiesel source. This alga produces hydrocarbon when subjected to stress conditions like the deficiency of certain nutrients, high temperature, excessive light intensities, and pH changes [10,11]. To cope with these conditions, *Botryococcus* may produce hydrocarbons. Some algae species, such as *Neochloris*, *Nannochloropsis*, and *Chlorella*, store more than 50% dry weight (dw) oil in the stationary phase [12] and scientists consider these microalgae to be alternative biodiesel sources [4,13,14]. Also, the biochemical content of seaweeds varies according to their type and environmental factors. Brown and green seaweeds have 50% dw of carbohydrates, while the ratio increases to 60% dw for green seaweeds [15]. The carbohydrate content of seaweeds includes saccharides such as ulvan, fucoidan, alginate, agar, laminaran, mannitol, rhamnose, fucose, and uronic acids [5]. These saccharides allow seaweed to be used as a source of bioethanol and biogas. The general opinion was that seaweeds are not suitable to be used as a source of biodiesel since they contain less than 5% dw [16,17]. However, the total lipid content of *Dictyotales* ordo, which belongs to the brown algae class, varies between 11% and 20% dw [17]. Scientists stated that seaweeds could be a good candidate for biodiesel source by producing algal biomass with high lipid content [6,17].

Algae are considered by some to be the most crucial alternative energy source for the future, and the biochemical composition plays a crucial role in their suitability as a potential biofuel feedstock. In this chapter, the biochemical compounds of algae and the influence of cultivation conditions on those compounds and, consequently, biofuel production are given significance.

3.2 Lipids and fatty acids in algae

3.2.1 Lipid groups in algal cells

Lipids consist of FAs and their derivatives like esters or amides. Lipids have numerous roles in the physiology of algae and are classified into two main groups as polar and nonpolar lipids. The polar lipids are structural lipids, and nonpolar lipids are storage lipids. Glycolipids and phospholipids are polar lipids; hydrocarbons, monoglycerides, diglycerides, triglycerides, and pigments are nonpolar lipids [12]. The ratio of these two types of lipids varies according to the species, the growth stages, culture conditions, and geographical location [11,12]. FAs are principally used to synthesize structural lipids via esterification under normal conditions and constitute about 5%−20% dw [18]. These lipids possess long chains of FAs, and they can be transformed into polyunsaturated FAs (PUFAs). Also, PUFAs can be divided into two classes: short-chain polyunsaturated FAs and long-chain polyunsaturated FAs. Long-chain PUFAs play a crucial role in membrane systems to protect the cell and to help in the separation of intracellular organelles. PUFAs such as docosahexaenoic acid and eicosapentaenoic acid provide the membrane fluidity that is essential for metabolic processes, and take part in different intracellular membrane fusion events [19]. Some algae species like *Phaeodactyllum tricornutum* and *Trepacantha barbata* are rich in PUFAs (Table 3.1).

However, under unfavorable conditions, the metabolic pathway shifts toward triacylglycerol (TAG) in lipid metabolism [18]. For this purpose, methods such as nutrient salts and trace elements, the addition of CO_2, temperature, salinity, pH, light sources, organic carbon sources (such as glucose, hormones, growth factors), low-dose ionizing radiation applications, low-dose cold atmospheric plasma pressure exposure, and using lipid-free microalgal waste and waste glycerol are used to enrich the algae lipid contents [4,20–22]. The growth slows down in the early stationary phase and organic carbon can also be converted into nonpolar lipids. TAGs are shaped by joining three FAs into a glycerol backbone in a dehydration reaction, and these are the most compact form of known energy storage so far [23]. TAGs are synthesized in the algal cells when energy input exceeds the energy consumption, thus saving the cell from oxidation by removing excess energy and electrons [5,18]. FAs are synthesized to form various types of lipids [18]. Palmitic acid (C16:0) and oleic acid (C18:1) are the primary FAs in algae (Table 3.1).

3.2.2 Triacylglycerol synthesis

TAG biosynthesis in algae occurs via glycerol pathways; one is the acetyl CoA-dependent pathway (de novo FA synthesis), and another is the acyl CoA-independent pathway (Fig. 3.1). TAGs are synthesized from fatty acyl CoA and glycerol-3-phosphate (G3P). The dihydroxyacetone phosphate, which is a glycolytic intermediate product, is first reduced to G3P. Later, G3P is acylated with G3P acyltransferase to generate monoacylglycerol-3-phosphate (MG3P). MG3P reacts with another acyl CoA to form phosphatidic acid (PA). Diacylglycerol is formed by removing the phosphate group from PA, which reacts with a third acyl CoA molecule to form TAG. Adenosine triphosphate (ATP) does not play a role in the synthesis of TAGs. Instead, energy is provided by breaking the high-energy trioser bond between the acyl form and CoA [35].

The formation of malonyl CoA (MCoA) from acetyl CoA is an irreversible reaction with acetyl CoA carboxylase. The biotin prosthetic group is found in the domain unit of the acetyl CoA carboxylase enzyme in plant cells. The reaction takes place in two steps.

In the first step, biotin-bound carbon dioxide, which is called biotinyl carbon dioxide ($B-CO_2$), is formed. This $B-CO_2$ transfers its CO_2 to acetyl CoA, and thus MCoA occurs. In all organisms, long-chain FAs are synthesized by FA synthase enzymes in an iterative chain of reactions. The oxidation of FAs occurs in three steps. In the first step, acetyl CoA is formed by removing two carbon units from the carboxyl end of FAs. In the second step of FA oxidation, the acetyl groups of acetyl CoA enter the citric acid cycle in the mitochondrial matrix or glyoxylate cycle in the peroxisome. The first two steps of FA oxidation,

TABLE 3.1 The minimum and maximum values of total lipid and fatty acid constituents in some micro- and macroalgal species.

Species	Total lipid (%)	$C_{12:0}$	$C_{13:0}$	$C_{14:0}$	$C_{15:0}$	$C_{16:0}$	$C_{17:0}$	$C_{18:0}$	$C_{20:0}$	$C_{15:1}$	$C_{16:1}$	$C_{17:1}$
Chlorophyceae												
Scenedesmus obliquus	19.25–36.00	–	–	0.08–2.24	0.51–0.81	21.60–27.05	0.68–0.97	–	–	0.84–2.75	0.13–5.02	–
Clamydonomas reinhardtii	4.94–14.81	–	0.51–5.06	0.00–0.49	–	25.70–32.1	–	4.40–10.02	–		3.38–7.66	
Neochloris oleoabundance	11.00–52.00	3.43–5.50	–	–	–	14.46–25.47	–	–	–	–	4.01	3.83–9.86
Trebouxiophyceae												
Botryococcus branuii	25.00–29.22	4.86–6.68	2.89–7.11	1.84–2.69	0.87–1.07	34.84–38.37	4.55–5.01	4.00–4.86	0.03–0.09	0.92–1.22	9.21–9.64	–
Chlorella sp.	4.09–15.53	–	–	0.46–2.93	0.63–0.91	27.05–69.77	0.56–2.65	1.39–6.16	–	–	1.15–2.84	–
Cyanophyceae												
Arthrospira platensis	15.24–26.71	–	–	1.03–1.96	–	22.26–31.55	–	3.08–8.57	1.64–3.95	–	3.08–8.57	–
Bacillariophyceae												
Phaeodactyllum tricornutum	2.79–53.04	–	–	1.62–8.30	–	3.11–24.55	–	0.11–3.52	0.13–0.59	–	0.92–2.67	–
Phaeophyceae												
Padina pavonica	1.82–3.01	1.18–1.64	4.28–6.84	4.84–7.66	0.95–3.98	42.61–48.64	0.58–0.71	1.62–6.24	0.94–1.74	0.85–3.28	1.54–2.34	0.34–0.67
Trepacantha barbata	1.57–3.57	–	0.19–0.75	6.37–7.57	0.35–0.57	27.02–31.33	5.96–8.77	0.80–1.27	0.23–0.39	1.67–2.50	4.87–39.00	3.55–5.75
Florideophyceae												
Kappaphycus alvarezii	2.60–3.00	–	–	1.06–1.28	0.10	36.36–36.65	–	21.58–25.94	–	–	7.19–9.23	–
Palmaria palmata	0.40–1.80	–	–	5.98–22.51	–	20.83–53.35	–	0.40–2.62	–	–	–	–
Ulvophyceae												
Ulva rigida	2.30–5.59	0.01–5.59	–	0.38–3.30	0.00–0.33	30.21–44.02	–	0.08–1.27	0.00–0.80	–	2.90–10.27	0.00–1.71

Species	$C_{18:1}$	$C_{18:2}$	$C_{18:3}$	$C_{20:1}$	$C_{20:3}$	$C_{20:4}$	$C_{20:5}$	$C_{22:1}$	$C_{22:6}$	T_{SFA}	T_{UFA}	References
Chlorophyceae												
Scenedesmus obliquus	8.00−15.17	13.38−20.73	36.31−41.78	−	−	−	−	−	−	23.57−29.86	70.14−76.43	[24]
Clamydonomas reinhardtii	17.96−29.92	4.64−12.10	3.80−16.88	−	−	−	−	−	−	33.34−44.75	55.25−66.66	[25]
Neochloris oleoabundans	3.17−22.79	15.56−35.09	8.07−24.52	−	−	−	−	−	−	14.46−25.97	−	[26]
Trebouxiophyceae												
Botryococcus branuii	12.71−14.45	1.55−2.44	2.50−5.08	0.20−0.51	0.02−0.27	0.02−0.09	0.01−0.07	0.05−0.19	−	65.27−67.04	32.95−35.09	[27]
Chlorella sp.	1.47−5.96	2.86−20.37	10.13−30.20	−	−	−	−	−	−	29.79−77.44	22.56−70.21	[28]
Cyanophycae												
Arthrospira platensis	11.17−17.32	11.37−12.73	13.99−19.16	−	−	−	−	−	5.42−1.372	30.99−43.49	56.51−69.01	[29]
Bacillariophyceae												
Phaeodactyllum tricornutum	0.13−2.12	0.09−1.65	−	0.02−0.68	−	−	0.13−9.27	0.05−1.16		6.97−32.51	1.78−32.67	[30]
Phaeophyceae												
Padina pavonica	0.92−4.28	0.50−5.70	0.14−0.39	0.03−0.30	0.19−0.81	0.05−0.18	0.48−1.22	0.11−0.18	5.99−8.84	77.98−85.59	14.41−22.02	[31]
Trepacantha barbata	20.16−25.45	2.61−5.60	2.88−5.76	0.07−0.61	7.71−12.12	0.22−0.62	1.97−2.89	−	0.00−3.52	45.94−53.74	46.26−54.06	[32]
Florideophyceae												
Kappaphycus alvarezii	3.16−3.23	0.37−0.50	0.13−0.19	−	0.75−1.29	4.50−5.95	7.05−10.29	−	−	59.32−63.75	36.25−40.68	[33]
Palmaria palmata	3.30−9.04	0.34−1.28	0.11−1.36	−	−	0.02−0.10	1.76−59.69	−	0.01−0.08	22.13−57.82	42.18−77.87	[34]
Ulvophyceae												
Ulva rigida	16.18−52.59	1.52−4.56	0.16−10.13	−	−	0.00−0.18	0.00−0.27	−	2.30−5.33	31.88−70.10	29.90−64.84	[16]

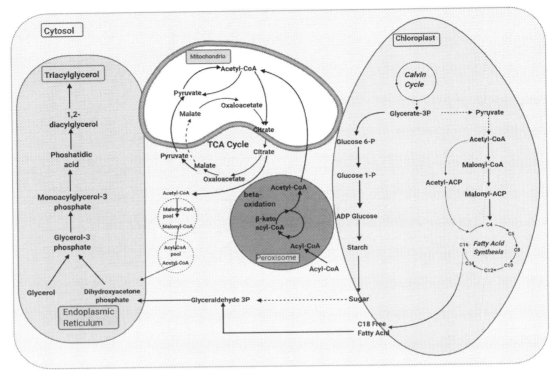

FIGURE 3.1 Simplified scheme of triacylglycerol synthesis in algae. *Source: Created with biorender.com.*

oxidation-reduction (REDOX) reactions, involve the loss or gain of electrons. Most redox reactions are metabolically driven by the breakdown of carbon-hydrogen bonds, resulting in electron loss or electron gain through the bond formation. These electrons act as NAD^+ and FAD^+ electron acceptors and form $NADH + H^+$ and $FADH_2$ by breaking up this electron transfer by hemolytic or heterolytic bonds. In the third step, NADH and $FADH_2$ are transferred to the respiratory chain in the mitochondria and produce ATP [35].

3.2.3 Factors affecting triacylglycerol synthesis

FA composition and TAGs synthesis in algae are affected by several growth conditions [5]. Factors that have an effect include nutrients, trace metals, carbon, temperature, light intensity and sources, salinity, pH, night illumination using monochromatic light-emitting diodes (LEDs), pretreatment with low-dose cold atmospheric pressure plasma (CAPP), and cultivation on wastes such as lipid-free microalgal biomass and waste glycerol [1,7,11,16,31,36]. Generally, the carbon fixation process in algae cells during photosynthesis may be influenced by all these factors. Thus TAGs and FA composition, which determine the efficiency of biofuels production, can be altered [3].

3.2.3.1 Nutrients and trace metals

Nitrogen is a vital element for forming amino acids, pigments, and enzymes, and it mainly affects the FA composition of primary production [3,12]. The biosynthesis and accumulation of lipids and TAGs are enhanced in algal culture under nitrogen starvation conditions [27,30,37,38]. The effects of nitrogen concentrations on biomass and lipid production of algae have been investigated in many studies. Ruangsomboon [27] stated that lipid content and lipid yield of *B. braunii* could be increased by applying a two-step culture. In the first step, a nitrogen sufficient medium can be used to increase lipid yield for 30 days, and then a nitrogen-deficient medium can be used for the next step. They also reported that using potassium nitrate as a nitrogen source increased the oil content of *Botryococcus branuii*. Yodsuwan et al. [30] found that nitrogen concentration has an inverse relationship between the specific growth rate and lipid accumulation of *Phaeodactylum tricornutum*. Avila-Leon et al. [26] reported that the lipid and FA composition of *Neochloris oleoabundans* was influenced by nitrogen starvation and stressing agents such as glycerol, sodium chloride, and sodium thiosulfate. Li et al. [28] observed that the saturated FA content of *Chlorella* sp. increased under nitrogen starvation and blue LED light conditions.

Phosphorus, which has an important function in metabolic processes, is another essential nutrient for algae growth. It plays a crucial role in signal transmission, energy transfer, and cellular activities. In its deficiency, it increases lipid accumulation of algae [12]. Studies have shown that phosphorus starvation enhanced the total lipid content in, *P. tricornutum*, *Chaetoceros* sp., *Pavlova lutheri*, and *Isocrysis galbana* [39]. However, lipid content decreases in *Chlorella* sp. in phosphate starvation conditions [28].

In addition, sulfur is one of the critical components for algae, and it is essential for lipid metabolism, the electron transport chain, and protein biosynthesis. Sulfur limitation decreases cell volume, growth rates, and photosynthetic activity, but it increases total lipid content in *Chlamydomonas reinhardtii* [40].

Magnesium also has a vital role in algae. The starvation of this element inhibits cell division. Ren et al. [41] investigated the effects of iron, magnesium, and calcium on the biomass and lipid accumulation of *Scenedesmus* sp., and they stated that lipid synthesis in this alga were increased by iron, magnesium, and calcium with the addition of EDTA in the culture. Calcium also regulates the lipid synthesis calcium signals in microalgae. They conclude that the total lipid content of *Scenedesmus* sp. were increased fourfold by adding calcium to the culture. Gorain et al. [42] studied the effects of Mg^{2+}, Ca^{2+}, and Na^+ on increasing lipid accumulation in *Chlorella vulgaris* and *Scenedesmus obliquus*. They found that calcium deficiency increased the lipid contents of both microalgae.

The trace metals, which are used in small quantities, are crucial for the metabolic functions of algae. The most critical trace metals are iron, manganese, zinc, cobalt, nickel, and copper, and among them, iron is the most important due to its effect on gene expression and metabolism regulation in algae [12]. Sun et al. [43] investigated the effect of nitrogen deficiency, light intensity, and iron on the FA profile on *N. oleoabundans*. They found that the light intensity, iron concentration, and nitrogen levels affected the TAGs and carbohydrate synthesis of alga. Other trace metals, such as copper, cadmium, and zinc, can affect both the algal biomass and lipid production. Sibi et al. [44] worked on copper stress on *Chlorella* species, and their results showed that copper stress promoted lipid production of alga.

3.2.3.2 Carbon sources

Carbon is an essential nonmineral nutrient, and it influences lipid production in algae cells [3]. The lipid content of algae varies depending on the carbon sources (inorganic or organic) used. Several studies have focused on the effect of inorganic and organic carbon sources on lipid and FA compositions of algae [1,14,24,45]. Ji et al. [24] found that the lipid synthesis of *S. obliquus* was enhanced with high CO_2 concentrations. On the other hand, excess CO_2 levels may destroy the lipid synthesis of algae [19].

Plant hormones, growth factors, vitamins, and glucose are among the organic carbon sources. Plant hormones affect the protein, lipid, and polysaccharide synthesis in algae. Growth regulators such as gibberellic acid, salicylic acid, jasmonic acid, and hormones such as melatonin and growth hormones are among the components that increase the lipid synthesis of algae. Du et al. [45] studied the effect of growth regulators on the lipid content of green alga, *Chlorella pyrenodosa*. As a result of the study, it was reported that 1-naphthaleneacetic acid and gibberellic acid growth regulators increased the lipid content of alga. El-Sheekh et al. [14] observed that the kinetin hormone stimulates the lipid production of *Desmodesmus intemedius*. Also, γ-aminobutyric acid promoted the growth and lipid production of *Monoraphidium* sp. cultures exposed to Cd^{2+} stress [22].

Glucose is also used for biodiesel production from algae. A study in which glucose was used as an organic carbon source showed that the lipid and MUFA content of *Auxenochlorella protot_hecoides* significantly increased with glucose addition [46]. Ak et al. [16] cultured the green seaweed *Ulva rigida* in five different sodium acetate concentrations. According to their results, saturated FA levels of *U. rigida* cultured in different sodium acetate concentrations were remarkably high. El-Sheekh et al. [7] also reported that the growth rate and lipid content of the green alga *C. vulgaris* positively affects sodium acetate.

Waste glycerol is also used in algae cultivation as an organic carbon source to reduce production cost [1,36]. Xu et al. [36] found that the lipid content of *S. obliquus* increases waste glycerol. In addition, the FA content of *S. obliquus* could be enhanced by waste glycerol and lipid-free algal hydrolysate combinations [1]. Chen et al. [13] reported that crude glycerol promotes lipid production in *Thraustochytrium* sp. BM2.

3.2.3.3 Temperature

Temperature is a key factor that influences the lipid production and formation of FAs [3,19]. Converti et al. [47] reported that the total lipid content of *C. vulgaris* decreased with increasing temperature, whereas the total lipid content of *Nannochloropsis oculata* increased with increasing temperature. The FA content of algae is also affected by temperature changes. Chaisutyakorn et al. [48] studied the effects of temperature on *Nannochloropsis* sp., *Tetraselmis suecica*, and *Chaetoceros* sp. They found that the saturated and monosaturated FA contents of *Nannochloropsis* sp. and *T. suecica* decreased with increasing temperature. They also reported that temperature did not change the FA composition of *Chaetoceros* sp.

3.2.3.4 Light intensity and sources

The lipid content and FA composition of algae influence light intensities, light sources, and wavelengths [19,21,43,49]. Previous studies have shown that low light intensity promotes polar lipid formation in algae, whereas higher light intensity reduces the polar lipid content [43]. The total lipid content of *N. oleoabundans* increased slightly with increasing light intensity but the FA profile did not change [43]. Takeshita et al. [50] studied the effect of high light intensity on lipid synthesis of *Parachlorella beijerinckii, Chlorella kessleri, Chlorellavulgaris, Chlorellaemersonii, Chlorella sorokiniana,* and *Chlorellaviscosa.* They stated that very high light intensity affects the FA composition of these six species, and they concluded that the types of FAs synthesized could be changed according to the species and strain. However, the high light intensity does not affect the TAG content of some microalgae species. Breuer et al. [51] reported that the light intensity did not affect the TAG synthesis of *S. obliquus.*

Also, a few works have been carried out on the effect of different wavelengths on total lipid and FA composition of algae. Total lipid content of *N. oleoabundans* was elevated by applying white LED lights [26]. Li et al. [28] searched the effect of LED wavelength and light intensity on the lipid composition of *Chlorella* sp. Their results showed that blue LEDs could prompt monounsaturated FA content for biodiesel production. Abomohra et al. [21] examined the effect of LEDs on the growth and biodiesel yield of the green microalga *S. obliquus.* At the end of the study, they determined that the blue—red LEDs combination significantly increased lipid production and biodiesel recovery. Also, LEDs' light affects the FA composition of seaweeds. Öztaşkent and Ak [32] reported that the LEDs influenced the monounsaturated and polyunsaturated FA levels of *Treptacantha barbata.*

3.2.3.5 *Salinity*

Salinity is also a crucial factor that affects the lipid and FA composition of algae. Salinity stress is one of the preferred methods to increase lipid content [3]. The lipid content of *Arthrospira platensis* increased with increasing salt concentration [29]. Avila-León [26] reported that the total oil content of *N. oleoabundans* varies according to the sodium chloride concentration in the environment. They also observed that increasing salinity levels from 1 to 22.2 mM enhanced the lipid synthesis of alga. Pugkaew et al. [52] investigated the effects of salinity changes on lipid and FA composition of *T. suecica.* They found that lipid content and FA compositions of alga did not affect the salinity changes. Also, El Maghraby and Fakhry [31] reported the same results for *Ulva linza, Jania rubens,* and *Padina pavonica.* Meanwhile, in a study by Renaud and Parry [53], *Isochrysis* sp. and *N. oculata* were cultivated with salt concentrations. The highest total FA amounts were found to be 35 ppt for *N. oculata* and 30 ppt for *Isochrysis* sp. Zhu et al. [54] found the highest lipid content at a salinity between 0.9% and 3.6% in *Schizochytrium limacinum.* However, they also reported that the FA composition of *S. limacium* changed at 0% salinity.

3.2.3.6 *pH*

pH affects the solubility of CO_2 and essential nutrients, so it is vital for the metabolism of algae. Researchers have reported different optimum pH values for different algae species. According to Ochoa-Alfaro [25], the highest lipid productivity was achieved at 7.8 for *C. reinhardtii.* Rai et al. [55] found that lipid accumulation was increased at pH 8 in *Chlorella* sp. However, Qiu et al. [56] recorded the maximum lipid accumulation for *C. sorokiniana* at pH 6.

3.2.3.7 *Low-dose rate of ionizing radiation*

Low-dose ionizing radiation applications are a promising alternative route for enhanced lipid production by algae. Gamma (γ) rays are high-energy electromagnetic ionized rays. Even low doses of these rays increase lipid synthesis by stimulating algal cells. Studies have shown that low doses of radiation increase the physiological activities of algae and plant cells. It has been determined that low-dose γ-radiation applied to *A. platensis* (a blue-green algae) increases the lipid content of the alga [20]. Jeong et al. [57] found that the lipid content of *T. suecica, Dunaliella tertiolecta, P. tricornutum,* and *Nannochloropsis oceanica* increased with low-dose γ-radiation. Tale et al. [58] reported that the lipid content of two different *C. sorokiniana* strains was enhanced with the low-dose γ-radiation they applied. Similar results were also reported for *Arthrospirafusiformis* [59].

3.2.3.8 *Low-dose cold atmospheric pressure plasma*

CAPP generates excited ions, atoms, and molecules. The effects of this application affect the cells of organisms. CAPP promotes seed replication and causes cell division [60]. The low dose of CAPP is a new method to increase the lipid content of algae. This method was first used by Almarashi et al. [61] to create physical stress on *C. vulgaris* for biodiesel production. As a result of the study, it was determined that the lipid content of algae that can be used in biodiesel production with CAPP increased.

3.3 Carbohydrates

3.3.1 Carbohydrates in algae

Photosynthetic organisms synthesize carbohydrates, which are essential molecules for all living cells. The general formula of carbohydrates can is $(CH_2O)n$, where n lies between 20 and 3000 [62]. They serve as essential structural components and as energy sources, that is, lipids and saccharides, in organisms. A saccharide is described as a group that contains sugars, starch, or cellulose. The saccharides are classified into four groups: monosaccharides, disaccharides, oligosaccharides, and polysaccharides.

Monosaccharides are either aldehydes or ketones that have one or more hydroxyl groups. Carbon atoms to which hydroxyl groups are attached are usually chiral centers, and stereoisomerism is common among monosaccharides [35]. They are the monomeric units of oligosaccharides and polysaccharides and, these molecules have more than one saccharide unit. Monosaccharides with four carbon atoms are called tetroses, those with five carbon atoms are called pentoses, those with six carbon atoms are called hexoses, and those with seven carbon atoms are called heptoses [35]. The carbonyl group of monosaccharides is covalently bonded along the chain with the oxygen of a hydroxyl group. Carbon atoms are numbered beginning from the reactive end of the molecule, the CHO (aldose) or "C" double bonded "O" (ketose) end of the molecule. Each carbon atom is then numbered in order through the end of the chain. The position of the OH group is important when numbering stereoisomers with more than three carbon atoms. The position of the OH group next to the last carbon atom determines whether it is an L or D stereoisomer. placed on the right are called D-sugars. In contrast, all sugars with hydroxyl groups in the highest chiral carbon atoms rotate plane-polarized light anticlockwise.

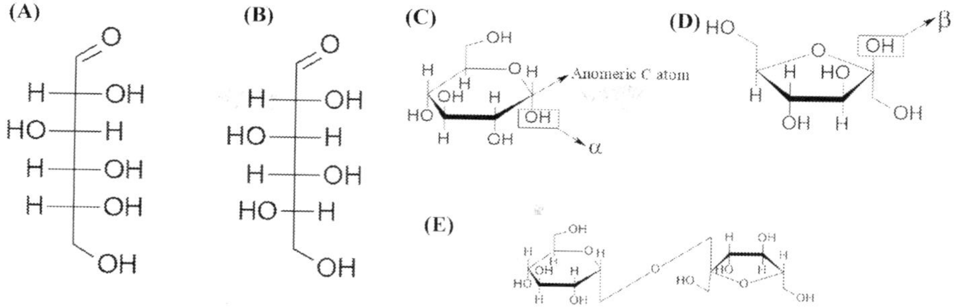

FIGURE 3.2 Structure of the main sugars: (A) D-glucose; (B) L-gluose; (C) α-D-glucopyranose; (D) 3-D-fructo-furanose; (E) O-α-D-glucopyranosyl (1→2)-3-D-fructofuranoside. *Source: Created with ChemSkecth.*

Carbonyl groups of aldehydes are reactive, and the oxygen atom of a hydroxyl group produces a hemiacetal. The reaction of the carbonyl group of a monosaccharide molecule with one of the hydroxyl groups for hemiacetal formation occurs intramolecularly (Fig. 3.2A and B). A six-membered ring is called a pyranose ring, and the oxygen atom of the hydroxyl group at C5 position reacts leading to ring formation. Monosaccharides can also occur in a five-membered ring called the furanose ring (Fig. 3.2C−E). The pyranose ring is more stable than the furanose ring, and as a consequence, most of the free pentoses, hexoses, and heptoses occur in pyranose rings. Several oligosaccharides and polysaccharides possess furanose rings [63]. When the hydroxyl group locates below the ring with D-sugars, it is called the alpha form, and the structure is described α-D-glucopyranose. The configuration with the newly shaped hydroxyl group at C1 position located above the ring is called the beta position, and the structure is termed β-D-glucopyranose. The new chiral carbon atom is called the anomeric carbon atom, and it can be classified as alpha (α) and beta (β) anomers. L-Series are the mirror image of the D-series. The anomeric hydroxyl group is situated in the alpha anomer and down in the beta anomer in L-series [35,63].

Disaccharides consist of two monosaccharides linked covalently by a glycosidic linkage shaped when a hydroxyl group on one sugar reacts with another anomeric carbon. The position of anomeric carbon atoms might be designated α- or β- or a combination of the two. There is no exact specification for the size range of oligosaccharides [63]. An oligosaccharide is a carbohydrate form built by the addition of three to 20 units of monosaccharides.

Polysaccharides are macromolecules of linked monosaccharide units and have a high molecular mass dependent on the number of its composing units, such as glucose, fructose, and glyceraldehyde [35]. Among these groups, polysaccharides are the most preferred form in bioethanol production [64]. Most of the remaining carbohydrate is transformed into polysaccharides which are also known as polymers of sugars [63]. The polysaccharides are categorized according to their biological functions, like structural polysaccharides and energy storage [65].

3.3.2 Carbohydrates synthesis in algae

Algal respiration encompasses the prevalent glycolytic and oxidative pentose phosphate pathways. Carbohydrates are storage products for energy and carbon in algae which are synthesized in dark reactions of photosynthesis. The tricarboxylic acid cycle, oxidative pentose phosphate pathway, glycolysis, and oxidative phosphorylation, which are part of dark reactions, are all essential for the production of proteins, nucleic acids, structural polysaccharides lipids. Organic carbon skeletons can also be generated in the light phase of photosynthesis in the oxidative pentose phosphate cycle. Besides, ATP and NADPH from the oxidative pentose phosphate cycle and oxidative phosphorylation pathway can be synthesized via light-dependent reactions [66].

The conversion of β-D-galactose to glucose-1-phosphate, which is often not suitable for other sugar types the sugar-like galactose and fructose converted first into the intermediate glycolytic pathway is synthesized by the action of four enzymes that constitute the Leloir pathway (Fig. 3.3). To complete the pathway, UDP-galactose is converted to UDP-glucose by UDP-galactose 4-epimerase. [67].

The low-molecular-weight compounds of algae play crucial roles in osmoregulation, while other compounds such as mannitol represent essential storage carbohydrates. Storage polysaccharides are found in red, brown, and green seaweeds, and cyanophytes [68,69]. Most green algae store starch, which is a mixture of branched and unbranched molecules [68].

Microalgae like *Spirulina*, *Chlorella*, *Chlamydomonas*, *Dunaliella*, and *Scenedesmus*, have a plentiful amount of starch, glycogen cellulose, or fermentable biomass content, which are used in bioethanol production [28,69,70]. Seaweeds like *Laminaria*, *Gracilaria*, *Gelidium*, *Ulva*, and *Enteromorpha* may also be utilized in bioethanol production by converting their storage compounds to fermentable sugars (Table 3.2). Furthermore, each seaweed class produces a range

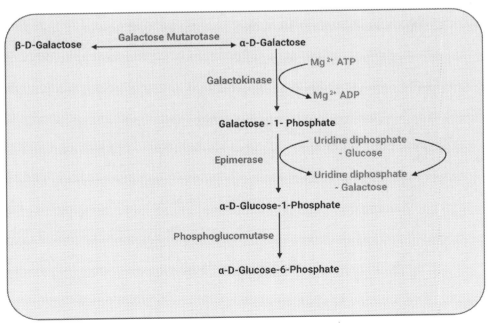

FIGURE 3.3 The Leloir pathway. Galactose and other similar sugar derivatives are converted into glucose by the epimerase enzyme. *Source: Created with biorender.com.*

TABLE 3.2   The total carbohydrate contents of microalgae and macroalgae.

Species	Total carbohydrate % dw	References
Chlorophyceae		
Scenedesmus obliquus	23.90	[24]
Scenedesmus sp.	34.48	[43]
Acutodesmus obliquus	19.77	[71]
Chlamydomonad reinhardtii	22.50	[70]
Chlamydomonas sp.	56.35	[43]
Neochloris oleoabundans	44.16	[43]
Eustigmatophyceae		
Nannochloropsis sp.	29.20	[72]
Nannochloropsis gaditana	21.30	[73]
Trebouxiophyceae		
Chlorella sp.	75.90	[28]
Chlorodentrophyceae		
Tetraselmis suecica	24.00	[49]
Tetraselmis sp.	22.00	[74]
Coccolithophyceae		
Iscochrysis galbana	10.64	[37]
Cyanophyceae		
Spirulina platensis	67.50	[69]
Porphyridiophyceae		
Porphyridium purpureum	33.00	[75]
Florideophyceae		
Gracilaria crassa	17.07	[76]
Gelidiella cornea	36.29	[77]
Gelidiella acerosa	16.62	[76]
Eucheuma isiforme	25.89	[77]
Phaeophyceae		
Sargassum filipendula	3.73	[77]
Dictyotha dichotoma	33.35	[78]
Padina gymnospora	1.86	[77]
Padina pavonica	14.73	[76]

(*Continued*)

TABLE 3.2 (Continued)

Species	Total carbohydrate % dw	References
Chlorophyceae		
Ulva pertusa	48.20	[79]
Ulva reticulata	15.37	[76]
Ulva rigida	65.93	[80]
Enteromorpha compressa	44.57	[78]
Enteromorpha intestinalis	34.28	[78]
Codium isthmocladum	16.72	[77]
Caulerpa racemosa	3.60	[77]
Halimeda tuna	17.12	[76]

of distinct compounds [68]. The carbohydrates of most green seaweeds are starch, sucrose, cellulose, and ulvan but the members of the genus *Acetabularia* often store inulin. Red algae primarily store carrageenan, agar, cellulose, lignin, and floridean starch. Brown seaweeds contain plenty of sugars such as laminarin, mannitol, alginate, cellulose, and fucoidan [68]. Table 3.2 summarizes the total carbohydrate content of some microalgae and seaweeds.

3.3.3 Factors affecting carbohydrate synthesis

The life cycle of algae and growth conditions can affect the carbohydrate synthesis in algae. Nutrient and trace metals, carbon sources, light sources and light intensity, temperature, salinity, and pH can influence the carbohydrate contents in algae [24,69,73,81].

3.3.3.1 *Nutrients and trace metals*

Nitrogen is a sine qua non nutrient for algal growth, and under nitrogen starvation, most algae can synthesize a large proportion of carbohydrates. Sun et al. [43] observed that nitrogen starvation caused an increase in the carbohydrate content of *N. oleoabundans*. They also found that the required period of nitrogen starvation for carbohydrate accumulation was shorter than for lipid accumulation. In contrast, Li et al. [28] observed that nitrogen starvation did not elevate the carbohydrate content of *Chlorella* sp. Similar results were also reported for *Isochrysis galbana* [37]. Also, Cyanobacteria can accumulate carbohydrates under nitrogen starvation conditions. According to De Philippis et al. [82], the glycogen was accumulated up to 70% dw in *Spirulina maxima* under nitrogen starvation conditions.

Phosphorus is also a crucial element in organic molecules. Its deficiency can change biochemical composition, especially carbohydrate synthesis. The highest carbohydrate content of *Porphyridium purpureum* was achieved at 0.14 g/L KH_2PO_4 concentration [75]. Li et al. [28] reported that phosphorus starvation enhanced the carbohydrate synthesis of *Chlorella* sp.

Sulfur is involved in the structure of polysaccharide, sulfolipids, and proteins, and the deficiency of this element affects the carbon metabolism of algae. Microalgae store a small amount of sulfur, and its deficiency affects them very rapidly. Ball et al. [83] observed that sulfur deficiency increased the carbohydrate content of *C. reinhardtii*. Arad [84] found that the polysaccharide content of red microalgae *Porphyridium* sp. under sulfate deficiency was reduced to half.

Iron is essential for photosynthetic components, and its deficiency or addition affects the carbohydrate synthesis of algae. Iron deficiency caused a decrease in the carbohydrate production of *Chaetoceros brevis* [85]. Sun et al. [43] searched the effects of Fe^{3+} on carbohydrate accumulation by *N. oleoabundans*. According to their results, carbohydrate content did not change with the addition of Fe^{3+}.

Hanifzadeh et al. [86] concluded that micronutrient optimization was essential to maximize biomass, lipid, and carbohydrate productivity of *C. sorokiniana* str. SLA-0. Ball et al. [83] found that manganese and potassium starvation increased the carbohydrate content of *C. reinhardtii*. Mohy El-Din [87] found that copper and lead affected the carbohydrate content of *Spirulina platensis*, and also observed that concentrations of the tested metals higher than 1 mg/L did not decrease the carbohydrate content. However, other studies showed that the copper toxicity decreased carbohydrate content in *D. tertiolecta* and *T. suecica* [88]. Also, selenium increased the carbohydrate content of *Gracilaria lemaneiformis* and *Ulva fasciata* [89].

3.3.3.2 Carbon sources

Some algae can use carbon in both inorganic and organic forms, and these sources affect their carbohydrate content. Carbohydrate is the first assimilatory product of carbon, and excess carbon is mobilized to lipids in algae. As a consequence, both are positively correlated [80]. Ji et al. [24] reported that the carbohydrate content of *S. obliquus* improved with 10% CO_2 in a culture medium. Izumo et al. [90] observed an inverse relationship between the carbohydrate content of *C. kessleri* and the carbon dioxide concentration in the environment. Algae can also use organic carbon sources under mixotrophic or heterotrophic conditions. For instance, Choix et al. [91] reported that *Chlorella* spp. can increase the carbohydrate content under heterotrophic culture conditions. In addition, Abreu et al. [92] showed that *C. vulgaris* synthesized more carbohydrates in mixotrophic cultivation than in phototrophic cultivation. Besides, the carbohydrate content of *C. kessleri* is enhanced with glucose [93]. Also, glycerol stimulates the carbohydrate production of *S. obliquus* [36]. Plant hormones also affect the carbohydrate production of algae. The carbohydrate content of *Acutodesmus obliquus* increased with cytokinin supplementation [71].

There are very few studies on the use of vitamins, which are organic carbon sources, to increase the carbohydrate content of algae. Ruangsomboon [10] reported that the addition of thiamine increases the hydrocarbon content in *B. braunii*, and the maximum carbohydrate content is achieved with the addition of cobalamin.

3.3.3.3 Temperature

The temperature also influences the carbohydrate content of algae. The optimum temperature varies depending on the species. Li et al. [70] reported that below the optimum temperature levels, the carbohydrate content of *C. reinhardtii* decreased.

72   3. Biochemical compounds of algae: sustainable energy sources for biofuel production

Araújo and Garcia [94] observed that the carbohydrate content of *Chaetoceros* cf. *wighamii* was high when they applied low temperature. However, Ogbonda et al. [95] increased the culture temperatures from 25°C to 40°C, and they found that the carbohydrate content of *Spirulina* sp. also slightly increased.

3.3.3.4 *Light sources and light intensity*

Light is the essential factor for both micro- and macroalgae. The source and intensity of light affect the biochemical composition of algae. LEDs are preferred light sources for increasing the carbohydrate content of algae. Li et al. [81] reported that red light with phosphorus starvation enhanced carbohydrate production of *Chlorella* sp. [81]. Li et al. [70] found that combined red–orange light at 24°C enhanced the carbohydrate content of *C. reinhardtii*. The mixed blue–red LED light also affected the polysaccharide content of *Ulva pertusa* [79]. Also, light intensity affects the carbohydrate accumulation of algae. Carbohydrate synthesis of *N. oleoabundans* was elevated by increasing the light intensity up to 100 μmol/photon m^2/s under nitrogen starvation conditions [43]. However, a further increase in light intensity did not support carbohydrate synthesis.

Onay [73] showed that the carbohydrate content of *Nannochloropsis gaditana* increased with increasing light intensity. Also, it was stated that carbohydrate and lipid contents of *Rhodymenia pseudopalmata* increased when cultured in continuous light-saturated conditions [96]. In contrast, Fábregas et al. [72] found that the carbohydrate content of *Nannochloropsis* sp. decreased with increasing light intensity. Ho et al. [38] determined that the carbohydrate content of *S. obliquus* increased until photoinhibition occurred at 420 μmol photon m^{-2}s^{-1}. Li et al. [28] observed that the carbohydrate content of *Ulva prolifera* did not influence the light and dark phase, and thalli continued to synthesize cell constituents at night.

3.3.3.5 *Salinity*

Salinity is also a crucial factor that affects both microalgae and seaweeds. The green alga *D. tertiolecta* adapted to high salt concentrations by synthesizing glycerol [97]. Also, it was reported that the carbohydrate content of *Nannochloropsis* sp. and *Tetraselmis* sp. increased at 30 ppt salinity [74]. However, Pliego-Cortés et al. [96] observed that the carbohydrate content of *R. pseudopalmata* tended to increase with decreasing salinity concentrations. Zheng et al. [98] concluded that the carbohydrate content of *U. prolifera* was enhanced at salinity levels of 35 ppt. In contrast, Pugkaew et al. [52] stated that salinity changes did not affect the lipid and FA composition of *T. suecica*.

3.3.3.6 *pH*

Another critical factor that seriously affects the carbohydrate content of algal cultures is pH. The optimum pH range in algae cultivation is 7.5–9. The highest carbohydrate contents were found for *Dunaliella bardawil* [99], *Nannochloropsis* sp. and *Tetraselmis* sp. [74], at pH 7.5 and pH 9 for *Chlorella ellipsoidea* [99].

3.3.3.7 *Low-dose gamma radiation*

Low-dose gamma radiation stimulates carbohydrate synthesis in algae [20,59]. Abomohra et al. [20] found that low-dose gamma radiation stimulated carbohydrate

Handbook of Algal Biofuels

content, and atmospheric pressure plasma generated excited ions, atoms, and molecules in *A. platensis*. Similar results were reported for *A. fusiformis* [59] and *C. vulgaris*[100].

3.4 Proteins

Proteins consist of countless small units called amino acids, and are an important class of biological macromolecules. Twenty types of amino acids are used in protein production. Among these 20 amino acids, only eight are essential amino acids. The other 12 amino acids are classified as nonessential amino acids. The essential amino acids are leucine, isoleucine, lysine, phenylalanine, methionine, threonine, valine, and tryptophan. However, histidine and arginine are also considered essential for children. Amino acids in the structure of proteins are in the form of L-amino acids. They are covalently joined by peptide bonds called polypeptides.

Moreover, protein consists of 100 - 1000 peptide bonds. Proteins are structures made up of 100 of individual covalent bonds. Each protein has a specific function, and each has a different three-dimensional structure [35].

The protein concentration of algae ranges from 5% to 70% of dry basic. The protein content varies depending on environmental conditions and species. Algae protein contains all amino acids, such as alanine, aspartic acids, arginine, glycine, glutamic acid, and proline. They need nitrogen compounds to synthesize proteins [27]. Nitrogen is a fundamental element in many molecules such as DNA, RNA, protein, and chlorophyll for algae. The limitation or starvation of these compounds changes protein, lipid, and carbohydrate synthesis [21,26].

When the studies on the protein contents of algae were examined, we concluded that a decrease in protein synthesis and growth in algae promote the polysaccharide and lipid accumulation. This situation shows us that the cultivation of such algae is suitable for biofuel production.

3.5 Conclusion

Algae are widely used in numerous areas, such as in the food, feed, medicine, fertilizer, chemical, and cosmetic industries, due to the various chemical compounds algae can produce. Ninety-six percent of the aquaculture products obtained by aquaculture are algae, and research into the biochemical composition of these photosynthetic organisms has gained momentum in the last 50 years. In recent years, sustainable resources have been used to meet the increasing energy needs. Scientists consider algae as a third-generation clean energy source due to their rapid growth capacity and high lipid and carbohydrate content. The metabolic pathways of the algal lipids and carbohydrates used in biofuel sources show differences according to species, environmental factors, and culture conditions. In recent years, scientists have focused on biomass production to evaluate high biomass yield and improve these valuable compounds.

In this chapter, we have presented a review on lipid, carbohydrates, and protein synthesis by algae and the individual effects of nutrients (nitrogen, phosphorus, sulfur, magnesium, calcium), trace metals, carbon, temperature, light intensity, salinity, and pH on the lipid, carbohydrate, and protein composition of algae. According to the literature, the metabolic pathways of algae influence the growth conditions. As a consequence, the cultivation techniques for biofuel sources, standardization, and optimization have gained importance. Further research should focus on the biomass production needed to evaluate biomass composition manipulation techniques for biofuel yield.

References

[1] A.E.-F. Abomohra, H. Eladel, M. El-Esawi, S. Wang, Q. Wang, Z. He, et al., Effect of lipid-free microalgal biomass and waste glycerol on growth and lipid production of *Scenedesmus obliquus*: innovative waste recycling for extraordinary lipid production, Bioresour. Technol. 249 (2018) 992–999.

[2] T. Suganya, M. Varman, H.H. Masjuki, S. Renganathan, Macroalgae and microalgae as a potential source for commercial applications along with biofuels production: a biorefinery approach, Renew. Sustain. Energy Rev. 55 (2016) 909–941.

[3] A. Juneja, R.M. Ceballos, G.S. Murthy, Effects of environmental factors and nutrient availability on the biochemical composition of algae for biofuels production: a review, Energies 6 (9) (2013) 4607–4638.

[4] A.E.-F. Abomohra, A.W. Almutairi, A close-loop integrated approach for microalgae cultivation and efficient utilization of agar-free seaweed residues for enhanced biofuel recovery, Bioresour. Technol. 317 (2020) 124027.

[5] K. Kumar, S. Ghosh, I. Angelidaki, S.L. Holdt, D.B. Karakashev, M.A. Morales, et al., Recent developments on biofuels production from microalgae and macroalgae, Renew. Sustain. Energy Rev. 65 (2016) 235–249.

[6] A.E.-F. Abomohra, A.H. El-Naggar, A.A. Baeshen, Potential of macroalgae for biodiesel production: screening and evaluation studies, J. Biosci. Bioeng. 125 (2) (2018) 231–237.

[7] M.M. El-Sheekh, S.F. Gheda, A.E.-K.B. El-Sayed, A.M. Abo Shady, M.E. El-Sheikh, M. Schagerl, Outdoor cultivation of the green microalga *Chlorella vulgaris* under stress conditions as a feedstock for biofuel, Environ. Sci. Pollut. Res. 26 (18) (2019) 18520–18532.

[8] A. Meinesz, How Life Bega—Evolution's Three Geneses, The University of Chicago Press, Chicago, IL, 2008.

[9] J. Seckbach, Algae and cyanobacteria in extreme environments, Cellular Origin, Life in Extreme Habitats and Astrobiology, Vol. 11, Springer, Dordrecht, 2007.

[10] S. Ruangsomboon, P. Sornchai, N. Prachom, Enhanced hydrocarbon production and improved biodiesel qualities of *Botryococcus braunii* KMITL 5 by vitamins thiamine, biotin and cobalamin supplementation, Algal Res. 29 (2018) 159–169.

[11] T. Manchanda, R. Tyagi, D.K. Sharma, Application of nutrient stress conditions for hydrocarbon and oil production by *Botryococcus braunii*, Biofuels 10 (3) (2019) 271–277.

[12] B. Sajjadi, W.-Y. Chen, A.A.A. Raman, S. Ibrahim, Microalgae lipid and biomass for biofuel production: a comprehensive review on lipid enhancement strategies and their effects on fatty acid composition, Renew. Sustain. Energy Rev. 97 (2018) 200–232.

[13] C.-Y. Chen, M.-H. Lee, Y.K. Leong, J.-S. Chang, D.-J. Lee, Biodiesel production from heterotrophic oleaginous microalga *Thraustochytrium* sp. BM2 with enhanced lipid accumulation using crude glycerol as alternative carbon source, Bioresour. Technol. 306 (2020) 123113.

[14] M.M. El-Sheekh, H.M. Eladel, A.E.-F. Abomohra, M.G. Battah, S.A. Mohamed, Optimization of biomass and fatty acid productivity of *Desmodesmus intermedius* as a promising microalga for biodiesel production, Energy Sources Part A (2019) 1–14.

[15] K. Sudhakar, R. Mamat, M. Samykano, W.H. Azmi, W.F.W. Ishak, T. Yusaf, An overview of marine macroalgae as bioresource, Renew. Sustain. Energy Rev. 91 (2018) 165–179.

[16] İ. Ak, C. Öztaşkent, Y. Özüdoğru, T. Göksan, Effect of sodium acetate and sodium nitrate on biochemical composition of green algae *Ulva rigida*, Aquac. Int. 23 (1) (2015) 1–11.

[17] B.J. Gosch, M. Magnusson, N.A. Paul, R. de Nys, Total lipid and fatty acid composition of seaweeds for the selection of species for oil-based biofuel and bioproducts, GCB Bioenergy 4 (6) (2012) 919–930.

[18] Q. Hu, M. Sommerfeld, E. Jarvis, M. Ghirardi, M. Posewitz, M. Seibert, et al., Microalgal triacylglycerols as feedstocks for biofuel production: perspectives and advances, Plant. J. 54 (4) (2008) 621–639.

References

[19] H. Alishah Aratboni, N. Rafiei, R. Garcia-Granados, A. Alemzadeh, J.R. Morones-Ramírez, Biomass and lipid induction strategies in microalgae for biofuel production and other applications, Microb. Cell Fact. 18 (1) (2019) 178.

[20] A.E.-F. Abomohra, W. El-Shouny, M. Sharaf, M. Abo-Eleneen, Effect of gamma radiation on growth and metabolic activities of *Arthrospira platensis*, Braz. Arch. Biol. Technol. 59 (2016).

[21] A.E.-F. Abomohra, H. Shang, M. El-Sheekh, H. Eladel, R. Ebaid, S. Wang, et al., Night illumination using monochromatic light-emitting diodes for enhanced microalgal growth and biodiesel production, Bioresour. Technol. 288 (2019) 121514.

[22] Y. Zhao, X. Song, D.-b Zhong, L. Yu, X. Yu, γ-Aminobutyric acid (GABA) regulates lipid production and cadmium uptake by *Monoraphidium* sp. QLY-1 under cadmium stress, Bioresour. Technol. 297 (2020) 122500.

[23] K. Spilling, J. Seppälä, Photobiology and lipid metabolism in algae, in: R. Gordon, J. Seckbach (Eds.), The Science of Algal Fuels, Cellular Origin, Life in Extreme Habitats and Astrobiology, in: 25, Springer, Dordrecht, 2012.

[24] M.-K. Ji, H.-S. Yun, J.-H. Hwang, E.-S. Salama, B.-H. Jeon, J. Choi, Effect of flue gas CO_2 on the growth, carbohydrate and fatty acid composition of a green microalga *Scenedesmus obliquus* for biofuel production, Environ. Technol. 38 (16) (2017) 2085−2092.

[25] A.E. Ochoa-Alfaro, D.E. Gaytán-Luna, O. González-Ortega, K.G. Zavala-Arias, L.M.T. Paz-Maldonado, A. Rocha-Uribe, et al., pH effects on the lipid and fatty acids accumulation in *Chlamydomonas reinhardtii*, Biotechnol. Prog. 35 (6) (2019) e2891.

[26] I.A. Avila-León, M.C. Matsudo, L.S. Ferreira-Camargo, J.N. Rodrigues-Ract, J.C.M. Carvalho, Evaluation of *Neochloris oleoabundans* as sustainable source of oil-rich biomass, Braz. J. Chem. Eng. 37 (1) (2020) 41−48.

[27] S. Ruangsomboon, Effects of different media and nitrogen sources and levels on growth and lipid of green microalga *Botryococcus braunii* KMITL and its biodiesel properties based on fatty acid composition, Bioresour. Technol. 191 (2015) 377−384.

[28] D. Li, Y. Yuan, D. Cheng, Q. Zhao, Effect of light quality on growth rate, carbohydrate accumulation, fatty acid profile and lutein biosynthesis of *Chlorella* sp. AE10, Bioresour. Technol. 291 (2019) 121783.

[29] R.N. Bhakar, R. Kumar, S. Pabbi, Total lipids and fatty acid profile of different *Spirulina strains* as affected by salinity and incubation time. Vegetos-, Int. J. Plant. Res. 26 (2s) (2013) 148−154.

[30] N. Yodsuwan, S. Sawayama, S. Sirisansaneeyakul, Effect of nitrogen concentration on growth, lipid production and fatty acid profiles of the marine diatom *Phaeodactylum tricornutum*, Agric. Nat. Resour. 51 (3) (2017) 190−197.

[31] D.M. El Maghraby, E.M. Fakhry, Lipid content and fatty acid composition of Mediterranean macro-algae as dynamic factors for biodiesel production, Oceanologia 57 (1) (2015) 86−92.

[32] C. Öztaşkent, İ. Ak, Effect of LED light sources on the growth and chemical composition of brown seaweed *Treptacantha barbata*, Aquac. Int. 29 (1) (2021) 193−205.

[33] Y.S. Yong, W.T.L. Yong, S.E. Ng, A. Anton, S. Yassir, Chemical composition of farmed and micropropagated *Kappaphycus alvarezii* (Rhodophyta, Gigartinales), a commercially important seaweed in Malaysia, J. Appl. Phycol. 27 (3) (2015) 1271−1275.

[34] O.G. Mouritsen, C. Dawczynski, L. Duelund, G. Jahreis, W. Vetter, M. Schröder, On the human consumption of the red seaweed dulse (*Palmaria palmata* (L.) Weber & Mohr), J. Appl. Phycol. 25 (6) (2013) 1777−1791.

[35] D.L. Nelson, M.M. Cox, Lehninger's Principles of Biochemistry, Freeman and Company, New York, 2008.

[36] S. Xu, M. Elsayed, G.A. Ismail, C. Li, S. Wang, A.E.-F. Abomohra, Evaluation of bioethanol and biodiesel production from *Scenedesmus obliquus* grown in biodiesel waste glycerol: a sequential integrated route for enhanced energy recovery, Energy Convers. Manag. 197 (2019) 111907.

[37] J.P. Fidalgo, A. Cid, E. Torres, A. Sukenik, C. Herrero, Effects of nitrogen source and growth phase on proximate biochemical composition, lipid classes and fatty acid profile of the marine microalga *Isochrysis galbana*, Aquaculture 166 (1) (1998) 105−116.

[38] S.-H. Ho, C.-Y. Chen, J.-S. Chang, Effect of light intensity and nitrogen starvation on CO_2 fixation and lipid/carbohydrate production of an indigenous microalga *Scenedesmus obliquus* CNW-N, Bioresour. Technol. 113 (2012) 244−252.

[39] K.I. Reitan, J.R. Rainuzzo, Y. Olsen, Effect of nutrient limitation on fatty acid and lipid content of marine microalgae1, J. Phycol. 30 (6) (1994) 972−979.

[40] M. Giordano, L. Prioretti, Sulphur and algae: metabolism, ecology and evolution, in: M.A. Borowitzka (Ed.), The Physiology of Microalgae, in: 6, Springer International Publishing, Switzerland, 2016, pp. 185−209.

[41] H.-Y. Ren, B.-F. Liu, F. Kong, L. Zhao, G.-J. Xie, N.-Q. Ren, Enhanced lipid accumulation of green microalga *Scenedesmus* sp. by metal ions and EDTA addition, Bioresour. Technol. 169 (2014) 763−767.

[42] P.C. Gorain, S.K. Bagchi, N. Mallick, Effects of calcium, magnesium and sodium chloride in enhancing lipid accumulation in two green microalgae, Environ. Technol. 34 (13−14) (2013) 1887−1894.

[43] X. Sun, Y. Cao, H. Xu, Y. Liu, J. Sun, D. Qiao, et al., Effect of nitrogen-starvation, light intensity and iron on triacylglyceride/carbohydrate production and fatty acid profile of *Neochloris oleoabundans* HK-129 by a two-stage process, Bioresour. Technol. 155 (2014) 204−212.

[44] G. Sibi, T.S. Anuraag, G. Bafila, Cooper stress on cellular contents and fatty acid profiles in *Chlorella* species, Line J. Biol. Sci. 14 (3) (2014).

[45] H. Du, F. Ahmed, B. Lin, Z. Li, Y. Huang, G. Sun, et al., The effects of plant growth regulators on cell growth, protein, carotenoid, PUFAs and lipid production of *Chlorella pyrenoidosa* ZF strain, Energies 10 (11) (2017) 1696.

[46] I. Krzemińska, M. Oleszek, Glucose supplementation-induced changes in the *Auxenochlorella protothecoides* fatty acid composition suitable for biodiesel production, Bioresour. Technol. 218 (2016) 1294−1297.

[47] A. Converti, A.A. Casazza, E.Y. Ortiz, P. Perego, M. Del Borghi, Effect of temperature and nitrogen concentration on the growth and lipid content of *Nannochloropsis oculata* and *Chlorella vulgaris* for biodiesel production, Chem. Eng. Process: Process. Intensif. 48 (6) (2009) 1146−1151.

[48] P. Chaisutyakorn, J. Praiboon, C. Kaewsuralikhit, The effect of temperature on growth and lipid and fatty acid composition on marine microalgae used for biodiesel production, J. Appl. Phycol. 30 (1) (2018) 37−45.

[49] F. Abiusi, G. Sampietro, G. Marturano, N. Biondi, L. Rodolfi, M. D'Ottavio, et al., Growth, photosynthetic efficiency, and biochemical composition of *Tetraselmis suecica* F&M-M33 grown with LEDs of different colors, Biotechnol. Bioeng. 111 (5) (2014) 956−964.

[50] T. Takeshita, S. Ota, T. Yamazaki, A. Hirata, V. Zachleder, S. Kawano, Starch and lipid accumulation in eight strains of six *Chlorella* species under comparatively high light intensity and aeration culture conditions, Bioresour. Technol. 158 (2014) 127−134.

[51] G. Breuer, P.P. Lamers, D.E. Martens, R.B. Draaisma, R.H. Wijffels, Effect of light intensity, pH, and temperature on triacylglycerol (TAG) accumulation induced by nitrogen starvation in *Scenedesmus obliquus*, Bioresour. Technol. 143 (2013) 1−9.

[52] W. Pugkaew, M. Meetam, K. Yokthongwattana, N. Leeratsuwan, P. Pokethitiyook, Effects of salinity changes on growth, photosynthetic activity, biochemical composition, and lipid productivity of marine microalga *Tetraselmis suecica*, J. Appl. Phycol. 31 (2) (2019) 969−979.

[53] S.M. Renaud, D.L. Parry, Microalgae for use in tropical aquaculture II: effect of salinity on growth, gross chemical composition and fatty acid composition of three species of marine microalgae, J. Appl. Phycol. 6 (3) (1994) 347−356.

[54] L. Zhu, X. Zhang, L. Ji, X. Song, C. Kuang, Changes of lipid content and fatty acid composition of *Schizochytrium limacinum* in response to different temperatures and salinities, Process. Biochem. 42 (2) (2007) 210−214.

[55] M.P. Rai, T. Gautom, N. Sharma, Effect of salinity, pH, light intensity on growth and lipid production of microalgae for bioenergy application, OnLine J. Biol. Sci. 15 (4) (2015).

[56] R. Qiu, S. Gao, P.A. Lopez, K.L. Ogden, Effects of pH on cell growth, lipid production and CO_2 addition of microalgae *Chlorella sorokiniana*, Algal Res. 28 (2017) 192−199.

[57] D.H. Jeong, M.H. Jeong, S.K. Jeong, K. Yang, W.S. Jo, Effect of continuous exposure to low-dose-rate gamma irradiation on cell growth and lipid accumulation of marine microalgae, Aquac. Int. 25 (2) (2017) 589−601.

[58] M.P. Tale, R. devi Singh, B.P. Kapadnis, S.B. Ghosh, Effect of gamma irradiation on lipid accumulation and expression of regulatory genes involved in lipid biosynthesis in *Chlorella* sp, J. Appl. Phycol. 30 (1) (2018) 277−286.

[59] H.A. Alghanmi, Evaluating the effect of different doses of gamma radiation on carbohydrates, proteins, and lipids content of *Arthrospira fusiformis*, Eurasian J. Biosci. 14 (1) (2020) 869−876.

[60] J. Tong, R. He, X. Zhang, R. Zhan, W. Chen, S. Yang, Effects of atmospheric pressure air plasma pretreatment on the seed germination and early growth of *Andrographis paniculata*, Plasma Sci. Technol. 16 (3) (2014) 260−266.

[61] J.Q.M. Almarashi, S.E. El-Zohary, M.A. Ellabban, A.E.-F. Abomohra, Enhancement of lipid production and energy recovery from the green microalga *Chlorella vulgaris* by inoculum pretreatment with low-dose cold atmospheric pressure plasma (CAPP), Energy Convers. Manag. 204 (2020) 112314.

References

[62] R. Harun, J.W.S. Yip, S. Thiruvenkadam, W.A.W.A.K. Ghani, T. Cherrington, M.K. Danquah, Algal biomass conversion to bioethanol — a step-by-step assessment, Biotechnol. J. 9 (1) (2014) 73—86.

[63] J.N. BeMiller, Carbohydrate Chemistry for Food Scientists, Elsevier, New York, 2018.

[64] M. Suutari, E. Leskinen, K. Fagerstedt, J. Kuparinen, P. Kuuppo, J. Blomster, Macroalgae in biofuel production, Phycol. Res. 63 (1) (2015) 1—18.

[65] M.J. Stadnik, M.Bd Freitas, Algal polysaccharides as source of plant resistance inducers, Trop. Plant. Pathol. 39 (2) (2014) 111—118.

[66] J.A. Raven, J. Beardall, Carbohydrate metabolism and respiration in algae, in: A.W.D. Larkum, S.E. Douglas, J.A. Raven (Eds.), Photosynthesis in Algae. Advances in Photosynthesis and Respiration, Vol. 14, Springer, Dordrecht, 2003.

[67] H.M. Holden, I. Rayment, J.B. Thoden, Structure and function of enzymes of the leloir pathway for galactose metabolism, J. Biol. Chem. 278 (45) (2003) 43885—43888.

[68] C.L. Hurd, P.J. Harrison, K. Bischof, C.S. Lobban, Seaweed Ecology and Physiology, second ed., Cambridge University Press, Cambridge, 2014.

[69] X. Li, W. Li, J. Zhai, H. Wei, Effect of nitrogen limitation on biochemical composition and photosynthetic performance for fed-batch mixotrophic cultivation of microalga *Spirulina platensis*, Bioresour. Technol. 263 (2018) 555—561.

[70] X. Li, J. Manuel, S. Slavens, D.W. Crunkleton, T.W. Johannes, Interactive effects of light quality and culturing temperature on algal cell size, biomass doubling time, protein content, and carbohydrate content, Appl. Microbiol. Biotechnol. 105 (2) (2021) 587—597.

[71] N. Renuka, A. Guldhe, P. Singh, F.A. Ansari, I. Rawat, F. Bux, Evaluating the potential of cytokinins for biomass and lipid enhancement in microalga *Acutodesmus obliquus* under nitrogen stress, Energy Convers. Manag. 140 (2017) 14—23.

[72] J. Fábregas, A. Maseda, A. Domínguez, A. Otero, The cell composition of *Nannochloropsis* sp. changes under different irradiances in semicontinuous culture, World J. Microbiol. Biotechnol. 20 (1) (2004) 31—35.

[73] M. Onay, Enhancing carbohydrate productivity from *Nannochloropsis gaditana* for bio-butanol production, Energy Rep. 6 (2020) 63—67.

[74] H. Khatoon, N. Abdu Rahman, S. Banerjee, N. Harun, S.S. Suleiman, N.H. Zakaria, et al., Effects of different salinities and pH on the growth and proximate composition of *Nannochloropsis* sp. and *Tetraselmis* sp. isolated from South China Sea cultured under control and natural condition, Int. Biodeterior. Biodegrad. 95 (2014) 11—18.

[75] G. Su, K. Jiao, Z. Li, X. Guo, J. Chang, T. Ndikubwimana, et al., Phosphate limitation promotes unsaturated fatty acids and arachidonic acid biosynthesis by microalgae *Porphyridium purpureum*, Bioprocess. Biosyst. Eng. 39 (7) (2016) 1129—1136.

[76] K. Manivannan, G. Thirumaran, G.K. Devi, P. Anantharaman, T. Balasubramanian, Proximate composition of different group of seaweeds from Vedalai coastal waters (Gulf of Mannar): Southeast coast of India, Middle East. J. Sci. Res. 4 (2) (2009) 72—77.

[77] D. Robledo, Y. Freile, Pelegrín, chemical and mineral composition of six potentially edible seaweed species of Yucatán, Bot. Marina 40 (1—6) (1997) 301—306.

[78] S. Gokulakrishnan, K. Raja, G. Sattanathan, J. Subramanian, Proximate composition of bio potential seaweeds from mandapam South East coast of India, Int. Lett. Nat. Sci. 45 (2015) 49—55.

[79] B. Le, J.-A. Shin, M.-G. Kang, S. Sun, S.H. Yang, G. Chung, Enhanced growth rate and ulvan yield of *Ulva pertusa* using light-emitting diodes (LEDs), Aquac. Int. 26 (4) (2018) 937—946.

[80] N. Balar, P. Sharnagat, P. Kumari, V.A. Mantri, Variation in the proximate composition of edible marine macroalga *Ulva rigida* collected from different coastal zones of India, J. Food Sci. Technol. 56 (10) (2019) 4749—4755.

[81] Y. Li, M. Zheng, J. Lin, S. Zhou, T. Sun, N. Xu, Darkness and low nighttime temperature modulate the growth and photosynthetic performance of *Ulva prolifera* under lower salinity, Mar. Pollut. Bull. 146 (2019) 85—91.

[82] R. De Philippis, C. Sili, M. Vincenzini, Glycogen and poly-3-hydroxybutyrate synthesis in *Spirulina maxima*, Microbiology 138 (8) (1992) 1623—1628.

[83] S.G. Ball, L. Dirick, A. Decq, J.C. Martiat, R. Matagne, Physiology of starch storage in the monocellular alga *Chlamydomonas reinhardtii*, Plant. Sci. 66 (1) (1990) 1—9.

[84] S. Arad, Y. Lerental, O. Dubinsky, Effect of nitrate and sulfate starvation on polysaccharide formation in *Rhodella reticulata*, Bioresour. Technol. 42 (2) (1992) 141—148.

[85] T. van Oijen, M.A. van Leeuwe, W.W.C. Gieskes, H.J.W. de Baar, Effects of iron limitation on photosynthesis and carbohydrate metabolism in the Antarctic diatom *Chaetoceros brevis* (Bacillariophyceae), Eur. J. Phycol. 39 (2) (2004) 161–171.

[86] M. Hanifzadeh, E.C. Garcia, S. Viamajala, Production of lipid and carbohydrate from microalgae without compromising biomass productivities: role of Ca and Mg, Renew. Energy 127 (2018) 989–997.

[87] S. Mohy El Din, Effect of copper and lead on growth and some metabolic activities of cyanobacterium *Spirulina platensis* (Nordstedt), Egypt. J. Bot. 57 (3) (2017) 445–456.

[88] C.Y. Lim, Y.H. Yoo, M. Sidharthan, C.W. Ma, I.C. Bang, J.M. Kim, et al., Effects of copper (I) oxide on growth and biochemical compositions of two marine microalgae, J. Env. Biol. 27 (3) (2006) 461–466.

[89] Z. Liu, Q. Wang, D. Zou, Y. Yang, Effects of selenite on growth, photosynthesis and antioxidant system in seaweeds, *Ulva fasciata* (Chlorophyta) and *Gracilaria lemaneiformis* (Rhodophyta), Algal Res. 36 (2018) 115–124.

[90] A. Izumo, S. Fujiwara, Y. Oyama, A. Satoh, N. Fujita, Y. Nakamura, et al., Physicochemical properties of starch in *Chlorella* change depending on the CO_2 concentration during growth: comparison of structure and properties of pyrenoid and stroma starch, Plant. Sci. 172 (6) (2007) 1138–1147.

[91] F.J. Choix, L.E. de-Bashan, Y. Bashan, Enhanced accumulation of starch and total carbohydrates in alginate-immobilized *Chlorella* spp. induced by *Azospirillum brasilense*: II. Heterotrophic conditions, Enzyme Microb. Technol. 51 (5) (2012) 300–309.

[92] A.P. Abreu, B. Fernandes, A.A. Vicente, J. Teixeira, G. Dragone, Mixotrophic cultivation of *Chlorella vulgaris* using industrial dairy waste as organic carbon source, Bioresour. Technol. 118 (2012) 61–66.

[93] X.-Y. Deng, C.-Y. Xue, B. Chen, P.K. Amoah, D. Li, X.-L. Hu, et al., Glucose addition-induced changes in the growth and chemical compositions of a freshwater microalga *Chlorella kessleri*, J. Chem. Technol. Biotechnol. 94 (4) (2019) 1202–1209.

[94] A. Eggert, Seaweed Responses to Temperature, in: B. Wiencke, K. Bischof (Eds.), Seaweed Biology, in: 219, Springer-Verlag, Berlin, 2012, pp. 47–66.

[95] K.H. Ogbonda, R.E. Aminigo, G.O. Abu, Influence of temperature and pH on biomass production and protein biosynthesis in a putative *Spirulina* sp, Bioresour. Technol. 98 (11) (2007) 2207–2211.

[96] H. Pliego-Cortés, E. Caamal-Fuentes, J. Montero-Muñoz, Y. Freile-Pelegrín, D. Robledo, Growth, biochemical and antioxidant content of *Rhodymenia pseudopalmata* (Rhodymeniales, Rhodophyta) cultivated under salinity and irradiance treatments, J. Appl. Phycol. 29 (5) (2017) 2595–2603.

[97] G. Arun, Osmoregulation in *Dunaliella*, Part II: photosynthesis and starch contribute carbon for glycerol synthesis during a salt stress in *Dunaliella tertiolecta*, Plant. Physiol. Biochem. 45 (9) (2007) 705–710.

[98] M. Zheng, J. Lin, S. Zhou, J. Zhong, Y. Li, N. Xu, Salinity mediates the effects of nitrogen enrichment on the growth, photosynthesis, and biochemical composition of *Ulva prolifera*, Environ. Sci. Pollut. Res. 26 (19) (2019) 19982–19990.

[99] Z.I. Khalil, M.M.S. Asker, S. El-Sayed, I.A. Kobbia, Effect of pH on growth and biochemical responses of *Dunaliella bardawil* and *Chlorella ellipsoidea*, World J. Microbiol. Biotechnol. 26 (7) (2010) 1225–1231.

[100] M.A. Toghyani, F. Karimi, S.A. Hosseini Tafreshi, D. Talei, Two distinct time dependent strategic mechanisms used by *Chlorella vulgaris* in response to gamma radiation, J. Appl. Phycol. 32 (3) (2020) 1677–1695.

CHAPTER

4

Algal physiology and cultivation

Kushi Yadav, Reetu, Shrasti Vasistha and Monika Prakash Rai

Amity Institute of Biotechnology, Amity University, Noida, India

4.1 Introduction

Algae are photosynthetic autotrophic organisms that have a unique ability to survive in a variety of water bodies, such as lakes, rivers, ponds, and seas. They utilize atmospheric CO_2 and produce oxygen through photosynthesis using sunlight. There is a huge diversity of algae on the basis of different properties and characteristics. Broadly based on size, algae are divided into two categories: microalgae, which are unicellular; and seaweed (macroalgae), which are multicellular. Both are tolerant of different environmental factors [1]. Algae have enormously important roles in upcoming industries due their ability to produce a variety of value-added products, including lipids, proteins, carbohydrates, carotenoids, and vitamins. These high value-added products make algae cultivation economical and sustainable [1]. Algae cultivation and biomass productivity are influenced by various environmental factors [2]. To make an economically viable bioproducts from algae it is necessary to enhance biomass productivity by optimizing the environmental factors [3]. This chapter focuses on the effects of environmental factors like pH, temperature, light, carbon dioxide, salinity, etc. directly on algae biomass productivity, which all affect the cost of algae-mediated products. Algae cultivation is an area of interest for industries and farms due to the enormous potential to generate a variety of high value-added products with huge industrial importance [4]. Obtaining value-added products from algae is an innovative approach which creates a competitive advantage in the agriculture industry that primarily contributes to products such as biodiesel with other value-added products [5–7]. To enhance biomass, it is essential to understand the different cultivation systems of both microalgae and seaweeds. In this chapter, different aspects of algae cultivation systems are discussed, such as open ponds, photobioreactors, and so on. To make algae-mediated products economically viable it is necessary to obtain more than one product, along with the use of integrated cultivation technology. Algae biomass is used by various industries to generate additional products apart from biofuel such as liquid

Handbook of Algal Biofuels
DOI: https://doi.org/10.1016/B978-0-12-823764-9.00016-9

© 2022 Elsevier Inc. All rights reserved.

biofertilizer, biostimulants, proteins, animal feed, and starch [8]. Algae cultivation technologies need to improve before initiating large-scale cultivation. As there are different challenges and advantages with different algae cultivation systems it is mandatory to find economical and environmentally sustainable systems for commercialization [9]. To make the production of algae-mediated high-value-added products economical for the present global market it is necessary to make the production technically viable and to maximize algae biomass productivity [10]. Instead of the use of freshwater and costly nutrients, wastewater is highly preferred for algae cultivation but there are certain challenges with regard to wastewater, which include recovering nutrients like nitrogen and phosphorous to avoid eutrophication and eliminating fertilizers. In general, 1 kg algae biomass is produced per cubic meter of sewage with average nutrient content. On average 100,000 people produce up to 50,000 per cubic meter of sewage, which could yield 50,000 kg/day of microalgae biomass [11]. It is necessary to perform technoeconomic analysis to understand the economic and environmental viability of the products produced. This analysis provides specific information about the different aspects of large-scale production, in both economic and environmental terms. This chapter discusses the potential for algae to treat wastewater by phycoremediation and the cost-effective production of algae-mediated value-added products.

4.2 Photosynthetic efficiency of algae

Microalgae are primitive plants and these microcellular plants perform photosynthesis to generate food and energy. Like other plants these microalgae have chloroplasts containing thylakoids, where the light reaction of adenosine triphosphate/nicotinamide adenine dinucleotide phosphate (ATP/NADPH synthesis) of photosynthesis takes place, whereas the dark reaction (carbohydrate synthesis) occurs in the stroma. A complete set of machinery—the photosynthetic unit (PSU)—is required for the metabolic process. PSU is part of the lipid membrane of the thylakoid comprising the pigment attributed reaction centers for the absorption of the light various electron carriers, pigments, and proteins necessary for ATP and NADPH generation [12,13]. At the resting stage, in a very quick reaction PSU absorbs the Z photons and gets excited by the chlorophyll a molecule [14]. Immediately after the excitation, to reduce the deteriorating effects of the excess energy absorbed, the small amount of the energy is dissipated as fluorescence by the chlorophyll molecule. The remaining energy is quenched in different ways in order to protect the proteins present in the photosystem, ultimately increasing the photosynthetic efficiency [15].

Further photocatalysis of water molecules also occurs, where the released electrons are accepted by the tyrosine-containing protein of the photosystem. After the release and acceptance of electrons by the successive acceptors and donors the electrons are finally taken up by $NADH^+$ in order to generate NADPH. Additionally, during the electron transport several protons are pumped through the thylakoid membrane across the ATP synthase producing ATP [15]. Hence, the output of the linear electron transport chain is the energy molecules ATP and NADPH which are further utilized in the dark reaction for synthesizing carbohydrates.

Microalgae biomass can be exploited for biofuels production and food/feed. Various value-added coproducts, like carotenoids and fatty acids, can be also be obtained which have specific therapeutic applications [16–20]. Lately all the major research is focused around alleviating the photosynthetic efficiency of the microalgal cells which would lead to enhanced biomass and as aforementioned this biomass can be utilized in disparate ways. Several strategies like abiotic stress, nanotechnology, and genetic engineering have been employed for increasing the photosynthetic efficiency, and thus increasing the overall biomass. In a recent study, gold nanoparticles (AuNPs) of the size 5 nm were administered at a concentration of 12 mg/L to a culture of *Chlorella zofingiensis*. This study reported a significance increase in the carotenoid production, as well as enhanced photosynthesis that was attributed to quick linear transport of the electrons [21]. Another study investigated the impact of polystyrene (PS) microplastics on *Chlamydomonas reinhardtii*, and an inhibitory effect at a concentration of 100 mg/L was observed [22]. The effect of salinity stress was monitored for *Scenedesmus obliquus* XJ002, and a 2.5-fold increase in lipid was observed, whereas there was a negative impact on the biomass [23].

Photosynthetic efficiency is also influenced by the cultivation system used for culturing the microalgae [24], for example, whether open ponds or closed photobioreactors are used. In these systems microalgal cells are cultivated at higher concentrations and are exposed to sunlight. Due to the high optical density, sunlight is not distributed evenly over the entire cultivation unit, leading to two conditions [25]. Firstly, there will be cells receiving limited sunlight and hence these cells will not have adequate energy and will perform photosynthesis inefficiently. Secondly, some cells will receive a high amount of sunlight, and these cells may experience photodamage of the proteins associated with the photosynthesis [25]. In both cases the overall photosynthetic efficiency and cell biomass is hampered, and therefore uniform exposure of the microalgae cells to the light is a prerequisite for efficient energy production.

Algae cells consist of light-harvesting complexes (LHC) or antenna in the thylakoid membrane that are responsible for the photosynthetic process. The recent genetic engineering, RNAi approach revolves around targeting the genes associated with the LHC [26–28]. In a parallel concept, the regulation of antenna proteins genes by NAB1 repressor was done cotranslationally or posttranslationally [29]. Another study targeted the overexpression of the nuclear-encoded, cbbX-homologous, candidate Ribulose-1,5-bisphosphate carboxylase-oxygenase RuBisCO activase in *Nannochloropsis oceanica*. An increase of approximately 28% was observed in the photosynthetic efficiency, and lipid productivity was enhanced by 41% [30]. Hence, photosynthetic efficiency may be affected by several factors and current research is continuing to unravel the solutions for increasing photosynthetic efficiency.

4.3 Factors influencing growth and biochemical composition

Many factors can affect the growth, biomass, and total lipid content of the microalgae cells. Various abiotic parameters can alter the cellular composition either by creating stress or by stimulating the synthesis of valuable metabolites. Microalgae are capable of adapting

4. Algal physiology and cultivation

and thriving in the stringent conditions caused by the changes in different factors like temperature, pH, CO_2, etc., although sometimes the changes may lead to a decline in the biomass, thus leading to an upsurge in the total cost of the disparate biorefineries [31]. Table 4.1 gives a brief summary of different factors affecting microalgae cells.

TABLE 4.1 Factors influencing different microalgal species.

Factor	Microalgae	Conditions	Observation	References
pH	Scenedesmus obliquus	8.0	—	[32]
	Chlorella sp.	8.0	Lipid content increase by 23%	[33]
	Chlorella vulgaris	6.0, 7.0, 9.0	—	[34]
	Botryococcus braunii	6.5	Increase in lipid	[35]
	Pavlova lutheri	8.0	35% lipid content	[36]
Light	Scenedesmus obliquus	11,100 lux	0.863 g/L	[37]
	Chlorella vulgaris	2700 lx light	Total lipid 19%	[38]
	Isochrysis galbana	Shorter light/dark regime	PUFAs increased	[39]
	Desmodesmus sp.	300 μE/m^{2}s	1.1 g/L biomass	[40]
Temperature	Dunaliella salina	22°C	—	[41]
	Heterochlorella luteoviridis	22°C	40.7% of PUFAs	[42]
	Tetraselmis subcordiformis	>20°C	Neutral lipids and PUFAs were decreased	[43]
	Nannochloropsis oculata	>30°C	Neutral lipids decreased and polar lipids increased	[43]
Salinity	Dunaliella salina	88–117 ppt	—	[44]
	Scenedesmus obliquus XJ002	0.20 M NaCl	2.52-fold higher lipid content	[45]
	Nannochloropsis salina	34 ppt	—	[46]
CO_2	Botryococcus braunii	10% CO_2	26.55 mg$_{dw}$/L/d biomass produced; 5.51 mg/L/d—about 21% lipid productivity	[47]
	Chlorella vulgaris	10%	104.76 mg$_{dw}$/L/d Biomass productivity; 6.91 mg/L/d lipid productivity	[47]
	Scenedesmus obliquus	15%	—	[48]
Nitrogen	Scenedesmus sp.	15 g/m^3	Protein synthesis enhanced	[49]
	Isochrysis galbana	72 mg/L	Fivefold increase in the concentration of PUFAs	[50]

Handbook of Algal Biofuels

Microalgae cells are photoautotrophic in nature, making light an important factor affecting the physiology of the cells. Light intensity higher than the required level might lead to photodamage, causing overproduction of the lipid. A similar study was done on *Chlorella* sp., at a light intensity of 320 $\mu E/m^2s$ the highest lipid was accumulated [51]. An improved lipid yield of 92.89% was observed in *Chlorella vulgaris* at a light intensity of 560 $\mu E/m^2s$ [52]. pH is a factor affecting enzyme activity and all the vital activities inside the microalgal cells [45]. Like in *Chlorella* sp. triacyl glycerides were less utilized with an increase in the pH [53].

Similarly, the optimum temperature required for microalgal growth is in the range of 16−27°C. Below and above this range the algae growth is either inhibited or ultraslow [45]. In a study on *Chlorella vulgaris*, lipid productivity was 14.7% at a temperature of 25°C, whereas it accumulated only 5.9% of lipid at a high temperature of 30°C. Other than the abiotic stress, nutrient imbalance may also trigger noticeable changes in the physiology of the microalgal cells. Because of the high photosynthetic activity at the start of the microalgae growth the internal carbon exceeds the C:N ratio causing protein synthesis. But later in the liquid culture nitrogen is consumed and lipid production is favored [54]. In a study on *Nannochloropsis oculata* an increase in the growth was observed by the supplementation of trace elements like Fe^{3+}, Zn^2, Mn^{2+}, etc. [55]. Therefore the parameters affecting the growth, biomass, lipid productivity, and biochemical composition of the microalgal cells can be optimized and further explored for benefits at a commercial scale.

4.4 Microalgae cultivation system

The presence of elevated photosynthesis, growth rate, and carbon dioxide sequestering ability support microalgae cultivation. The most advantageous factor in microalgae cultivation is wastewater management, where agricultural, dairy, municipal, and various other industrial wastes are utilized as a source of nitrogen and phosphorous during cultivation. Microalgae cultivation is not complicated and is environmentally friendly. Microalgae have a considerable lipid content, which makes them suitable candidates for biofuel production, as shown in Fig. 4.3. Microalgae growth and lipid accumulation depends upon various aspects, including concentration of both micronutrients and macronutrients, temperature, intensity of light, photoperiod, carbon dioxide levels, and so on. The temperature for the most satisfactory results of microalgae cultivation is considered to be 20−30°C. The amount of unsaturated fatty acid is reduced with an increase in temperature due to physiological adaptations [56]. Various biodiesel attributes, such as liquefying point, boiling point, amount of iodine and cetane, along with lubricant effect, are affected by the variety of fatty acid precursor [56,57]. Precursors like palmitic acid (C16:0), palmitoleic acid (C16:1), stearic acid (C18:0), oleic acid (C18:1), linoleic acid (C18:2), and linolenic acid (C18:3) are considered as standard for the production of quality biodiesel [58]. There are certain microalgae species which modify their lipid biosynthesis process to elevate the level of lipid production, specifically triacylglycerol under certain deficient conditions, as shown in Fig. 4.3. The selection of algae species from the wide variety of microalgae species and the optimization of all factors influencing microalgal cultivation allow the attainment of environmentally and economically sustainable of biofuel of sufficient quantity and quality. The selected species undergo a wide range of experiments to optimize the cultivation strategy for both increased biomass and biofuel production. Microfluidic chip is a technique used for various diverse applications involving ecotoxicology screening, cell reorganization, lipid analysis, sorting, cell viability, cultivation under different conditions, and measuring the amount of self-secreted molecules like ethanol and lactate [59,60]. Biofuel and biomass production along with phytoremediation ability also depend upon the type of the cultivation system as they affect various other factors.

Broadly, there are two major types of photoautotrophic systems: open and closed systems. The closed systems include photobioreactors, as shown in Fig. 4.1. The major advantages of photobioreactors are that they are extremely adjustable in terms of light, with a high intensity light attainable, and they allow proper stirring and a well-controlled environment for the microalgae [61,62]. There are a variety of photobioreactor types, such as flat plate, tubular, or columns. Tubular reactors are widely used but bubble column and airlift photobioreactors cultivate better biomass yield [63]. The major disadvantages with these systems are that they are not economically viable, and are susceptible to overheating, excessive oxygen elevation, cell damage, and biofouling. Scaling up is also difficult in practical terms, as discussed in Table 4.2 [70]. In contrast open pond cultivation systems are not so well-controllable but they are economically

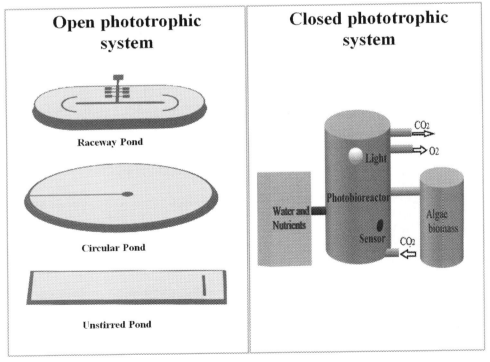

FIGURE 4.1 Types of microalgae phototrophic cultivation system.

TABLE 4.2 Comparison of different types of cultivation systems depending on various growth parameters for microalgae production.

Parameters	Closed cultivation system	Open cultivation system	Hybrid cultivation system	References
Contamination risk	Less	More	Less	[64]
Light required	More	Less	Moderate	[65]
Water evaporation	Less	More	Moderate	[66]
Biomass quality	Reproducible	Variable	Reproducible	[67]
Product quality	High	Low	Moderate	[65]
Input energy requirement	More	Less	Moderate	[66]
Operation type	Batch	Batch	Continuous	[68]
Area required	Moderate	More	More	[67]
Controllability	More	Less	Moderate	[69]
Economical viability	Low	High	Moderate	[67]
Atmospheric CO_2 utilization	Low	High	Moderate	[68]

sustainable as environmental carbon dioxide is directly used by microalgae and a rotating arm is provided for regular stirring [71]. Of the open pond cultivation forms, the widely used types are raceway pond, the circular pond tank, the closed pond, and the shallow big pond. The main drawbacks of open cultivation related to microalgae are contagion by bacteria or other microalgae species, along with evaporation, uncontrolled temperature, large space, inferior light intensity, and atmospheric carbon dioxide diffusion [65,70]. Both open and closed cultivation systems have their own advantages and disadvantages, but when both of them are combined they form a hybrid cultivation system which provides magnificent biomass yield and elevated nutrient removal [72]. The hybrid cultivation system initially involves microalgae culture in a closed photobioreactor which for further cultivation is transferred to an open cultivation system to collaboratively provide good biomass yield in an economically sustainable manner. This system is acceptable at a commercial level [73]. There are various major types of metabolic pathways for microalgae cultivation which include autotrophic, heterotrophic, mixotrophic, and photoheterotrophic. The autotrophic pathway is a purely photosynthetic pathway so it can convert inorganic carbon into organic form in light [74]. In contrast in a heterotrophic pathway the direct organic carbon is provided but in the absence of light, whereas in a mixotrophic pathway microalgae growth can be autotrophic as well as mixotrophic depending upon the food provided. The photoheterotrophic pathway includes both organic carbon and the presence of light. Among all, mixotrophic metabolism is most considered as it provides the highest biomass and lipid yield [75].

4.5 Artificial growth of seaweeds

Seaweed cultivation methods are broadly divided into two categories, that is, wild seaweeds and aquaculture seaweeds. Drift seaweeds are the main natural source for the wild seaweed cultivation method. In contrast the aquaculture seaweeds method includes land-based cultivation, which occurs in ponds and tanks and ocean-based cultivation that involves the sea as source. Among all these methods, land-based method is more preferred for macroalgal cultivation for various species such as *Ulva rotundata, Monostroma,* and *Laminaria digitata* [69,76]. In some cases, for example, *Eucheuma, Undaria,* and *Kappaphycus,* sea-mediated macroalgae cultivation is preferred [77]. Overall macroalgae cultivation is done in a sequential manner, including experimental plant setup containing an incubation unit along with flourishing parts [78].

Seawater filtration, an appropriate air and light supply, a chiller unit with tank, and storage are some essential parts of an incubator. A few factors on which macroalgae cultivation directly depend are funds, place of cultivation, seaweed species (usually native species are preferred), depth of water, nutrient availability, water turbidity and temperature, water current, irradiance, and photoperiod [79]. To improve the productivity and reduce the challenges in macroalgae cultivation for industrial production it is crucial to recognize the relation between the factors that affect cultivation and intrinsic physiological response. Various studies were conducted to develop more understanding about physiological aspects, which leads to elevated biomass production, such as in the brown seaweed *Saccharina latissima*. *S. latissima* is grown under stress conditions, which include thermal shocks and thallus dissection which may enhance meiospores, and to avoid fertilization red light is provided, whereas white light increases the fertilization of meiospores [80]. To obtain juvenile sporophytes, it is necessary to maintain seeded ropes in mass cultivation; this takes 8 months, thus making the hatchery phase expensive and problematic. In contrast to conventional methods, using textile sheets provides a double yield of biomass because the sheet acts as a physical substratum for meiospores, gametophytes, and sporophytes with a 4-month duration. However, for effective juvenile development the appropriate twine must be chosen for the sporophytic seeding technique, as shown in Fig. 4.2 [80]. Further

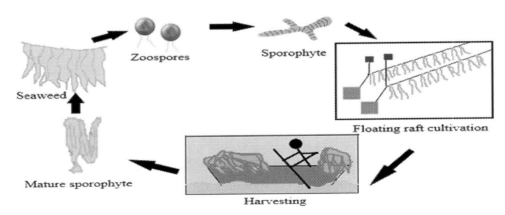

FIGURE 4.2   Artificial cultivation of seaweeds.

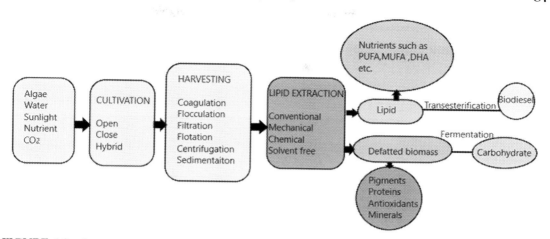

FIGURE 4.3 Overview of algae-mediated biodiesel and coproducts synthesis.

development was undertaken on the breeding method specifically for *Palmaria palmata* cultivation on the germination, maceration, and agitation process in order to increase seedling level on the physical substratum; this included tetraspore libration and maintenance in order to yield the seedling [81]. Various conventional process such as off-bottom, rafts, and long-line ropes are used for the extraction of commercially valuable products, such as agar and carrageenan, from various propagules including *Gracilaria*, *Kappaphycus*, and *Eucheuma* [82]. Using different binding methods propagules are bonded on ropes that are anchored to stakes in substrata or are suspended from floats. The feasibility of the industrialization and elevated production of carrageenophyte is increased by replacing the ropes by utilizing a tubular net similar to that already used in mollusc farm [82]. During cultivation on a muddy base tubular nets are advantageous as they enhance seawater flow, hold seedlings on the raft, save time by managing cultivator, and circumvent environmental invasiveness [83].

It is crucial to investigate both physiochemical and biological parameters prior to the selection of macroalgae species and the site of cultivation, because there are certain factors such as light intensity, temperature, wave velocity, current intensity, and seaward high waves that affect the growth of algae differently [84]. Instead of costly nutrients, macroalgae can be grown on wastewater, which makes macroalgae cultivation ecologically and economically beneficial. Macroalgae have the potential to uptake nutrients like nitrogen and phosphorus from wastewater, which ultimately leads to wastewater treatment, macroalgae cultivation, and cost reduction. Macroalgae species that have been reported to have been grown in wastewater are *Gracilaria edulis*, *Gracilaria changii*, *Ulva pertusa*, and *Cladophora* [85–87]. (Fig. 4.3).

4.6 Cost analysis of algae cultivation

The systems of algae cultivation are largely categorized into open raceway ponds (OPRs) and closed photobioreactors (PBRs) [88]. The ORPs require low capital and

88   4. Algal physiology and cultivation

TABLE 4.3   The annual operating expense for 1 year in open raceway ponds and photobioreactors (M$s) [90].

Parameter	Open raceway ponds	Photobioreactors
CO_2	3.14	3.14
Nutrients	6.32	6.32
Labor	9.64	13.88
Hydrogen	1.48	1.48
Wastewater	2.79	0.19
Utilities	0.97	1.06
Maintenance, tax, and insurance cost	7.18	25.54
Sum of operating expense	31.53	51.59

operating cost, along with a minimal energy requirement for mixing the cultures. However, the PBRs are effective as they can be operated under controlled environment and also they can be designed according to specific requirements [89]. The PBRs are closed systems which occupy less space and hence the light is properly utilized by the cultures. The techniques used for microalgae cultivation majorly determine the economic viability of the process. Researchers have determined that OPRs are more preferred economically as they offer energy balance in a better way than PBRs (Table 4.3). The production costs of microalgae biomass in OPRs, horizontal PBRs, and flat plate PBRs were estimated to be 4.95, 4.16, and 5.96 €/kg, respectively, for a 100-hectare area [91]. Chisti et al. [92] evaluated the price per gallon for 100,000 kg of biomass to be $2.95 and $3.80 in PBRs and OPRs, respectively. Richardson et al. [93] determined that OPRs offers the production of biomass at low cost compared to PBRs. The production cost of 10 MG/yr of biomass was estimated for OPRs to be 12.74 $/gallon, whereas for PBRs the cost was 32.57 $/gallon. In addition, the selling price of lipid from microalgae was determined to be 8.52 $/gallon for OPRs and 18.10 $/gallon for PBRs [90]. In contrast, recent literature has reported an increase in lipid productivity was obtained in PBRs in compared to OPRs. The lipid content of 35% was observed in *Nannochloropsis* sp. when cultivated in PBRs [94]. Studies have revealed that the cultivation of microalgae in PBRs offers a lower risk of net cash income (NCI) by 26% for producing lipids in contrast to OPRs [95].

4.7 Integrated cultivation system

The major bottleneck for the mass cultivation of microalgae is the expensive culture media used for microalgae growth which adds cost to the entire process [95]. It has been estimated that the annual worldwide consumption of freshwater is approx. 3700 billion m^3, as per 2014 statistics [96]. Hence, in order to reduce the cost, an integrated approach has been considered to be a subject of significant research in this area. The microalgae grown in wastewater has been considered to be a practical solution to replace synthetic media.

Handbook of Algal Biofuels

Wastewater consists of organic and inorganic nutrients like carbon, nitrogen, and phosphorous, which are essential for microalgae growth [97]. The biomass obtained from cultivating microalgae in wastewater can be further converted into biofuel and other by-products. The wastewater offers various advantages for microalgae biomass production, such as (1) low-cost media; (2) provides sufficient nutrients for microalgae growth; (3) offers integrated benefits, such as treating wastewater and boosting biomass; and (4) sustains huge biomass and biofuel production [98]. Various wastewaters have been explored for microalgae cultivation, including primary wastewater, secondary wastewater, textile wastewater, domestic wastewater, palm mill effluents, and agro industrial wastewater [99]. In addition, *Chlorella* sp., *Chlorosarcinopsis* sp., and *Scenedesmus* sp. have shown significant results for efficient nutrient uptake and the removal of toxic pollutants from wastewater [100]. The underlying mechanism for the uptake of ammonium, nitrate, and phosphorus by microalgae in wastewater is represented in Eq. (4.1):

$$H_2O + CO_2 + NH_3 + PO_4 + light + microalgae \rightarrow C_{106}H_{263}O_{110}N_{16}P + C_6H_{12}O_6 + O_2 \quad (4.1)$$

Chlorella and *Scenedesmus* sp. have been reported to facilitate the complete breakdown of organic pollutants from the palm oil industry. Simultaneously a growth rate of 0.53 g/d was observed when *Scenedesmus* sp. was cultivated in textile wastewater [101]. Singh et al. [102] studied the potential of the microalgae *Chlorella* sp., *Chlorococcum* sp., and *Neochloris* sp. in pharmaceutical effluents. Recently, researchers have focused on the binary culture process for improved phytoremediation of wastewater through microalgae. Binary cultures (microalgae−microalgae) have been studied primarily due to their improved performance for wastewater treatment and the increase in biomass [99]. Table 4.4. summarizes the various wastewaters used for microalgae cultivation with improved biomass.

TABLE 4.4 Comparative studies of different microalgae grown in wastewater.

Type of wastewater	Microalgae	Nutrient removal (%)	Biomass production (g/L)	References
Primary treated	*Chlorosarcinopsis* sp.	TN: 87 TP: 82 TOC: 97	3.43	[97]
Secondary treated		TN: 85 TP: 81 TOC: 95	2.14	
Aquaculture	*Nannochloris oculata*	TN: 78 PO$_4$: 14 NO$_2$: 84	—	[103]
Piggery	*Scenedesmus* sp.	TN: 24 TP: 60	—	
Primary	*Galdieria sulphuraria*	NH$_3$: 6.26 P: 1.41	—	[104]
Pharmaceutical	*Neochloris* sp.	COD: 95	0.52	[102]
	Chlorococcum sp.	COD: 90	0.129	

(Continued)

90   4. Algal physiology and cultivation

TABLE 4.4   (Continued)

Type of wastewater	Microalgae	Nutrient removal (%)	Biomass production (g/L)	References
Primary	*Chlorella* sp.	COD: 67 NH_3: 100 PO_4: 97	0.419	[105]
Starch containing textile	*Scenedesmus* sp.	COD: 71	1.4	[106]
Molasses	*Scenedesmus sp*	NH_3: 76 PO_4: 64	3.2	[107]
Dairy manure	*Chlorella sp*	TN: 82	–	[106]
Aquaculture	*Scenedesmus obliquus*	TN: 86 TP: 82	0.0426	[107]
Tannery	*Chlorella sorokiniana*	COD: 80 $N-NH_4$: 74 $P-PO_4^3$: 93	–	[108]
	Chlorella variabilis	COD: 84 $N-NH_4$: 68 $P-PO_4^3$: 93	–	
	Scenedesmus sp.	COD: 66 $N-NH_4$: 47 $P-PO_4^3$: 70	–	
Raw sewage	*Chlorella sorokiniana*	COD: 69 $N-NH_4$: 86 $P-PO_4^3$: 68	1.31	[109]
	Scenedesmus sp.	COD: 76 $N-NH_4$: 98 $P-PO_4^3$: 98	1.06	
Anaerobic digestion effluent	*Acutodesmus obliquus*	COD: 45 $N-NH_4$: 7 $P-PO_4^3$: 84	1.1	[110]
	Parachlorella kessleri	COD: 39 $N-NH_4$: 15 $P-PO_4^3$: 84	1	
Landfill leachate and urban wastewater	*Desmodesmus* sp.	COD: 64 $N-NH_4$: 79 $P-PO_4^3$: 43	1.2	[111]
	Scenedesmus obliquus	COD: 67 $N-NH_4$: 82 $P-PO_4^3$: 41	1.3	

4.8 Conclusion

Algae have tremendous potential for the productin of products like biofuel, food, and
high-value-added compounds. This chapter focuses on a variety of environmental factors
that affect algae growth, which are necessary to understand when designing commercially
viable algae technologies. Environmental factors directly affect algae growth, and thus

ultimately affect biomass productivity. Hence, it is necessary to optimize such conditions to obtain maximum biomass productivity, which directly influences the algae-mediated value-added products. The algae cultivation system needs to be improved by introducing wastewater instead of costly media to make production economically and environmentally viable. Technoeconomic analysis is a powerful tool to identify the cost-effectiveness of the process by focusing on the type of targeted industry for which biomass is produced and to provide the input and output energy balance ratios specifically in large-scale production. To make algae biomass products cost-effective various advancements in cultivation and processing technologies are required for the complete utilization of algae biomass. Hence, more technology-based research is required to increase algae productivity and the industrialization of algae-mediated products in a cost-effective and environment-friendly manner.

Acknowledgments

The corresponding author (MPR) expresses her gratitude to the Mission innovation unit, Department of Biotechnology, New Delhi (INDIA) for financial support [file no. BT/PR31218/PBD/26/771/2019]. Authors KY, Reetu, and MPR are thankful to Amity University Uttar Pradesh, Noida for providing the required facilities.

References

[1] P.L. Show, J.S. Tan, S.Y. Lee, K.W. Chew, M.K. Lam, J.W. Lim, et al., A review on microalgae cultivation and harvesting, and their biomass extraction processing using ionic liquids, Bioengineered 11 (2020) 116−129.
[2] S.P. Slocombe, J.R. Benemann, Microalgal Production for Biomass and High-Value Products, CRC Press, Boca Raton, FL, 2016, pp. 13−26.
[3] J. Chen, Y. Wang, J.R. Benemann, X. Zhang, H. Hu, S. Qin, Microalgal industry in China: challenges and prospects, J. Appl. Phycol. 28 (2016) 715.
[4] E.C. Odjadjare, T. Mutanda, A.O. Olaniran, Potential biotechnological application of microalgae: a critical review, Crit. Rev. Biotechnol, Early Online (2015) 1−16.
[5] A. Thrassou, D. Vrontis, H. Chebbi, D. Yahiaoui, A preliminary strategic marketing framework for new product development, J. Transnatl. Manag. 17 (1) (2012) 21−44.
[6] G. Tardivo, A. Thrassou, M. Viassone, F. Serravalle, Value co-creation in the beverage and food industry, Br. Food J. 119 (2017) 2359−2372.
[7] A. Thrassou, D. Vrontis, S. Bresciani, The agile innovation pendulum: a strategic marketing multicultural model for family businesses, Int. Stud. Manag. Organ. 48 (2018) 105−120.
[8] K.N. Ingle, E. Vitkin, A. Robin, Z. Yakhini, D. Mishori, A. Golberg, Macroalgae biorefinery from *Kappaphycus alvarezii*: conversion modeling and performance prediction for India and Philippines as examples, BioEnergy Res. 11 (2018) 22−32.
[9] A. Golberg, A. Liberzon, Modeling of smart mixing regimes to improve marine biorefinery productivity and energy efficiency, Algal Res. 11 (2015) 28−32.
[10] K. Balina, F. Romagnoli, D. Blumberga, Seaweed biorefinery concept for sustainable use of marine resources, Energy Procedia 128 (2017) 504−511.
[11] I.T.D. Cabanelas, Z. Arbib, F.A. Chinalia, C.O. Souza, J.A. Perales, P.F. Almeida, et al., From waste to energy: microalgae production in wastewater and glycerol, Appl. Energy 109 (2013) 283−290.
[12] F.J.R. Taylor, U. Pollingher, Ecology of dinoflagellates, in: F.J.R. Taylor (Ed.), The Biology of Dinoflagellates, Blackwell Scientific, 407, Publications, Oxford, 1987, p. 429.
[13] C. Zonneveld, Photoinhibition as affected by photoacclimation in phytoplankton: a model approach, J. Theor. Biol. 193 (1) (1998) 115−123.
[14] F. Rubio, F. Camacho, J.M. García Camacho, Y. Fernández Sevilla, Chisti, E. Molina Grima, A mechanistic model of photosynthesis in microalgae, Biotechnol. Bioeng. 81 (4) (2003) 459−473.

[15] P. Müller, X.-P. Li, K.K. Niyogi, Non-photochemical quenching, a response to excess light energy, Plant. Physiol. 125 (4) (2001) 1558–1566.

[16] M. Bilal, T. Rasheed, I. Ahmed, H.M.N. Iqbal, High-value compounds from microalgae with industrial exploitability—a review, Front. Biosci. (Scholar Ed.) 9 (2017) 319–342.

[17] I.-S. Ng, S.-I. Tan, P.-H. Kao, Y.-K. Chang, J.-S. Chang, Recent developments on genetic engineering of microalgae for biofuels and bio-based chemicals, Biotechnol. J. 12 (10) (2017) 1600644.

[18] E. Koutra, C.N. Economou, P. Tsafrakidou, M. Kornaros, Bio-based products from microalgae cultivated in digestates, Trends Biotechnol. 36 (8) (2018) 819–833.

[19] Y. Maeda, T. Yoshino, T. Matsunaga, M. Matsumoto, T. Tanaka, Marine microalgae for production of biofuels and chemicals, Curr. Opin. Biotechnol. 50 (2018) 111–120.

[20] N. Renuka, A. Guldhe, R. Prasanna, P. Singh, F. Bux, Microalgae as multi-functional options in modern agriculture: current trends, prospects and challenges, Biotechnol. Adv. 36 (4) (2018) 1255–1273.

[21] X. Li, H. Sun, X. Mao, Y. Lao, F. Chen, Enhanced photosynthesis of carotenoids in microalgae driven by light-harvesting gold nanoparticles, ACS Sustain. Chem. Eng. 8 (20) (2020) 7600–7608.

[22] S. Li, P. Wang, C. Zhang, X. Zhou, Z. Yin, T. Hu, et al., Influence of polystyrene microplastics on the growth, photosynthetic efficiency and aggregation of freshwater microalgae *Chlamydomonas reinhardtii*, Sci. Total. Environ. 714 (2020) 136767.

[23] X. Ji, J. Cheng, D. Gong, X. Zhao, Y. Qi, Y. Su, et al., The effect of NaCl stress on photosynthetic efficiency and lipid production in freshwater microalga—*Scenedesmus obliquus* XJ002, Sci. Total. Environ. 633 (2018) 593–599.

[24] J.H. De Vree, R. Bosma, M. Janssen, M.J. Barbosa, R.H. Wijffels, Comparison of four outdoor pilot-scale photobioreactors, Biotechnol. Biofuels 8 (1) (2015) 215.

[25] G. Perin, A. Bellan, A. Bernardi, F. Bezzo, T. Morosinotto, The potential of quantitative models to improve microalgae photosynthetic efficiency, Physiol. Plant. 166 (1) (2019) 380–391.

[26] Z. Perrine, S. Negi, R.T. Sayre, Optimization of photosynthetic light energy utilization by microalgae, Algal Res. 1 (2) (2012) 134–142.

[27] S. Cazzaniga, L. Dall'Osto, J. Szaub, L. Scibilia, M. Ballottari, S. Purton, et al., Domestication of the green alga *Chlorella sorokiniana*: reduction of antenna size improves light-use efficiency in a photobioreactor, Biotechnol. Biofuels 7 (1) (2014) 157.

[28] G. Perin, A. Bellan, A. Segalla, A. Meneghesso, A. Alboresi, T. Morosinotto, Generation of random mutants to improve light-use efficiency of *Nannochloropsis gaditana* cultures for biofuel production, Biotechnol. Biofuels 8 (1) (2015) 161.

[29] J. Beckmann, F. Lehr, G. Finazzi, B. Hankamer, C. Posten, L. Wobbe, et al., Improvement of light to biomass conversion by de-regulation of light-harvesting protein translation in *Chlamydomonas reinhardtii*, J. Biotechnol. 142 (1) (2009) 70–77.

[30] L. Wei, Q. Wang, Y. Xin, Y. Lu, J. Xu, Enhancing photosynthetic biomass productivity of industrial oleaginous microalgae by overexpression of RuBisCO activase, Algal Res. 27 (2017) 366–375.

[31] I. Pancha, K. Chokshi, S. Mishra, Enhanced biofuel production potential with nutritional stress amelioration through optimization of carbon source and light intensity in *Scenedesmus* sp. CCNM 1077, Bioresour. Technol. 179 (2015) 565–572.

[32] J. Yang, B. Li, C. Zhang, H. Luo, Z. Yang, pH-associated changes in induced colony formation and growth of *Scenedesmus obliquus*, Fund. Appl. Limnol. 187 (3) (2016) 241–246.

[33] M.P. Rai, T. Gautom, N. Sharma, Effect of salinity, pH, light intensity on growth and lipid production of microalgae for bioenergy application, OnLine, J. Biol. Sci. 15 (4) (2015) 260.

[34] T. Mathimani, L. Uma, D. Prabaharan, Formulation of low-cost seawater medium for high cell density and high lipid content of *Chlorella vulgaris* BDUG 91771 using central composite design in biodiesel perspective, J. Clean. Prod. 198 (2018) 575–586.

[35] J. Jin, C. Dupré, J. Legrand, D. Grizeau, Extracellular hydrocarbon and intracellular lipid accumulation are related to nutrient-sufficient conditions in pH-controlled chemostat cultures of the microalga *Botryococcus braunii* SAG 30.81, Algal Res. 17 (2016) 244–252.

[36] S.M.U. Shah, C.C. Radziah, S. Ibrahim, F. Latiff, M.F. Othman, M.A. Abdullah, Effects of photoperiod, salinity and pH on cell growth and lipid content of *Pavlova lutheri*, Ann. Microbiol. 64 (1) (2014) 157–164.

[37] E. Sforza, B. Gris, C. de Farias Silva, Tomas Morosinotto, A. Bertucco, Effects of light on cultivation of *Scenedesmus obliquus* in batch and continuous flat plate photobioreactor, Chem. Eng. 38 (2014) 211.

References

[38] B. Durairaj, S. Muthu, T. Xavier, Antimicrobial activity of *Aspergillus niger* synthesized titanium dioxide nanoparticles, Adv. Appl. Sci. Res. 6 (1) (2015) 45–48.

[39] K.K. Sharma, H. Schuhmann, P.M. Schenk, High lipid induction in microalgae for biodiesel production, Energies 5 (5) (2012) 1532–1553.

[40] J.C. Nzayisenga, X. Farge, S.L. Groll, A. Sellstedt, Effects of light intensity on growth and lipid production in microalgae grown in wastewater, Biotechnol. Biofuels 13 (1) (2020) 1–8.

[41] Z. Wu, P. Duangmanee, P. Zhao, N. Juntawong, C. Ma, The effects of light, temperature, and nutrition on growth and pigment accumulation of three *Dunaliella salina* strains isolated from saline soil, Jundishapur J. Microbiol. 9 (1) (2016).

[42] T. Menegol, A.B. Diprat, E. Rodrigues, R. Rech, Effect of temperature and nitrogen concentration on biomass composition of *Heterochlorella luteoviridis*, Food Sci. Technol. 37 (No. SPE) (2017) 28–37.

[43] L. Wei, X. Huang, Z. Huang, Temperature effects on lipid properties of microalgae *Tetraselmis subcordiformis* and *Nannochloropsis oculata* as biofuel resources, Chin. J. Oceanol. Limnol. 33 (1) (2015) 99–106.

[44] T. Vo, D. Tran, Effects of salinity and light on growth of *Dunaliella isolates*, J. Appl. Environ. Microbiol. 2 (5) (2014) 208–211.

[45] V. Ananthi, K. Brindhadevi, A. Pugazhendhi, A. Arun, Impact of abiotic factors on biodiesel production by microalgae, Fuel 284 (2021) 118962.

[46] M.L. Bartley, W.J. Boeing, A.A. Corcoran, F.O. Holguin, T. Schaub, Effects of salinity on growth and lipid accumulation of biofuel microalga *Nannochloropsis salina* and invading organisms, Biomass Bioenergy 54 (2013) 83–88.

[47] C. Yoo, S.-Y. Jun, J.-Y. Lee, C.-Y. Ahn, H.-M. Oh, Selection of microalgae for lipid production under high levels carbon dioxide, Bioresour. Technol. 101 (1) (2010) S71–S74.

[48] S.P. Singh, Priyanka Singh, Effect of CO_2 concentration on algal growth: a review, Renew. Sustain. Energy Rev. 38 (2014) 172–179.

[49] L.-L. Zhuang, Y. Azimi, D. Yu, Y.-H. Wu, H.-Y. Hu, Effects of nitrogen and phosphorus concentrations on the growth of microalgae *Scenedesmus*. LX1 in suspended-solid phase photobioreactors (ssPBR), Biomass Bioenergy 109 (2018) 47–53.

[50] M.J. Zarrinmehr, O. Farhadian, F.P. Heyrati, J. Keramat, E. Koutra, M. Kornaros, et al., Effect of nitrogen concentration on the growth rate and biochemical composition of the microalga, Isochrysis galbana, Egypt. J. Aquat. Res. 46 (2) (2020) 153–158.

[51] X. Guo, G. Su, Z. Li, J. Chang, X. Zeng, Y. Sun, et al., Light intensity and N/P nutrient affect the accumulation of lipid and unsaturated fatty acids by *Chlorella* sp, Bioresour. Technol. 191 (2015) 385–390.

[52] M. Hultberg, H.L. Jönsson, K.-J. Bergstrand, A.S. Carlsson, Impact of light quality on biomass production and fatty acid content in the microalga *Chlorella vulgaris*, Bioresour. Technol. 159 (2014) 465–467.

[53] L. Peng, C.Q. Lan, Z. Zhang, C. Sarch, M. Laporte, Control of protozoa contamination and lipid accumulation in *Neochloris oleoabundans* culture: effects of pH and dissolved inorganic carbon, Bioresour. Technol. 197 (2015) 143–151.

[54] A. Concas, M. Pisu, G. Cao, A novel mathematical model to simulate the size-structured growth of microalgae strains dividing by multiple fission, Chem. Eng. J. 287 (2016) 252–268.

[55] X. Dou, X.-H. Lu, M.-Z. Lu, L.-S. Yu, R. Xue, J.-B. Ji, The effects of trace elements on the lipid productivity and fatty acid composition of *Nannochloropis oculata*, J. Renew. Energy 4 (2013) 1–11.

[56] M. Fadhlaoui, V. Laderriere, I. Lavoie, C. Fortin, Influence of temperature and nickel on algal biofilm fatty acid composition, Environ. Toxicol. Chem. 8 (2020) 1566–1577.

[57] J. Van Wagenen, T.W. Miller, S. Hobbs, P. Hook, B. Crowe, M. Huesemann, Effects of light and temperature on fatty acid production in *Nannochloropsis salina*, Energies 5 (2012) 731–740.

[58] Y. Zheng, T. Li, X. Yu, P.D. Bates, T. Dong, S. Chen, High-density fed-batch culture of a thermotolerant microalga *chlorella sorokiniana* for biofuel production, Appl. Energy 108 (2013) 21–287.

[59] S. Abalde-Cela, A. Gould, X. Liu, E. Kazamia, A.G. Smith, C. Abell, High-throughput detection of ethanol-producing *cyanobacteria* in a microdroplet platform, J. R. Soc. Interface 12 (2015) 106.

[60] P. Hammar, S.A. Angermayr, S.L. Sjostrom, J. Van Der Meer, K.J. Hellingwerf, E.P. Hudson, et al., Single-cell screening of photosynthetic growth and lactate production by *cyanobacteria*, Biotechnol. Biofuels 8 (2015) 193.

[61] O.K. Lee, E.Y. Lee, Sustainable production of bioethanol from renewable brown algae biomass, Biomass Bioenergy 92 (2016) 70–75.

[62] L. Zhang, B. Zhang, X. Zhu, H. Chang, S. Ou, H. Wang, Role of bioreactors in microbial biomass and energy conversion, Green. Energy Technol. (2018) 33−78.

[63] İ. Ekİn, Types of microalgae cultivation photobioreactors and production process of microalgal biodiesel as alternative fuel, Acta Biologica Turc. 47 (2020) 777−780.

[64] J. Sen Tan, S.Y. Lee, K.W. Chew, M.K. Lam, J.W. Lim, S.H. Ho, et al., A review on microalgae cultivation and harvesting, and their biomass extraction processing using ionic liquids, Bioengineered, 2020.

[65] Solovchenko, A., & Chekanov, K. Production of carotenoids using microalgae cultivated in photobioreactors, in: Production of Biomass and Bioactive Compounds Using Bioreactor Technology, 2014, pp. 63−91.

[66] R.R. Narala, S. Garg, K.K. Sharma, S.R. Thomas-Hall, M. Deme, Y. Li, et al., Comparison of microalgae cultivation in photobioreactor, open raceway pond, and a two-stage hybrid system, Front. Energy Res. (2016).

[67] J.S. Chang, P.L. Show, T.C. Ling, C.Y. Chen, S.H. Ho, C.H. Tan, et al., Photobioreactors, in: Current Developments in Biotechnology and Bioengineering: Bioprocesses, Bioreactors and Controls, 2017.

[68] O.M. Adeniyi, U. Azimov, A. Burluka, algae biofuel: current status and future applications, Renew. Sustain. Energy Rev. (2018).

[69] C. Praeger, M.J. Vucko, L. McKinna, R. de Nys, A. Cole, Estimating the biomass density of macroalgae in land-based cultivation systems using spectral reflectance imagery, Algal Res. 50 (2020) 102009.

[70] S. Bhadra, P.A. Salam, N.K. Sarker, Microalgae-based biodiesel production in open raceway ponds using coal thermal flue gas: a case of West Bengal, India. Environ. Qual. Manag. 29 (2020) 27−36.

[71] O.M. Adeniyi, U. Azimov, A. Burluka, Algae biofuel: current status and future applications, Renew. Sustain. Energy Rev. 90 (2018) 316−335.

[72] S.A. Razzak, S.A.M. Ali, M.M. Hossain, H. deLasa, Biological CO_2 fixation with production of microalgae in wastewater — a review, Renew. Sustain. Energy Rev. 76 (2017) 379−390.

[73] C. Song, X. Han, Y. Qiu, Z. Liu, S. Li, Y. Kitamura, Microalgae carbon fixation integrated with organic matters recycling from soybean wastewater: effect of pH on the performance of hybrid system, Chemosphere. 248 (2020) 126094.

[74] S. Daliry, A. Hallajisani, J. Mohammadi Roshandeh, H. Nouri, A. Golzary, Investigation of optimal condition for *Chlorella vulgaris* microalgae growth, Glob. J. Environ. Sci. Manag. 3 (2017) 217−230.

[75] M. Scarsella, G. Belotti, P. De Filippis, M. Bravi, Study on the optimal growing conditions of *Chlorella vulgaris* in bubble column photobioreactors, Chem. Eng. Trans. 20 (2010).

[76] H. Taher, S. Al-Zuhair, A.H. Al-Marzouqi, Y. Haik, M.M. Farid, A review of enzymatic transesterification of microalgal oil-based biodiesel using supercritical technology, Enzyme Res. (2011) 1−5.

[77] M. Min, L. Wang, Y. Li, M.J. Mohr, B. Hu, W. Zhou, et al., Cultivating *chlorella sp.* in a pilot-scale photobioreactor using centrate wastewater for microalgae biomass production and wastewater nutrient removal, Appl. Biochem. Biotechnol. 165 (2011) 123−137.

[78] W.H. Chen, B.J. Lin, M.Y. Huang, J.S. Chang, Thermochemical conversion of microalgal biomass into biofuels: a review, Bioresour. Technol. 184 (2015) 314−327.

[79] S. Aravind, P.S. Kumar, N.S. Kumar, N. Siddarth, Conversion of green algal biomass into bioenergy by pyrolysis. A review, Environ. Chem. Lett. 5 (2020) 1−21.

[80] P.D. Kerrison, G. Twigg, M. Stanley, D. De Smet, G. Buyle, A. Martínez Pina, et al., Twine selection is essential for successful hatchery cultivation of *Saccharina latissima*, seeded with either meiospores or juvenile sporophytes, J. Appl. Phycol 31 (2019) 3051−3060.

[81] P.S. Schmedes, M.M. Nielsen, J.K. Petersen, Improved *Palmaria palmata* hatchery methods for tetraspore release, even settlement and high seedling survival using strong water agitation and *macerated propagules*, Algal Res. 40 (2019) 101494.

[82] L. Hayashi, S. de J. Cantarino, A.T. Critchley, Challenges to the future domestication of seaweeds as cultivated species: understanding their physiological processes for large-scale production, Adv. Bot. Res. 95 (2020) 57−83.

[83] M. Pereira, D. Matias, F. Pereira, C. Reis, M.F. Simões, P. Rijo, Antimicrobial screening of *Plectranthus madagascariensis* and *P. neochilus* extracts, J. Biomed. Biopharm. Res. 12 (2015) 127−138.

[84] R.C. Rajak, S. Jacob, B.S. Kim, A holistic zero waste biorefinery approach for macroalgal biomass utilization: a review, Sci. Total. Environ. 716 (2020) 137067.

[85] S. Mawi, S. Krishnan, M.F.M. Din, N. Arumugam, S. Chelliapan, Bioremediation potential of macroalgae *Gracilaria edulis* and *Gracilaria changii* co-cultured with shrimp wastewater in an outdoor water recirculation system, Environ. Technol. Innov. 17 (2020) 100571.

References

[86] A. Vadiveloo, E.G. Nwoba, N.R. Moheimani, Viability of combining microalgae and macroalgae cultures for treating anaerobically digested piggery effluent, J. Environ. Sci. (China) 82 (2019) 132–144.

[87] J.M. Valero-Rodriguez, S.E. Swearer, T. Dempster, R. de Nys, A.J. Cole, Evaluating the performance of freshwater macroalgae in the bioremediation of nutrient-enriched water in temperate environments, J. Appl. Phycol 32 (2020) 641–652.

[88] R. Kumar, A.K. Ghosh, P. Pal, Synergy of biofuel production with waste remediation along with value-added co-products recovery through microalgae cultivation: a review of membrane-integrated green approach, Sci. Total. Environ. 698 (2020) 134169.

[89] F.A. Fernández, J.M.F. Sevilla, E.M. Grima, Costs analysis of microalgae production, Biofuels from Algae, Elsevier, 2019, pp. 551–566.

[90] R. Davis, A. Aden, P.T. Pienkos, Techno-economic analysis of autotrophic microalgae for fuel production, Appl. Energy 88 (2011) 3524–3531.

[91] N. Norsker, M.J. Barbosa, M.H. Vermue, R.H. Wijffels, Microalgal production—a close look at the economics, Biotechnol. Adv. 29 (2011) 24–27.

[92] Y. Chisti, Biodiesel from microalgae, Biotechnol. Adv. 25 (2007) 294–306.

[93] J.W. Richardson, M.D. Johnson, J.L. Outlaw, Economic comparison of open pond raceways to photo bioreactors for profitable production of algae for transportation fuels in the Southwest, Algal Res. 1 (2012) 93–100.

[94] J.W. Richardson, M.D. Johnson, X. Zhang, P. Zemke, W. Chen, Q. Hu, A financial assessment of two alternative cultivation systems and their contributions to algae biofuel economic viability, Algal Res. 4 (2014) 96–104.

[95] W.S. Chai, W.G. Tan, H.S.H. Munawaroh, V.K. Gupta, S.H. Ho, P.L. Show, Multifaceted roles of microalgae in the application of wastewater biotreatment: a review, Environ. Pollut. (2021) 116236.

[96] G.S. Diniz, A.F. Silva, O.Q.F. Araújo, O.Q.F.R.M. Chaloub, The potential of microalgal biomass production for biotechnological purposes using wastewater resources, J. Appl. Phycol. 29 (2017) 821–832.

[97] S. Vasistha, A. Khanra, M.P. Rai, Influence of microalgae-ZnO nanoparticle association on sewage wastewater towards efficient nutrient removal and improved biodiesel application: an integrated approach, J. Water Process. Eng. (2020) 101711.

[98] A. Shahid, S. Malik, H. Zhu, J. Xu, M.Z. Nawaz, S. Nawaz, et al., Cultivating microalgae in wastewater for biomass production, pollutant removal, and atmospheric carbon mitigation; a review, Sci. Total. Environ. 704 (2020) 135303.

[99] H. Al-Jabri, P. Das, S. Khan, M. Thaher, M.A. Quadir, Treatment of Wastewaters by Microalgae and the Potential Applications of the Produced Biomass—a Review, Water 13 (2021) 27.

[100] H.B. Hariz, M.S. Takriff, Palm oil mill effluent treatment and CO_2 sequestration by using microalgae—sustainable strategies for environmental protection, Environ. Sci. Pollut. Res. (2017) 1–32.

[101] S. Bhattacharya, S.K. Pramanik, P.S. Gehlot, H. Patel, T. Gajaria, S. Mishra, et al., Process for preparing value-added products from microalgae using textile effluent through a biorefinery approach, ACS Sustain. Chem. Eng. 5 (2017) 10019–10028.

[102] A. Singh, S.B. Ummalyma, D. Sahoo, Bioremediation and biomass production of microalgae cultivation in river water contaminated with pharmaceutical effluent, Bioresour. Technol. 307 (2020) 123233.

[103] Q. Emparan, R. Harun, M.K. Danquah, Role of phycoremediation for nutrient removal from wastewaters: a review, Appl. Ecol. Environ. Res. 17 (2019) (2019) 889–915.

[104] D. Tchinda, S.M. Henkanatte-Gedera, I.S.A. Abeysiriwardana-Arachchige, H.M.K. Delanka- Pedige, S.P. Munasinghe-Arachchige, Y. Zhang, et al., Single-step treatment of primary effluent by Galdieria sulphuraria: removal of biochemical oxygen demand, nutrients, and pathogens, Algal Res. 42 (2019) 101578.

[105] L. Evans, S.J. Hennige, N. Willoughby, A.J. Adeloye, M. Skroblin, T. Gutierrez, Effect of organic carbon enrichment on the treatment efciency of primary settled wastewater by Chlorella vulgaris, Algal Res. 24 (2017) 368–377.

[106] C. Lin, M.T. Nguyen, C. Lay, Starch-containing textile wastewater treatment for biogas and microalgae biomass production, J. Clean. Prod. 168 (2017) 331–337.

[107] S. Yang, J. Xu, Z. Wang, L. Bao, E.Y. Zeng, Cultivation of oleaginous microalgae for removal of nutrients and heavy metals from biogas digestates, J. Clean. Prod. 164 (2017) 793–803.

[108] M. Nagi, M. He, D. Li, T. Gebreluel, B. Cheng, C. Wang, Utilization of tannery wastewater for biofuel production: new insights on microalgae growth and biomass production, Sci. Rep. 10 (2020) 1−14.

[109] S.K. Gupta, F.A. Ansari, A. Shriwastav, N.K. Sahoo, I. Rawat, F. Bux, Dual role of *Chlorella* sorokiniana and *Scenedesmus obliquus* for comprehensive wastewater treatment and biomass production for bio-fuels, J. Clean. Prod. 115 (2016) 255−264.

[110] E. Koutra, G. Grammatikopoulos, M. Kornaros, Microalgal post-treatment of anaerobically digested agro-industrial wastes for nutrient removal and lipids production, Bioresour. Technol. 224 (2017) 473−480.

[111] A. Hernández-García, S.B. Velásquez-Orta, E. Novelo, I. Yáñez-Nogueza, Wastewater-leachate treatment by microalgae: biomass, carbohydrate and lipid production, Ecotoxicol. Environ. Saf. 174 (2019) 435−444.

CHAPTER 5

Genetic manipulation of microalgae for enhanced biotechnological applications

Ashutosh Pandey[1,2], Gaurav Kant[1], Shadma Afzal[1], Manish Pratap Singh[1], Nand Kumar Singh[1], Sanjay Kumar[3] and Sameer Srivastava[1]

[1]Department of Biotechnology, Motilal Nehru National Institute of Technology Allahabad, Prayagra, India [2]Department of Biotechnology, IMS Engineering College (Affilliated to Dr. APJ Abdul Kalam Technical University, Lucknow) Ghaziabad, India [3]School of Biochemical Engineering, Indian Institute of Technology (BHU) Varanasi, Varanasi, India

5.1 Introduction

Microalgae have been widely reported for their biotechnological application in the commercial production of many potential bioactive molecules. Microalgae are capable of producing different types of metabolites such as carotenoids, vitamins, carbohydrates, proteins, and lipids that are used for animal as well as human consumption [1,2]. In various environmental conditions, they have distinctive photosystems and greater adaptability, making them an important player in the overproduction of the desired commodity. Over the past few decades microalgae have been genetically manipulated using various approaches such as the classical genetic modification approach (insertional mutagenesis and targeted gene alteration) and modern novel approach (riboswitches-based gene regulation for efficient transgene expression) [3], along with the screening of efficient clones through luciferase reporters coupled with promoters and the application of inducible chloroplast genes for metabolic reconstruction of the lipid pathways [4–7].

Novel genetic modification techniques primarily relay upon the genome availability of a particular microalgal strain, metabolic pathways encoded by genes, and the capability to engineer them. Recent sequencing techniques have been easing the sequencing

Handbook of Algal Biofuels
DOI: https://doi.org/10.1016/B978-0-12-823764-9.00019-4

© 2022 Elsevier Inc. All rights reserved.

process for the whole genome as well as the targeted gene sequence of the microalgae [8–10]. The capability of the reverse genetic approach to engineer a strain has been limited and has faced different bottlenecks. For example, targeted gene disruption and insertional inactivation of genes via mutagenesis have been reported in *Phaeodactylum tricornutum* [11], *Nannochloropsis* sp. [12,13], and *Chlamydomonas reinhardtii* [14], but they are difficult to replicate in other strains, being very laborious and requiring expert screening of clones. This impedes greatly the testing of gene and pathway functions based on hypotheses and hampers the production of reasonable strains of microalgae [15].

This chapter highlights the key strategies for the genetic alteration of microalgae, the methods used for gene modification, and a detailed discussion of genetically modified microalgae used for the development of triacylglycerol (TAG) and high-value-added compounds. Specifically, the application and use of CRISPR technology for selective genetic engineering in some algae species is summarized. As CRISPR technology continues to progress, it is predicted that the CRISPR/Cas9 method will revolutionize future solutions for effective and rigorous genomic manipulation to produce useful microalgae resources in an efficient and economically feasible manner.

5.2 Genetic modification of algae for the generation of energy and value-added metabolites

5.2.1 Enhancement of bioenergy products

Microalgae harboring genetically engineered lipid biosynthesis genes/pathway are more suitable to be used as a biodiesel feedstock. This includes the enhancement of a single or several genes involved in the lipid production pathway or the blocking of the alternative/competitive pathway. Sequentially there are three significant steps in the lipid production pathway: malonyl-CoA synthesis, acyl chain elongation, and TAG formation. The three existing competitive pathways of lipid synthesis are β-oxidation of fatty acids, phospholipid biosynthesis, and the conversion of phosphoenolpyruvate to oxaloacetate. In the first step of lipid biosynthesis Acetyl-CoA carboxylase (ACC) plays a key role, as it catalyzes the carboxylation of acetyl-CoA to form malonyl-CoA, the first intermediate product of the fatty acid elongation pathway [16,17]. ACC has three activity subunits composed from several polypeptides encoded by various genes, comprising biotin carboxylase, biotin carboxyl carrier protein, and carboxyl transferase activity [18]. These three operation subunits assemble a completely functioning ACC with a certain unique difference between species [18]. To date, several studies on the overexpression of ACC from different species in diverse organisms have concluded that, due to the increased malonyl-CoA pool, the rate of fatty acid biosynthesis increased, however, the lipid content was not significantly improved [19,20]. It is suggested that in certain species, the ACC is not the rate-determining step, as the expression of ACC exceeds at some level [21]. After malonyl-CoA is synthesized from acetyl-CoA by ACC, a number of reactions are catalyzed by fatty acid synthase (FAS) for the production of fatty acids [17]. FASs include malonyl/acetyltransferase, acyl carrier protein (ACP), ketoacyl synthase (KS), ketoacyl reductase (KR), dehydrase (DH), enoyl reductase (ER), and thioesterase (TE) [22]. First, malonyl-CoA:ACP transacetylase catalyzes the reaction to malonyl-CoA by adding an acyl carrier protein (ACP) and releases the intermediate product of malonyl-CoA-ACP that can enter the elongation cycle of fatty acids [21].

A sequence of enzyme-based Claisen condensation by KS, KR, DH, and ER will take place during the fatty acid elongation cycle to form the final product palmitic-ACP or stearic-ACP [17,21]. Then, TE hydrolyzes the thioester bonds of palmitic-ACP or stearic-ACP and produces palmitic acid or stearic acid. KS is stimulated to increase fatty acid accumulation, resulting in changes in cell physiology, lipid profile, decrease in cell growth rate, and rate of lipid synthesis [18,21]. On the other hand, due to the inherent properties of fatty acid synthase (FAS), it is difficult to increase the development of fatty acids by changing the direction. This is because the enzyme is composed of several subunits and a change on any subunit will influence the efficiency of the whole FAS [23]. In addition, desaturases and elongases have been incorporated into fatty acid synthesis to obtain longer or unsaturated fatty acids, which use palmitic acid or stearic acid as substrates [20,21,23]. Triacylglyceride (TAG) formation is the final stage of lipid synthesis in microalgae. In the glycerol phosphate-based pathway glycerol-3-phosphate is converted into phosphatidate (PA) by glycerol phosphate acyltransferase (GPAT) and acylglycerol phosphate acyltransferase (AGPAT) [18,23,24]. Phosphatidic acid phosphatase (PAP) catalyzes the dephosphorylation of PA to generate diacylglycerol. Ultimately diacylglycerol acyltransferase (DGAT), a key enzyme in the TAG synthesis pathway, carries out the esterification of diacylglycerol to TAG [25−27]. Studies on the overexpression of DGAT in microbes as well as plants have been documented, which indicate that doing so could increase TAG accumulation [28]. Successful expression of DGAT gene in *Scenedesmus obliquus* increased the lipid content by 128% [21]. The esterification of DGAT-catalyzed diacylglycerol is therefore the rate-determining stage in lipid synthesis. Alternatively, blocking the competing pathways of lipid synthesis can also increase lipid production. Among the three competitive pathways, the β-oxidation pathway is the easiest competitive pathway. It separately breaks down the fatty acid in cytosol or mitochondria and peroxisome for prokaryotes or eukaryotes [28−30]. Studies show that lipid production could also be effectively improved by the direct knockout of genes related to the β-oxidation pathway or by indirect inhibition of the β-oxidation pathway by reducing the transport system of acyl-CoA [31]. Another competitive pathway is the pathway of phospholipid biosynthesis, where phosphatidate is transformed into CDP-diacylglycerol and enters the pathway of phospholipid biosynthesis to shape the cell membrane instead of the formulation of TAG [32,33]. However, inhibiting the phospholipid biosynthesis pathway produces abnormally long fatty acids because this inhibition affects the cell physiology as there is a lack of phospholipids for cell membrane formation [32,33]. The reaction which transforms phosphoenolpyruvate into oxaloacetate is another example. The essential metabolite in the biolipid synthesis pathway to be converted into pyruvate or oxaloacetate is phosphoenolpyruvate (PEP). Consequently, several studies have shown that the inhibition of PEPC activity can increase the lipid content by killing the PEPC gene [20,21,34].

5.2.2 Enhancement of carbohydrate accumulation in microalgae

Carbohydrates could be converted to a variety of biofuels, including ethanol, butanol, H_2, lipids, and/or methane. They are stored in microalgae in several ways. The phyla Rhodophyta, Chlorophyta, Glaucophyta, and Dinophyta store carbohydrates as linear α-1,4-and branched α-1,6-glycosidic linkages [35]. Moreover, in green algae, starch is synthesized and stored within the chloroplast, and stored in the cytoplasm or in the periplastidial space [36]. The key rate-limiting step of starch synthesis is identified as ADP-glucose pyrophosphorylase (AGPases). AGPases from *E. coli* (*glgC16*) or the recombinant *rev6* have been shown to successfully increase starch content in other plants [37]. These could be used as target genes to be expressed in microalgae with no AGPase activity, such as the *C. reinhardtii sta1* or *sta6* mutant. Starch-synthesizing enzymes could be an alternative technique for increasing microalgal starch. Another alternative genetic engineering method has been also tested for starch accumulation. Since starch can be degraded by hydrolytic and/or phosphorolytic mechanisms, any method to stop this mechanism would ultimately increase the accumulation of starch. Hydrolytic starch degradation requires an enzyme capable of hydrolyzing semicrystalline carbohydrates at the surface of the insoluble starch granule. In *Arabidopsis thaliana*, α-amylase (AMY3) is shown to possess the ability to participate in starch degradation. A homologous protein like AMY3 is found in *Chlamydomonas. reinhardtii* [35], The degradation of starch stored in plastids is stimulated by the phosphorylation of glucose residues at the root of amylopectin by glucan-water dikinases (GWD) [38]. The C-3 position in the glucan can also be phosphorylated by the phosphoglucan water dikinases (PWD). These phosphorylations are involved in degradation of starch granules [39]. In a seminal work in *A. thaliana*, the loss of the GWD (*sex1* phenotype) resulted in four to six times higher starch levels than those in wild-type leaves [40], while disruptions in PWD resulted in a less severe phenotype having starch accumulation but not at the level as reported by Yu et al. [40]. The effect of the loss of GWD and PWD suggests that these genes could be a promising target for a starch accumulation phenotype in microalgae (Table 5.1).

5.2.3 Other value-added compound production

Carotenoids are a diverse group of pigments that are commonly found and synthesized in green algae. Carotenoids play a key role in efficient photosynthesis. As for humans, carotenoids have indeed been known to be used to treat certain diseases or used as a nutrient supplement [49]. As a result, carotenoids are now marketed as an attractive commodity in the field of wellness food. Due to their high productivity and high growth rate microalgae are now used to produce and extract the carotenoids in large quantities [50]. The carotenoid biosynthesis pathway has been extensively studied and characterized, hence in order to enhance the production of carotenoids, screening of high carotenoid-producing microalgae and metabolic engineering approaches have been also applied [49,51]. The carotenoid synthesis pathway is a well-characterized pathway with phytoene synthase (PSY) catalyzing the first step of carbon flux towards the production of phytoene. This is considered to be the rate-determining step in the pathway [52]. Subsequently the phytoene is converted to lycopene by phytoene desaturase (PDS), ζ-carotene desaturase (ZDS), and carotene

TABLE 5.1 Genetic studies conducted to enhance lipid biosynthesis through genetic and metabolic engineering in different kinds of microalgae.

Microalgae	Enzyme	Genes	Molecular approach/ transformation technique	Targeted pathway	Results	References
Chlamydomonas reinhardtii	Steroyl-acyl carrier protein (ACP) desaturase	CrFAB₂	Overexpression of functional gene (alteration in fatty acid composition)/ glass bead	Fatty acid biosynthesis	Increase total fatty acid content by 28%, 2.4-fold increases in oleic acid (18:1), and slight increase in linoleic acid (18:2).	[1]
Chlamydomonas reinhardtii	Diacylglycerol acyltransferase	CrDGAT2−4	RNAi silencing	Terminal step of fatty acid biosynthesis	24.3% Increase in lipid content	[41]
		CrDGAT2−1, CrDGAT2−5			Decrease in lipid biosynthesis by 16%−24% and 24%−34% respectively	
Chlamydomonas reinhardtii	Citrate synthase	CrCIS	Blocking of competitive pathway	Citric acid synthesis	Decrease in mRNA level, Increase expression of triacylglycerol (TAG) biosynthesis pathway related genes by 209%−266%, increase TAG accumulation by 169.5%.	[8]
Chlamydomonas reinhardtii	Phosphoenol pyruvate carboxylase	CrPEPC	Blocking of competitive pathway/	Oxaloacetate acid synthesis	Lipid content (TAG) increased by 20%, Increased expression of TAG biosynthesis related gene i.e., DGAT/Phosphatidate phosphatase	[42]
Chlamydomonas reinhardtii	Acetyl CO-A Carboxylase, enoyl ACP reductase, phosphatidyl- glycerophosphate synthase, monogalactosyldiacylglycerol, sulfolipid synthase.	accD, ENR1, PGP1, MGD1, SQD2	Overexpression of Dof-type transcription factor/ *Agrobacterium* mediated transformation	Fatty acid and glycerolipid biosynthesis	Total lipid content increased by around twofold	[43]

(Continued)

TABLE 5.1 (Continued)

Microalgae	Enzyme	Genes	Molecular approach/ transformation technique	Targeted pathway	Results	References
Scenedesmus obliquus	Starchless mutant	–	Starchless mutant (slm1)	Fatty acid biosynthesis	Increase TAG content from 45% ± 1% to 57% ± 0.2% Decrease the biomass productivity	[25]
Chlorella ellipsoidea	Acetyl Co-A carboxylase	ACCase	Overexpression of transcription factor GmDof4	Fatty acid biosynthesis	Lipid content increased by 46.4% to 52.9% under mixotrophic culture condition. Significantly unregulated ACCase. Significant increase in C18:1/ C18:2/C18:3 and C16:0.	[44]
Phaeodactylum tricornutum	Diacylglycerol acyltransferase	DGAT2	Overexpression of functional gene/ Electroporation	Terminal step of fatty acid biosynthesis	Increase neutral lipid content by 35%. Alter fatty acid profile. Increase PUFA (EPA) by 76.2%.	[45]
Thassiosira pseudonana 1A6/1B1	Lipase	Thaps3_264297	Antisense knockdown/RNAi technology/ microparticle bombardment	Catabolism of fatty acids	Increase lipid content by 2.4–3.3-fold	[46]
Phaeodactylum trocornutum	Mallic enzyme	PtME	Overexpression of functional gene	Decarboxylation of malate to pyruvate	Lipid content increased by 2.5-fold (57.8% w/w)	[3]
Chlorella pyrenoidosa	NADP-malic enzyme	ptME	Overexpression of functional gene	Decarboxylation of malate to pyruvate	Increased neutral lipid content by upto 3.2-fold.	[47]
Phaeodactylum tricornutum	Glycerol-3-phosphate acyltransferase	ptGPAT	Overexpression of functional gene/ electroporation	TAG biosynthesis pathway	Increased neutral lipid content by twofold (42.6% w/w)	[48]

cis-trans isomerase [52–54]. Lycopene, being an important intermediate is generally targeted to increase carotenoids biosynthesis. The PSY and PDS were identified as the candidate genes for carotenoids biosynthesis [55]. Overexpression of the PSY gene was attempted in some microalgae, such as *Chlamydomonas reinhardtii, Haematococcus pluvialis*, and *Phaeodactylum tricornutum*, which led to an increase in carotenoids production [56,57]. Similarly, PDS gene expression has also been manipulated to enhance carotenoids production in *Chlamydomonas reinhardtii, Haematococcus pluvialis*, and *Chlorella zofingiensis* [28,58].

Various lycopene cyclases have been reported to influence carotenoid synthesis. However, reports regarding the regulation of cyclase and hydrolase by genetic engineering for microalgae are limited in numbers. It has also been reported that the deprivation of nitrogen may enhance the expression of cyclase [49]. Additionally, β-carotene and zeaxanthin could be oxidized to form more high-value biocompounds, such as canthaxanthin, violaxanthin, or astaxanthin in specific microalgae, that is, *Haematococcus pluvialis* and *Chlorella zofingiensis* [59]. In this pathway, β-carotene oxygenase (BKT) is one of the important enzymes that could be manipulated into other model microbes to produce astaxanthin. In a very recent report, BKT was successfully introduced into *Chlamydomonas reinhardtii* to produce astaxanthin [54]. This field of research generally opts for the enhancement of carotenoids production by changing culture conditions, by introducing some nutritional or abiotic stress over genetic engineering methods. The enhancement of carotenoid production by genetic engineering techniques is still limited and needs more attention. Moreover, the codon optimization for all genes expressed in different microalgae remains a significant challenge to be overcome for the optimized expression of the desired genes [60]. The high growth rate, high protein content, and FDA-approval for human use are some of the advantages of *Chlorella* species. In a previous study, lutein accumulation was shown to be upto 42.0 mg/L in *Chlorella sorokiniana* [61]. Similarly, *Chlorella pyrenoidosa* and *Chlorella vulgaris* were optimized for the accumulation of proteins upto 57% and 58%, respectively, while the average protein content in *Chlorella* sp. was estimated to be near 50% in most previous studies [62]. A successful genetic modification of *Chlorella zofingiensis* was reported by Liu et al. for enhanced carotenoid accumulation [63–65]. There are various reports where *Chlorella* sp. have been used as model microalgae for the expression of heterologous proteins. In a seminal work, the human growth hormone was first reported in genetically modified *Chlorella vulgaris* for pharmaceutical use [66]. However, the yield was quite low and genetically modified *Chlorella* sp. are still difficult to adapt for industry at the current time [67]. Some scientific reports on transgenic microalgae to produce versatile biocompounds are summarized in Table 5.2.

5.3 Advance methods used in genome editing

5.3.1 RNA interference

RNAi is considered as a regulatory phenomenon in the eukaryotic gene regulation via dsRNA functioning on the transcriptional and translational repression. It is considered to have enormous ability for the regulation of different cellular events and metabolic processes applied in strain improvements. RNAi technology is reported to be better than

TABLE 5.2 Summary of biochemical production by transgenic microalgae.

Algal strain	Target gene	Genetic method	Observations	References
Chlamydomonas reinhardtii	PSY gene from *Chlorella zofingiensis*	Overexpression	Contents of the carotenoids were 2.0- and 2.2-fold increased	[61]
Haematococcus pluvialis	Modified PDS gene from *Haematococcus pluvialis* (L504R)	Overexpression	43-fold higher resistance to the bleaching herbicide norflurazon and increase in astaxanthin production by about 35%	[68]
Chlamydomonas reinhardtii	Modified PDS gene from *Chlamydomonas reinhardtii* (L505F)	Overexpression	27.7-fold higher resistance to the herbicide norflurazon and increased carotenoids production by about 20%	[69]
Chlorella zofingiensis	Modified PDS gene from *Chlorella zofingiensis*	Overexpression	Produced 32.1% more total carotenoids and 54.1% more astaxanthin	[61]
Scenedesmus obliquus CPC2	DGTT1 gene from *Chlamydomonas reinhardtii*	Overexpression	The biomass concentration and biomass productivity were increased 20.0% and 232.6%, respectively	[70]
Chlamydomonas reinhardtii	PSY gene from *Dunaliella salina*	Overexpression	Increased content of carotenoids by 25%−160%	[52]
Synechococcus elongates PCC7942	Carbonic anhydrase	Overexpression	Carbon dioxide fixation increased 41%	[71]
Synechococcus sp. *PCC 7942*	Adh and pdc genes from *Zymomonas mobilis* with the *Escherichia coli* lac and CI-PL temperature inducible promoter		Increase in ethanol productivity	[72,73]

antisense technology in terms of stability, accuracy, and efficiency. RNAi was first discovered in *Caenorhabditis elegans*. It is basically a reverse genetic approach to elaborate the role of genes and their activity specific for an mRNA molecule [74,75]. Genome-wide analysis of the microalgae provides an overview for the function and regulatory mechanism of the many predicted genes that have been still unexplored. The major functional elements of RNAi machinery show that these are frequently distributed among most of the algal lineages but form the evolutionary evidence that it was obvious that RNAi-mediated silencing of the genes have been completely or partially absent in some algal species with small nuclear genomes [76]. In most algal species high-throughput sequencing has identified composite RNA molecules including small interfering RNA and microRNA but their functional role in regulatory mechanism remains poorly explored [77].

RNAi technology may be considered as a reliable tool for targeted silencing of genes in microalgae. The first attempt at RNAi-mediated silencing was performed on *Chlamydomonas reinhardtii* using a cascade of tandem IR transgenes that uniformly regulate the transcriptional silencing of the gene with a selectable phenotype mediated by RNAi silencing [78]. RNAi-mediated transcriptional silencing have been also reported in some diatoms and the findings support the fact that in diatoms functional gene silencing cascades are present and modify the pathways through transcriptional silencing and translational modifications. RNAi silencing was performed on *Phaeodactylum tricornutum* diatom with targeted genes and successfully knocked down the GUS reporter gene as well as DPH1 (endogenous phytochrome) and CPF1 (cryptochrome) genes [79]. RNAi-based silencing was also performed on *D. salina* for GAPDH gene using the recombinant plasmid p7NBTFIR with a homologous hairpin loop sequence (hpRNA) of GAPDH. The output was investigated using real-time PCR and about 41.2%−67.4% of transcription suppression was reported compared to wild-type *Dunaliella salina*, and microscopic observations showed the loss of motility in transformed cells [80]. Individual *Chlamydomonas* sp. miRNAs may have the potential to control the expression of several target genes via relaxed base coupling.

In *Chlamydomonas* sp. cre-miR1174.2 miRNA is functional and capable of initiating site-specific cleavage of the target sequence. It was also reported that hybridization of a nucleotide at the 5′ region of a miRNA with a target sequence has a better inhibitory transcriptional response, while the 3′ pairing does not have any silencing effect [81]. The modulatory role of endogenous miRNA has been established by the in silico approach on microalgae and has identified about 39 precursor miRNAs that generate 45 AGO3-associated miRNA sequences. The inhibitory potential of these miRNAs depends upon the extent of the candidate binding site on the transcripts [82]. High-throughput sequencing of *Coccomyxa subellipsoidea* C-169 (a good strain for biofuel production) genome reveals the presence of total 124 miRNAs displaying 384 genes as a potential target. These target genes are mainly grouped into six major classes on the KEGG pathway and are significantly enriched in C5-branched dibasic acid metabolism, leucine, and isoleucine biosynthesis, pantothenate and CoA biosynthesis, butanoate metabolism, and alpha-linolenic acid metabolism. Analysis of these putative target genes may help in the bioengineering of the *Coccomyxa subellipsoidea* C-169 for better biofuel production [83]. RNAi-based silencing was also performed on some industrial strains of microalgae, such as *Nannochloropsis* sp., which is a model organism for industrial microalgal oil production. The experimental design achieved about 40%−80% success rate and about 62%−83% carbonic anhydrase (CA) gene was silenced by RNAi, while at the same time the photosynthetic oxygen evolution rate of the transformed cells was improved to 68%−100% at pH 6 compared to the wild-type strain Fig. 5.1, 5.2 and 5.3.

5.3.2 Zinc-finger nucleases for targeted genome editing

Zinc figure nucleases (ZFNs) were developed for targeted genome-editing using sequence-specific nucleases that could program themselves for precise sequence recognition. ZFNs have the DNA binding domain at their N terminal and a nonspecific DNA binding domain at the C terminal of the endonucleases. They have the specificity on the template strand and cover about three base pairs for their binding. Two Fok1 endonucleases get dimerized and initiate the double-stranded breaks followed by the nonhomologous or homologous recombination to

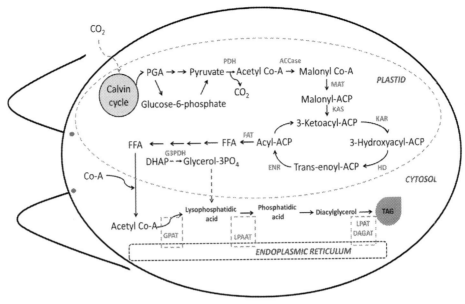

FIGURE 5.1 Metabolic pathway of microalgal triacylglycerol/lipid biosynthesis shown in black and enzymes shown in red.

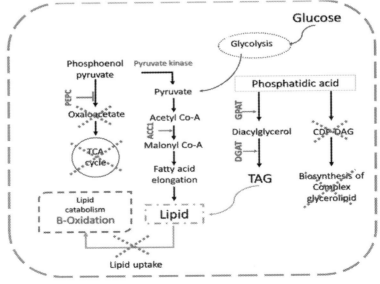

FIGURE 5.2 Showing the multichannel pathway of lipid accumulation in microalgal cell. Inhibiting the activity of phosphoenolpyruvate carboxylase leads to the conversion of phosphoenolpyruvate to pyruvate by the activity of pyruvate kinase; overexpression of enzyme glycerol phosphate acyltransferase and diacylglycerol acyltransferase leads to the transformation of substrate phosphatidic acid into triacylglycerol; and blocking the uptake of lipid promotes the higher accumulation of lipid inside the cell.

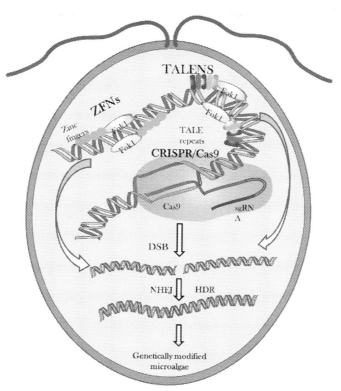

FIGURE 5.3 Pictorial presentation of genome editing using Zinc figure nucleases, Transcription activator-like effector nucleases, and CRISPR/Cas9.

repair the breaks. To date a number of higher and lower organisms have been exposed to ZFNs-based site-directed genome modification [84]. There are only few reports available in the literature on the ZFNs-mediated genomic editing in microalgae. It was attempted to modify the photoreceptor gene COP3 and COP4 genes in model microalgae (*C. reinhardtii*) using engineered ZFNs and the resulting clones showed 5%–15% success rate in gene disruption [85,86]. To build unique ZFNs with strong specificity and affinity against target sites is the most difficult challenge for the advance of ZFN technology. This necessitates the requirement for the proper validation of the ZFNs using various tools and reports before plaining any crucial study [75].

5.3.3 CRISPR/Cas9

Cellular metabolic pathway engineering is a highly effective and emerging approach for increasing the yield of desired products by editing the pathway at genetic level, and offers a stable and promising tool for strain improvement [87]. Nowadays, some powerful genome-editing tools such as RNAi are frequently used to manipulate the role of a particular gene in a targeted metabolic pathway by applying the basic and simple gene-silencing method.

However, partial suppression and RNAi transgene silencing are the major limitations of the RNAi-based gene-silencing method of genome editing [88]. In comparison, the developing field of genome engineering offers a new and stable avenue to alter the function of a targeted gene with respect to improving strain quality as well as the productivity of the final product by bypassing the limitations of RNAi [89]. The emerging field of genome editing includes CRISPR/cas9 technology, which provides a simple, accurate, and efficient process for genome modification. This approach uses a small guide RNA molecule that directs the Cas9 protein subunit of CRISPR to target the specific loci in the genome of organism. A modified version of Cas9 protein lacking nuclease activity (dCas9) can tightly regulate the expression of target genes of a metabolic pathway on the association with CRISPR module. The activity of a specific target gene of a metabolic pathway can be stimulated or repressed, depending on the nature of effector molecules associated with the dCas9 protein, therefore the expression profile of a single or two or more genes of a specific metabolic pathway can be activated or deactivated by the association of guide RANs for every targeted gene in the dCas9 protein mutants of the CRISPR/Cas9 module [90]. The flexibility in the wide application of the CRISPR/Cas9 module toward genome modification makes it a reliable, stable, powerful, and promising genome-editing tool for metabolic pathway engineering. In microalgal systems, there is a rising demand for the CRISPR/Cas9 approach because it has significant scope in microalgal strain modification to improve and amplify the yield of biofuel production, microalgal lipid content, and other value-added products like nutraceuticals. In contrast with other genome-editing approaches, one of the most significant merits of applying the SRISPR/Cas9 approach in microalgal systems is the ease of multiplexing, which is totally different from the conventional methods of mutagenesis and RNAi-mediated gene-silencing techniques, and provides a simple and less complicated approach for modifying a targeted metabolic pathway. In the field of lipid biosynthesis engineering, it has been suggested that the CRISPR/Cas9 approach of genome editing can increase the lipid content in the oleaginous microalgae by blocking the several competitive metabolic pathways that directly or indirectly inhibit the biosynthesis pathway of lipids, including the starch biosynthesis, lipid degradation, and beta oxidation pathways. Apart from gene knockout or gene silencing, a dCsa9 mutant of CRISPR/Cas9 system plays an important role in activating stress-associated elements of the lipid biosynthesis pathway under nonstress condition. Indigenous species of microalgae have some limitations for the industrial production of biofuels for commercial application including suboptimal lipid content and light harvesting efficiency. Therefore solving these restrictions can improve the large-scale production of biofuels without altering the physiology of microalgae; hence introducing a truncated LHC in microalgal strain is associated with susceptibility against photodamage [91]. Therefore it is very important to have a specific system inside the cell that can detect and stimulate an effective response in the presence of any chemical compounds and the fluctuation in light intensity. In this way the development of a dCas9 mutant of the CRISPR module can be stimulated in the presence of chemicals and light, providing a tool for the conditional modulation for optimal growth of algal biomass without changing the normal cellular metabolic processes of the cell [92]. With wide applications, this system has some cytotoxic effects in several microalgal species, which restrict the large-scale utilization of this system. However, it has been reported that in various cyanobacterial species the cytotoxic effect of Cas9 nuclease of CRISPR/Cas9 module can be reduced by replacing the Cas9 protein with the Cas12a variant [93]. The recent development of variants of CRISPR module has extended

5.3 Advance methods used in genome editing

the application of genetic-editing tools in the field of microalgal genome modification to provide an optimized and standard platform for better biofuel production as well as other value-added product formation. Jiang and coworkers successful described the use of Cas9 protein of CRISPR system in microalgae *C. reinhardtii* to create different alterations in four distinct genes within 24 h of transformation [94]. They have suggested that the transformation efficiency was very low, getting only a single viable colony while performing sixteen distinct transformation, due to the virulent nature of Cas9.

Researchers have suggested that the CRISPR/Cas9 approach of genome engineering can create a stable mutant with the desired expression of genes in *P. tricornutum* [95]. Apart from *Streptococcus pyogenes*-derived Cas9 nuclease, other nucleases, including Cpf1 (Cas12a), (derived from *Acidominococcus* (AsCpf1), *Lachnospiraceae* (LbCpf1), or *Francisella novicida* (FnCpf1)), also offer a great and efficient tool for microalgal genome engineering. These nucleases use only a single guide RAN molecule for multiple targeted loci [96–98]. It has been reported that the use of Cpf1 nuclease with the CRISPR module (CRISPR/Cpf1) can increase the chance of homologous end-joining in green microalgae *Chlamydomonas* with 10% transformation efficiency [99], whereas CRISPR with Cas9 nuclease activity (CRISPR/Cas9) has 0.02% nonhomologous end-joining efficiency [96,97]. CRISPRi improved the production of lipid (94%) through a successful knockdown of PEPC in *Chlamydomonas renhardtii*.

5.3.4 Transcription activator-like effector nucleases

Transcription activator-like effector nucleases (TALENs) are a specific type of restriction enzyme composed of two different domains with different activities. Two different domains are the DNA recognition domain and the DNA cleavage domain, in which a TAL effector protein molecule is used by the DNA recognition domain of TALEN to bind a specific stretch of DNA sequence by modifying two important amino acids in the polypeptide chain of TAL protein [100]. The nature of the nonspecific DNA cleavage domain of TALEN is like the restriction enzyme Fok1 but it can modify its activity and specificity at the time of genome modification. In this strategy, after the activity of the DNA recognition domain of TALEN, the targeted sequence of the DNA molecule is cleaved by the activity of DNA cleavage domain of TALENs, and then the genome modification process can start by the recruitment of HR or nonhomologous end-joining (NHEJ). In this approach, the nonspecific DNA cleavage domain is required for target specific cleavage of recognized loci within the genome of the organism, guided by the highly conserved DNA binding domain derived from transcription activator-like effectors (TALEs) [101]. The single nucleotide recognition nature of each conserved DNA binding domain of TALE makes this approach quite simple compared with other genome modification approaches. Several examples are available for the successful use of TALEN approach in the field of microalgal genome editing. The aforementioned approach was first described by Daboussi, et al. in microalgae *Phaeodactylum tricornutum*, wherein they altered the expression of a specific gene (UDP-glucose pyrophosphorylase) of carbohydrate metabolic pathway, leading to improved cytosolic lipid content by upto 45-fold [11]. Another application of TALENs approach was also described in microalgae *Phaeodactylum tricornutum*, in which the authors successfully knocked out the function of the urease gene, which was further

Handbook of Algal Biofuels

confirmed and validated by several experiments such as lack of cell growth in the presence of urea in growth media and by several molecular techniques including PCR, Southern blot, and Western blot analysis. The TALEN strategy of gene alteration has been also applied to silence the function of uridine monophosphate synthase gene and developed a knockout mutant for uridine monophosphate synthase in *Phaeodactylum tricornutum* for investigating the role of uridine monophosphate synthase gene as a strong cytosolic positive selectable marker [147]. A mutated form of this approach lacking nuclease activity (TALEs) serving as a transcriptional activator recently has been applied to overexpress the activity of two arylsulfatases and HLA3 protein in microalgae *Chlamydomonas reinhardtii* [102]. The TALEN-mediated genome modification approach becomes an emerging tool of the molecular field which is continuously exploited by several researchers for investigating the function of a specific gene of targeted loci through NHEJ by destroying the activity of red/far-red light-sensing phytochromes [103].

5.4 Future perspectives of genetic engineering in microalgae

A variety of bioproducts are being produced by using algae as miniature bioreactors such as cosmetics, feed, pigments, pharmaceuticals, algae-based biomaterials, nutraceuticals, and other chemicals. Annually, a million tons of algae are harvested by the algal biotechnology industries. A prevailing driving force in such industries is the enticing opportunity for the utilization of genetically enhanced organisms or engineered novel strains of microalgae. In recent years, data availability at the molecular level of microalgae has achieved immense scientific attention. Genetic engineering comprises cutting-edge tools which exploit the genetic information at cellular level along with molecular biology procedures to boost productivity. Jagadevan et al. [104] reported that genetic engineering methods reduce the cost of microalgal biofuels to 15%−20% as well as also increase investment in the manufacturing of other highly valued compounds and bioproducts. Topical developments in the field of the genetic manipulation of microalgae have turned out to be promising in terms of opening new areas of probable application for these organisms as efficient integrated biorefineries. As compared to natural organisms, transgenic algae promise a considerably wider field of application. By means of additionally acquired physiological capabilities and new biochemical reactions the transgenics open the door for molecular farming and improvements in algal bioproducts. The results of a recent report showed that with a combination of unmodified microalgal strains with transgenic ones, the cost dropped by 85%. The development of Aquaculture Raceway Integrated Design (ARID), that is, an open pond cultivation system that reduces cost by 16%, is also emphasized. The system is economical as is utilizes less energy and provides an improved yield. The transgenic microalgal strains have several advantages, such as quicker growth, cost-efficiency, higher cell densities, amplified valuable compounds, and an enhanced robustness under economic growth conditions. Gujjala et al. [105] stated that the opportunities of genetic engineering tools on microalgae are limitless and these approaches can be used for creating and modifying metabolic pathways. Metabolic and genetic engineering together can also be used for the enhanced production of numerous secondary metabolites, and exclusively carotenoids in microalgae [106]. Glycoengineering is one of the approaches that has enormous potential for the creation of nonimmunogenic biopharmaceuticals in microalgae [107]. Various omics approaches are now being used in biotechnology, bioengineering, and other allied fields. Omics tools provide an estimation of predicted results and offer a close insight into the genome under study [108]. These approaches are aimed at enhancing microalgal volumes to yield valued metabolites along with creating specific and robust transgenic strains for increased photosynthetic efficiency, CO_2 fixation, lipid content, biomedical applications, and biomass productions [2]. Profound information of the microalgal genome is obligatory for its application, hence algomics is applied. Algomics is an application of genomic and post-genomic tools to make a microalga renewable, sustainable, and economically viable for bioenergy and other products [108]. Using algomics various interpretations can be made in different aspects of genome sequences, genome editing, and its effect (using CRISPR-Cas9 system). Furthermore, the study of existing metabolic pathways, and their modification can also be construed. In order to make economically modest microalgal products there is a persistent need for genetic engineering tools and integrated omics approaches for the development of high-performance transgenic microalgal strains. However, there are some limitations in these approaches. Gene-silencing mechanisms of foreign genes can lessen the frequency of expression in transformed strains. To counter this the use of special constitutive promotors (*HSP70A* promoter) can be undertaken in the heterologous DNA. This can counter the transcriptional silencing of downstream promoters in a transgene set as these promoters can induce a constitutively open chromatin state. Also, phenotypic screening of the transformants can be performed to overcome the problem of gene silencing. Charoonnart et al. [109] reported an improved tactic to use promoters which are firmly controlled and have a large dynamic range. These can be further regulated by effective codon optimization and used to express a desired gene. Another limitation is the intricate algal cell wall and organelle structure which also cause hindrances in sequencing and understanding of the algal genome [110]. Another challenge is the lack of efficient chloroplast and mitochondria transformation systems. Any algal transformation scheme starts with the choice of apposite target organism. Microalgae that have a short life cycle and high cell densities that can be cultivated in inexpensive culture media under definite ecological conditions are favored. The chloroplast transformation process has only been studied in a few useful microalgal species, such as *Chlamydomonas reinhardtii*, *Dunaliella tertiolecta*, and *Phaeodactylum tricornutum* [111]. Currently, molecular data on microalgae are limited and genetic engineering approaches are offered only for limited algal species. Some of the available websites and algal databases are presented in Table 5.3. However, further positive and efficient studies are needed for additional whole genome data of microalgae. Relatively few microalgal genome sequencings projects have been completed (Table 5.4) and others are ongoing or in planning, enabling genetic engineering. These projects are aiming to produce the genomic, transcriptomic, and proteomic information in different microalgal species. Cheng et al. [143] reported that a phylodiverse genome sequencing project (10KP) is estimated to be completed by 2023. The project is designed to sequence the genome of around 1000 green algae and upto 3000 protists. However, it is essential to interpret the full annotation of genes and their networking of the metabolic cascades, in order to harness the full potential of microalgae [144]. Regardless of the disparity in the inter- and intraspecies omics datasets, an inclusive usage of system biology approach could be done, and the conserved factors can be extracted to predict the outcomes [145].

Another challenge for the use of genetically modified microalga is that GMO genetically modified organism (GMO) products are unfavorable for human use. Also, there is a risk

TABLE 5.3 Completely sequenced algal genome.

Algal strain	Pedigree	Accession number	Genome size (Mb)	Organization	NCBI Link	References
Chlamydomonas reinhardtii CC-503	Chlorophyta; Chlorophyceae; Volvocales; Chlamydomonadaceae	NW_001843471.1	111.1	Joint genome institute	https://www.ncbi.nlm.nih.gov/nuccore/NW_001843471.1	[112]
Bathycoccus prasinos (RCC 1105)	Chlorophyta; Prasinophyceae; Mamiellales; Mamiellaceae	PRJNA231566	15.07	–	–	[113]
Dunaliella salina CCAP 19/18	Chlorophyta; Chlorophyceae; Volvocales; Dunaliellaceae	NSFN00000000	343.704	Joint genome institute	https://www.ncbi.nlm.nih.gov/nuccore/NSFN00000000.1	[114]
Micromonas commoda (RCC299)	Chlorophyta; Prasinophyceae; Mamiellales; Mamiellaceae		21.11	–	–	[115]
Phaeodactylum tricornutum CCAP1055/1	Heterokontophyta; Bacillariophyceae; Naviculales; Phaeodactylaceae	NC_011669	27.5	Diatom consortium	https://www.ncbi.nlm.nih.gov/nuccore/NC_011669	[116]
Micromonas pusilla CCMP1545	Chlorophyta; Prasinophyceae; Mamiellales; Mamiellaceae	PRJNA422663	21.96	–	–	[117]
Nannochloropsis gaditana CCMP526	Heterokontophyta; Eustigmatophyceae; Eustigmatales; Monodopsidaceae	NW_005803952	33.98	Colorado school of mines	https://www.ncbi.nlm.nih.gov/nuccore/NW_005803952.1	[35]
Ostreococcus tauri (RCC4221)	Chlorophyta; Prasinophyceae; Mamiellales; Mamiellaceae		13.03	–	–	[118]
Nannochloropsis salina CCMP1776	Heterokontophyta; Eustigmatophyceae; Eustigmatales; Monodopsidaceae	AFGQ01002567	24.4	Chinese academy of sciences	https://www.ncbi.nlm.nih.gov/nuccore/AFGQ01002567	[119]
Ostreococcus lucimarinus (CCE9901)	Chlorophyta; Prasinophyceae; Mamiellales; Mamiellaceae	13.2	–	–	–	[120]
Chlorella sorokiniana ASM313072v1	Chlorophyta; Trebouxiophyceae; Chlorellales; Chlorellaceae	PKFC00000000	58.53	Los alamos national lab	https://www.ncbi.nlm.nih.gov/nuccore/PKFC00000000.1	[121]
Chlorella vulgaris strain NJ-7 NJ-7_scaffold00001	Chlorophyta; Trebouxiophyceae; Chlorellales; Chlorellaceae	VATV01000001	39.08	Chinese academy of sciences	https://www.ncbi.nlm.nih.gov/nuccore/VATV01000001.1	[75]
Coccomyxa subellipsoidea C-169	Chlorophyta; Chlorophyceae; Chlorococcales; Coccomyxaceae	GSE76638; PRJNA428141	48.83	–	–	[122]
Tetradesmus obliquus	Chlorophyta; Chlorophyceae; Sphaeropleales; Scenedesmceae	FNXT00000000	157.946	Laboratory of systems and synthetic biology	https://www.ncbi.nlm.nih.gov/nuccore/FNXT00000000.1	[123]

5.4 Future perspectives of genetic engineering in microalgae

TABLE 5.4 Some databases and websites used in algal biotechnology.

Databases/websites	Link	References
pico-PLAZA	http://bioinformatics.psb.ugent.be/pico-plaza/	[124]
Alga-PrAS (algal protein annotation suite)	http://alga-pras.riken.jp/	[125]
Phytozome	phytozome.jgi.doe.gov	[126]
KEGG (kyoto encyclopedia of genes and genomes)	https://www.genome.jp/kegg/	[127]
The greenhouse	greenhouse.lanl.gov	[128]
BioCyc database collection	https://biocyc.org/	[129]
Pathway tools	http://pathwaytools.org/	[130]
MetaCyc	https://metacyc.org/	[130]
AlgaeBase	https://www.algaebase.org/	[131]
COBRA (constraint-based reconstruction and analysis)	https://opencobra.github.io/	[132]
BLAST (basic local alignment search tool)	https://blast.ncbi.nlm.nih.gov/Blast.cgi	[133]
EXPASY	https://www.expasy.org/	[134]
JGI (Joint Genome Institute)	https://mycocosm.jgi.doe.gov/phycocosm/home	[135]
Alganaut	https://alganaut.uts.edu.au/	[136]
DiatomCyc	http://www.diatomcyc.org/	[137]
Algae resource database	https://shigen.nig.ac.jp/	[138]
DIDI (diatoms image database of India)	http://indianalgae.co.in	[139]
Algaepath	http://algaepath.itps.ncku.edu.tw/	[140]
dEMBF (database of enzymes for microalgal biofuel feedstock)	http://bbprof.immt.res.in/embf/index.php?page = home	[141]
dEMBF v2.0 (database of enzymes for microalgal biofuel feedstock)	http://bbprof.immt.res.in/embf/	[142]

of the interaction of genetically modified species with wild types which could result in the development of unknown species. Kumar et al. [145] reported that precise genome-editing technologies, for instance, nontransgenic and marker-free CRISPR, can reform the microalgal bioengineering methods for the creation of non-GMO algal products. This is likely to increase the biosafety and improve the regulatory matters linked with the usage of genetically modified microalgae. The probable hazards have prompted scientists to arm these genetically engineered microalgae with physiognomies that will limit their endurance in case of the release of these microalgae in the environment [146]. There is a need for a road map for a regulatory regime that would cover the commercial/industrial usage of genetically modified microalgae. A proper environmental assessment is hence required to prevent the genetically engineered microalgae from evolving. Drastic change is needed in the existing regulations for the use of genetically modified microalgae. Simultaneously, revolution is also needed in academic and industrial community regarding the regulations for the usage of genetically modified strains. Besides an improvement in the economical competitivity of an algal-derived product, the creation of a competent extraction method or the use of entire cells is desired. Selectable marker genes, reporter genes, promoters, transformation methods, and other genetic techniques are readily accessible. Altogether these hold the potential for the development of microalgae as a next-generation renewable reserve. Undeniably, alongside all the progress, bioprospection for novel and robust microalgal strains with industrial practicality must move forward and continue [145].

5.5 Concluding remarks

In the fields of algal bioengineering, genetic molecular techniques have made tremendous development. The ability to manipulate, replicate, and alter DNA, RNA, proteins, and organisms through transformation techniques, cloning, and gene-editing tools has enabled the creation of powerful biological insights and applications. As is already evident, the future of molecular techniques lies in the development of more powerful editing tools, the simplification of high-performance techniques, and the implementation of more automated techniques. The latter point is particularly important when it comes to considering applications, such as algal biofuels. Various algal species are expected to be created to meet the requirements of environmental standards. Approaches for genetic modification may play a major role in the elucidation and characterization of physiological variations between species. The continued growth of tools for gene manipulation is important for algal biology.

Acknowledgments

Authors (Ashutos Pandey, Manish Pratap Singh, and Shadma Afzal) and Gaurav Kant would like to give their sincere thanks to the Ministry of Human Resource and Development (MHRD) New Delhi, India and Department of Biotechnology (DBT) New Delhi, India, respectively for Ph.D. fellowship.

References

[1] K. Hwangbo, J.-W. Ahn, J.-M. Lim, Y.-I. Park, J.R. Liu, W.-J. Jeong, Overexpression of stearoyl-ACP desaturase enhances accumulations of oleic acid in the green alga Chlamydomonas reinhardtii, Plant. Biotechnol. Rep. 8 (2014) 135−142. Available from: https://doi.org/10.1007/s11816-013-0302-3.

[2] A. Pandey, A. Gupta, A. Sunny, S. Kumar, S. Srivastava, Multi-objective optimization of media components for improved algae biomass, fatty acid and starch biosynthesis from Scenedesmus sp. ASK22 using desirability function approach, Renew. Energy. 150 (2020) 476−486. Available from: https://doi.org/10.1016/j.renene.2019.12.095.

[3] J. Xue, Y.-F. Niu, T. Huang, W.-D. Yang, J.-S. Liu, H.-Y. Li, Genetic improvement of the microalga Phaeodactylum tricornutum for boosting neutral lipid accumulation, Metab. Eng. 27 (2015) 1−9. Available from: https://doi.org/10.1016/j.ymben.2014.10.002.

[4] C. Bogen, A. Al-Dilaimi, A. Albersmeier, J. Wichmann, M. Grundmann, O. Rupp, et al., Reconstruction of the lipid metabolism for the microalga Monoraphidium neglectum from its genome sequence reveals characteristics suitable for biofuel production, BMC Genomics 14 (2013) 926. Available from: https://doi.org/10.1186/1471-2164-14-926.

References

[5] F. Liu, S.J. Pang, N. Xu, T.F. Shan, S. Sun, X. Hu, et al., Ulva diversity in the Yellow Sea during the large-scale green algal blooms in 2008–2009: Ulva green tide in the Yellow Sea, Phycol. Res. 58 (2010) 270–279. Available from: https://doi.org/10.1111/j.1440-1835.2010.00586.x.

[6] J. Neupert, D. Karcher, R. Bock, Generation of Chlamydomonas strains that efficiently express nuclear transgenes, Plant. J. 57 (2009) 1140–1150. Available from: https://doi.org/10.1111/j.1365-313X.2008.03746.x.

[7] N. Shao, R. Bock, A codon-optimized luciferase from Gaussia princeps facilitates the in vivo monitoring of gene expression in the model alga Chlamydomonas reinhardtii, Curr. Genet. 53 (2008) 381–388. Available from: https://doi.org/10.1007/s00294-008-0189-7.

[8] X. Deng, J. Cai, X. Fei, Effect of the expression and knockdown of citrate synthase gene on carbon flux during triacylglycerol biosynthesis by green algae Chlamydomonas reinhardtii, BMC Biochem. 14 (2013) 38. Available from: https://doi.org/10.1186/1471-2091-14-38.

[9] P.P. Jutur, A.A. Nesamma, Genetic engineering of marine microalgae to optimize bioenergy production, Handbook of Marine Microalgae, Academic Press, 2015, pp. 371–381. Available from: https://doi.org/10.1016/B978-0-12-800776-1.00024-8.

[10] Q. Wang, Y. Lu, Y. Xin, L. Wei, S. Huang, J. Xu, Genome editing of model oleaginous microalgae Nannochloropsis spp. by CRISPR/Cas9, Plant. J. 88 (2016) 1071–1081. Available from: https://doi.org/10.1111/tpj.13307.

[11] F. Daboussi, S. Leduc, A. Maréchal, G. Dubois, V. Guyot, C. Perez-Michaut, et al., Genome engineering empowers the diatom Phaeodactylum tricornutum for biotechnology, Nat. Commun. 5 (2014) 3831. Available from: https://doi.org/10.1038/ncomms4831.

[12] O. Kilian, C.S.E. Benemann, K.K. Niyogi, B. Vick, High-efficiency homologous recombination in the oil-producing alga Nannochloropsis sp, Proc. Natl. Acad. Sci. 108 (2011) 21265–21269. Available from: https://doi.org/10.1073/pnas.1105861108.

[13] G. Perin, A. Bellan, A. Segalla, A. Meneghesso, A. Alboresi, T. Morosinotto, Generation of random mutants to improve light-use efficiency of Nannochloropsis gaditana cultures for biofuel production, Biotechnol. Biofuels. 8 (2015) 161. Available from: https://doi.org/10.1186/s13068-015-0337-5.

[14] J.A.E. Nelson, P.A. Lefebvre, Chapter 73 Transformation of Chlamydomonas reinhardtii, Methods Cell Biol, Elsevier, 1995, pp. 513–517. Available from: https://doi.org/10.1016/S0091-679X(08)60854-7.

[15] M. Hlavova, Z. Turoczy, K. Bisova, Improving microalgae for biotechnology—from genetics to synthetic biology, Biotechnol. Adv. 33 (2015) 1194–1203. Available from: https://doi.org/10.1016/j.biotechadv.2015.01.009.

[16] S.O. Duke, F.E. Dayan, Bioactivity of herbicides, Compr. Biotechnol, Elsevier, 2011, pp. 23–35. Available from: https://doi.org/10.1016/B978-0-08-088504-9.00273-7.

[17] L.M. Salati, A.G. Goodridge, Fatty acid synthesis in eukaryotes, New Compr. Biochem, Elsevier, 1996, pp. 101–127. Available from: https://doi.org/10.1016/S0167-7306(08)60511-6.

[18] J.L. Harwood, Plant lipid metabolism, Plant Biochem, Elsevier, 1997, pp. 237–272. Available from: https://doi.org/10.1016/B978-012214674-9/50007-2.

[19] P.K. Sharma, M. Saharia, R. Sriversustava, S. Kumar, L. Sahoo, Tailoring microalgae for efficient biofuel production, Front. Mar. Sci. 5 (2018) 382. Available from: https://doi.org/10.3389/fmars.2018.00382.

[20] J.L. Blatti, J. Michaud, M.D. Burkart, Engineering fatty acid biosynthesis in microalgae for sustainable biodiesel, Curr. Opin. Chem. Biol. 17 (2013) 496–505. Available from: https://doi.org/10.1016/j.cbpa.2013.04.007.

[21] J.J. Thelen, J.B. Ohlrogge, Metabolic engineering of fatty acid biosynthesis in plants, Metab. Eng. 4 (2002) 12–21. Available from: https://doi.org/10.1006/mben.2001.0204.

[22] I.I.G.S. Verwoert, K.H. van der Linden, M.C. Walsh, H.J.J. Nijkamp, A.R. Stuitje, Modification of Brassica napus seed oil by expression of the Escherichia coli fabH gene, encoding 3-ketoacyl-acyl carrier protein synthase III, Plant. Mol. Biol. 27 (1995) 875–886. Available from: https://doi.org/10.1007/BF00037016.

[23] H.S. Sul, C.M. Smas, D. Wang, L. Chen, Regulation of fat synthesis and adipose differentiation, Prog. Nucleic Acid Res. Mol. Biol, Elsevier, 1998, pp. 317–345. Available from: https://doi.org/10.1016/S0079-6603(08)60896-X.

[24] X.-M. Sun, L.-J. Ren, Q.-Y. Zhao, X.-J. Ji, H. Huang, Enhancement of lipid accumulation in microalgae by metabolic engineering, Biochim. Biophys. Acta BBA - Mol. Cell Biol. Lipids 1864 (2019) 552–566. Available from: https://doi.org/10.1016/j.bbalip.2018.10.004.

[25] G. Breuer, L. de Jaeger, V.P.G. Artus, D.E. Martens, J. Springer, R.B. Draaisma, et al., Superior triacylglycerol (TAG) accumulation in starchless mutants of Scenedesmus obliquus: (II) evaluation of TAG yield and productivity in controlled photobioreactors, Biotechnol. Biofuels. 7 (2014) 70. Available from: https://doi.org/10.1186/1754-6834-7-70.

116 5. Genetic manipulation of microalgae for enhanced biotechnological applications

[26] W.-J. Tan, Y.-C. Yang, Y. Zhou, L.-P. Huang, L. Xu, Q.-F. Chen, et al., Diacylglycerol acyltransferase and Diacylglycerol kinase modulate triacylglycerol and phosphatidic acid production in the plant response to freezing stress, Plant. Physiol. 177 (2018) 1303–1318. Available from: https://doi.org/10.1104/pp.18.00402.

[27] M. Zhang, J. Fan, D.C. Taylor, J.B. Ohlrogge, *DGAT1* and *PDAT1* acyltransferases have overlapping functions in *Arabidopsis* triacylglycerol biosynthesis and are essential for normal pollen and seed development, Plant. Cell 21 (2009) 3885–3901. Available from: https://doi.org/10.1105/tpc.109.071795.

[28] Q. Liu, R.M.P. Siloto, R. Lehner, S.J. Stone, R.J. Weselake, Acyl-CoA:diacylglycerol acyltransferase: molecular biology, biochemistry and biotechnology, Prog. Lipid Res. 51 (2012) 350–377. Available from: https://doi.org/10.1016/j.plipres.2012.06.001.

[29] W. Banaś, A. Sanchez Garcia, A. Banaś, S. Stymne, Activities of acyl-CoA:diacylglycerol acyltransferase (DGAT) and phospholipid:diacylglycerol acyltransferase (PDAT) in microsomal preparations of developing sunflower and safflower seeds, Planta 237 (2013) 1627–1636. Available from: https://doi.org/10.1007/s00425-013-1870-8.

[30] E.B. D'Alessandro, N.R. Antoniosi Filho, Concepts and studies on lipid and pigments of microalgae: a review, Renew. Sustain. Energy Rev. 58 (2016) 832–841. Available from: https://doi.org/10.1016/j.rser.2015.12.162.

[31] G. Musso, M. Cassader, E. Paschetta, R. Gambino, Bioactive lipid species and metabolic pathways in progression and resolution of nonalcoholic steatohepatitis, Gastroenterology 155 (2018) 282–302 e8. Available from: https://doi.org/10.1053/j.gastro.2018.06.031.

[32] G.M. Carman, G.-S. Han, Regulation of phospholipid synthesis in yeast, J. Lipid Res. 50 (2009) S69–S73. Available from: https://doi.org/10.1194/jlr.R800043-JLR200.

[33] R. Coleman, Enzymes of triacylglycerol synthesis and their regulation, Prog. Lipid Res. 43 (2004) 134–176. Available from: https://doi.org/10.1016/S0163-7827(03)00051-1.

[34] T.G. Dunahay, E.E. Jarvis, S.S. Dais, P.G. Roessler, Manipulation of microalgal lipid production using genetic engineering, Appl. Biochem. Biotechnol. 57–58 (1996) 223–231. Available from: https://doi.org/10.1007/BF02941703.

[35] R. Radakovits, R.E. Jinkerson, S.I. Fuerstenberg, H. Tae, R.E. Settlage, J.L. Boore, et al., Draft genome sequence and genetic transformation of the oleaginous alga Nannochloropsis gaditana, Nat. Commun. 3 (2012) 686. Available from: https://doi.org/10.1038/ncomms1688.

[36] P. Deschamps, I. Haferkamp, C. d'Hulst, H.E. Neuhaus, S.G. Ball, The relocation of starch metabolism to chloroplasts: when, why and how, Trends Plant. Sci. 13 (2008) 574–582. Available from: https://doi.org/10.1016/j.tplants.2008.08.009.

[37] D.M. Stark, K.P. Timmerman, G.F. Barry, J. Preiss, G.M. Kishore, Regulation of the amount of starch in plant tissues by ADP glucose pyrophosphorylase, Science 258 (1992) 287–292. Available from: https://doi.org/10.1126/science.258.5080.287.

[38] G. Ritte, M. Heydenreich, S. Mahlow, S. Haebel, O. Kötting, M. Steup, Phosphorylation of C6- and C3-positions of glucosyl residues in starch is catalysed by distinct dikinases, FEBS Lett. 580 (2006) 4872–4876. Available from: https://doi.org/10.1016/j.febslet.2006.07.085.

[39] S.C. Zeeman, T. Delatte, G. Messerli, M. Umhang, M. Stettler, T. Mettler, et al., Starch breakdown: recent discoveries suggest distinct pathways and novel mechanisms, Funct. Plant. Biol. 34 (2007) 465. Available from: https://doi.org/10.1071/FP06313.

[40] T.-S. Yu, H. Kofler, R.E. Häusler, D. Hille, U.-I. Flügge, S.C. Zeeman, et al., The arabidopsis sex1 mutant is defective in the R1 protein, a general regulator of starch degradation in plants, and not in the chloroplast hexose transporter, (n.d.) 13.

[41] X.-D. Deng, B. Gu, Y.-J. Li, X.-W. Hu, J.-C. Guo, X.-W. Fei, The roles of acyl-CoA: diacylglycerol acyltransferase 2 genes in the biosynthesis of triacylglycerols by the green algae Chlamydomonas reinhardtii, Mol. Plant. 5 (2012) 945–947. Available from: https://doi.org/10.1093/mp/sss040.

[42] X. Deng, J. Cai, Y. Li, X. Fei, Expression and knockdown of the PEPC1 gene affect carbon flux in the biosynthesis of triacylglycerols by the green alga Chlamydomonas reinhardtii, Biotechnol. Lett. 36 (2014) 2199–2208. Available from: https://doi.org/10.1007/s10529-014-1593-3.

[43] A. Ibáñez-Salazar, S. Rosales-Mendoza, A. Rocha-Uribe, J.I. Ramírez-Alonso, I. Lara-Hernández, A. Hernández-Torres, et al., Over-expression of dof-type transcription factor increases lipid production in Chlamydomonas reinhardtii, J. Biotechnol. 184 (2014) 27–38. Available from: https://doi.org/10.1016/j.jbiotec.2014.05.003.

Handbook of Algal Biofuels

References

[44] J. Zhang, Q. Hao, L. Bai, J. Xu, W. Yin, L. Song, et al., Overexpression of the soybean transcription factor GmDof4 significantly enhances the lipid content of Chlorella ellipsoidea, Biotechnol. Biofuels. 7 (2014) 128. Available from: https://doi.org/10.1186/s13068-014-0128-4.

[45] Y.-F. Niu, M.-H. Zhang, D.-W. Li, W.-D. Yang, J.-S. Liu, W.-B. Bai, et al., Improvement of neutral lipid and polyunsaturated fatty acid biosynthesis by overexpressing a type 2 diacylglycerol acyltransferase in marine diatom Phaeodactylum tricornutum, Mar. Drugs. 11 (2013) 4558–4569. Available from: https://doi.org/10.3390/md11114558.

[46] E.M. Trentacoste, R.P. Shrestha, S.R. Smith, C. Gle, A.C. Hartmann, M. Hildebrand, et al., Metabolic engineering of lipid catabolism increases microalgal lipid accumulation without compromising growth, Proc. Natl. Acad. Sci. 110 (2013) 19748–19753. Available from: https://doi.org/10.1073/pnas.1309299110.

[47] J. Xue, L. Wang, L. Zhang, S. Balamurugan, D.-W. Li, H. Zeng, et al., The pivotal role of malic enzyme in enhancing oil accumulation in green microalga Chlorella pyrenoidosa, Microb. Cell Fact. 15 (2016) 120. Available from: https://doi.org/10.1186/s12934-016-0519-2.

[48] Y.-F. Niu, X. Wang, D.-X. Hu, S. Balamurugan, D.-W. Li, W.-D. Yang, et al., Molecular characterization of a glycerol-3-phosphate acyltransferase reveals key features essential for triacylglycerol production in Phaeodactylum tricornutum, Biotechnol. Biofuels. 9 (2016) 60. Available from: https://doi.org/10.1186/s13068-016-0478-1.

[49] C. Zhang, Biosynthesis of carotenoids and apocarotenoids by microorganisms and their industrial potential, in: L.Q. Zepka, E. Jacob-Lopes, V.V.D. Rosso (Eds.), Prog. Carotenoid Res, InTech, 2018. Available from: https://doi.org/10.5772/intechopen.79061.

[50] J. Fiedor, K. Burda, Potential role of carotenoids as antioxidants in human health and disease, Nutrients 6 (2014) 466–488. Available from: https://doi.org/10.3390/nu6020466.

[51] L.A.C. Cardoso, S.G. Karp, F. Vendruscolo, K.Y.F. Kanno, L.I.C. Zoz, J.C. Carvalho, Biotechnological production of carotenoids and their applications in food and pharmaceutical products, in: D.J. Cvetkovic, G.S. Nikolic (Eds.), Carotenoids, InTech, 2017. Available from: https://doi.org/10.5772/67725.

[52] I. Couso, M. Vila, H. Rodriguez, M.A. Vargas, R. León, Overexpression of an exogenous phytoene synthase gene in the unicellular alga Chlamydomonas reinhardtii leads to an increase in the content of carotenoids, Biotechnol. Prog. 27 (2011) 54–60. Available from: https://doi.org/10.1002/btpr.527.

[53] N. Arrach, R. Fernández-Martín, E. Cerdá-Olmedo, J. Avalos, A single gene for lycopene cyclase, phytoene synthase, and regulation of carotene biosynthesis in Phycomyces, Proc. Natl. Acad. Sci. United States A. 98 (2001) 1687–1692. Available from: https://doi.org/10.1073/pnas.021555298.

[54] D.K. Saini, S. Pabbi, A. Prakash, P. Shukla, Synthetic biology applied to microalgae-based processes and products, Handb. Microalgae-Based Process. Prod, Elsevier, 2020, pp. 85–98. Available from: https://doi.org/10.1016/B978-0-12-818536-0.00004-X.

[55] C. Shang, W. Wang, S. Zhu, Z. Wang, L. Qin, M.A. Alam, et al., The responses of two genes encoding phytoene synthase (Psy) and phytoene desaturase (Pds) to nitrogen limitation and salinity up-shock with special emphasis on carotenogenesis in Dunaliella parva, Algal Res. 32 (2018) 1–10. Available from: https://doi.org/10.1016/j.algal.2018.03.002.

[56] F. Bohne, H. Linden, Regulation of carotenoid biosynthesis genes in response to light in Chlamydomonas reinhardtii, Biochim. Biophys. Acta BBA - Gene Struct. Expr. 1579 (2002) 26–34. Available from: https://doi.org/10.1016/S0167-4781(02)00500-6.

[57] W.-R. Lin, S.-I. Tan, C.-C. Hsiang, P.-K. Sung, I.-S. Ng, Challenges and opportunity of recent genome editing and multi-omics in cyanobacteria and microalgae for biorefinery, Bioresour. Technol. 291 (2019) 121932. Available from: https://doi.org/10.1016/j.biortech.2019.121932.

[58] P.T. Tran, M.N. Sharifi, S. Poddar, R.M. Dent, K.K. Niyogi, Intragenic enhancers and suppressors of phytoene desaturase mutations in Chlamydomonas reinhardtii, PLoS One 7 (2012) e42196. Available from: https://doi.org/10.1371/journal.pone.0042196.

[59] J.I. Galarza, J.A. Gimpel, V. Rojas, B.O. Arredondo-Vega, V. Henríquez, Over-accumulation of astaxanthin in Haematococcus pluvialis through chloroplast genetic engineering, Algal Res. 31 (2018) 291–297. Available from: https://doi.org/10.1016/j.algal.2018.02.024.

[60] C. Wang, Z. Hu, C. Zhao, X. Mao, Isolation of the β-carotene ketolase gene promoter from Haematococcus pluvialis and expression of ble in transgenic Chlamydomonas, J. Appl. Phycol. 24 (2012) 1303–1310. Available from: https://doi.org/10.1007/s10811-011-9781-1.

[61] B.F. Cordero, I. Obraztsova, I. Couso, R. Leon, M.A. Vargas, H. Rodriguez, Enhancement of lutein production in Chlorella sorokiniana (Chorophyta) by improvement of culture conditions and random mutagenesis, Mar. Drugs. 9 (2011) 1607–1624. Available from: https://doi.org/10.3390/md9091607.

[62] E.W. Becker, Micro-algae as a source of protein, Biotechnol. Adv. 25 (2007) 207–210. Available from: https://doi.org/10.1016/j.biotechadv.2006.11.002.

[63] J. Liu, Batch cultivation for astaxanthin analysis using the green microalga Chlorella zofingiensis under multitrophic growth conditions, in: C. Barreiro, J.-L. Barredo (Eds.), Microb. Carotenoids, Springer, New York, NY, 2018, pp. 97–106. Available from: https://doi.org/10.1007/978-1-4939-8742-9_5.

[64] J. Liu, X. Mao, W. Zhou, M.T. Guarnieri, Simultaneous production of triacylglycerol and high-value carotenoids by the astaxanthin-producing oleaginous green microalga Chlorella zofingiensis, Bioresour. Technol. 214 (2016) 319–327. Available from: https://doi.org/10.1016/j.biortech.2016.04.112.

[65] J. Liu, Z. Sun, H. Gerken, Z. Liu, Y. Jiang, F. Chen, Chlorella zofingiensis as an alternative microalgal producer of astaxanthin: biology and industrial potential, Mar. Drugs. 12 (2014) 3487–3515. Available from: https://doi.org/10.3390/md12063487.

[66] J.-F. Arnaud, F. Viard, M. Delescluse, J. Cuguen, Evidence for gene flow via seed dispersal from crop to wild relatives in *Beta vulgaris* (Chenopodiaceae): consequences for the release of genetically modified crop species with weedy lineages, Proc. R. Soc. Lond. B Biol. Sci. 270 (2003) 1565–1571. Available from: https://doi.org/10.1098/rspb.2003.2407.

[67] B. Yang, J. Liu, Y. Jiang, F. Chen, *Chlorella* species as hosts for genetic engineering and expression of heterologous proteins: Progress, challenge and perspective, Biotechnol. J. 11 (2016) 1244–1261. Available from: https://doi.org/10.1002/biot.201500617.

[68] J. Steinbrenner, G. Sandmann, Transformation of the green alga Haematococcus pluvialis with a phytoene desaturase for accelerated astaxanthin biosynthesis, Appl. Environ. Microbiol. 72 (2006) 7477–7484. Available from: https://doi.org/10.1128/AEM.01461-06.

[69] J. Liu, H. Gerken, J. Huang, F. Chen, Engineering of an endogenous phytoene desaturase gene as a dominant selectable marker for Chlamydomonas reinhardtii transformation and enhanced biosynthesis of carotenoids, Process. Biochem. 48 (2013) 788–795. Available from: https://doi.org/10.1016/j.procbio.2013.04.020.

[70] C.-Y. Chen, A.-L. Kao, Z.-C. Tsai, T.-J. Chow, H.-Y. Chang, X.-Q. Zhao, et al., Expression of type 2 diacylglycerol acyltransferse gene *DGTT1* from *Chlamydomonas reinhardtii* enhances lipid production in *Scenedesmus obliquus*, Biotechnol. J. 11 (2016) 336–344. Available from: https://doi.org/10.1002/biot.201500272.

[71] P.-H. Chen, H.-L. Liu, Y.-J. Chen, Y.-H. Cheng, W.-L. Lin, C.-H. Yeh, et al., Enhancing CO_2 bio-mitigation by genetic engineering of cyanobacteria, Energy Environ. Sci. 5 (2012) 8318. Available from: https://doi.org/10.1039/c2ee21124f.

[72] J. Dexter, P. Fu, Metabolic engineering of cyanobacteria for ethanol production, Energy Environ. Sci. 2 (2009) 857. Available from: https://doi.org/10.1039/b811937f.

[73] J. Dexter, P. Armshaw, C. Sheahan, J.T. Pembroke, The state of autotrophic ethanol production in Cyanobacteria, J. Appl. Microbiol. 119 (2015) 11–24. Available from: https://doi.org/10.1111/jam.12821.

[74] R.F. Ketting, The many faces of RNAi, Dev. Cell. 20 (2011) 148–161. Available from: https://doi.org/10.1016/j.devcel.2011.01.012.

[75] M. Fayyaz, K.W. Chew, P.L. Show, T.C. Ling, I.-S. Ng, J.-S. Chang, Genetic engineering of microalgae for enhanced biorefinery capabilities, Biotechnol. Adv. 43 (2020) 107554. Available from: https://doi.org/10.1016/j.biotechadv.2020.107554.

[76] D.H. Kim, J.J. Rossi, Overview of gene silencing by RNA interference, Curr. Protoc. Nucleic Acid. Chem. 36 (2009). Available from: https://doi.org/10.1002/0471142700.nc1601s36.

[77] C. Zhang, S. Lin, Initial evidence of functional siRNA machinery in dinoflagellates, Harmful Algae 81 (2019) 53–58. Available from: https://doi.org/10.1016/j.hal.2018.11.014.

[78] M. Schroda, RNA silencing in Chlamydomonas: mechanisms and tools, Curr. Genet. 49 (2006) 69–84. Available from: https://doi.org/10.1007/s00294-005-0042-1.

[79] V. De Riso, R. Raniello, F. Maumus, A. Rogato, C. Bowler, A. Falciatore, Gene silencing in the marine diatom Phaeodactylum tricornutum, Nucleic Acids Res. 37 (2009) e96. Available from: https://doi.org/10.1093/nar/gkp448.

Handbook of Algal Biofuels

References

119

[80] Y. Jia, L. Xue, H. Liu, J. Li, Characterization of the glyceraldehyde-3-phosphate dehydrogenase (GAPDH) gene from the halotolerant alga Dunaliella salina and inhibition of its expression by RNAi, Curr. Microbiol. 58 (2009) 426−431. Available from: https://doi.org/10.1007/s00284-008-9333-3.

[81] T. Yamasaki, A. Voshall, E.-J. Kim, E. Moriyama, H. Cerutti, T. Ohama, Complementarity to an miRNA seed region is sufficient to induce moderate repression of a target transcript in the unicellular green alga *Chlamydomonas reinhardtii*, Plant. J. 76 (2013) 1045−1056. Available from: https://doi.org/10.1111/tpj.12354.

[82] A. Voshall, E.-J. Kim, X. Ma, E.N. Moriyama, H. Cerutti, Identification of AGO3-associated miRNAs and computational prediction of their targets in the green alga Chlamydomonas reinhardtii, Genetics 200 (2015) 105−121. Available from: https://doi.org/10.1534/genetics.115.174797.

[83] R. Yang, G. Chen, H. Peng, D. Wei, Identification and characterization of MiRNAs in Coccomyxa subellipsoidea C-169, Int. J. Mol. Sci. 20 (2019) 3448. Available from: https://doi.org/10.3390/ijms20143448.

[84] T. Gaj, C.A. Gersbach, C.F. Barbas, ZFN, TALEN, and CRISPR/Cas-based methods for genome engineering, Trends Biotechnol. 31 (2013) 397−405. Available from: https://doi.org/10.1016/j.tibtech.2013.04.004.

[85] I. Sizova, M. Fuhrmann, P. Hegemann, A Streptomyces rimosus aphVIII gene coding for a new type phosphotransferase provides stable antibiotic resistance to Chlamydomonas reinhardtii, Gene 277 (2001) 221−229. Available from: https://doi.org/10.1016/S0378-1119(01)00616-3.

[86] A. Greiner, S. Kelterborn, H. Evers, G. Kreimer, I. Sizova, P. Hegemann, Targeting of photoreceptor genes in *Chlamydomonas reinhardtii* via Zinc-finger nucleases and CRISPR/Cas9, Plant. Cell 29 (2017) 2498−2518. Available from: https://doi.org/10.1105/tpc.17.00659.

[87] I. Ng, S. Tan, P. Kao, Y. Chang, J. Chang, Recent developments on genetic engineering of microalgae for biofuels and bio-based chemicals, Biotechnol. J. 12 (2017) 1600644. Available from: https://doi.org/10.1002/biot.201600644.

[88] A. Banerjee, C. Banerjee, S. Negi, J.-S. Chang, P. Shukla, Improvements in algal lipid production: a systems biology and gene editing approach, Crit. Rev. Biotechnol. 38 (2018) 369−385. Available from: https://doi.org/10.1080/07388551.2017.1356803.

[89] S.Y. Gan, C.A. Maggs, Random mutagenesis, and precise gene editing technologies: applications in algal crop improvement and functional genomics, Eur. J. Phycol. 52 (2017) 466−481. Available from: https://doi.org/10.1080/09670262.2017.1358827.

[90] A. Piatek, Z. Ali, H. Baazim, L. Li, A. Abulfaraj, S. Al-Shareef, et al., RNA-guided transcriptional regulation in planta via synthetic dCas9-based transcription factors, Plant. Biotechnol. J. 13 (2015) 578−589. Available from: https://doi.org/10.1111/pbi.12284.

[91] T. de Mooij, M. Janssen, O. Cerezo-Chinarro, J.H. Mussgnug, O. Kruse, M. Ballottari, et al., Antenna size reduction as a strategy to increase biomass productivity: a great potential not yet realized, J. Appl. Phycol. 27 (2015) 1063−1077. Available from: https://doi.org/10.1007/s10811-014-0427-y.

[92] L.R. Polstein, C.A. Gersbach, A light-inducible CRISPR-Cas9 system for control of endogenous gene activation, Nat. Chem. Biol. 11 (2015) 198−200. Available from: https://doi.org/10.1038/nchembio.1753.

[93] M.I.S. Naduthodi, M.J. Barbosa, J. van der Oost, Progress of CRISPR-Cas based genome editing in photosynthetic microbes, Biotechnol. J. 13 (2018) 1700591. Available from: https://doi.org/10.1002/biot.201700591.

[94] W. Jiang, A.J. Brueggeman, K.M. Horken, T.M. Plucinak, D.P. Weeks, Successful transient expression of Cas9 and single guide RNA genes in Chlamydomonas reinhardtii, Eukaryot. Cell. 13 (2014) 1465−1469. Available from: https://doi.org/10.1128/EC.00213-14.

[95] M. Nymark, A.K. Sharma, T. Sparstad, A.M. Bones, P. Winge, A CRISPR/Cas9 system adapted for gene editing in marine algae, Sci. Rep. 6 (2016) 24951. Available from: https://doi.org/10.1038/srep24951.

[96] K. Baek, D.H. Kim, J. Jeong, S.J. Sim, A. Melis, J.-S. Kim, et al., DNA-free two-gene knockout in Chlamydomonas reinhardtii via CRISPR-Cas9 ribonucleoproteins, Sci. Rep. 6 (2016) 30620. Available from: https://doi.org/10.1038/srep30620.

[97] S.-E. Shin, J.-M. Lim, H.G. Koh, E.K. Kim, N.K. Kang, S. Jeon, et al., CRISPR/Cas9-induced knockout and knock-in mutations in Chlamydomonas reinhardtii, Sci. Rep. 6 (2016) 27810. Available from: https://doi.org/10.1038/srep27810.

[98] D.C. Swarts, M. Jinek, Cas9 vs Cas12a/Cpf1: Structure-function comparisons and implications for genome editing, Wiley Interdiscip. Rev. RNA. 9 (2018) e1481. Available from: https://doi.org/10.1002/wrna.1481.

[99] A. Ferenczi, D.E. Pyott, A. Xipnitou, A. Molnar, Efficient targeted DNA editing and replacement in *Chlamydomonas reinhardtii* using Cpf1 ribonucleoproteins and single-stranded DNA, Proc. Natl. Acad. Sci. 114 (2017) 13567−13572. Available from: https://doi.org/10.1073/pnas.1710597114.

Handbook of Algal Biofuels

[100] D. Carroll, Genome engineering with targetable nucleases, Annu. Rev. Biochem. 83 (2014) 409–439. Available from: https://doi.org/10.1146/annurev-biochem-060713-035418.

[101] J.K. Joung, J.D. Sander, TALENs: a widely applicable technology for targeted genome editing, Nat. Rev. Mol. Cell Biol. 14 (2013) 49–55. Available from: https://doi.org/10.1038/nrm3486.

[102] H. Gao, D.A. Wright, T. Li, Y. Wang, K. Horken, D.P. Weeks, et al., TALE activation of endogenous genes in Chlamydomonas reinhardtii, Algal Res. 5 (2014) 52–60. Available from: https://doi.org/10.1016/j.algal.2014.05.003.

[103] A.E. Fortunato, M. Jaubert, G. Enomoto, J.-P. Bouly, R. Raniello, M. Thaler, et al., Diatom phytochromes reveal the existence of far-red-light-based sensing in the ocean, Plant. Cell 28 (2016) 616–628. Available from: https://doi.org/10.1105/tpc.15.00928.

[104] S. Jagadevan, A. Banerjee, C. Banerjee, C. Guria, R. Tiwari, M. Baweja, et al., Recent developments in synthetic biology and metabolic engineering in microalgae towards biofuel production, Biotechnol. Biofuels. 11 (2018) 185. Available from: https://doi.org/10.1186/s13068-018-1181-1.

[105] L.K.S. Gujjala, S.P.J. Kumar, B. Talukdar, et al., Biodiesel from oleaginous microbes: opportunities and challenges, Biofuels 10 (2019) 45–59. Available from: https://doi.org/10.1080/17597269.2017.1402587.

[106] A.N. Yadav, D. Kour, K.L. Rana, N. Yadav, B. Singh, V.S. Chauhan, et al., Metabolic engineering to synthetic biology of secondary metabolites production, New Future Dev. Microb. Biotechnol. Bioeng, Elsevier, 2019, pp. 279–320. Available from: https://doi.org/10.1016/B978-0-444-63504-4.00020-7.

[107] L. Barolo, R.M. Abbriano, A.S. Commault, J. George, T. Kahlke, M. Fabris, et al., Perspectives for glycoengineering of recombinant biopharmaceuticals from microalgae, Cells 9 (2020) 633. Available from: https://doi.org/10.3390/cells9030633.

[108] M.J.M.F. Reijnders, R.G.A. van Heck, C.M.C. Lam, M.A. Scaife, V.A.P.M. dos Santos, A.G. Smith, et al., Green genes: bioinformatics and systems-biology innovations drive algal biotechnology, Trends Biotechnol. 32 (2014) 617–626. Available from: https://doi.org/10.1016/j.tibtech.2014.10.003.

[109] P. Charoonnart, S. Purton, V. Saksmerprome, Applications of microalgal biotechnology for disease control in aquaculture, Biology 7 (2018) 24. Available from: https://doi.org/10.3390/biology7020024.

[110] L.R. Dahlin, M.T. Guarnieri, Recent advances in algal genetic tool development, Curr, Biotechnol. 5 (2016) 192–197. Available from: https://doi.org/10.2174/2211550105666160127230814.

[111] Y.M. Dyo, S. Purton, The algal chloroplast as a synthetic biology platform for production of therapeutic proteins, Microbiology 164 (2018) 113–121. Available from: https://doi.org/10.1099/mic.0.000599.

[112] J. Shrager, C. Hauser, C.-W. Chang, E.H. Harris, J. Davies, J. McDermott, et al., *Chlamydomonas reinhardtii* genome project. a guide to the generation and use of the cDNA information, Plant. Physiol. 131 (2003) 401–408. Available from: https://doi.org/10.1104/pp.016899.

[113] H. Moreau, B. Verhelst, A. Couloux, E. Derelle, S. Rombauts, N. Grimsley, et al., Gene functionalities and genome structure in Bathycoccus prasinos reflect cellular specializations at the base of the green lineage, Genome Biol. 13 (2012) R74. Available from: https://doi.org/10.1186/gb-2012-13-8-r74.

[114] K. Liolios, K. Mavromatis, N. Tavernarakis, N.C. Kyrpides, The genomes on line database (GOLD) in 2007: status of genomic and metagenomic projects and their associated metadata, Nucleic Acids Res. 36 (2007) D475–D479. Available from: https://doi.org/10.1093/nar/gkm884.

[115] J. Guo, S. Wilken, V. Jimenez, C.J. Choi, C. Ansong, R. Dannebaum, et al., Specialized proteomic responses and an ancient photoprotection mechanism sustain marine green algal growth during phosphate limitation, Nat. Microbiol. 3 (2018) 781–790. Available from: https://doi.org/10.1038/s41564-018-0178-7.

[116] C. Bowler, A.E. Allen, J.H. Badger, J. Grimwood, K. Jabbari, A. Kuo, et al., The Phaeodactylum genome reveals the evolutionary history of diatom genomes, Nature. 456 (2008) 239–244. Available from: https://doi.org/10.1038/nature07410.

[117] A.Z. Worden, J.-H. Lee, T. Mock, P. Rouze, M.P. Simmons, A.L. Aerts, et al., Green evolution and dynamic adaptations revealed by genomes of the marine picoeukaryotes micromonas, Science 324 (2009) 268–272. Available from: https://doi.org/10.1126/science.1167222.

[118] E. Derelle, C. Ferraz, S. Rombauts, P. Rouze, A.Z. Worden, S. Robbens, et al., Genome analysis of the smallest free-living eukaryote Ostreococcus tauri unveils many unique features, Proc. Natl. Acad. Sci. 103 (2006) 11647–11652. Available from: https://doi.org/10.1073/pnas.0604795103.

[119] P. Bohutskyi, S. Chow, B. Ketter, M.J. Betenbaugh, E.J. Bouwer, Prospects for methane production and nutrient recycling from lipid extracted residues and whole Nannochloropsis salina using anaerobic digestion, Appl. Energy. 154 (2015) 718–731. Available from: https://doi.org/10.1016/j.apenergy.2015.05.069.

References

[120] B. Palenik, J. Grimwood, A. Aerts, P. Rouzé, A. Salamov, N. Putnam, et al., The tiny eukaryote *Ostreococcus* provides genomic insights into the paradox of plankton speciation, Proc. Natl. Acad. Sci. 104 (2007) 7705–7710. Available from: https://doi.org/10.1073/pnas.0611046104.

[121] H.N. Dawson, R. Burlingame, A.C. Cannons, Stable transformation of chlorella: rescue of nitrate reductase-deficient mutants with the nitrate reductase gene, Curr. Microbiol. 35 (1997) 356–362. Available from: https://doi.org/10.1007/s002849900268.

[122] G. Blanc, I. Agarkova, J. Grimwood, A. Kuo, A. Brueggeman, D.D. Dunigan, et al., The genome of the polar eukaryotic microalga Coccomyxa subellipsoidea reveals traits of cold adaptation, Genome Biol. 13 (2012) R39. Available from: https://doi.org/10.1186/gb-2012-13-5-r39.

[123] M.J. Wynne, J.K. Hallan, Reinstatement of *Tetradesmus* G. M. Smith (Sphaeropleales, Chlorophyta): reinstatement of *Tetradesmus* G. M. Smith (Sphaeropleales, Chlorophyta), Feddes Repert. 126 (2015) 83–86. Available from: https://doi.org/10.1002/fedr.201500021.

[124] K. Vandepoele, M. Van Bel, G. Richard, S. Van Landeghem, B. Verhelst, H. Moreau, et al., pico-PLAZA, a genome database of microbial photosynthetic eukaryotes: genomics update, Environ. Microbiol. 15 (2013) 2147–2153. Available from: https://doi.org/10.1111/1462-2920.12174.

[125] A. Kurotani, Y. Yamada, T. Sakurai, Alga-PrAS (algal protein annotation suite): a database of comprehensive annotation in algal proteomes, Plant. Cell Physiol. (2017) pcw212. Available from: https://doi.org/10.1093/pcp/pcw212.

[126] P.J. Maughan, M.A.S. Maroof, G.R. Buss, [No title found], Mol. Breed. 6 (2000) 105–111. Available from: https://doi.org/10.1023/A:1009628614988.

[127] M. Kanehisa, KEGG: kyoto encyclopedia of genes and genomes, Nucleic Acids Res. 28 (2000) 27–30. Available from: https://doi.org/10.1093/nar/28.1.27.

[128] C.R. Steadman Tyler, B.T. Hovde, H.E. Daligault, X.L. Zhang, Y. Kunde, B.L. Marrone, et al., High-quality draft genome sequence of the green alga Tetraselmis striata (Chlorophyta) generated from PacBio sequencing, Microbiol. Resour. Announc. 8 (2019) MRA.00780-19, e00780-19. Available from: https://doi.org/10.1128/MRA.00780-19.

[129] P.D. Karp, Expansion of the BioCyc collection of pathway/genome databases to 160 genomes, Nucleic Acids Res. 33 (2005) 6083–6089. Available from: https://doi.org/10.1093/nar/gki892.

[130] P.D. Karp, The MetaCyc database, Nucleic Acids Res. 30 (2002) 59–61. Available from: https://doi.org/10.1093/nar/30.1.59.

[131] M.D. Guiry, G.M. Guiry, L. Morrison, F. Rindi, S.V. Miranda, A.C. Mathieson, et al., AlgaeBase: an on-line resource for algae, Cryptogam. Algol. 35 (2014) 105–115. Available from: https://doi.org/10.7872/crya.v35.iss2.2014.105.

[132] A. Ebrahim, J.A. Lerman, B.O. Palsson, D.R. Hyduke, COBRApy: constraints-based reconstruction and analysis for python, BMC Syst. Biol. 7 (2013) 74. Available from: https://doi.org/10.1186/1752-0509-7-74.

[133] S.F. Altschul, W. Gish, W. Miller, E.W. Myers, D.J. Lipman, Basic local alignment search tool, J. Mol. Biol. 215 (1990) 403–410. Available from: https://doi.org/10.1016/S0022-2836(05)80360-2.

[134] P. Artimo, M. Jonnalagedda, K. Arnold, D. Baratin, G. Csardi, E. de Castro, et al., ExPASy: SIB bioinformatics resource portal, Nucleic Acids Res. 40 (2012) W597–W603. Available from: https://doi.org/10.1093/nar/gks400.

[135] H. Nordberg, M. Cantor, S. Dusheyko, S. Hua, A. Poliakov, I. Shabalov, et al., The genome portal of the department of energy joint genome institute: 2014 updates, Nucleic Acids Res. 42 (2014) D26–D31. Available from: https://doi.org/10.1093/nar/gkt1069.

[136] J. Ashworth, P.J. Ralph, An explorable public transcriptomics compendium for eukaryotic microalgae, Syst. Biol. (2018). Available from: https://doi.org/10.1101/403063.

[137] M. Fabris, M. Matthijs, S. Rombauts, W. Vyverman, A. Goossens, G.J.E. Baart, The metabolic blueprint of Phaeodactylum tricornutum reveals a eukaryotic Entner-Doudoroff glycolytic pathway: the Phaeodactylum tricornutum metabolic blueprint, Plant. J. 70 (2012) 1004–1014. Available from: https://doi.org/10.1111/j.1365-313X.2012.04941.x.

[138] S. Lee, P.E. Jung, Y. Lee, Publicly-funded biobanks and networks in East Asia, SpringerPlus 5 (2016) 1080. Available from: https://doi.org/10.1186/s40064-016-2723-2.

[139] L.K. Pandey, K.K. Ojha, P.K. Singh, C.S. Singh, S. Dwivedi, E.A. Bergey, Diatoms image database of India (DIDI): A research tool, Environ. Technol. Innov. 5 (2016) 148–160. Available from: https://doi.org/10.1016/j.eti.2016.02.001.

[140] H.-Q. Zheng, Y.-F. Chiang-Hsieh, C.-H. Chien, B.-K. Hsu, T.-L. Liu, C.-N. Chen, et al., AlgaePath: comprehensive analysis of metabolic pathways using transcript abundance data from next-generation sequencing in green algae, BMC Genomics 15 (2014) 196. Available from: https://doi.org/10.1186/1471-2164-15-196.

[141] N. Misra, P.K. Panda, B.K. Parida, B.K. Mishra, dEMBF: a comprehensive database of enzymes of microalgal biofuel feedstock, PLOS One 11 (2016) e0146158. Available from: https://doi.org/10.1371/journal.pone.0146158.

[142] S. Sahoo, S.R. Mahapatra, B.K. Parida, P.K. Narang, S. Rath, N. Misra, et al., dEMBF v2.0: an updated database of enzymes for microalgal biofuel feedstock, Plant. Cell Physiol. 61 (2020) 1019–1024. Available from: https://doi.org/10.1093/pcp/pcaa015.

[143] S. Cheng, M. Melkonian, S.A. Smith, S. Brockington, J.M. Archibald, P.-M. Delaux, et al., 10KP: a phylodiverse genome sequencing plan, GigaScience 7 (2018). Available from: https://doi.org/10.1093/gigascience/giy013.

[144] C. Fajardo, M. Donato, R. Carrasco, G. Martínez-Rodríguez, J.M. Mancera, F.J. Fernández-Acero, Advances and challenges in genetic engineering of microalgae, Rev. Aquac. 12 (2020) 365–381. Available from: https://doi.org/10.1111/raq.12322.

[145] G. Kumar, A. Shekh, S. Jakhu, Y. Sharma, R. Kapoor, T.R. Sharma, Bioengineering of microalgae: recent advances, perspectives, and regulatory challenges for industrial application, Front. Bioeng. Biotechnol. 8 (2020) 914. Available from: https://doi.org/10.3389/fbioe.2020.00914.

[146] A.A. Snow, V.H. Smith, Genetically engineered algae for biofuels: a key role for ecologists, BioScience 62 (2012) 765–768. Available from: https://doi.org/10.1525/bio.2012.62.8.9.

[147] M. Serif, G. Dubois, A. Finoux, M. Teste, D. Jallet, F. Daboussi, One-step generation of multiple gene knockouts in the diatom Phaeodactylum tricornutum by DNA-free genome editing, Nature Communications 9 (1) (2018) 1–10. Available from: https://doi.org/10.1038/s41467-018-06378-9.

CHAPTER 6

The current status of various algal industries

Ahmed E. AlProl[1] and Marwa R. Elkatory[2]

[1]National Institute of Oceanography and Fisheries, Hurghada, Egypt [2]Advanced Technology and New Materials Research Institute (ATNMRI), City of Scientific Research and Technological Applications (SRTA-City), Alexandria, Egypt

6.1 Introduction

Algae are simple organisms with chlorophyll that comprise groups of cells in colonies that are not fundamentally very interrelated and they are polyphyletic in nature [1]. Algae are being used for a long period of time due to their high biomass production in different extreme habitats as compared to cereal-based crops. Algae can be cultivated with brackish or saltwater on marginal soil, and thus do not compete with traditional agricultural resources. Natural algae products for human use, such as food and medical treatments, have long been widely explored. Due to their many advantages over various crops, algae are known as third-generation biofuels. Microalgae have high carbon and lipid concentration, and are rapidly growing [2] as an energy-production biomass of the future [3,4]. More energy is needed than ever in the world today. The main factor causing the rising energy demand is population growth. The human population is estimated to overstep 10 billion by 2050. A new energy is therefore required that is cleaner and more sustainable. Egypt has faced population growth and increased energy demand throughout the years. The use of algae for energy production in the modern decades has received a lot of attention. Algae can be clean and sustainable as a source of energy that can be converted into biofuel. In the form of cofiring, bio-oil pyrolysis, biodiesel, bioethanol, and biomethane fermentation, marine biomass of algae could be used as a feedstock to harvest various fuels, including bioelectricity.

To date, due to its restricted cultivation systems, the costs for the development of algae biofuels are very high. But over time, new technologies for algae cultivation are being established ranging from tropical to moderate climatic conditions in various

124 6. The current status of various algal industries

climate zones. Based on archeological evidence from Chile 14,000 years BP, algae organisms have been in the animal diet for many centuries [5]. Out of the 23.8 million tons, sea algae consumed in the 2012, the FAO [6] reported that 38% was consumed by humans and known by them as sea algae (such as nori/laver, kelp), without the use in food and beverage of any additional hydrocolloid intake (e.g., agar, alginates, carrageenan). The human consumption of algal foods varies by nation, with Japanese diets representing a recent annual per capita consumption ranging from 9.6 g day^{-1} in 2014 to 11.0 g day^{-1} in 2010 [7]. Marine algae are categorized into micro/macro algae and are extremely useful for industrial wastewater treatment. Most of the products of algae have economical value and they are commonly used since they are a good source of pharmaceuticals, fibers, vitamins, minerals, lectins, antioxidants, pigments, polysaccharides, proteins, steroids, halogenated compounds, and polyunsaturated fatty acids and lipids.

The present chapter is designed to provide information on various applications for algae such as biofuels, food, polymer products, the pharmaceutical industries, and wastewater treatment, as well as on industrial development and various products for industrial use for algae.

6.2 The algae industry

Algae have become one of the long-term, most promising sources of sustainability for heat, fruit, feed, and other biomass and oil coproducts. The many and varied benefits associated with how and where they grow make them so attractive. It is estimated that per unit area the oil yield from algae ranges from 20,000 to 80,000 liters per acre annually, 7—31 times more than the next best cultivation of palm oil. The algae industry offers a wide range of products, valued at US$ 5.5—6 billion per year.

Food products contribute nearly $5 billion. The industry uses 7.5—8 million tons of wet algae annually. This is either obtained from natural (wild) algae or from (farmed) crops. Algae production has increased rapidly, because the supply from natural resources has been overwhelmed by demand. Commercial harvesting takes place in the waters of many countries from cold to temperate to tropical zones in both the Northern and Southern Hemisphere [8,9].

6.2.1 Commercial production

World algae production comes from two sources: first, wild stock production; second, aquaculture (mariculture, soil cultivation, and agriculture). FAO statistics show that the production from the collection of wild stocks was stable in 2003—12, at over 1 million tons (wet weight).

The highest producers in 2012 were Chile (436.035 tons; 39% of total world production), China (257.640 tons; 23%), Norway (98.514 tons; 9%), Japan (98.079 tons), France (41.229 tons; approximately 4%), Ireland (29.500 tons; 2.73%) and Iceland (18.079 tons; 2%). There were 24 other countries contributing less than 1% each. Since 2003, with the

Handbook of Algal Biofuels

exception of 2007, Chile has consistently been the leading producer. Three countries (Chile, Norway, and Japan) posted only 1 year's production in 10 years from 2003 to 2012. Aquaculture provides the bulk of algae worldwide. The output of mariculture aquatic algae, valued at nearly $ 6 billion, reached 24.9 million tons in 2012, around 88% (21 million tons) was brown, red, and green macroalgae. The aquaculture recorded by FAO [6] accounted for approximately 96% or 23.8 million tons. FAO data show a steady rise in red macroalgae with stable production of brown macroalgae of about 8% all year-round from 2003 to 2012. For this reason, it ranged from 4%−12%. China is the leading provider, although the trend is declining, with 50% of global production over a 10-year period (2003−12). In the period from 2003 to 2006 the Philippines ranked second, generating 9%−10%, after which Indonesia overtook it. Algae companies have had many commercial achievements in the United States in recent years. In mid-2011, Sapphire Energy, for example, broke ground with a 300-acre (1 million gallons) facility for algae biofuels the Columbus, New Mexico project. Phycal, Inc. was awarded a United States DOE grant to finance its Hawaiian Electric Company purchase agreement to supply power-producing algae fuel. Solazyme has been awarded a contract with the United States Navy to supply algal biofuels.

6.3 Main chemical compounds and bioactive compounds in algae

Bioactive compounds are materials containing functional properties in the animal body which are physiologically active. There is much interest in the creation and processing of various biocompounds which could probably serve as working ingredients, for example, phycocyanines, polyphenols, carotenoids, fatty acids, and polyinsaturated materials [10]. Microalgal strains are considered rich sources of antioxidants compounds, with possible application in medications, food, and cosmetics [11]. Microalgae isolates can block UV irradiation, and can produce pigments, lipids, and polysaccharides with antioxidant activity, such as propionate dimethylsulfonium and mycosporine amino acids [12]. Polysaccharides are a high-value-added product class with applications in the fields of cosmetics, pharmaceuticals, and foods. The sulfate esters of microalgal polysaccharides are known as sulfated polysaccharides and have particular medical uses. A blue photosynthetic pigment, which belongs to the category of large quantities of phycobiliproteins in cyanobacteria, rhodiophitis, and cryptophytes [13], is one good example of such compounds.

Marine macroalgae with low lignin, cellulose, and lipid content have been characterized [14]. Marine brown algae (particularly *Ascophylum nodosum*) may be rich in degradable polyphenols and may inhibit anaerobic digestion. Seaweeds replicate in several different ways; they can be reproduced by the combination of male and female gametes. The algae also expand and break into several small parts [15]. Brown algae are used for alginate development. Alginate is a main polysaccharide in brown macroalgae and has carboxylated groups, comprising mannuronic (M) and guluronic (G) acids. Alginic acid salts with monovalent ions, while salts with divalent or multipurpose metal ions and acid itself are insoluble.

Alginates are used to make thickening agents and stabilizers for foods and cosmetics [14]. The sulfonic acid of fucoidane is the second richest functional acid group in brown algae. Blocks of L-fucose are a key constituent of Fucoidan branched sulfate esters. Fucoidane acid hydrolysis also produces varying amounts of D-galactose, D-xylose, and uronic acid [16]. Red algae have been employed for the treatment of malaria, as well as for antifouling and antibiotic applications [15].

For the removal of pollutants, the good metal binding capacity of algae has been attributed to the presence of carboxyl and sulfate groups in algal cell wall polysaccharides that can act as binding sites for the removal of metal ions.

6.4 Energy production

The identification of potential renewable energy sources has recently gained traction [17]. Many countries currently use fossil energy sources, for example, wind, biomass, waste, solar, hydro, and geothermal energy [18]. The IEA recently said that waste and fuel energy has a greater alternative fuel capacity than other renewable sources (International Energy Agency 2010). Fuel biofuel (biodiesel, bioethanol, and biogas) is currently recognized for its potential in sustainable energy production as an alternative and green renewable fuel [19,20] (Fig. 6.1).

6.4.1 Biofuels

Biofuel products can refer to all types of liquid, solid, or gaseous fuels that can be produced from raw materials as renewable sources. The forms and amounts of biomass, energy type, and economic returns of the product are key elements of any conversion process [21]. Biofuel production has recently increased around the world, especially with regard to bioethanol

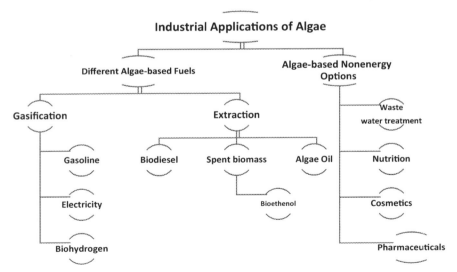

FIGURE 6.1 Some industrial application of algae and the energy production pathway.

production from sugar crops and cereals such as sorghum, sugar cane, and sugar beet. Agricultural crops are categorized as first generation biofuels owing to their application as food or fodder, however the race among food and fuel produces enough biofuel for the overall consumption of fuel. With the combined benefit of high productivity of biomass, wastewater treatment, annual production, algae chemical composition, and algal oil content, algae cultivation techniques can be controlled by changing algae. The third form of biofuel, which is considered an alternative energy source for fossil fuel, does not have the drawbacks of both the first- and second-generation biofuels products [22].

6.4.1.1 Bioethanol production

Because of high algae productivity, the variable chemical composition, diversity, and high photosynthetic rate of species of algae, the production of bioethanol from algae is of unusual importance. Because of the high content of polysaccharides, carbohydrates, and cellulose, algae are the best source for generating bioethanol. In the processing of algae bioethanol, fermentation and gasification are widely used in two processes [23]. Commercial ethanol production occurs on a large scale by a fermentation process from sugar and starch crops in various countries. The biomass is moistened, and by various methods the starch is converted into sugar. The microalgal starch is taken from the cells using acid, alkaline, enzymes, yeast (*Saccharomyces cerevisiae*).

Biofuel production or microalgae biomass can be produced by two main phases involving downstream and upstream techniques. Differing cultivation technologies are used in the upstream process in order to increase the quality and the amount of biomass, while the downstream stage emphasizes the technologies of harvest and sustainable biofuel production. Kim et al. [24] developed bioethanol using *Chlamydomonas* sp. (a psychrophilic microalgae). This study produced the maximum yield of 0.15 g FAME/g KNM0029C after incubation in a 20-L photobioreactor. Residual biomass was pretreated for the production of bioethanol and the yields were compared by various techniques. After sonication, pretreatment with enzymes (amyloglucosidase) resulted in the maximum bioethanol production (0.2 g/g residual biomass). Brown macroalgae is a probable feedstock for ethanol manufacture thanks to its content of high carbohydrate and easy cultivation [25], which can be altered for fermentation to ethanol. There have also been reports of red algae fermentation that has turned acid agar into sugars, but the ethanol output is low, with a maximum theoretical output of upto 45%. Algal fermentation yields 0.08 and 0.12 kg^{-1} kg of dry algae according to the genera of the algae and the different pretreatments [26]. Bioethanol production from microalgae was studied by Bertucco [27]. The saccharification and hydrothermic liquid and flash hydrolysis of biomass by acids or enzymes and alternative methods were discussed. Since glucose is the main monosaccharide in microalgal biomass, high rates of hydrolysis and fermentation can be achieved (more than 80% of the efficiency as a sum of these two processes). The process' sustainability could be improved by anaerobic digestion for vinasse recycling and CO_2 emission reduction by bioethanol and biogas production. The possible production of bioethanol from the algae *Ulva lactua* was evaluated by El-Sayed et al. [28] using yeast fermentation. The Plackett-Burman experimental design and immobilization method with assisted solid materials were used to optimize the bioethanol process. Experimental results showed that bioethanol production has a substantial effect on the concentration of sugar, pH, and inoculum scale.

Oluwatosin et al. [29] investigated the feasibility of pretreated and hydrolyzed bioethanol produced by dilute sulfuric acid and the commercially available enzymes from brown macroalgae, such as *Laminaria digitate* and *A. nodosum*. Glucose and rhamnose were, respectively, the prevalent sugars. Yoza and Masutani [30] found that upto 49% of seaweeds biomass acid pretreatment can produce sugars, while 20% of sugar can be released by enzyme hydrolysis based on its dry mass. Ueno et al. [31] investigated the ability of *Chlorococcum littorale* marine green alga in dark fermented ethanol and approximately 27% of the algae starch was used within 24 h at 25°C.

6.4.1.2 Biodiesel

The study of biodiesel production has been extensively recognized in modern years and it is yielded from oilseed crops, such as palm oil, soybean oil, and rapeseed oil [32]. The transesterification method is the conversion of raw compounds, such as lipids (triacyleglycerols/free fatty acids) to nontoxic (ecofriendly) biodiesel. During the transesterification process, crude oil of a catalyst reacts with an alcohol (usually methanol) and fatty acid methyl esters are made as the final product with glycerol. The application of acid catalyst has been viewed as advantageous, as reported by Meng et al. [33], however an alkali catalyst is recycled commercially owing to its fast nature upto 400 times compared to acid catalyst [34]. Some criteria should be taken into consideration for the production of cost effective biodiesel and calculation of feedstock, such as the quality and utilization of by-products and the type of raw material used [35].

Chlorella protothecoides and *Chlorella vulgaris* are two chief strains, whose high oil content has been investigated by numerous studies for biodiesel production, for example, Gao et al. [36]; Wahlen et al. (2011) and Cao et al. [37] studied the direct production of biodiesel from algae of *Chlorella pyrenoidosa* via a transesterification process. The results showed that in the optimal environment (100 mg dry weight equivalent microalgae, 8 mL n-hexane, 4 mL methanol, 0.5 M H_2SO_4, 120°C, 3 h reaction contact time) the production of biodiesel reached 92.5%.

6.4.1.3 Biomethane Production

Nowadays, the international production of biomethane from seaweed is gaining significance. An anaerobic digester comprises synergistic microbial populations, which convert algal organic substances (protein, carbohydrate, and lipid) to carbon dioxide and methane gas. Methane gas is broadly used as a chemical feedstock and fuel.

Diagne et al. (2018) examined the use of *Ulva lactuca* and *Codium tomentosum* as green macroalgae as a source of biogas. The results showed that *U. lactuca* and *C. tomentosum* realized a heating value of 2151 kWh/t VS. Moreover, the methane potential of algae was twice (216.4 L CH_4/kg VS) that of cow manure (100.3 L CH_4/kg VS).

Cardeña et al. (2016) studied the improvement of methane gas production from different algae cultures via the ozonation pretreatment process. In this work, the use of ozone to develop the production of methane from three algae sources was examined. The methane produced in every case could be augmented to a different extent (6%−66%).

Wang et al. [38] studied the thermal treatment of microalgae for the improved production of biomethane using the *Chlorella* strain. The production of methane from untreated algal biomass was 155 mL/g VS, but at 70 and 90°C for 0.5 h, improved the production of methane by 48%. While the thermal treatment at 121°C for 0.3 h yielded 322 mL/g VS.

6.4 Energy production

González-Fernández et al. [39] obtained 88% enhancement of methane after applying ultrasound with untreated *Scenedesmus* algae. This study reported that the production of methane was 153.5 mL CH_4 g^{-1} CO Din, which is lower than that obtained after treatment with ozonation conditions (259.6 mL CH_4 g^{-1} CO Din).

Passos et al. [40] achieved 169.9 mL CH_4 g^{-1} VS methane production with 61% as a result of improvement of methane generation from algal biomass by thermal pretreatment.

6.4.2 Biobutanol production

Butanol production is used as source for fuel in transportation and has been recommended as a possible candidate for biofuel production owing to its high energy density and low vapor pressure [41]. In butanol production bacteria are used to digest sugar and starch, and can utilize the cellulose present in alga. Therefore the production of butanol can be as economical as ethanol production [36]. Several *Clostridium* strains are able to yielding ethanol, acetone, and butanol via anaerobic fermentation by consuming pentoses and hexoses [42]. Butanol has been manufactured from anaerobic fermentation of *U. lactuca* via *Clostridium* sp, however the butanol produced is lower, upto 0.16 g butanol g^{-1}, than the ethanol production in similar conditions [35].

U. lactuca algae used as a feedstock for anaerobic fermentation with *Clostridium* sp. was treated with boiling water, and then underwent enzymatic hydrolysis via cellulases for the manufacture of ethanol, acetone, and butanol, producing 0.35 g (ethanol, acetone, and butanol) g^{-1} sugar [42].

Ulva sp. extracted from Jamaica Bay, New York City, was recycled to yield butanol at an experimental scale via the saccharification of sugars [41].

Gao et al. [36] produced butanol from microalgae after the extraction of ionic liquid. The study compared the suitability of ionic liquid extracted algae (ILEA) and hexane extracted algae (HEA) for acetone, butanol, and ethanol (ABE) production. The highest butanol titer (8.05 g L^{-1}) was achieved with the fermentation of the acid hydrolysates of HEA. However, it required detoxification to support product formation after acid hydrolysis, while ILEA did not. Direct ABE fermentation of ILEA and HEA (without detoxification) starches resulted in a butanol titers of 4.99 and 6.63 g L^{-1}, respectively, which significantly simplified the LEA to butanol process.

6.4.3 Bio-oil

Bio-oils are manufactured through thermochemical conversion—a process which converts algal biomass into bio-oil along with gas and char in the absence of oxygen and at very high temperature.

The process can be divided into thermochemical liquefaction and pyrolysis [43]. The pyrolysis technique is achieved at a high temperature from 350°C to 530°C to produce gaseous, solid, and liquid parts. The liquid part is made of a nonaqueous and an aqueous phase dubbed tar or bio-oil and the algal biomass is dried. In the thermochemical liquefaction technique, wet algal biomass is considered at a high pressure of about 10 MPa and a low temperature of about 300°C. The bio-oil comprises different organic substances accumulated as proteins, lipids, and carbohydrates in algal biomass. Compared to lipids existing in algae, the quantity of the yield is high. Several microalgae have been examined to

Handbook of Algal Biofuels

130   6. The current status of various algal industries

produce bio-oil via thermal liquefaction or pyrolysis [44], for example, the development of various hydrocarbons through the pyrolysis of *Dunaliella* strain biomass.

Bio-oil yields from microalgae have been stated to be upto 41% for *Spirulina* sp. [45], approximately 24%−45% for *Scenedesmus* microalgae [46], and get to 49% for *Desmodesmus* [47].

Bio-oil produced from macroalgae has been reported to reach 23% via the liquefaction method [48]. However, in *Laminaria saccharina* 63% energy restoration was calculated by Anastasakis and Ross [49].

6.4.4 Biohydrogen

Nowadays, a much interest is being paid to the production of biohydrogen, which is hydrogen yielded through photosynthetic organisms for renewable energy. Hydrogen ions can be manufactured by algal biomass at certain conditions. Biohydrogen production at huge scales is not possible owing to the low concentration of algal biomass and the high cost of the technique [50].

Three pathways are recommended for the production of hydrogen: (1) direct photolysis, (2) indirect photolysis; and (3) ATP driven pathway.

Direct photolysis occurs when manufactured oxygen and hydrogen are removed. Water splitting and photosynthesis are linked in this technique, which leads to the production of hydrogen and oxygen. Additionally, the hydrogenase enzyme used in this technique is oxygen sensitize. Indirect methods are commonly favored. In a sulfur-limited and anaerobic, and starch-controlled environment, the cell walls of algal biomass are transformed into hydrogen.

Cyanobacteria are the chief manufacturers of biohydrogen by the biological method, with nitrogenase enzymes and hydrogenase acting as catalysts [51].

6.4.5 Advantage of the algal biomass for biodiesel and bioethanol production

There are numerous advantages of algal biomass biodiesel and bioethanol production, such as:

1. Biodiesel and bioethanol production by algae does not compete with food production in water and on land: freshwater and maritime algae may be utilized.
2. The carbohydrate content, such as starches and sugars, in cells of algae is high; the carbohydrates can be fermented to yield bioethanol.
3. Algae have no lignin and low hemicellulose levels, which allow an increased hydrolysis efficiency [52], thus decreasing the production cost.
4. Large amounts of CO_2 gas are removed from the atmosphere and power plants by the usage of suitable technology options [53].
5. Algae grow quickly and can be grown in various water environments, for example, wastewater, freshwater, and saline water [54].
6. Efficient photosynthetic mechanism [55].
7. Algae can purify wastewater, removing high concentrations of nitrogen, phosphorus, and heavy metals from wastewater.
8. Algae can yield 10 times more biodiesel and bioethanol per each unit area of ground compared to a typical terrestrial plant [56].

Handbook of Algal Biofuels

6.4.6 Challenges facing algae biomass for biofuel production

1. The energy demand and high cost of reaping unicellular algae.
2. Evolvement broad-scale protocols and methodology [57].
3. There are presently no industrial facilities for producing biofuel from algal biomass.
4. Analysis of the special effects of numerous factors that modify the quantity of lipid content productivities by microorganisms; and enhancement of numerous modifications to both downstream processes (lipid extraction, conversion techniques, and biodiesel) and upstream (drying biomass, microalgae culture, and harvesting) [18].
5. The study of energy balance is not obviously addressed, as reported from the thermodynamic analysis [18].

6.5 Algae-based nonenergy field

The use of algal biomass is practically endless for many products, due to the variations in chemical composition and the large diversity of the various cultivation processes. These days, the option of algal biomass-based products is narrow owing to the untapped nature of this natural source. There are several influences for this: nonawareness around these products, economic limit, and endorsement of novel products via regulating authorities [58], and the lack of investments for founding large-scale facilities.

The FAO [59] reported that macroalgae and some microalgal species can be grown at commercial scale at the sea coastal area to provide biomass for production of biofuel, specialized chemicals for food processing, cosmetics, pharmaceuticals, soil additives, biofertilizers, animal fodder, and other end products as shown in Table 6.1.

6.5.1 Pharmaceuticals

Algal biomass is a rich resource of naturally active substances containing primary and secondary metabolites with different bioactive substances, which can be candidates for the pharmaceutical field [67]. The bioactive substances in algae can be used for different applications, such as nutrients, proteins, and vaccines, which require very high costs to obtain from other plant sources and animals, or may not be available [68]. Pharmaceutical products from algae are characterized by different bioactive agents and their high value.

Omega 3 polyunsaturated fatty acids compounds are very important for body metabolism in humans. Algae have been scanned as a source of vitamins and vitamin precursors, such as tocopherol, ascorbic acid and riboflavin.

Chlorophyts and *Bryophyta* algae are recycled for the production of these essential fatty acids. Different unicellular algal biomass, such as *Chlorella vulgaris* and *Chlamydomonas pyrenoidosa*, have antibacterial activity against numerous pathogens such as Gram-positive and Gram-negative bacteria from growth media and cell extracts. Few studies of in antifungal activities from extracts of diatoms and green algae have been reported. Blue—green macroalgae and microalgae such as *Prymnesium parvum* and *Ochromonas* species are able to yield toxic compounds with immense potential for pharmaceutical products [69].

132   6. The current status of various algal industries

TABLE 6.1 Products of some species of algae and their application.

Product	Use/application	Algae	References
Glycerin	Pharmaceuticals, paints, industry bulk chemical	By-product of biodiesel production	Subhadra and Edwards [60]
Phycobiliproteins	Biomedical uses, (fluorescent markers), food coloring, pharmaceuticals, cosmetics	*Arthrospira* sp. (*Chloroxybacteria*), *Porphyra* sp. *and Rhodella* sp. (*Rhodophyta*), *potentially* and *Glaucophyta*	Milledge [61]
ß-carotene3	Food coloring, functional	*Dunaliella* sp.	Ben-Amotz [62]; Milledge [61]
Agarose3	Biotechnological	*Rhodophyta*	Radmer [63]
Alginates3	Food, medical uses, paper	*Laminaria* sp. and *Macrocystis* sp. (*Phaeophyceae*) and *Ascophyllum* sp.	Radmer [63]
Vitamin B12	Health food and neutraceuticals,	(*Chlorophyta*), and (*Chloroxybacteria*)	Harun et al. [54]
Provitamin A3	Foods	*Arthrospira platensis* and (*Chloroxybacteria*)	Harun et al. [54]
Phycocyanins	Nutraceuticals, cosmetics	*Spirulina platensis*	Priyadarshani and Biswajit [64]
Fatty acids	Cosmetics and pharmaceuticals	*Odontella aurita*	Priyadarshani and Biswajit [64]
Ascorbic acid	Foods	*Chlorella vulgaris*	Priyadarshani and Biswajit [64]
Astaxanthin	Pharmaceuticals and feed additives	*Haematococcus pluvialis*/Chlorophyta	Del Campo et al. [65]
Carotenoids, Lutein	Foods and food supplement,	*Muriellopsis* sp./Chlorophyta	Blanco et al. [66]

Nostoc species comprise a broad spectrum of polyunsaturated fatty acids that contain vital fatty acids, for example, eicosapentaenoic acid, linoleic, octadecatetraenoic, α-linolenic, and γ-linolenic [70].

Chlorella species, a green alga, is rich in minerals, chlorophyll, polysaccharides, proteins, B vitamins, especially B12, and necessary amino acids with molecular components of 23% (w/w) of carbohydrate, 53% (w/w) of protein, 5% (w/w) of minerals, oligo elements, and 9% (w/w) of lipids. Concentrations of these nutrients can be altered via the manipulation of the culture environments [71].

6.5.2 Medicinal uses of algae drugs

Different types of algal biomass have various medicinal properties, which can be used for different treatments. Some algae and their medicinal cures are as follows [72]:

Handbook of Algal Biofuels

1. *Cyanobacteria* species are used to produce extracellular and intracellular metabolites with a wide range of biological activities, such as antiviral, antibacterial, and antifungal activity[73].
2. *Corallina* strain: pesticides can be extracted.
3. *Grateloupia* strain: can be used as a treatment to lower blood sugar
4. *Enteromorpha* species: can be recycled to treat parasitic disease, hemorrhoids, fever reducyion capacity, goiter, coughing, and bronchitis.
5. *Sargassum* strain: can be recycled to treat edema; reduce inflammation; cervical lymphadenitis; induce urination; and as a source of iodine and potassium.
6. *Acetabularia* strain: can be used for the treatment of edema and urinary diseases.
7. *Gelidium* sp.: can be extracted as agar.

6.5.3 Antiviral activity of algal compounds

Viruses have been the cause of mass pandemic and epidemic outbreaks of possibly dangerous and deadly infections, for example, hepatitis, influenza, HIV, etc.

There is an urgent necessity for the discovery of antiviral substances. Some have been extracted from algal bioactive substances. For example, from more than 600 cultures of Cyanobacteria algae, various extracts were examined for inhibition of cellular contaminations such as HSV-2 and HIV-1. However, the success rate was 10%.

In recent years, several studies have been carried out to discover cytotoxic metabolites or new antibiotics of microalgae. For example, Abdo et al. [74] studied the antiviral activity of freshwater algae. In this study some species of algal were isolated from the Nile river of Egypt and their biological antiviral activity in drinking water was studied. These species were identified as *Cosmarium leave* (green algae) and *Chroococcus turgidus, Spirulina platensis, Oscillatoria limnetica*, and *Anabaena sphaerica* (green, Cyanobacteria, blue). They were cultivated by a photobioreactor and purified. A dry algal strain was extracted with MeOH to provide methanol extracts for freshwater.

Talyshinsky et al. [75] studied the antiviral activity (retroviruses) of red microalgal polysaccharides. Microalgal polysaccharides significantly inhibited the production of retroviruses (murine leukemia virus- MuLV) and cell transformation by murine sarcoma virus (MuSV-124) in cell culture.

Furthermore, the enhanced antiviral activity due to sulfur-comprising polysaccharides has been exposed through *Porphyridium* (a red algae) blocking the adsorption process of virions against HSV-2 and HSV-1 [76].

Also, *S. platensis* algae displayed antiviral activity against different viruses such as measles, herpes simplex, and human cytomegalo viruses through blocking their entry owing to the existence of sulfur-containing polysaccharides [77].

6.5.4 Algae as a source of antioxidant properties

Algal biomass includes water-soluble antioxidants, such as vitamins, polyphenols, and phycobiliproteins. Antioxidants assist in the prevention of cancer growth via the regression of premalignant lesions [78]. Numerous studies have found that several algal species inhibit oxidative damage through the method of active oxygen and scavenging free

134

radicals, which aid in cancer prevention. Antioxidants help to counter different diseases, such as inflammations, chronic disorders, and cardiovascular diseases. Algal biomass comprises a wide variety of bioactive substances, which are very well used commercially [79].

Osman [80] isolated the sulfate polysaccharides from aquatic algae to release radical scavenging activities. Also, the polyphenols and (pholorotannins) that originate in aquatic algae have antioxidant properties.

6.5.5 Anticancer activity of algal substances

Different bioactive substances may inhibit cancer. More than 1000 compounds have been selected from cyanobacteria and used for antineoplastic activity, due to the presence of leukemia inhibitors. A success rate of 7% was reported by Patterson et al. [81].

Aquatic microalgae comprising numerous extracts were investigated by mechanism-dependent assays using different enzymes like protein tyrosine kinase and protein kinase C [82]. *Poteriochromonas malhamensis* algae has shown the ability to prevent enzyme activity (protein tyrosine kinase), owing to the presence of the novel drug chlorosulfo lipid.

Several cyanobacteria are promising manufacturers of bioactive substances that can destroy cancer cells via a method of apoptosis, affecting cell signaling through the activation of enzymes of protein kinase C [83]. These drugs are more effective against lung carcinoma and lymphocytic leukemia [84]. Cryptophycin as a metabolite has high anticancer activity and could be extracted from Nostoc ATCC 53789 as reported by Schwartz et al. [85],

James and Thomas [86] assessed the anticancer effect of limnic algal biomass extract of karunya algae culture collection 2 and algae extract of karunya algae culture collection 23 in the MG-63 cell line. Analysis of GC-Ms showed that the substances with potential therapeutic anticancer activity effects are stigmast-4-en-3-one, epoxygedunin, and quercetin.

6.5.6 Pigments and carotenoids

Many pigments related to the incidence of light originate in algae. When extracted from micro/microalgae, they have marketable uses as cosmetic ingredients and natural food colorants. Also, carotenoids produced from microalgae have different commercial applications, for example, *Dunaliella* algae produce β-carotene as a vitamin supplement, canthaxanthan, lutein; zeaxanthin for coloring of chicken skin and medicinal use; and astaxanthin to provide a red color for some species of fish, such as salmon [64]. There are a number of algae which contain appreciable amounts of carotene. β-Carotene is known as a food colorant or color enhancer in fish flesh, egg yolks, and can improve the health and fertility of grain-fed cattle.

Del Campo et al. [65] report that halophytic *Dunaliella* sp. are used in numerous commercial facilities for the production of β − carotene in China, Australia, and the United States. It can account for 14% of its whole dry weight.

One plant is run by Cognis Nutrition and Health (about 800 hectares) and yields β-carotene products and powder of *Dunaliella* for human and animal feed uses. Bixin, zeaxanthin, lutein, canthaxanthin, and lycopene are other carotenoids produced, however in much smaller quantities (Spolaore et al. 2006). *Haematococcus pluvialis* is a green microalgae

Handbook of Algal Biofuels

that can produce carotenoids, such as astaxanthin. It can be manufactured in major amounts with 1.5%–4% of dry biomass. The synthetic equivalent (95% of the market) is utilized in aquaculture and has an estimated commercial value of US$ 200 Mio with an average price of US$ 2500 kg^{-1}.

6.6 Cosmetics

In recent decades microalgas have been used to produce different cosmetics, for example, hair care products, antiaging cream, sun protectants, and rejuvenating products.

Seaweeds can improve various successful defense methods against free radicals and reactive oxygen's producing compounds in cosmetic against the dangerous effects of UV radiation, which are still used on the consumer market [87]. Some microalgae produced secondary metabolites that can aid in avoiding skin damage, defects, inhibit inflammation, quicken healing, treat seborrhea, and retain skin moisture [88]. Any algae styles synthesized materials that absorb UV radiation and prevent action against Melanin formation with certain chemical structures [89].

At present, there is growth in the production of carotenoids and chlorophyll by *S. platensis*, *C. vulgaris*, and *Nostoc*, when cultivated in the presence of UV radiation, as reported by Sharma and Sharma [90].

Sun and Chavan [91] note that an extract from *Fucus vesiculosus* is used to minimize the existence of dark circles under the eye by hemo-oxygenase stimulation. Furthermore, the antioxidant properties of the algal extract may improve the appearance of eye bags and promote collagen development, helping to minimize wrinkles and fine lines.

6.6.1 Sunscreen

Cosmetics consisting of algae peptides are present in many skin and face cares products such as skin lotion, milky lotion, face lotion, cream, rincing, shampoo, hair caresolutions, hair restoring agents and body soap goods [92].

An extract from *Chlorogloeopses* spp. is used to avoid UVB or UVA damage to the keratinous tissues, creating free radicals through exposure to UV, and preventing the development of photoaging, cutaneous shrinkage, and wrinkles [93]. Some forms of protein that preserve moisture and have a high viscosity are present in some plants such as *Porphyra*, *Chlorella* sp., and *Spirulina* sp, [89].

Squalene extract, produced from *Thraustochytrium* and *Aurantiochytrium*, is used in the cosmetics field to improve the properties of ideal skin becauss it is nontoxic, nonsensitizing, and nonirritating, and can stimulate emollient and antistatic actions in moisturizing creams [94].

6.6.2 Whitening

Nowadays, some cosmetic products use whitening agents that inhibit tyrosinase enzyme to prevent pigmentation of skin and stimulate bleaching. Pure extract of *Nannochloropsis oculata* species contains an antityrosinase material and zeaxanthin that are used in creams [95].

6.6.3 Hair care

Some marine algae produce commercial compounds used in sun protection and antiaging, for example, extracts of *Chlorococcum* sp., *Thalassiosira* sp., *Monodus* sp., and *Chaeloceros* sp. act to stop hair damage as they can modify melanogenesis recovery and stimulate melanocyte proliferation, keratinocyte differentiation, and the development of hair follicles [96].

Chlorella sp. have the ability to soften and increase flexibility of both hair and skin [97].

6.7 Food ingredients and polymers

Macroalgae are grown and harvested for phycocolloids, such as alginates, carrageenans, agar, and other minor products. These products are some of the most desirable products that can be gelled and thickened by algae. These polymers are found in cell walls or in the cells with storage purposes.

6.7.1 Alginate

The abundance of sulfated polysaccharides in their cell walls is a characteriztic of marine algae. Alginate is the largest brown alga polysaccharide comprising 14%−40% dry weight [98]. Alginate's advantages include the potential to consume contaminants, lower cholesterol intake, and change the colonic bacterial profiles [99]. Alginates' metal chelating capabilities make them useful scavengers of toxic components in the human intestine; this chelating often can contribute to important di- or polyvalent metals' nutritional deficiencies [99]. Also, dietetic alginates offer a feeling of satiety, so weight reduction measures have been tested, although their effectiveness in this role remains uncertain [100]. Alginate has a chelating property that makes it a potential candidate for food and pharmaceutical applications [101]. Alginate is used to scale cotton yarn in the textiles industry and because of its gelling properties it is of great importance. In total alginate's market value in 2003 was around US$ 213 million.

6.7.2 Carragenans

Carrageenan is a mixture of linear polymers of sulfated galactans which constitute cell wall material of marine red algae. Mostly, the chain consists of alternating units of 3-linked-β-D-galactopyranose (G-unit) and 4-linked-α-D-galactopyranose (D-unit) or 4-linked 3,6-anhydrogalactose (DA-unit) [9]. They are removed with hot water from cell walls.

The majority of carrageenans are now manufactured from the cell walls of different Eucheuma and *Kappaphycus alvarezii* species, as well as from the *Chondrus crispus* (Irish Moos) and the *Gigartina stellata*. Most of the algae for carrageenan production are farmed, since the natural resources cannot meet the demand. *Eucheuma* sp. and *K. alvarezii* are most commonly grown in the Philippines, Indonesia, and Tanzania on a set, off-bottom line or floating rafts.

Carrageenes are used as emulsion and suspension stabilizers in the dairy, garment, and pharmaceutical industries. Algae carrageenan is a water-soluble polysaccharide that is used in different foods as an emulsifying and stabilizing agent. Carrageenans, because of their thickening effect, are used for various foods such as jelly, preserves, desserts, and meat products. Several therapeutic applications have also been investigated for carrageans, for example, as antiviral, antitumor, and anticoagulant agents. In 2003, the development of carrageenan was around US\$ 300 million [9].

6.7.3 Agar

Agars and carrageenans are 1,3-alpha-1,4-beta-galactans from the cell walls of red algae, substituted by zero (agarose), one (kappa-), two (iota-), or three (lambda-carrageenan) sulfate groups per disaccharidic monomer [63]. The agars are collected like carrageenans with hot water. Agar comes from macroalgae; the most common raw material for agar production is upplied by the *Gelidium* and *Gracilaria* genera. *Gelidium* to produce commercial agar is obtained from the wild, while in Chile, China, and Indonesia, *Gracilaria* has been grown on lines of ropes or nets in protected ocean bays, or in earth ponds [9]. Agar is commonly used in agricultural products like frozen food, desserts, fruit juices, and in industrial applications such as paper sizing, textile printing, and molecular biology. As agarose it is used for the manufacture of capsules, tablets, anticoagulants, and in a variety of biomedical sector sectors [102]. Agars are used as stabilizers and gelling agents in emulsions and suspensions. Approximately 90% of the agar produced is used for food and the other 10% for bacteriological and other biotechnological usage. Some 7630 tons of agar were produced in 2001 for a market value of approximately US\$ 137 million [9].

6.7.4 Aquaculture feed

Various algae are currently used as feed for the growth of various fish [103]. *Chlorella, Tetraselmis, Pavlova, Phaeodactylum, Nannochloropsis, Skeletonema,* and *Thalassiosiraare* the most widely used algae for aquaculture feeding.

Like *Spirulina*, *Chlorella* has the GRAS certificate from the FDA and can therefore be used, with good hygiene and manufacturing practise, as a food without risk to human health [104]. *Cryptonemia* and *Hypneacervicorn*, protein-rich microalgae, were tested in the diet of shrimps [105]. Over the last few decades, hundreds of microalgae have been studied as food, but fewer than 20 in aquaculture are currently of major importance.

6.8 Algae industrial companies

As a possible candidate for transforming biomass to beneficial compounds such as biofuels and others, researchers are looking at algae. Algal biofuels are a growing sector with many recent start-ups, including Craig Venter Synthetic Genomics, and big-time investors such as Bill Gates. However, bioefuels remain a competitive fuel with some difficulties. Today's algae industry focuses on bringing scientific and technical advancements in the processing of

138 6. The current status of various algal industries

algae products. The key reasons for these efforts are the algae's high per acre yield and its aptitude for generating various items, from low-value substances to large quantities of commodities such as fuels and feedstuffs. The number of algae research projects and companies in the United States has risen by more than one order of magnitude in recent years, with many, if not most, focused on the production of biofuels. The overall investments in this field (both private and public) are over US\$ 2 billion dollars. This pattern of growth is taking place worldwide. Many businesses believe their commercial production is just a few years away. Table 6.2 lists some companies that are involved in the marine algae industry.

6.9 Wastewater treatment by marine algae

Industrial wastewater now contains huge amounts of heavy metals, dyes, hydrocarbons, and colorants. The mobility of the heavy metals, dyes, hydrocarbons, and dyes, and their accumulation in the food chain are a major environmental problem in this area. Furthermore, these contaminants are nondegradable and thus permanent, causing environmental and health problems [111]. Many materials have been documented to function efficiently for the removal of contaminants from industrial effluents (e.g., bark, chitosan, wool, cotton, clay, fungi, alginate, dead biomass and waste fruit residues) [112]. Algae are used to eliminate dangerous organic and inorganic pollutants such as heavy metals, dyes,

TABLE 6.2 Major algae-based companies.

Company name	Origin	Production field	Major products	References
Algenol	Southwest Florida, USA	Ethyl alcohol production from seaweeds	Ethanol, gasoline, jet and diesel	pr-canada.net.
Taiwan Chlorella Manufacturing	Taiwan	Chlorella production	Addition of chlorella to foods as pasta, cookies. chlorella develops the quality of nutritional of diet.	http://www. taiwanchlorella. com
Solix Algredients	Fort Collins, USA	algae cultivation	Solasta astaxanthin extract (*Haematococcus pluvialis*), solmega DHA omega-3	[107]
TerraVia Holdings, Inc.	USA	food manufacturers; personal care producers	Solazyme; AlgaVia; AlgaWise; AlgaWise Ultra omega-9	Food Navigator-USA.com [108]
Sapphire Energy	Columbus, New Mexico, USA	Energy produces (from cultivation to production)	Green Crude	[109]
Aurora biofuels	California, USA	biofuels; pharmaceutical, nutrition	A2 Omega-3 - · A2 Feed— · A2 Fuel · A2 Protein	[110]

Handbook of Algal Biofuels

hydrocarbon, and dyes in the wastewater recycling plant in the pollution control sector. Different algae have been used for pollutant removal and have been studied as a biosorbent. The existence of polysaccharides, proteins, or lipids at the surface of their cell walls containing amino, hydroxyl, carboxyl, and sulfate groups can serve as binding sites for metals and have shown high metal binding abilities [113].

6.9.1 Removal of heavy metals and dyes

Research on bioremediation of heavy metals and complex dyes by algae has been carried out in recent years in both batch and in continuous mode tests, for example, with live cells, dead cells, pretreated dead cells, and immobilized metal and metal complex dyes. However, due to the lack of comprehensive knowledge of the various aspects of bioremediation in different modes of operation, the functional application of the process is still minimal. There has been extensive discussion of the algae involved in metal removal and metal dyes, systems, and removal methods, along with metal recovery processes and the various technoeconomic problems to consider for the process to be scaled up.

The major sources of heavy metal-contaminated wastewater (e.g., Cr, Co, Ni, Cu, Zn, Cd, Hg, Pb, and Me) are industries like electromagnetic plate, metal plate, leather tanning, chromatic ore processing, batteries, printing pigments, oils, photographic materials, automotive, wood preservation, alloys, and fabrics and textiles [114]. The effluents of these factories release metals and dyes into the water [115]. In general, the large-scale factories are fitted with effluent treatment plants before effluent disposal. However, effluent treatment is not always done in small and medium-sized enterprises for economic reasons and thus effluent disposal is of great concern. Various study groups, such as Ajjabi and Chouba [116] and Gupta and Rastogi [117], have conducted biosorption experiments with different forms of algae (green, red, and brown) for the removal of heavy metal in aqueous solutions.

A systematic biosorption method of heavy metals by *Gelidium sesquipedale* algae was attempted by Vilar et al. [118]. The biosorption method was designed and optimized gradually in order to characterize the biosorbent, to determine the balance, to calculate the kinetic biosorption/desorption and to determine the pervasiveness of a real reactor. Deng et al. [119] studied the the biosorption of Cu(II) and Pb(II) from the solution by *Cladophora fasciculis* and the effect of initial pH, high metal concentration on the biosorbtion process. The biosorption was reportedly pH-dependent, thus pH may affect both the biomass and metal chemistry protonation of the functional groups. The cell wall ligands are protonated in acidic environments and therefore metal cations cannot approach them as repulsive forces are in place. With pH increases, the metal ions will be exposed to more complex with amino, phosphorus, and carboxylic groups, since they have a negative charge which results in an attractive force. *Sargassum latifolium* algae was further researched by Al prol et al. [120] as an eco-friendly material for treating aqueous solution toxic nickel(II) and lead(II) ions. The goal of this study was to estimate the effectiveness of *S. latifolium* brown algae as alternative low-cost adsorbent for the removal of Ni^{2+} and Pb^{2+} ions from metal solution by adsorption. The highest metal removal was made with pH 7 and 6 of the initial metal at a concentration of 10 mg/L, a temperature of 25°C and 30°C, at 5 and 60 min of each contact time for Ni^{2+} and Pb^{2+} ions, and optimum 2.5 g. During the study of investigation data and model parameters, different isothermic

140   6. The current status of various algal industries

modeling, namely Langmuir, Freundlich, Temkin, and Dubin—Radushkevich, were used. The Langmuir isotherm was measured to be 0.276 and 0.171 for full removal capacities (Qmax).

The cadmium removal of aqueous solution with *U. lactuca*, a marine green algae was studied by Ghoneim et al. [121]. The aim of this study was to assess the efficiency of sea algae to extract metals from the aqueous solution. *U. lactuca*, collected from the Suez Bay intertidal region, in the northern part of the Red Sea, was utilized to minimize aqueous cadmium. The results showed an optimum pH level of 5.5; with approximately 0.1 g of *U. lactuca*. In aqueous solutions, *U. lactuca* was adequate to extract 99.2% of 10 mg/L Cd^{2+} at 30°C. The obtained results well matched with the isotherms Langmuir and Freundlich with capacity of 29.1 mg/g. Table 6.3 summarizes a significant literature review for algae metals removal, all of which showed that algae are very capable of binding/accumulating heavy metals. Hopefully algae can be used to extract heavy metal ions from contaminated wastewaters in order to achieve fruitful results.

6.9.2 Removal of nutrients

The most valuable nutrients for algae are nitrogen (N) and phosphorous (P). These can be supplied as a basic, readily available, but significant cost factor, in agricultural fertilizer

TABLE 6.3 Literature review of heavy metals removal by marine algae.

Algae	Metal/metal complex dye	Process/condition	Findings	References
Spirulina platensis	Copper	Batch process, pH: 7, (100 mg/L of initial cd concentration): (37°C): (90 min)	Percentage removal: 90.6%	Al-Homaidan et al. [122]
Utricularia aurea	Cd(II), Pb(II)	Batch, 30 min for lead and 90 min at pH 4	87% for lead and 78% for cadmium	Yoonaiwong et al. [123]
Pelvetia canaliculata	Cu(II), Zn(II)	Batch system, pH 4	Particular uptake = 2.4: 2.4 (mEq g)	Girardi et al. [124]
Ulothrix Zonata	Cu, Zn, Cd, Pb	Batch, 25°C, pH of 4—5, contact time of 60 min and adsorbent level of 1.5 g/L	Removal of Cu, Zn, Pb(II), and Cd(II), was 98.2%, 96%, 98.4% and 94.7% respectively	Malakootian et al. [125]
Anabaena sphaerica	Cd, Pb	Batch system, time: 60 min, pH: 4	Specific uptake 111.1—121.95 mg/g	Abdel-Aty et al. [126]
Galerina vittiformis	Zn, Pb, Cr, Cu, Cd	Batch process 50 mg/kg of metals	Biosorption possible	Damodaran et al. [127]
Erythrodontium barteri	Cu(II)	PH 5°C—27°C Cu concentration of 100 mg/L, 300 min	97.50% of the metal ion	Adesola Babarinde et al. [128]
Green macroalgae	Co, pb, Cd	Batch system, pH: 2—6, 10°C—40°C: 4 h	adsorption-capacity = 0.3494, 0.2942, 0.3587 mmol/g	Bulgariu and Bulgariu [129]

Handbook of Algal Biofuels

[130]. The cheaper sources of these nutrients are available with many choices. Aresta et al. [131], mentioned wastewater effluent from fish, whereby 84%−96% N and 72%−87% P were extracted from piggery wastewater by cultivating algae, thereby reducing eutrophication. Nutrient recycling is another choice in the process depending on the technology chosen for treatment. For example, nutrient recycling after anaerobic digestion or after gasification [23]. Microalgae are also used in household wastewater tertiary treatment in ripening pools, or in municipal wastewater systems on a small to medium scale. Mainly the optional designs involve relatively low dimensions and high-quality algal ponds that are fine, relying on mechanical mixing to optimize algal production and biological oxygen demand. The surface growth is supported by relatively deep ponds.

6.9.3 Algae as a monitor of water quality

Over the decades several scientists have identified the use of algal bioassays as indicative species of water quality in response to environmental disturbances and their use [132]. In 1959, Palmer released a composite of *Euglena*, *Oscillatoria*, *Chlamydomonas*, *Scenedesmus*, *Chlorella*, *Nizschia*, and *Navicula* which could be used as a water pollution indicator, whereas the presence of different organisms such as Lemanea, Stigeoclonium and certain species of Micrasterias, Staurastrum, Pinnularia Meridion and Surirella would indicate that the water sample would be considered unpolluted.

U. lactuca blooms suggest excess nitrogen eutrophication in estuarine waters [133] which is associated with nonpoint source pollution and point source contamination. Nonetheless, macroalgae production and harvesting can eliminate nutrients and thus can be used to reduce eutrophication [134]. The sulfur content of *U. lactuca* can be upto 5%. The effect is large amounts of anaerobic digestion of hydrogen sulfide (H_2S). The "rotten egg" scent of H_2S occurs on the low tide when the blooms form in the bay of long shallow estuaries with eutrophication [133].

6.10 Conclusion

Algae are a very broad and varied group. Algae can be cultivated almost anywhere, even in sewage or saltwater, they require no fertile soil or food crops, and they require less energy to process. In addition to high lipid levels, increasing densities, and the presence of useful items, algae have much faster growth rates than terrestrial cultures.

This chapter shows that algae appear to be the only source of renewable biofuels that are capable of meeting the global demand for transport fuels. The production of algae under controlled conditions gives a good yield of oil. The oil extracted from microalgae produce more oil than any other oilseeds. Microalgae can be converted to bioethanol, biodiesel, bio-oil, biomethane, and biohydrogen by biochemical and thermochemical techniques. Also, this chapter focuses specifically on applications of the algal biomass-based nonenergy field like pharmaceuticals, pigments, carotenoids, cosmetics, food ingredients, polymers, and wastewater treatment.

Some primary perspectives for the future are reported in this chapter:

142   6. The current status of various algal industries

1. Continue to extend our awareness of these algal species that they are engineered to grow a new energy industry.
2. Enhance cost-effective and reliable construction.
3. In their field of exploration, the full use of algae and the exploration of their various applications are needed.
4. Preserve the integrity of algal culture by eliminating predatory and other species pollution.
5. Regulate temperature fluctuations and reduce evaporation water losses.
6. Boost resource usage and increase bio refinery efficiency.
7. Production of precious co-products.

References

[1] M.S. Montasser, et al., A novel eco-friendly method of using red algae (Laurencia papillosa) to synthesize gold nanoprisms, J. Nanomed. Nanotechnol. 7 (2016) 383.

[2] Castellanos, C., Batch and continuous studies of Chlorella vulgaris in photo-bioreactors, Unpublished master's thesis, London, The University of western Ontario, 2013.

[3] O. Fenton, D. Uallachain, Agricultural nutrient surpluses as potential input sources to grow third generation biomass (microalgae): a review, Algal Res. 1 (2012) 49–56.

[4] M.A. Hassaan, A. Pantaleo, F. Santoro, M.R. Elkatory, G. De Mastro, A.E. Sikaily, et al., Techno-economic analysis of ZnO nanoparticles pretreatments for biogas production from barley straw, Energies 13 (19) (2020) 5001.

[5] T. Dillehay, C. Ramirez, M. Pino, M. Collins, J. Rossen, J. Pinot-Navarro, Monte Verde: seaweed, food, medicine and the peopling of South America, Sci 320 (2008) 784–789.

[6] FAO, The State of the World Fisheries and Aquaculture, FAO, Rome, 2014, p. 223.

[7] MHLW, The National Health and Nutrition Survey in Japan, 2004–2014. The Ministry of Health, Labour and Welfare. http://www.mhlw.go.jp/bunya/kenkou/kenkou_eiyou_chosa.htm, 2014.

[8] R. Amirante, G. Demastro, E. Distaso, M.A. Hassaan, A. Mormando, A.M. Pantaleo, et al., Effects of ultrasound and green synthesis ZnO nanoparticles on biogas production from Olive Pomace, Energy Procedia 148 (2018) 940–947.

[9] McHugh, D.J., A guide to the seaweed industry. Rome, FAO. FAO Fisheries,Technical Paper No. 44, 2003.

[10] M. Plaza, S. Santoyo, L. Jaime, R. García-Blairsy, M. Herrero, F. Señoráns, Screening for bioactive compounds from algae, J. Pharm. Biomedi. Anal. 51 (2) (2010) 450–455.

[11] H. Li, K. Cheng, C. Wong, K. Fan, F. Chen, Y. Jiang, Evaluation of antioxidant capacity and total phenolic content of different fractions of selected microalgae, Food Chem. 102 (3) (2007) 771–776.

[12] T. Mata, A. Martins, N. Caetano, Microalgae for biodiesel production and other applications: a review, Renew. Sustain. Energy Rev. 14 (1) (2010) 217–232.

[13] A. Gupta, J. Sainis, Isolation of C-phycocyanin from Synechococcus sp., (Anacystisnidulans BD1), J. Appl. Phycol. 22 (3) (2010) 231–233.

[14] G. Jard, H. Marfaing, H. Carrère, J. Delgenes, J. Steyer, C. Dumas, French Brittany macroalgae screening: composition and methane potential for potential alternative sources of energy and products, Bioresou. Techno. 06 (2013) 114. Available from: https://doi.org/10.1016/j.biortech.

[15] Werner, A., Clarke, D., Kraan, S., Strategic review of the feasibility of seaweed aquaculture in Ireland, NDP Marine RTDI desk study series REFERENCE: DK/01/008. ISSN: 1649 5063, 2004.

[16] W. Mackie, R.D. Preston, Cell wall and intercellular region polysaccharides, in: W.D.P. Stewart (Ed.), Algal Physiology and Biochemistry, Blackwell Scientific Publications, Oxford, UK, 1974, pp. 58–64.

[17] K. Lum, J. Kim, X. Lei, Dual potential of microalgae as a sustainable biofuel feedstock and animal feed, J. Anim. Sci. Biotech. 4 (1) (2013) 53.

[18] M.K. Lam, K.T. Lee, Microalgae biofuels: a critical review of issues, problems and the way forward, Biotech. Adv 30 (3) (2012) 673–690.

References

[19] J.K. Pittman, A. Dean, O. Osundeko, The potential of sustainable algal biofuel production using wastewater resources, Bioreso. Techno. 102 (1) (2011) 17–25.

[20] M.A. Hassaan, A. Pantaleo, L. Tedone, M.R. Elkatory, R.M. Ali, A.E. Nemr, et al., Enhancement of biogas production via green ZnO nanoparticles: experimental results of selected herbaceous crops, Chem. Eng. Commun. (2019) 1–14.

[21] Wout, R., Greenwell, H., Davies, D., Theodorou, M., Methods of ensiling algae, ensiled algae and uses of ensiled algae, WO2013045931-A1, 2013.

[22] L. Brennan, P. Owende, Biofuels from microalgae-a review of technologies for production, processing, and extractions of biofuels and co-products, Renew. Sustain. Energy Rev. 14 (2) (2010) 557–577.

[23] T. Minowa, S. Sawayama, A novel microalgal system for energy production with nitrogen cycling, Fuel. 78 (1999) 12–13.

[24] E. Kim, S. Kim, H. Choi1, S. Han, Co-production of biodiesel and bioethanol using psychrophilic microalga Chlamydomonas sp. KNM0029C isolated from Arctic sea ice, Biotechn. Biofuels 13 (20) (2020) 1–13.

[25] K.A. Jung, S.R. Lim, Y. Kim, J. Park, Potentials of macroalgae as feedstocks for biorefinery, Bioresour. Technol. 135 (2013) 182–190.

[26] Roesijadi, G., Copping, A.E., Huesemann, M.H., Foster, J., Benemann, J.R., Techno-economic feasibility analysis of offshore seaweed farming for bioenergy and biobased products; PNNL-19944,Washington, DC, United States Department of Energy, 2010.

[27] A. Alberto Bertucco, Bioethanol from microalgal biomass: a promising approach in biorefinery Carlos Eduardo de Farias Silva, Brazi. Arch. Bio. Technol 62 (2019) 1–14.

[28] W.M.M. El-Sayed, H.A.H. Ibrahim, U.M. Abdul-Raouf, M.M. El-Nagar, Evaluation of bioethanol production from Ulva lactuca by Saccharomyces cerevisiae, J. Biotechnol. Biomater. 6 (2016) 226.

[29] O. Oluwatosin, A. Joseph, B. Heike, W. Graeme, Ethanol production from brown seaweed using non-conventional yeasts, Bioethanol 2 (2016) 134–145.

[30] B.A. Yoza, E.M. Masutani, The analysis of macroalgae biomass found around Hawaii for bioethanol production, Environ. Technol. 34 (2013) 1859–1867.

[31] Y. Ueno, N. Kurano, S. Miyachi, Ethanol production by dark fermentation in the marine green alga, Chlorococcum littorale, J. Ferment. Bioeng. 86 (1998) 38–43.

[32] Argonne National laboratory (ANl), Life-cycle assessment of energy and greenhouse gas effects of soybean-derived biodiesel and renewable fuels, 2008.

[33] X. Meng, J. Yan, X. Xu, L. Zhang, Q. Nie, et al., Biodiesel production from oleaginous microorganisms, Renew. Energy 34 (2009) 1–5.

[34] X. Miao, Q. Wu, Biodiesel production from heterotrophic microalgal oil, Bioresour. Technol. 97 (2006) 841–846.

[35] Nikolaison, L., Dahl, J., Bech, K.S., Bruhn, A., Rasmussen, M.B., Energy production from macroalgae, in: Proceedings of the 20th European Biomass Conference, Milan, Italy, 18–22 June 2012, 2012.

[36] K. Gao, V. Orr, L. Rehmann, Butanol fermentation from microalgae-derived carbohy-drates after ionic liquid extraction, Bioresour. Technol. 206 (2016) 77–85.

[37] H. Cao, Z. Zhang, X. Wu, X. Miao, Direct biodiesel production from wet microalgae biomass of Chlorella pyrenoidosa through in situ Transesterification, Bio. Med. Res. Inter. (2013) 1–7.

[38] M. Wang, E. Lee, M.P. Dilbeck, M. Liebelt, Q. Zhang, Thermal pretreatment of microalgae for biomethane production: experimental studies, kinetics and energy analysis, J. Chem. Tech. Biotechnol. 92 (2016) 399–407.

[39] C. González-Fernández, B. Sialve, N. Bernet, J. Steyer, Comparison of ultrasound and thermal pretreatment of Scenedesmus biomass on methane production, Bioresour. Technol. 110 (2012) 610–616.

[40] F. Passos, J. García, I. Ferrer, Impact of low temperature pretreatment on the anaerobic digestion of microalgal biomass, Bioresour. Technol. 138 (2013) 79–86.

[41] T. Potts, J. Du, M. Paul, P. May, R. Beitle, J. Hestekin, The production of butanol from Jamaica bay macro algae, Environ. Prog. Sustain. Energy 31 (2012) 29–36.

[42] V. Wal Hetty, L. Bram, B. Sperber, R. Robert, W.B. Bakker, Production of acetone, butanol, and ethanol from biomass of the green seaweed Ulva lactuca, Bioresour. Technol. 128 (2013) 431–437.

[43] A. Demirbas, Mechanisms of liquefaction and pyrolysis reactions of biomass, Energ. Convers. Manag. 41 (2000) 633–646.

[44] W.M. Peng, Q.Y. Wu, P.G. Tu, N.M. Zhao, Pyrolytic characteristics of microalgae as renewable energy source determined by thermogravimetric analysis, Bioresou. Techno. 80 (2001) 1–7.

[45] U. Jena, K.C. Das, Comparative evaluation of thermochemical liquefaction and pyrolysis for bio-oil production from microalgae, Energy Fuels 25 (2011) 5472–5482.

[46] D.R. Vardon, B.K. Sharma, G.V. Blazina, K. Rajagopalan, T.J. Strathmann, Thermochemical conversion of raw and defatted algal biomass via hydrothermal liquefaction and slow pyrolysis, Bioresour. Technol. 109 (2012) 178–187.

[47] L.G. Alba, C. Torri, C. Samori, J. Van der Spek, D. Fabbri, Hydrothermal treatment of microalgae: evaluation of the process as conversion method in an algae biorefinery concept, Energy Fuels 26 (2012) 642–657.

[48] F. Murphy, G. Devlin, R. Deverell, K. McDonnell, Biofuel production in Ireland-An approach to 2020 targets with a focus on algal biomass, Energies. 6 (2013) 6391–6412.

[49] K. Anastasakis, A.B. Ross, Hydrothermal liquefaction of the brown macro-alga laminaria saccharina: effect of reaction conditions on product distribution and composition, Bioresour. Technol. 102 (2011) 4876–4883.

[50] N. Saifuddin, P. Priatharsini, Developments in bio-hydrogen production from algae: a review, Res. J. Appl. Sci. Eng. Technol. 12 (2016) 968–982.

[51] Q.Q. Guan, P.E. Savage, C.H. Wei, Gasification of alga nannochloropsis sp in supercritical water, J. Supercrit. Fluids. 61 (2012) 139–145.

[52] F.S. Eshaq, M. Ali, M. Mohd, Production of bioethanol from next generation feed-stock alga Spirogyra species. Inter, J. Eng. Sci. Techno. 3 (2) (2011) 1749–1755.

[53] V. Vasudevan, R. Stratton, M. Pearlson, G. Jersey, A. Beyene, J. Weissman, et al., Environmental performance of algal biofuel technology options, Environ. Sci. Techno. 46 (4) (2012) 2451–2459.

[54] R. Harun, M. Danquah, G. Forde, Microalgal biomass as a fermentation feedstock for bioethanol production, J. Chem. Techno. Biotech. 85 (2) (2010) 199–203.

[55] S.A. Scott, M.P. Davey, J.S. Dennis, I. Horst, C.J. Howe, D.J. Lea-Smith, et al., Biodiesel from algae: challenges and prospects, Curr. Opin. Biotech. 21 (3) (2010) 277–286.

[56] P.M. Schenk, S.R. Thomas-Hall, E. Stephe, U. Marx, J. Mussgnug, C. Posten, et al., Second generation biofuels: high-efficiency microalgae for biodiesel production, Bio Energ. Res. J. 1 (1) (2008) 20–43.

[57] R. Halim, B. Gladman, M.K. Danquah, P.A. Webley, Oil extraction from microalgae for biodiesel productn, Bioresour. Technol 102 (1) (2011) 178–185.

[58] M. Edwards, Green Algae Strategy - End Biowar I and Engineer Sustainable Food and Biofuels, Lu Press, Tempe, Arizona, USA, 2008.

[59] Food and Agriculture Organization of the United Nations. https://www.FAO.org/home/en/ (accessed 30.11.16).

[60] B.G. Subhadra, M. Edwards, Coproduct market analysis and water footprint of simulated commercial algal biorefineries, Appl. Energy. 88 (2011) 3515–3523.

[61] J. Milledge, Commercial application of microalgae other than as biofuels: a brief review, Revi. Environ. Sci. Biotech. 10 (2011) 31–41.

[62] A. Ben-Amotz, Industrial production of microalgal cell-mass and secondary products-major industrial species: Dunaliella, Handbook of Microalgal Culture, Blackwell Publishing Ltd, 2007.

[63] R.J. Radmer, Algal diversity and commercial algal products, Bio. Sci. 46 (1996) 263–270.

[64] I. Priyadarshani, R. Biswajit, Commercial and industrial applications of micro algae—a review, J. Algal Biomass Utln 3 (2012) 89–100.

[65] J. Del Campo, M. Garcia-Gonzales, M. Guerrero, Outdoor cultivation of microalgae for carotenoid production: current state and perspectives, Appl. Micro. Biotech. 74 (2007) 1163–1174.

[66] A. Blanco, J. Moreno, J. Del Campo, J. Rivas, J. Guerrero, Outdoor cultivation of lutein-rich cells of Muriellopsis sp in open ponds, Appl. Microbio. Biotechnol 73 (2007) 1259–1266.

[67] M.A. Rania, M.T. Hala, Antibacterial and antifungal activity of cynobacteria and green microalgae evaluation of medium components by plackett-burman design for antimicrobial activity of spirulina platensis, Glob. J. Biotechnol. Biochem. 3 (2008) 24–31.

[68] M.A. Borowitzka, Micro-algal Biotechnology, Cambridge University Press, Cambridge, 1988, p. 477.

[69] H. Katircioglu, Y. Beyatli, B. Aslim, Z. Yuksekdag, T. Atici, Screening for antimicrobial agent production in fresh water, Internet J. Microbiol. 2 (2006) 187–197.

[70] M. Wang, Y. Xu, G. Jiang, L. Li, T. Kuang, Membrane lipids and their fatty acid composition in Nostoc flagelliforme cells, Acta. Bot. Sin. 42 (12) (2000) 1263–1266.

Handbook of Algal Biofuels

References

[71] J. Costa, M. Morais, Microalgae for food production, in: C.R. Soccol, A. Pandey, C. Larroche (Eds.), Fermentation Process Engineering in the Food Industry, Taylor & Francis, Boca Raton, USA, 2013, p. 486.

[72] M. Cemile, E. Çigdem, The effects of oxidative stress and some of the popular antioxidants on reproductive system: a mini review, J. Nutr. Foo. Sci. 6 (2016) 464.

[73] N. Noaman, A. Fattah, M. Khaleafa, S. Zaky, Factors affecting antimicrobial activity of Synechococcus leopoliensis, Microbiol. Res. 159 (2004) 395—402.

[74] S. Abdo, M. Hetta, W. El-Senousy, R. Salah El Din, G. Ali, Antiviral activity of freshwater algae, J. Appl. Pharma. Sci. 2 (2012) 21—25.

[75] M. Talyshinsky, Y. Souprun, M. Huleihel, Anti-viral activity of red microalgal polysaccharides against Retroviruses, Cancer Cell Inter. 2 (2002) 8.

[76] M. Huleihel, V. Ishamu, J. Tal, S. Arad, Antiviral effect of red microalgal polysaccharides on Herpes simplex and Varicella zoster viruses, J. Appl. Phycol. 13 (2001) 127—134.

[77] S. Ayehunie, A. Belay, T.W. Baba, R.M. Ruprecht, Inhibition of HIV-1 replication by an aqueous extract of Spirulina platensis, J. Aquir. Immun. Defic. Syndr. Hum. Retrovirol. 18 (1998) 7—12.

[78] P. Bisen, Nutritional therapy as a potent alternate to chemotherapy against cancer, J. Cancer Sci. Ther. 8 (6) (2016) 168.

[79] A. Özgür, Y. Tutar, Heat shock proteins: an important target for development of next generation cancer drugs, J. Pharma. Sci. (2016).

[80] E. Osman, Effects of celecoxib or omega-3 fatty acids alone and in combination with risperidone on the behavior and brain biochemistry using amphetamine-induced model of schizophrenia in rats, J. Pharma. Rep. 1 (2016) 1—16.

[81] G. Patterson, K. Baker, C. Baldwin, C. Bolis, F. Caplan, Antiviral activity of cultured blue-green algae Cyanophyta sp, J. Phycol. 29 (1993) 125—130.

[82] W.H. Gerwick, M.A. Roberts, P.J. Proteau, J.L. Chen, Screening cultured marine microalgae for anticancer type activity, J. Appl. Phycol. 6 (1994) 143—149.

[83] N. Boopathy, K. Kathiresan, Anticancer drugs from marine flora: an overview, J. Oncol. 4 (2010) 1—18.

[84] Furusawa, E., Moore, R., Mynderse, J., Norton, T., Patterson, G.M.L., New purified culture of Scytonema pseudohofmanni ATCC 53141 is used to produce scytophycins A, B, C, D and E, which are potent cytotoxins and antineoplastic agents, USA Patent Number 5281533, 1994.

[85] R.E. Schwartz, C.F. Hirsch, D.F. Sesin, J.E. Flor, M. Chartrain, Pharmeaceuticals from cultured algae, J. Ind. Microbiol. 5 (1990) 13—123.

[86] J. James, J. Thomas, Anticancer activity of microalgae extract on human cancer cell line (MG-63), Asian J. Pharm. Clin. Res. 12 (1) (2019) 139—142.

[87] L. Gouveia, A. Batista, I. Sousa, A. Raymundo, N.M. Bandarra, Microalgae in novel food products, in: K.N. Papadopoulos (Ed.), Food Chem. Res. Dev, Nova Scien, M. B. Algal Resea., 25, 2017, pp. 483—487.

[88] S. Kim, Y. Ravichandran, S. Khan, Y. Kim, Prospective of the cosmeceuticals derived from marine organisms, Biotech. Biopro. Eng. 13 (2008) 511—523.

[89] Hagino, H., Saito, S., Use of algal proteins in cosmetics, 03029218.9, (2010), 2015.

[90] R. Sharma, V.K. Sharma, Effect of ultraviolet-B radiation on growth and pigments of Chlorella vulgaris, J. Indian. Bot. Soc. 94 (2015) 81—88.

[91] Sun, Y., Chavan, M., Cosmetic compositions comprising marine plants, 14/077 934, 2014.

[92] Z. Draelos, New treatments for restoring impaired epidermal barrier permeability: skin barrier repair creams, Clin. Dermatol. 30 (2012) 345—348.

[93] O'connor, C., Skill, S.C., Llewellyn, C.A., Topical composition, PCT/GB2011/051138, 2011.

[94] Pora, B., Qian, Y., Caulier, B., Comini, S., Looten, P., Segueilha, L., Method for the preparation and extraction of squalene from microalgae, 14 (118), 641, 2014.

[95] S. Babitha, S.E. Kim, Effect of marine cosmeceuticals on the pigmentation of skin, Mar. Cosmeceuticals Trends Prospect, CRC Press, New York, 2012, pp. 63—65.

[96] Zanella, L., Pertile, P., Massironi, M., Massironi, M., Caviola, E., Extracts of microalgae and their application, 13/883, 193, 2014.

[97] Brooks, G. Franklin, S., 2013. *Cosmetic compositions comprising microalgal components.* U.S. Patents, US20110250178A1.

[98] J. Ramberg, E. Nelson, R. Sinnott, Immunomodulatory dietary polysaccharides: a systematic review of the literature, Nutria. J. 9 (2010) 54.

[99] I.A. Brownlee, A. Allen, J. Pearson, P. Dettmar, M. Havler, M.R. Atherton, et al., Alginate as a source of dietary fiber, Crit. Rev. Food. Sci. Nutr. 45 (2005) 497—510.

[100] N. Yavorska, Sodium alginate a potential tool for weight management: effect on subjective appetite, food intake, and glycemic and insulin regulation, J. Undergrad. Life Sci. 6 (2012) 66—69.

[101] L.M. Colla, C.O. Reinehr, C. Reichert, J.A.V. Costa, Production of biomass and nutraceutical compounds by Spirulina platensis under different temperature and nitrogen regimes, Bioresour. Technol. 98 (2007) 1489—1493.

[102] C. Ververis, K. Georghiou, Cellulose, hemicelluloses, lignin and ash content of some organic materials and their suitability for use as paper pulp supplements, Bioresour. Technol. 98 (2007). 296—230.

[103] Y.C. Chen, Immobilized isochrysis galbana (Haptophyta) for long-term storage and applications for feed and water quality control in clam (Meretrix lusoria) cultures, J. Appl. Phycol. 15 (2003) 439—444.

[104] J.A. Costa, E. Radmann, V. Cerqueira, G. Santos, M. Calheiros, Fatty acids from the microalgae Chlorella vulgaris and Chlorella minutissima grown under different conditions, Alimentos Nutr. Araraquara 17 (4) (2006) 429—436.

[105] R.L. Da Silva, J.M. Barbosa, Seaweed meal as a protein source for the white shrimp Lipopenaeus vannamei, J. Appl. Phycol. 21 (2008) 193—197.

[106] Pr-canada.net. Algenol partners with lee county as commission votes to approve incentive funding for Florida-based.

[107] http://www.solixalgredients.com.

[108] Food Navigator-USA.com. Solazyme rebrands as Terra Via, raises $28 m as part of plan to focus on food, nutrition, personal care: 'We're redefining the future of food. Retrieved 2017.

[109] http://www.sapphireenergy.com/locations/green-crude-farm.html.

[110] http://en.openei.org/wiki/Aurora-BioFuels-Inc.

[111] T. Akar, S. Tunali, Biosorption performance of Botrytis cinerea fungal by-products for removal of Cd (II) and Cu (II) ions from aqueous solutions, Miner. Eng. 18 (2005) 1099—1109.

[112] Q. Yu, J.T. Matheickal, P. Yin, P. Kaewsarn, Heavy metal uptake capacities of common marine macro algal biomass, Water Res. 33 (1999) 1534—1537.

[113] S.K. Das, A.K. Guha, Biosorption of chromium by *Termitomyces clypeatus*, Colloids Surf. B 60 (2007) 46—54.

[114] L.N. Du, Y.Y. Yang, G. Li, S. Wang, X.M. Jia, Y.H. Zhao, Optimization of heavy metal-containing dye acid black 172 decolorization by Pseudomonas sp. DY1 using statistical designs, Int. Biodeterior. Biodegrade. 64 (7) (2010) 566—573.

[115] S. Ilyas, J.C. Lee, B.S. Kim, Bioremoval of heavy metals from recycling industry electronic waste by a consortium of moderate thermophiles: process development and optimization, J. Clean. Prod. 70 (2014) 194—202.

[116] L. Ajjabi, L. Chouba, Biosorption of Cu^{2+} and Zn^{2+} from aqueous solutions by dried marine green macroalga Chaetomorpha linum, J. Environ. Manage. 90 (2009) 3485—3489.

[117] V.K. Gupta, A. Rastogi, Biosorption of lead from aqueous solutions by green algae Spirogyra species: kinetics and equilibrium studies, J. Hazard. Mater. 152 (2008) 407—414.

[118] V.J.P. Vilar, C.M.S. Botelho, R.A.R. Boaventura, Modeling equilibrium and kinetics of metal uptake by algal biomass in continuous stirred and packed bed adsorbers, Adsorption. 13 (2007) 587—601.

[119] L. Deng, Y. Su, H. Su, X. Wang, X. Zhu, Biosorption of copper (II) and lead (II) from aqueous solutions by nonliving green algae Cladophora fascicularis: equilibrium, kinetics and environmental effects, Adsorption. 2 (2006) 267—277.

[120] A. Al prol, M. El-Metwally, A. Adel, Sargassum latifolium as eco-friendly materials for treatment of toxic nickel (II) and lead (II) ions from aqueous solution, Egy. J. Aqua. Biol. Fish. 23 (5) (2019) 285—299.

[121] M.M. Ghoneim, H.S. El-Desoky, K.M. El-Moselhy, A. Amer, E.H. Abou El-Naga, L. Mohamedein, et al., Removal of cadmium from aqueous solution using marine green algae Ulva lactuca, Egy. J. Aqua. Res. 40 (2014) 235—242.

[122] A.A. Al-Homaidan, H.J. Al-Houri, A.A. Al-Hazzani, G. Elgaaly, N. Moubayed, Biosorption of copper ions from aqueous solutions by Spirulina platensis biomass, Arab. J. Chem. 7 (1) (2014) 57—62.

[123] W. Yoonaiwong, P. Kaewsarn, p Reanprayoon, Biosorption of lead and cadmium ions by non-living aquatic macrophyte, Utricularia aurea, Sustain. Environ. Res. 21 (6) (2011) 369—374.

[124] F. Girardi, et al., Marine macroalgae Pelvetia canaliculata (Linnaeus) as natural cation exchanger for metal ions separation: a case study on copper and zinc ions removal, Chem. Eng. J. 247 (2014) 320−329.

[125] M. Malakootian, M. Ahmadian, S. Gh, Equilibrium and kinetic modeling of heavy metals biosorption from three different real industrial wastewaters onto Ulothrix Zonata algae, Austr. J. Bas. App. Sci. 5 (12) (2011) 1030−1037.

[126] A.M. Abdel-Aty, N.S. Ammar, H.H. Abdel Ghafar, R.K. Ali, Biosorption of cadmium and lead from aqueous solution by fresh water alga Anabaena sphaerica biomass, J. Adv. Res. 4 (4) (2013) 367−374.

[127] D. Damodaran, K. Vidya Shetty, B. Raj Mohan, Effect of chelaters on bioaccumulation of Cd(II), Cu(II), Cr (VI), Pb(II) and Zn(II) in Galerina vittiformis from soil, Int. Biodeterior. Biodegrad. 85 (2013) 182−188.

[128] N. Adesola Babarinde, O. Oyesiku, O. Dairo, Isotherm and thermodynamic studies of the biosorption of copper (II) ions by Erythrodontium barteri, Inter. J. Phy. Sci. 2 (11) (2007) 300−304.

[129] D. Bulgariu, L. Bulgariu, Equilibrium and kinetics studies of heavy metal ions biosorption on green algae waste biomass, Bioresour. Technol. 103 (1) (2012) 489−493.

[130] Y. Chisti, Response to Reijnders: do biofuels from microalgae beat biofuels from terrestrial plants, Trends Biotechnol. 26 (7) (2008) 351−352.

[131] M. Aresta, A. Dibenedetto, G. Barberio, Utilization of macro-algae for enhanced CO_2 fixation and biofuels production: development of a computing software for an LCA study, Fuel Process. Technol. 86 (14−15) (2005) 1679−1693.

[132] Mohamed, N.A., Application of algal ponds for wastewater treatment and algal production, M.Sc. Thesis, Fac. of Sci. (Cairo Univ.) Bani-Sweef Branch, 1994.

[133] E. Allen, J. Browne, J.D. Murphy, Evaluation of biomethane potential from co-digestion of nitrogenous substrates, Environ. Technol. (2013). Available from: https://doi.org/10.1080/09593330.2013.806564.

[134] A. Hughes, K. Black, I. Campbell, J. Heymans, K. Orr, M. Stanley, et al., Comments on prospects for the use of macro-algae for fuel in Ireland and UK: an overview of marine management issues, Mar. Policy 38 (2013) 554−556.

CHAPTER 7

Algal biomass as a promising tool for CO_2 sequestration and wastewater bioremediation: an integration of green technology for different aspects

Reda M. Moghazy, Sayeda M. Abdo and Rehab H. Mahmoud

Water Pollution Research Department, National Research Centre, Cairo, Egypt

7.1 Introduction

The increasing population and human behavior has increased the shortage of drinking water which has led to an extreme release of wastewater into aquatic environments. These issues have contributed to the pollution of freshwater bodies and eutrophication [1]. Algal blooms are produced mainly due to eutrophication, which also causes the invasion of water resources by aquatic plants, oxygen consumption, the loss of many key species, and the deterioration of the aquatic ecosystems [2]. In addition, the increasing atmospheric emissions of carbon dioxide have negative environmental impacts, such as increasing the greenhouse effect and acidification of the ocean. Therefore community economies have to decrease these emissions. Moreover, since this greenhouse gas phenomenon presents a long time ago [3], new techniques to reduce CO_2 and flue gas emissions in the air are required [4]. CO_2 sequestering and storage (CCS) techniques are the most important part of these procedures for the emissions of greenhouse gases [5,6]. This is a promising solution to transform carbon dioxide into organic matter through biological processes. Photosynthesis has long been perceived as an approach to capture anthropogenic CO_2 and algae have been well-known as quickly growing photosynthetic organisms whose CO_2 capturing rates can be a lot higher than those of higher plants. Also they can proficiently take up and utilize dissolved inorganic forms (CO_2 and HCO_3) of carbon in an aquatic environment [7]. Algae have the ability to sequester CO_2 and its subsequent utilization rather than being strictly utilized for CCS. The fixation of carbon by microalgae is embedded and transformed into biomass, particularly lipids and carbohydrates, which in turn could be useful

Handbook of Algal Biofuels
DOI: https://doi.org/10.1016/B978-0-12-823764-9.00015-7

© 2022 Elsevier Inc. All rights reserved.

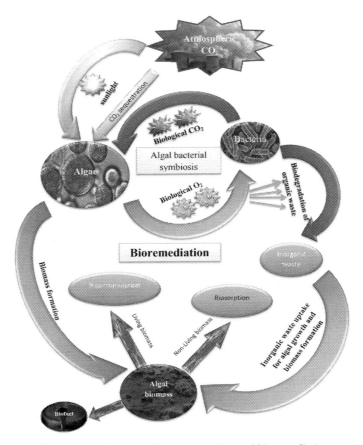

FIGURE 7.1 Biomass utilization in relation to CO_2 sequestration and bioremediation.

for different applications, such as bioenergy, bioremediation tools for wastewater treatments [8–10], foods, or fine chemicals. Cultivation of photoautotrophic microorganisms is considered to be one possible strategy for biological wastewater treatment because they capture CO_2 during growth. The nutrients remaining in the wastewater, rather than being waste, are utilized as an algal nutrient that accumulates effectively inside the cell [11]. For this purpose, this chapter illustrates the role of algal biomass in CO_2 sequestration, where the algal cell contains approximately 45%–50% carbon content, and the use of algae in different bioremediation approaches and their utilization in different aspects such as biodiesel production Fig. 7.1.

7.2 Strategies of carbon dioxide sequestration

CO_2 sequestration is defined as the process of reducing the amount of CO_2 that is discharged into the atmosphere. Different approaches for CO_2 storage could be taken into consideration, and can be categorized into:

7.2.1 Nonbiological methods

These include ocean storage, mineralization, and geological storage. Carbon dioxide storage approaches should have the following specifications: (1) minimal environmental impact; (2) safe storage; (3) indefinite storage; and (4) verifiable storage [12].

7.2.2 Biological sequestration

Natural sequestration is achieved biologically and performed by living organisms including many microorganisms and plants, leading to carbon fixation and storage by different biological processes. Almost all sequestration processes take place biologically, although nowadays a few artificial strategies are also an integral part of this. This mainly includes the sequestration taking place in the ocean through terrestrial sequestration and fertilization [13].

7.2.3 Phytosequestration

Phytosequestration refers to the atmospheric capture of CO_2 by photosynthetic organisms. There are many benefits to phytosequestration in the carbon sequestration strategy. Jansson et al. reported that 99.9% of the C present in the world's biota is contributed by microbial biomass and vegetation [14].

7.3 Carbon dioxide biosequestration using microalgae

CO_2 is the main carbon source that can be utilized by microalgae. Gases discharged from waste during combustion are vital CO_2 sources that can be applied on large-scale production systems of microalgae, because they usually contain a volume fraction of CO_2 from 5% to 15%. The CO_2 transformed into the algal biomass form can be used as fertilizer, feed, food, or fuel. Transformation processes release CO_2 back into the environment through respiration, digestion, and biomass combustion. Thus it would assume logically that microalgal biomass is not also a simple CO_2 reservoir but also that cells are a biological postcombustion tool for capturing gaseous CO_2 that is discharged from power plants. CO_2 resulting from microalgae biosequestration can initiate valued biomass of algal production and simultaneously lower the greenhouse gas emissions. However, little advance has been made in decreasing its cost. The most important benefit of the sequestration of CO_2 by the microalgal cells is its growth, which is rapid and widespread. Also, the production of biofuel from microalgal biomass is highly biodegradable and free of toxic materials or sulfur [15]. Different species of microalgae are found to be tolerant of the emitted CO_2 concentrations of flue gas. A few microalgal groups are tolerant of high CO_2 emission, such as *Arthrospira* (formerly *Spirulina*) spp., *Chlorella* spp., *Botryococcus braunii*, *Scenedesmus dimorphus*, and *Nannochloropsis oculate* [16]. *Scenedesmus* spp. under high concentrations of CO_2 (10%–80%) exhibits a higher tolerance response than *Chlorella* spp.; these two species of algal are adapted to grow in lower concentrations of carbon dioxide (10%–30%). However, *Scenedesmus* spp. growth was suppressed under the concentration of CO_2 reaching 100% [17]. A mutant algal species of *Chlorella* spp. (strain KR-1) showed a tolerant ability to grow under levels of CO_2 as high as 70% [18]. Flue gases emitted from a hot stove, a power plant in a steel plant, and a coke oven were used for *Chlorella* sp. (strain MFT-15) cultivation. This strain could utilize CO_2, SO_2, and NOX concentrations in the different flue streams at a good rate. Another study reported the excellent tolerance of *S. dimorphus* against high gas emissions levels. this alga could tolerate CO_2 concentrations of 10–20% and NO concentrations of 100–500 ppm, while the maximum SO_2 concentration tolerated was 100 ppm [19]. Different fixation rates of CO_2 are achieved by various species of microalgae or even by the same mutant strains. Five different microalgal species showed significant differences in the reduction rate of CO_2 for mass cultivation as follows: *B. braunii* SAG-30.81, *Spirulina platensis*, *Dunaliella tertiolecta* SAD-13.86LEB-52, *Chlorella vulgaris* LEB-104, and *Chlorococcum littorale* gave biomass production rates of 496.98, 318.61, 272.4, 251.64, and 1000 mg/L/day, respectively, [20]. The productivity of *C. vulgaris* reached up to 260 mg/L/h during the fixation of CO_2 discharged from flue gas [21], while *Chlorella sorokiniana* give a yield of 330 mg/L after 96 h of the flue gas exposure containing CO_2 the biomass [22]. *Scenedesmus obliquus* (mutant strain WUST4) can be grown to gain a high concentration of biomass (0.922 g/L) when exposed to flue gas containing under 10% CO_2 concentrations, while the nonmutant *strain* under the higher concentration of CO_2 (20%) gave a lower biomass concentration (0.653 g/L) [23].

Approximately 45%–50% of the content of algae cells is carbon, thus algae require a continual intake of carbon. With a low supplementary carbon source, the algal growth will quickly become limited. CO_2 is the most common carbon source added to autotrophic cultures, usually by bubbling through sparging stones or perforated pipes, or by floating gas exchangers or hollow fiber membranes.

CO_2 is consumed and O_2 is released into the liquid through photosynthesis. Photooxidative damage to chlorophyll can be caused by high O_2 levels which will inhibit photosynthesis and reduce productivity. In open pond systems, O_2 mass transfer is easy due to there being a large surface area exposed to O_2, but in closed pond systems, such as photobioreactors, there are additional instruments known as gas exchange chambers that aid gas exchange to reduce the dissolved O_2 levels.

CO_2 is also often necessary to maintain a steady pH in the algal culture system. CO_2 will be converted into carbonic acid when dissolved into the liquid phase so that pH will decrease. As algae consume the carbonic acid the pH rises, so the pH can be controlled very effectively by controlling CO_2 levels [24].

7.4 Bioremediation: an ecofriendly approach for wastewater treatment

The bioremediation process is the use of the metabolic activity of microorganisms to treat the contaminated sources, that is, ground, surface, or wastewater, soils, sediments, and air in the environment. The bioremediation process includes mineralization and detoxification of waste to convert it into inorganic compounds such as CO_2, methane, and water. The process of bioremediation includes both phytoremediation and microbial remediation. The use of microalgae, macroalgae, and cyanobacteria for the reduction of CO_2 from the air and treating the nutrients and xenobiotics from wastewater is called phycoremediation. Phycoremediation is considered one of the best sustainable mechanism to remediate the hazardous compounds located in the environment. The employment of algae, bacteria, and fungi for remediation purposes is also called microbial remediation. The multiple steps involved are dependent on different enzymatic reactions [25].

7.5 Algae-based wastewater treatment plants

Microalgae are photosynthetic unicellular autotrophs. With the aid of the light, algae transform H_2O and CO_2 into carbohydrates and oxygen and supply energy for algal growth. Moreover, phosphorus and nitrogen are utilized as organic nutrients for algae cell proliferation. The capability of oxygen production and the take-up of nutrients from water highlights microalgal biomass cultivation as a cost-effective technology for wastewater treatment [26].

The wastewater treatment process using microalgae is carried out in closed or open cultivation systems. Wastewater treatment plants designed in an open system, such as artificial ponds, natural lagoons, and reservoirs, have a simple construction and maintenance, and thus are the favored option for microalgae-based wastewater treatment [27]. Nonetheless, the treatment operation is difficult to control. Closed systems, which mostly are photobioreactors, are more controllable throughout the operation but are costly to set up. The photosynthetic active radiation (PAR) is considered the main factor for algae productivity, with the optimal range of photon flux being between 30 and 400 $\mu mol/m^2 s$. Providing extra CO_2 can increase the algae biomass productivity; it hinders the development of an alkaline pH level. Relevant pH values for high algal productivity are recorded to be between 7 and 9 [28].

7.5.1 Algae-based municipal wastewater treatment process

Municipal wastewater treatment is usually designed to dispose of hazardous organic wastes by subjecting them to biodegradation, which is achieved by different microorganisms such as bacteria. During the biodegradation process, the organic waste is converted into smaller molecules (CO_2, PO_4, NH_3, etc.), and requires a continuous oxygen supply. The oxygen supply process is high-cost and requires a considerable workforce and expertise. The cultivation of microalgae in wastewater treatment ponds and tanks can overcome this problem. During photosynthesis, the algae release O_2 which is considered as a continued source of oxygen for biodegradation. Nutrient removal is the main role of algae-based municipal wastewater treatment systems (removal of phosphorus and nitrogen). In addition, algal play an important role in pathogen removal throughout the photosynthesis process, increasing the pH because of the simultaneous ejection of CO_2 and H+ ions [29], and bicarbonate uptake when the algae consume carbon. Rose et al. [30] reported the killing of 100% of most pathogenic bacteria, including *Escherichia coli*, and viruses at a pH of 9.2 for 24 h. Parhad and Rao [31] also found that with a pH higher than 9.2, *E. coli* could not live in wastewater. The added value is the produced biomass that can be used as biofuel feedstock [9,32].

154

7. Algal biomass

Two different systems have been recently used for municipal wastewater treatment. These systems can be integrated into secondary treatment stages: waste stabilization pond systems (WSPs) and high rate algal ponds (HRAP).

7.5.1.1 Waste stabilization pond systems

Waste stabilization pond systems allow the symbiotic growth of heterotrophic bacteria and microalgae. The algae provide oxygen as a photosynthetic by-product; the provided oxygen is then consumed by the aerobic bacteria that are utilized for biooxidation of the organic waste into inorganic nutrients utilized by microalgae. CO_2 is the end product of this process, which in turn is fixed into carbon inside the algal cell via photosynthesis [33].

7.5.1.2 High rate algal pond systems

High rate algal pond systems is an integration of the intensified oxidation ponds and an algal reactor. HRAP is paddlewheel-mixed, shallow, open raceway pond for providing highly efficient wastewater treatment. This differs from conventional oxidation ponds in that HRAP is a result of a deep algal photosynthesis machine that provides oxygen saturation to initialize the aerobic treatment conditions and the consumption of wastewater nutrients into algal biomass [33].

7.5.2 Effluent from industrial wastewater treatment plants and microalgae

Industrial wastewater treatment depends on the immobilization of biomass and is an efficient procedure of retaining biomass during the wastewater treatment process. Microalgal immobilization and embedding into polymers such as chitosan, cellulose, alginate, or carrageenan, has been reported by various authors [34,35].

After the removal of industrial material ions that are loaded into microbial biomass, it is possible to recover valuable elements again, such as gold and silver.

7.6 Algal bacterial symbiosis system for wastewater treatment: role and effect of carbon dioxide

In recent years, the algal bacterial symbiosis (ABS) system has been gaining relevance in biomass production coupled with pollutant removal [36] Such microalgae—bacteria associations can be used as potent biological systems with the capability to work effectively under fluctuating growth conditions and nutrient stress due to their various metabolic activities and their ability to adapt to different environments. Naturally, most of the bacteria and microalgae build their microecosystems where they affect the growth of each other in many ways [37]. Furthermore, certain bacteria have been reported to ease the accumulation of algal cells, thereby facilitating the harvesting process of the microalgal biomass [38]. Thus such microalgae—bacteria associations can be used for efficient wastewater bioremediation along with symbiotically grown biomass [39].

7.6.1 Microalgae—bacteria symbiosis mechanism

In this symbiotic system, microalgae cells can fix CO_2 and nutrients for the photosynthesis process, producing huge quantities of O_2 [40]. Furthermore, aerobic bacteria can adopt O_2 which helps in the degradation of the organic carbons into inorganic carbon for microalgae growth [41]. Inorganic nutrients present in wastewater, especially phosphorus and nitrogen, can be absorbed to simulate the algae growth and maintain high cell levels [42]. Hence, these complex relationships between bacteria and microalgae in the ABS reaction system can eliminate the pollutants in wastewater while continuing their growth [43].

7.6.2 Impact of Microalgae—bacteria system on the production of algal biomass and associated compounds

A great effort has been exerted to collect the axenic cultures of algal monocultures for biomass production improvement. Recently, the interactions between microalgae and microorganisms have been shown to have extensive possible applications in aquaculture, so that it has been recognized that the enhancement of the productivity of algal biomass enriched with valuable compounds, such as carbohydrates and lipids, can be exploited in many applications. In this respect, algae-bacterial ineteraction plays a great role in algal growth by enhancement of their chemotaxis, motility, quorum sensing systems, type IV secretion systems, and growth promoters synthesis [44]. It was reported that the bacterium *Halomonas* sp. simulates the algal growth of *Porphyrdium purpureum* and *Amphidinium operculatum*, for vitamin B12 production (Cyanocobalamin) [45]. *Halomonas* sp. initiated the cyanocobalamin production when the culture was fed with Fucidin, a commercial algal extract. This finding suggests a vitamin exchange for carbon fixation in the bacteria [45]. Also, it was shown that both the vitamin B1 and B12 required for the algal growth of the marine dinoflagellate alga *Lingulodinium polyedrum* can be achieved by the presented bacterial consortium in the same culture [46]. Moreover, it is found that the heterotrophic bacterium *Mesorhizobium* sp. promotes the biomass productivity of the vitamin B12-dependent alga, *Lobomonas rostrata* [47]. *Chlamydomonas reinhardtii* encodes for a B12-independent methionine synthase (MetE) so it does not need vitamin B12 for growth. The presence of *C. reinhardtii* with *Mesorhizobium* sp. leads to a reduction in MetE expression, indicating that vitamin B12 can be transported from the B12-independent algae by the bacterium [47]. In 1995 it was reported that an oligotrophic and halophilic bacterium, *Halomonas* sp., enhanced the iron available to *Dunaliella bardawil* through the ability of this bacterium to give siderophores which increase the solubility of Fe, thus enhancing its availability to *Dunaliella* and promoting its survival under insufficient Fe conditions [48].

7.6.3 Microalgal—bacteria relation and production of biofuel

Biomass and biofuels are key assets for a sustainable, secure, and efficient energy systems [49]. Microalgal cells utilized the sunlight energy to transform CO_2 into lipids, carbohydrates, and proteins, thus the production of algal biomass can be assisted by CO_2 biofixation, this is demonstrated where algal biomass of weight of 1 kg of dry can utilize about 1.83 kg of CO_2 [50]. Microalgae have great ability to synthesize a wide range of biofuels because the algal cells contain a high content of energy storage molecules, such as lipids and carbohydrates [51]. In addition, algal protein plays vital role in human nutrition, animal feed, and aquaculture [52]. The stored carbohydrates in microalgae help in hydrogen production, while photosynthate redirected to lipids may be converted into biodiesel. The remaining algal biomass could be further fermented into biogas or ethanol by yeasts or by anaerobic bacteria [53].

7.6.3.1 Biodiesel

After the trans-esterification process triacylglycerol (TAG) and other lipids produced by microalgae are converted into fatty acid methyl esters, which are the precursors to biodiesel [54]. Different algal species are screened for high lipid content and have been applied for biodiesel production [55]. The lipid profile of *C. vulgaris* in many studies shows high lipid concentrations, which makes it one of the most common strains applied in the production of biodiesel. Concerning the ABS system, various species of *Pseudomonas* were found in association with *C. vulgaris* during its growth in the open pond [56]. Moreover, coculturing of the marine microalga *Tetraselmis striata* with both of the bacterial strains *Pelagibaca bermudensis* and *Stappia* sp. exhibited high lipids content and fast algal growth, so the microalga *T. striata* has become a good candidate for biodiesel production [57]. Also, a successful symbiotic relationship between the green alga *Auxenochlorella protothecoides* and *E. coli* was studied. This association led to a two- to sixfold increase in algal biomass productivity and the neutral lipid concentration was found to be doubled in comparison with the axenic growth; this increase was owing to the supply by *E. coli* of degradation products and thiamine derivatives to *A. protothecoides* [58].

7.6.3.2 Biohydrogen

Microalgae can produce biohydrogen through the biophotolysis process [59], in which protons and electrons are extracted from water and transferred via ferredoxin/NADPH to hydrogenase. Electrons are accepted by hydrogenase enzymes from reduced ferredoxin and utilized for the reduction of protons into molecular biohydrogen [60]. The hydrogenase enzymes are more sensitive to oxygen so severe anaerobic conditions are required for the production of the biohydrogen process by microalgal cell, which is where the role of the symbiosis system comes. For instance, the bacterial strains *Rhodococcus* sp. and *Leifsonia* sp. demonstrated a significant enhancement of the production of biohydrogen in the microalga *Chlamydomonas* due to the oxygen consumption during the bacterial respiration, which is necessary for the initiation of a Fe-dependent hydrogenase enzyme in *Chlamydomonas* [61]. Also, the high biohydrogen productivity was observed through the artificial symbiosis between the hydrogenase-deficient *E. coli* and *Chlamydomonas* [61]. The sunlight energy can also be transformed into bioelectricity by symbiotic collaboration between microalgae that is characterized by photosynthetic activity and heterotrophic bacteria—this form is called microalgal fuel cells—without the external addition of exogenous nutrients or organics [62].

7.6.3.3 Biogas and bioethanol

Algal biomass residue can be used in bioethanol or biogas production. Biogas production can be carried out through an anaerobic digestion process in which microalgal biomass is incubated with anaerobic microbes that transform the carbon of algal cells and other organic compounds into CO_2, organic acids, and methane gas (CH_4) [63]. Many factors affect the productivity of biogas, one of them is the microalgae used, as the relative ratio of cellular lipids, proteins, and carbohydrates may affect the efficiency of anaerobic bacteria [64]. Another is resistance factor of the algal cell walls against protease, which may hinder their degradation. Fortunately, there are about nine bacterial strains able to damage the rigid cell walls of the algae *B. braunii* and *Nannochloropsis gaditana* [65]. Microalgal biomass can produce bioethanol through the fermentation process. The achievement of bioethanol production by the microalga *C. reinhardtii* depends to great extent on a complex symbiotic relationship, as the process starts with amylase enzymatic hydrolysis from the bacterium strain *Bacillus licheniformis* and then comes fermentation by brewer's yeast *Saccharomyces cerevisiae* [66].

7.7 Biosorption and bioaccumulation

7.7.1 Biosorption

Biosorption is a concept that refers to loading the pollutant ion (sorbent) into the biological material (biosorbent). The biosorbent laden with pollutant ions is then recovered and reused again, and the pollutant ion is recovered by eluent (desorbing agent). It is essential to select the desorbing agent very carefully, so that all adsorbent ions are recovered from the biosorbent by a low solution volume without destroying the eluent; this would sustain its biosorption ability and allow reuse several times in numerous biosorption cycles. The highly loaded biosorbent can be then remediated by conventional treatment techniques which are capable of metal recovery, such as electrolysis. The prefix "bio" in biosorption indicates that the sorbent originates from a biological source [67].

The biosorption process is characterized by widely available and easily used biosorbents, low operation costs, low sewage sludge quantity disposal, and low COD concentration of wastewater. There are similarities in the operation of the biosorption process to the conventional adsorption or ion-exchange processes, but it differs from them in that the sorbent originates from biological material. Biosorbents allow selective pollutant removal and can be regenerated and reused, particularly in low concentration effluents. The limitation faced by the biosorption is a shorter lifetime of biosorbents as compared with conventional sorbents [68]. This limitation is overcome by the immobilization or embedding of the algal biomass in polymeric substances, such as chitosan, cellulose, carrageenan, or alginate. Biosorbent materials are renewable sorbents, such as algae, plant, bacteria, fungi, or animal origin [68]. Biosorbents can be classified according to their cost-effectiveness as high- and low-cost sorbents. Low-cost sorbents include the biomass located naturally which can be directly collected from nature (e.g., seaweeds) and waste or by-products from industry (e.g., from fermentation processes by yeasts) [69]. High-cost sorbents include biosorbent materials which are specially prepared for biosorption purposes. These materials should have a high characterization of biosorption and should be easily recovered and reused. The new biomaterials used for biosorption should be characterized to determine their biosorption ability. Different conditions affect the removal of the capacity of biosorption (expressed in either mg/g or meq/g), such as temperature, pH, the sorbate concentration, and sorbent dose. The biosorption mechanism was explained by different processes: precipitation and surface complexation, ion exchange, or physical adsorption. Recently, the ion-exchange process was confirmed as the only dominating role. Since it is found that the mechanism was ion-exchange, protons were addressed by the completion between pH and metal cations on the binding sites, so that pH is the most strong effective condition in the operation that influences the biosorption process [70]. The other factors essential in the biosorption process include the concentration and type of the sorbate and the biosorbent, the biosorption mode (in case of suspension or immobilized state of biomass), and the existence of other anions and cations. The processes of deprotonation or protonation of the binding sites of sorbent ions are determined by pH and thus influence the availability of the binding site to the sorbate. The release of pollutant ions from the binding site is achieved by lowering of pH. This is one of the roles of pH that is used for the pollutant cations recovery and desorption of the biosorbent. There is a similarity between organic and inorganic cations in the pH effect where they both are bound to functional groups that bear negative charges exposed on the cellular surfaces. There is a similarity in the chemical composition of a given group in the cell wall of organisms so that biosorption properties are similar. Seaweeds are considered as excellent biosorbents, because most seaweeds, such as brown (*Sargassum* sp.) and green (*Ulva* sp.), contain carboxyl and hydroxyl groups in their alginate in cell walls [71]. The bad properties of the sorption process are found in red algae that contain carrageenan, exposing sulfonate groups and hydroxyl. Generally, the biosorption process can be defined as an ion exchange process that takes place between sorbate ions and the binding sites of many functional groups located on the cell wall surface: carboxyls, phosphoryls, sulfonates, aminos, amidos, and imidazoles [72]. Other techniques that can characterize the biosorption properties include spectroscopic techniques, that is, IR, electron microscopy (scanning, and transmission), electron dispersive spectroscopy, Raman, X-ray photoelectron spectroscopy, nuclear magnetic resonance, and X-ray diffraction analysis. Also, the stability of biosorption affinity between the sorbate and the functional group can be detected and this enables the description of how the biomass can adsorb different sorbates. There are different factors that affect the biosorption efficiency, such as pH, the biosorbent type (specially the cell wall composition), the presence of other competing ions (both anions and cations), and temperature, [73]. Generally, an increase of pH toward more alkaline values causes the deprotonation of the binding sites of biosorbent ions exposed by cellular surfaces. But decreasing the value of pH to become more acidic initiates the competition of protons and positively charged pollutant ions. However, these rules refer only to cations. Since the reverse of the biosorption process, that is, decreasing pH value, causes deprotonation. This feature is used in the regeneration (recovery and reuse) of the biosorbents. Another mechanism can be explained at high pH, where metal complexes solubility would decrease [73]. It is important to explain the biosorption mechanism and factors affecting the efficiency of the optimum operation conditions for the biosorption process itself and also for the biosorbent regeneration in order to achieve the highest removal performance. It is required to describe the pH effect to give an explanation for solution chemistry (redox, complexation, hydrolysis, precipitation, and reactions) which is affected by this factor [74].

Many functional groups responsible for pollutant ions binding by ions exchange contain a carbon source that is formed by the CO_2 sequestration process, such as the carboxylic ($-COOH$), amide ($CONH_2$), or carbonyl ($R-C=O$) groups. These functional groups are affected by pH values due to protonation and deprotonation which affects the biosorption or desorption of pollutant ions of the sorbate on the biosorption surfaces. Table 7.1 shows a comparison of the uptake of pollutants by different algal biomass.

7.7.2 Bioaccumulation

Bioaccumulation is the process of the sorbate accumulating metabolically inside the cell, which occurs in two steps. The first step is the same as the biosorption, and this step is rapid. The second one is slower, taking place by the metabolic transport of pollutant ions inside the cells by the active transport system. The bioaccumulation process is in nonequilibrium [83].

The process is more complicated than the biosorption process itself due to it requiring the metabolic activity of the cell. It proceeds by the cultivation of algal cells in the presence of sorbate ions. Algae sequester the carbon dioxide as the carbon source for algal biomass growth while the nutrient utilized in the algal growth medium is bicarbonate. The following reaction takes place in the vicinity of the cell wall:

$$HCO_3^- \rightarrow CO_2 + OH^-$$

TABLE 7.1 Comparison of the biosorption uptake of different pollutants by different algal biomass.

Biomass	Pollutant	Uptake capacity (q_{max} mg/g)	References
Carolina	MB	55	[75]
Chaetophora elegans		143	[76]
		143	
Chlamydomonas variabilis		18.3	[77]
Ulva fasciata		244	[78]
Sargassum dentifolium		66.6	
Wastewater algal biomass		21.8	[79]
Cladophora glomerata	Pb (II)	26.5	[80]
	Cu (II)	15.5	
Ulva fasciata	Cu (II)	125	[75]
Sargassum dentifolium		250	
Cystoseira indicia		103.093	[81]
Sargassum sp.		87.05	[82]

160   7. Algal biomass

Thus in the vicinity of the surface of the cell, the hydroxide ion concentration increases. This leads to hydroxide precipitation of toxic pollutants and this supports the processes of wastewater treatment. In other cases, various metabolic activity contributes to the removal of the toxic ions: such as extracellular metabolic transformations (biomineralization, and biotransformations), and the creation of insoluble phosphates or sulfides that are removed by the cell [84]. The removal of Mn, Fe, and Pb ions takes place by this mechanism. When considering the operational aspects, the bioremoval of pollutant ions is usually performed by growing algal cells in batch culture systems [83]. The bioaccumulation process is characterized in that there is no need to harvest the algal biomass again from the environment after treatment. Also, a further step of the treatment processes is reduced: collection, drying, processing, and storage [85]. The bioaccumulation process of pollutants is strongly influenced by different operational conditions, especially the presence of toxic pollutants in the growth medium of algae which can inhibit growth of the cells and the subsequent bioaccumulation process. This is considered a severe defect of this bioaccumulation process due to making it difficult to remediate a high-loaded solution by pollutants. This bioaccumulation application is performed in most conventional municipal and industrial wastewater treatment plants that depend on living organisms.

In bioaccumulation, pollutant ions are transported throughout algal cell walls and membranes. Inside the algal cells, these ions are attached to intracellular algal structures. different processes are concluded into the mechanism of the bioaccumulation process: including intracellular accumulation and reduction or oxidation reactions [86]. This bioaccumulation process is very complicated and depends on different factors (mostly, similar to the factors affecting the organism cultivation): the growth medium composition (the medium, in this case, refers to wastewater), temperature, pH, the existence of other pollutant ions (which can be suppressors of algal growth, as well), or other surfactants, inhibitors, etc. Toxic pollutant ions form complex structures with the algal cell membrane. These structures cause damage its integrity and impair its function [86]. Also, a high concentration of pollutants causes an accumulation effect that may be responsible for the morphological and physiological changes of algal cells [87]. Toxic metal ions interact with sulfydryl groups of enzymes causing the inhibition of cell growth. On the other hand, this process is sustained by the formation of metallothioneins (proteins with low molecular weight), which are rich with thiol groups (e.g., cysteine), attaching them in the inactive biological form and thus excluded from metabolic reactions. Algal species isolated from polluted environments exhibit efficient bioaccumulation. Discoloration of waste effluents is essential due to the photosynthetic process being repressed in the contaminated waters. Due to dyes being compounds that are not easily biodegradable, they are useful in either bioaccumulation or biosorption processes [85]. The dye type and concentration are the most effective parameter of the bioaccumulation process, and pH has also been investigated. The best bioaccumulation efficiency was attained at low concentrations of dyes [88].

7.7.2.1 Metal bioaccumulation induction to lipid production

Numerous metals, such as Cu, Mn, Fe, Zn, and Ni, are considered as important micronutrients for different organisms in many metabolic processes [89] as they have a vital function as precursors in many vitamins, some structural proteins in cell membranes, and catalytic cofactors for various metalloenzymes [90]. It is found that a high load of metals

Handbook of Algal Biofuels

7.7 Biosorption and bioaccumulation

bioaccumulating into the algal cell can induce reactive oxygen species (ROS) production, chlorophyll production inhibition [91], and negative disruption of cell proliferation. The useful result of these effects is the accumulation of lipid inside the microalgal cell. In various studies, heavy metal bioaccumulation stress has led to the induction of lipid content production in some microalgal cells. Ren et al. investigated the effects of Mg^{2+}, Ca^{2+}, and Fe^{3+} at concentration of $0-0.73$, $0-0.98$, and $0-0.12$ g/L, respectively, on the biological accumulation of the lipid of the *Scenedesmus* sp. It was found that lipid productivity and the total content of lipid reached 29.7% and 28.2%, respectively, with the addition of EDTA during algal cultivation [92]; this clarifies that the main pathways of metabolic activity related to the formation and breakdown of the lipid in *Scenedesmus* sp. could be affected by Fe^{3+}, Ca^{2+}, and Mg^{2+}. These heavy metal ions have many roles in controlling the pathway of metabolic action in microalgal cells and their lipid bioaccumulation processes [93]. For example, Ca^{2+} is considered to be a universal messenger that initiates the signaling and developmental stimuli of the environment [94]. Also, Mg^{2+} has been recognized as an essential ion in signaling to perform the mediation and activation of many biochemical activities in the cell, such as carbon fixation regulation in chloroplasts through the process of the Calvin cycle. Furthermore, it is proved that as Mg^{2+} increases it could induce acetyl-CoA carboxylase enzyme (fatty acid synthesis regulator), which has a role in increasing the content of lipids in the cells [94]. Battah et al. studied the influence of Mn^{2+} and Co^{2+} on the percentage of lipid content of green alga *C. vulgaris* [95], using $MnCl_2$ solution at different concentrations (2, 10, and 12 μM). The results show that with the increase of concentration the lipid content percentage significantly increased by 14%, 16%, and 15%, respectively. They also proved that the addition of different concentrations of $CoNO_3$ induced up to 25% more lipid content than the controls. An additional finding showed that the addition of five different Fe^{3+} concentrations increases the total lipid content of *C. vulgaris* up to 56.6% in the culture medium. Einicker-Lamas et al. found that Cd^{+2} induces the total lipid content in *Euglena gracilis*. It is reported that different combinations of Cd^{+2} concentration affect *C. vulgaris* by increasing acetone mobile polar lipids, phospholipids, and TAGs [96]. Bioaccumulation can be categorized into two types:

7.7.2.2 Biocoagulation

Biocoagulation refers to the biological coagulation of dye molecules through the extracellular biopolymer's surface liberated by the algal cell during the metabolic conversion of dyes. These biopolymers are located as surface long-chain of different functional groups and have excellent coagulation properties. The dye molecules settle down when adsorbed on these polymers [97]. The dye's biological decolonization mechanism by flocculation in the case of algae may be due to the interaction between these extracellular polymers that have high molecular weights and the polymers which attach physically or electrostatically and accordingly bridge the diffusion into a three-dimensional matrix of sufficient magnitude to decrease under quiescent conditions [97].

7.7.2.3 Biodegradation/bioconversion

Biodegradation is commonly mediated by the reduced enzymes, such as azo dye reductase, which reduce azo dyes by cleaving the azo linkage ($-N{=}N-$) present in

7. Algal biomass

such dyes, forming aromatic amines that are then further reduced by the Azo reductase enzyme of algae.

$$R_1 - N{=}N - R_2 + 4e - + 4H^+ R_1 - NH_2 + R_2 - NH$$

The chemical composition of the dyes and the algal cell control the rate of reduction [98]. Laccase and polyphenol oxidase enzymes have also been documented in the degradation of dyes. In a study, the blue−green alga *Oscillatoria curviceps* BDU92191 was found to degrade Acid Black 1 dye as a source of nitrogen using nitrogen-assimilating enzymes such as nitrite reductase and glutamine synthase. El-Sheekh et al. [99] revealed that the azo reductase enzyme activity was increased by the addition of dyes in the algal strains. Besides, the results showed that some functional groups are biodegraded. This study also indicated that p-amino benzene addition acts as the inducer for the azo reductase enzyme. The azo reductase induction phenomenon has been documented by other researchers as well [100].

7.8 Conclusion

Algal biomass production is a nascent stage for several technologies. Algae can utilize CO_2 as the main carbon source through a various metabolic processes inside the algal cell. The fixation of carbon by microalgae is embedded and transformed into biomass, hence CO_2 is a precursor to the formation of carbohydrates, lipids, and proteins. These constituents are considered as an axis core for many valuable applications driven from algae, especially biofuel production, human nutrition, animal feed and as bioremediation tools for wastewater treatments.

References

[1] M.G. de Morais, J.A.V. Costa, Biofixation of carbon dioxide by Spirulina sp. and Scenedesmus obliquus cultivated in a three-stage serial tubular photobioreactor, J. Biotechnol. 129 (2007) 439−445. Available from: https://doi.org/10.1016/j.jbiotec.2007.01.009.

[2] J. Ruiz, Z. Arbib, P.D. Álvarez-Díaz, et al., Photobiotreatment model (PhBT): a kinetic model for microalgae biomass growth and nutrient removal in wastewater, Env. Technol. 34 (2013) 979−991.

[3] B.C., O'Neill, M. Oppenheimer, Dangerous climate impacts and the Kyoto Protocol, Science (2002). 80.

[4] R.A. Pielke, An idealized assessment of the economics of air capture of carbon dioxide in mitigation policy, Environ. Sci. Policy 12 (2009) 216−225. Available from: https://doi.org/10.1016/j.envsci.2009.01.002.

[5] Durand B., Carbon Dioxide Capture and Storage, 2005.

[6] C. Gough, State of the art in carbon dioxide capture and storage in the UK: an experts' review, Int. J. Greenh. Gas. Control. 2 (2008) 155−168. Available from: https://doi.org/10.1016/S1750-5836(07)00073-4.

[7] G. Amoroso, D. Sültemeyer, C. Thyssen, H.P. Fock, Uptake of HCO_3- and CO_2 in cells and chloroplasts from the microalgae Chlamydomonas reinhardtii and Dunaliella tertiolecta, Plant. Physiol. 116 (1998) 193−201.

[8] H.S. Doma, R.M. Moghazy, R.H. Mahmoud, Environmental factors controlling algal species succession in High Rate Algal Pond, Egypt. J. Chem. (2020).

[9] H.M. El-Kamah, S.A. Badr, R.M. Moghazy, Reuse of wastewater treated effluent by lagoon for agriculture and aquaculture purposes, Aust. J. Basic. Appl. Sci. 5 (2011).

[10] R.M. Abdelhameed, E. Alzahrani, A.A. Shaltout, R.M. Moghazy, Development of biological macroalgae lignins using copper based metal-organic framework for selective adsorption of cationic dye from mixed dyes, Int. J. Biol. Macromol. (2020). Available from: https://doi.org/10.1016/j.ijbiomac.2020.10.157.

[11] A. Brar, M. Kumar, V. Vivekanand, N. Pareek, Photoautotrophic microorganisms and bioremediation of industrial effluents: current status and future prospects, 3 Biotech. 7 (2017) 18.

References

[12] K.S. Lackner, S. Brennan, Envisioning carbon capture and storage: expanded possibilities due to air capture, leakage insurance, and C-14 monitoring, Clim. Change 96 (2009) 357–378.

[13] P. Nogia, G.K. Sidhu, R. Mehrotra, S. Mehrotra, Capturing atmospheric carbon: biological and nonbiological methods, Int. J. Low-Carbon Technol. 11 (2016) 266–274. Available from: https://doi.org/10.1093/ijlct/ctt077.

[14] C. Jansson, S.D. Wullschleger, U.C. Kalluri, G.A. Tuskan, Phytosequestration: carbon biosequestration by plants and the prospects of genetic engineering, Bioscience 60 (2010) 685–696. Available from: https://doi.org/10.1525/bio.2010.60.9.6.

[15] A. Demirbas, Biodiesel from oilgae, biofixation of carbon dioxide by microalgae: a solution to pollution problems, Appl. Energy (2011). Available from: https://doi.org/10.1016/j.apenergy.2010.12.050.

[16] W.Y. Cheah, P.L. Show, J.S. Chang, et al., Biosequestration of atmospheric CO_2 and flue gas-containing CO_2 by microalgae, Bioresour. Technol. (2015).

[17] N. Hanagata, T. Takeuchi, Y. Fukuju, et al., Tolerance of microalgae to high CO_2 and high temperature, Phytochemistry (1992). Available from: https://doi.org/10.1016/0031-9422(92)83682-O.

[18] K.D. Sung, J.S. Lee, C.S. Shin, S.C. Park, Isolation of a new highly CO_2 tolerant fresh water Microalga Chlorella sp. KR-1, Renew. Energy 16 (1999) 1019–1022. Available from: https://doi.org/10.1016/S0960-1481(98)00362-0.

[19] Y. Jiang, W. Zhang, J. Wang, et al., Utilization of simulated flue gas for cultivation of Scenedesmus dimorphus, Bioresour. Technol. 128 (2013) 359–364.

[20] E.B. Sydney, W. Sturm, J.C. de Carvalho, et al., Potential carbon dioxide fixation by industrially important microalgae, Bioresour. Technol. 101 (2010) 5892–5896.

[21] Larsson M., Lindblom J., Algal flue gas sequestration and wastewater treatment: an industrial experiment, Master Sci Thesis Stock Sweden, 2011.

[22] A.M. Lizzul, P. Hellier, S. Purton, et al., Combined remediation and lipid production using Chlorella sorokiniana grown on wastewater and exhaust gases, Bioresour. Technol. 151 (2014) 12–18. Available from: https://doi.org/10.1016/j.biortech.2013.10.040.

[23] F.-F. Li, Z.-H. Yang, R. Zeng, et al., Microalgae capture of CO_2 from actual flue gas discharged from a combustion chamber, Ind. Eng. Chem. Res. 50 (2011) 6496–6502. Available from: https://doi.org/10.1021/ie200040q.

[24] K. Kumar, C.N. Dasgupta, B. Nayak, et al., Development of suitable photobioreactors for CO_2 sequestration addressing global warming using green algae and cyanobacteria, Bioresour. Technol. 102 (2011) 4945–4953.

[25] M.M. El-Sheekh, A.A. Farghl, H.R. Galal, H.S. Bayoumi, Bioremediation of different types of polluted water using microalgae, Rend. Lincei. (2016). Available from: https://doi.org/10.1007/s12210-015-0495-1.

[26] A. Lavrinovičs, T. Juhna, Review on challenges and limitations for algae-based wastewater treatment, Constr. Sci. (2018). Available from: https://doi.org/10.2478/cons-2017-0003.

[27] C.U. Ugwu, H. Aoyagi, H. Uchiyama, Photobioreactors for mass cultivation of algae, Bioresour. Technol. (2008).

[28] M.T. Ale, M. Pinelo, A.S. Meyer, Assessing effects and interactions among key variables affecting the growth of mixotrophic microalgae: PH, inoculum volume, and growth medium composition, Prep. Biochem. Biotechnol. (2014). Available from: https://doi.org/10.1080/10826068.2013.812562.

[29] H.J. Fallowfield, N.J. Cromar, L.M. Evison, Coliform die-off rate constants in a high rate algal pond and the effect of operational and environmental variables, Water Sci. Technol. 34 (1996) 141–147. Available from: https://doi.org/10.1016/S0273-1223(96)00831-1.

[30] Rose P.D., Dunn K.M., Maart B.A., Shipin O., Integrated Algal Ponding Systems and the Treatment of Saline Wastewaters, 2002.

[31] N.M. Parhad, N.U. Rao, Effect of pH on survival of Escherichia coli, J. Water Pollut. Control. Fed. (1974) 980–986.

[32] H.M. Elkamah, H.S. Doma, S. Badr, et al., Removal of fecal coliform from HFBR effluent via stabilization pond as a post treatment, Res. J. Pharm. Biol. Chem. Sci. 7 (2016).

[33] Santhanam N., Oilgae guide to algae-based wastewater treatment, Tamilnadu Home Algal Energy, 2009.

[34] R.M. Moghazy, A. Labena, S. Husien, et al., Neoteric approach for efficient eco-friendly dye removal and recovery using algal-polymer biosorbent sheets: characterization, factorial design, equilibrium and kinetics, Int. J. Biol. Macromol. 157 (2020) 494–509. Available from: https://doi.org/10.1016/j.ijbiomac.2020.04.165.

[35] E.S. Mansor, A. Labena, R.M. Moghazy, A.E. Abdelhamid, Advanced eco-friendly and adsorptive membranes based on Sargassum dentifolium for heavy metals removal, recovery and reuse, J. Water Process. Eng. 37 (2020) 101424. Available from: https://doi.org/10.1016/j.jwpe.2020.101424.

[36] Y. Shen, J. Gao, L. Li, Municipal wastewater treatment via co-immobilized microalgal-bacterial symbiosis: microorganism growth and nutrients removal, Bioresour. Technol. (2017). Available from: https://doi.org/10.1016/j.biortech.2017.07.041.

[37] S.R. Subashchandrabose, B. Ramakrishnan, M. Megharaj, et al., Consortia of cyanobacteria/microalgae and bacteria: biotechnological potential, Biotechnol. Adv. 29 (2011) 896—907.

[38] R.J. Powell, R.T. Hill, Mechanism of algal aggregation by Bacillus sp. strain RP1137, Appl. Env. Microbiol. (2014). Available from: https://doi.org/10.1128/AEM.00887-14.

[39] R. Muñoz, B. Guieysse, Algal-bacterial processes for the treatment of hazardous contaminants: a review, Water Res. (2006).

[40] L. Liu, H. Fan, Y. Liu, et al., Development of algae-bacteria granular consortia in photo-sequencing batch reactor, Bioresour. Technol. (2017). Available from: https://doi.org/10.1016/j.biortech.2017.02.025.

[41] B. Zhang, P.N.L. Lens, W. Shi, et al., Enhancement of aerobic granulation and nutrient removal by an algal—bacterial consortium in a lab-scale photobioreactor, Chem. Eng. J. (2018). Available from: https://doi.org/10.1016/j.cej.2017.11.151.

[42] N. Abdel-Raouf, A.A. Al-Homaidan, I.B.M. Ibraheem, Microalgae and wastewater treatment, Saudi J. Biol. Sci. (2012).

[43] S. Bharte, K. Desai, The enhanced lipid productivity of Chlorella minutissima and Chlorella pyrenoidosa by carbon coupling nitrogen manipulation for biodiesel production, Env. Sci. Pollut. Res. (2019). Available from: https://doi.org/10.1007/s11356-018-3757-5.

[44] H. Luo, M.A. Moran, Evolutionary ecology of the marine roseobacter clade, Microbiol. Mol. Biol. Rev. (2014). Available from: https://doi.org/10.1128/mmbr.00020-14.

[45] M.T. Croft, A.D., Lawrence, E. Raux-Deery, et al., Algae acquire vitamin B12 through a symbiotic relationship with bacteria, Nature (2005). Available from: https://doi.org/10.1038/nature04056.

[46] R. Cruz-López, H. Maske, The vitamin B1 and B12 required by the marine dinoflagellate Lingulodinium polyedrum can be provided by its associated bacterial community in culture, Front. Microbiol. (2016). Available from: https://doi.org/10.3389/fmicb.2016.00560.

[47] E. Kazamia, H. Czesnick, T.T.V. Nguyen, et al., Mutualistic interactions between vitamin B12-dependent algae and heterotrophic bacteria exhibit regulation, Environ. Microbiol. (2012). Available from: https://doi.org/10.1111/j.1462-2920.2012.02733.x.

[48] S.A. Amin, D.H. Green, M.C. Hart, et al., Photolysis of iron-siderophore chelates promotes bacterial-algal mutualism, Proc. Natl. Acad. Sci. USA (2009). Available from: https://doi.org/10.1073/pnas.0905512106.

[49] A.T. McCurdy, A.J. Higham, M.R. Morgan, et al., Two-step process for production of biodiesel blends from oleaginous yeast and microalgae, Fuel (2014). Available from: https://doi.org/10.1016/j.fuel.2014.07.099.

[50] L. Brennan, P. Owende, Biofuels from microalgae—a review of technologies for production, processing, and extractions of biofuels and co-products, Renew. Sustain. Energy Rev. (2010).

[51] N. Elhalawany, M.E. El-Naggar, A. Elsayed, et al., Polyaniline/zinc/aluminum nanocomposites for multifunctional smart cotton fabrics, Mater. Chem. Phys. 249 (2020) 123210. Available from: https://doi.org/10.1016/j.matchemphys.2020.123210.

[52] Hayes M., Skomedal H., Skjånes K., et al., Microalgal proteins for feed, food and health, in: Microalgae-Based Biofuels and Bioproducts: From Feedstock Cultivation to End-Products, 2017.

[53] L. Zhu, The combined production of ethanol and biogas from microalgal residuals to sustain microalgal biodiesel: a theoretical evaluation, Biofuels Bioprod. Bioref. (2014). Available from: https://doi.org/10.1002/bbb.1442.

[54] S.A. Scott, M.P. Davey, J.S. Dennis, et al., Biodiesel from algae: challenges and prospects, Curr. Opin. Biotechnol. (2010).

[55] A. Pugliese, L. Biondi, P. Bartocci, F. Fantozzi, Selenastrum capricornutum a new strain of algae for biodiesel production, Fermentation (2020). Available from: https://doi.org/10.3390/FERMENTATION6020046.

[56] Z. Guo, Y.W. Tong, The interactions between Chlorella vulgaris and algal symbiotic bacteria under photoautotrophic and photoheterotrophic conditions, J. Appl. Phycol. (2014). Available from: https://doi.org/10.1007/s10811-013-0186-1.

[57] J. Park, B.S. Park, P. Wang, et al., Phycospheric native bacteria Pelagibaca bermudensis and Stappia sp. Ameliorate biomass productivity of Tetraselmis striata (KCTC1432BP) in co-cultivation system through mutualistic interaction, Front. Plant. Sci. (2017). Available from: https://doi.org/10.3389/fpls.2017.00289.

References

[58] B.T. Higgins, I. Gennity, S. Samra, et al., Cofactor symbiosis for enhanced algal growth, biofuel production, and wastewater treatment, Algal Res. (2016). Available from: https://doi.org/10.1016/j.algal.2016.05.024.

[59] F. Khosravitabar, Microalgal biohydrogen photoproduction: scaling up challenges and the ways forward, J. Appl. Phycol. (2020).

[60] S.I. Allakhverdiev, V. Thavasi, V.D. Kreslavski, et al., Photosynthetic hydrogen production, J. Photochem. Photobiol. C. Photochem. Rev. (2010).

[61] G. Lakatos, Z. Deák, I. Vass, et al., Bacterial symbionts enhance photo-fermentative hydrogen evolution of Chlamydomonas algae, Green. Chem. (2014). Available from: https://doi.org/10.1039/c4gc00745j.

[62] R.H. Mahmoud, S.M. Abdo, F.A. Samhan, et al., Biosensing of algal-photosynthetic productivity using nanostructured bioelectrochemical systems, J. Chem. Technol. Biotechnol. (2020). Available from: https://doi.org/10.1002/jctb.6282.

[63] P.E. Wiley, J.E. Campbell, B. McKuin, Production of biodiesel and biogas from algae: a review of process train options, Water Environ. Res. (2011). Available from: https://doi.org/10.2175/106143010x12780288628615.

[64] A.M. Illman, A.H. Scragg, S.W. Shales, Increase in Chlorella strains calorific values when grown in low nitrogen medium, Enzyme Microb. Technol. (2000). Available from: https://doi.org/10.1016/S0141-0229 (00)00266-0.

[65] C. Muñoz, C. Hidalgo, M. Zapata, et al., Use of cellulolytic marine bacteria for enzymatic pretreatment in microalgal biogas production, Appl. Environ. Microbiol. (2014). Available from: https://doi.org/10.1128/AEM.00827-14.

[66] C.E. de Farias Silva, A. Bertucco, Bioethanol from microalgae and cyanobacteria: a review and technological outlook, Process. Biochem. (2016).

[67] K. Chojnacka, Biosorption and bioaccumulation—the prospects for practical applications, Environ. Int. 36 (2010) 299−307. Available from: https://doi.org/10.1016/j.envint.2009.12.001.

[68] C. White, S.C. Wilkinson, G.M. Gadd, The role of microorganisms in biosorption of toxic metals and radionuclides, Int. Biodeterior. Biodegrad. 35 (1995) 17−40.

[69] J. Wang, C. Chen, Biosorption of heavy metals by Saccharomyces cerevisiae: a review, Biotechnol. Adv. 24 (2006) 427−451. Available from: https://doi.org/10.1016/j.biotechadv.2006.03.001.

[70] S. Schiewer, B. Volesky, Biosorption processes for heavy metal removal, Environmental Microbe-metal Interactions, American Society of Microbiology, 2000, pp. 329−362.

[71] T.A. Davis, B. Volesky, A. Mucci, A review of the biochemistry of heavy metal biosorption by brown algae, Water Res. 37 (2003) 4311−4330. Available from: https://doi.org/10.1016/S0043-1354(03)00293-8.

[72] Volesky B., Schiewer S., Biosorption, metals, in: Encyclopedia of Bioprocess Technology, 2002.

[73] G.M. Naja, B. Volesky, Treatment of metal-bearing effluents: removal and recovery, Heavy Met. Env. (2009) 247−292.

[74] A.E. Ofomaja, Y.-S. Ho, Effect of pH on cadmium biosorption by coconut copra meal, J. Hazard. Mater. 139 (2007) 356−362. Available from: https://doi.org/10.1016/j.jhazmat.2006.06.039.

[75] El-wakeel S.T., Moghazy R.M., Labena A., Algal biosorbent as a basic tool for heavy metals removal; the first step for further applications, 2508 (2019) 75−87.

[76] F. Mikati, M. El Jamal, Biosorption of methylene blue on chemically modified Chaetophora Elegans algae by carboxylic acids, J. Sci. Ind. Res. (India) (2013).

[77] R.M. Moghazy, Activated biomass of the green microalga Chlamydomonas variabilis as an efficient biosorbent to remove methylene blue dye from aqueous solutions, Water SA 45 (2019) 20−28.

[78] R.M. Moghazy, A. Labena, S. Husien, Eco-friendly complementary biosorption process of methylene blue using micro-sized dried biosorbents of two macro-algal species (Ulva fasciata and Sargassum dentifolium): full factorial design, equilibrium, and kinetic studies, Int. J. Biol. Macromol. 134 (2019) 330−343. Available from: https://doi.org/10.1016/j.ijbiomac.2019.04.207.

[79] R.M. Moghazy, S. Abdo, The efficacy of microalgal biomass collected from high rate algal pond for dyes biosorption and biofuel production, Res. J. Chem. Env. 22 (2018) 54−60.

[80] E. Yalçin, K. Çavuşoğlu, M. Maraš, M. Biyikoğlu, Biosorption of lead (II) and copper (II) metal ions on Cladophora glomerata (L.) Kütz. (Chlorophyta) algae: effect of algal surface modification, Acta Chim. Slov. (2008).

[81] M. Akbari, A. Hallajisani, A.R. Keshtkar, et al., Equilibrium and kinetic study and modeling of Cu (II) and Co (II) synergistic biosorption from Cu (II)-Co (II) single and binary mixtures on brown algae C. indica, J. Environ. Chem. Eng. 3 (2015) 140−149.

Handbook of Algal Biofuels

166 7. Algal biomass

[82] J.P. Chen, L. Yang, Chemical modification of Sargassum sp. for prevention of organic leaching and enhancement of uptake during metal biosorption, Ind. Eng. Chem. Res. 44 (2005) 9931–9942.

[83] Z. Aksu, G. Dönmez, The use of molasses in copper(II) containing wastewaters: effects on growth and copper(II) bioaccumulation properties of Kluyveromyces marxianus, Process. Biochem. 36 (2000) 451–458. Available from: https://doi.org/10.1016/S0032-9592(00)00234-X.

[84] Lloyd J.R., Bioremediation of metals; the application of micro-organisms that make and break minerals. interactions 2: M2, 2002.

[85] Z. Aksu, G. Dönmez, Combined effects of molasses sucrose and reactive dye on the growth and dye bioaccumulation properties of Candida tropicalis, Process. Biochem. (2005). Available from: https://doi.org/10.1016/j.procbio.2004.09.013.

[86] P. Yilmazer, N. Saracoglu, Bioaccumulation and biosorption of copper(II) and chromium (III) from aqueous solutions by Pichia stipitis yeast, J. Chem. Technol. Biotechnol. (2009). Available from: https://doi.org/10.1002/jctb.2088.

[87] M.-I. de Silóniz, L. Balsalobre, C. Alba, et al., Feasibility of copper uptake by the yeast Pichia guilliermondii isolated from sewage sludge, Res. Microbiol. 153 (2002) 173–180. Available from: https://doi.org/10.1016/S0923-2508(02)01303-7.

[88] Z. Aksu, Reactive dye bioaccumulation by Saccharomyces cerevisiae, Process. Biochem. (2003). Available from: https://doi.org/10.1016/S0032-9592(03)00034-7.

[89] D. Kar, P. Sur, S.K. Mandal, et al., Assessment of heavy metal pollution in surface water, Int. J. Environ. Sci. Technol. (2008). Available from: https://doi.org/10.1007/BF03326004.

[90] G. Ndeezi, J.K. Tumwine, B.J. Bolann, et al., Zinc status in HIV infected Ugandan children aged 1–5 years: a cross sectional baseline survey, BMC Pediatr. (2010). Available from: https://doi.org/10.1186/1471-2431-10-68.

[91] F. Van Assche, H. Clijsters, Effects of metals on enzyme activity in plants, Plant. Cell Environ. (1990).

[92] H.Y. Ren, B.F. Liu, F. Kong, et al., Enhanced lipid accumulation of green microalga Scenedesmus sp. by metal ions and EDTA addition, Bioresour. Technol. (2014). Available from: https://doi.org/10.1016/j.biortech.2014.06.062.

[93] Z.Y. Liu, G.C. Wang, B.C. Zhou, Effect of iron on growth and lipid accumulation in Chlorella vulgaris, Bioresour. Technol. (2008). Available from: https://doi.org/10.1016/j.biortech.2007.09.073.

[94] L. Huang, J. Xu, T. Li, et al., Effects of additional Mg^{2+} on the growth, lipid production, and fatty acid composition of Monoraphidium sp. FXY-10 under different culture conditions, Ann. Microbiol. (2014).

[95] M. Battah, Y. El-Ayoty, A.E.F. Abomohra, et al., Effect of Mn^{2+}, Co^{2+} and H_2O_2 on biomass and lipids of the green microalga Chlorella vulgaris as a potential candidate for biodiesel production, Ann. Microbiol. (2015). Available from: https://doi.org/10.1007/s13213-014-0846-7.

[96] M. Einicker-Lamas, G. Antunes Mezian, T. Benevides Fernandes, et al., Euglena gracilis as a model for the study of Cu^{2+} and Zn^{2+} toxicity and accumulation in eukaryotic cells, Environ. Pollut. (2002). Available from: https://doi.org/10.1016/S0269-7491(02)00170-7.

[97] S.V. Mohan, Y.V. Bhaskar, J. Karthikeyan, Biological decolourisation of simulated azo dye in aqueous phase by algae Spirogyra species, Int. J. Environ. Pollut. (2004). Available from: https://doi.org/10.1504/IJEP.2004.004190.

[98] P. Sharma, H. Kaur, M. Sharma, V. Sahore, A review on applicability of naturally available adsorbents for the removal of hazardous dyes from aqueous waste, Environ. Monit. Assess. (2011).

[99] M.M. El-Sheekh, M.M. Gharieb, G.W. Abou-El-Souod, Biodegradation of dyes by some green algae and cyanobacteria, Int. Biodeterior. Biodegrad. (2009). Available from: https://doi.org/10.1016/j.ibiod.2009.04.010.

[100] L. Jinqi, L. Houtian, Degradation of azo dyes by algae, Environ. Pollut. (1992). Available from: https://doi.org/10.1016/0269-7491(92)90127-V.

Handbook of Algal Biofuels

CHAPTER 8

Application of halophilic algae for water desalination

Shristy Gautam[1] and Dhriti Kapoor[2]

[1]Department of Molecular Biology and Genetic Engineering, School of Bioengineering and Biosciences, Lovely Professional University, Phagwara, India [2]Department of Botany, School of Bioengineering and Biosciences, Lovely Professional University, Phagwara, India

8.1 Introduction

Water scarcity is now one of the world's biggest crises that has impacted many facets of human health, economic growth, and the stability of the environment. Water desalination has been employed to resolve this problem. It is the process of the removal of salt and other minerals from salty waters and encompasses a wide range of approaches from conventional distillation to reverse osmosis. Although 3/4 of the Earth's surface is covered with water, more than 97% of the water is in the form of saltwater that cannot be used specifically for drinking or agriculture [1]. The treatment of high salt wastewater is difficult for professionals in wastewater treatment. Because of the large amount of salt, saltwater can neither be incorporated into surface water nor into a general washing system without pretreatment [2]. High levels of salt can dramatically affect water quality and the composition of the organisms in the receiving water body or can cause significant harm to the treatment processes of wastewater, especially biologically, but also physically and chemically. Insufficiently treated saline wastewater effluent can also induce salinization or sodification in the recipient water system and its vicinity by replacing Ca^{2+} and Mg^{2+} with Na^+ [2]. Consequently, before application into surface waters or general wastewater systems, salt content of saline wastewaters must be decreased.

Many well-known physiochemical approaches to the desalination of the seawater are available, for example, reverse osmosis, forward osmosis, multistage flash distillation, electrodialysis, vapor compression distillation, and multiple effect distillation. These energy-intensive systems have detrimental environmental consequences that need to be taken into account, measured, and mitigated. These environmental consequences consist of water intake impingement and entrainment, which lead to a new source of death in the marine ecosystem [3]. The release of greenhouse gases from reverse osmosis desalination plants and air contaminants from thermoelectricity also exacerbates the change of environment. New technology for water recovery and improvements in water infrastructure for water communities will partly mitigate the need. These methods, however, require long-term investment and are facing increasingly chronic water issues in many developing nations [4]. It is, therefore important for the support of urban, rural, and agriculture prosperity and environmental conservation to boost water supplies at an affordable price. The interesting biological systems render very effective segregation of living organisms without sacrificing their habitat, compared with the artificial desalination process. This has provided biologists and engineers with ample inspiration for a large number of research works to find robust new ways of desalinizing water with lower energy usage, a reduction in costs, and mitigation of the environmental impacts.

Many studies have shown that microalgae can be used for treating wastewater of different origin [5]. Some microalgae species can also tolerate or decrease water and salt levels. Algae as a group encompass a large variety of salt tolerances. Some species can only withstand millimolar salt concentration, while others are able to live in saturated salt. In relation to salinity adaptation, the algae can be divided generally into halotolerant and halophile; the latter require salt to maximize growth and the former have reaction mechanisms allowing for their presence in saline medium. Changes in the salt content affect algae in three ways: (1) osmotic stress, which directly influences the water potential of the cell; (2) ionic (salt) stress induced by inevitable ion absorption or release, which is also an acclimatization; and (3) changes in cellular ion ratios due to selective membrane permeability. Many algae can be adapted by biochemical strategies such as osmolytes production and accumulation or functioning of the Na^+/K^+ pump system for changes in salinity [6−8]. The activity of the Na^+/K^+ pump helps to improve the osmotic potential and thus affects turgor pressure and cell volume [9]. Ion concentrations are increased in all cell components of the algae exposed to high salt content, but selective uptake of organelles can change ionic proportions [10]. The micro green algae *Dunaliella salina* is an example of a common salt-excluding algae since it is able to live in 3 M NaCl [11] concentration due to the extreme ability to block salt. A Na^+/H^+ antiporter is responsible for this ability on the vacuole (tonoplast) and $H^+/ATPase$ pumps at the plasma membrane. Therefore the halophilic microalgae *D. salina* has been able to grow with an optimal salinity level of 100%−125% in a wide salinity range of 50%−250%. Microalgal cells, with a relatively high level of salinity, withstand osmotic stress by using the majority of their energy in order to sustain cell pressure at a certain level, instead of growth [12]. In addition, increasing NaCl levels induces Na^+ channel blockers to reduce the cell size and the yield of biomass [13]. *D. salina* KSA-HS022, however, showed a 5.8-fold improved productivity of biomass. Slow adaptation can give microalgal cells enough time to withstand stress by producing osmoregulatory substances as a result of a gradual salinity rise [14]. A number of low-molecular-weight organic solutes have been detected that contribute to osmoregulation, including fatty acids, monosaccharides, and free amino acids. Oxidative stress-mediated enzymes may also be developed to resolve the stress of salinity [15]. Some halophilic algae not only tolerate salt, but can also concentrate many times as much salt as the water in which they live [16]. Algae that inhabit estuaries and saline inland zones with high salt and brackish marsh environments can be selected significantly to tolerate salt. This provides a large reservoir of algae, supplying many applicants for the treatment of saline water in inventive biotechnology systems [17].

The concept of biodesalination is fresh and the research is still in its infancy. Macrophyte algae secrete and absorb the salts from the solution [18]. *Pheridia tenuis* sequesters salt in its vacuoles and removes salt from water; if left in a bucket that contains seawater the water will become salt-free [17]. Salts are consumed by halophytes, in which salt is accumulated up a higher osmotic gradient. It has been recorded that the system has high salinity removals between 13% (for the first day) and 63% (for the sixth day) and that with increased salinity the removal is increased. A very interesting study on the isolation of halophilic microalgae was carried out in Egypt [19], where it was observed that halophile algae (*Scenedesmus* sp., *Chlorella vulgaris*) grown in a photobioreactor (PBR) successfully sequestered salt from brackish water and seawater under carefully controlled conditions, thus allowing their possible utilization in biological desalination. The biological mechanism and factors influencing salinity reduction have been studied. CO_2 was used for algae growth, and useful lipids from algae were extracted as an end-product which could be used to produce biofuel and provide energy for a desalination process [20]. The halophilic microalgae reported from saline and alkaline lakes include *Anabaenopsis arnoldii, Chloroflexus, Synechococcus* sp., *Chroococcus, D. salina, Spirulina* sp., *Arthrospira platensis*, etc. A few algae, including *Spirulina* sp. and *D. salina*, produce β-carotene, glycerol, biofuel, essential omega-fatty acid, etc. and are used as food supplement sources. A number of interesting applications in biotechnology were identified by halophile microorganisms. They may be less "fashionable" in biotechnology research than thermophiles, but the number of applications already exploited or established is still remarkable [21]. It has been shown that salinity tolerance for green microalgae can either be increased or decreased depending on the culture media composition [20]. Desalination by using algae for the removal of salts of saline water and water processing for various applications is a new idea and in previous research it was used and tested for industrial wastewater treatment, which decreased algae costs to a minimum while retaining performance without any reductions. The results obtained in the past studies were encouraging and effective for the desalination of marine wastewater and have achieved success, gaining the achieval of a removal efficiency of upto 95% until the prices became reasonably affordable for future use for different purposes [22] (Fig. 8.1). This chapter explains the use of halophilic algae which are not only salt tolerant but also capable of absorbing and concentrating the salt solutes at higher levels than the water they are in. In addition, the use of the halophytic technologies with the development of potential biofuels from algae balances the need for a sustainable approach to clean, affordable water and renewable energy.

8.2 Marine environment

The hypersaline environments are also called thalassohaline environments and are caused by the evaporation of marine water which leaves mineral depositions dominated primarily by sodium chloride (NaCl) leading to concentrations exceeding 300 Practical Salinity Units [23,24]. The salts and the pH composition of the region are essentially similar to seawater [25]. Water with 3% or higher salinity is designated as salt water. Brackish

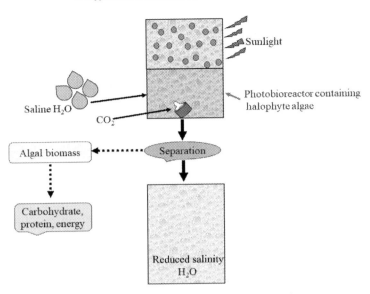

FIGURE 8.1 Schematic representation of water desalination by halophyte algae.

rivers, sea, coastal waters, salt-lakes, and salt marshes are known as salt water. The most frequent dissolved ions in salt water and soil that have significant physiological functions in most living organisms are sodium and chloride. However, high sodium in cells can cause protein damage and interfere with metabolism [1,26]. Living organisms therefore use various sodium balance strategies, including blocking Na^+ from entering the cytosol (salt strength) via the cell membrane and a reduction of cytosol Na^+ (salt tolerance) concentration through intricate transport systems [27,28].

A lot of microorganisms, including algae, bacteria, and fungi, have been reported to live in salterns [29]. Some algal species are classified as halophiles in salt water, lakes, crystallizing ponds, and microbial beds in hypersaline conditions and are optional rather than obligatory halophyte species [30]. Halophilic algal species are polyextremophilic and include such species as *Chlamydomonas nivalis* (psychrophilic green algae) and *Cyanidium caldarium* (thermophilic red algae). The most prominent species of halophilic algal is *Dunaniella salina*, a green alga, dominating the population of the algae in hypersaline environments. The *Amphora* sp. MUR 258 diatom has evolved well over a very broad range of high salinity rates (7%−12% NaCl), which confirms that the hypersaline algae need a high salt level environment of at least 7% NaCl over seawater for optimal growth. Whilst many studies on diatoms are undertaken in the flora of salt lakes, very few studies are carried out on their salinity tolerance and development. Clavero et al. [31] analyzed 34 benthic diatom strains with a total salinity range of 0.5%−17.5% isolated from thalassic hypersaline lakes, with the five most halotolerant taxa having total salts rising upto 15%, and the growth of *Amphora coffeaeformis* at a rate of upto 7% NaCl. The high salinity limit for diatoms is about 15%, presumably because of their osmoregulation capability and in algae this limit tends to be set by the form of osmoregulatory compatible solution. The precise mechanism of osmoregulation in halophile diatoms, including *Amphora* sp. MUR 258, also needs clarity [32].

The following are some of the most famous saltwater habitats:

Dead Sea: the Dead Sea is a characteristic example of the thalassohaline climate. The lake is around 320 m deep, the temperature of the water is 21°C–36°C, and the salt level is 78% NaCl. The pH is very low, and the most abundant ions are Na^+, Cl^-, and Mg^{2+} [33].

Great Salt Lake: the Great Salt Lake is the biggest salt lake in the western hemisphere in Utah, United States. This is a medium depth (~10 m maximum) thalassohaline lake in a desert of salt. In contrast to the Dead Sea, on the alkaline side the Great Salt Lake has a relatively high pH with 33% NaCl [33].

Solar Lake: the Solar Lake is also a strongly hypersaline lake with a shallow depth of 4–6 m, situated on the Sinai coast of the Gulf of Aqaba. Sunlight reaches the bottom of the lake (thus giving the name of the Solar Lake), with a high rate of evaporation in sediment and water as well as with intensive and complex microbial interactions. The lake's water gets completely oxygenated in the summer, but stratifies in the fall. Due to high evaporation rates the lake salinity increases to 20% NaCl during the summer.

Natrun valley: in the arid area of Central Northern Egypt, the Natrun valley or Wadi El Natrun is located below sea level. The northern part of the valley has eight seasonal, sometimes completely dry, hypersaline lakes in it. The lakes of Natrun Valley have also a high evaporation rate like Solar Lake. Water sources are provided from underground drainage of water from the Great Nile River via the burdi swamps. The salinity near the valley sediment ranges from 3.1% to 8.6% NaCl [33].

La Mala valley: the region of La Mala lies 780 m above sea level with a slope of 2%. The field is thalassohaline and the surface area of the soil is furnished with salt water. Well water is the other water resource for the ecosystem. The water supply to the well's chloride is smaller than the ions in the water Mg^{2+}, Ca^{2+}, and K^+. The water supply has an average salinity of 18% NaCl.

In recent decades, the salinization of both terrestrial and aquatic environments has become a rising global problem. Apart from the rising amount of primary salinization due to climate change, secondary salinization is increasingly highlighted by the increasing volume of saline residues. Salt wastewater is highly variable in composition and concentration and is depending on the sources. Over 20% of world agriculture is afflicted by primary salinization, excessive irrigation or rainfall may convert saline farmland into saline wastewater sources [34,35]. A growing amount of salt wastewater (inequitable irrigation performed, inadequate drains, fertilization) is leading to agricultural activities anyway. Nutrients (nitrogen and phosphorus) and non or barely degradable organic (pesticides and herbicides) chemicals may be present in agricultural wastewater alongside different salts [36–38].

8.3 Isolation of halophilic microalgae

Different techniques for isolating halophilic microalgae are well-known, including streak plating technique, micropipette washing technique, centrifugal washing technique, etc. The isolation of algal species should be carried out carefully. However, by using the micropipette washing procedure, the pure culture will not be free from contamination. The centrifuge washing technique with streaking technique should be used in order to free isolated microalgae from contamination. Halophilic microalgae can be isolated from

172   8. Application of halophilic algae for water desalination

the fluvial ecosystem by using various media, such as Johnson medium, BG 11 medium, BBM medium, Chu-10 medium, and F/2 medium.

Johnson medium: Unicellular halophilic microalgal organisms have been isolated using enriched cultures. In order to enrich the culture, a mixture of microorganisms in water samples have been centrifuged for 10 min at $3000 \times g$. Pellets were produced in 50 mL of modified Johnson medium [39] at salinity of 50% using NaCl. The glass tubes were continuously blown with filtered air, supplemented by 3% carbon dioxide at 0.2 vvm air aeration rate. At 25°C and 90 μmol photon/m^2/s cultures were incubated with white-light-emitting diodes at a 16:8 h light:dark interval [14].

BG 11 medium: The algae culture was inoculated into 250 mL BG 11 medium consisting of ferric ammonium citrate (0.006 g/L); dipotassium hydrogen phosphate (0.0314 g/L); magnesium sulfate (0.036 g/L); calcium chloride dehydrate (0.0367 g/L); sodium carbonate (0.020 g/L); disodium magnesium EDTA (0.001 g/L); sodium nitrate (1.5 g/L); citric acid (0.056 g/L); and the final pH was adjusted to 8. Enriched algae culture incubation was conducted at 30°C, with a photoperiod of 200 lux at a 16:8 light:dark cycle. At 5-day intervals growth was observed. Serial dilution technique and antibiotic combinations have been used for purification of the axenic algal culture [40].

Bold's Basal Medium (BBM): Microalgae were grown on enhanced BBM consisting of NaNO$_3$ (0.250 g/L), NaCl (0.025 g/L), K$_2$HPO$_4$ (0.075 g/L), KH$_2$PO$_4$ (0.175 g/L), H$_3$BO$_3$ (0.011 g/L), EDTA (0.050 g/L), KOH (0.031 g/L), Co (NO$_3$)$_2$ · 6H$_2$O (0.490 mg/L), CuSO$_4$ (1 mg/L), MnCl$_2$ · 4H$_2$O (1.440 mg/L), ZnSO$_4$ · 7H$_2$O (8.820 mg/L), MoO$_3$ (0.710 mg/L), MgSO$_4$ · 7H$_2$O (0.075 g/L), CaCl$_2$ · 2H$_2$O (0.025 g/L), FeSO$_4$ · 7H$_2$O (5 mg/L), and KOH adjusted to a pH 8.0. The algae were separated by the streak plate technique. A loop full of algae colony was streaked for invert culture in BBM. When the algal colonies grew, they were streaked up again until single colonies appeared microscopically on the plate [41].

Chu-10 medium: Algae have been cultivated under laboratory conditions in modified Chu No. 10 nutrient solution. Samples of water collected were inoculated aseptically into a sterilized, liquid, modified Chu10 medium consisting of Ca(NO$_3$)$_2$ (0.04 g/L), K$_2$HPO$_4$ (0.01 g/L), MgSO$_4$0.7H$_2$O (0.025 g/L), Na$_2$CO$_3$ (0.02 g/L), Na$_2$SiO$_3$ (0.025 g/L), ferric citrate (0.003 g/L), and citric acid (0.003 g/L) prepared using seawater. The algal culture is cultivated at room temperature and illuminated with two cool, white fluorescent lamps that radiate around 20 μE/m/^{2}/s for a 16:8 light:dark regime [42].

F/2 medium: Algal colonies, which were formed by streaking technique and micropipetting technique, were transferred to a liquid f/2 medium [43] consisting of NaNO$_3$ (0.075 g/L), NaH$_2$PO$_4$.H$_2$O (0.005 g/L), microelement stock solution (1 mL/L), vitamin solution (1 mL/L), microelement stock solution [FeCl$_3$0.6H$_2$O (3.150 g), Na$_2$EDTA (4.160 g), MnCl$_2$0.4H$_2$O (0.180 g), CoCl$_2$0.6H$_2$O (0.010 g), CuSO$_4$0.5H$_2$O (0.010 g), ZnSO$_4$.H$_2$O (0.022 g), Na$_2$MoO$_4$0.2H$_2$O (0.006 g)], vitamin solution [Biotin (Vitamin H) (0.5 mg), thiamine HCl (Vitamin B1) (100 mg), cyanocobalamin (Vitamin B12) (0.5 mg)], and the pH was adjusted to 8.0 for growth. The liquid culture was preserved in 70 μmol photons/m^2/s at a light:dark interval of 12:12 h at 25°C and was manually shaken once daily. Algae were initially isolated by streaking on agar plates. The growth medium used for agar plating consisted of charcoal filtrated seawater with a salinity modified by NaCl for salinity of environmental water samples enriched with 75 g/L NaNO$_3$ and 5 g/L NaH$_2$PO$_4$ · H$_2$O. 10 mg/L GeO$_2$ was applied to inhibit diatomic growth on the medium of half of the agar layer. At 30 μmol photons/m^2/s, the plates induced a light:dark interval of 12:12 h at 25°C and the presence of microalgae or

Handbook of Algal Biofuels

other microorganisms was verified periodically. Heavy bacterial contamination plates were again streaked to isolate microalgae cells on new plates. On the new plates of agar, enriched with twice as much nitrate and phosphate, small colonies could be observed and selected colonies were reconstructed after 17 incubation days [44].

8.4 Efficiency of microalgae and seaweeds for water desalination

The desalination of water has developed into a major industrial technique to offer freshwater for various societies and commercial trades that performs as a significant component of increasing the economy of a country [45]. Desalination by the use of algae for the elimination of different salts from saline H_2O and H_2O processing for use in a variety of functions is a novel perception and has been introduced and evaluated for the treatment of contaminated water [46]. The green algae *Scenedesmus* spp. were studied to examine their elimination ability of different nutrients in saline H_2O. Elimination efficacy was studied in the output of the reactor for total dissolved solids, Na, Cl^-, and PO_4^{2-}, that were found to be about 97%, whereas the elimination efficacy of both Nitrate and SO_4^{2-} were found to be about 93%. This demonstrates the advantage of this method to generate appropriate potable H_2O from seawater [47]. *Scendesmus* algae species development was efficaciously enhanced in saline H_2O as it absorbs salts and utilizes them for their growth processes and eliminates salts from brine H_2O [46]. *C. vulgaris'* ability to desalinize water was assessed using three dissimilar methods, that is, static, continuous flow, and a photobioreactor MDC. It was found that under these processes, the salt removal ability of *C. vulgaris* significantly improved [48]. *Nannochloropsis oculata* and *Dunaliella tertiolecta* significantly improved biomass formation, and also were helpful in the removal of salt ions such as nitrogen, phosphate, and fluoride. Hence, this strategy is recognized as a beneficial technique to reduce the cost of salt removal by making profits from biofuel formation with good ecological paybacks, for instance, CO_2 extenuation and improved wastewater treatment [49].

The nutrient depriving capacity of nine microalgal species from the genera *Monoraphidium*, *Desmodesmus*, *Chlorella*, *Chlorococcum*, and *Scenedesmus* were tested. It was shown that almost all species have the potential to reduce conductivity and degrade noteworthy levels of chloride and nutrients, Thus it was reported that these freshwater algal species are promising assets to recover water quality through bioremediation practices in salt-rich wastewater bodies [2]. Furthermore, only small variations were noted in the desalination potential of *Scenedesmus* sp. and *C. vulgaris* strains [20]. *Scenedesmus obliquus* successfully removed higher NaCl concentration from brackish water and it was concluded that the salt exclusion improved with the increase in the salinity in brackish water, which was possibly because of the greater NaCl gradient from the bulk solution to the algae surface [50]. *P. tenuis* is recognized to sequester salt inside the vacuoles of its cells, the various salts from water when it is cultivated in bucket containing seawater [17]. *S. obliquus* removed salt from water in which 30% of salt exclusion was observed after 16 days of incubation and it was also found that *S. obliquus* significantly enhanced the proportion of desalination and elimination of NaCl as the salinity of the water was elevated [51]. *S. obliquus* considerably removed K^+, Na^+, Ca^{2+}, and Mg^{2+} from saline alkaline water and this ability of *S. obliquus* improved as the salinity concentration increased [52].

Handbook of Algal Biofuels

8.5 Economic feasibility

In spite of the presence of many varieties of microalgae, primarily chlorophyta, two blue–green algae, and diatoms are cultured at the large scale in various cultivation practices and have been recognized to increase the nutritional value of traditional food [53]. The green single-celled alga, *D. salina* is a major microalgal species, which is presently grown on a commercial level and it is day by day gaining the increasing attention of researchers due to its efficient feasibility. *D. salina* uptakes carotenoids at higher salt concentrations, and it is the richest source of β-carotene, with an optimum yield of β-carotene of around 24% NaCl and is manufactured on a large scale at the industrial level [54]. *D. salina* also is capable of growing in high-salt concentration water bodies by the uptake of glycerol to maintain proper osmotic level [55]. Microalgae such as *Chlorella* also are utilized for biomass production because of their maximum yield, suitable biochemical components, global spread, and domination over extra algal moieties in exposed culture processes [56,57]. They also have promising lipid content, which allows efficacious production of biofuels from microalgae [58–60]. The wide variety and richness of macroalgae, as well as the presence of large amount of carbohydrates, are significant bases for the production of renewable carbon, which in turn forms sustainable fuels and chemicals. It is also recognised as a beneficial moiety for the generation of biofuels such as biohydrogen and bioethanol; however, effective industrial distribution of macroalgae-based biorefineries is dependent on their commercial feasibility [61] (Fig. 8.2).

Among these macroalgae, brown algae accumulate relatively high lipids, which is recognized as the one of the major constituents for the generation of biofuels and biomass

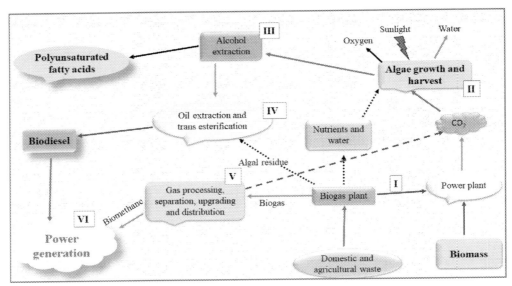

FIGURE 8.2 Diagrammatic representation showing the use of biomass for algae cultivation and domestic/agriculture waste for biodiesel production.

[62]. Microalgae are successful in comparison to conventional biofuel methods due to their ability to fix CO_2 from commercial outlet gases, their potential to grow in extremely briny water, the degradation of contaminants; and enhanced levels of lipids, sugars, and proteins [63,64]. Algal species present in the sea are more suitable for cultivation in salt water as compared to freshwater types because these algal species are acclimatized to the ecosystem with higher salt concentrations. Microalgae in community sewage disposal waste and agronomic manure also are utilized as a moiety for significant mineral nutrients [65–67]. Growing sea algae, that is, *N. oculata* and *D. tertiolecta*, in the waste products might be a better strategy to eliminate contaminants and can also create biomass for biofuel generation. Therefore this technique can be recognized as a beneficial technique to reduce the costs of salt removal by making profits from biofuel generation with helpful ecological payback, for example, CO_2 fixation and clean wastewater [49]. Fuels produced using algae are better than first generation biofuels, because they do not require land and eliminate food demand problems. They are also used for biodiesel production due to their low lifetime greenhouse gases release, as algae biomass changes environmental CO_2 via photosynthesis into a bioplant asset which is ultimately discharged back to the environment through microbes [68,69].

8.6 Role of algal biofuel in desalination process

The use of algae to minimize salinity and provide the required growth of algae has a much lower demand for energy when the algae are harvested for biofuel or protein. A cost-effective and efficient algal cell separation from water is necessary for biodesalination. The efficient use of biodesalination technologies will provide a reliable, low-cost supply of water to drink and farm, while generating valuable bio-based fuel as an energy by-product for desalination [20]. When saline microalgae are exposed to salt stress, the fatty acids in the membrane start to denature. In response to unfavorable conditions, microalgae increase lipid synthesis [70]. Many researchers have explored the possibility of lipid accumulation to protect salt pressure. Furthermore, other studies indicate that an increase in salinity affects even the composition of intracellular lipids. Concentrations of saturated and monounsaturated fatty acids (palmitic acids), have increased in *Dunaliella*, whereas polyunsaturated fatty acids (linoleic acid) decrease due to high salt stress [13]. It has also been found that the fatty acid compositions of polar lipids change. *Dunaliella* is considered to be a good feedstock for the production of biodiesel. *Dunaliella* sp. fatty acids are also methylated to produce biodiesel [71]. The effects on lipid development of varying NaCl concentration have been well studied [59,72,73] and salt stress is an effective way of promoting algal lipid content. Kaewkannetra et al. [74] found a higher content of lipid in *S. obliquus* under salt stress, and a maximum accumulation of 0.3 M sodium chloride was observed. Salama et al. [75] indicated that in BB medium modified with 0.025 M sodium chloride, the highest biomass and lipid contents of *S. obliquus* were obtained. Pancha et al. [76] observed that when *Scenedesmus* sp. was treated with 0.4 M sodium chloride, the highest lipid content was found. Therefore it would be a good way to minimize costs, reduce freshwater use, and encourage the lipid content of algae if brackish water were to be used as the source of water in which to grow algae. Biological desalination and lipid

176   8. Application of halophilic algae for water desalination

development can be accomplished simultaneously by means of brackish water. The further optimization of the biodesalination process and the use of algal for desalination and biofuels might make this technology attractive and sustainable.

8.7 Conclusion

The rapid population growth, increase in per capita consumption, and rapid development generates an urgent demand for water. Halophytic technology offers a safe option for clean and safe water. Biodesalination has reached a promising stage in the emerging sector. It has now become a separate field of research that can change the development of freshwater through conventional approaches to desalination methods. A new concept is desalination focused on the use of algae to extract salt from salt water and to refine water for different purposes. Research has shown that the use of algae reduces costs to a minimum, without decreasing efficiency. The findings were encouraging and optimistic with respect to marine desalination and succeeded. The method proceeds to achieve removal efficiency of upto 95% until the price for potential uses for different purposes is relatively affordable. This would open the door to a new path in which the issue of water desalination can be solved by minimizing the overall production cost.

References

[1] H.T. El-Dessouky, H.M. Ettouney, Fundamentals of Salt Water Desalination, Elsecvier, 2002.

[2] A. Figler, D. Dobronoki, K. Márton, S.A. Nagy, I. Bácsi, Salt tolerance and desalination abilities of nine common green microalgae isolates, Water 11 (12) (2019) 2527.

[3] S. Lattemann, T. Höpner, Environmental impact and impact assessment of seawater desalination, Desalination 220 (1–3) (2008) 1–15.

[4] K. Watkins, Human development report 2006-beyond scarcity: power, poverty and the global water crisis, UNDP Human Development Reports, 2006.

[5] N. Abdel-Raouf, A. Al-Homaidan, I. Ibraheem, Microalgae and wastewater treatment, Saudi J. Biol. Sci. 19 (3) (2012) 257–275.

[6] N. von Alvensleben, K. Stookey, M. Magnusson, K. Heimann, Salinity tolerance of Picochlorum atomus and the use of salinity for contamination control by the freshwater cyanobacterium Pseudanabaena limnetica, PLoS One 8 (5) (2013) e63569.

[7] A.F. Talebi, M. Tabatabaei, S.K. Mohtashami, M. Tohidfar, F. Moradi, Comparative salt stress study on intracellular ion concentration in marine and salt-adapted freshwater strains of microalgae, Not. Sci. Biol. 5 (3) (2013) 309–315.

[8] I. Slama, C. Abdelly, A. Bouchereau, T. Flowers, A. Savoure, Diversity, distribution and roles of osmoprotective compounds accumulated in halophytes under abiotic stress, Ann. Bot. 115 (3) (2015) 433–447.

[9] N. Erdmann, M. Hagemann, Salt acclimation of algae and cyanobacteria: a comparison, Algal Adaptation to Environmental Stresses, Springer, 2001, pp. 323–361.

[10] G. Demetriou, C. Neonaki, E. Navakoudis, K. Kotzabasis, Salt stress impact on the molecular structure and function of the photosynthetic apparatusâ€"the protective role of polyamines, Biochim. Biophys. Acta 1767 (4) (2007) 272–280.

[11] R. Taheri, M. Shariati, Study of the inhibitory effect of the media culture parameters and cell population to increase the biomass production of Dunaliella tertiolecta, Prog. Biol. Sci. 3 (2) (2013) 123–133.

[12] T. Ishika, P.A. Bahri, D.W. Laird, N.R. Moheimani, The effect of gradual increase in salinity on the biomass productivity and biochemical composition of several marine, halotolerant, and halophilic microalgae, J. Appl. Phycol. 30 (3) (2018) 1453–1464.

Handbook of Algal Biofuels

References

[13] M. Takagi, T. Yoshida, Effect of salt concentration on intracellular accumulation of lipids and triacylglyceride in marine microalgae *Dunaliella* cells, J. Biosci. Bioeng. 101 (3) (2006) 223−226.

[14] A.E.-F. Abomohra, A.H. El-Naggar, S.O. Alaswad, M. Elsayed, M. Li, W. Li, Enhancement of biodiesel yield from a halophilic green microalga isolated under extreme hypersaline conditions through stepwise salinity adaptation strategy, Bioresour. Technol. 310 (2020) 123462.

[15] G. Kan, C. Shi, X. Wang, Q. Xie, M. Wang, X. Wang, et al., Acclimatory responses to high-salt stress in Chlamydomonas (Chlorophyta, Chlorophyceae) from Antarctica, Acta Oceanol. Sin. 31 (1) (2012) 116−124.

[16] M.A. Khan, D.J. Weber, Ecophysiology of High Salinity Tolerant Plants, Springer Science & Business Media, 2006.

[17] N.P. Yensen, Halophyte Uses for the Twenty-First Century, Ecophysiology of High Salinity Tolerant Plants, Springer, 2008, pp. 367−396.

[18] M. Omar, D. Balla, Drainage water purification in saline detention ponds with duckweeds, in: Proceedings of the Twenty-Third European regional conference,"Progress in Managing Water for Food and Rural Development" Lviv, Ukraine, 2009.

[19] A.E.-F. Abomohra, M. El-Sheekh, D. Hanelt, Screening of marine microalgae isolated from the hypersaline Bardawil lagoon for biodiesel feedstock, Renew. Energy 101 (2017) 1266−1272.

[20] E. Sahle-Demessie, A.A. Hassan, A. El Badawy, Bio-desalination of brackish and seawater using halophytic algae, Desalination 465 (2019) 104−113.

[21] A. Oren, Biotechnological applications and potentials of halophilic microorganisms, Halophilic Microorg. Environ. (2002) 357−388.

[22] M. El Nadi, F. El Sergany, R.G.H. Water, Desalination by Algea, ASU J. Civ. Eng. 2 (2010) 105−114.

[23] A. Oren, Diversity of halophilic microorganisms: environments, phylogeny, physiology, and applications, J. Ind. Microbiol. Biotechnol. 28 (1) (2002) 56−63.

[24] N. Gunde-Cimerman, P. Zalar, S. de Hoog, A. Plemenitaš, Hypersaline waters in salterns−natural ecological niches for halophilic black yeasts, FEMS Microbiol. Ecol. 32 (3) (2000) 235−240.

[25] C. Gostinčar, M. Lenassi, N. Gunde-Cimerman, A. Plemenitaš, Fungal adaptation to extremely high salt concentrations, Advances in Applied Microbiology, Elsevier, 2011, pp. 71−96.

[26] J. Hall, Guyton and Hall Textbook of Medical Physiology: Enhanced E-book, Elsevier Health Sciences, 2010.

[27] S. Shabala, A. Moreno, Y. Hariadi, A. Mackay, Y. Tian, J. Bose, Halophytes: what makes them special? Revealing ionic mechanisms of salinity tolerance, in: Proceedings of the Eighteenth International Botanical Congress, 2011, p. Sym030.

[28] S. Shabala, R. Munns, Salinity stress: physiological constraints and adaptive mechanisms, Plant Stress. Physiol. 1 (1) (2012) 59−93.

[29] N. Gunde-Cimerman, P. Zalar, U. Petrovic, M. Turk, T. Kogej, G. De Hoog, et al., Fungi in salterns, in: A. Ventosa (Ed.), Halophilic Microorganisms, Springer-Verlag, Berlin, 2004.

[30] B.P. Burns, F. Goh, M. Allen, B.A. Neilan, Microbial diversity of extant stromatolites in the hypersaline marine environment of Shark Bay, Aust. Environ. Microbiol. 6 (10) (2004) 1096−1101.

[31] E. Clavero, M. Hernández-Mariné, J.O. Grimalt, F. Garcia-Pichel, Salinity tolerance of diatoms from thalassic hypersaline environments, J. Phycol. 36 (6) (2000) 1021−1034.

[32] I. Indrayani, N.R. Moheimani, K. de Boer, P.A. Bahri, M.A. Borowitzka, Temperature and salinity effects on growth and fatty acid composition of a halophilic diatom, Amphora sp. MUR258 (Bacillariophyceae), J. Appl. Phycol. (2020) 1−11.

[33] B.J. Javor, Hypersaline Environments: Microbiology and Biogeochemistry, Springer Science & Business Media, 2012.

[34] J. Li, L. Pu, M. Han, M. Zhu, R. Zhang, Y. Xiang, Soil salinization research in China: advances and prospects, J. Geogr. Sci. 24 (5) (2014) 943−960.

[35] H. Zhang, G. Zhang, X. Lü, D. Zhou, X. Han, Salt tolerance during seed germination and early seedling stages of 12 halophytes, Plant Soil. 388 (1−2) (2015) 229−241.

[36] D.-M. Jiang, X.-Y. Wang, M.-H. Liu, G.-F. Lu, Study on agricultural structure and non-point source pollution: a case in Dapu Town of Yixing City, Ecol. Econ. 2 (3) (2006) 270−280.

[37] M.G. Nia, H. Rahimi, T. Sohrabi, A. Naseri, H. Tofighi, Potential risk of calcium carbonate precipitation in agricultural drain envelopes in arid and semi-arid areas, Agric. Water Manage. 97 (10) (2010) 1602−1608.

[38] B. Sun, L. Zhang, L. Yang, F. Zhang, D. Norse, Z. Zhu, Agricultural non-point source pollution in China: causes and mitigation measures, Ambio 41 (4) (2012) 370−379.

Handbook of Algal Biofuels

[39] M.K. Johnson, E.J. Johnson, R.D. MacElroy, H.L. Speer, B.S. Bruff, Effects of salts on the halophilic alga *Dunaliella viridis*, J. Bacteriol. 95 (4) (1968) 1461–1468.

[40] H.H. Bhatt, R. Pasricha, V.N. Upasani, Isolation and characterization of a halophilic cyanobacterium Euhalothece SLVH01 from Sambhar salt lake, India, Int. J. Curr. Microbiol. Appl. Sci. 5 (2) (2016) 215–224.

[41] C. Liu, J. Liu, S. Hu, X. Wang, X. Wang, Q. Guan, Isolation and identification of a halophilic and alkaliphilic microalgal strain, PeerJ 7 (2019) e7189.

[42] S. Balakrishnan, P. Santhanam, S. Jeyanthi, M. Divya, M. Srinivasan, Preliminary screening of halophilic microalgae collected from different salt pans of Tuticorin, southeast coast of India, Proceedings of the Zoological Society, Springer, 2019, pp. 90–96.

[43] R.R. Guillard, J.H. Ryther, Studies of marine planktonic diatoms: I. Cyclotella nana Hustedt, and *Detonula confervacea* (Cleve) Gran, Can. J. Microbiol. 8 (2) (1962) 229–239.

[44] S. Fon-Sing, M. Borowitzka, Isolation and screening of euryhaline *Tetraselmis* spp. suitable for large-scale outdoor culture in hypersaline media for biofuels, J. Appl. Phycol. 28 (1) (2016) 1–14.

[45] M. Shatat, S.B. Riffat, Water desalination technologies utilizing conventional and renewable energy sources, Int. J. Low-Carbon Technol. 9 (1) (2014) 1–19.

[46] R. El Sergany, M. El Fadly, A. El, Nadi, brine desalination by using algae ponds under nature conditions, Am. J. Environ. Eng. 4 (4) (2014) 75–79.

[47] M. El Nadi, I. Waheb, S. Saad, Using continuous flow algae ponds for water desalination, El Azhar Univ., Faculty Eng., CERM Civil Eng. 33 (4) (2011).

[48] B. Kokabian, U. Ghimire, V.G. Gude, Water deionization with renewable energy production in microalgae-microbial desalination process, Renew. Energy 122 (2018) 354–361.

[49] S.A. Shirazi, J. Rastegary, M. Aghajani, A. Ghassemi, Simultaneous biomass production and water desalination concentrate treatment by using microalgae, Desalin. Water Treat. 135 (2018) 101–107.

[50] J. Wei, L. Gao, G. Shen, X. Yang, M. Li, The role of adsorption in microalgae biological desalination: salt removal from brackish water using *Scenedesmus obliquus*, Desalination 493 (2020) 114616.

[51] X. Gan, G. Shen, B. Xin, M. Li, Simultaneous biological desalination and lipid production by *Scenedesmus obliquus* cultured with brackish water, Desalination 400 (2016) 1–6.

[52] Z. Yao, C. Ying, J. Lu, Q. Lai, K. Zhou, H. Wang, et al., Removal of K +, Na +, Ca^{2+}, and Mg^{2+} from saline-alkaline water using the microalga Scenedesmus obliquus, Chin. J. Oceanol. Limnol. 31 (6) (2013) 1248–1256.

[53] G. Randrianarison, M.A. Ashraf, Microalgae: a potential plant for energy production, Geol. Ecol. Landsc. 1 (2) (2017) 104–120.

[54] P. DasSarma, J.A. Coker, V. Huse, S. DasSarma, Halophiles, industrial applications, Encyclopedia of Industrial Biotechnology: Bioprocess, Bioseparation, Cell and Technology, Wiley, 2009, pp. 1–43.

[55] H. Kageyama, R. Waditee-Sirisattha, Y. Tanaka, T. Takabe, Osmoprotectant and sunscreen molecules from halophilic algae and cyanobacteria, Algal Green Chemistry, Elsevier, 2017, pp. 1–16.

[56] Z.-Y. Wu, C.-L. Shi, X.-M. Shi, Modeling of lutein production by heterotrophic *Chlorella* in batch and fed-batch cultures, World J. Microbiol. Biotechnol. 23 (9) (2007) 1233–1238.

[57] S. Huo, Z. Wang, S. Zhu, W. Zhou, R. Dong, Z. Yuan, Cultivation of *Chlorella zofingiensis* in bench-scale outdoor ponds by regulation of pH using dairy wastewater in winter, South China, Bioresour. Technol. 121 (2012) 76–82.

[58] R. Halim, M.K. Danquah, P.A. Webley, Extraction of oil from microalgae for biodiesel production: a review, Biotechnol. Adv. 30 (3) (2012) 709–732.

[59] M. El-Sheekh, A.E.-F. Abomohra, D. Hanelt, Optimization of biomass and fatty acid productivity of *Scenedesmus obliquus* as a promising microalga for biodiesel production, World J. Microbiol. Biotechnol. 29 (5) (2013) 915–922.

[60] V. Skorupskaite, V. Makareviciene, D. Levisauskas, Optimization of mixotrophic cultivation of microalgae *Chlorella* sp. for biofuel production using response surface methodology, Algal Res. 7 (2015) 45–50.

[61] N.M. Konda, S. Singh, B.A. Simmons, D. Klein-Marcuschamer, An investigation on the economic feasibility of macroalgae as a potential feedstock for biorefineries, BioEnergy Res. 8 (3) (2015) 1046–1056.

[62] G. Roesijadi, S.B. Jones, L.J. Snowden-Swan, Y. Zhu, Macroalgae as a Biomass Feedstock: A Preliminary Analysis, Pacific Northwest National Lab. (PNNL), Richland, WA, 2010.

[63] H.K. Reddy, T. Muppaneni, J. Rastegary, S.A. Shirazi, A. Ghassemi, S. Deng, ASI: hydrothermal extraction and characterization of bio-crude oils from wet chlorella sorokiniana and dunaliella tertiolecta, Environ. Prog. Sustain. Energy 32 (4) (2013) 910–915.

References

[64] J. Rastegary, S.A. Shirazi, T. Fernandez, A. Ghassemi, Water resources for algae-based biofuels, J. Contemp. Water Res. Educ. 151 (1) (2013) 117−122.

[65] C.J. Ridley, B.M. Parker, L. Norman, B. Schlarb-Ridley, R. Dennis, A.E. Jamieson, et al., Growth of microalgae using nitrate-rich brine wash from the water industry, Algal Res. 33 (2018) 91−98.

[66] M. Jämsä, F. Lynch, A. Santana-Sánchez, P. Laaksonen, G. Zaitsev, A. Solovchenko, et al., Nutrient removal and biodiesel feedstock potential of green alga UHCC00027 grown in municipal wastewater under Nordic conditions, Algal Res. 26 (2017) 65−73.

[67] M. Taziki, H. Ahmadzadeh, M.A. Murry, S.R. Lyon, Nitrate and nitrite removal from wastewater using algae, Curr. Biotechnol. 4 (4) (2015) 426−440.

[68] T.M. Mata, A.A. Martins, N.S. Caetano, Microalgae for biodiesel production and other applications: a review, Renew. Sustain. Energy Rev. 14 (1) (2010) 217−232.

[69] J. Pickett, D. Anderson, D. Bowles, T. Bridgwater, P. Jarvis, N. Mortimer, et al., Sustainable Biofuels: Prospects and Challenges, The Royal Society, London, 2008.

[70] M.L. Bartley, W.J. Boeing, A.A. Corcoran, F.O. Holguin, T. Schaub, Effects of salinity on growth and lipid accumulation of biofuel microalga *Nannochloropsis salina* and invading organisms, Biomass Bioenergy 54 (2013) 83−88.

[71] S. Rasoul-Amini, P. Mousavi, N. Montazeri-Najafabady, M.A. Mobasher, S.B. Mousavi, F. Vosough, et al., Biodiesel properties of native strain of *Dunaliella salina*, Int. J. Renew. Energy Res. (IJRER) 4 (1) (2014) 39−41.

[72] K. Minas, E. Karunakaran, T. Bond, C. Gandy, A. Honsbein, M. Madsen, et al., Biodesalination: an emerging technology for targeted removal of Na + and Cl − from seawater by cyanobacteria, Desalin. Water Treat. 55 (10) (2015) 2647−2668.

[73] L. Pirastru, F. Perreault, F.L. Chu, A. Oukarroum, L. Sleno, R. Popovic, et al., Long-term stress induced by nitrate deficiency, sodium chloride, and high light on photosystem II activity and carotenogenesis of green alga *Scenedesmus* sp, Botany 90 (11) (2012) 1007−1014.

[74] P. Kaewkannetra, P. Enmak, T. Chiu, The effect of CO_2 and salinity on the cultivation of *Scenedesmus obliquus* for biodiesel production, Biotechnol. Bioprocess. Eng. 17 (3) (2012) 591−597.

[75] E.-S. Salama, H.-C. Kim, R.A. Abou-Shanab, M.-K. Ji, Y.-K. Oh, S.-H. Kim, et al., Biomass, lipid content, and fatty acid composition of freshwater *Chlamydomonas mexicana* and *Scenedesmus obliquus* grown under salt stress, Bioprocess. Biosyst. Eng. 36 (6) (2013) 827−833.

[76] I. Pancha, K. Chokshi, R. Maurya, K. Trivedi, S.K. Patidar, A. Ghosh, et al., Salinity induced oxidative stress enhanced biofuel production potential of microalgae *Scenedesmus* sp. CCNM 1077, Bioresour. Technol. 189 (2015) 341−348.

CHAPTER 9

Biofuel versus fossil fuel

Ahmed I. EL-Seesy[1,2,*], Mostafa E. Elshobary[3,4,*] and Zhixia He[1]

[1]Institute for Energy Research, Jiangsu University, Zhenjiang, P.R. China
[2]Department of Mechanical Engineering, Benha Faculty of Engineering, Benha University, Benha, Egypt [3]School of Food and Biological Engineering, Jiangsu University, Zhenjiang, P.R. China [4]Department of Botany and Microbiology, Faculty of Science, Tanta University, Tanta, Egypt

9.1 Introduction

The world's population has continued to grow over the past 50 years, which has led directly to a substantial increase in primary energy expenditure. The global energy consumption increased by more than 5.5% throughout the last 40 years, which is the highest growth percentage in this period [1,2] and is expected to reach 25% in 2040 [3]. Henceforth, the world is facing two challenges: global energy demands and environmental pollution, which could endanger the survivability of humankind [4–6]. The global energy dilemma has been growing day by day over the past few decades due to the diminishing traditional fuel resources, such as fossil fuels. The massive use of fossil fuels, mostly as liquids such as biodiesel, for transportation and power generation has released high CO_2 emissions into the atmosphere. The transport sector is assessed to account for 30% of global energy use and is responsible for about 15% of human-made CO_2, as well as 31% of global man-derived O_3 emissions [7,8]. There is an urgent need to minimize their emissions to avoid the adverse effects of global warming and air pollution—more than 4 million people die each year from air pollution [9]. Although renewable and sustainable electricity is planned to increase in the close future in the transportation sector, liquid fuels are still dominant and necessary [10], whereas electricity generation makes up only 4.5% of the global demand in 2017 [11]. Also, the dependence on conventional fuels has led to fluctuations in the price of mineral oil.

---

* Authors Ahmed I. EL-Seesy and Mostafa E. Elshobary contributed equally to this chapter.

Additionally, the dependence on conventional fuels has led to variations in the price of mineral oil. The rise in the price of mineral oil has increased the load on consumers, businesses, and investors, as it creates intense struggle in other countries that use biofuels or alternative energy sources. The growing future energy demands, coupled with concerns about environmental hazards, places greater emphasis on the fabrication of clean liquid fuels, known as biofuels, as appropriate unconventional energy sources. In this regard, the usage of biofuels is attractive and more scalable than other substitutes. Biofuels are manufactured from various biomass feedstocks, which are renewable feedstocks if the biomass multiplying cycle is continued. Biofuel is a clean, renewable energy source that assures a country's economy while committing to a green environment [12].

Biofuels, that is, biogas, bioalcohols (bioethanol and biobutanol), biodiesel, bio-oil, and biohydrogen, are most leading renewable energy sources. The sustainable production of biofuels is a valuable means for reducing climate change, improving economies, especially in the least developed countries, and boosting energy security for the whole world [1,2]. There are several kinds of algal biofuels involving solid biofuel (burning of biomass), liquid biofuels (bioethanol, biobutanol, biooil, and biodiesel), dry biofuel (bioelectricity), or gaseous biofuels (biogas, biomethane, syngas, and biohydrogen). These biofuel types are obtained by various processes such as fermentation, anaerobic digestion, pyrolysis, gasification, lipid transesterification, and liquefaction [12].

Biogas is typically a mixture of methane (CH_4) and CO_2 that is used to produce heat and/or electricity or is liquefied into methanol and compressed directly into car fuel [3,4]. Various sources of biomass can be involved in biogas production, including energy crops, agricultural wastes, food wastes, animal and human feces, industrial wastes, and microalgae. Even though each biomass resource has its character, ability, and practicality in the production of biogas, since the 1950s microalgae seem to have received a lot of interest in this field [5–7].

Bioalcohol is an entirely nonpolar, long hydrocarbon, that has properties that are very comparable to gasoline's properties, and its vaporization heat is superior to that of gasoline, thus making it ideal for operating in gasoline vehicles without modifying or altering the combustion engine [13]. Bio-oil is a mixture formed by breaking down biomacromolecules, such as carbohydrates, proteins, and lipids, into an organic liquid phase. Bio-oil can be normally stored or transported and has lower nitrogen and sulfur contents, as well as high gross heating values [14,15].

Biodiesel sounds very exciting for numerous reasons; it is biodegradable and nontoxic and can replace diesel fuel without reducing the operation and performance of the engine. Also, biodiesel has a high flash point, proper lubrication, and high cetane number. Moreover, it has physicochemical properties incredibly near to fossil diesel fuel, enabling its use either in a pure or combined state with conventional diesel fuel [16,17]. Therefore it provides a dual solution to the crises of fossil fuel depletion and environmental degradation. Generally, biodiesel can be classified into three distinctive creations. The first group depends on edible high oleaginous crops as a feedstock (e.g., soybean, sunflowers, etc.). It is economically scalable and has been produced on an industrial scale. However, there is a big debate about its production. The use of arable land to compete with food security could lead to starvation and raises the risk of price increases that makes their social acceptance more difficult [18]. Furthermore, its impact on greenhouse gas emissions and net energy output is meager [19–21]. Thus the second generation used nonedible plants, biomass residues, and waste biofuels [18]. However, the dependence on arable land is the key drawback of most second-generation biofuels. This challenge is definitely averted by investigating the third-generation of biofuels that are developed primarily from microalgae biomass [22].

Photosynthesis is the primary process in biofuel generation, relying on the conversion of solar energy to chemical energy via the synthesis of high calorific compounds, including carbohydrates and lipids. These compounds are considered the primary source of renewable biofuels, such as bioethanol and biodiesel [19−21]. Photosynthetic efficiency (PhE) differs according to the organisms. The maximum PhE in the terrestrial planets can range from 4.6% to 6% [23], while algae oils signify the greatest PhE amongst all photosynthetic organisms of upto 9% PhE [24]. Compared to the higher plants, algae, including microalgae and macroalgae, exhibit many advantages, including a more rapid growth rate, not being restricted to a specific season, can be cultivated in barren lands using wastewater, and do not compete with food security [4,25,26,82,83]. Additionally, microalgae can produce significantly higher areal energy yields of 30−100 times per hectare of land used, which is feasible with microalgae biomass, for example, the lipid content of *Nannochloropsis salina* was found to reach 37.7% of cellular dry weight [27]. With regard to macroalgae, *Ulva intestinalis* showed the greatest annual lipid productivity of 3.0 g/m^2 year with lipid content of 6.13% of cellular dry weight [5]. *D. fasciola* showed a lipid yield of 4.92% dry weight [82].

There have been a number of articles published on utilizing algal biodiesel in *compression-ignition* (CI) engines in order to evaluate its possibility as a renewable source. Meanwhile, there is a shortage of articles describing CI engine performance and emission features, including the utilization of microalgae or macroalgae biofuels. Furthermore, there are several conflicting findings among these articles. Thus this chapter aims to perform a vital assessment of state-of-the-art in this exact subject that indicates the utilization and comparison of biofuels with diesel fuel through the published articles and point out areas for additional investigation.

9.2 Mechanisms of fossil fuel and biofuel production

It is commonly recognized that the conventional fuel is a fuel created by biological activities, that is, anaerobic decay of buried dead animals and others. It comprises great proportions of carbon and includes petroleum, coal, and natural gas [28]. It varies from volatile substances with small carbon-to-hydrogen proportions (like methane) to liquids (like petroleum), to nonvolatile substances that are very carbon-rich, like anthracite coal. On the other hand, biofuel production differs according to its type of biofuel and its feedstocks. Moreover, some biofuels need pretreatment steps before conversion to biofuel, such as liquid and gaseous biofuels. Pretreatment is a critical step in the conversion of biomass to its simple components that can be easily converted to biofuel. All biofuel mechanisms and pretreatment methods are listed in Table 9.1.

9.3 Algal biomass

Algal biomass includes two main groups: microalgae and macroalgae; both of them can be grown or harvested from natural sources [36]. Microalgae can be planted either in a

184

9. Biofuel versus fossil fuel

TABLE 9.1 Mechanisms of different types of biofuel and pretreatment methods.

Biofuel	Biomass pretreatment	Mechanism	Remarks	References
Biogas	Three pretreatments can be used to improve biogas production including mechanical (shaking, ultrasound, and microwave), thermal, and hydrothermal, to alter biomass through breaking down cell walls, reducing particle sizes, and degrading biopolymers.	Anaerobic digestion	Biochemical constituents of biomass are digested through sequential biochemical reactions due to the coomensalism carried out by a variety of microbes in the digester.	[26]
Bioalcohol	Physical, chemical, physicochemical, biological, and enzymatic are commonly the main pretreatment methods in the fermentation of lignocellulosic biomass or cellulosic microalgal biomass.	Fermentation	Fermentation is carried out using microorganisms, such as bacteria and yeast, which are able to generate alcohols from different carbohydrate types, including hexoses or pentose.	[29−31]
Bioethanol	Hydrolysis is based on breaking down the cell and hydrolysis of the biomass chemically or via enzymes in photobioreactors.			[18,32]
Bio-oil		Pyrolysis	Pyrolysis is the thermal degradation of dry biomass by heat (low or high) in the absence of oxygen, which results in the production of various products including bio-oils, solid residues (biochar), and gaseous products.	[14]
		Hydrothermal liquefaction	The process that converts wet biomass into liquid fuels in the presence of water or water-containing solvent/cosolvent and a catalyst.	
Biodiesel	Lipid extraction	Transesterification	The use of alkali catalysis, such as hydroxide (KOH), sodium hydroxide (NaOH), or sodium methoxide (CH_3ONa) with alcohol (methanol or ethanol) to covert crude oil to biodiesel.	[18,33]
		Esterification	A significant quantity of acid (5%−25%) and high alcohol:fatty acids molar ratio (20:1−40:1) are necessary.	[18,32]
		Enzymatic conversion	Lipase, an enzyme, has a good ability to catalyze the conversion of oils to biodiesel in the form of a solution or immobilized form.	[18,34]
		Noncatalytic conversion	Comprises supercritical conversion, microwave-assisted conversion, or ultrasound-assisted conversion.	[18,35]

Handbook of Algal Biofuels

closed or open-air culture scheme. The most common open scheme is raceway ponds, while photobioreactors (PBRs) are the most famous closed culture system. A raceway pond is composed of a shallow ~1-m-deep twist recirculation canal, in which microalgae are mixed and circulated by a paddle wheel in the open air. PBRs are closely controlled culture systems made of transparent material that allows light to pass through under the computerized system control [37,38]. These two systems vary in economic feasibility and biomass productivity. The raceway systems are cheap to make and work, while the quality and productivity of the biomass are higher in the PBRs compared to the raceway [36]. Several types of PBRs have been made, such as tubular, plate, bubble, Christmas, foil, horizontal, and porous PBRs [37,39]. Disappointingly, most trials of building cost-effective PBRs have so far failed, which is attributed to its complication and exclusive building materials [40,41]. Therefore hybrid systems were investigated using both systems to maximize the advantages and minimize the drawbacks in two stages. The hybrid cultivation scheme is the sophisticated approach in the algal culture where the cultivation medium is transferred from PBR to an open system. Firstly, PBRs are chosen to reduce contamination and lipid productivity and increase algal biomass. Then, the culture is transferred to the open system to reduce the production cost and to promote carbohydrate and lipids accumulation [39]. Moreover, some supplements could improve the biomass and lipid productivity, such as carbon-rich compounds or carbohydrates, and wastes, such as dairy products in heterotrophic systems. Using this approach can further reduce the costs of microalgal production and harvesting, as well as enable the treatment of various categories of wastewater such as municipal, industrial, and agricultural wastewaters [25,42]. On the other hand, naturally occurring microalgae may flourish, especially in eutrophicated water bodies [43]. In algal blooms, the huge naturally appearing microalgal biomass can achieve many grams of fresh cells per liter, turning the surface water green, red, or brown. There are numerous microalgal species known to cause blooming globally [44]. When microalgae are grown, forming a biofilm, the algal cells are naturally growing, accumulated, and easily collected. This can lead to a reduction in the cultivation and harvesting costs as well as the total biofuel production cost [45].

Macroalgae or seaweeds also can be cultivated or grow naturally. In 2012 the annual production of global farmed and captured seaweeds was estimated in 33 countries to be 23.8 million wet tons, which accounted to about US$ 5.5−6 billion. *Laminaria japonica* was the utmost cultured at 6.8 million tons [46]. Seaweeds are innovative products that can be utilized in the feed, food, cosmetics, chemicals, and pharmaceutical industries [47]. When the main target outcome is bioenergy, the cultivation of seaweeds is unlikely to be commercially feasible [48,49]. Therefore several pieces of research have been conducted to make it possible by depending on the remediation ability of such biomass [50,51]. Macroalgae can be utilized as nutrients-removing organisms in highly eutrophicated seawater that is characterized by high levels of N, P, CO_2, and a low content of dissolved O_2 [51,52]. Consequently, there is a significant interest in eliminating a large amount of C, N, and P surplus nutrients with comprehensive macroalgae cultivation [51,53]. Biomass produced by such cultivations could be used for bioenergy production, as well as high-value products [5,54]. In general, the extreme natural growth or blooming of microalgae and seaweeds has been detected due to the excessive eutrophication of the water body [43,55]. The drift and subsequent degradation of this biomass resource are deemed to be a real

186   9. Biofuel versus fossil fuel

pollution challenge, which can be resolved via the utilization of such biomass for bioenergy production [56,57].

9.4 Algal biodiesel and other types of physicochemical properties

Table 9.1 illustrates the comparison between diesel fuel, algal oil, algal biodiesel, and the American Society for Testing and Materials (ASTM) standard for biodiesel. It can be seen that the kinematic viscosity for algal oil is about 10 times that of diesel fuel, while algal biodiesel is almost comparable to conventional fuel and within the ASTM standards. However, there are certain drawbacks that are occasionally reduced but are essential when biodiesel is transported to the final customers, where microalgae and algae generate biodiesel mostly comprising polyunsaturated acids [58], which perform inadequately compared with the mainstream alternative. The level of unsaturation influences the oxidative stability, ignition quality, and cold flow features [59].

9.5 Combustion and emission parameters

Biodiesel has promising properties that allows it to be a fitting alternative to conventional fuel in CI engines via cutting-edge techniques, including an advanced injection system and boost power. Nevertheless, each kind of biodiesel has its own unique physical and chemical properties that are attributed to each having a varying fatty acid methyl ester (FAME) features. These highlights are satisfactory for the assessment of the aspects of biodiesels depending on their FAME performance. The algal oil and algal biodiesel are considered to be an encouraging alternative source to replace or reduce the dependency on conventional fuel. As a result, academics have made enormous efforts on enhancing algal oil exploitation as an alternative fuel in diesel engines [60–65]. The utilization of pure algal oil, algal biodiesel, and their blends with conventional fuel in CI engines is demonstrated in Table 9.1. These investigations have shown that the brake-specific fuel consumption (bsfc) was increased, while the brake thermal efficiency (BTE) was reduced using algal biodiesel or a blend with conventional fuel, which was attributed to its lower energy content. Regarding engine emissions, the soot, unburned hydrocarbons (UHC), and carbon monoxide (CO) quantities were diminished, whereas the oxides of nitrogen (NO_x) intensity was increased with the addition of algal biodiesel. Concerning combustion parameters, the cylinder pressure, heat release rate (HRR), and ignition delay (ID) depended on the engine running terms.

Over the past few decades, there have been substantial endeavors undertaken on the utilization of numerous metal oxide nanoparticles, including ZnO, TiO, Al_2O_3, and CeO_2, to augment the ignition behavior of diesel and biodiesel and to reduce diesel engine emissions [66–74]. Therefore scientists have staged massive effort for enhancing the application of nanoparticles as additives by exploring the engine aspects of diesel–algae biodiesel combinations [60,61]. Details on the supplementation of nanoparticles with diesel–algae biodiesel combinations on the CI engine performance and emissions are listed in Table 9.2. These investigations have shown that bsfc is reduced and the BTE is enhanced with the

Handbook of Algal Biofuels

TABLE 9.2 Evaluation of numerous properties of diesel, algal oil, and algal biodiesel with ASTM biodiesel standards [59,66,75–77].

Properties	Diesel fuel (ASTM D7675)	Algae oil	Microalgae *Spirulina* biodiesel	*Jatropha* biodiesel	Waste cooking oil biodiesel	Biodiesel (ASTM D6751)
Appearance	Clear	Translucent	Translucent	Translucent	Translucent	Translucent
Color	Yellowish	Yellow/greenish	Greenish yellow			Greenish yellow
Kinematic viscosity, 40 C (cSt)	1.9 to 4.1	33.74	4.41	5.26	5.21	1.9–6
Acid no. (mg KOH per g)	Max 0.5	0.2	0.374	–	–	Max 0.5
Calorific value (MJ/kg)	40–45	35.8	37.5	40.63	38.11	39
Flash point (C)	>62	220	115	–	–	Min 100
Cetane number	45–60	–	37–72	52	49	35–50
Density (kg/L)	0.84–0.9	0.897	0.864	0.883	0.879	0.86–0.9

addition nanoparticles. Also, the smoke opacity, UHC, and CO concentrations are lowered, whereas the NO_x emission is increased. Moreover, there are a few experiments that have researched the impacts of combining oxygenated fuel like diethyl ether (DEE) with diesel–algae biodiesel mixtures [78] (Tables 9.3 and 9.4).

9.6 Conclusions

The utilization of algae for biodiesel creation generates numerous benefits in environment-friendly and land use terms. The oil separated from macroalgae or microalgae biomass has unique characteristics like high unsaturated concentration. Hence, the evaluation of their usage in CI engines is extremely valuable. The strain choice impacts the chemical composition which in turn affects engine performance and emission features. There is a significant discrepancy in the number of articles published regarding algae, biodiesel, economic evaluation, and the association with engine performance. *Chlorella* is one of the most studied microalgae in terms of diesel engine assessment, which is associated with its global accessibility. A wide variety of mixtures up to B50 but also pure biodiesel have been examined. The reduction of BTE and increase of bsfc are mostly noted. The mixture giving findings nearest to diesel fuel is B20. The soot, CO, and UHC levels can be reduced when using algal biodiesel, but an increase in NO_x level is commonly mentioned, owing to the higher temperatures in the combustion chamber. Also, the use of nanoparticle additives with diesel–algae biodiesel blends has shown promising results in terms of enhancing the bsfc and exhaust emissions, although the NO_x level is increased. The

TABLE 9.3 Burning and emission features of diesel engines run by algae biodiesel–diesel combinations.

Engine	Base fuel	Running condition	Proportions of algae	Performance results	Burning results	Emission results	References
Single cylinder, 499.56cc	Diesel–water emulsion.	Constant speed and load.	5%,7.5%, 10%, and 20% vol.	Reduced brake thermal efficiency (BTE); increased bsfc.	Enhanced p and heat release rate (HRR); reduced ignition delay (ID).	Diminished NO_x, and increased CO.	[63]
Single cylinder, 553cc	Diesel fuel–microalgae *Spirulina*.	Variable compression ratio; constant and load.	5% and 10% vol for each of five components (biocrude paraffin, xylene, cyclopentanone, dioctylphthalate, and butanol).	Reduced BTE; increased bsfc.	Diminished p, HRR; enlarged ID.	Enlarged NO_x; sowered PM.	[60]
Single cylinder, 552.64cc	Diesel fuel–microalgae *Spirulina*.	Constant speed with variable loads.	20%,40%, 60%, 80% and 100% vol.	Diminished BTE; increased bsfc.	Diminished p, HRR, and enlarged ID.	Lowered soot, NO_x, and PM.	[61]
Turbocharged-six-cylinder, 5900cc	Diesel fuel–microalgae fuel components.	Constant speed with high load.	5% and 10% vol. for each 5 components (biocrude paraffin, xylene, cyclo-pentanone, dioctylphthalate and butanol).	Dropped BTE; enlarged bsfc.	Dropped p, HRR, and ID.	Enlarged NO_x; lowered CO, PM, and total particle number (PN).	[62]
Single cylinder, 661.11 cc	Diesel–algal biodiesel.	Variable compression ratio; constant and load.	5%, 10%, 15%, 20%, 25%, 30%, 35%, 40%, 45%, and 50vol.	———	———	Diminished UHC and CO; enlarged NO_x.	[65]
Single cylinder, 661 cc	Diesel fuel, algal biodiesel, and algal oil.	Constant speed with different loads.	100%	Diminished BTE; enlarged bsfc.	Diminished p; enlarged ID.	Lowered NO_x, UHC, CO, and smoke opacity.	[75]

TABLE 9.4 Burning and emission aspects of CI engines powered by algae biodiesel—diesel combinations with nanoparticles and oxygenated additives.

Engine	Base fuel	Working situation	Proportions of algae	Percentages of additives	Performance findings	Burning findings	Emission findings	Reference
Single cylinder, 661 cc	Diesel—*Botryococcus braunii* algae biodiesel and CuO_2 nanoparticles.	Different speed and loads.	20% vol.	25, 50, 75, and 100 ppm.	Diminished bsfc and enhanced brake thermal efficiency (BTE)	—	—	[79]
Single-cylinder, 661 cc	Diesel fuel/*Naviculla* sp. algae oil biodiesel with MgO nanoparticles.	Constant speed with different loads.	20% vol.	25, 50, and 100 ppm.	Diminished bsfc and enhanced BTE.	—	Enlarged NO_x; diminished CO, UHC, and smoke opacity.	[80]
Single-cylinder, 661 cc	Diesel fuel/*Naviculla* sp. algae oil biodiesel with La_2O_3 nanoparticles.	Constant speed and load.	20%	50 and 100 ppm.	Diminished bsfc and enhanced BTE.	—	Lowered the total particulate matter and soot, heightened NO_x.	[81]
Single-cylinder, 661 cc	Diesel fuel/microalgae biodiesel—oxygenated additive diethyl ether (DEE)—antioxidant *N,N′*-Diphenyl-*p*-phenylenediamine (DPPD).	EGR; constant speed and load.	80%, 85%, and 90%.	10, 15 and 20% (DEE); 1000 and 2000 ppm (DPPD).	Diminished bsfc and improved BTE.	—	Lowered the particulate emission UHC, CO, and NO_x.	[78]

evaluation of biodiesel generated from macroalgae or microalgae is an area that is not very well studied, and to date various references and articles have created conflicting findings or inadequate research, as this review has revealed. Thus further investigations are needed to fill this research gap in the published papers related to nanoparticles and oxygenated additives with pure algal oil and algal biodiesel blended with higher alcohols.

Acknowledgments

This book chapter was sponsored by the National Natural Science Foundation of China (51876083, 51776088, and 52050410330).

References

[1] F. Alam, A. Date, R. Rasjidin, S. Mobin, H. Moria, A. Baqui, Biofuel from algae-Is it a viable alternative? Proc. Eng. 49 (2012) 221−227.

[2] C.S. Jones, S.P. Mayfield, Algae biofuels: versatility for the future of bioenergy, Curr. Opin. Biotechnol. 23 (2012) 346−351.

[3] M.L.N.M. Carneiro, F. Pradelle, S.L. Braga, M.S.P. Gomes, A.R.F.A. Martins, F. Turkovics, et al., Potential of biofuels from algae: comparison with fossil fuels, ethanol and biodiesel in Europe and Brazil through life cycle assessment (LCA), Renew. Sustain. Energy Rev. 73 (2017) 632−653.

[4] M.E. Elshobary, H.M. Zabed, J. Yun, G. Zhang, X. Qi, Recent insights into microalgae-assisted microbial fuel cells for generating sustainable bioelectricity, Int. J. Hydrog. Energy 46 (2020) 3135−3159.

[5] M.E.H. Osman, A.M. Abo-Shady, M.E. Elshobary, M.O. Abd El-Ghafar, A. Abomohra, Screening of seaweeds for sustainable biofuel recovery through sequential biodiesel and bioethanol production, Environ. Sci. Pollut. Res. 27 (2020) 32481−32493.

[6] V.K. Gupta, M.G. Tuohy (Eds.), Biofuel Technologies: Recent Developments, Springer, 2013.

[7] G. Perin, P.R. Jones, Economic feasibility and long-term sustainability criteria on the path to enable a transition from fossil fuels to biofuels, Curr. Opin. Biotechnol. 57 (2019) 175−182.

[8] J. Fuglestvedt, T. Berntsen, G. Myhre, K. Rypdal, R.B. Skeie, Climate forcing from the transport sectors, Proc. Natl. Acad. Sci. 105 (2008) 454−458.

[9] WHO, World Health Organisation. Ambient air pollution: health impacts <http://www.who.int/airpollution/ambient/health-im> (2018) (accessed 18.12.20).

[10] B. Dudley, BP statistical review of world energy, BP Stat. Rev., London, 2018 (accessed 06.08.18).

[11] I. International Energy Agency, World energy outlook. <https://www.iea.org/weo2017/> 2017 (accessed 12.08.20).

[12] A. Abomohra, M.E. Elshobary, Biodiesel, Bioethanol, and Biobutanol Production from Microalgae, Springer, Singapore, Singapore, 2019. Available from: https://doi.org/10.1007/978−981-13−2264-8.

[13] V. Hönig, M. Kotek, J. Mařík, Use of butanol as a fuel for internal combustion engines, Agron. Res. 12 (2014) 333−340.

[14] C. Yang, R. Li, B. Zhang, Q. Qiu, B. Wang, H. Yang, et al., Pyrolysis of microalgae: a critical review, Fuel Process. Technol. 186 (2019) 53−72.

[15] A.E.F. Abomohra, M. Elsayed, S. Esakkimuthu, M. El-Sheekh, D. Hanelt, Potential of fat, oil and grease (FOG) for biodiesel production: a critical review on the recent progress and future perspectives, Prog. Energy Combust. Sci. 81 (2020) 100868.

[16] A.E. Atabani, M.M. El-Sheekh, G. Kumar, S. Shobana, Edible and Nonedible Biodiesel Feedstocks: Microalgae and Future of Biodiesel, Clean Energy for Sustainable Development, Elsevier Inc, 2017, pp. 507−556.

[17] S. Basumatary, Yellow oleander (Thevetia peruviana) seed oil biodiesel as an alternative and renewable fuel for diesel engines: a review, Int. J. ChemTech Res. 7 (2015) 282−340.

[18] I. Abdelsalam, M. Elshobary, M.M. Eladawy, M. Nagah, Utilization of multi-tasking non-edible plants for phytoremediation and bioenergy source—a review, Phyton (B. Aires) 88 (2019) 69−90.

References

[19] A. Abomohra, M.E. Elshobary, Biodiesel, bioethanol and biobutanol production from microalgae biomass, in: M. Alam, Z. Wang (Eds.), Microalgae Biotechnology for Development of Biofuel and Wastewater Treatment, Springer, Singapore, 2019, pp. 293–321. Available from: https://doi.org/10.1007/978-981-13-2264-8_13. Microalgae Biotechnol. Dev. Biofuel Waste Water Treat., Springer Singapore Press, 2019: pp. 293–321.

[20] A. Benoist, D. Dron, A. Zoughaib, Origins of the debate on the life-cycle greenhouse gas emissions and energy consumption of first-generation biofuels—a sensitivity analysis approach, Biomass Bioenergy 40 (2012) 133–142.

[21] P. Collet, A. Hélias, L. Lardon, M. Ras, R.-A. Goy, J.P. Steyer, Life-cycle assessment of microalgae culture coupled to biogas production, Bioresour. Technol. 102 (2011) 207–214.

[22] A. Singh, S.I. Olsen, A critical review of biochemical conversion, sustainability and life cycle assessment of algal biofuels, Appl. Energy. 88 (2011) 3548–3555.

[23] X.G. Zhu, S.P. Long, D.R. Ort, What is the maximum efficiency with which photosynthesis can convert solar energy into biomass? Curr. Opin. Biotechnol. 19 (2008) 153–159.

[24] P.S. Shukla, E.G. Mantin, M. Adil, S. Bajpai, A.T. Critchley, B. Prithiviraj, *Ascophyllum nodosum*-based biostimulants: sustainable applications in agriculture for the stimulation of plant growth, stress tolerance and disease management, Front. Plant Sci. 10 (2019) 655.

[25] M.E. Elshobary, A.M. Abo-Shady, H.M. Khairy, D. Essa, H.M. Zabed, X. Qi, et al., Influence of nutrient supplementation and starvation conditions on the biomass and lipid productivities of *Micractinium reisseri* grown in wastewater for biodiesel production, J. Environ. Manage. 250 (2019) 109529.

[26] H.M. Zabed, S. Akter, J. Yun, G. Zhang, Y. Zhang, X. Qi, Biogas from microalgae: technologies, challenges and opportunities, Renew. Sustain. Energy Rev. 117 (2020) 109503.

[27] M. Ashour, M.E. Elshobary, R. El-shenody, A. Kamil, A.E. Abomohra, Evaluation of a native oleaginous marine microalga *Nannochloropsis oceanica* for dual use in biodiesel production and aquaculture feed, Biomass Bioenergy 120 (2019) 439–447.

[28] G. Mutezo, J. Mulopo, A review of Africa's transition from fossil fuels to renewable energy using circular economy principles, Renew. Sustain. Energy Rev. 137 (2021) 110609.

[29] T. Yoshida, Y. Tashiro, K. Sonomoto, Novel high butanol production from lactic acid and pentose by *Clostridium saccharoperbutylacetonicum*, J. Biosci. Bioeng. 114 (2012) 526–530.

[30] C.K. Phwan, H.C. Ong, W.H. Chen, T.C. Ling, E.P. Ng, P.L. Show, Overview: comparison of pretreatment technologies and fermentation processes of bioethanol from microalgae, Energy Convers. Manage. 173 (2018) 81–94.

[31] L. Kexun, L. Shun, L. Xianhua, An overview of algae bioethanol production, Int. J. Energy Res. 38 (2014) 965–977.

[32] S. Kang, J. Fu, G. Zhang, From lignocellulosic biomass to levulinic acid: a review on acid-catalyzed hydrolysis, Renew. Sustain. Energy Rev. 94 (2018) 340–362.

[33] H.M. Zabed, S. Akter, J. Yun, G. Zhang, F.N. Awad, X. Qi, et al., Recent advances in biological pretreatment of microalgae and lignocellulosic biomass for biofuel production, Renew. Sustain. Energy Rev. (2019) 105–128.

[34] J.M. Marchetti, V.U. Miguel, A.F. Errazu, Possible methods for biodiesel production, Renew. Sustain. Energy Rev. 11 (2007) 1300–1311.

[35] B. Bharathiraja, M. Chakravarthy, R.R. Kumar, D. Yuvaraj, J. Jayamuthunagai, R.P. Kumar, et al., Biodiesel production using chemical and biological methods—a review of process, catalyst, acyl acceptor, source and process variables, Renew. Sustain. Energy Rev. 38 (2014) 368–382.

[36] M. Dębowski, M. Zieliński, A. Grala, M. Dudek, Algae biomass as an alternative substrate in biogas production technologies—review, Renew. Sustain. Energy Rev. 27 (2013) 596–604.

[37] M.R. Tredici, Mass production of microalgae: photobioreactors, Handb. Microalgal Cult. Biotechnol. Appl. Phycol. 1 (2004) 178–214.

[38] Y. Chisti, Biodiesel from microalgae, Biotechnol. Adv. 25 (2007) 294–306.

[39] R. Ganesan, S. Manigandan, M.S. Samuel, R. Shanmuganathan, K. Brindhadevi, N.T. Lan Chi, et al., A review on prospective production of biofuel from microalgae, Biotechnol. Reports. 27 (2020) e00509.

[40] S.S. Oncel, Microalgae for a macroenergy world, Renew. Sustain. Energy Rev. 26 (2013) 241–264.

[41] E. Stephens, I.L. Ross, J.H. Mussgnug, L.D. Wagner, M.A. Borowitzka, C. Posten, et al., Future prospects of microalgal biofuel production systems, Trends Plant Sci. 15 (2010) 554–564.

[42] J.B.K. Park, R.J. Craggs, A.N. Shilton, Wastewater treatment high rate algal ponds for biofuel production, Bioresour. Technol. 102 (2011) 35–42.

[43] M.E. Elshobary, D.I. Essa, A.M. Attiah, Z.E. Salem, X. Qi, Algal community and pollution indicators for the assessment of water quality of Ismailia canal, Egypt, Stoch. Environ. Res. Risk Assess. 34 (2020) 1089–1103.

[44] HKSAR, Agricultural, fisheries and conservation department, HKSAR. Hong Kong Red Tide Information Network. <https://www.afcd.gov.hk/english/fisheries/hkredtide/redtide.html> (accessed 24.08.20).

[45] L.B. Christenson, R.C. Sims, Rotating algal biofilm reactor and spool harvester for wastewater treatment with biofuels by-products, Biotechnol. Bioeng. 109 (2012) 1674–1684.

[46] S.M. Phang, A.T. Critchley, P.O. Ang, Advances in seaweed cultivation and utilization in Asia, in: Asian Fish. Forum 2004 Penang, Malaysia, University of Malaya Maritime Research Centre, 2006.

[47] M.E.H. Osman, A.M. Abu-Shady, M.E. Elshobary, A. Aboshady, M.E. Elshobary, The seasonal fluctuation of the antimicrobial activity of some macroalgae collected from Alexandria coast, Egypt, in: B. Annous, J. Gurtler (Eds.), Salmonella - Distribution, Adaptation, Control Measures and Molecular Technologies, InTech, Rijeka, 2012, pp. 173–186.

[48] J.K. Pittman, A.P. Dean, O. Osundeko, The potential of sustainable algal biofuel production using wastewater resources, Bioresour. Technol. 102 (2011) 17–25.

[49] A. Bruhn, J. Dahl, H.B. Nielsen, L. Nikolaisen, M.B. Rasmussen, S. Markager, et al., Bioenergy potential of *Ulva lactuca*: biomass yield, methane production and combustion, Bioresour. Technol. 102 (2011) 2595–2604.

[50] J.C. Sanderson, M.J. Dring, K. Davidson, M.S. Kelly, Culture, yield and bioremediation potential of *Palmaria palmata* (Linnaeus) Weber & Mohr and *Saccharina latissima* (Linnaeus) C.E. Lane, C. Mayes, Druehl & G.W. Saunders adjacent to fish farm cages in northwest Scotland, Aquaculture 354–355 (2012) 128–135.

[51] C.G. Golueke, W.J. Oswald, H.B. Gotaas, Anaerobic digestion of algae, Appl. Microbiol. 5 (1957) 47–55.

[52] Y.F. Yang, X.G. Fei, J.M. Song, H.Y. Hu, G.C. Wang, I.K. Chung, Growth of *Gracilaria lemaneiformis* under different cultivation conditions and its effects on nutrient removal in Chinese coastal waters, Aquaculture 254 (2006) 248–255.

[53] A.H. Buschmann, F. Cabello, K. Young, J. Carvajal, D.A. Varela, L. Henríquez, Salmon aquaculture and coastal ecosystem health in Chile: analysis of regulations, environmental impacts and bioremediation systems, Ocean. Coast. Manage. 52 (2009) 243–249.

[54] I.K. Chung, Y.H. Kang, C. Yarish, P.K. George, J.A. Lee, Application of seaweed cultivation to the bioremediation of nutrient-rich effluent, Algae 17 (2002) 187–194.

[55] P. Morand, Coastal eutrophication and excessive growth of macroalgae, Recent. Res. Dev. Environ. Biol. 1 (2004) 395–449.

[56] P. Morand, X. Briand, R.H. Charlier, Anaerobic digestion of ulva sp. 3. Liquefaction juices extraction by pressing and a technico-economic budget, J. Appl. Phycol. 18 (2006) 741–755.

[57] G. Migliore, C. Alisi, A.R. Sprocati, E. Massi, R. Ciccoli, M. Lenzi, et al., Anaerobic digestion of macroalgal biomass and sediments sourced from the *Orbetello lagoon*, Italy Biomass Bioenergy 42 (2012) 69–77.

[58] T.V. Ramachandra, M. Durga Madhab, S. Shilpi, N.V. Joshi, Algal biofuel from urban wastewater in India: scope and challenges, Renew. Sustain. Energy Rev. 21 (2013) 767–777.

[59] R. Piloto-Rodríguez, Y. Sánchez-Borroto, E.A. Melo-Espinosa, S. Verhelst, Assessment of diesel engine performance when fueled with biodiesel from algae and microalgae: an overview, Renew. Sustain. Energy Rev. 69 (2017) 833–842.

[60] U. Rajak, P. Nashine, T.N. Verma, A. Pugazhendhi, Performance, combustion and emission analysis of microalgae *Spirulina* in a common rail direct injection diesel engine, Fuel 255 (2019) 115855.

[61] U. Rajak, P. Nashine, T.N. Verma, Effect of *spirulina* microalgae biodiesel enriched with diesel fuel on performance and emission characteristics of CI engine, Fuel 268 (2020) 117305.

[62] F.M. Hossain, M. Nurun Nabi, R.J. Brown, Investigation of diesel engine performance and exhaust emissions of microalgae fuel components in a turbocharged diesel engine, Energy Convers. Manage. 186 (2019) 220–228.

[63] X. Wei, P. Hellier, F. Baganz, Impact on performance and emissions of the aspiration of algal biomass suspensions in the intake air of a direct injection diesel engine, Energy Convers. Manage. 205 (2020) 112347.

[64] B. Wang, P.G. Duan, Y.P. Xu, F. Wang, X.L. Shi, J. Fu, et al., Co-hydrotreating of algae and used engine oil for the direct production of gasoline and diesel fuels or blending components, Energy 136 (2017) 151–162.

[65] M. Yadav, S.B. Chavan, R. Singh, F. Bux, Y.C. Sharma, Experimental study on emissions of algal biodiesel and its blends on a diesel engine, J. Taiwan. Inst. Chem. Eng. 96 (2019) 160–168.

[66] A.I. EL-Seesy, H. Hassan, S. Ookawara, Performance, combustion, and emission characteristics of a diesel engine fueled with Jatropha methyl ester and graphene oxide additives, Energy Convers. Manage. 166 (2018).

[67] A.I. EL-Seesy, H. Kosaka, H. Hassan, S. Sato, Combustion and emission characteristics of a common rail diesel engine and RCEM fueled by n-heptanol-diesel blends and carbon nanomaterial additives, Energy Convers. Manage. 196 (2019).

[68] A.I. EL-Seesy, H. Hassan, Combustion characteristics of a diesel engine fueled by biodiesel-diesel-n-butanol blend and titanium oxide additives, in: Energy Procedia, 2019.

[69] A.I. EL-Seesy, H. Hassan, Investigation of the effect of adding graphene oxide, graphene nanoplatelet, and multiwalled carbon nanotube additives with n-butanol-Jatropha methyl ester on a diesel engine performance, Renew. Energy 132 (2019).

[70] A.I. El-Seesy, A.M.A. Attia, H.M. El-Batsh, The effect of Aluminum oxide nanoparticles addition with Jojoba methyl ester-diesel fuel blend on a diesel engine performance, combustion and emission characteristics, Fuel 224 (2018).

[71] M. Nour, A.I. EL-Seesy, A.K. Abdel-Rahman, M. Bady, Influence of adding aluminum oxide nanoparticles to diesterol blends on the combustion and exhaust emission characteristics of a diesel engine, Exp. Therm. Fluid Sci. 98 (2018).

[72] A.I. El-Seesy, H. Hassan, S. Ookawara, Effects of graphene nanoplatelet addition to jatropha biodiesel–diesel mixture on the performance and emission characteristics of a diesel engine, Energy 147 (2018).

[73] A.I. El-Seesy, A.K. Abdel-Rahman, M. Bady, S. Ookawara, Performance, combustion, and emission characteristics of a diesel engine fueled by biodiesel-diesel mixtures with multi-walled carbon nanotubes additives, Energy Convers. Manage. 135 (2017).

[74] A.I. El-Seesy, A.K. Abdel-Rahman, M. Bady, S. Ookawara, The influence of multi-walled carbon nanotubes additives into non-edible biodiesel-diesel fuel blend on diesel engine performance and emissions, in: Energy Procedia, 2016.

[75] S.S. Satputaley, D.B. Zodpe, N.V. Deshpande, Performance, combustion and exhaust emissions analysis of a diesel engine fuelled with algae oil and algae biodiesel, Mater. Today Proc. 5 (2018) 23022–23032.

[76] A.I. EL-Seesy, Z. He, H. Hassan, D. Balasubramanian, Improvement of combustion and emission characteristics of a diesel engine working with diesel/jojoba oil blends and butanol additive, Fuel 279 (2020) 118433.

[77] M.S. Gad, A.I. EL-Seesy, A. Radwan, Z. He, Enhancing the combustion and emission parameters of a diesel engine fueled by waste cooking oil biodiesel and gasoline additives, Fuel 269 (2020).

[78] V. Krishna Kolli, S. Gadepalli, J. Deb Barma, M. Krishna Maddali, S. Barathula, N. Kumar reddy Siddavatam, Establishment of lower exhaust emissions by using EGR coupled low heat loss diesel engine with fuel blends of microalgae biodiesel-oxygenated additive DEE-antioxidant DPPD, Therm. Sci. Eng. Prog. 13 (2019) 100401.

[79] T. Dharmaprabhakaran, S. Karthikeyan, M. Periyasamy, G. Mahendran, Combustion analysis of CuO_2 nanoparticle addition with blend of botryococcus braunii algae biodiesel on CI engine, Mater. Today Proc. 33 (2020) 2874–2876.

[80] J. Arunprasad, T. Elango, Performance and emission characteristics of engine using *Naviculla* Sp. algae oil methyl ester with MgO nanoparticles, in: Materials Today: Proceedings, 2020.

[81] J. Arunprasad, T. Elango, Performance and emission analysis of engine using Naviculla Sp. Algae oil methyl ester with La_2O_3 nanoparticles, Mater. Today Proc. 33 (2020) 3177–3181.

[82] M. Elshobary, R. El-Shenody, A. Abomohra, Sequential biofuel production from seaweeds enhances the energy recovery: A case study for biodiesel and bioethanol production, Int. J. Energy Res. 45 (4) (2020) 6457–6467. Available from: https://doi.org/10.1002/er.6181.

[83] D. Essa, A. Abo-Shady, H. Khairy, Abomohra A., M. Elshobary, Potential cultivation of halophilic oleaginous microalgae on industrial wastewater, Egypt. J. Bot. 58 (2) (2018) 205–216. Available from: https://doi.org/10.21608/ejbo.2018.809.1054.

Algae for biodiesel production

Mohammadhosein Rahimi, Fateme Saadatinavaz and Mohammadhadi Jazini

Department of Chemical Engineering, Isfahan University of Technology, Isfahan, Iran

10.1 Introduction

With the growing population and development of different industries, energy consumption has been progressively growing during recent decades. Although fossil fuels meet the energy-based human needs, they are expected to diminish in a few years [1]. However, their consumption causes many environmental problems by releasing different toxic gases, especially CO_2, to the atmosphere, leading to global warming [2]. Accordingly, using renewable energy resources has recently attracted worldwide attention due to their well-known benefits, specifically reducing the carbon footprint on the planet, since CO_2 levels are expected to rise by 80% between 2007 and 2030 [3,4]. Biomass-derived biofuels, divided into liquid, gaseous, and solid fuels, are of substantial importance and they are one of the main preferred renewable, sustainable energy sources. Amongst these biofuels, biodiesel is very promising for use as a substitute to fossil fuels due to its advantages such as low toxicity, no release of pollutants (e.g., sulfates, and aromatic compounds), and its biodegradability [5]. In comparison with diesel, biodiesel could be used either as an alternative or blended with diesel without any specific modifications in the engine [6]. Furthermore, increasing the biodiesel application in the transport section, due to its similar properties to diesel and lower emissions, will significantly minimize the environmental problems. Up to now, several feedstocks have been utilized for biodiesel production, such as edible/nonedible oils and wastes [7–9].

The availability and type of feedstock are essential issues in the production of different generations of biodiesel. Also, choosing an appropriate feedstock for biodiesel production is highly important, while about 75% of the biodiesel production cost is related to the feedstock [10]. Biodiesel is produced from the transesterification of renewable oil feedstocks and a short-chain alcohol, mostly methanol, either in the presence of a catalyst or not [11]. Therefore biodiesel is generally divided into three different generations according to the raw materials used for its production [12]. Among different feedstocks for biodiesel production, some edible vegetable oils such as palm, sunflower, soya bean, rapeseed, cottonseed, kernel, coconut, groundnut, and canola oil are considered to be the most common used feedstocks for biodiesel production, and they are known as first-generation feedstocks [4]. The use of such materials in biodiesel production is hugely questionable as it should be noted that fuel production should not compete with human food reserves. Additionally, lots of water would need to be used to cultivate such feedstocks in addition to the large area of arable land occupied [13]. Therefore some countries have banned using these feedstocks for biodiesel production [14]; however, second- and third-generation feedstocks have been considered for biodiesel production as they include nonedible feedstocks [12]. Accordingly, some feedstocks such as nonedible vegetable oils, animal fats, waste oils, and greases have been considered for the production of second-generation biodiesel. These feedstocks have partially solved the first-generation feedstocks' problems as they do not compete with the food and water supply [15,16].

Algae are a wide range of photosynthetic organisms that produce extensive amounts of lipids, proteins, carbohydrates, and value-added materials (e.g., carotenoids, vitamins, etc.), and are considered to be third-generation feedstocks for biodiesel production. Compared to higher plants, the algae growth rate is faster. At the same time, less water is needed for their cultivation, they can grow in different habitats, such as wastewater, their potential for CO_2 remediation may be higher, no arable land is needed for their cultivation, they require less area compared to higher plants, there is no or very little lignin in their structure, they do not compete with the human food supply, and they can produce high-value products. Thus the possibility of using these materials for biodiesel production may help make the process more economical [17]. Algal species are classified according to their morphology, physiology, and phylogeny. Commonly, microalgae and macroalgae/ seaweeds are known as algae, which can be distinguished according to their dimensions. Microalgae are unicellular with dimensions under 0.4 mm, while macroalgae are multicellular entities ranging upto tens of meters [18,19]. Also, seaweeds, as their name suggest, are mostly located in marine habitats and consist of stem, shoot, and roots [17]. Furthermore, there are many differences in the biochemical composition of micro- and macroalgae. Microalgae may contain even upto 50%−70% of proteins and more than 30% of lipids; however, these fractions may change in some stress conditions such as high light and/or nutrients manipulation [20]. In turn, macroalgae water content is very high (upto 90% fresh weight), and their highest component is carbohydrates, followed by proteins at between 25%−77%, and 5%−43% dry matter, respectively [21]. Due to the high lipid content of microalgae, they have attracted more attention for biodiesel production. However, researchers have done many investigations on the production of biodiesel from seaweeds. Macroalgae have some advantages over microalgae as their harvest is more easily achieved, and they can generate more net energy [22].

This chapter aims to give an overview of the production of biodiesel using algal biomass. In this regard, the lipid content and fatty acid profiles of different algal species have been summarized. In addition, the effect of environmental conditions on the lipid content, productivity, and fatty acid profiles of algal species has been reviewed. Furthermore, different transesterification methods of algal biomass have been described, as well as different characteristics of biodiesel. Finally, the economic feasibility of algal biodiesel has been highlighted.

10.2 Lipids in algal biomass

Algae cells contain different concentrations of lipids, carbohydrates, proteins, and nucleic acids, while their contents depend on the species and growth conditions. Within the algal cells, lipids are the most energy-rich components with the value of 37.6 kJ/g [23], making them suitable for biodiesel production. The ratio of these lipids classes is highly dependent on the growth phase of algae [9,24]. In general, there are two major lipid classes in algal biomass, known as polar and nonpolar. Polar lipids (e.g., phospholipids and glycolipids) are also known as structural lipids, and are involved in protecting the cells and maintaining their shape. In contrast, the main function of nonpolar lipids (also known as neutral lipids) is energy storage. Monoacylglycerols, diacylglycerols (DAGs), triacylglycerols (TAGs), and free fatty acids (FFAs) are known as neutral lipids. In addition to these major classes, different amounts of other types of lipids such as hydrocarbons, terpenoids, pigments, sphingolipids, etc., are available in different algal species [25]. Fatty acids are displayed as C, followed by two numbers, where the first number indicates the number of carbon atoms, and the second number shows the number of double bonds. Fatty acids are classified as saturated (SFAs) and unsaturated fatty acids (UFAs) due to the presence or absence of double bonds. UFAs are subsequently classified into monounsaturated fatty acids (MUFAs), with a single double bond, and polyunsaturated fatty acids (PUFAs), with two or more bonds. Comparing polar and neutral lipids, the accumulation of neutral lipids is a priority for biodiesel production; however, excessive amounts of PUFAs may destructively alter the biodiesel properties [26]. Accordingly, lipid content and productivity of algal species play a vital role in the production of biodiesel. As reported in the literature, macroalgae dry matter accounts for approximately 10%–15%, while most of it is carbohydrates (about 60%); however, the lipids (approximately 0.3%–6% of dry matter) could be used for biodiesel production [27]. In contrast, microalgae species contain higher lipid contents, while this content in some species may exceed 80% of the microalgal biomass [28].

10.2.1 Lipids biosynthesis in algal biomass

Lipid biosynthesis in algae happens in the chloroplast, where CO_2 is converted to acetyl-CoA leading the metabolism to produce FAs [29]. However, compared to higher plants, the regulation of lipids biosynthesis in algae (exclusively biosynthetic pathways of FAs and TAGs) is not well understood, while understanding these characteristics could be useful in enhancing either lipid production or strain improvements. Although the research on the metabolism and lipid biosynthesis of algal species using some promising microalgae (e.g., *Chlamydomonas reinhardtii*, *Ostreococcus tauri*, etc.) is ongoing, it is generally accepted that the biosyntheses of FAs and TAGs in algae are similar to higher plants, due to the similarity of isolated enzymes and/or genes from algae and such plants [30].

The lipid biosynthesis in the plants has been defined in a superior way in comparison with algae. Accordingly, the pathway of de novo FAs synthesis by algae is considerably derived from the similar characteristics of plants, while many similarities about FAs synthesis has been investigated in different algal species and plants [31–33]. The de novo synthesis of FAs occurs in the stroma of all eukaryotic species, as explained in detail previously [25,30]. Briefly, the first step for de novo FA synthesis is the production of malonyl-CoA as a result of ATP-dependent carboxylation of acetyl-CoA, catalyzed by acetyl-CoA carboxylase (ACCase). An acetyl-CoA pool is present in the chloroplast to activate carboxylic acid metabolites. It also is made as a result of reactions inside the mitochondria and peroxisomes [34]. The next step is the conversion of malonyl-CoA to a malonyl-acyl carrier protein (ACP) in the presence of malonyl-CoA:ACP malonyltransferase. After this step, the produced malonyl-ACP, the central carbon donor for further FA synthesis, is ligated to an acetyl-CoA molecule leading the metabolism to the formation of 3-ketoacyl-ACP in the presence of ketoacyl-ACP synthase (KAS) and the release of a CO_2 molecule. The product of the previous step (3-ketoacyl-ACP) is then reduced, dehydrated, and reduced again by ketoacyl-ACP reductase, hydroxyacyl-ACP dehydrase, and enoyl-ACP reductase, respectively, to form a 6-carbon-ACP. Finally, C16 and/or C18 FAs are produced as a result of this cycle (Fig. 10.1).

The utilization of algal oils in different industrial applications led the researchers to figure out the metabolism of TAGs. Generally, algae contain small amounts of TAGs during normal growth conditions. It is well-established that this content could be enhanced significantly during stress conditions, such as nutrient manipulation, high temperature, and high light intensity. Accordingly, the biosynthesis of TAGs in algae occurs via the Kennedy pathway as a result of four reactions (Fig. 10.1). In a nutshell, during the first two steps, the produced FAs in the chloroplast are transferred sequentially from CoA to positions 1 (first step) and 2 (second step) of glycerol-3-phosphate (G3P). These two reactions are catalyzed by glycerol-3-phosphate acyltransferase, and lyso-phosphatidic acid acyltransferase, respectively, resulting in the formation of phosphatidic acid (PA). Following these reactions, DAG is released as a result of PA dephosphorylation, catalyzed by PA phosphatase. Finally, another FA is placed in

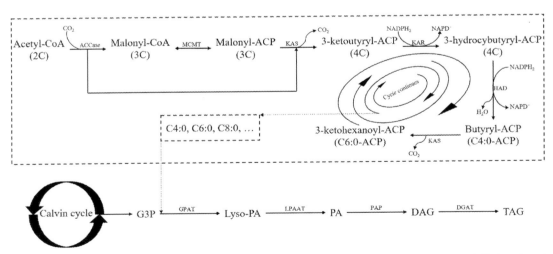

FIGURE 10.1 Simplified de novo fatty acid and TAG pathways in chloroplast. *ACCase*, acetyl-CoA carboxylase; *MCMT*, malonyl-CoA,ACP malonyltransferase; *KAS*, ketoacyl-ACP synthase; *KAR*, ketoacyl-ACP reductase; *HAD*, hydroxyacyl-ACP dehydrase; *G3P*, glycerol-3-phosphhate; *GPAT*, glycerol-3-phosphate acyltransferase; *Lyso-PA*, lysophosphatidic acid; *LPAAT*, lyso-phosphatidic acid acyltransferase; *PA*, phosphatidic acid; *PAP*, phosphatidic acid phosphatase; *DAG*, diacylglycerol; *DGAT*, diacylglycerol acyltransferase; and *TAG*, triacylglycerol.

10.2 Lipids in algal biomass

position 3 of DAG (the last vacant position). The last reaction is catalyzed by diacylglycerol acyltransferase (DGAT) resulting in TAG synthesis. It is noteworthy that during this pathway, PA and DAG could be directly used as precursors for the polar lipids (e.g., phospholipids, phosphatidylcholine, galactolipids, etc.) biosynthesis.

10.2.2 Lipid content and fatty acid profiles of algal biomass

It is estimated that there are about 300,000 algae species around the world [17]. Accordingly, the biochemical properties of thousands of algae from different taxonomic groups have been investigated so far. Due to these studies, several algal species have been suggested to be oleaginous and applicable for biodiesel production. However, the lipid content of algal species varies depending on the species and growth conditions [35]. The lipid content of some algal species is presented in Table 10.1. While the lipid content of the species is of significant importance,

TABLE 10.1 Oil content of some algal species.

Algal species	Lipid content (% dry weight)	References	Algal species	Lipid content (% dry weight)	References
Microalgae			Microalgae		
Ankistrodesmus falcatus	59.9	[36]	*Scenedesmus quadricauda*	18	[37]
Ankistrodesmus gracilis	7.9–20.5	[38]	*Tetraselmis suecica*	12.9	[37]
Botryococcus braunii	25–75	[39]	Macroalgae		
Chaetoceros calcitrans	40	[37]	*Acanthophora spicifera*	10–12	[40]
Chaetoceros muelleri	43.4	[37]	*Catenella repens*	8	[41]
Chlorella emersonii	29	[42]	*Caulerpa peltata*	11.42–12.69	[40,43]
Chlorella minutissima	31	[42]	*Caulerpa racemose*	9–10.5	[40]
Chlorella pyrenoidosa	19.97–44.07	[44]	*Chaetomorpha aerea*	8.5	[40]
Chlorella sorokiniana	10	[42]	*Cladophora fascicularis*	15.7	[40]
Chlorella vulgaris	26–47.53	[45–47]	*Cladophora glomerata*	18	[48]
Chlorella zofingiensis	64.7	[49]	*Dictyota dichotoma*	6.87	[22]
Chlorococcum pamirum	30.2–65.3	[50]	*Enteromorpha compressa*	11.45	[40]

(Continued)

Handbook of Algal Biofuels

200

10. Algae for biodiesel production

TABLE 10.1 (Continued)

Algal species	Lipid content (% dry weight)	References	Algal species	Lipid content (% dry weight)	References
Dunaliella bioculata	8	[51]	Geledium pusillum	9.7	[41]
Dunaliella salina	6	[51]	Gracilaria vermiculophylla	9.1	[22]
Ellipsoidion parvum	17	[52]	Hypnea valentiae	9.6—11.6	[40]
Euglena acus	5.78	[53]	Laurencia papillosa	8.9—10.8	[40]
Euglena gracilis	14—20	[51]	Padina boryana	11.37	[22]
Nannochloropsis oceanica	54.3	[54]	Padina tetrastromatica	8.15	[55]
Parachlorella kessleri	31—66	[49]	Ulva fasciata	9.85	[8]
Porphyridium purpureum	11	[42]	Ulva intestinalis	11.63	[22]
Scenedesmus abundans	36.9—67.59	[56]	Ulva lactuca	9.6—12.5	[22,40]
Scenedesmus dimorphus	16—40	[51]	Ulva linza	8.3	[22]
Scenedesmus obliquus	19	[52]	Ulva reticulate	8.5	[40]

the lipid productivity indicates the production capacity of biodiesel by algae. In this regard, it has been reported that the higher lipid species growth rate is slower than the lower lipid species [57]. In addition, the methods utilized to enhance the lipid content in some species may simultaneously decrease the biomass production, causing a reduction in the overall lipid productivity. This is why the lipid productivity of algal species is considered to be their capacity for biodiesel production [26]. Accordingly, for biodiesel production, considering suitable environmental conditions for high growth and accumulation of considerable amounts of lipids is of significant importance alongside choosing an appropriate algal species.

In addition to the lipid content of algal cells, the composition of the lipids is of significant importance, as the biodiesel characteristics highly depend on the fatty acid composition of algae. As an example, the biodiesel produced from algal oil with high PUFAs content exhibits good cold-flow properties while high amounts of SFAs in the oil enhance cetane number (CN) and the oxidative stability (OS) of biodiesel [11]. In this regard, the ability of algal strains to generate a high quality of fatty acid compositions is an important factor as well as the high lipid content and productivity. The fatty acid profiles of some algal species used for biodiesel production are summarized in Table 10.2.

TABLE 10.2 Fatty acid profile of selected algal species.

Fatty acid	C12	C14	C15	C16	C16:1	C16:2	C16:3	C18	C18:1	C18:2	C18:3	C20:5	Others	SFAs	MUFAs	PUFAs	References
Microalgae																	
Amphora sp.	–	–	–	28.61	38.16	–	–	12.66	–	3.86	4.55	–	–	41.28	38.16	8.42	[87]
Batrachospermum sirodotia	–	–	0.68	17.24	4.31	3.24	1.70	–	15.82	4.90	–	–	1.47[a]	–	–	–	[88]
B. braunii	–	0.73	–	7.17	–	–	–	1.59	77.22	5.16	5.34	–	1.26[b]	9.85	79.61	10.54	[89]
Botryococcus terribilis	0.62	–	–	35.22	–	–	–	3.12	39.74	5.02	7.22	–	2.81[c]	43.15	44.29	12.56	[89]
C. muelleri	–	16.50	0.71	28.84	39.48	2.17	1.37	1.92	0.96	0.79	0.51	2.28	4.47[d]	47.98	40.44	11.59	[38]
Chlamydomonas sp.	0.8	1	–	29.1	6.2	1.5	1.3	5	20.1	6.5	12.6	–	5.6[b], 9.5[e]	–	–	–	[9]
C. emersonii	–	–	–	14.75	–	–	–	9.80	17.01	9.04	29.32	–	2.74[c]	24.55	17.01	38.37	[87]
C. vulgaris	–	–	–	14.55	1.18	–	–	10.51	23.62	13.80	32.10	–	–	25.06	24.80	46.90	[87]
C. pamirum	–	1.55	–	25.77	10.29	–	–	28.84	13.86	10.28	7.4	–	1.07[f], 0.65[g], 0.23[h]	56.40	25.92	17.68	[51]
D. salina	–	–	–	16.33	1.03	–	–	6.43	19.57	6.76	27.70	–	2.28[c]	22.77	22.89	34.47	[87]
Dunaliella sp.	–	–	–	9.19	0.80	–	–	4.27	22.51	3.84	44.31	–	1.42[c]	13.47	24.74	48.15	[87]
Dinoflagellate	–	6.01	–	16.65	3.35	–	–	–	2.10	–	–	2.89	44.98[a], 2.65[h], 18.27[i]	25.31	5.45	66.14	[90]
Kirchneriella lunaris	–	1.92	–	37.16	0.67	–	13.95	–	18.73	–	–	–	17.71[a]	–	–	–	[88]
Lyngbya kuetzingii	–	–	–	24.98	2.86	2.68	2.05	1.60	21.52	12.11	18.38	2.38	5.99[a]	–	–	–	[88]
N. oceanica	–	4.57	0.63	33.08	26.71	–	–	3.67	18.16	1.73	–	5.80	5.03[d], 0.62[i]	42.57	57.43	57.43	[55]
Nannochloropsis oculata	–	4.9	–	18.4	26.1	–	–	–	3.0	6.2	–	34.0	7.4[d]	23.3	29.1	47.6	[91]

(Continued)

TABLE 10.2 (Continued)

Fatty acid	C12	C14	C15	C16	C16:1	C16:2	C16:3	C18	C18:1	C18:2	C18:3	C20:5	Others	SFAs	MUFAs	PUFAs	References
Nannochloropsis sp.	–	5.37	–	28.83	32.93	–	–	0.98	21.16	2.24	–	6.33	–	35.18	54.09	8.57	[90]
Pediastrum tetras	0.4	1.8	–	29.7	1	0.3	4.2	2.8	33.9	8.2	11.7	0.6	1.8[e], 1.5[j]	–	–	–	[9]
P. tricornutum	–	8.7	–	18.7	26.4	5.6	9.3	–	2.1	5.1	–	20.9	3.2[j]	27.4	28.5	44.2	[91]
Scenedesmus incrassatulus	–	0.79	–	24.93	3.73	–	–	4.17	45.36	6.60	5.97	–	1.08[b], 0.93[f], 3.10[j]	–	–	–	[92]
S. obliquus	–	6.75	–	20.26	42.06	1.84	–	1.51	5.11	0.83	–	2.11	–	–	–	–	[88]
Scenedesmus sp.	–	–	–	18.42	2.31	3.26	–	3.43	49.64	11.30	8.26	–	–	21.85	51.95	22.82	[90]
Selenastrum capricornutum	3.81	–	–	45.62	5.96	–	5.00	–	4.27	11.75	–	–	–	–	–	–	[88]
Spirulina platensis	1.14	2.52	–	41.21	3.39	–	–	1	4.11	12.64	17.79	–	3.90[k], 3.07[l]	–	–	–	[93]
Tetraselmis sp.	–	–	–	25.4	4.9	–	–	–	16.2	10.0	22.4	2.4	1.5[c], 4.2[e], 13.0[j]	25.4	22.6	52.0	[91]

Fatty acid	C12	C13	C14	C14:1 n-5	C15	C16	C16:1	C16:1 n-7	C16:1 n-9	C16:1 n-10	C16:2 n-6	C16:4 n-3	C17	C18	C18:1 n-9	C18:2 n-6	C18:3 n-3	References
Macroalgae																		
C. peltata	7.28	4.16	3.02	–	–	36.82	–	–	–	–	–	–	–	4.58	2.31	18.19	7.03	[44]
Caulerpa prolifera	–	1.04	6.48	1.98	3.40	11.72	–	9.29	–	–	–	–	5.13	9.57	13.01	7.13	10.11	[94]
Chaetomorpha linum	–	–	12.06	–	–	10.55	3.63	–	–	–	–	–	–	1.18	33.33	18	0.28	[95]
Cladophora albida	0.21	–	12.48	–	0.56	33.04	–	13.90	–	–	2.46	–	0.26	1.28	13.30	15.54	–	[96]
Derbesia tenuissima	–	–	1.02	0.40	0.46	9.84	–	1.74	0.28	–	0.41	0.40	0.24	0.53	1.76	1.93	9.46	[97]
Dictyochloropsis splendida	–	–	13.04	–	–	81.14	–	–	–	–	–	–	1.01	–	0.26	4.39	0.15	[98]
D. dichotoma	–	–	15.42	–	0.97	24.75	–	15.49	–	–	0.44	–	–	2.85	8.49	5.55	–	[96]
Dictyota spiralis	–	–	14.00	–	0.73	21.69	–	19.58	–	–	–	–	0.23	2.43	9.47	6.05	–	[96]

															SFAs	MUFAs	PUFAs	References
E. compressa	—	—	2.16	—	—	70.26	3.71	—	—	—	—	—	—	2.95	2.38	18.54	—	[99]
Oedogonium sp.	—	—	0.76	0.43	0.50	8.59	—	1.68	0.30	—	0.93	1.43	0.27	0.36	1.24	2.17	12.84	[97]
P. boryana	—	—	5.53	—	—	37.04	—	3.66	0.82	1.83	—	—	—	4.53	20.42	2.32	1.10	[22]
Taonia atomaria	—	—	7.07	—	0.56	25.41	—	8.09	—	—	—	—	0.18	1.04	8.09	10.08	—	[96]
U. fasciata	—	1.82	9.46	—	2.37	45.18	—	—	—	—	—	—	3.08	10.50	15.92	2.78	—	[8]
U. intestinalis	—	—	9.96	—	—	28.54	—	—	0.46	2.10	—	0.25	—	0.47	18.45	5.64	—	[22]
U. lactuca	—	—	1.20	—	—	32.94	—	—	0.22	6.83	—	0.36	—	0.94	11.04	4.10	8.81	[22]

Fatty acid (Continued)	C18:3 n-6	C18:4 n-3	C20	C20:1 n-9	C20:3 n-6	C20:4 n-6	C20:5 n-3	C22	C22:1 n-9	C24	Others	SFAs	MUFAs	PUFAs	References
Macroalgae															
C. peltata	—	—	0.95	—	—	—	—	8.27	—	—	5.04[m], 2.35[n]	—	—	—	[44]
Caulerpa prolifera	—	2.32	1.31	—	—	4.09	5.14	—	2.55	—	3.55[o], 2.36[p]	38.66	61.53*		[94]
Chaetomorpha linum	—	—	0.21	0.83	—	—	—	—	—	0.95	0.28[q]	—	—	—	[95]
Cladophora albida	—	—	0.38	0.18	—	1.37	2.02	0.75	0.35	1.08	0.86[r]	50.03	27.73	22.24	[96]
Derbesia tenuissima	0.87	0.96	0.24	0.22	1.46	0.38	1.15	0.91	—	1.38	0.73[s], 0.22[t], 4.92[u], 0.26[v]	34.6	12.2	53.2	[97]
Dictyochloropsis splendida	—	—	—	—	—	—	—	—	—	—		95.19	4.81*		[98]
D. dichotoma	2.63	—	1.98	0.31	2.60	11.46	6.57	—	—	—		45.98	24.28	29.74	[96]
Dictyota spiralis	3.38	—	1.12	0.29	2.63	18.40	—	—	—	—		40.20	29.34	30.46	[96]
E. compressa	—	—	—	—	—	—	—	—	—	—		—	—	—	[99]
Oedogonium sp.	1.39	2.58	0.21	0.21	0.43	0.32	1.13	—	—	0.31	0.83[s], 0.66[t], 6.05[u], 0.84[v], 0.30[w]	23.5	10	66.4	[97]

(Continued)

TABLE 10.2 (Continued)

Fatty acid	C12	C14	C15	C16	C16:1	C16:2	C16:3	C18	C18:1	C18:2	C18:3	C20:5	Others	SFAs	MUFAs	PUFAs	References
P. boryana	0.13	0.99	1.95	—	—	—	—	—	—	—	—			49.05	26.72	4.54	[22]
Taonia atomaria	1.71	—	0.74	0.30	2.14	18.64	13.55	0.48	0.85	—	0.84[r]			35.47	17.34	47.19	[96]
U. fasciata	—	—	2.79	—	—	—	—	2.14	—	—	1.54[x],0.67[y]			—	—	—	[8]
U. intestinalis	—	—	10.97	—	—	—	—	—	—	—	0.73[f]			49.94	21.74	5.88	[22]
U. lactuca	—	8.19	8.88	—	—	—	—	—	—	1.40	—			45.35	18.10	23.46	[22]

*Total unsaturated fatty acids,
[a]C22:6,
[b]C17:1,
[c]C20:1,
[d]C20:4,
[e]C18:4,
[f]C14:1,
[g]C24:1,
[h]C24,
[i]C22:5,
[j]C17:0,
[k]C8,
[l]C15:1,
[m]C16:2,
[n]C21,
[o]C22:2n-6,
[p]C22:4n-6,
[q]C24:1n-9,
[r]C22:6n-3,
[s]C15:1n-5,
[t]C16:2n-4,
[u]C16:3n-3,
[v]C16:3n-6,
[w]C20:2n-6,
[x]C10:1,
[y]C16:1n-12.

10.2.3 Effect of environmental conditions on algal lipids

Culture conditions remarkably affect the biological properties of algal species. Stress conditions, such as high light irradiation and nutrient manipulation (nutrient depletion, limitation, and starvation), highly affect the biochemical composition and growth characteristics of different species. Many researchers have investigated the effect of nitrogen, phosphorus, carbon, salinity, pH, temperature, light, etc., on the growth and biochemical composition of a wide range of algal species [35,71,72].

10.2.3.1 Nutrients effect (nitrogen, phosphorus, carbon, and micronutrients)

As discussed, nutrients manipulation is a common method for modifying the lipid content of algal species. In general, algae species are different in their nutritional needs, although nitrogen, phosphorus, sulfur, and carbon are essential macronutrients for metabolism, growth, and reproduction of different species [73,74]. In this regard, nitrogen manipulation is widely used to increase the lipid content and other algal species compounds, such as carotenoids [20,75]. Nitrogen is known as a vital element for the growth of algal species. Accordingly, when the nitrogen in the culture is depleted, the cell growth stops as a result of a photosynthetic efficiency decrease, but the photosynthesis energy could be employed by lipid biosynthesis pathways to accumulate considerable amounts of lipids [23]. Generally, during the nitrogen limitation conditions, the carbohydrates content increased as well as the lipids, while the protein content decreased. However, it is noteworthy that these variations in microalgae are species-specific. For example, the effect of nitrogen starvation on the growth and biochemical characteristics of different microalgal species was investigated previously [20]. According to this study results, the protein content of five species (*Scenedesmus obliquus*, *Dunaliella tertiolecta*, *Chlorella zofingiensis*, *Phaeodactylum tricornutum*, and *Neochloris oleoabundans*) decreased during nitrogen-deficient cultivation. In addition, the lipid content of *S. obliquus*, *C. zofingiensis*, *P. tricornutum*, and *N. oleoabundans* species was significantly increased. Besides an increase was observed in the carbohydrate contents of *S. obliquus*, *D. tertiolecta*, *C. zofingiensis*, and *P. tricornutum*. It is noteworthy that the biomass concentration of the studied species increased during the nitrogen starvation conditions, however, these increments were lower in comparison with the nitrogen sufficient cultures. Accordingly, the highest increase in biomass concentration was for *S. obliquus* and *C. zofingiensis* during the nitrogen starvation conditions. These changes may result from the accumulation of storage compounds such as TAGs, while nitrogen stress is a popular method to enhance TAGs (20%−50% dry weight) induction [30,76]. The reason for the increase in TAGs is the activation of DGAT (Section 10.2.1) as a result of nitrogen starvation, converting acyl-CoA to TAGs [77]. Although the biomass concentration of some species may increase during the nitrogen starvation conditions, the overall lipid efficiency of most species is diminished during this phase of growth which indicates the necessity of an appropriate correlation between growth and lipid accumulation. In other words, the best algal species to be used as feedstock for biodiesel production are the ones that accumulate lipids and simultaneously grow well with less damage to the growth rate during stress conditions. In addition to biomass and lipid content of algal species, nitrogen stress has been reported to modify some species' fatty acid composition. Generally, this kind of stress increases C18:0 while decreasing C16:0 [23]. However, these changes are strain-specific, as in some species the amount of SFAs does not change significantly, while a reduction of some SFAs has been recorded in some cases [36,78].

Phosphorus is another nutrient considered a substantial factor that directly influences the growth and lipid accumulation of algal cells. This component also plays a vital role in several metabolisms, such as photosynthesis, energy transfer, and signaling pathways [79,80]. During the nutrient sufficient cultivation, the cells can store phosphorus as polyphosphate granules and utilize these stored compounds when the nutrients are insufficient [79]. Generally, phosphorus limitation conditions cause a reduction in cell division rates, resulting in a slight decrease in the photosynthetic rates, leading the metabolism to accumulate carbon in the form of TAGs rich in SFAs and MUFAs [71]. Furthermore, in addition to lipids, depending on the species, microorganisms are able to accumulate carbohydrates or both lipids and carbohydrates, in phosphorus-deficient environments [81,82]. Phosphorus limitation also has been reported to be effective on the lipid composition of algal biomass, while in some cases this stress condition caused a significant increase in UFAs [71]. In addition to manipulating nitrogen or phosphorus, researchers have investigated the effect of both nitrogen and phosphorus concentrations on growth characteristics and lipid accumulation of different species. Accordingly, the effect of nitrogen and/or phosphorus limitation was investigated for the growth and lipid accumulation of *Senedesmus* sp. LX1 [83]. The results suggested that the highest lipid accumulation of 53% and 30% were obtained during phosphorus and nitrogen limitation conditions. However, the lipid accumulation rate and productivity were not at their highest rate. Furthermore, the highest lipid productivity was obtained when the initial nitrogen concentration was 25 mg/L and the nitrogen to phosphorus ratio (N/P) was controlled at 20:1. In addition, an efficient removal of nitrogen and phosphorus was obtained when N/P was between 5:1 to 8:1.

As another essential macronutrient for algal species growth and metabolism, carbon could be utilized in the form of CO_2 and organic carbons (e.g., glucose, glycerol, acetate, etc.) [84]. A carbon source could be effective on both lipid content and composition of algal species [85]. For instance, the optimum concentration of CO_2 for *S. obliquus* growth was reported to be 9%, while the lipid content and productivity of this microalga was maximum when 12% CO_2 purged the cultures [86]. The effect of different concentrations of CO_2 (2%—15%) on biomass and lipid production of *Nannochloropsis oculate* was investigated [87]. The optimum CO_2 concentration was 2%, while the highest lipid content, biomass, and lipid productivity were obtained when 2% CO_2 purged the cultures. Similar parameters were investigated for *Chlorella* sp. when between 0.03% and 15% CO_2 purged the cultures [88,89], and the optimum CO_2 concentration was reported to be 2%—3%. In addition to CO_2, different organic carbon sources were reported to enhance the growth and lipid content of algal species [13,90,91]. As an example, the lipid content of *S. obliquus* increased by 5% (dry weight) when 2.5 g/L waste glycerol was added to the cultures [13]. In another investigation, heterotrophic cultivation of *C. zofingiensis* using glucose, mannose, and fructose resulted in an increase in the biomass of this microalga [84]. In the case of lipid composition, it has been reported that the accumulation of SFAs was induced in the presence of high CO_2 concentrations (2%−10%), whereas the production of UFAs was enhanced when the CO_2 concentration is low (less than 2%) [85].

In addition to macronutrients, algae need trace amounts of micronutrients, such as iron, zinc, magnesium, copper, calcium, manganese, etc., for growth and metabolism, while they significantly affect the lipid and carbohydrate content of various algae as well as their growth rate [72,92]. In addition, the fatty acid profile of algal species could be modified as a result of these nutrients manipulation [23]. However, the effect of these micronutrients' concentration on the growth and biochemical composition of different species is considerably different. For instance, iron is known as one of the most impressive elements in photosynthesis. Generally, Fe^{3+} in the media could increase the exponential growth phase duration resulting in the final cell concentration enhancement [93]. Also, the presence of this micronutrient has a significant effect on the enhancement of lipid content in different species (e.g., *C. sorokiniana*, *C. vulgaris*, etc.) [93,94]. It is noteworthy that Fe^{3+} in the media should be at the optimum; otherwise, it has an inhibitory effect on the growth and lipid production. Magnesium is another essential micronutrient for growth of algae. The presence of Mg^{2+} in the media in high concentrations may enhance ACCase activity, resulting in an increase of the production of cells and algal lipids (e.g., *Monoraphidium* sp.) [95,96]. As well as Fe^{3+} and Mg^{2+}, the presence of Ca^{2+} in cultures has been reported to be effective for lipid content enhancement; however, the effect of excessive amounts of Ca^{2+} on biomass concentration is negligible [95]. Other micronutrients, such as zinc, copper, cadmium, and manganese, were reported to be effective for algal growth and lipid accumulation [97,98].

10.2.3.2 *Other environmental conditions (salinity, light, pH, and temperature)*

In addition to the effect of nutrient manipulation on lipid accumulation and the composition of algal species, other environmental conditions affect lipid accumulation of algal species and their growth. In this regard, salinity stress is a common method for lipid accumulation enhancement during the two-stage cultivation of different algae. Accordingly, during the first stage, algae are cultivated in a salt-free medium (optimum conditional growth of algae) to considerably produce biomass, and then salt stress is employed for the enhancement of lipid accumulation (second stage). While the cells could be damaged by salt, the utilization of enough salt can enhance lipid biosynthesis [99].

Similar to other environmental conditions, salinity tolerance differs in different species of algae. Accordingly, growth characteristics and lipid accumulation of *C. sorokiniana* were investigated when cultivated in different concentrations of NaCl (0−30 g/L) [100]. In this study, the highest biomass concentration and productivity were obtained during the salinity stress when the concentration of NaCl was adjusted to 10 g/L. Additionally, the utilization of 20 g/L NaCl for this stress resulted in the considerable accumulation of lipids as well as their productivity. Other algal strains, such as *S. obliquus* [101], *Chlamydomonas mexicana* [102], *C. vulgaris* [103], etc., were reported to synthesize considerable amounts of lipids during salinity stress. In another investigation, the fatty acid composition of *C. mexicana* and *S. obliquus* was improved in the presence of 50 mM NaCl [104].

208 10. Algae for biodiesel production

Generally, light is a vital source for photosynthetic microorganisms like algae, considerably affecting their growth and biochemical composition. While the requirements of algal species are different, increasing the light intensity has been suggested to enhance the photosynthetic rate of algae [25], except during the initial stage of cultivation, which leads to photoinhibition due to the low cell densities [105]. Commonly, when algae were cultivated in the presence of low and high light intensities, the formation of polar and neutral lipids happened in the algal cells, respectively [30]. Besides, moderate light intensity and duration are essential to obtain the optimum conditions. In this regard, the growth and biomass production is highly influenced by the light/dark cycle as well as the biochemical composition of algae [106]. According to previous investigations, increasing the light/dark cycle will increase photosynthetic efficiency [107,108]. However, to enhance lipids concentration, moderate light/dark cycles are required, while shorter or longer cycles may lower the content of lipids [105].

The growth rate and composition of algal species are also affected by the pH of cultures. Regulations in the pH alter the absorption and availability of nutrients (e.g., carbon, iron, etc.), which affect the growth rate indirectly. In addition, pH directly affects microalgae growth as the optimal pH of different microalgal species is strain-specific and narrow [85]. pH is also an important factor that affects the lipid content and composition of different microalgae [109,110]. As an example, biomass and lipid productivities of *Chlorella* sp. and *T. suecica* are optimum at pH 7 and 7.5, respectively [111]. The microalga *Scenedesmus* sp. was also found to grow well and produce high amounts of lipids in basic conditions (pH > 9) [112].

During the cultivation of algal species, the temperature is an important factor as it can change the composition and growth of these microorganisms. Commonly, these strains can tolerate different temperatures according to their species [113]. Accordingly, they are categorized into three groups of psychrophilic (17°C), mesophilic (20°C–25°C), and thermophilic (40°C) strains based on their optimal growth temperature. However, the temperature variations may elevate or reduce the lipid content in different species [114,115]. For example, the lipid content of *N. oculate* enhanced twofold when the temperature increased from 20°C to 25°C, while an increase in the temperature from 25°C to 30°C caused a significant reduction in the lipid content of *C. vulgaris* [78]. In the case of *Scenedesmus* sp., by increasing the temperature from 20°C to 25°C the lipid content decreased by 10% dry weight [116]. In addition, the SFAs content was increased by a temperature increase and vice versa. Also, it has been reported that increasing the temperature (higher than the optimum temperature) may enhance the concentration of MUFAs as well as decreasing the amount of PUFAs [23].

10.3 Different methods of transesterification

Microalgae species appear to be potential feedstocks for biodiesel production due to their high potential for lipids production. Besides, they have been proposed to meet

Handbook of Algal Biofuels

international demands as the only renewable source of biodiesel as an alternative for fossil fuels in the transportation sector [117]. As reported previously, microalgae are categorized as low, medium, and high oil species, which respectively contain about 30%, 50%, and 70% dry weight oil content [29]. Accordingly, these species are able to provide between 58,000 to about 137,000 L oil/ha yearly, which subsequently leads to produce about 51,000–121,000 kg biodiesel/ha in a year. As estimated, the oil content and biodiesel production of microalgae are many times higher than other feedstocks (e.g., soybean, sunflower, jatropha, palm oil, coconut, etc.). In addition, these species require less land compared to other agricultural feedstocks [29]. The ability of macroalgae for bioenergy production has not been well studied in comparison with microalgae. Nevertheless, some macroalgal species lipid content is more than 10% dry weight, and these are considered to be suitable candidates for biodiesel production [22]. In general, the beneficial use of algae requires the selection of the appropriate species.

There are several methods for biodiesel production, such as pyrolysis, transesterification, micro-emulsification, etc., while the most selective method is transesterification. The reaction between TAGs/FFAs in the oil and a short-chain alcohol as a solvent is known as transesterification. The products of this reaction are FAMEs and glycerol. It is worth noting that the biodiesel produced by transesterification is favorable for industrial production and its properties are similar to diesel [4]. In general, there are four different methods for transesterification based on the stages implemented during this process (Fig. 10.2). Also, transesterification could be classified according to the catalyst (acidic, basic, enzymatic, etc.) used in the process.

Conventionally, biodiesel is produced using two methods. The first method consists of three steps as follows the harvested algal biomass should be dried. Then the lipids are extracted from dried biomass. Finally, the extracted lipids converted to biodiesel by transesterification method (Fig. 10.2, process 1). In the second method, the wet algal biomass could be utilized in lipid extraction without drying (Fig. 10.2, process 2). As it is obvious, performing conventional methods require lipid extraction before transesterification. In contrast, another method is employed for transesterification known as in situ (or direct) transesterification. The main difference between conventional and in situ

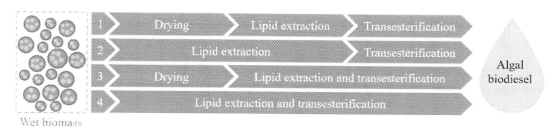

FIGURE 10.2 Production of biodiesel using algal cells via conventional (1 and 2), and in situ (3 and 4) transesterification.

transesterification is the stage of lipid extraction. While in the latter, the lipid extraction happens simultaneously with transesterification when the biomass is exposed to an appropriate catalyst and a solvent under high temperature. According to Fig. 10.2, in situ transesterification could be performed either after drying the harvested biomass (process 3) or directly without drying (process 4). According to previous research, more than 80% of total energy for biodiesel production is related to the drying of algal biomass [118]. In addition, the combination of lipid extraction and transesterification to a single process will reduce the costs of energy consumption and equipment installation, resulting in the reduction of whole process cost [119]. Also, as reported in previous investigations, production of 1 kg biodiesel from dry biomass requires 107.3 MJ of energy, while the same from wet algae requires 42.3 MJ [120]. Accordingly, among different transesterification methods, wet in situ transesterification (Fig. 10.2 process 4) is preferred due to reducing the operating and capital costs. However, the biggest challenge of wet in situ transesterification is the presence of water during the reaction [121,122]. In this regard, the catalyst activity could be diminished, the reaction of alcohol and catalyst with lipid molecules could be negatively affected, and when the water content is high the process could be completely inhibited.

In order to achieve sufficient biodiesel production, a suitable alcohol and a catalyst are essential. Lots of research has been undertaken into producing biodiesel using different alcohols, such as methanol, butanol, propanol, and ethanol, but methanol is the most widely used alcohol due to its physical and chemical advantages as well as its low price [123]. In addition, in order to enhance alcohol solubility, a catalyst is needed during the transesterification. To date, many base catalysts (e.g., sodium hydroxide, sodium methoxide, and potassium hydroxide), acid catalysts (e.g., sulfuric acid, phosphoric acid, and hydrochloric acid), and enzymatic catalysts (e.g., lipases) have been extensively used for biodiesel production. Base catalysts are not appropriate for lipid samples containing high FFAs [124], while soap and water formation is possible due to FFAs and base reaction. However, this kind of catalysts' reaction rate was reported to be significantly faster than acid catalysts [123,125]. In contrast, the yield of acid catalysts is high and could be effective in the presence of high FFAs concentration [126]. Accordingly, the base catalysis is preferred due to its fast reaction rate. In addition to these two types of catalysts, enzymes such as lipases can play an influential role in lipid mixtures containing high amounts of FFAs. Enzymes are advantageous over acid and base catalysts while they can catalyze several bioconversion reactions, are more environmentally friendly, by-products removal is easier, and there are no side reactions using the enzymes [127]. However, using enzymes as catalysts involves some disadvantages like the high price of enzymes and the longer duration of the reaction compared to other catalysts. Therefore due to the presence of alcohols such as methanol during the transesterification, the operational life of the enzymes is short; however, the immobilization of enzymes may enhance their stability [127]. Recently, nanocatalysts (e.g., CaO, $Ca(OCH_3)_2$, $KF/CaO-Fe_3O_4$, etc. [128−130]) have attracted scientists' attention due to their high activity, stability, and reusability [131]. Another advantage of this kind of catalyst is their high opposition to saponification and efficient volume to surface ratio [130].

Many researchers have investigated the effect of different catalysts on the transesterification of algal biomass [132–134]. For instance, the biodiesel yield of 86.1% was reported when the lipids of *Spirulina maxima* were used for transesterification in the presence of KOH as catalyst [135]. Also, the maximum yield of 90.6% and 98.11% was reported for biodiesel production using *E. compressa* and *C. peltata* oil as a result of base transesterification using NaOH [43,70]. In another study, after utilizing KOH solution with pumice powder as a catalyst for the transesterification of *C. vulgaris* lipids, the biodiesel's yield was 85% [134]. The production of biodiesel using wet biomass of *Nannochloropsis gaditana* was performed using acetyl chloride as a catalyst where 100% of saponified lipids were transferred to FAMEs [136]. The wet biomass of *Chlamydomonas* sp. JSC4 (68.7wt.% water content) was applied to direct transesterification, resulting in the conversion of 97.2% of the extracted oil to biodiesel [137]. In the case of enzymatic transesterification performed on the biomass of *C. vulgaris* (water content of 86%–91%) using *Burkholderia* lipase as catalyst, more than 90% biodiesel conversion was achieved [133]. As an example of a nanocatalyst for transesterification, calcium methoxide was utilized as catalyst for biodiesel production using *Nannochloropsis* sp., and a 96% yield of FAME was achieved in this research [129].

10.4 Biodiesel characteristics

There are some advantages of biodiesel utilization such as less CO_2 emissions, high cetane number, easy biodegradability, more efficient combustion, etc., which make biodiesel an attractive fuel to be used in diesel engines. However, the utilization of this fuel has some drawbacks (e.g., high viscosity, lower calorific value (CV), higher NOx emissions, etc.), so that it is necessary to thoroughly check its properties before using it, following international standards. To be a suitable alternative for petroleum diesel, the produced biodiesel's physicochemical properties must be within the worldwide accepted standard specifications. The main standards for biodiesel characteristics have been developed by the American Society for Testing and Materials (ASTM), and the European Committee for Standardization (EN). Also, other standards of biodiesel have been extensively summarized previously [138].

As defined in previous sections, biodiesel is a mixture of different FAMEs, mostly consisting of C16 and C18 fatty acids. Furthermore, the biodiesel characteristics are determined by the type of fatty acids. In other words, the fatty acid profiles directly affect the properties of biodiesel. For instance, an increase of chain length results in the increase of some biodiesel characteristics such as cetane number, CV, viscosity, and melting point [139], whereas the mentioned characteristics decrease by increasing the degree of unsaturation (DU). The properties of biodiesel from algal feedstocks have not been extensively investigated. Therefore some characteristics of biodiesel are discussed in this section. In addition, different characteristics of biodiesel derived from some algae species are summarized in Table 10.3.

TABLE 10.3 Properties of biodiesel obtained from different algal species.

Property	FP	CN	D	AN	IN	CP	CFPP	CV	KV 40°C	SV	OS	DU	LCSF	References
Unit	°C	–	kg/m^3	mg KOH/g	g I$_2$/100 g	°C	°C	MJ/kg	mm^2/s	mg KOH	h	wt.%	wt.%	
ASTM	130*	47*	880	0.5[+]	–	−3 to (−12)	+5[+]	–	1.9–6.0	370[+]	–	–	–	[138,140]
EN	101*	51*	860–900	0.5[+]	120[+]	–	–	35*	3.5–5.0	–	3*	–	–	[138,140]
Microalgae														
Aphanothece microscopica	–	51.3–55.8	–	–	65.4–79.9	–	(−4.6)–24.9	–	–	225.1	–	65.3–70.6	3.8	[141,142]
Batrachospermum sirodotia	–	59.5	875	–	50.1	13.27	–	39.4	4.89	–	–	–	–	[59]
Chlorella marina	98	–	971	0.474	–	–	−10	43	4.8	–	–	–	–	[143]
Chlorella minutissima	–	84	880	–	74	–	−6	44	3.8	145	12	75	3.1	[144]
Chlorella protothecoides	115	–	–	0.374	–	–	−11	41	5.2	–	–	–	–	[145]
Chlorella sp.	179	–	883	0.37	97.12	–	–	37.06	4.73	–	6.8	–	–	[146]
Chlorella vulgaris	–	54.1–56.7	880	–	65.0–110.4	2.44	4.5	40.8	4.38	217.8	–	74.1	6.7	[59,141]
Coelastrella sp. M-60	–	55.5	–	–	84.9	–	44.3	–	–	194.5	–	91.9	19.3	[147]
Crypthecodinium cohnii	95	46.5	912	0.14	–	16.1	–	39.86	5.06	–	–	–	–	[148]
Dinoflagellate	–	–	878	0.44	–	–	–	39.84	3.74	–	1	–	–	[61]
Dunaliella tertiolecta	–	52.2	–	–	83.8	–	−8.4	–	–	220.8	–	98.0	2.6	[141]
Isochrysis sphaerica	–	61.4	874	–	29.3	17.02	–	38.9	5.07	–	–	–	–	[59]
Kirchneriella lunaris	–	51.7	882	–	137.3	−2.38	–	41.5	4.15	–	–	–	–	[59]

Species													Ref.
Lyngbya kuetzingii	–	52.0	882	–	134.0	−1.79	–	41.4	4.18	–	–	–	–	[59]
Micractinium sp. M-13	–	62.9	–	–	53.4	–	4.0	–	–	190.6	–	57.8	3.9	[147]
Nannochloropsis oculata	–	46.7	889	–	–	–	−10.7	40.3	3.92	–	–	–	–	[62]
Nannochloropsis salina	95	–	992	0.32	–	–	−10	40	3.2	–	–	–	–	[143]
Nannochloropsis sp.	–	–	854	0.46	–	–	–	39.81	5.76	–	1.93	–	–	[61]
Navicula sp.	–	56.7	878	–	81.8	7.58	–	40.2	4.62	–	–	–	–	[59]
Phaeodactylum tricornutum	–	48.3−55.1	863−879	–	58.7−99.2	4.47	−12.3−(−10.6)	40.6	2.95−4.47	266.1	–	52.7	1.3	[59,62,141]
Phromidium sp.	–	54.6	–	–	74.5	–	4.4	–	–	217.9	–	76.3	6.6	[141]
Scenedesmus incrassatulus	–	62	803	–	–	9	−4	41	3.78	–	19	–	–	[63]
Scenedesmus obliquus	–	51.77−57.7	877	0.42	68.2−98.69	9.65	4.9−20.8	37.67−39.9	4.72	217.5	3.54	68.7	11.9	[59,141,149]
Scenedesmus sp.	–	–	852	0.52	–	–	–	39.76	4.15	–	5.42	–	–	[61]
Selenastrum capricornutum	–	59.6	875	–	49.0	13.49	–	39.4	4.90	–	–	–	–	[59]
Spirulina platensis	130	–	860	0.45	–	–	–	41.36	5.66	–	–	–	–	[64]
Staursatrum sp.	–	61.3	874	–	30.3	16.84	–	38.9	5.06	–	–	–	–	[59]
Tetraselmis sp.	–	47.3	876	–	–	–	−8.5	40.9	3.70	–	–	–	–	[62]

(Continued)

TABLE 10.3 (Continued)

Property	FP	CN	D	AN mg	IN	CP	CFPP	CV	KV 40°C	SV mg	OS	DU	LCSF	References
Unit	°C	–	kg/m³	KOH/g	g I_2/100 g	°C	°C	MJ/kg	mm²/s	KOH	h	wt.%	wt.%	
Macroalgae														
Caulerpa peltata	178	57.9	885	0.40	–	3	–	–	4.25	–	–	–	–	[43]
Caulerpa racemosa	155	58.23	868	0.13	–	–4	–	–	4.30	–	–	–	–	[150]
Cladophora glomerata	110	–	892	0.94	76.24	–	–	33.6	3.8	–	–	–	–	[48]
Enteromorpha compressa	166	58.5	878	0.43	–	3	–	–	4.35	–	–	–	–	[70]
Melanothamnus afaqhusainii	135	–	880	0.50	51	1	–	42.8	–	–	–	–	–	[151]
Padina boryana	–	59.49	–	–	50.53	13.20	–	39.43	4.89	–	–	–	–	[22]
Padina tetrastromatica	158	57.5	883	0.50	–	–	–	38.5	4.92	–	–	–	–	[55]
Ulva fasciata	–	55.6	840	13.73 14[a]	95.86–91	–	–	38.2–38.98	35.2[b]	189.69	–	–	–	[8,152]
Ulva intestinalis	–	59.70	–	–	48.18	13.62	–	39.37	4.91	–	–	–	–	[22]
Ulva lactuca	–	54.90	–	14.27	97.22–101.62	4.03	–	40.64	4.45	212.9	–	–	–	[22,153]
Ulva linza	–	80.2	840	12[a]	90	–	–	38.37	–	–	–	–	–	[152]
Ulva reticulate	–	33.0	850	14[a]	85	–	–	33.5	–	–	–	–	–	[152]
Ulva rigida	–	78.4	860	13[a]	88	–	–	36.7	–	–	–	–	–	[152]
Ulva tubulosa	–	16.2	840	13[a]	89	–	–	33.4	–	–	–	–	–	[152]

*Minimum.

† Maximum.

[a]In these studies acid value reported as the mL of H_3PO_4/g.

[b]The kinematic viscosity in this study reported is cSt at 40°C.

AN, acid number; ASTM, American society for testing and materials; CN, cetane number; CP, cloud point; CFPP, cold filter plugging point; CV, calorific value; D, density; DU, degree of unsaturation; EN, European committee for standardization; FP, flash point; IN, iodine number; KV, kinematic viscosity; LCSF, long chain saturation factor; OS, oxidation stability; SV, saponification value.

10.4.1 Boiling point, flash point, and calorific value

The boiling point (BP) is the temperature at which the vapor pressure is equal to the surrounding pressure. In addition, the BP is known as a feature to indicate a fuel's volatility, so that the higher the BP, the lower the volatility. Furthermore, the type of bonds between the components affects the BP.

The flash point (FP) is defined as the lowest temperature at which the fuel vapors ignite near the combustion source. In the field of storage and transit, biodiesel is preferable to petroleum diesel, while the FP of biodiesel (more than 150°C) is higher than that of petroleum diesel (55°C−65°C) [138]. This makes the biodiesel safe as a transport fuel. Therefore some algal species, such as *Chlorella* sp., *S. platensis*, *C. peltata*, etc., meet both ASTM and EN standards (Table 10.3).

The value of released energy by the consumption of a unit amount of fuel is called the CV or heating value. The CV of biodiesel is lower than the diesel fuel [154]. In addition, the energy of biodiesel is about 10% less than diesel due to the presence of high amounts of oxygen in biodiesel. Therefore the energy level of diesel blends is reduced. Accordingly, CV is severely affected by the biodiesel degree of unsaturation.

10.4.2 Cetane number, acid number, iodine number, and sulfur content

The ignition characteristics of biodiesel in the engine are illustrated by cetane number [141]. The ignition delay phase of fuel is directly affected by this feature. The length of the fatty acid chain and saturation degree positively improve the biodiesel cetane number. Biodiesel's cetane number is higher as it contains more oxygen, and its combustion sufficiency is better [155]. As depicted in Table 10.3 the cetane number of most algal-derived biodiesel meets the ASTM and EN standards.

Acid number (AN) (or acid value) indicates the value of FFAs in the biodiesel sample. The higher the FFA content, the higher the acid number, which may cause corrosion in the engine [156]. The acid number value is described by mg KOH/g, showing the required amount of KOH to neutralize it. As illustrated in Table 10.3, many of the studied algal species meet the standard limitations.

Iodine number (IN) (or iodine value) indicates the biodiesel degree of unsaturation by determining the amount of iodine (in grams) that can be combined with 100 g of biodiesel [157]. During the contact of fuel and air, the iodine number is used to figure out the oxidation tendency and degree of unsaturation of fuel. The iodine number in most biodiesel derived from algal species is below 110, while the EN standard is 120 g I_2/100 g (Table 10.3).

The amount of sulfur in the fuel determines the amount of sulfur oxide produced during combustion. It is noteworthy that biodiesel produced from vegetable oils contains a lower content of sulfur [158].

10.4.3 Cloud point, cold filter plugging point, and pour point

The temperature at which the precipitation occurs as a result of crystal formation is known as the cloud point (CP). It is also defined as the lowest temperature at which the fuel could be ignited efficiently. The CP of biodiesel derived from different feedstocks is different due to the differences in their FAs content [159].

The minimum temperature at which a certain amount of fuel passes through a standard filter for cooling under certain conditions is defined as the cold filter plugging point (CFPP) [139]. These two parameters (CP and CFPP) are used to describe the highest filterability of the fuel. Accordingly, biodiesel efficiency at low temperature is not suitable due to wax creation and flow reduction.

The lowest temperature at which a liquid fuel loses its flow characteristics is known as pour point (PP) [156]. The parameters of CP, CFPP, and PP are directly related to each other.

In comparison with MUFAs and PUFAs, the SFAs CP and pour point is higher, showing a high value of CFPP, suggesting that the cold-flow properties of biodiesel are considerable when the fuel contains less SFAs [26]. These mentioned characteristics may vary between different countries according to the weather conditions.

10.4.4 Kinematic viscosity and density

Viscosity is an important physical parameter used to describe the flow capability of fuel. Kinematic viscosity (KV) is a significant parameter that evaluate biodiesel quality. The biodiesel viscosity is higher than fossil fuels due to its chemical structure [159,160]. The lower the viscosity, the smaller the droplets, resulting in better fuel supply to the combustion chamber. Approximately the biodiesel of all mentioned algal species is in the range of the ASTM and EN standards (Table 10.3). However, in some cases, this parameter may not meet the standards due to long-chain fatty acid contents and the high degree of unsaturation [29].

Density (D) is a fuel characteristic that determines the appropriate amount of fuel to be used in the combustion chamber. This parameter also affects characteristics such as the amount of energy and air—fuel (A/F) ratio. Different factors such as biodiesel production process, FAME profile, and type of feedstock affect the viscosity of biodiesel.

10.4.5 Oxidation stability

In addition to biodiesel's mentioned characteristics, the quality of this fuel could be evaluated using its oxidation. This is one of the most important factors and is known as oxidation stability (OS). The presence of some components, such as metals and water, influences the oxidation of biodiesel. In addition, the presence of USFAs and double bonds causes oxidation as a result of reactions with oxygen [139].

Therefore biodiesel is more susceptible to oxidation than petroleum diesel. The OS of biodiesel was considerable when *S. incrassatulus* and *Chlorella* sp. were used as feedstock for biofuels production (Table 10.3).

10.4.6 Water and sediment content

The presence of water and sediment in the biodiesel determines the purity of biodiesel. The presence of water in biodiesel decreases its calorific value and causes corrosion in the engine, while may cause clogging in fuel lines.

10.5 Economic feasibility

In comparison with terrestrial plants (first- and second-generation feedstocks), algae cultivation is easier and faster; they do not need arable land, the productivity of algal biomass per land area is higher, they can use wastewater as nutrient and simultaneously reduce the hazardous material of the wastewater, and they also can capture CO_2 and convert it to valuable products (lipids, proteins, and carbohydrates). Due to these benefits, algal species produce a considerable amount of lipids, while their biodiesel conversion efficiency is high in different species [43,70,136]. In addition, the biodiesel production from algal oils is at least nine times higher than other edible and nonedible plants, while they occupy between 0.1 to 0.2 m^2 land for production of 1 kg biodiesel [29]. As previously estimated, total biodiesel amounts of 8.2 and 12 t/h/y could be obtained when *C. vulgaris* and *S. obliquus* are used as feedstock [11]. However, the commercialization of third-generation biodiesel production may face many difficulties as the large-scale production of algal biomass is still challenging. Accordingly, biomass cultivation and harvesting cost more than other operation stages [11]. One way to make algal production more economically viable may be the possibility of refining algal biomass to different products as well. In simple terms, all fractions of algal biomass should be utilized as valuable compounds. However, to commercialize algal cultivation and biodiesel production, many more investigations are needed to enhance the energy and resource efficiency.

10.6 Conclusions and perspectives

In this chapter, different characteristics of algal species have been extensively reviewed. As a renewable and sustainable energy, algal biodiesel could be suitable as an alternatives to petroleum diesel. However, large-scale cultivation technology and the commercialization of biodiesel production are in their infancy due to the operational costs. In this regard, scientists have undertaken several investigations to attempt to lower the costs of biodiesel production. For instance, the lipid extraction from wet biomass has recently attracted researchers' attention in order to reduce the total operational cost. Also, other aspects of biodiesel production require further optimization. In this regard, the growth and cultivation of algal species and the production of biofuels from these feedstocks require more attention.

References

[1] M.K. Yesilyurt, M. Aydin, Experimental investigation on the performance, combustion and exhaust emission characteristics of a compression-ignition engine fueled with cottonseed oil biodiesel/diethyl ether/diesel fuel blends, Energy Convers. Manage. 205 (2020) 112355.

[2] T. Nematian, Z. Salehi, A. Shakeri, Conversion of bio-oil extracted from *Chlorella vulgaris* microalgae to biodiesel via modified superparamagnetic nano-biocatalyst, Renew. Energy 146 (2020) 1796–1804.

[3] G.Y. Yew, S.Y. Lee, P.L. Show, Y. Tao, C.L. Law, T.T.C. Nguyen, et al., Recent advances in algae biodiesel production: from upstream cultivation to downstream processing, Bioresour. Technol. Rep. 7 (2019) 100227.

[4] D. Singh, D. Sharma, S.L. Soni, S. Sharma, P.K. Sharma, A. Jhalani, A review on feedstocks, production processes, and yield for different generations of biodiesel, Fuel 262 (2020) 116553.

[5] A. Ali, A. Qadir, M. Kuddus, P. Saxena, M. Zainul, Abdin, production of biodiesel from algae: an update, in: L.M.T. Martínez, O.V. Kharissova, B.I. Kharisov (Eds.), Handbook of Ecomaterials, Springer International Publishing, Cham, 2019, pp. 1953−1964.

[6] S. Saravanan, N. Krishnamoorthy, Investigation on reduction in consequences of adding antioxidants into the algae biodiesel blend as a CI engine fuel, Fuel 276 (2020) 117993.

[7] L. Zhang, K.C. Loh, A. Kuoki, Y.W. Tong, Microbial biodiesel production from industrial organic wastes by oleaginous microorganisms: current status and prospects, J. Hazard. Mater. 402 (2020) 123543.

[8] T. Veeranan, R. Kasirajaan, B. Gurunathan, R. Sahadevan, A novel approach for extraction of algal oil from marine macroalgae *Ulva fasciata*, Renew. Energy 127 (2018) 64−73.

[9] C.D. Calixto, J.K. da Silva Santana, V.P. Tibúrcio, L.F.B.L. Pontes, C.F. Costa Sassi, M.M. Conceição, et al., Productivity and fuel quality parameters of lipids obtained from 12 species of microalgae from the northeastern region of Brazil, Renew. Energy 115 (2018) 1144−1152.

[10] T.M.I. Mahlia, Z.A.H.S. Syazmi, M. Mofijur, A.E.P. Abas, M.R. Bilad, H.C. Ong, et al., Patent landscape review on biodiesel production: technology updates, Renew. Sust. Energ. Rev. 118 (2020) 109526.

[11] A.E.F. Abomohra, W. Jin, R. Tu, S.F. Han, M. Eid, H. Eladel, Microalgal biomass production as a sustainable feedstock for biodiesel: current status and perspectives, Renew. Sust. Energy Rev. 64 (2016) 596−606.

[12] A.E.F. Abomohra, M. Elsayed, S. Esakkimuthu, M. El-Sheekh, D. Hanelt, Potential of fat, oil and grease (FOG) for biodiesel production: a critical review on the recent progress and future perspectives, Prog. Energy Combust. Sci. 81 (2020) 100868.

[13] S. Xu, M. Elsayed, G.A. Ismail, C. Li, S. Wang, A.E.F. Abomohra, Evaluation of bioethanol and biodiesel production from *Scenedesmus obliquus* grown in biodiesel waste glycerol: a sequential integrated route for enhanced energy recovery, Energy Convers. Manag. 197 (2019) 111907.

[14] Y. Ma, Z. Gao, Q. Wang, Y. Liu, Biodiesels from microbial oils: opportunity and challenges, Bioresour. Technol. 263 (2018) 631−641.

[15] P. Andreo-Martínez, V.M. Ortiz-Martinez, N. Garcia-Matinez, A.P. de los Rios, F.J. Hernandez-Fernandez, Production of biodiesel under supercritical conditions: state of the art and bibliometric analysis, Appl. Energy 264 (2020) 114753.

[16] T. Nematian, M. Barati, Nanobiocatalytic processes for producing biodiesel from algae, in: M. Rai, A.P. Ingle (Eds.), Sustainable Bioenergy, Elsevier, 2019, pp. 299−326. Chapter 11.

[17] T. Suganya, M. Varman, H.H. Masjuki, S. Renganathan, Macroalgae and microalgae as a potential source for commercial applications along with biofuels production: a biorefinery approach, Renew. Sust. Energ. Rev. 55 (2016) 909−941.

[18] R.G. Saratale, G. Kumar, R. Banu, A. Xia, S. Periyasamy, G.D. Saratale, A critical review on anaerobic digestion of microalgae and macroalgae and co-digestion of biomass for enhanced methane generation, Bioresour. Technol. 262 (2018) 319−332.

[19] B. Jin, P. Duan, Y. Xu, F. Wang, Y. Fan, Co-liquefaction of micro-and macroalgae in subcritical water, Bioresour. Technol. 149 (2013) 103−110.

[20] G. Breuer, P.P. Lamers, D.E. Martens, R.B. Draaisma, R.H. Wijffels, The impact of nitrogen starvation on the dynamics of triacylglycerol accumulation in nine microalgae strains, Bioresour. Technol. 124 (2012) 217−226.

[21] G. Pablo, J.S. Gomes-Dias, C.M.R. Rocha, A. Romaní, G. Garrote, L. Domingues, Recent trends on seaweed fractionation for liquid biofuels production, Bioresour. Technol. 299 (2020) 122613.

[22] A.E.F. Abomohra, A.H. El-Naggar, A.A. Baeshen, Potential of macroalgae for biodiesel production: screening and evaluation studies, J. Biosci. Bioeng. 125 (2018) 231−237.

[23] B. Sajjadi, W.Y. Chen, A.A.A. Raman, S. Ibrahim, Microalgae lipid and biomass for biofuel production: a comprehensive review on lipid enhancement strategies and their effects on fatty acid composition, Renew. Sust. Energ. Rev. 97 (2018) 200−232.

[24] S.K. Sinha, A. Gupta, R. Bharalee, Production of biodiesel from freshwater microalgae and evaluation of fuel properties based on fatty acid methyl ester profile, Biofuels 7 (1) (2016) 69−78.

[25] Y. Li-Beisson, J.J. Thelen, E. Fedosejevs, J.L. Harwood, The lipid biochemistry of eukaryotic algae, Prog. Lipid Res. 74 (2019) 31−68.

[26] S. Deshmukh, R. Kumar, K. Bala, Microalgae biodiesel: a review on oil extraction, fatty acid composition, properties and effect on engine performance and emissions, Fuel Process. Technol. 191 (2019) 232−247.

References

[27] H. Chen, D. Zhou, G. Luo, S. Zhang, J. Chen, Macroalgae for biofuels production: progress and perspectives, Renew. Sustain. Energy Rev. 47 (2015) 427–437.

[28] Y. Chisti, Biodiesel from microalgae beats bioethanol, Trends Biotechnol. 26 (3) (2008) 126–131.

[29] L. Gouveia, A.C. Oliveira, R. Congestri, L. Bruno, A.T. Soares, R.S. Menezes, et al., Biodiesel from microalgae, in: C.G. Fernandez, R. Muñoz (Eds.), Microalgae-Based Biofuels and Bioproducts, Elsevier, 2017, pp. 235–258. Chapter 10.

[30] Q. Hu, M. Sommerfeld, E. Jarvis, M. Ghirardi, M. Posewitz, M. Seibert, et al., Microalgal triacylglycerols as feedstocks for biofuel production: perspectives and advances, Plant. J. 54 (4) (2008) 621–639.

[31] Y. Li-Beisson, F. Beisson, W. Riekhof, Metabolism of acyl-lipids in *Chlamydomonas reinhardtii*, Plant. J. 82 (3) (2015) 504–522.

[32] S.S. Merchant, J. Kropat, B. Liu, J. Shaw, J. Warakanont, TAG, You're it! *Chlamydomonas* as a reference organism for understanding algal triacylglycerol accumulation, Curr. Opin. Biotechnol. 23 (3) (2012) 352–363.

[33] B. Liu, C. Benning, Lipid metabolism in microalgae distinguishes itself, Curr. Opin. Biotechnol. 24 (2) (2013) 300–309.

[34] R. Leonardi, Y.M. Zhang, C.O. Rock, S. Jackowski, Coenzyme A: back in action, Prog. Lipid Res. 44 (2–3) (2005) 125–153.

[35] P. Singh, S. Kumari, A. Guldhe, R. Misra, I. Rawat, F. Bux, Trends and novel strategies for enhancing lipid accumulation and quality in microalgae, Renew. Sustain. Energy Rev. 55 (2016) 1–16.

[36] P. Singh, A. Guldhe, S. Kumari, I. Rawat, F. Bux, Investigation of combined effect of nitrogen, phosphorus and iron on lipid productivity of microalgae *Ankistrodesmus falcatus* KJ671624 using response surface methodology, Biochem. Eng. J. 94 (2015) 22–29.

[37] X.W. Wang, J.R. Liang, C.S. Luo, C.P. Chen, Y.H. Gao, Biomass, total lipid production, and fatty acid composition of the marine diatom *Chaetoceros muelleri* in response to different CO_2 levels, Bioresour. Technol. 161 (2014) 124–130.

[38] T.B. Fioresi, L.H.S. Tavares, Cultivo de *Ankistrodesmus gracilis* (Chlorophyta) em laboratório à base de esterco suíno, Biotemas 21 (1) (2008) 7–16.

[39] T.M. Mata, A.A. Martins, N.S. Caetano, Microalgae for biodiesel production and other applications: a review, Renew. Sustain. Energy Rev. 14 (1) (2010) 217–232.

[40] M.P. Sudhakar, B. Ramesh Kumar, T. Mathimani, K. Arunkumar, A review on bioenergy and bioactive compounds from microalgae and macroalgae-sustainable energy perspective, J. Clean. Prod. 228 (2019) 1320–1333.

[41] N. Burman, G.G. Satpati, S. SenRoy, N. Khatoon, R. Sen, S. Kanjilal, et al., Mapping algae of Sundarban origin as lipid feedstock for potential biodiesel application, J. Algal Biomass Util. 3 (2) (2012) 42–49.

[42] M.J. Griffiths, S.T. Harrison, Lipid productivity as a key characteristic for choosing algal species for biodiesel production, J. Appl. Phycol. 21 (5) (2009) 493–507.

[43] S. Tamilarasan, R. Sahadevan, Ultrasonic assisted acid base transesterification of algal oil from marine macroalgae *Caulerpa peltata*: optimization and characterization studies, Fuel 128 (2014) 347–355.

[44] X. Wen, Y. Geng, Y. Li, Enhanced lipid production in *Chlorella pyrenoidosa* by continuous culture, Bioresour. Technol. 161 (2014) 297–303.

[45] S.J. Lee, B.D. Yoon, H.M. Oh, Rapid method for the determination of lipid from the green alga *Botryococcus braunii*, Biotechnol. Tech. 12 (7) (1998) 553–556.

[46] X.Y. Deng, K. Gao, M. Addy, D. Li, R.C. Zhang, Q. Lu, et al., Cultivation of *Chlorella vulgaris* on anaerobically digested swine manure with daily recycling of the post-harvest culture broth, Bioresour. Technol. 247 (2018) 716–723.

[47] Y. Ye, Y. Huang, A. Xia, Q. Fu, Q. Liao, W. Zeng, et al., Optimizing culture conditions for heterotrophic-assisted photoautotrophic biofilm growth of *Chlorella vulgaris* to simultaneously improve microalgae biomass and lipid productivity, Bioresour. Technol. 270 (2018) 80–87.

[48] M. Yuvarani, D. Kubendran, A.R.S. Aathika, P. Kathik, M.P. Prekuma, V. Karthikeyan, et al., Extraction and characterization of oil from macroalgae *Cladophora glomerata*, Energy Sources A Recovery, Util. Environ. Eff. 39 (23) (2017) 2133–2139.

[49] T. Takeshita, I.N. Ivanov, K. Oshima, K. Ishii, H. Kawamoto, S. Ota, et al., Comparison of lipid productivity of *Parachlorella kessleri* heavy-ion beam irradiation mutant PK4 in laboratory and 150-L mass bioreactor, identification and characterization of its genetic variation, Algal Res. 35 (2018) 416–426.

[50] P. Feng, Z. Deng, Z. Hu, Z. Wang, L. Fan, Characterization of *Chlorococcum pamirum* as a potential biodiesel feedstock, Bioresour. Technol. 162 (2014) 115−122.

[51] A. Demirbas, Use of algae as biofuel sources, Energy Convers. Manage. 51 (12) (2010) 2738−2749.

[52] A.E.F. Abomohra, M. Wagner, M. El-Sheekh, D. Hanelt, Lipid and total fatty acid productivity in photoautotrophic fresh water microalgae: screening studies towards biodiesel production, J. Appl. Phycol. 25 (4) (2013) 931−936.

[53] L.U. Abubakar, A. Mutie, E.U. Kenya, Characterization of algae oil (oilgae) and its potential as biofuel in Kenya, J. Appl. Phytotechnol. Environ. Sanitation 1 (4) (2012) 147−153.

[54] Y. Xiao, J. Zhang, J. Cui, X. Yao, Y. Feng, et al., Simultaneous accumulation of neutral lipids and biomass in *Nannochloropsis oceanica* IMET1 under high light intensity and nitrogen replete conditions, Algal Res. 11 (2015) 55−62.

[55] V. Ashokkumar, M.R. Salim, Z. Salam, P. Sivakumar, C.T. Chong, S. Elumalai, et al., Production of liquid biofuels (biodiesel and bioethanol) from brown marine macroalgae *Padina tetrastromatica*, Energy Convers. Manage. 135 (2017) 351−361.

[56] S. Mandotra, P. Kumar, M.R. Suseela, P.W. Ramteke, Fresh water green microalga *Scenedesmus abundans*: a potential feedstock for high quality biodiesel production, Bioresour. Technol. 156 (2014) 42−47.

[57] P.T. Vasudevan, M. Briggs, Biodiesel production—current state of the art and challenges, J. Ind. Microbiol. Biotechnol. 35 (5) (2008) 421.

[58] A.F. Talebi, S.K. Mohtashami, M. Tabatabaei, M. Tohidfar, A. Bagheri, M. Zeinalabedini, et al., Fatty acids profiling: a selective criterion for screening microalgae strains for biodiesel production, Algal Res. 2 (3) (2013) 258−267.

[59] M. Song, H. Pei, W. Hu, G. Ma, Evaluation of the potential of 10 microalgal strains for biodiesel production, Bioresour. Technol. 141 (2013) 245−251.

[60] I.A. Nascimento, S.S.I. Marques, I.T.D. Cabanelas, S.A. Pereira, J.I. Druzian, C.O. de Souza, et al., Screening microalgae strains for biodiesel production: lipid productivity and estimation of fuel quality based on fatty acids profiles as selective criteria, Bioenergy Res. 6 (1) (2013) 1−13.

[61] L. Chen, T. Liu, W. Zhang, X. Chen, J. Wang, Biodiesel production from algae oil high in free fatty acids by two-step catalytic conversion, Bioresour. Technol. 111 (2012) 208−214.

[62] K.N. Gangadhar, H. Pereira, H.P. Diogo, R.M.B. Dos Santos, B.L.A.P. Devi, R.B.N. Prasad, et al., Assessment and comparison of the properties of biodiesel synthesized from three different types of wet microalgal biomass, J. Appl. Phycol. 28 (3) (2016) 1571−1578.

[63] M.T. Arias-Peñaranda, E. Cristiani-Urbina, C. Montes-Horcasitas, F. Esparza-García, G. Torzillo, R.O. Cañizares-Villanueva, *Scenedesmus incrassatulus* CLHE-Si01: a potential source of renewable lipid for high quality biodiesel production, Bioresour. Technol. 140 (2013) 158−164.

[64] P. Nautiyal, K. Subramanian, M. Dastidar, Production and characterization of biodiesel from algae, Fuel Process. Technol. 120 (2014) 79−88.

[65] M.M. Ismail, G.A. Ismail, M.M. El-Sheekh, Potential assessment of some micro- and macroalgal species for bioethanol and biodiesel production, Energy Sources A Recovery Util. Environ. Effects (2020) pp. 1−17.

[66] F. Borghini, L. Lucattini, S. Focardi, S. Focardi, S. Bastianoni, Production of bio-diesel from macro-algae of the Orbetello lagoon by various extraction methods, Int. J. Sustain. Energy 33 (3) (2014) 695−703.

[67] H. Pereira, L. Barreira, F. Figueiredo, L. Custódio, C. Vizetto-Duarte, C. Polo, et al., Polyunsaturated fatty acids of marine macroalgae: potential for nutritional and pharmaceutical applications, Mar. Drugs 10 (9) (2012) 1920−1935.

[68] N. Neveux, M. Magnusson, T. Maschmeyer, R. de Nys, N.A. Paul, Comparing the potential production and value of high-energy liquid fuels and protein from marine and freshwater macroalgae, Glob. Change Biol. Bioenergy 7 (4) (2015) 673−689.

[69] A.E.M.M. Afify, E.A. Shalaby, S.M. Shanab, Enhancement of biodiesel production from different species of algae, Grasas Aceites 61 (4) (2010) 416−422.

[70] T. Suganya, N.N. Gandhi, S. Renganathan, Production of algal biodiesel from marine macroalgae *Enteromorpha compressa* by two step process: optimization and kinetic study, Bioresour. Technol. 128 (2013) 392−400.

[71] E. Spijkerman, A. Wacker, Interactions between P-limitation and different C conditions on the fatty acid composition of an extremophile microalga, Extremophiles 15 (5) (2011) 597.

References

221

[72] M.M. Hanifzadeh, E.C. Garcia, S. Viamajala, Production of lipid and carbohydrate from microalgae without compromising biomass productivities: role of Ca and Mg, Renew. Energy 127 (2018) 989–997.

[73] S. Aslan, I.K. Kapdan, Batch kinetics of nitrogen and phosphorus removal from synthetic wastewater by algae, Ecol. Eng. 28 (1) (2006) 64–70.

[74] H.A. Aratboni, N. Rafiei, R. Garcia-Granados, A. Alemzadeh, J.R. Morones-Ramirez, Biomass and lipid induction strategies in microalgae for biofuel production and other applications, Microb. Cell Fact. 18 (1) (2019) 178.

[75] K.J. Mulders, J.H. Janssen, D.E. Martens, R.H. Wijffels, P.P. Lamers, Effect of biomass concentration on secondary carotenoids and triacylglycerol (TAG) accumulation in nitrogen-depleted *Chlorella zofingiensis*, Algal Res. 6 (2014) 8–16.

[76] N. Poddar, R. Sen, G.J.O. Martin, Glycerol and nitrate utilisation by marine microalgae *Nannochloropsis salina* and *Chlorella* sp. and associated bacteria during mixotrophic and heterotrophic growth, Algal Res. 33 (2018) 298–309.

[77] M. Takagi, K. Watanabe, K. Yamaberi, T. Yoshida, Limited feeding of potassium nitrate for intracellular lipid and triglyceride accumulation of *Nannochloris* sp. UTEX LB1999, Appl. Microbiol. Biotechnol. 54 (1) (2000) 112–117.

[78] A. Converti, A.A. Casazza, E.Y. Ortiz, P. Perego, M. Del Borghi, Effect of temperature and nitrogen concentration on the growth and lipid content of *Nannochloropsis oculata* and *Chlorella vulgaris* for biodiesel production, Chem. Eng. Process. 48 (6) (2009) 1146–1151.

[79] P. Feng, Z. Xu, L. Qin, M.A. Alam, Z. Wang, S. Zhu, Effects of different nitrogen sources and light paths of flat plate photobioreactors on the growth and lipid accumulation of *Chlorella* sp. GN1 outdoors, Bioresour. Technol. 301 (2020) 122762.

[80] F. Chu, J. Cheng, X. Zhang, Q. Ye, J. Zhou, Enhancing lipid production in microalgae *Chlorella* PY-ZU1 with phosphorus excess and nitrogen starvation under 15% CO_2 in a continuous two-step cultivation process, Chem. Eng. J. 375 (2019) 121912.

[81] G. Markou, I. Chatzipavlidis, D. Georgakakis, Carbohydrates production and bio-flocculation characteristics in cultures of *Arthrospira* (*Spirulina*) *platensis*: improvements through phosphorus limitation process, Bioenergy Res. 5 (4) (2012) 915–925.

[82] I. Khozin-Goldberg, Z. Cohen, The effect of phosphate starvation on the lipid and fatty acid composition of the fresh water eustigmatophyte *Monodus subterraneus*, Phytochemistry 67 (7) (2006) 696–701.

[83] L. Xin, H. Hong-Ying, G. Ke, S. Ying-Xue, Effects of different nitrogen and phosphorus concentrations on the growth, nutrient uptake, and lipid accumulation of a freshwater microalga *Scenedesmus* sp, Bioresour. Technol. 101 (14) (2010) 5494–5500.

[84] N. Sun, Y. Wang, Y.T. Li, J.C. Huang, F. Chen, Sugar-based growth, astaxanthin accumulation and carotenogenic transcription of heterotrophic *Chlorella zofingiensis* (Chlorophyta), Process. Biochem. 43 (11) (2008) 1288–1292.

[85] D. Cheng, Q. He, Assessment of environmental stresses for enhanced microalgal biofuel production—an overview, Front. Energy Res. 2 (2014) 26.

[86] H.H. Abd El Baky, G.S. El-Baroty, A. Bouaid, M. Martinez, J. Aracil, Enhancement of lipid accumulation in *Scenedesmus obliquus* by optimizing CO_2 and Fe^{3+} levels for biodiesel production, Bioresour. Technol. 119 (2012) 429–432.

[87] S.Y. Chiu, C.Y. Kao, M.T. Tsai, S.C. Ong, C.H. Chen, C.S. Lin, Lipid accumulation and CO_2 utilization of *Nannochloropsis oculata* in response to CO_2 aeration, Bioresour. Technol. 100 (2) (2009) 833–838.

[88] A.B. Fulke, S.N. Mudliar, R. Yadav, A. Shekh, N. Srinivasan, R. Ramanan, et al., Bio-mitigation of CO_2, calcite formation and simultaneous biodiesel precursors production using *Chlorella* sp, Bioresour. Technol. 101 (21) (2010) 8473–8476.

[89] S.Y. Chiu, C.Y. Kao, C.H. Chen, T.C. Kuan, S.C. Ong, C.S. Lin, Reduction of CO_2 by a high-density culture of *Chlorella* sp. in a semicontinuous photobioreactor, Bioresour. Technol. 99 (9) (2008) 3389–3396.

[90] L. Liu, Y. Zhao, X. Jiang, X. Wang, W. Liang, Lipid accumulation of *Chlorella pyrenoidosa* under mixotrophic cultivation using acetate and ammonium, Bioresour. Technol. 262 (2018) 342–346.

[91] G.B. Leite, K. Paranjape, A.E.M. Abdelaziz, P.C. Hallenbeck, Utilization of biodiesel-derived glycerol or xylose for increased growth and lipid production by indigenous microalgae, Bioresour. Technol. 184 (2015) 123–130.

[92] H. Ermis, U. Guven-Gulhan, T. Cakir, M. Altinbas, Effect of iron and magnesium addition on population dynamics and high value product of microalgae grown in anaerobic liquid digestate, Sci. Rep. 10 (1) (2020) 1–12.

Handbook of Algal Biofuels

[93] M. Wan, X. Jin, J. Xia, J.N. Rosenberg, G. Yu, Z. Nie, et al., The effect of iron on growth, lipid accumulation, and gene expression profile of the freshwater microalga *Chlorella sorokiniana*, Appl. Microbiol. Biotechnol. 98 (22) (2014) 9473–9481.

[94] Z.Y. Liu, G.C. Wang, B.C. Zhou, Effect of iron on growth and lipid accumulation in *Chlorella vulgaris*, Bioresour. Technol. 99 (11) (2008) 4717–4722.

[95] H.Y. Ren, B.F. Liu, F. Kong, L. Zhao, G.J. Xie, N.Q. Ren, Enhanced lipid accumulation of green microalga *Scenedesmus* sp. by metal ions and EDTA addition, Bioresour. Technol. 169 (2014) 763–767.

[96] L. Huang, J. Xu, T. Li, L. Wang, T. Deng, X. Yu, Effects of additional Mg^{2+} on the growth, lipid production, and fatty acid composition of *Monoraphidium* sp. FXY-10 under different culture conditions, Ann. Microbiol. 64 (3) (2014) 1247–1256.

[97] J. Kropat, A. Hong-Hermesdorf, D. Casero, P. Ent, M. Castruita, M. Pellegrini, et al., A revised mineral nutrient supplement increases biomass and growth rate in *Chlamydomonas reinhardtii*, Plant. J. 66 (5) (2011) 770–780.

[98] M. Hanikenne, S.S. Merchant, P. Hamel, Transition metal nutrition: a balance between deficiency and toxicity, in: E.H. Harris, D.B. Stern, G.B. Witman (Eds.), The *Chlamydomonas* Sourcebook, second ed., Elsevier, 2009, pp. 333–399. Chaper 10.

[99] C.H. Su, L.J. Chien, J. Gomes, Y.S. Lin, Y.K. Yu, J.S. Liou, et al., Factors affecting lipid accumulation by *Nannochloropsis oculata* in a two-stage cultivation process, J. Appl. Phycol. 23 (5) (2011) 903–908.

[100] L. Zhang, H. Pei, S. Chen, L. Jiang, Q. Hou, Z. Yang, et al., Salinity-induced cellular cross-talk in carbon partitioning reveals starch-to-lipid biosynthesis switching in low-starch freshwater algae, Bioresour. Technol. 250 (2018) 449–456.

[101] P. Kaewkannetra, P. Enmak, T. Chiu, The effect of CO_2 and salinity on the cultivation of *Scenedesmus obliquus* for biodiesel production, Biotechnol. Bioprocess. Eng. 17 (3) (2012) 591–597.

[102] E.S. Salama, R.A.I. Abou-Shanab, J.R. Kim, S. Lee, S.H. Kim, S.E. Oh, et al., The effects of salinity on the growth and biochemical properties of *Chlamydomonas mexicana* GU732420 cultivated in municipal wastewater, Environ. Technol. 35 (12) (2014) 1491–1498.

[103] X. Duan, G.Y. Ren, L.L. Liu, W.X. Zhu, Salt-induced osmotic stress for lipid overproduction in batch culture of *Chlorella vulgaris*, Afr. J. Biotechnol. 11 (27) (2012) 7072–7078.

[104] E.S. Salama, H.C. Kim, R.A.I. Abou-Shanab, M.K. Ji, Y.K. Oh, S.H. Kim, et al., Biomass, lipid content, and fatty acid composition of freshwater *Chlamydomonas mexicana* and *Scenedesmus obliquus* grown under salt stress, Bioproc. Biosyst. Eng. 36 (6) (2013) 827–833.

[105] S. Wahidin, A. Idris, S.R.M. Shaleh, The influence of light intensity and photoperiod on the growth and lipid content of microalgae *Nannochloropsis* sp, Bioresour. Technol. 129 (2013) 7–11.

[106] I. Krzemińska, B. Pawlik-Skowrońska, M. Trzcińska, J. Tys, Influence of photoperiods on the growth rate and biomass productivity of green microalgae, Bioproc, Biosyst. Eng. 37 (4) (2014) 735–741.

[107] J.U. Grobbelaar, Upper limits of photosynthetic productivity and problems of scaling, J. Appl. Phycol. 21 (5) (2009) 519–522.

[108] C. Yeesang, B. Cheirsilp, Effect of nitrogen, salt, and iron content in the growth medium and light intensity on lipid production by microalgae isolated from freshwater sources in Thailand, Bioresour. Technol. 102 (3) (2011) 3034–3040.

[109] Q. Zhang, T. Wang, Y. Hong, Investigation of initial pH effects on growth of an oleaginous microalgae *Chlorella* sp. HQ for lipid production and nutrient uptake, Water Sci. Technol. 70 (4) (2014) 712–719.

[110] B. Skrupski, K.E. Wilson, K.L. Goff, J. Zou, Effect of pH on neutral lipid and biomass accumulation in microalgal strains native to the Canadian prairies and the Athabasca oil sands, J. Appl. Phycol. 25 (4) (2013) 937–949.

[111] N.R. Moheimani, Inorganic carbon and pH effect on growth and lipid productivity of *Tetraselmis suecica* and *Chlorella* sp (Chlorophyta) grown outdoors in bag photobioreactors, J. Appl. Phycol. 25 (2) (2013) 387–398.

[112] R. Gardner, P. Peters, B. Peyton, K.E. Cooksey, Medium pH and nitrate concentration effects on accumulation of triacylglycerol in two members of the chlorophyta, J. Appl. Phycol. 23 (6) (2011) 1005–1016.

[113] M. Ras, J.P. Steyer, O. Bernard, Temperature effect on microalgae: a crucial factor for outdoor production, Rev. Environ. Sci. Biotechnol. 12 (2) (2013) 153–164.

[114] C.J. Zhu, Y. Lee, T. Chao, Effects of temperature and growth phase on lipid and biochemical composition of *Isochrysis galbana* TK1, J. Appl. Phycol. 9 (5) (1997) 451–457.

References

[115] Z.Y. Du, C. Benning, Triacylglycerol accumulation in photosynthetic cells in plants and algae, in: Y. Nakamura, Y.L. Beisson (Eds.), Lipids in Plant and Algae Development, Springer, 2016, pp. 179–205. Chapter 8.

[116] L. Xin, H. Hong-Ying, Z. Yu-Ping, Growth and lipid accumulation properties of a freshwater microalga *Scenedesmus* sp. under different cultivation temperature, Bioresour. Technol. 102 (3) (2011) 3098–3102.

[117] Y. Chisti, Biodiesel from microalgae, Biotechnol. Adv. 25 (3) (2007) 294–306.

[118] L. Lardon, A. Hélias, B. Sialve, J.P. Steyer, O. Bernard, Life-cycle assessment of biodiesel production from microalgae, Environ. Sci. Technol. 43 (17) (2009) 6475–6481.

[119] B. Kim, H.Y. Heo, J. Son, J. Yang, Y.K. Chang, J.H. Lee, et al., Simplifying biodiesel production from microalgae via wet in situ transesterification: a review in current research and future prospects, Algal Res. 41 (2019) 101557.

[120] P.D. Patil, H. Reddy, T. Muppaneni, A. Mannarswamy, T. Schuab, F.O. Holguin, et al., Power dissipation in microwave-enhanced in situ transesterification of algal biomass to biodiesel, Green Chem. 14 (3) (2012) 809–818.

[121] V. Kumar, M. Muthuraj, B. Palabhanvi, A.K. Ghoshal, D. Das, Evaluation and optimization of two stage sequential in situ transesterification process for fatty acid methyl ester quantification from microalgae, Renew. Energy 68 (2014) 560–569.

[122] C.Z. Liu, S. Zheng, L. Xu, F. Wang, C. Guo, Algal oil extraction from wet biomass of *Botryococcus braunii* by 1, 2-dimethoxyethane, Appl. Energy 102 (2013) 971–974.

[123] K.A. Salam, S.B. Velasquez-Orta, A.P. Harvey, A sustainable integrated in situ transesterification of microalgae for biodiesel production and associated co-product-a review, Renew. Sust. Energ. Rev. 65 (2016) 1179–1198.

[124] H. Fukuda, A. Kondo, H. Noda, Biodiesel fuel production by transesterification of oils, J. Biosci. Bioeng. 92 (5) (2001) 405–416.

[125] S. Saka, D. Kusdiana, Biodiesel fuel from rapeseed oil as prepared in supercritical methanol, Fuel 80 (2) (2001) 225–231.

[126] G.H. Huang, F. Chen, D. Wei, X.W. Zhang, G. Chen, Biodiesel production by microalgal biotechnology, Appl. Energy 87 (1) (2010) 38–46.

[127] N. Pragya, K.K. Pandey, P.K. Sahoo, A review on harvesting, oil extraction and biofuels production technologies from microalgae, Renew. Sustain. Energy Rev. 24 (2013) 159–171.

[128] S. Siva, C. Marimuthu, Production of biodiesel by transesterification of algae oil with an assistance of nano-CaO catalyst derived from egg shell, Int. J. Chemtech Res. 7 (4) (2015) 2112–2116.

[129] S.H. Teo, A. Islam, Y.H. Taufiq-Yap, Algae derived biodiesel using nanocatalytic transesterification process, Chem. Eng. Res. Des. 111 (2016) 362–370.

[130] S. Hu, Y. Guan, Y. Wang, H. Han, Nano-magnetic catalyst $KF/CaO-Fe_3O_4$ for biodiesel production, Appl. Energy 88 (8) (2011) 2685–2690.

[131] G. Baskar, R. Aiswarya, Trends in catalytic production of biodiesel from various feedstocks, Renew. Sustain. Energy Rev. 57 (2016) 496–504.

[132] B. Kim, H. Im, J.W. Lee, In situ transesterification of highly wet microalgae using hydrochloric acid, Bioresour. Technol. 185 (2015) 421–425.

[133] D.T. Tran, C.L. Chen, J.S. Chang, Effect of solvents and oil content on direct transesterification of wet oil-bearing microalgal biomass of *Chlorella vulgaris* ESP-31 for biodiesel synthesis using immobilized lipase as the biocatalyst, Bioresour. Technol. 135 (2013) 213–221.

[134] A.P. Cercado, F. Ballesteros Jr, S. Capareda, Ultrasound assisted transesterification of microalgae using synthesized novel catalyst, Sustain. Environ. Res. 28 (5) (2018) 234–239.

[135] M.A. Rahman, M.A. Aziz, R.A. Al-khulaidi, N. Sakib, M. Islam, Biodiesel production from microalgae *Spirulina maxima* by two step process: optimization of process variable, J. Radiat. Res. Appl. Sci. 10 (2) (2017) 140–147.

[136] M.D. Macías-Sánchez, A. Robles-Medina, E. Hita-Peña, M.J. Jiménez-Callejón, L. Estéban-Cerdán, P.A. González-Moreno, et al., Biodiesel production from wet microalgal biomass by direct transesterification, Fuel 150 (2015) 14–20.

[137] C.L. Chen, C.C. Huang, K.C. Ho, P.X. Hsiao, M.S. Wu, J.S. Chang, Biodiesel production from wet microalgae feedstock using sequential wet extraction/transesterification and direct transesterification processes, Bioresour. Technol. 194 (2015) 179–186.

[138] A.E. Atabani, A.S. Silitonga, H.C. Ong, T.M.I. Mahlia, H.H. Masjuki, I.A. Badruddin, et al., Non-edible vegetable oils: a critical evaluation of oil extraction, fatty acid compositions, biodiesel production, characteristics, engine performance and emissions production, Renew. Sustain. Energy Rev. 18 (2013) 211–245.

[139] G. Knothe, Dependence of biodiesel fuel properties on the structure of fatty acid alkyl esters, Fuel Process. Technol. 86 (10) (2005) 1059–1070.

[140] D. Singh, D. Sharma, S.L. Soni, S. Sharma, D. Kumari, Chemical compositions, properties, and standards for different generation biodiesels: a review, Fuel 253 (2019) 60–71.

[141] E.C. Francisco, D.B. Neves, E. Jacob-Lopes, T.T. Franco, Microalgae as feedstock for biodiesel production: carbon dioxide sequestration, lipid production and biofuel quality, J. Chem. Technol. Biotechnol. 85 (3) (2010) 395–403.

[142] E. Jacob-Lopes, T.T. Franco, From oil refinery to microalgal biorefinery, J. CO_2 Util. 2 (2013) 1–7.

[143] A. Muthukumar, S. Elayaraja, T. Ajithuar, S. Kumaresan, T. Balasubramanian, Biodiesel production from marine microalgae *Chlorella marina* and *Nannochloropsis salina*, J. Pet. Technol. Altern. Fuels 3 (5) (2012) 58–62.

[144] N. Arora, A. Patel, P.A. Pruthi, K.M. Poluri, V. Pruthi, Utilization of stagnant non-potable pond water for cultivating oleaginous microalga *Chlorella minutissima* for biodiesel production, Renew. Energy 126 (2018) 30–37.

[145] H. Xu, X. Miao, Q. Wu, High quality biodiesel production from a microalga *Chlorella protothecoides* by heterotrophic growth in fermenters, J. Biotechnol. 126 (4) (2006) 499–507.

[146] V. Makarevičienė, S. Lebedevas, P. Rapalis, M. Gumbyte, V. Skorupskaite, J. Žaglinskis, Performance and emission characteristics of diesel fuel containing microalgae oil methyl esters, Fuel 120 (2014) 233–239.

[147] R. Karpagam, K.J. Raj, B. Ashokkumar, P. Varalakshmi, Characterization and fatty acid profiling in two fresh water microalgae for biodiesel production: lipid enhancement methods and media optimization using response surface methodology, Bioresour. Technol. 188 (2015) 177–184.

[148] M.A. Islam, M.M. Rahman, K. Heimann, M.N. Nabi, Z.D. Ristovski, A. Dowell, et al., Combustion analysis of microalgae methyl ester in a common rail direct injection diesel engine, Fuel 143 (2015) 351–360.

[149] A. Guldhe, B. Singh, I. Rawat, K. Permaul, F. Bux, Biocatalytic conversion of lipids from microalgae *Scenedesmus obliquus* to biodiesel using *Pseudomonas fluorescens* lipase, Fuel 147 (2015) 117–124.

[150] M. Balu, K. Lingadurai, P. Shanmugam, K. Raja, N.B. Teja, V. Vijayan, Biodiesel production from *Caulerpa racemosa* (macroalgae) oil, Indian, J. Mar. Sci. 49 (04) (2020) 616–621.

[151] A.M. Khan, M.S. Hussain, Production of biofuels from marine macroalgae *Melanothamnus afaqhusainii* and *Ulva fasciata*, J. Chem. Soc. Pak. 37 (2) (2015) 371–379.

[152] G. Sivaprakash, K. Mohanrasu, V. Ananthi, M. Jothibasu, D.D. Nguyen, B. Ravindran, et al., Biodiesel production from *Ulva linza*, *Ulva tubulosa*, *Ulva fasciata*, *Ulva rigida*, *Ulva reticulate* by using Mn_2ZnO_4 heterogenous nanocatalysts, Fuel 255 (2019) 115744.

[153] T. Suganya, S. Renganathan, Optimization and kinetic studies on algal oil extraction from marine macroalgae *Ulva lactuca*, Bioresour. Technol. 107 (2012) 319–326.

[154] A.S. Ramadhas, S. Jayaraj, C. Muraleedharan, Biodiesel production from high FFA rubber seed oil, Fuel 84 (4) (2005) 335–340.

[155] M. Lapuerta, O. Armas, J. Rodriguez-Fernandez, Effect of biodiesel fuels on diesel engine emissions, Prog. Energy Combust. Sci. 34 (2) (2008) 198–223.

[156] A.E. Atabani, A.S. Silitonga, I.A. Badruddin, T.M.I. Mahlia, H.H. Masjuki, S. Mekhilef, A comprehensive review on biodiesel as an alternative energy resource and its characteristics, Renew. Sustain. Energy Rev. 16 (4) (2012) 2070–2093.

[157] G. Knothe, Analyzing biodiesel: standards and other methods, J. Am. Oil Chem. Soc. 83 (10) (2006) 823–833.

[158] M. Balat, H. Balat, Progress in biodiesel processing, Appl. Energy 87 (6) (2010) 1815–1835.

[159] M. Bhuiya, M.G. Rasul, M.M.K. Khan, N. Ashwath, A.K. Azad, Prospects of 2nd generation biodiesel as a sustainable fuel—Part 2: properties, performance and emission characteristics, Renew. Sustain. Energy Rev. 55 (2016) 1129–1146.

[160] G. Knothe, K.R. Steidley, Kinematic viscosity of biodiesel components (fatty acid alkyl esters) and related compounds at low temperatures, Fuel 86 (16) (2007) 2560–2567.

C H A P T E R

11

Eco-friendly biogas production from algal biomass

Mohamed A. Hassaan[1], Marwa R. Elkatory[2], Ahmed El Nemr[1] and Antonio Pantaleo[3]

[1]National Institute of Oceanography and Fisheries, Cairo, Egypt [2]Advanced Technology and New Materials Research Institute (ATNMRI), City of Scientific Research and Technological Applications (SRTA-City), Alexandria, Egypt [3]Agriculture and Environmental Sciences Department, Bari University, Bari, Italy

11.1 Introduction

A discussion of fuel, food, and indirect land use transition (ILUC) drives the reasoning for processing algae biogas. The ethics for energy and not for food are questionable if we are to use our scarce resources of arable land (0.2 ha of arable land per capita worldwide). Agricultural land and our seas and oceans draw bioenergy off algae. Algae can be used to clean water with overloaded nutrients (e.g., associated with salmon farming), while CO_2 from power plants can be captured in microalgae [1–6]. Regarding seaweeds, many species can be collected and differentiated, for example, in color, in several ways. The genetic variation between the green algae *Ulva lactuca* and the brown marine algae *Fucus* is greater than the difference between *U. lactuca* and an oak tree. *U. lactuca* contains a great deal of sulfur and normally has a ratio of C:N under 10, making it particularly difficult for monodigestion. The C:N ratio and the corresponding real biomethane yields usually increase from the winter season to the summer season and reach a C:N ratio of more than 20 in late summer. This does not apply to brown seaweed, for instance, *Laminaria*. Seaweed can be collected as a residue or grown in aquaculture (e.g., *Laminaria* sp., in combination with salmon farms). Seaweeds can be collected as residues (such as the algal bloom associated with the green algae *U. lacuca*); brown seaweed (such as *Fucus* sp. or *Ascophylum nodosum*). The size of seaweed associated with aquaculture is potentially very important in a large sustainable biofuel industry [1–3].

Handbook of Algal Biofuels
DOI: https://doi.org/10.1016/B978-0-12-823764-9.00023-6

© 2022 Elsevier Inc. All rights reserved.

Many types of microalgae can also be isolated from different habitats. The culturing can occur in open ponds or in closed photobioreactors (costly in relations of financial expenditure, energy input, and operation). The C:N ratio in microalgae appears to be lower than that in marine algae, but the composition varies considerably, depending on the species and the nutrient availability. The ambition is to increase the production of lipids for esterification in biodiesel production [1]. Lipids often produce high biogas levels, but anaerobic digestion (AD) cannot be stabilized in microalgae with excess lipid concentrations. The great benefit of AD is that it does not need a pure culture and does not require the production of a particular compound (such as lipid for biodiesel) [1–12]. The cost of processing microalgae biomass can be substantially decreased by both of these benefits. In order to capture CO_2 produced by power plants, microalgae can be used. The microalgae can be digested for biogas, but when combusted this emits CO_2. The benefits of capturing CO_2 from fossil fuels are therefore more substantial than sequestering CO_2 through the work of original fossil combustible. The scale of large carbon trapping ponds or photobioreactors is very extraordinary. The input of energy in harvesting, mixing, and converting microalgae into biogas is extremely important and can be of a scale that uses extra energy than in the biogas phase.

Pretreatment refers to downstream processing phases in algae that are sufficient for improving biofuel production on an industrial scale after harvest and prior to AD or fermentation. In recent years, various pretreatment systems have been developed in order to increase the availability of lignocellulosic materials to degrade and boost the viscosity and flow of substrates [8–12]. This pattern was driven by an increase in gas production, improvement in process stability, or an accelerated rate of decomposition. Pretreatment techniques can be divided into physical, biochemical, and chemical concepts, but they are also used together. In this chapter, an overview is given of the input substrates for biogas plants, pretreatment technologies, and intrinsic inhibitor devices or inhibitors developed after pretreatments that can reduce biogas production.

The understanding of biofuels in the last decade has been seriously impaired, in particular first-generation biofuels. In 2008 a big rise in the cost of crops used to grow biofuel came about, leading to food unrest in some developing countries. This began the debate on food fuels, which took place with regard to new policies and regulations [1,13,14].

In Europe for example, Biofuel Directive (2003) (2003/30/EC) specified that by 2010, biofuels should be 5.75% of the market of transport fuel (by energy content). However, following the food fuel debates, the Renewable Energy Directive (RED, 2009), which sets out 10% of transport energy to be renewable by 2020, put more emphasis upon renewable energy than biofuels. This made a shift in approach possible, for example by using electric vehicles.

Biofuels of the third generation do not have to be grown on farmland. Biofuels of the typical third generation are usually from algae that have lower area requirements compared to land crops such as maize, rape or grass [15]. The microalgae offer higher production per hectare than crop plants [16]. There is a broad variety of findings in literature, partly because computations are based on laboratory or pilot data. The third-generation transport biofuel of the future is mooted to be from both microalgae and macroalgae. However the commercial production of algal biofuels is currently limited. In scientific publications, sustainable algal bioenergy pathways are not well known (or agreed upon). Jard et al. [13] find that biogas production from seaweed can be commercialized because even complex carbohydrates can become biogas.

11.2 Structural and chemical composition of seaweeds

11.2.1 Moisture and salt content

The water content of seaweeds is greater than that of several land crops [12,17,18]. The higher value for heating the seaweed (HHV) in terms of high ash content is smaller than the terrestrial energy crops [19]; however, higher calorific values in demineralized algae [20] were achieved. Seaweed has an additionally greater content of salt (sodium chloride) than land plants, with 15% unwashed *Sargassum muticum* dry weight of salt [18].

11.2.2 Structure composition

There are structural variations between green, red, and brown algae [12,21]. In brief, the main skeletal cell component in brown algae is cellulose and xylane, whereas it is cellulose and mannan in red and green algae [12,21,22]. Notably, only crystalline xylan and mannan have been found structurally in marine green and red algae [21]. Polysaccharides are microfibrils which have different structural configurations with flat bands of cellulose and mannans, whereas xylene is a helix [21]. The microfibrils are either organized in structure or randomly scattered through each layer according to the species [23]. In addition, these microfibrils are connected to the polysaccharide matrix including different polysaccharides of the sulfate or carboxylic form (Table 11.1) [29]. For example, a role for linking the cellulosic backbone was suggested in sulfated fucane [30]. Proteins similar to sulfated fucans and phenols were also found in brown algae [30]. A significant factor for the rigidity of the cell wall structure is the relation between the phenols and the alginates [30]. Furthermore as discussed here, phenols can be inhibitory in biofuel microorganisms.

11.2.3 Polysaccharides

The appropriateness of diverse categories of treatment for seaweed handling can vary with the chemical composition of the algae [31,32]. There are many similarities between the different seaweed groups determined with 107 forms of chemical profiling, and the water-soluble and insoluble properties of green and red algae were more similar, compared with brown algae that possess more distinct properties [33]. However, their sugar

TABLE 11.1 Sugars and polysaccharides in red, green, and brown seaweed [12].

Seaweed	Polysaccharides	Sugars	References
Red	Xylans, mannans, agar, carrageenan, agaropectin, cellulose	3,6-anhydro-D-galactose, glucose, D-galactose, D-fructose	[21,24,25]
Green	Cellulose, hemicellulose, ulvan, xylopyranose, starch, xyloglucan, glucopyranose, glucuronan	Rhamnose, galactose, glucose, xylose, uronic acids	[24,26]
Brown	Alginates, cellulose, fucoidan, laminaran	Glucuronate, sulfated fucose, mannitol, glucose, guluronate, mannuronate	[24,27,28]

228   11. Eco-friendly biogas production from algal biomass

and amino acid compositions are clearly differentiated [24]. The polysaccharides and sugars are unique in various forms of algae (Table 11.1). Wei et al. [24] stressed the structures of these polysaccharides. The general sulfur content of marine algae is high because there are sulfate polysaccharides with diverse amounts of different polysaccharide SO_4 groups, often varying in diverse phyla and forms [34,35]. Brown algae contain sulfate fucans [36], red algae contain sulfate galactans (agar and carrageen), and green algae contain sulfate xylo arabinogalactans [20,21,29].

The biochemical composition of the polysaccharides of algae influences their structural characteristics [37], which are defined as egg boxes, double-helix carrageens, and perforated ulcers [21,38]. The impact of pretreatment and biofuel outputs are likely to be influenced by such variations. The average methane production rates for AD are just 23%–28% of the theoretical methane capacity for alginic acid, which is considerably lower than that of cellulose and AD and its sodium salt [39].

11.2.4 Chemical composition variability

Owing to sea currents, light intensity, and temperature variations, the composition of the algae varies between species, seasons, and geographical location [27]. For example, the quality of amino acid in August in *Saccharina latissima* was nearly twice as high as in June, and the mineral and ash content also increased [40]. At various times of the year, polysaccharides like fucoidane also show different branching degrees, sulfurization, and the chain length [27]. The seaweed also developed more cell wall polysaccharides in the presence of higher heavy metal content [41]. Phenolic compounds also differ spatially and temporarily in algae, which are possible inhibitors of AD and fermentation [42].

11.3 Anaerobic digestion

AD occurs as a natural biochemical process under organic conditions, in which organic matter is mainly converted by oxidation into biogas (CO_2 and CH_4) in the absence of oxygen. This phase takes place naturally in the higher intestinal tract (e.g., sediments, damp areas, marshes, paddy fields, etc.). Here electron acceptors, including dioxygen, nitrate and sulfate, are depleted from environments. The methane formed in this case is replaced by carbon dioxide [11,43–45]. AD's high degree of capacity to minimize organic matter relative to aerobic degradation is an important attribute. Furthermore, the energy transfer in CH_4 formed during the digestion process makes the process cost efficient [46]. Solid residues from anaerobic decomposition can be used as organic soil fertilizer [47–51].

11.3.1 Microbiology of anaerobic digestion

AD is a process in which microbiological anaerobic decomposition produces energy and creates CH_4 through the metabolization of organic material in an oxygen-free environment [44,45,52]. Although the AD method has a general microbiology that is well known, the mechanisms and relationships are not fully understood just yet within the different

Handbook of Algal Biofuels

AD process microorganisms. [11,44]. The method can be divided into the four steps and phases. The microorganisms comprise their own characteristic group (Fig. 11.1).

11.3.1.1 Hydrolysis

In AD, the hydrolysis process is the first and limited step in which insoluble biopolymers break down to soluble organic complexes such as oligomers and monomers. This may occur by hydrolysis and small fermentation processes, and this part of the process occurs without the formation of methane. In the case of the most difficult biological large size and morphology, a pretreatment step should be included [43,44]. More complex biopolymers are not available for intracellular metabolism.

11.3.1.2 Acidogenesis

Due to a broad range of optional anaerobes and anaerobes, soluble compounds formed by hydrolysis are degraded in the acid-forming step. The oxidation of these compounds contributes to carbon dioxide, hydrogen gas, and organic acid production. The same bacterium can, however, carry out hydrolytic, fermentative, and acidic operation [43,44].

11.3.1.3 Acetogenesis

The third stage in AD is acetogenesis, where fermented products by the use of acetogenic bacteria are transformed to acetate, CO_2, and H_2, especially fatty acids and alcohols. These bacteria are called acetogens and are obligatory sources of H_2. Later those methanogens are used for the processing of CH_4 [43–45]. H_2 concentration is a key factor regulating the metabolism of acetate and methane formation. The formation of biogas from fermentation products only occurs if the hydrogen concentration is less than the feasible thermodynamic concentration, therefore, it is difficult to detect H_2 in the developed biogas. Meanwhile, methanogens' biological function includes constant H_2 delivery from the redox response [45,54]. The link between acetogens, methanogens, and a mechanism called interspecies are the syntrophic pass of hydrogen or electron flow interspecies [55].

11.3.1.4 Methanogenesis

During this level, the system greatly affects microorganisms. Each specific type of methanol is controlled by various factors, including volatile fatty acids (VFA), hydrogen stress (hydrogen pressure), buffering capacity, liquid bicarbonate concentration, carbon dioxide (pH) gas phase concentration, NH_4, toxic substances, nutrient availability, and other environmental factors, such as light and temperature [45,46,53,56,57].

Methanogenesis has three primary pathways: acetotrophic (metabolized by acetate), hydrogenotrophic (metabolized by H_2/CO_2), and methylotrophic (methylated one-carbon compound metabolized). AD consists of a set of steps taken to turn raw products into methane, carbon dioxide, and bacterial biomass by separate classes of bacteria (Fig. 11.2). Biogas formed in algae AD usually comprises 50−70% hydrogen, <2% hydrogen sulfide, <3.5% methane [58−60], and 30%−45% carbon dioxide. AD is mainly the facultative form of biomass with large water content, for instance seaweed; it bears a high humidity biomass without any energy disadvantages occurring as a result of dewatering and drying [61,62]. It is an infrastructure/engineering process that is reasonably straightforward.

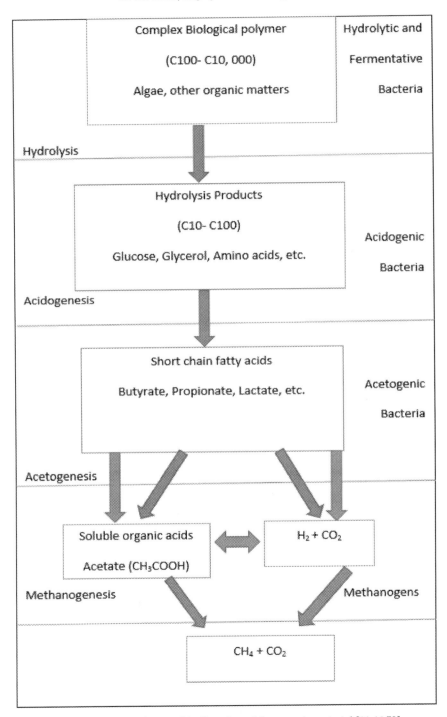

FIGURE 11.1 The different stages of anaerobic digestion of the organic material [11,46,53].

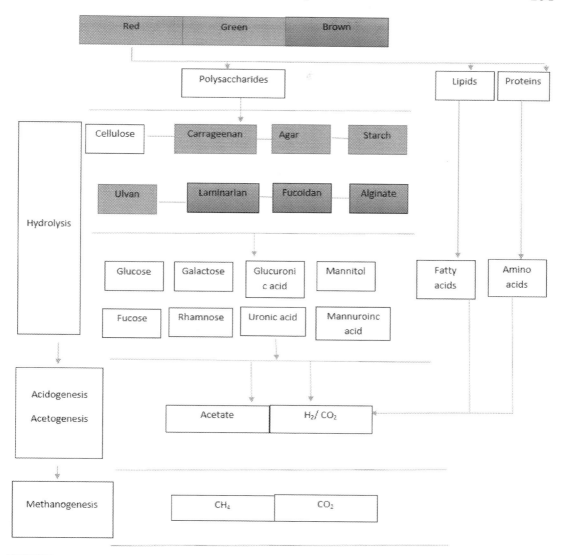

FIGURE 11.2 The main stages for biogas production in different seaweeds [5].

AD often uses all the organic carbon content in algae; thus biogas systems produce more crop power per hectare than liquid biofuels [63]. Tokyo Gas recently showed that 20 m^3 methane was generated by 1 ton of algae [64]. At present, seaweed biofuel AD could be the most suitable for industry [5,65].

Mainly acetate (Eq. 11.1) or formate (Eq. 11.2) is transformed in methanogenesis to CO_2 and CH_4 as final products for organic matter AD. CO_2 and H_2 in their metabolic process (Eq. 11.3) by methanogens can be used to generate H_2O an CH_4. Certain chemolithotrophic substances can use carbon monoxide (Eq. 11.4) in the development of CH_4 methanogens. Methylotrophic methanogens specifically generate CH_4 from methyl group substrate ($-CH_3$, for example), not by CO_2 (Eq. 11.5). Archaea can oxidize acetate to H_2 and CO_2 which is the cause of CH_4 as the acetate-used methanogens are blocked by ammonia, sulfides, etc. [43,44,52,66,67].

$$CH_3COOH \rightarrow CH_4 + CO_2 \tag{11.1}$$

$$H_2COOH \rightarrow CH_4 + CO_2 \tag{11.2}$$

$$CO_2 + 4H_2 \rightarrow CH_4 + 2H_2O \tag{11.3}$$

$$4CO + H_2O \rightarrow CH_4 + 3CO_2 \tag{11.4}$$

$$3CH_3OH + 3H_2 \rightarrow 3CH_4 + 3H_2O \tag{11.5}$$

11.3.2 Anaerobic digestion of seaweed

Through its ILUC policies the European commission has recently released a recommendation to decrease the use of food-based biofuels by 10%−5%. The significance and technological difficulties associated with the advancement of technology were indicated by studies undertaken in the early 1970s on the creation of macroalgae biogas [45,68−71]. Analysis from a limited number of research groups has already shown that seaweeds are a viable and important catalyst for growth in the biofuels industry and have provided the basis for research, as well as an increase in public and private funding [72]. These studies have found that the technology required more testing to meet its maximum potential in its earlier stages of growth and process optimization [11].

11.3.3 Optimization of anaerobic digestion

Possible ways to optimize AD in a specific sustenance range from ecological and microbial interactions in the reactor to improved reactor performance, substrate combination, the addition of opposite nutrients, and/or the pretreatment of the substrates to make AD more comfortable. In this analysis many methods were employed to increase the production of biogas from the AD of algae [11].

11.3.4 Anaerobic digestion process parameters

11.3.4.1 Temperature and digester configuration

The temperature setup and digester setup are key elements that influence the efficient start and maintenance of the AD process, including other operational parameters. In general, the AD method can be carried out in three major areas: psychrophilic (15°C−25°C), mesophilic (35°C−37°C), and thermophilic (45°C−60°C). AD has become the subject of most applications and research activities in the region of 30°C−45°C. It is well known that once the maximum growth temperature is achieved, the decomposition rate for a certain substrate increases [11]. The AD will influence the metabolic behavior of the microbial consortia, resulting in a decrease in the kinetics of the reactor for temperatures below and above the optimal growth temperature [53]. The optimal AD temperature of the algae *Laminaria digitata* has been defined. The psychrophilic $(20°C \pm 2°C)$, mesophilic $(35°C \pm 2°C)$, and thermophilic $(45°C \pm 2°C)$ conditions were used for the analysis [11]. Anaerobic digesters are constructed in a variety of ways to handle a given substratum successfully. Batch or continuous feeding reactors may be run with mixed with media and/or single- or double-phase systems that enable the development or suspension of growth.

11.3.4.2 Codigestion

Codigestion is a method that, in order to boost biogas production, combines AD of a main substratum with minor amounts of single or extra substrates. This technique is known for optimizing the nutrient content of a mixture and reducing the influence of inhibitory substrate compounds in the AD phase [11,49].

11.4 Pretreatments

Pretreatment is a metabolic mechanism of biogas conversion. During AD, the first steps of hydrolysis and acidogenesis are also assumed to be rate limiting because of the hydrolysis ability of the substratum. Strengthening of complex biopolymers, damage to cell walls, and removal of chemicals from polymers are the main elements of various pretreatments [73]. The next step would also make organic material easier for microorganisms to degrade [74]. To boost degradability overall, various technologies are used and loosely grouped into: [73,74]:

- Physical, such as mechanical, ultrasonic, steam, grinding, and thermal.
- Chemical, such as acid, solvents, alkali, and oxidizing agents.
- Enzymatic or biological.
- A combination of the above.

The effects of a pretreatment of a given medium depend not only on the pretreatment process but also on substrate characteristics [74]. While pretreatment to boost the AD of different substrates has been widely studied and a lot of technologies in commercial plants now are used, only a few studies have targeted the use of macroalgae pretreatments [13,75,76].

11.4.1 Physical treatment

11.4.1.1 Mechanical treatment

The physical structure of seaweeds is specifically influenced by mechanical treatment to facilitate the access of hydrolyzing agents to polysaccharides. Tables 11.2 shows the mechanical treatments on seaweed prior to the processing of methane. Different scholars approach biomass in various ways, allowing simple comparisons of shifts in yields. This covers all the inoculum/substrate (I/S) ratios and biochemical measuring methods (BMP) capable of impacting the final gas volume [82]. The methane outputs of the feedstock and

TABLE 11.2 Methane yields following mechanical pretreatment of seaweed [12].

Algae	Pretreatment	I/S ratio; Source	BMP method	CH$_4$ yield (mL/g VS)	Change (%)	BI[1] (%)	References
Gracilaria vermiculophylla	WM: Washed and macerated; UM: unwashed and macerated; WDM: washed, dried and macerated. *Control*: without maceration.	4:1; brewery WWT plant	Glass vials with rubber stopper, 37°C, 28 days	UM: 338 WM: 481 WDM: 349	+ 14.6 + 11.9 + 7.7	– – –	[77]
G. vermiculophylla	M: Freshwater washed and macerated. *Control*: chopped (2 × 2 cm) without maceration.	6:1 in 500 mL; with cattle manure, lab reactor	Bottles with rubber stoppers, 53°C, 34 days	147	+ 11.4	–	[76]
Chaetomorpha linum	M: Freshwater washed and macerated. *Control*: chopped (2 × 2 cm) without maceration.	6:1 in 500 mL; with cattle manure, lab reactor	Bottles with rubber stoppers, 53°C, 34 days	195	+ 17.5	–	[76]
Ulva lactuca	M: Freshwater washed and macerated. *Control*: chopped (2 × 2 cm) without maceration.	6:1 in 500 mL; with cattle manure, lab reactor	Bottles with rubber stoppers, 53°C, 34 days	255	+ 67.7	–	[76]
Saccharina latissima	M: Freshwater washed and macerated. *Control*: chopped (2 × 2 cm) without maceration.	6:1 in 500 mL; with cattle manure, lab reactor	Bottles with rubber stoppers, 53°C, 34 days	333	– 2.1	–	[76]
Laminaria spp.	Ball milled, unwashed, and dried at 80°C for 1 day, with particle size of 1–2 mm. *Control*: cut and unwashed only.	1:1.33 in 400 mL; WWT plant	Bottle sealed with adaptor attached to gas measuring device (GMD), 38°C, 25 days, shaken manually.	1 mm: 241 2 mm: 260	– 26.5 – 20.7	– –	[78]

Species	Pretreatment	I/S; inoculum	Conditions	mL/g VS	BI (%)		Ref.
Fucus vesiculosus	WC; Washed and chopped (<5 mm); UWC: unwashed and chopped; WNC: washed and not chopped. Control: NWC: not washed or chopped	5:1 in 60 mL; WWT plant	Serum bottles, gas measured with syringe, 37°C, 30 days.	W + C: 81.1 UW + C: 67.3 W + NC: 73.1	+ 574.3 + 493.6 + 527.5	25 21 23	[79]
Laminaria spp.	CUWB: Cut without washing and beaten, 76 μm gap for 10 min. Control: CUW: cut and unwashed only.	1:1.33 in 400 mL; WWT plant	Bottle with adaptor attached to GMD, 38°C, 25 days, shaken manually	335	+ 2.1	–	[78]
Laminaria spp.	CUWB: Cut without washing and beaten, 76 μm gap for 15 min. Control: CUW: cut and unwashed only.	1.2:1 in 400 mL; WWT plant	Bottle attached to GMD, 38°C, 14 days, shaken manually	240	+ 8.6	–	[80]
Ascophylum nodosum	CUWB: Cut without washing and beaten, 76 μm gap for 15 min. Control: CUW: cut and unwashed only.	3:1 in 400 mL; WWT plant	Bottle attached to GMD, 38°C, 14 days, shaken manually	240	+ 8.6	–	[80]
U. lactuca	Washed and dried for 1 day. Control: Unwashed (−20°C) and grinded: 10−15 mm.	3:1 in 400 mL; seaweed and grass, dairy slurry.	Bioprocess AMPTS II system, 37°C, 30 days.	221	+ 33.9	55	[81]

I/S, inoculum/substrate; Biodegradability index (BI[1]) based on Buswell equation.

the BMP monitoring time [60,75], with average values of 60% CH_4, 40% CO_2, and other trace gases are seen, rather than the yields of biogas, if possible due to varying methane composition (less than 40% to ~70%) [72].

In order to demonstrate the utility of the specific pretreatment method in comparison with theoretical results the biodegradability index (BI) is used to compare more easily the efficacy of pretreatment [83]. The methane production after pretreatment is computed by dividing the theoretical output given by the authors and presented as a percentage. Related to the percentage change, the change in the BI is often used to show the efficacy of pretreatment compared to untreated control algae.

11.4.1.2 Size reduction

Biomass pieces after being well ground are widely used to increase the surface area to the appropriate size to enhance hydrolysis and to aid fermentation in order to convert complex carbohydrates into simple sugars (Fig. 11.2) [84]. However for lignocellulosic terrestrial plants, the same friction strategies may not evoke the anticipated surface area increase in the algae: for example, friction of *Laminaria digitata* that has flat blades did not increase its surface area significantly [85]. Wet mechanical milling of *L. digitate* using cutting disks did not improve glucose release [86]. Experimental studies have shown a number of effects on methane production from chopped or decomposed samples of different algal species (Table 11.2) [76]. The increase in unique methane production from maceration in *Gracilaria vermiculophylla* has been reported to be 8—16% [77]. The rise in methane yield was also observed with the smaller particulate scale of *Laminaria* spp. [76]. On the other hand, Montingelli et al. [80] observed decreased yields of methane after the milling of *Laminaria* spp. due to complex hydrolysis rates assisted by greater surface areas, which were attributable to the inhibitory effect of the aggregation of VFA and successive pH decrease during the acidogenesis process (Fig. 11.2). It was suggested that methane yields could be improved during AD by a combination of smaller and larger particle sizes [76].

11.4.1.3 Thermal treatment

Treatment with brown seaweed *Nizimuddinia zanardini* produced upto 84% of constituents, for example, mannitol and hemicelluloses at 121°C, and contributed to a 22% rise in methane production in contrast to untreated seaweed (Table 11.3) [89]. Thermal pretreatment is able to release sugar or remove polysaccharides from seaweed. The rise in bioethanol was attributed to an increase in exposed fibers and eroding of seaweed surfaces, allowing for higher enzyme degradation. Any marine structures, such as the cortex of the alga, were not impaired however by the treatment of autoclaves, suggesting the significance of the structural composition of the sea algae and the correct pretreatment procedure for effectively hydrolyzing these components [12].

11.4.1.4 Microwave pretreatment

Due to its very high moisture level, microwave pretreatment is appropriate for algae, facilitating rapid changes in pressure and temperature within the cells. This allows the rupture of the cell wall and enables a surface area increase for subsequent processes of bioethanol or biogas (Fig. 11.2) [91]. Kostas et al. [92] have studied the processes and the use of microwave pretreatment biomass in bioenergy development. The pretreatment of

TABLE 11.3 Methane yields following thermal pretreatment of seaweed (Maneein et al. [12]).

Algae	Pretreatment	I/S ratio; source	BMP method	CH4 yield (mL/g VS)	Change (%)	References
Ulva spp.	Freshwater rinsed, blended into slurry 1. Thermal: 90°C with no chemical, 2. HCl: 0.1 M, 90°C 3. NaOH: 0.1 M, 90°C Each treatment was magnetically stirred (10 min), oven (6 h), shaken 1 min every 30 min. *Control*: untreated slurry only	Unknown (70 mL inoculum and 35 mL slurry); sewage sludge digester	Bottles with rubber stopper, 35°C, 30 days. Shaken manually intermittently.	Thermal: 293.0 0.1 M HCl: 284.8 0.1 M NaOH: 251.3	+ 15.8 + 12.7 − 0.7	[87]
Saccharina latissima	Defrosted, shredded into slurry. steam exploded 130°C or 160°C, 10 min. *Control*: untreated slurry only	7:1 in 700 g; sewage treatment plant.	Bottles with rubber stopper, shaker (90 rpm, 37°C). Refed day 67, biogas shown: day 119.	130°C: 268 160°C: 260	+ 20.2 + 16.6	[88]
Nizimuddinia zanardini	Washed and dried (40°C, 1 day); hammer milled to <1 mm. 5% seaweed, 121°C, 0.5 h. *Control*: untreated.	Unknown; WWT plant	Bottles closed with rubber stopper, 37°C, 40 days.	143	+ 22	[89]
Laminaria spp.	Cut seaweed, Freshwater immersed, microwaved (560 W), held for 30 s. *Control*: cut unwashed seaweed.	1:1.33 in 400 mL; WWT plant	Bottles sealed with adaptor attached to GMD, 38°C, 25 days, shaken daily.	244	− 25.6	[78]
Fucus vesiculosus	Cut and grounded (mortar and pestle) microwaved (700 W), for 3 min *Control*: not microwave	1:3; WWT plant	Bottles with rubber stopper and metal cap, 37°C, 22 days. Shaken daily.	146.9	+ 92.3	[90]

I/S, Inoculum/substrate.

microwaves for high-value materials and polysaccharides, such as carrageenan, agar, and fucoidan from seaweed, has mainly been used [93]. When compared for *Undaria pinnatifida*, the microwave was more effective in removing polysaccharides and fucoidane, as well as in reducing molecular weight compounds [94]. Methane yields will then be obtained provided that the required microorganisms that can use the compounds are present. Another gain is their fast heating period that could stabilize and reduce the degradation of sugar at higher temperatures causing lower concentrations of inhibitory substances, such as furfural compounds, formic acid, and levulinic acid (sometimes produced when heated in standard inductive heat) [89]. When *Fucus vesiculosus* was treated with microwaves the production of methane also grew by 92% compared with untreated algae [90]. This is a 10% rise in BI compared to washing and cutting which has the BMP process and the same *F. vesiculosus* harvest time. However, microwave pretreatment, where it was kept for 30 s at boiling point, was ineffective for *Laminaria* spp. [78]. This resulted in a 27% fall in methane yields in comparison with untreated algae.

11.4.1.5 *Ultrasonic treatment*

Ultrasound is less acceptable as a pretreatment technology for anaerobic digester liquid wastewater than posttreatment [10]. The ultrasound frequency of more than 20 kHz is used to shape and then implode cavities or fluid-free bubbles, resulting in shockwaves, called cavitation. These forces allow microbial cell walls in the fluid to destruct. Generally, this technique is used in wastewater sludge treatment, but the effect of the ultrasonic substrate on biogas production is very low. Ultrasonic therapy only breaks down microbiological biomass, rather than inputs [10].

11.4.1.6 *Electrokinetic disintegration*

For a range of processes in contemporary biotechnology, electrical fields are used. Electrokinetic degradation is primarily used for the treatment of wastewater sludge [10]. The principal inhibitor of successful sewage slug AD is the presence of flocculents and additives composed of negative charged molecules in ionic cation [95]. The electric field is used to disrupt the ionic connections of the drainage sludge and thereby separates the floccs [95]. Electrical fields are also likely to interrupt the microbial cells by altering the cell membrane load. The effect on lignocellulosic material, if any, is not clear. Some German companies manufacture electromotor in which the loop is supplied with an electrode in a pipe that normally covers a voltage of around 30 kV [96,97]. A rise of about 20% was claimed for the biogas yield from the sewage sludge [97]. A study conducted at the Bavarian state research center for agriculture showed no substantial growth in biogas output from agricultural waste [98]; it is also alleged that this system will improve biogas yield from agricultural residues [96]. With regard to ultrasonic dissociation, anaerobic digestion of effluent or pretreatment of a substrate comparable to waste sludge may be more suitable for postapplication treatment.

11.4.1.7 *Extrusion*

Biomass is permitted to demonstrate compression, heat, and shear strength during extrusion pretreatment. It causes physical destruction and chemical modifications to the cells during extruder passage [4]. The extruder consists of individual or double tubing, which spin into a thick barrel, fitted with a temperature regulator. As a biomass material moves through the barrel, friction and intense shearing allow the temperature and pressure to rise. If the finishing line leaves, there is a pressure release from the biomass content that induces structural modifications in the harvested biomass that make digestion smoother in the next phase [4].

Maroušek [99] analyzed extrusion factors of pelleted hay for maximum cumulative biogas output by announcing an improvement of some 33%, in the maximum biogas production of 405 m^3/ton TS (52.3% methane), in optimum pressure conditions of 1.3 MPa, reaction time of 7 min, and 8% dry matter. In order to increase biogas production from an agricultural portion of municipal solid waste. Novarino and Zanetti [100] used extrusion preprocessing to produce an 800 L/kg VS of biogas with an estimated content of 60% methane.

11.4.2 Chemical treatment

Improvements in hydrolysis and solubilization of seaweed throughout alkaline and acid pretreatments make them widely researched for AD methods [101]. As pretreatment approaches for aquatic algae, other chemicals have also been researched, including ionic liquid, "organosolv," and sodium-chlorite. In the hydrolysis of polysaccharides of algae, chemical properties play a significant part [102].

11.4.2.1 Alkali or acidic treatment

Alkalis, most often NaOH, have been introduced to induce fiber swelling and increase pore size, thus allowing sugars to be released from the cell walls to enhance effective enzyme hydrolysis, or eventual fermentation [12,103]. Hydrochloric acid (HCl), H_2SO_4, and low pH flue gas condensates have been investigated for acid pretreatment (Table 11.4). Hemicellulose and other carbohydrates, such as laminarin, are considered to be hydrolysis acids [101,105]. The hydrolysis of cell walls of marine algae releases their cell content [105]. Acids were, however, criticized as a danger hazard (particularly acids), a great acid recycling expense, and process equipment requirements for acid-resistant processing [105].

After the pretreatment with NaOH or HCl the benefits of the improved solubility of algae in biogas yields were much reduced (Table 11.4). *F. vesiculosus* pretreatment with 0.2 M HCl (80−12 h), increasing methane production by approximately 2.5 times relative to untreated seabed, was 1.6 times greater than hydrothermal (80−24 h) [106]. In a related method with *Ulva* spp. using HCl or NaOH of 0.1 M, and NaOH at 60°C, 75°C, and 90°C in contrast to thermal pretreatment alone, methyl outputs were not increased (Table 11.3) [87]. However, both studies used different approaches for evaluating BMP, so that it was impossible to specifically compare the two experiments. Furfural compound concentrations, especially following thermal−chemical treatment for lignocellulosic biomass, were not inhibitory in cellulose AD [107], implying that if the appropriate algae are picked, thermochemical algae hydrolysis may be useful when biogas rates are increased.

TABLE 11.4 Methane yields following thermochemical pretreatment of seaweed [12].

Algae	Pretreatment	I/S ratio; source	BMP method	CH$_4$ yield (mL/g VS)	Change (%)	References
Gracilaria vermiculophylla	Algae washed, macerated then: 1. 0.1, 0.3, 0.5 g NaOH g^{-1} seaweed (20°C, 55°C, 90°C, 1 bar, 3.5 bar, 6 bar, 60 and 90 mins) 2. Control: Untreated	4:1; brewery WWT	Glass vials with rubber stopper, 37°C, 24 days	353–380	−21 to −26.6	[77]
Palmaria palmata	Dried (40°C), chopped (2 × 2 cm) then: 1. 0.04 g NaOH/g TS (50 g L-1): 20°C–70°C, 24 h; 160°C, 0.5 h. 0.02 g HCl g^{-1} TS, 160°C, 0.5 h. 2. Control: untreated	2:1 in 400 mL; sugar WWT	Glass vials with rubber stopper, 35°C, 60 days.	20°C–70°C: 362–365 160°C NaOH: 282 160°C HCl: 268	+ 17.5–18.5 − 8.4 − 13	[13]
Ulva spp.	Washed, sun dried (1–2 weeks) then 1. 0.04 g NaOH g^{-1} TS (20°C, 24 h); 0.04 g HCl g^{-1} TS (150°C, 0.5 h) 2. Control: untreated	2:1 in 400 mL; sugar wastewater industry	Glass vials with rubber stopper, 35°C until no gas production.	NaOH: 148HCl: 77	+ 12.1 − 41.7	[103]
Ulva spp.	Freshwater rinsed, blended to slurry *then* 1. 500 mL slurry, no chemical; 0.01 M HCl; 0.1 M NaOH. All 90°C, 6 h, manual shaking every 0.5 h, 1 min. 2. Control: untreated	Sewage sludge digester	Bottles with rubber stopper, 35°C, 30 days, shaken manually intermittently.	Only thermal: 293.0 0.1 M HCl: 284.8 0.1 M NaOH: 251.3	+ 15.8 + 12.7 − 0.7	[87]
Laminaria digitata	Freshwater rinsed, dried (75°C, 24 h), milled then: 1. 20% solids loading, 2.5% citric acid (CA); 6% citric acid; 1% lactic acid (LA), autoclaved (120°C, 1 h, 1 atm) 2. Control: untreated	2:1 in 30 mL; bovine slurry adapted to seaweed.	Serum bottles (pH 7.3–7.5) with rubber stopper, 35°C, 32 days.	2.5% CA: 237 6% CA: 69 LA: 161	+ 3.9 − 69.7 − 29.4	[104]

11.4.2.2 Peroxide treatment

Thermal processes are used in this pretreatment system to improve disturbance by hydroxyl radicals of the seaweed crystalline structure and hydrogen bonds [12,108]. A higher conversion concentrations of cellulose to glucose from alkaline and the acid pretreatments in seaweed were obtained with peroxide treatment by 88.1%, followed by a pretreatment (100°C, 30 min, 0.018% H_2O_2, and $FeSO_4$ 11.9 mM) [109], followed by a step to radical hydroxyl reaction [109]. In addition, four distinct species of marine algae have been treated, leading to a rise in their glucose production compared to unprocessed marine algae, suggesting an increase in enzyme digestibility [108].

11.4.2.3 Oxidative

The wet air oxidation is a pretreatment solution for the full breakdown of organic compounds into CO_2 and H_2O, and increased interaction between O_2 and organic matter. High temperature conditions (and consequently high pressure) are important to achieve this [110]. The resulting high pressure is needed both to sustain the high temperature and to help to increase the DO concentration and thus the oxidation rate. The wet air oxidation of the complex biomethanated effluent distillery has been used by Chandra et al. [111] to improve its biodegradability. They registered an improvement in pretreated effluent biogas upto 2.8 times the untreated wastewater with a CH_4 concentration of upto 64.14%.

11.4.2.4 Ozonation

In oxidizing substrates, ozone (O_3) is a strong oxidant. In different feedstock lignin can be depleted. It reacts to convert the proteins, lipids, polysaccharides, and other intractable compounds into biodegradable fragments. Anaerobic microorganisms can efficiently access and assimilate the effective dissolution of cell walls and the release of more soluble and easily biodegradable organics [4,112]. Goel et al. [113] investigated the AD of ozone excess sludge pretreated throughout long-standing lab-scale reactor operations. They found that the pretreatment of O_3 is effective to partly solubilize the sludge solids and to increase anaerobic degradability. The degree of solubility and effectiveness of digestion are contingent on the dose of O_3. The AD performance at 0.05 g O_3/g TS has risen to about 59% in contrast to 31% in the control run.

11.4.3 Biological pretreatment

Biological pretreatment is a simplified form of pretreatment technology, often referred to as preacidification or multistage fermentation, which distinguishes the initial stages of AD (hydrolysis and acidogenesis) from acetogenesis and methanogenesis [10]. For this form of pretreatment, a two-stage digestion system is popular. The principle of digestion is analogous to the various chambers of ruminant digestive systems in individual vessels. The first digester's pH value should range from 4 to 6 and thus inhibit methane production [10,114]. The VFA accumulate in this inhibition. The preacidification gas emitted during this process has high CO_2 and H_2 concentrations. H_2 development correlates with fatty acid production and is an important predictor for determining the preacidification step. At pH 6, H_2 production is essentially high and then stops, while H_2 production is initially

242   11. Eco-friendly biogas production from algal biomass

lower at pH 4 but stays longer and thus H_2 production becomes greater overall [10,115]. It is a significant factor in the amount of H_2 production. In constant fermentation testing, Antonopoulou et al. [116] demonstrated that the H_2 level was 35%–40% v/v of the preacidification stage's total gas. The decrease rates of substrates in AD are strongly optimistic for microbiological pretreatment. In overall, hemicellulose, cellulose, and starch decomposing enzymes function best at temperatures between pH 4 and 6 and between 30°C and 38°C, which increases the rate of degradation in this step by ensuring the optimum environment for hydrolytic enzymes, particularly for degradation of the carbohydrates [10]. At a hydraulic retention time of roughly 30 days, Liu et al. [115] attained an extra biogas yield of 21%. This was attributed to accelerated oxidation by increased activity of the hydrolyte enzyme. The methane concentration in biogas is also a beneficial consequence of this pretreatment process. CO_2 is produced in the preacidification stage in addition to H_2 and VFA development. CO_2 can be present as carbonate ion CO_3^{2-} at higher pH levels; as bicarbonate HCO_3; at neutral pH as and as CO_2 in acidic conditions. Because of the low pH, most carbonate is volatile and CO_2 released by a preacidification stage into the hydrolysis steam [10]. This means a higher CH_4 concentration in the gas phase is present for the methanogenesis step. In a two-phase method digesting grass silage, Nizami et al. [117] obtained biogas of 71% methane. In a wet single-stage system, the same grass silage developed a 52% methane biogas. Several plant constructors, ranging from constant to lots preacidification systems, provide preacidification systems in large-scales biogas plants.

11.4.4 Nanoparticles treatment

Actually, nanoparticles (NPs) are commonly used in marketed items. The use of NPs in development is known to be great, providing excellent physiochemical properties (e.g., optical, magnetic, and electrical). Additives have been a popular approach to improving AD performance [118]. Various research has discussed the addition of numerous forms of NPs to enhance the yield of biogas and to boost AD [3,17,118,119]. Mu et al. [120] said the effect on methane production was minimized by chemical ZnO NPs alone. Furthermore the reduction of ZnO NP (lower than 6 mg/g TS) doses has a slight to no effect on the yield of CH_4. Hassaan et al. [3] mentioned that NPs can improve the AD procedure and motivate slurry digestion, which enhances the production of biogas. The biogas output increase from durum wheat is accomplished with a low concentration of ZnO NPs (5 and 10 μg/mL), while a higher ZnO NPs concentration (upto 20 μg/mL) achieves an inhibitory effect. Wang et al. [121] reported that the lower (1.3 and 4.6 mg/mL) Fe^{2+} concentrations were observed to increase AD, while the higher (3.3 and 9.8 μg/mL) Ag^+ and Mg^{2+} concentrations were observed to decrease AD. Comparing the AD of sludge induced by zerovalent Fe NPs (nZVI), MgO NPs or Ag NPs, with that encouraged by the conforming quantities of Fe^{2+}, Ag^+, and Mg^{2+}, it was discovered that the released Fe^{2+}, Ag^+, and Mg^{2+} were mainly accountable for the improvement and/or inhibition influences of nZVI, Ag NPs, and MgO NPs [121] in contrast to the levels induced by Fe^{2+}, Ag^+, and Mg^{2+}. Toxicity can be achieved by the interaction between NPs and the bacterial cell wall [122]. When it is accurate, higher levels of metal NPs are necessary to surround and shield bacterial cells

Handbook of Algal Biofuels

from their surroundings, leaving little possibility of nutrients being ingested in order to proceed with the process of life. This may explain why higher levels of ZnO NPs in our work inhibit and cut biogas yield by 50%.

11.4.5 Inhibitor removal

There are a variety of inhibitors, including heavy metals, phenolics, and salt, involved in reducing ethanol yields and methane fermentation. The elimination of these inhibitors before fermentation as part of the pretreatment phase not only can AD increase yields of biofuels, but it can also produce high-value by-products [12].

11.5 The challenges of biogas production from algae

A high conversion of organic material to methane is one of the main factors for accomplishing a confident energy stability in algal AD; increasing the CH_4 production of sea algae AD has been shown to be a critical parameter in enhancing the energy stability of processes and minimizing greenhouse gas emissions [39,123]. Real-world biogas yields are considerably less than the theoretical limit from digestion in many algae. Seaweeds show less than 50% of the standard methane yield [124,125] (~ 200 L CH_4 g^{-1} VS) of the common commercial feedstocks [126]. For example, the *S. muticum* methane potential is only ~ 0.13 L CH^4 g^{-1} VS and below 27% of the theoretical maximum yield of methane [13,17]. However the "anaerobic digestibility," with a theoretical methane capacity ranging from 19%−81%, is a considerable variability between algae species [63]. However, Spain had a CH_4 yield of around 50% theoretically [127] in microalgal biomass generated in a wastewater treatment system called HRAP in Barcelona. The yield of methane from microalgae can differ broadly, relying on circumstances and species, from 0.024 to 0.6 L CH_4 g^{-1} VS [128,129]. Again for the same species the methane yield reported can differ significantly, for example, 0.063 [130] and 0.323 L CH_4 g^{-1} VS [130] were recorded as the AD methane yield of *Dunaliella* [131].

The reasons for low methane yields in many species of algae are widely debated but include:

1. Composition and structure of the cell wall;
2. Polysaccharides is not completely hydrolyzed;
3. Polyphenols;
4. Organic sulfur compounds;
5. Additional antimicrobials and toxins;
6. Ratio of C:N
7. concentrations of heavy metals

The bacteria in question are susceptible to the chemical composition of the feedstock when AD emits methane [132−134]. The structure and compounds of different types and species of algae are substantially different [58].

11.6 Conclusion

Seaweed biomass can serve as a feedstock for the production of biofuels, but is currently mainly harvested for high-value natural products and food. However, the overall value of algae biomass in terms of its ability to generate biofuels would require adequate pretreatment for algae. A variety of technologies for biogas substrates pretreatment based on a variety of principles are available. The bioavailability of the substrate increases if the substrate composition and pretreatment technology are balanced suitably. This could increase the efficiency of gas yield and decomposition rates in AD of the biogas reactors. When considering the energy and cost balance, pretreatment energy demand is determined by the technology. In most situations, low-energy demand pretreatment procedures provide less advantage than pretreatment with higher energy inputs, but not always. In other words, the degradation rate and resulting biogas yield are decreased. Greater gas outcomes are due to deterioration and greater availability of recalcitrant materials for lignocellulose complexes. In this section, a variety of methods are available for studying the structure of the algae cell wall differently, improved access to marine polymers for initial hydrolysis, and elimination/inhibition of the development of compounds, such as acids and phenols that could inhibit subsequent microbial metabolism. The investments in recalcitrant pillar pretreatment are currently large because of the high process engineering expenditure. Nevertheless, new substrates for biogas processing can be provided if these costs are reduced to a relatively high level. Finally, it is possible to establish pretreatments for algae that optimize the sustainability of biofuel production at the lowest costs with a sound understanding of the history and sources of sea algae and their composition and biomass.

References

[1] J.D. Murphy, B. Drosg, E. Allen, J. Jerney, A. Xia, C. Herrmann, A Perspective on Algal Biogas, IEA Bioenergy, 2015, pp. 5–35.

[2] M.A. Hassaan, A. Pantaleo, L. Tedone, M.R. Elkatory, R.M. Ali, A.E. Nemr, et al., Enhancement of biogas production via green ZnO nanoparticles: experimental results of selected herbaceous crops, Chem. Eng. Commun. (2019) 1–14.

[3] M.A. Hassaan, A. Pantaleo, F. Santoro, M.R. Elkatory, G. De Mastro, A.E. Sikaily, et al., Techno-economic analysis of ZnO nanoparticles pretreatments for biogas production from barley straw, Energies 13 (19) (2020) 5001.

[4] T. Karuppiah, V.E. Azariah, Biomass pretreatment for enhancement of biogas production, Anaerobic Digestion, IntechOpen, 2019.

[5] J.J. Milledge, B.V. Nielsen, P.J. Harvey, The inhibition of anaerobic digestion by model phenolic compounds representative of those from *Sargassum muticum*, J. Appl. Phycol. 31 (1) (2019) 779–786.

[6] J.J. Milledge, B.V. Nielsen, S. Maneein, P.J. Harvey, A brief review of anaerobic digestion of algae for bioenergy, Energies 12 (6) (2019) 1166.

[7] R. Amirante, G. Demastro, E. Distaso, M.A. Hassaan, A. Mormando, A.M. Pantaleo, et al., Effects of ultrasound and green synthesis ZnO nanoparticles on biogas production from Olive Pomace, Energy Procedia 148 (2018) 940–947.

[8] Green algae as a substrate for biogas production - cultivation and biogas potentials. Master's programme Science for Sustainable Development Master's Thesis, 30 ECTS credits (2010). http://liu.diva-portal.org/smash/get/diva2:343859/FULLTEXT01.pdf.

[9] L. Nikolaisen, P. Daugbjerg Jensen, K. Svane Bech, Energy production from marine biomass (*Ulva lactuca*), 2011. https://backend.orbit.dtu.dk/ws/portalfiles/portal/12709185/Ulva_lactuca.pdf.

[10] G. Bochmann, L.F. Montgomery, Storage and pre-treatment of substrates for biogas production, The Biogas Handbook, Woodhead Publishing, 2013, pp. 85–103.

References

[11] C.H.V. Ramirez, Biogas Production From Seaweed Biomass: A Biorefinery Approach (Doctoral dissertation, Doctoral thesis), Institute of Technology, Sligo, Ireland, 2017.

[12] S. Maneein, J.J. Milledge, B.V. Nielsen, P.J. Harvey, A review of seaweed pre-treatment methods for enhanced biofuel production by anaerobic digestion or fermentation, Fermentation 4 (4) (2018) 100.

[13] G. Jard, H. Marfaing, H. Carrère, J.P. Delgenès, J.P. Steyer, C. Dumas, French Brittany macroalgae screening: composition and methane potential for potential alternative sources of energy and products, Bioresour. Technol. 144 (2013) 492–498.

[14] J. Benemann, Microalgae for biofuels and animal feeds, Energies 6 (11) (2013) 5869–5886.

[15] R. Davis, A. Aden, P.T. Pienkos, Environmental life cycle comparison of algae to other bioenergy feedstocks, Environ. Sci. Technol. 44 (5) (2010) 813–1819. doi: 10.1021/es902838n; A.F. Clarens, E.P. Resurreccion, M.A. White, L.M. Colosi, Use of Algae in a Landfill Leachate Treatment System (Ph.D. thesis), Drexel University, 2017.

[16] S.S. Oncel, Microalgae for a macroenergy world, Renew. Sustain. Energy Rev. 26 (2013) 241–264.

[17] J.J. Milledge, P.J. Harvey, Potential process 'hurdles' in the use of macroalgae as feedstock for biofuel production in the British Isles, J. Chem. Technol. Biotechnol. 91 (8) (2016) 2221–2234.

[18] J.J. Milledge, P.J. Harvey, Ensilage and anaerobic digestion of *Sargassum muticum*, J. Appl. Phycol. 28 (5) (2016) 3021–3030.

[19] A.B. Ross, J.M. Jones, M.L. Kubacki, T. Bridgeman, Classification of macroalgae as fuel and its thermochemical behaviour, Bioresour. Technol. 99 (14) (2008) 6494–6504.

[20] L.M. Díaz-Vázquez, A. Rojas-Pérez, M. Fuentes-Caraballo, I.V. Robles, U. Jena, K.C. Das, Demineralization of *Sargassum* spp. macroalgae biomass: selective hydrothermal liquefaction process for bio-oil production, Front. Energy Res. 3 (2015) 6.

[21] B. Kloareg, R.S. Quatrano, Structure of the cell walls of marine algae and ecophysiological functions of the matrix polysaccharides, Oceanogr. Mar. Biol. 26 (1988) 259–315.

[22] B. Darcy-Vrillon, Nutritional aspects of the developing use of marine macroalgae for the human food industry, Int. J. Food Sci. Nutr. 44 (1993) S23–S35.

[23] C.L. Hurd, P.J. Harrison, K. Bischof, C.S. Lobban, Seaweed Ecology and Physiology, second ed., Cambridge University Press, Cambridge, 2014.

[24] Y. Date, K. Sakata, J. Kikuchi, Chemical profiling of complex biochemical mixtures from various seaweeds, Polym. J. 44 (8) (2012) 888–894.

[25] E.J. Yun, I.G. Choi, K.H. Kim, Red macroalgae as a sustainable resource for bio-based products, Trends Biotechnol. 33 (5) (2015) 247–249.

[26] C. Bobin-Dubigeon, M. Lahaye, F. Guillon, J.L. Barry, D.J. Gallant, Factors limiting the biodegradation of *Ulva* sp. cell-wall polysaccharides, J. Sci. Food Agric. 75 (3) (1997) 341–351.

[27] L.E. Rioux, S.L. Turgeon, M. Beaulieu, Characterization of polysaccharides extracted from brown seaweeds, Carbohydr. Polym. 69 (3) (2007) 530–537.

[28] J.S. Jang, Y. Cho, G.T. Jeong, S.K. Kim, Optimization of saccharification and ethanol production by simultaneous saccharification and fermentation (SSF) from seaweed, Saccharina japonica, Bioprocess. Biosyst. Eng. 35 (1–2) (2012) 11–18.

[29] A. Synytsya, J. Čopíková, W.J. Kim, Y. Park, Cell wall polysaccharides of marine algae, in: S.K. Kim (Ed.), Springer Handbook of Marine Biotechnology, Springer, Berlin/Heidelberg, 2015, pp. 543–590.

[30] E. Deniaud-Bouët, N. Kervarec, G. Michel, T. Tonon, B. Kloareg, C. Hervé, Chemical and enzymatic fractionation of cell walls from Fucales: insights into the structure of the extracellular matrix of brown algae, Ann. Bot. 114 (6) (2014) 1203–1216.

[31] J.A. Gallagher, L.B. Turner, J.M. Adams, S. Barrento, P.W. Dyer, M.K. Theodorou, Species variation in the effects of dewatering treatment on macroalgae, J. Appl. Phycol. 30 (4) (2018) 2305–2316.

[32] A.R. Cabrita, M.R. Maia, I. Sousa-Pinto, A.J. Fonseca, Ensilage of seaweeds from an integrated multi-trophic aquaculture system, Algal Res. 24 (2017) 290–298.

[33] F. Wei, K. Ito, K. Sakata, Y. Date, J. Kikuchi, Pretreatment and integrated analysis of spectral data reveal seaweed similarities based on chemical diversity, Anal. Chem. 87 (5) (2015) 2819–2826.

[34] R.M. Rodriguez-Jasso, S.I. Mussatto, L. Pastrana, C.N. Aguilar, J.A. Teixeira, Chemical composition and antioxidant activity of sulphated polysaccharides extracted from *Fucus vesiculosus* using different hydrothermal processes, Chem. Pap. 68 (2) (2014) 203–209.

246   11. Eco-friendly biogas production from algal biomass

[35] E. Percival, The polysaccharides of green, red and brown seaweeds: their basic structure, biosynthesis and function, Br. Phycol. J. 14 (2) (1979) 103—117.

[36] O. Berteau, B. Mulloy, Sulfated fucans, fresh perspectives: structures, functions, and biological properties of sulfated fucans and an overview of enzymes active toward this class of polysaccharide, Glycobiology 13 (6) (2003) 29R—40R.

[37] A. Alves, R.A. Sousa, R.L. Reis, A practical perspective on ulvan extracted from green algae, J. Appl. Phycol. 25 (2) (2013) 407—424.

[38] A. Robic, C. Gaillard, J.F. Sassi, Y. Lerat, M. Lahaye, Ultrastructure of ulvan: a polysaccharide from green seaweeds, Biopolymers 91 (8) (2009) 652—664.

[39] J.J. Milledge, S. Heaven, Energy balance of biogas production from microalgae: effect of harvesting method, multiple raceways, scale of plant and combined heat and power generation, J. Mar. Sci. Eng. 5 (1) (2017) 9.

[40] S. Sharma, L. Neves, J. Funderud, L.T. Mydland, M. Øverland, S.J. Horn, Seasonal and depth variations in the chemical composition of cultivated *Saccharina latissima*, Algal Res. 32 (2018) 107—112.

[41] L.R. Andrade, R.N. Leal, M. Noseda, M.E.R. Duarte, M.S. Pereira, P.A. Mourão, et al., Brown algae overproduce cell wall polysaccharides as a protection mechanism against the heavy metal toxicity, Mar. Pollut. Bull. 60 (9) (2010) 1482—1488.

[42] E. Plouguerné, K. Le Lann, S. Connan, G. Jechoux, E. Deslandes, V. Stiger-Pouvreau, Spatial and seasonal variation in density, reproductive status, length and phenolic content of the invasive brown macroalga *Sargassum muticum* (Yendo) Fensholt along the coast of Western Brittany (France), Aquat. Bot. 85 (4) (2006) 337—344.

[43] T.D. Brock, M.T. Madigan, J.M. Martinko, and J. Parker. Biology of Microorganisms. London: Prentice Hall International (1994).

[44] M. Gerardi, The Microbiology of Anaerobic Digesters, New Jersey, USA, 2003.

[45] D.L. Klass, Biomass for Renewable Energy, Fuels, and Chemicals, Elsevier, 1998.

[46] D.P. Chynoweth, Environmental impact of biomethanogenesis, Environ. Monit. Assess. 42 (1996) 3—18.

[47] J. Fry, Methane Digesters for Fuel Gas and Fertilizer with Complete Instructions for Two Working Models, The New Alchemy Institute, MA, USA, 1973.

[48] C. Vaneeckhaute, E. Meers, E. Michels, G. Ghekiere, F. Accoe, F.M.G. Tack, Closing the nutrient cycle by using bio-digestion waste derivatives as synthetic fertilizer substitutes: a field experiment, Biomass Bioenergy 55 (2013) 175—189.

[49] J.B. Holm-Nielsen, T. Al Seadi, P. Oleskowicz-Popiel, The future of anaerobic digestion and biogas utilization, Bioresour. Technol. 100 (22) (2009) 5478—5484.

[50] T. Al Seadi, C. Lukehurst, Quality management of digestate from biogas plants used as fertiliser, IEA Bioenergy 37 (2012) 40.

[51] C.T. Lukehurst, P. Frost, T. Al Seadi, Utilisation of digestate from biogas plants as biofertiliser, IEA Bioenergy 2010 (2010) 1—36.

[52] D.F. Toerien, W.H.J. Hattingh, Anaerobic digestion I. The microbiology of anaerobic digestion, Water Res. 3 (6) (1969) 385—416.

[53] Y. Chen, J.J. Cheng, K.S. Creamer, Inhibition of anaerobic digestion process: a review, Bioresour. Technol. 99 (10) (2008) 4044—4064.

[54] V. Müller, Energy conservation in acetogenic bacteria, Appl. Environ. Microbiol. 69 (11) (2003) 6345—6353.

[55] T. Amani, M. Nosrati, S.M. Mousavi, Using enriched cultures for elevation of anaerobic syntrophic interactions between acetogens and methanogens in a high-load continuous digester, Bioresour. Technol. 102 (4) (2011) 3716—3723.

[56] M.S. Switzenbaum, E. Giraldo-Gomez, R.F. Hickey, Monitoring of the anaerobic methane fermentation process, Enzyme Microb. Technol. 12 (10) (1990) 722—730.

[57] D. Karakashev, D.J. Batstone, I. Angelidaki, Influence of environmental conditions on methanogenic compositions in anaerobic biogas reactors, Appl. Environ. Microbiol. 71 (1) (2005) 331—338.

[58] B.K. Tiwari, D.J. Troy, Seaweed sustainability—food and nonfood applications, Seaweed Sustainability, Academic Press, 2015, pp. 1—6.

[59] P. Peu, J.F. Sassi, R. Girault, S. Picard, P. Saint-Cast, F. Béline, et al., Sulphur fate and anaerobic biodegradation potential during co-digestion of seaweed biomass (*Ulva* sp.) with pig slurry, Bioresour. Technol. 102 (23) (2011) 10794—10802.

[60] C.H. Vanegas, J. Bartlett, Green energy from marine algae: biogas production and composition from the anaerobic digestion of Irish seaweed species, Environ. Technol. 34 (15) (2013) 2277—2283.

Handbook of Algal Biofuels

References

[61] M. Aresta, A. Dibenedetto, G. Barberio, Utilization of macro-algae for enhanced CO_2 fixation and biofuels production: development of a computing software for an LCA study, Fuel Process. Technol. 86 (14–15) (2005) 1679–1693.

[62] Y.N. Barbot, H. Al-Ghaili, R. Benz, A review on the valorization of macroalgal wastes for biomethane production, Mar. Drugs 14 (6) (2016) 120.

[63] E. Allen, D.M. Wall, C. Herrmann, A. Xia, J.D. Murphy, What is the gross energy yield of third generation gaseous biofuel sourced from seaweed? Energy 81 (2015) 352–360.

[64] M. Huesemann, G. Roesjadi, J. Benemann, F.B. Metting, Biofuels from microalgae and seaweeds, Biomass Biofuels (2010) 165–184.

[65] J. Lewis, F. Salam, N. Slack, M. Winton, L. Hobson, Product Options for the Processing of Marine Macro-Algae—Summary Report, The Crown Estates, 2011.

[66] J.G. Ferry, Methane from acetate, J. Bacteriol. 174 (17) (1992) 5489.

[67] U. Deppenmeier, V. Müller, G. Gottschalk, Pathways of energy conservation in methanogenic archaea, Arch. Microbiol. 165 (1996) 148–163.

[68] C.G. Golueke, W.J. Oswald, H.B. Gotaas, Anaerobic digestion of algae, Appl. Microbiol. 5 (1) (1957) 47.

[69] K.T. Bird, P.H. Benson. Seaweed Cultivation for Renewable Resources. New York: Elsevier, 1987.

[70] K.T. Bird, D.P. Chynoweth, D.E. Jerger, Effects of marine algal proximate composition on methane yields, J. Appl. Phycol. 2 (3) (1990) 207–213.

[71] R.A. Troiano, D.L. Wise, D.C. Augenstein, R.G. Kispert, C.L. Cooney, Fuel gas production by anaerobic digestion of kelp, ReReC 2 (1976) 171–176.

[72] A.D. Hughes, M.S. Kelly, K.D. Black, M.S. Stanley, Biogas from Macroalgae: is it time to revisit the idea? Biotechnol. Biofuels 5 (1) (2012) 86.

[73] P. Kumar, D.M. Barrett, M.J. Delwiche, P. Stroeve, Methods for pretreatment of lignocellulosic biomass for efficient hydrolysis and biofuel production, Ind. Eng. Chem. Res. 48 (8) (2009) 3713–3729.

[74] Y. Sun, J. Cheng, Hydrolysis of lignocellulosic materials for ethanol production: a review, Bioresour. Technol. 83 (1) (2002) 1–11.

[75] S. Tedesco, T.M. Barroso, A.G. Olabi, Optimization of mechanical pre-treatment of *Laminariaceae* spp. biomass-derived biogas, Renew. Energy 62 (2014) 527–534.

[76] H.B. Nielsen, S. Heiske, Anaerobic digestion of macroalgae: methane potentials, pre-treatment, inhibition and co-digestion, Water Sci. Technol. 64 (8) (2011) 1723–1729.

[77] J.V. Oliveira, M.M. Alves, J.C. Costa, Design of experiments to assess pre-treatment and co-digestion strategies that optimize biogas production from macroalgae *Gracilaria vermiculophylla*, Bioresour. Technol. 162 (2014) 323–330.

[78] M.E. Montingelli, K.Y. Benyounis, J. Stokes, A.G. Olabi, Pretreatment of macroalgal biomass for biogas production, Energy Convers. Manag. 108 (2016) 202–209.

[79] L. Pastare, I. Aleksandrovs, D. Lauka, F. Romagnoli, Mechanical pre-treatment effect on biological methane potential from marine macro algae: results from batch tests of *Fucus vesiculosus*, Energy Procedia 95 (2016) 351–357.

[80] M.E. Montingelli, K.Y. Benyounis, B. Quilty, J. Stokes, A.G. Olabi, Influence of mechanical pretreatment and organic concentration of Irish brown seaweed for methane production, Energy 118 (2017) 1079–1089.

[81] E. Allen, J. Browne, S. Hynes, J.D. Murphy, The potential of algae blooms to produce renewable gaseous fuel, Waste Manag. 33 (11) (2013) 2425–2433.

[82] F. Raposo, V. Fernández-Cegrí, M.A. De la Rubia, R. Borja, F. Béline, C. Cavinato, et al., Biochemical methane potential (BMP) of solid organic substrates: evaluation of anaerobic biodegradability using data from an international interlaboratory study, J. Chem. Technol. & Biotechnol. 86 (8) (2011) 1088–1098.

[83] M.R. Tabassum, A. Xia, J.D. Murphy, Comparison of pre-treatments to reduce salinity and enhance biomethane yields of *Laminaria digitata* harvested in different seasons, Energy 140 (2017) 546–551.

[84] P. Tsapekos, P.G. Kougias, I. Angelidaki, Biogas production from ensiled meadow grass; effect of mechanical pretreatments and rapid determination of substrate biodegradability via physicochemical methods, Bioresour. Technol. 182 (2015) 329–335.

[85] D. Manns, S.K. Andersen, B. Saake, A.S. Meyer, Brown seaweed processing: enzymatic saccharification of Laminaria digitata requires no pre-treatment, J. Appl. Phycol. 28 (2) (2016) 1287–1294.

[86] G.E. Symons, A.M. Buswell, The methane fermentation of carbohydrates1, 2, J. Am. Chem. Soc. 55 (5) (1933) 2028–2036.

[87] H. Jung, G. Baek, J. Kim, S.G. Shin, C. Lee, Mild-temperature thermochemical pretreatment of green macro-algal biomass: effects on solubilization, methanation, and microbial community structure, Bioresour. Technol. 199 (2016) 326–335.

[88] V. Vivekanand, V.G. Eijsink, S.J. Horn, Biogas production from the brown seaweed *Saccharina latissima*: thermal pretreatment and codigestion with wheat straw, J. Appl. Phycol. 24 (5) (2012) 1295–1301.

[89] P. Yazdani, A. Zamani, K. Karimi, M.J. Taherzadeh, Characterization of Nizimuddinia zanardini macro-algae biomass composition and its potential for biofuel production, Bioresour. Technol. 176 (2015) 196–202.

[90] F. Romagnoli, L. Pastare, A. Sabūnas, K. Bāliņa, D. Blumberga, Effects of pre-treatment on biochemical methane potential (BMP) testing using Baltic Sea *Fucus vesiculosus* feedstock, Biomass Bioenergy 105 (2017) 23–31.

[91] XXX.

[92] E.T. Kostas, D. Beneroso, J.P. Robinson, The application of microwave heating in bioenergy: a review on the microwave pre-treatment and upgrading technologies for biomass, Renew. Sustain. Energy Rev. 77 (2017) 12–27.

[93] A.T. Quitain, T. Kai, M. Sasaki, M. Goto, Microwave–hydrothermal extraction and degradation of fucoidan from supercritical carbon dioxide deoiled Undaria pinnatifida, Ind. Eng. Chem. Res. 52 (23) (2013) 7940–7946.

[94] A.T. Quitain, T. Kai, M. Sasaki, M. Goto, Supercritical carbon dioxide extraction of fucoxanthin from Undaria pinnatifida, J. Agric. Food Chem. 61 (24) (2013) 5792–5797.

[95] V.K. Tyagi, S.L. Lo, Application of physico-chemical pretreatment methods to enhance the sludge disintegration and subsequent anaerobic digestion: an up to date review, Rev. Environ. Sci. Bio/Technol. 10 (3) (2011) 215.

[96] Hugo Vogelsang Maschinenbau GmbH, Product Information. Hugo Vogelsang Maschinenbau GmbH, Oldenburg, 2011. <http://www.engineered-to-work.com/en/BioCrack_Productinformation.html> (accessed 20.11.20).

[97] Süd Chemie Elektrokinetische Desintegration, Trink- und Abwasserbehandlung, Sud-chemie AG, Moosburg, 2011. <http://www.s-cpp.com/scmcms/web/binary.jsp?nodeId = 6324&binaryId = 10100&preview = &disposition = inline&lang = de> (accessed 20.11.20).

[98] A. Lehner, M. Effenberger, A. Gronauer, Optimierung der Verfahrenstechnik landwirtschaftlicher Biogasanlagen. LfL, 2010.

[99] J. Maroušek, Finding the optimal parameters for the steam explosion process of hay, Rev. Técnica de la Facultad de Ingeniería. Univ. del. Zulia 35 (2) (2012) 170–178. 5495.

[100] D. Novarino, M.C. Zanetti, Anaerobic digestion of extruded OFMSW, Bioresour. Technol. 104 (2012) 44–50.

[101] C. Vanegas, A. Hernon, J. Bartlett, Influence of chemical, mechanical, and thermal pretreatment on the release of macromolecules from two Irish seaweed species, Sep. Sci. Technol. 49 (1) (2014) 30–38.

[102] S.G. Wi, H.J. Kim, S.A. Mahadevan, D.J. Yang, H.J. Bae, The potential value of the seaweed Ceylon moss (*Gelidium amansii*) as an alternative bioenergy resource, Bioresour. Technol. 100 (24) (2009) 6658–6660.

[103] XXX.

[104] XXX.

[105] P. Schiener, M.S. Stanley, K.D. Black, D.H. Green, Assessment of saccharification and fermentation of brown seaweeds to identify the seasonal effect on bioethanol production, J. Appl. Phycol. 28 (5) (2016) 3009–3020.

[106] XXX.

[107] F. Monlau, C. Sambusiti, A. Barakat, M. Quéméneur, E. Trably, J.P. Steyer, et al., Do furanic and phenolic compounds of lignocellulosic and algae biomass hydrolyzate inhibit anaerobic mixed cultures? A comprehensive review, Biotechnol. Adv. 32 (5) (2014) 934–951.

[108] F. Gao, L. Gao, D. Zhang, N. Ye, S. Chen, D. Li, Enhanced hydrolysis of *Macrocystis pyrifera* by integrated hydroxyl radicals and hot water pretreatment, Bioresour. Technol. 179 (2015) 490–496.

[109] L. Gao, D. Li, F. Gao, Z. Liu, Y. Hou, S. Chen, et al., Hydroxyl radical-aided thermal pretreatment of algal biomass for enhanced biodegradability, Biotechnol. Biofuels 8 (1) (2015) 1–11.

[110] P.J. Strong, B. McDonald, D.J. Gapes, Combined thermochemical and fermentative destruction of municipal biosolids: a comparison between thermal hydrolysis and wet oxidative pre-treatment, Bioresour. Technol. 102 (9) (2011) 5520–5527.

References

[111] T.S. Chandra, S.N. Malik, G. Suvidha, M.L. Padmere, P. Shanmugam, S.N. Mudliar, Wet air oxidation pretreatment of biomethanated distillery effluent: mapping pretreatment efficiency in terms color, toxicity reduction and biogas generation, Bioresour. Technol. 158 (2014) 135–140.

[112] A. Elliott, T. Mahmood, Pretreatment technologies for advancing anaerobic digestion of pulp and paper biotreatment residues, Water Res. 41 (19) (2007) 4273–4286.

[113] R. Goel, T. Tokutomi, H. Yasui, Anaerobic digestion of excess activated sludge with ozone pretreatment, Water Sci. Technol. 47 (2013) 207–214.

[114] R.K. Thauer, Biochemistry of methanogenesis: a tribute to Marjory Stephenson: 1998 Marjory Stephenson prize lecture, Microbiology 144 (9) (1998) 2377–2406.

[115] D. Liu, D. Liu, R.J. Zeng, I. Angelidaki, Hydrogen and methane production from household solid waste in the two-stage fermentation process, Water Res. 40 (11) (2006) 2230–2236.

[116] G. Antonopoulou, H.N. Gavala, I.V. Skiadas, K. Angelopoulos, G. Lyberatos, Biofuels generation from sweet sorghum: fermentative hydrogen production and anaerobic digestion of the remaining biomass, Bioresour. Technol. 99 (1) (2008) 110–119.

[117] A.S. Nizami, A. Orozco, E. Groom, B. Dieterich, J.D. Murphy, How much gas can we get from grass? Appl. Energy 92 (2012) 783–790.

[118] E. Abdelsalam, M. Samer, Y.A. Attia, M.A. Abdel-Hadi, H.E. Hassan, Y. Badr, Influence of zero valent iron nanoparticles and magnetic iron oxide nanoparticles on biogas and methane production from anaerobic digestion of manure, Energy 120 (2017) 842–853.

[119] H. Mu, Y. Chen, Long-term effect of ZnO nanoparticles on waste activated sludge anaerobic digestion, Water Res. 45 (17) (2011) 5612–5620.

[120] H. Mu, Y. Chen, N. Xiao, Effects of metal oxide nanoparticles (TiO_2, Al_2O_3, SiO_2 and ZnO) on waste activated sludge anaerobic digestion, Bioresour. Technol. 102 (22) (2011) 10305–10311.

[121] T. Wang, D. Zhang, L. Dai, Y. Chen, X. Dai, Effects of metal nanoparticles on methane production from waste-activated sludge and microorganism community shift in anaerobic granular sludge, Sci. Rep. 6 (1) (2016) 1–10.

[122] K.S. Siddiqi, A. ur Rahman, A. Husen, Properties of zinc oxide nanoparticles and their activity against microbes, Nanoscale Res. Lett. 13 (1) (2018) 1–13.

[123] S.P. Mayfield, Consortium for Algal Biofuel Commercialization (CAB-COMM) Final Report (No. EE0003373), Univ. of California, San Diego, CA, 2015.

[124] M. Alvarado-Morales, A. Boldrin, D.B. Karakashev, S.L. Holdt, I. Angelidaki, T. Astrup, Life cycle assessment of biofuel production from brown seaweed in Nordic conditions, Bioresour. Technol. 129 (2013) 92–99.

[125] J.L. Chen, R. Ortiz, T.W. Steele, D.C. Stuckey, Toxicants inhibiting anaerobic digestion: a review, Biotechnol. Adv. 32 (8) (2014) 1523–1534.

[126] H.H. Nguyen, S. Heaven, C. Banks, Energy potential from the anaerobic digestion of food waste in municipal solid waste stream of urban areas in Vietnam, Int. J. Energy Environ. Eng. 5 (4) (2014) 365–374.

[127] F. Passos, R. Gutiérrez, D. Brockmann, J.P. Steyer, J. García, I. Ferrer, Microalgae production in wastewater treatment systems, anaerobic digestion and modelling using ADM1, Algal Res. 10 (2015) 55–63.

[128] K.C. Tran, Anaerobic Digestion of Microalgal Biomass: Effects of Solid Concentration and Pre-Treatment (Doctoral dissertation), University of Southampton, 2017.

[129] K.P. Roberts, S. Heaven, C.J. Banks, Quantification of methane losses from the acclimatisation of anaerobic digestion to marine salt concentrations, Renew. Energy 86 (2016) 497–506.

[130] M.J. Fernández-Rodríguez, B. Rincón, F.G. Fermoso, A.M. Jiménez, R. Borja, Assessment of two-phase olive mill solid waste and microalgae co-digestion to improve methane production and process kinetics, Bioresour. Technol. 157 (2014) 263–269.

[131] J.H. Mussgnug, V. Klassen, A. Schlüter, O. Kruse, Microalgae as substrates for fermentative biogas production in a combined biorefinery concept, J. Biotechnol. 150 (1) (2010) 51–56.

[132] R. Samson, A. LeDuy, Improved performance of anaerobic digestion of *Spirulina maxima* algal biomass by addition of carbon-rich wastes, Biotechnol. Lett. 5 (10) (1983) 677–682.

[133] J.B.K. Park, R.J. Craggs, A.N. Shilton, Wastewater treatment high rate algal ponds for biofuel production, Bioresour. Technol. 102 (1) (2011) 35–42.

[134] C. González-Fernández, B. Sialve, N. Bernet, J.P. Steyer, Impact of microalgae characteristics on their conversion to biofuel. Part II: focus on biomethane production, Biofuels Bioprod. Bioref. 6 (2) (2012) 205–218.

CHAPTER 12

Algal biomass for bioethanol and biobutanol production

Marwa R. Elkatory[1], Mohamed A. Hassaan[2] and Ahmed El Nemr[2]

[1]Advanced Technology and New Materials Research Institute (ATNMRI), City for Scientific Research and Technological Applications (SRTA-City), Alexandria, Egypt [2]National Institute of Oceanography and Fisheries, Cairo, Egypt

12.1 Introduction

Fossil fuels, coal, petroleum, and gas are still major energy sources to date [1]; however, the growing use of fossil fuels has led to significant environmental concerns. Moreover, fossil fuels are known to be unverifiable because of their diminishing reserves. It is desperately important to pursue clean and green technologies, such as biofuels. Biofuels are carbon-neutral, help to limit gaseous pollutants emissions, and are perceived to have environmental advantages when compared with conventional fuels [2]. Both terrestrial and marine biomass are known as renewable sources of energy. With regard to renewable energy sources, aquatic biomass is the most eligible technique to be applied on a huge measure deprived of any environmental or economic sanctions [3]. Algae, both marine and freshwater, are among those that are not frequent in many environments for production [4]. The early studies focused primarily on the use of algae as feed for animals, bioenergy, and in aquaculture.

Also, as new bioresource for renewable energy production, algae have gained a great deal of attention. To one side from extra sources of biomass, algae have a great biomass revenue per unit of zone and light, can have a high content of oil or starch, and don't require freshwater or agronomic land, and nutrient supplies could be met either by wastewater or by seawater. A selection of organic fragments, mainly lipids and carbohydrates, are produced by algae, which can be used to obtain a fuel. Table 12.1 shows that algae (microalgae and macroalgae) have recently been identified as an attractive, alternative, sustainable source of biomass from food or cellulosic materials for the production of

252   12. Algal biomass for bioethanol and biobutanol production

TABLE 12.1 The major differences between microalgae and macroalgae.

Parameters	Microalgae	Macroalgae
Size	1–50 μm	Up to 60 m
Physical structure	Doesn't have leaves, roots, and stems	Macroalgae have a thallus and occasionally a foot and a stem
Energy density	High	Moderate
Biomass yield	High	Moderate
Cultivation	Photobioreactors or open ponds	Natural environments such as ocean
Harvesting	Difficult	Easy
Oil yield	Microalgae produce comparatively high amounts of lipids	Macroalgae produce only small amounts of lipids

biofuels [5]. The production of marine algae does not compete with the production of food, as algae do not require freshwater or arable land, but will use land depending on the selected cultivation method. CO_2 from industrial pollution may be used by algae as a source of carbon. In addition, per unit area, the biomass yield is higher than that of terrestrial biomass [6]. While the production of biofuels from microalgae has been intensively studied for a few years, there is only an interest in using macroalgae as feedstock for the production of biofuels. Macroalgae have the benefit of being much easier to harvest than microalgae, so that there is no issue with low feedstock concentration.

There are differences in the levels of hydrocarbons, lipids, oil, etc. in various species, for example., *Botryococcus braunii* produces abundant hydrocarbons, predominantly triterpene oils, which are almost 30%–40% of its dry weight (DW). *Botryococcus* oils have botryococcenes, alkadiene, and alkatriene oils as the principal hydrocarbons [7]. *B. braunii*'s main product is very close to fossil fuel complexes, and these useful photosynthetic algae bio-oils can produce a widespread variety of fuels, such as gasoline and diesel [8].

Algae are brilliant small plants that can turn biomass into energy to address the difficulties associated with fossil fuel exhaustion and the rise in emissions worldwide. Algae are also exclusive sources of a plethora of exceptional bioproducts, ranging from bioceuticals to antimicrobials, to UV protectants, in addition to biofuels (Fig. 12.1). In their early commercialization, focusing on improved methods of growth and harvest can assist. The improvised of downstream handling will expose innovative doors to the screening of new algal strains techniques for improving algal biofuel quality. The protection of our nonrenewable sources of energy and the atmosphere can be linked with biofuels as a route toward a better future.

For generating biofuels based on algae demands, a wider range of modern equipment for the broad application of biotechnological methods, more optimized cultural conditions, novel strains and consortiums and the key mechanisms for reflection are required [9]. The development of H_2 from algae is well-known, but further investigations are needed for commercial implementation. The algal biodiesel has obtained the most emphasis among the major biofuels to date, and is probably the only algal biofuel endeavor that has been

Handbook of Algal Biofuels

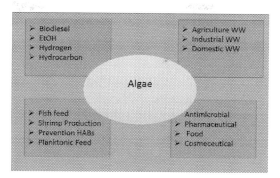

FIGURE 12.1 Different applications of algae.

upgraded to pilot or full scale, although the development of bioethanol using algae requires careful consideration and commitment.

This chapter discusses the potential for energy products (bioethanol and biobutanol), including both macro- and microalgae in freshwater and marine environments. The diversity of algal sources and their environments is also discussed in this chapter, along with the production systems shown for use, in addition to the cultivation of algal biomass.

12.2 Current biofuels status

Present biofuels development activities concentrate mainly on bioethanol, with other possible oils, such as biobutanol, becoming main players if technical progress is made. 1-Butanol has some properties that, when compared to C_2H_5OH, makes it a decent selection for fuel improvement. With a lower heating value of about 100,000 btu/gal, C_4H_9OH has a higher energy density as opposed to C_2H_5OH at 76,000 btu/gal. In his 1992 Buick test driven on B100 (100% C_4H_9OH), Ramey found that the gas mileage was 8% higher (depending on the amount of C_4H_9OH used) than with gasoline [10,11]. In 2007, Edmunds tested a flex-fuel Chevrolet Tahoe on both E85 (85% C_2H_5OH, 15% gasoline) and gasoline and found that the gas mileage when powered on E85 was 26% lower [11]. This is an additional benefit for C_4H_9OH as opposed to C_2H_5OH. C_4H_9OH can run without engine modifications in most cars, whereas C_2H_5OH needs comprehensive modifications at concentrations greater than 10%. This is due to C_4H_9OH being significantly less corrosive than C_2H_5OH. C_2H_5OH is extremely water-miscible and can absorb moisture from the air. If the C_2H_5OH/gasoline mixture's water content becomes too great, the fuel will separate into a water/C_2H_5OH layer at the bottom of the container.

Agrofuel was the first generation of biofuels to use maize, sugarcane, palm, sweet sorghum, and soybean as food stocks for the manufacture of specific cultivated plants (Table 12.2). Agrofuel was developed to provide bioethanol and derived botanical extracts to manufacture biodiesel by fermenting yeast with sugars or starch from vegetables [12]. Both the food and water industries are significantly impacted by these systems [13]. Biofuels of the second generation relied on nonfood plants such as silver grass, grass,

254   12. Algal biomass for bioethanol and biobutanol production

TABLE 12.2 The methodologies for developing better biofuel technology.

The biofuel generation	Biofuel raw material	Technology	Limits
First generation	It is derived from foodstuff harvests with decent glucose content, such as sugarcane. The octane rating is higher in these materials	Enzymatic	Foodstuff harvests abstracted from consumer to biofuel industries; food safety risks
Second generation	Contains cellulose, weeds, biomass waste, corn stalks, stalks of weed and wood	Uses liquid technologies to produce solid biomass biofuel. much as ruminants digest the grass they feed	Using broad soil mass for crops producing more fuel biomass, thus cutting the equilibrium crop yield
Third generation	Algae was used in much ongoing research as the third-generation biofuel raw material	Thermochemical, biochemical, and chemical	Unsteady fuel compared to first and second-generation fuels because of its extraordinary rate of unsaturation
Fourth generation	Solar and photobiological biofuels	Continuous investigation on solar energy transformation into fuel	Well-organized tools for improved practice of resources employed for fuel energy manufacture

Jatropha, and switchgrass [14]. Algal biofuel is the third generation of biofuels and minimizes land and water use and the extreme usage of toxic biocides [15].

Algal biomass can be used as a raw material in biofuel production via pyrolysis or for the generation of biogas and bioethanol by fermentation. Several requirements as mentioned below should be met by macro and microalgae for bioenergy production:

1. abundant;
2. simple to harvest;
3. capable of surviving among currents in the open ocean; and
4. manufactured at a cost equivalent to or less than existing sources.

12.3 Bioalcohols

As eco-friendly and renewable sources of energy, bioalcohols such as bioethanol and biobutanol are currently promoted globally. The extreme energy crisis in recent years has made the exploration of biofuels inevitable. The creation of new sources of renewable energy will help to make operations more competitive and viable and overcome the environmental barriers. Various bioalcohols, such as C_4H_9OH or C_2H_5OH, have been studied as alternatives to fossil fuels. Due to its lower operational costs and higher yields, bioethanol is the most researched bioalcohol [16]. The need for sustainable production of fuels and chemicals has prompted extensive research in the chemical industry to replace fossil fuel sources with renewable sources.

Handbook of Algal Biofuels

12.3.1 Bioethanol

Bioethanol is currently the most common biofuel in the world [17]. As an alternative to petroleum, it has many attractive features [18] and can help to enable a smoother conversion from oil to biological industries [19]. Bioethanol has appeared as a possible transport fuel and has been used to substitute MTBE, unlike other bioalcohols that are still under investigation (methyl tertiary butyl ether).

This work evaluates the environmental burden of bioethanol from onshore grown macroalgae based on literature evidence and preliminary studies. The study revealed that compared to fossil fuel and sugarcane bioethanol, an optimized device model enables the development of an environmentally effective biofuel. The findings of the life cycle assessment (LCA) were extremely reliant on the form of coproduct control chosen, as shifting coproduct control from energy distribution to replacement substantially decreased the environmental burden of the studied system.

12.3.2 Biobutanol

The United States currently uses around 360 million gallons of fuel annually (DOE/EIA-0383 2012). This statistic is projected to marginally reduce as higher fuel pump rates lead to less driving and thus a drop in requests. Nevertheless, demand for vehicle fuel could increase and offset decreased consumption in the United States in developed countries. The higher costs of gasoline are mirrored in high transport-based amounts for goods and facilities. Therefore the EU formally released the 2003/30/EC Biofuels Directive, which stipulated that 5.75% of all fossil fuels (gasoline and diesel) be supplemented by biofuels by 2010. Later, the Energy Independence and Security Act for 2007 was adopted by the US Congress, which expanded national clean energy requirements by 9 billion gallons by the year 2008 and eventually to 36 billion gallons by 2022, among other rules and goals. The by-product of the use of fossil fuels is a greenhouse gas called carbon dioxide. The increase in average global temperatures has been related to an increase in the atmospheric concentration of carbon dioxide. Around 1/4 of the CO_2 emissions produced by humans are from the use of fossil-derived vehicle fuel for vehicles and light trucks [20–22]. Therefore CO_2 neutral fuel alternatives to fossil fuel such as C_4H_9OH could have major consequences on the world economy and the environment.

Biobutanol (C_4H_9OH), due to its longer chain, has the primary advantage of higher heating value (similar to gasoline) than bioethanol. Biobutanol (Fig. 12.2) has a boiling point of 118°C. The industrial use of C_4H_9OH dates back to the First World War, when the lacquer industry started. C_4H_9OH is used in paints, plastics, and polymers with a market value of $5 million. C_4H_9OH demand is likely to rise at a rate of 3.2% pa [23].

There are a variety of benefits to C_4H_9OH over C_2H_5OH, and butanol's physical properties include:

1. higher density of energy;
2. the lower pressure of vapor;
3. less corrosive; and
4. lower water solubility.

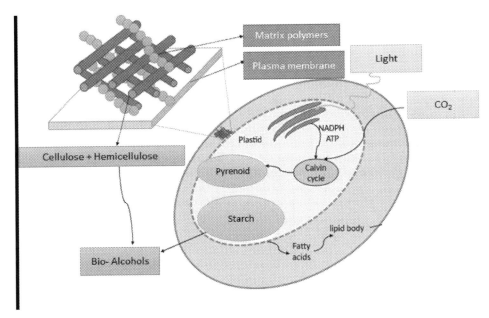

FIGURE 12.2 Various fuels resulting from microalgae.

Bioethanol and its blends may be transferred through standing pipes (different to C_2H_5OH) and can be used without modification in present machines [24]. David Ramey, of Butyl Fuel, LLC, demonstrated the possibility of using pure C_4H_9OH as a motor fuel through two extended journeys in the United States in 2005 and 2007 (http://www.butanol.com, accessed 13.01.21). In 2006 a joint venture was launched by British Petroleum and Dupont to further develop the fermentative output of C_4H_9OH and commercialize it.

12.4 Microalgae

In general, microalgae are microorganisms present in both freshwater and marine environments that are photosynthetic. Microalgae have been categorized on the basis of numerous features, for instance, pigmentation, the result of photosynthetic storage, photosynthetic membrane arrangements, and other morphological characteristics. There are currently four classes of microalgae: diatoms (*Bacillariophyceae*), blue—green algae (*Cyanophyceae*), green algae (*Chlorophyceae*), and golden algae (*Chrysophyceae*) [25]. In commercial production, the leading species of microalgae are *Isochrysis*, *Chaetoceros*, *Dunaliella*, *Chlorella*, and *Arthrospira* (*Spirulina*) [26]. *Chlorella* species, like other microalgae species, are able to alter nutrition modes from phototrophic to heterotrophic [8,27]. Mixotrophic growth may also be necessary for certain algae. The common constituents in microalgae are biomolecules such as proteins, carbohydrates, nucleic acids, and lipids [28].

There are many different forms of renewable biofuel that microalgae can supply (Table 12.3, Fig. 12.2). These comprise CH_4 that is formed by the anaerobic digestion (AD)

12.4 Microalgae

TABLE 12.3 Diverse types of microalgal feedstocks and their modification activities.

Feedstocks	Alteration method	Final product
<10% of total suspended solids in culture	AD	CH_4
	Direct combustion	H_2, alcohols OH, or alkanes
	Fermentation	Bioethanol
>15% total suspended solids slurry	Supercritical water gasification	Syngas
	Hydrothermal liquefaction	Hydrothermal liquefaction
Dried powder	Direct combustion	Heat
	Pyrolysis	Bio-oil

process of algae biomass, [29], biodiesel resulting from microalgal oil [30], and biohydrogen formed by a photobiological process [30]. The concept of generating microalgal biofuel is not a novel one [31], however taking into consideration the increasing cost of petroleum, it is now being taken seriously. The worry about global warming related to the use of fossil fuels is also generating serious interest [32].

The excellence or category of the end product relies on the accumulation of lipids or carbohydrates (Table 12.4). An integrated bioreactor system that lowers production costs and reduces land use with higher productivity will increase the economic value of biofuel. The use of a wastewater supply for the cultivation of microalgae can have a double advantage in commercial biofuel manufacture [48].

During the scalability process, the general cost of biofuel manufacture lies in the kinetic efficiency. The final cost of the generated fuel is determined by the comparative price of the closed and open cultivation systems. The open method could be simply scaled up compared to the closed system, but has the downside of reduced production of biomass. Therefore the choice of the required farming scheme and the cost of manufacture influences the excellence of the end product [48].

Each of the processes for the efficient production of biofuels should be applied to microalgal biomass. The steps include:

1. fermentation of carbohydrates into bioalcohols;
2. biomass thermochemical conversion or gasification;
3. triglyceride extraction accompanied by transesterification into biodiesel; and
4. manufacture of biogas through AD [49].

In order to increase the productivity of microalgae biofuels, numerous patents have been issued. The main emphasis of these patents is on the choice of the strains, procedures, and control [50].

Microalgae have also been researched for the production of bioethanol. The bioethanol yields from different strains of microalgae are shown in Table 12.5. High polysaccharide levels have been found to accumulate jointly as starch in the green *Chlorococum* spp. and in *Spirogyra* spp. [64]. Harun et al. confirmed that *Chlorococum* sp. is a green alga. Samples that are preextracted for lipids produce 60% more C_2H_5OH than these that remain as dried

Handbook of Algal Biofuels

258

12. Algal biomass for bioethanol and biobutanol production

TABLE 12.4 Total carbohydrates of some representative microalgal strains.

Microalgae species	Total carbohydrates (% per DW)	Reference
Arthrospira platensis	40.8	[33]
Dunaliella tertiolecta	50.6	[33]
Nannochloropsis sp.	56.8	[33]
Galdieria partita Sentz	50.1	[33]
Dunaliella salina Teod	69.7	[33]
Nostoc sp.	52.3	[33]
Cosmarium sp.	58.4	[33]
Chlorella vulgaris IAMC-534	37 (starch)	[34]
C. vulgaris CCAP 211/11B	55	[35]
C. vulgaris P12	41(starch)	[36]
C. vulgaris FSP-E	55 (starch)	[37]
Chlamydomonas reinhardtii UTEX 90	60	[38]
C. reinhardtii IAM C-238	55 (starch)	[39]
Chlorococum humicola	32.5	[40]
Chlorococcum sp. TISTR 8583	26 (starch)	[41]
S. acutiformis TISTR 8495	16.4 (starch)	[41]
S. obliquus CNW-N	51.8	[42]
Tetraselmis sp. CS-362	26	[43]
Chlorophyta strain SP2−3	70.35 ± 8.17	[44]
Spirogyra sp.	33−64	[45]
Chlorella sp. strain MI	57.84 ± 16.62	[44]
Porphyridium cruentum	40−57	[45]
Porphyridium cruentum	40−57	[46]
Dunaliella salina	55.5	[47]
Desmodesmus sp. strain FG	53.47 ± 14.18	[44]
Scenedesmus sp. strain SP2−9	52.93 ± 4.14	[44]
Chlorophyta strain C1C	52.85 ± 6.04	[44]
Chlorophyta strain C1	51.45 ± 4.45	[44]
Ankistrodesmus sp. strain LP1	51.25 ± 8.99	[44]
Chlorella vulgaris *(dry-ash-free)	51	[8]
Chlorella sorokiniana strain RP	49.8 ± 2.40	[44]

(*Continued*)

Handbook of Algal Biofuels

12.5 Biofuel production processes

TABLE 12.4 (Continued)

Microalgae species	Total carbohydrates (% per DW)	Reference
Ankistrodesmus sp. strain SP2−15	47.25 ± 9.54	[44]
Scenedesmus sp. strain SP1−20	46.11 ± 6.40	[44]
Scenedesmus sp. strain PL	45.66 ± 10.08	[44]
Chlorella sp. strain SP2−1	44.37 ± 2.23	[44]
Scenedesmus sp. strain RD	41.87 ± 13.15	[44]
Dunaliella bardawil	40.4	[47]

unharmed cells. This means that microalgae can be used as a method to increase their total cost-effective value for the making of C_2H_5OH and lipid biofuels from the same biomass. Bioethanol is likely to be a substitute fuel, but it is extremely necessary to guarantee that the development of this fuel is not impeded by the limitations of raw materials [64]. In this sense, the microalgae cell collecting period is shorter (1−10 days) than the other feedstock (1 or 2 times per year, harvesting time) and could therefore deliver ample supplies to meet the C_2H_5OH production requirements [65]. In addition, algae are capable of converting photons and can synthesize and absorb vast quantities of carbohydrate biomass from the cheapest source of raw materials for the production of bioethanol [66]. The *Chlamydomonas perigranulata* fermentation process to yield C_2H_5OH, C_4H_9OH, CH_5COOH, and CO_2 has been identified [67], demonstrating algal biomass' multiutility. Remarkably, Harun et al. [64] noted that more than 60% more C_2H_5OH was delivered by the lipid-extracted microalgae compared to the dried/intact microalgae, indicating the importance of spent biomass for C_2H_5OH manufacture.

12.5 Biofuel production processes

Biofuel production from microalgae begins with the large-scale production of biomass and the efficient extraction of an intermediate compound (triglycerides) to be processed into useful byproducts. The various steps of the processing of microalgae for biofuel development are discussed concisely below.

12.5.1 Biomass production

Artificially cultivating microalgae under optimum growing conditions will improve biomass revenue to the full level compared to ordinary circumstances. Using the ordinary resource for microalgal development, however, may decrease the charge of biofuel handling and marketable manufacture [68]. Using sunlight, the growth control influence and the light source may be made usable, but this can impact the production of biomass at a continuous pace. The main feature of phototropic microalgae is the assimilation of carbon dioxide. Lower CO_2 concentrations can be used for (about 150,000 ppm) [69]. In the

Handbook of Algal Biofuels

TABLE 12.5 Bioethanol production from different microalgal feedstocks showing the theoretical yield.

Strain	Hydrolysis treatment	Biomass load (g/L)	Sugar concentration (g/L)	Fermenting microorganism	C_2H_5OH (g/L)	C_2H_5OH (g/g biomass)	% of theoretical yield	Reference
Chlamydomonas reinhardtii	H_2SO_4	50	28.5	*Saccharomyces cerevisiae*	14.6	0.290	100	[51]
Spirogyra sp.	Enzymatic	50	12.5	*S. cerevisiae*	NA	0.080	78.4	[52]
Chlorococcum infusionum	NaOH (SHF) a	50	NA	*S. cerevisiae*	NA	0.260	NA	[53]
Chlorella vulgaris	H_2SO_4/ Enzymatic	5	5.5	*Escherichia coli*	1.7	0.400	61.0	[54]
Scenedesmus obliquus	H_2SO_4	500	63.2	*Kluyveromyces marxianus*	11.7	0.023	36.3	[55]
C. vulgaris	Enzymatic (SHF)a	20	7.8	*Zymomonas mobilis*	3.6	0.180	87.6	[37]
C. vulgaris	Enzymatic (SSF) a	20	NA	*Z. mobilis*	4.3	0.210	87.1	[37]
C. vulgaris	H_2SO_4 (SHF) a	50	23.6	*Z. mobilis*	11.7	0.230	96.7	[37]
S. obliquus CNW-N	H_2SO_4	40	16.0	*Z. mobilis*	8.6	0.210	99.8	[37]
S. abundans PKUAC12	H_2SO_4/Enzymatic	50	10.8	*S. cerevisiae*	4.7	0.100	85.5	[56]
Mychonastes afer PKUAC 9	H_2SO_4/Enzymatic	50	6.0	*S. cerevisiae*	2.8	0.060	92	[56]
C. reinhardtii	H_2SO_4	NA	NA	*S. cerevisiae*	8.7	0.150	86.0	[57]
C. vulgaris	Enzymatic	10	1.2	*S. cerevisiae*	0.6	0.070	89.0	[58]
Scenedesmus bijugatus b	H_2SO_4 (SSF)a	20	5.2	*S. cerevisiae*	NA	0.158	72.5	[59]
Chlamydomonas mexicana	Enzymatic (SHF)a	38	22.5	*S. cerevisiae*	8.5	0.410	72.0	[60]
C. mexicana	Enzymatic (SSF)a	38.1	22.5	*S. cerevisiae*	10.5	0.500	88.2	[60]
C. vulgaris	Hydrochloric acid (HCl)	100	10.15	*Brettanomyces custersii* H1−603	3.75	0.04	72.4	[61]

Senedesmus dimorphus b	Enzymatic (SSF)a	25	13.50	*S. cerevisiae*	7.34	0.26	95.59	[62]
Desmodesmus sp.	H_2SO_4 (SHF)a	100	55.3	*S. cerevisiae*	23.0	0.23	81.4	[44]
Chlorophyta sp.	H_2SO_4 (SHF)a	100	72.9	*S. cerevisiae*	23.6	0.24	63.7	[44]
Chlorophyta sp. C	H_2SO_4 (SHF)a	100	137.2	*S. cerevisiae*	61.2	0.31	87.4	[44]
Scenedesmus dimorphus	H_2SO_4 (SHF)a	18	7.7	*S. cerevisiae*	3.6	0.178	80.3	[63]
S. dimorphus	Enzymatic (SHF)a	18	7.6	*S. cerevisiae*	4.3	0.183	84.3	[63]
S. dimorphus	Untreated biomass (SSF)a	18	8.0	*S. cerevisiae*	4.5	0.181	81.2	[63]
S. dimorphus b	Organosolv-treated biomass (SSF)a	18	10.2	*S. cerevisiae*	6.8	0.266	91.3	[63]

262 12. Algal biomass for bioethanol and biobutanol production

processing unit, soluble carbonates such as $NaHCO_3$ and Na_2CO_3 and any external source of carbon dioxide from power plants are used [70]. Other inorganic complexes like Si, P, and N are also required [71]. Based on its nutrient requirements, there are three major biomass production mechanisms, for example:

- Photoautotrophic processing
- Heterotrophic production
- Mixotrophic processing
- Photoautotrophic biomass manufacture

The photosynthetic manner of development that uses inorganic salts and mixtures in the existence of sun or light source is photoautotrophic biomass production. This production mode is the cost-effectively feasible procedure that can be approved out as follows by two traditional methods.

12.5.1.1 The open pond production unit

One of the advantages of this system is the inexpensive and more convenient way of growing a particular form and type of plant. The open pond production unit (OPPU) may be a lagoon, a lake, or a natural artificial water system [72]. The shallow broad, closed, raceway and circular ponds tank are diverse categories of open pond systems (OPS) commonly operated in the development of great scale microalgae [73]. The raceway pond is the highly favored artificial pond system [74]. Biomass revenue production relies significantly on the position of the pond system [75]. In the OPS, the growth limiting factors include pH, DO, culture toxicity, temperature of O_2 accumulation, and light source strength. The other big problem in the mass processing of microalgae is the pollution in the open pond environment. Some microalgae, such as Chlorella in a great nutritional state, *Spirulina* in a higher basic state, *Dunaliella* in a higher saline situation, can only be grown in a particular condition to avoid the cross pollution from the open surroundings [76].

12.5.1.2 Closed photobioreactor production unit

The closed photobioreactor production unit (CPPU) will overwhelmed challenges in the OPPU, such as open environment cross pollution and maintaining a particular condition for microalgae growth with a systemized setup. Under defined conditions, a single algal species may be sustained at a continuous rate for an extended period. However, in contrast to the open method, the charge of manufacture is high [77]. The column photobioreactor, flat-plate photobioreactor, and tubular photobioreactor can be closed photobioreactors in industries. A photobioreactor in the gene consists of a series of similar tubes manufacture of glass or plastic [78]. These tubes of 0.1 m are arranged upright or horizontal to absorb the full sunlight [77]. For optimum gas exchange and efficiency, they are also given continuous mixing and agitation. The culture of algal is again distributed in liquid media for exchange of O_2 and CO_2 [79]. The pilot-scale development, thanks to its controllability, is strongly favored by these CPPU. This approach has a detrimental impact on the tube dimensional length restriction which will contribute to CO_2 loss, pH variation and O_2 build-up.

Handbook of Algal Biofuels

12.5.1.3 Hybrid two-stage production unit

The hybrid two-stage production unit combines OPPU and CPPU systems (more than 2 configurations) to minimize the drawbacks of both systems, such as decreased energy usage and improved efficiency and assignment [80]. The photobioreactor is used in the initial stage for the continuous growth of the managed atmosphere which will overcome open emission. The latter process uses the open approach to maximize lipid aggregation by putting cells under nutritional tension [81].

12.5.1.4 Heterotrophic biomass production

For its strength and nutrients, heterotrophic processing uses organic compounds such as glucose. This manner of manufacture of biomass is self-regulating of the resource of light. Therefore, for growth and mass production, greater surface area is not generally needed [82]. In the fermenter, the heterotrophic microalgae are normally cultured [79] However, due to the creation of organic carbon (OC) through photosynthesis, energy consumption is high. [77].

12.5.1.5 Mixotrophic biomass production

The microalgae both have characteristically traits in the processing of autotrophic and heterotrophic biomass while generating mixotrophic biomass. In mixotrophic biomass production, neither light intensity nor OC source is an important factor for growth. A major example of a mixotrophic biomass growth mode is the cyanobacterium (*Chlamydomonas reinhardtii*) and green algae (*Spirulina platensis*) [83]. These species use light for organic carbon for aerobic breathing and photosynthesis, thus minimizing the loss of biomass during the dim stage, as glucose use during the diurnal phases control development [83]. In contrast to photoautotrophic, the growth yield of the mixotrophic is higher but would be less than heterotrophic development of biomass. *Spirulina*, for example, exhibits greater mixotrophic development in the dark breathing period than others due to lower photo inhibition and a lack of biomass [84].

12.5.2 Biomass recovery/harvesting microalgal biomass

The recovery of biomass takes up approximately 20%−30% of the overall cost of production. A critical step in biofuel processing is the solid−liquid separation of microalgal biomass [85]. The cost-effective significance of the generated biomass is then calculated. The method of harvesting relies on the strain and size chosen for biomass and cell density production. Dissimilar methods require solid−liquid isolation:

12.5.2.1 Bulk harvesting

A concentration of 100−800 tons, the biomass isolated from the bulk medium provides a solid biomass of around 2%−7%. It depends on the process of harvesting and the original concentration of biomass. This combines the microalgal cells and thereby improves the procedures.

12.5.2.2 Filtration

Filtration is the process of physical recapture of microalgal cells which involves filtration of tangential flow, filtration of dead ends, microfiltration, and filtration of pressure, ultrafiltration, and vacuum filtration. For gathering bigger microalgae of $>70\,\mu m$, such as *Coelastrum* and *Spirulina*, this method is applicable. It is also operated for the retrieval of lesser $<30\,\mu m$ microalgae under applied pressure by the membrane filtration process. The commonly used membrane filtration methods, for instance, are ultrafiltration and microfiltration. The traditional filtration method for larger cell separation utilizes some diatomaceous filter aid or additional constituents such as cellulose to boost the performance of biomass repossession [86].

12.5.2.3 Flocculation

Flocculation is the practice that can improve the efficiency of biomass recovery, as exercised before any other harvesting techniques [87]. Moreover, Flocculation is the practice by which the adding of any cationic fragments for example multivalent chitosan, cations, and other polymers causes aggregation of the algal cells. Organic flocculants can avoid pH changes and have the benefit of using low flocculant concentrations to avoid environmental threats [88]. By adding cations, the negative charge of the algal cells is neutralized or diminished, thereby causing biomass aggregation. Through the bridging phase, the cells can also aggregate. Metal salts for example $Al_2(SO_4)_3$, $FeCl_3$ and $Fe_2(SO_4)_3$ are several examples of widely used flocculants [89].

12.5.2.4 Flotation

Flotation is the physical method that, naturally due to the aggregation of lipid content in cells, creates the algal cells drift above the liquid medium. The dispersion of the microsized air bubble on the liquefied medium which allows algal cells to be caught to the surface can also acquire flotation. This strategy takes advantage of the lack of chemical consumption. The technological viability, however, is fewer discussed [90].

12.5.3 The dewatering process and biomass extraction

The process of dehydration assists in enhancing the viability of the desired final product. Drum dehydrating, sun dehydrating, fluidized bed dehydrating, spray dehydrating and freeze-drying are different forms of dehydration process [91,92]. The time overwhelming, but ancient method used for the dehydration method is sun drying, while spray and freeze-drying are expensive but commonly used methods for drying high-value, large-scale items. In order to avoid differences in the lipid production and composition, the optimum drying temperature must be preserved. The higher the temperature, the greater the lipid composition and harvest changes would be. Conversely, when the biomass dried at 60°C, the lipid TAG content remains unchanged but changes the lipid yield.

12.5.4 Bioalcohols conversion

Theoretically practicable approaches involving chemical, biochemical and thermochemical alteration and direct combustion procedure include the conversion of biomass to

12.6 Bioethanol from microalgae

Currently, bioethanol is considered an alternative fossil fuel source. C_2H_5OH is considered in other developing countries as a replacement for petrol [93] because of its similar characteristics to gasoline. Currently, corn and sugarcane are the feedstock sources of C_2H_5OH. There is an issue with these feedstocks that are linked to the first-generation biofuel, that is, disagreements over food versus fuel. Therefore these feedstocks compete with the food chain and land use, restricting production in this way [94]. Basically, C_2H_5OH is made from sugar fermentation unconstrained from various biomass forms, for instance farming and organic waste crops [95].

Microalgae, because of their high carbohydrate content, may be used as an alternate feedstock to bioethanol. Cultivation, biomass processing, hydrolysis, fermentation, and C_2H_5OH recovery are the steps involved in the production of C_2H_5OH via microalgae (Table 12.5). The yeast cells are added to the sugar syrup for fermentation when the cells are broken down into sugar units after hydrolysis, and C_2H_5OH is then purified by distillation [95]. For bioethanol production, the microalgae *Chlorella vulgaris* has been tested and the findings have shown that the theoretical yield is 87.6% with a concentration of 11.7 g/L [37]. As a fuel, microalgal C_2H_5OH can be used; the released CO_2 can be reused for algal cultivation for the production of biomass that can be used in AD [37], thus reducing the impact of greenhouse gases. Nevertheless, the commercial-scale C_2H_5OH manufacturing process from algal biomass is still under development worldwide and further research is needed in this regard.

12.7 Biobutanol from microalgae

Carbohydrate residues of biodiesel-derived microalgae shall be taken into account. Study into carbohydrate residue of biodiesel-derived microalgae such as C_2H_5OH, C_4H_9OH, as well as biogas, has been undertaken [96–100]. In 1861 Louis Pasteur carried out the first fermentation process to form C_4H_9OH. Since the fermentation, acetone (C_3H_6O), C_2H_5OH, and C_4H_9OH have been modified to make acetone, butanol and ethanol (ABE) fermentation [101]. The processing of these solvents is therefore biologically known as ABE fermentation. Biobutanol is biomass-derived C_4H_9OH; it can be used as a combustion fuel in an internal gasoline motor. Since this hydrocarbon is fully nonpolar, it is identical to petrol, which is ideal for operation in petrol vehicles without requiring upgrades or mechanism modifications [102]. The vaporization heat for C_4H_9OH is marginally higher than that for gasoline [102]. Therefore C_4H_9OH vaporization rate is comparable to gasoline, so the C_4H_9OH blended gasoline did not demonstrate a cold start problem and can be used instead of gasoline as a 100% biobutanol fuel [103]. Biobutanol has the properties of high miscibility, low volatility, high energy content from 33.07 to 36.1 MJ/kg [104] and a density of 810 kg/m^3 [104].

Studies on the chemical, heat, and biological pretreatment of feedstock for the production of biobutanol have been conducted with microalgal biomass (Table 12.6). The thermal decomposition of microalgae at 108°C for ABE fermentation has already been examined in

266

12. Algal biomass for bioethanol and biobutanol production

TABLE 12.6 Assessment of the feedstock pretreatment circumstances and biobutanol creation.

Fermentation bacteria	Feedstock	Pretreatment technique	C_4H_9OH yield (g/g)	Reference
Clostridium beijerinckii NCIMB 8052	*Ulva lactuca*	H_2SO_4 and enzymes	0.35 acetone−butanol−ethanol (ABE) g/g sugar	[105]
Clostridium acetobutylicum ATCC 824	*Ulva lactuca*	H_2SO_4 and enzymes	0.08 ABE g/g sugar	[105]
C. acetobutylicum ATCC 824	*Ulva lactuca*	NaOH and enzymes	NA	[105]
Clostridium saccharoperbutylacetonicum N1−4	Wastewater algae	H_2SO_4	0.201	[106]
Clostridium saccharoperbutylacetonicum N1−4	Wastewater algae	H_2SO_4 and enzymes	0.249	[106]
C. saccharoperbutylacetonicum ATCC 27021	Wastewater algae	Thermal pretreatment with steam (121°C, 15 psig, 1 h)	0.13 g/g algae	[107]
C. saccharoperbutylacetonicum N1−4	Wastewater algae	H_2SO_4	NA	[107]
C. beijerinckii ATCC55025	*Ulva lactuca*	H_2SO_4	0.29 g/g sugar	[108]
C. saccharoperbutylicum ATCC27021	*Ulva lactuca*	H_2SO_4	NA	[108]
C. acetobutylicum B1787	*Arthrospira platensi a*	H_2SO_4	0.16	[33]
C. acetobutylicum B1787	*Arthrospira platensi b*	H_2SO_4	0.29	[33]
C. acetobutylicum B1787	*Nannochloropsis* sp.	H_2SO_4	0.18	[33]
C. acetobutylicum B1787	*Nannochloropsis* sp. *c*	H_2SO_4	0.12	[33]
C. acetobutylicum ATCC 824	De-oiled biomass of C.	2% H_2SO_4 + 2% NaOH	0.09 g/g carbohydrate	[96]
C. acetobutylicum ATCC 824	*Chlorella vulgaris* JSC-6	1% NaOH then 3% H2SO4	(or 0.24 g/g sugar)	[97]
C beijerinckii BA101 1	Liquefied corn starch	N/A	0.41 g ABE/g sugar	[109]
Clostridium beijerinckii BA101	Starch—based packing	Blending, Autoclaving at 121°C for 15 min	0.37 g ABE/g sugar	[110]

Handbook of Algal Biofuels

several species of microalgae [111]. Mechanical and thermal pretreatments typically result in cell disorder, but the yield of monosaccharides is not high. Chemical pretreatment may break cell walls into fermentable sugars and hydrolyze the microalgal carbohydrates, but during chemical pretreatment some inhibiting compounds are produced that affect cell growth and the development of C_4H_9OH in fermentation such as furfural, furfural 5-hydroxymete, acetic acid, oxalic acid, and formic acid (Table 12.6).

12.8 Macroalgae

Macroalgae are the most important part of aquatic habitats and are used to protect marine resources by avoiding pollution and eutrophication [112]. Macroalgae are classed as lower plants since they do not have roots, branches, or leaves. Instead, they are composed of a thallus and occasionally a foot and a stem. Some animals enclose structures filled with gas to assist in flexibility. They may grow very rapidly up to tens of meters in length [113]. Macroalgae vary in different ways, such as anatomy, ecophysiology, and longevity. They are categorized into *Phaeophyta* (brown), *Rhodophyta* (red), and *Chlorophyta* (green) algae according to their pigmentation [114]. Macroalgae grow on rocky substrates in their natural environment and form steady, multiple layers of constant vegetation, absorbing nearly all the photons available. Around 200 macroalgae species, of which about 10 are intensively cultivated, are used worldwide, for example, *Phaeophyta, Laminaria japonica, Undaria pinnatifida, Rhodophyta japonica, Eucheuma, Kappaphycus, Gracilaria, Porphyra, Chlorophyta, Monostromaa,* and *Enteromorpha* [113]. Macroalgae typically grow rapidly and can reach heights of up to 60 m in length [115]. Macroalgae growth rates greatly exceed those of land plants. For instance, the average productivity of brown algae biomass was almost $3.3-11.3$ kg $DW/m^2/year$ for uncultivated algae and up to 13.1 kg DW/m^2 over 7 months for cultivated algae, compared with $6.1-9.5$ kg fresh weight/$m^2/year$ for sugarcane, the most vigorous land plant (Table 12.7) [116]. They are available in natural water basins on a seasonal basis. A potential solution to the energy crisis is the production of macroalgae at sea, which doesn't require land or fertilizer. Macroalgae are primarily used to produce food and to extract hydrocolloids, and C_2H_5OH can be extracted from algae [117]. Macroalgal biomass comprises large amounts (at least 50%) of sugar that can be used in the manufacture of C_2H_5OH fuel [118−125].

Macroalgae such as *Ascophyllum nodosum, Enteromorpha intestinalis, Codium tomentosum, Fucucus Spiralis, Sargassum muticum, Saccorhiza polyschides, Ulva rigida,* and *Pelvetia canaliculate* have recently been reported [119] as options for biodiesel production, via triglycerides in a cross-esterification process [120,121]. The macroalgae content of water above the level of terrestrial (80%−85%) plants make them more suitable for microbial conversion than for straightforward combustion or thermochemical translation. Macroalgae, such as *Sargassum* spp., *Gelidium amansii, Gracilaria* spp., *Prymnesium parvum,* and *Laminaria* spp. are promising candidates for the manufacture of bioethanol [118,122]. The red algae, *Gelidium J.V. Lamouroux,* was used for the processing of waste products into bioethanol [123]. Also, the green algae *Ulva* spp., with a higher content of polysaccharides [124] have been used in the manufacture of C_2H_5OH and CH_4 [123]. Meinita [125] have lately reported bioethanol production from *Kappaphycus alvarezii.* Likewise [126], bioethanol manufacture from *Eucheuma* and *Hypnea* has also been reported. In mid-2008, research was started into the manufacture of bioethanol and

268

12. Algal biomass for bioethanol and biobutanol production

TABLE 12.7 The carbohydrate content of macroalgae.

Species	Phylum	Carbohydrates (%)
Caulerpa lentillifera	Green algae	38.7
Chaetomorpha linum	Green algae	54
Codium fragile	Green algae	58.7
Ulva fasciata	Green algae	43
Ulva lactuca	Green algae	54.3
Ulva pertusa	Green algae	52.3
Ulva rigida	Green algae	53
Enteromorpha	Green algae	64.9
Monostroma	Green algae	63.9
Laminaria japonica	Brown algae	51.3
Laminaria digitate	Brown algae	66.4
Alaria	Brown algae	39.8
Sargassum fulvellum	Brown algae	39.6
Sargassum polycystum	Brown algae	33.5
Sargassum vulgare	Brown algae	61.6
Saccharina latissimi	Brown algae	16.4
Padina	Brown algae	31.6
Cryptonemia crenulata	Red algae	47
Kappaphycus alvarezzi	Red algae	60.7
Eucheuma cottonii	Red algae	26.5
Gelidium amansi	Red algae	66
Hypnea charoides	Red algae	57.3
Hypnea musciformis	Red algae	39
Hydropuntia dentata	Red algae	39
Lomentaria hakodatensis	Red algae	40.4
Porphyra	Red algae	45.1
Rhodymenia	Red algae	44.6
Gracilaria	Red algae	61.75

biofertilizer from *K. alvarezii* in the lab using a combined two-product method [127]. Benjamin [128] in his American patent clarified the application of genetically modified green macroalgal (*Enteromorpha*) cells for bioethanol creation. Uchida and Murata [129] reported lactic acid and

Handbook of Algal Biofuels

C_2H_5OH fermentation by means of numerous kinds of green, red, and brown algae. Furthermore, Adams et al. [122] described the result of enzymatic pretreatment for their bioethanol manufacture in a brown alga, *Saccharina latissima* (*Laminaria saccharina*).

In fact, there is only limited knowledge available on the efficacy of these macroalgal carbohydrate processes [121], although some new breakthroughs in the development of C_2H_5OH from brown macroalgae have been produced [122]. Red algae generate high quantities of carbohydrates for the production of bioethanol. Despite the absence of lignin cross-linking in macroalgae, they develop in the water environment where they can grow erect [130]. Although macroalgae contain a low amount of lignin, they contain significant amounts of sugars that could be used for the production of bioethanol in the fermentation phase [118]. Nevertheless, the carbohydrate content is affected by the inclusion of agar, a polymer of galactose and galactopyranose, in some algae, such as marine red algae. Modern study has developed saccharification methods to discharge glucose from cellulose and galactose from agar through fermentation to generate C_2H_5OH [118].

Macroalgae is now known as a biomass that can be used in third-generation bioenergy production [130]. This biomass, with many carbohydrate-rich organisms and known to produce low or no lignin at all, is ideal for generating bioethanol [131]. It is also assumed that the production of this crop is sustainable. Other benefits of the use of nonfood macroalgae for the production of bioethanol include decreased competition for agricultural food and feed crops, high yields per region, and non-dependence on agricultural fertilizers, pesticides, agricultural land, or freshwater [121,122]. In addition, significant quantities of CO_2 are consumed by macroalgal development and preexisting markets for bioethanol macroalgae [132−134].

12.9 Aquaculture seaweed cultivation

Seaweed can be grown either in the ocean or in ponds or tanks. Depending on the species being cultivated, a broad range of techniques are used to cultivate seaweed. Several types of large-scale cultivation systems were planned and tested for open sea applications in the American Biomass Program [135,136]. In general, fragments of adult plants, juvenile plants, sporelings, or spores are seeded on ropes or other substrates in nurseries and the plants are grown to maturity in the sea. Some of the cultivation systems widely used are as follows.

12.9.1 Land-based cultivation systems

12.9.1.1 Tanks

Seaweed is grown in tanks that receive a constant aerated stream of seawater. The aeration provides the tanks with vigorous water movement, sending the algal thalli up and down in the water column in a circular pattern. This vigorous aeration allows for dissolved carbon dioxide to be quickly consumed, and the biomass is carried to the surface. The use of tanks will provide the highest efficiency per unit area per day and is more effective than any other form of agriculture. The performance of these systems depends to a great extent on the input of various energy types. By pumping more seawater, the temperature and salinity can also be manipulated. The pH of the tanks should be controlled in the range of 7.9—8.3, and the nutrient levels of the medium should be observed.

12.9.1.2 *Ponds*

Macroalgae can be grown in ponds as well. The low-cost operating system is a benefit of this strategy. In general, water exchange in the ponds is achieved with the use of tide gates. Owing to the absence of water movement in the ponds, yields are low. Tank systems are usually in smaller modules that allow a predictable maintenance schedule, several steps however, can be precisely controlled and managed to reduce the labor input, although this type of system has high operational costs. Ponds are larger, and a significant outbreak of epiphytes or other weedy species may be much harder to bring under control.

12.9.1.3 *Seaweed cultivation in the sea*

Aquaculturists produce most macroalgae in the sea. A number of farm structures are used to keep the marine algae in place. These include long lines to which seaweed strings, nets stretched out on frames, and ropes supported by poles are attached.

12.9.1.4 *Species-specific cultivation methods employed*

The methods operated to grow seaweed depend primarily on the species being cultivated, and biogeographical considerations. Generally, adult plant fragments, juvenile plants, or spore lings are planted on ropes in nurseries and the plants are then raised in the sea to maturity. The diverse life cycles of different kinds of seaweed mean that the thorough understanding of the species under cultivation is essential.

12.10 Bioethanol from macroalgae

It was predicted that cellulosic C_2H_5OH would play a major role in achieving US objectives in the 2007 Energy Independence and Protection Act for renewable biofuels [137—139]; but there is still no large-scale marketable manufacture of bioethanol from lignocellulosic constituents, despite comprehensive research and the availability of low-cost lignocellulosic biomass [138]. Some of the difficulties faced by straw bioethanol production is that lignin, present in many terrestrial sources of second-generation biofuel biomass, can inhibit the biodegradation of hemicellulose and cellulose by cellulase [140]. Macroalgae don't usually comprise large amounts of lignin and thus offer the possibility of more readily turning macroalgae components into sugars, but seaweed contains low amounts of glucose-containing polysaccharides. Therefore, C_2H_5OH would need to be generated from the additional constituents of seaweed carbohydrates, that is, mannitol, sulfated polysaccharides, alginate, carrageenan, and agar [131]. Conversely, some of the sugars delivered by the degradation of *Saccharomyces cerevisiae* seaweed polysaccharides such as rhamnose and xylose are poorly used by yeast. One of the main drawbacks of macroalgae as a bioethanol feedstock is the absence of tractable microorganisms that can effectively transform monosaccharides extracted from seaweed into C_2H_5OH [141]. Seaweed biomass hydrolysis, typically by the use of acids or enzymes, is necessary in order to release sugars into polysaccharides for fermentation [142]. The polysaccharides in

TABLE 12.8 Different species of different groups of macroalgae for bioethanol production [139].

Feedstock	Biomass potential	Reference
Gelidium sp.	C_2H_5OH	[143]
Kappaphycus alvarezii	C_2H_5OH	[144]
Ulva sp.	C_2H_5OH	[145]

brown, green, and red algae can be hydrolyzed by acid hydrolysis [5]. Saccharification and the application of fermentation inhibitor removal materials would add significant costs to the production of bioethanol from seaweed compared to more willingly fermented substrates, for example, sugars derived from sugarcane. Blue, red, and green algae were all fermented into C_2H_5OH in Table 12.8, but brown algae are advised to be the key feedstock for the development of bioethanol because they have great carbohydrate content and can be easily grown [142,146−148]. Consequently, the balance of C_2H_5OH produced from seaweed, like that from corn, can also be negligible. It has been estimated that the energy return on energy investment (EROI) of seaweed C_2H_5OH is comparable to corn C_2H_5OH at 1.788 [149]. Both Denmark [150] and Japan [148] have proposed large seaweed C_2H_5OH production facilities, but the economic and energy viability of these outlines is uncertain, and there appears to be no large-scale production of macroalgae C_2H_5OH to date [150]. In a brown seaweed report, Horn [123] stated that a profitable industrial process would require higher C_2H_5OH yields and the "total" use of all components of the seaweed biomass, including the processing of nonfuel items of higher value. With 86% of cars sold in Brazil in 2008 able to consume C_2H_5OH or a combination of C_2H_5OH and fossil fuel petroleum, bioethanol can be readily used in existing supply chains.

12.11 Biobutanol from macroalgae

While the cultivation of seaweed for biogas and bioethanol is being discovered in South America, Asia, and Europe, macroalgae biobutanol is attracting research interest and investment in the United States [5]. C_4H_9OH has been studied for about 100 years as a transport fuel and has been proposed as a biofuel with the potential not only to increase, but also to replace C_2H_5OH as a gasoline additive because of its low vapor pressure and higher energy density [108]. The production of C_4H_9OH from biomass may also be more energy efficient than C_2H_5OH, as certain bacteria used in the manufacture of C_4H_9OH digest cellulose as well as sugar and starch [150]. By anaerobic fermentation from a large diversity of sugars, both pentoses (C5) and hexoses (C6), a practice known as ABE fermentation [105], some *Clostridium* species are able to produce ABE, but C_4H_9OH has an inhibitory action to fermentation. C_4H_9OH with a yield of 0.16 g C_4H_9OH/g C5 and C6 sugars was produced by the fermentation of *Ulva lactuca* by the *Clostridium* strains, but the C_4H_9OH yield is less than that of C_2H_5OH produced under comparable conditions [151]. The *Clostridium* fermentation feedstock for the production of ABE with a yield of 0.35 g ABE/g sugar was used for the manufacture of *U. lactuca* solubilized by hot-water treatment followed by hydrolysis using commercial sugar cellulases [105].

While these findings demonstrate the ability of seaweed, *U. lactuca*, as an ABE fermentation substrate, ABE was proposed for the subsequent development of 1,2-propanediol (propylene glycol) in a seaweed biorefinery, instead of as a source of C_4H_9OH as a fuel, replacing the fossil fuel-derived product. In research on naturally occurring macroalgae (*Ulva*) from Jamaica Bay, New York City, C_4H_9OH was found to be produced from algal sugars on a pilot scale [108]. In a brown algae (*Saccharina*) sample, while C_4H_9OH fermentation of soluble organic matter extracted from acid was shown, yields were poor at 0.12 g/g of soluble solids extracted [146]. Mannitol and laminarin-derived glucose were found to be the main fermentation substrates, but abundant alginates were recalcitrant and did not undergo fermentation. Alginate is the largest organic fraction of brown algae such as *Laminaria hyperobea* [120]. It was concluded that there is still a need for major changes to make industrial-scale C_4H_9OH commercially viable from seaweed fermentation [146]. It has been found that the use of substrates will account for total direct cost of 60% in a typical industrial plant for ABE fermentation [152]. Thus cheaper and readily available brown macroalgae (*U. pinnatifida*, *Laminaria japonica*, *Ecklonia stolonifera*, *Hizikia fusiforme*, and *Sargassum fulvellum*) and red macroalgae (*Porphyra tenera* and *G. amansii*) were investigated for the *Clostridium saccharoperbutylacetonicum* N1–4 C_4H_9OH development. The research circumvented the determination of the development of C_4H_9OH by *C. saccharoperbutylacetonicum* N1–4. Instead, the reduction sugars consumed during fermentation were not calculated and the overall reduction sugar analysis was performed. Despite the absence of a specific sugar reduction review, the objective of determining the potential of macroalgae for the development of biobutanol was nevertheless developed and properly investigated. Moreover, to detoxify the pretreated hydrolysate of the best macroalgae substrate, the overliming and activated carbon process was also applied [139].

The results concluded that red algae fermentation (Table 12.9) did not surpass the brown macroalgae fermentation. Among the three brown macroalgae, the *Poryphyra* sp. fermentation was the largest, but it was about two times lower than the development of C_4H_9OH. The lowest C_4H_9OH produced from the red algae tested was from *G. amansii* fermentation, which could be accounted for by the production of higher inhibitor concentrations. Polyphenols, especially tannins, are the growth inhibitor compounds associated

TABLE 12.9 Brown and red alga fermentation for biobutanol production [139].

Species	Group	Reducing sugar, initial (g/L)	C_4H_9OH (g/L)	Solvent (g/L)[a]	C_4H_9OH yield (g/g)[b]
Undaria pinnatifida	Brown macroalgae	4.93 ± 0.15	1.39 ± 0.22	2.67	0.014
Laminaria japonica	Brown macroalgae	7.60 ± 0.55	5.51 ± 0.35	7.11	0.055
Ecklonia stolonifera	Brown macroalgae	12.36 ± 0.54	3.94 ± 0.45	5.48	0.040
Hizikia fusiforme	Brown macroalgae	12.61 ± 0.45	1.29 ± 0.25	2.61	0.013
Sargassum fulvellum	Brown macroalgae	8.86 ± 0.34	1.76 ± 0.4	3.14	0.018
Poryphyra sp.	Red macroalgae	6.48 ± 0.35	2.56 ± 0.32	2.87	0.025
Gelidium amansii	Red macroalgae	6.15 ± 0.55	0.62 ± 0.35	0.62	0.006

[a]*Sum of the final concentration of acetone, butanol, and ethanol concentration.*
[b]*Calculated from the final concentration of butanol divided by the initial concentration of macroalgae.*

with the composition of macroalgae [153]. Furthermore, excessive pretreatment might have decomposed the reducing sugars into nonutilizable and inhibitory compounds such as 5-hydroxymethylfurfural and other phenolic compounds [139].

12.12 Conclusion

Algae are desirable feedstock sources for the production of biofuels. Chlorophytes (including microalgae and macroalgae) are the main groups of algae that are used for energy production Carbohydrates are a major component of algae that can make a significant contribution to bioalcohols production (bioethanol and biobutanol). The cultivation method is also one of the important factors for the production of high-quality bioalcohols. In this chapter we illustrate various methods of planting, collecting, and handling. The key problems tend to be the selection of high carbohydrates with algal strains for high facilities, activities, and maintenance costs. In conclusion, algae are an excellent biomass source for the production of biofuels because of their potential for high-quality biofuel, excellent productivity, and a reduced environmental impact compared to other sources.

References

[1] E.A. Soliman, M.R. Elkatory, A.I. Hashem, H.S. Ibrahim, Synthesis and performance of maleic anhydride copolymers with alkyl linoleate or tetra-esters as pour point depressants for waxy crude oil, Fuel 211 (2018) 535–547.

[2] R.M. Ali, M. El Katory, M. Hassaan, K. Amer, A. El Geiheini, Highly crystalline heterogeneous catalyst synthesis from industrial waste for sustainable biodiesel production, Egypt. J. Chem. 63 (4) (2020) 15–16.

[3] M. Aresta, A. Dibenedetto, G. Barberio, Utilization of macro-algae for enhanced CO_2 fixation and biofuels production: development of a computing software for an LCA study, Fuel Process. Technol. 86 (14–15) (2005) 1679–1693.

[4] IEA Report, Carbon Dioxide Utilization: Evaluation of Specific Biological Processes Which Have the Capability of Directly Utilizing High Concentrations of Carbon Dioxide as Found in the Flue Gas Streams From Power Generation Plant, Chemical Society of Japan Publishers, 1994.

[5] N. Wei, J. Quarterman, Y.-S. Jin, Marine macroalgae: an untapped resource for producing fuels and chemicals, Trends Biotechnol. 31 (2) (2013) 70–77. Available from: https://doi.org/10.1016/j.tibtech.2012.10.009.

[6] K. Gao, K. McKinley, Use of macroalgae for marine biomass production and CO_2 remediation: a review, J. Appl. Phycol. 6 (1) (1994) 45–60. Available from: https://doi.org/10.1007/BF02185904.

[7] A. Banerjee, R. Sharma, Y. Chisti, U.C. Banerjee, *Botryococcus braunii*: a renewable source of hydrocarbons and other chemicals, Crit. Rev. Biotechnol. 2002 (22) (2002) 245–279.

[8] H. Xu, X.L. Miao, Q.Y. Wu, High quality biodiesel production from a microalga *Chlorella protothecoides* by heterotrophic growth in fermenters, J. Biotechnol. 126 (4) (2006) 499–507.

[9] M.V. Rodionova, R.S. Poudyal, I. Tiwari, R.A. Voloshin, S.K. Zharmukhamedov, H.G. Nam, et al., Biofuel production: challenges and opportunities, Int. J. Hydrogen Energy 42 (12) (2017) 8450–8461.

[10] K.R. Szulczyk, Which is a better transportation fuel-butanol or ethanol? Int. J. Energy Env. 1 (2010) 501–512.

[11] Edmunds, http://www.edmunds.com/fuel-economy/e85-versus-gasolinecomparison-test.html, 2007.

[12] J. Lü, C. Sheahan, P. Fu, Metabolic engineering of algae for fourth generation biofuels production, Energy Environ. Sci. 4 (2011) 2451–2466.

[13] E. Rosenthal, UN report describes risks of inaction on climate change, N. Y. Times (2007) 17.

[14] T.R. Brown, R.C. Brown, A review of cellulosic biofuel commercial-scale projects in the United States, Biofuels Bioprod. Biorefin 7 (2013) 235–245.

[15] A. Raheem, P. Prinsen, A.K. Vuppaladadiyam, M. Zhao, R. Luque, A review on sustainable microalgae based biofuel and bioenergy production: recent developments, J. Clean. Prod. 181 (2018) 42–59.

[16] R.A. Voloshin, M.V. Rodionova, S.K. Zharmukhamedov, T. Nejat Veziroglu, S.I. Allakhverdiev, Review: biofuel production from plant and algal biomass, Int. J. Hydrogen Energy 1 (2016) 1–17.

[17] K. Ullah, et al., Assessing the potential of algal biomass opportunities for bioenergy industry: a review, Fuel 143 (2014) 414–423.

[18] S. Akhlaghi, et al., Deterioration of automotive rubbers in liquid biofuels: a review, Renew. Sustain. Energy Rev. 43 (2015) 1238–1248.

[19] S.P.S. Chundawat, B. Venkatesh, B.E. Dale, Effect of particle size based separation of milled corn stover on AFEX pretreatment and enzymatic digestibility, Biotechnol. Bioeng. 96 (2) (2007) 219–231.

[20] A. Jaffe, K. Medlock, R. Soligo, The Status of World Oil Reserves: Conventional and Unconventional Resources in the Future Supply Mix, James A. Baker III Institute for Public Policy, Rice University, 2011.

[21] Y. Deng, S.S. Fong, Metabolic engineering of Thermobifida fusca for direct aerobic bioconversion of untreated lignocellulosic biomass to 1-propanol, Metab. Eng. 13 (5) (2011) 570–577. < http://www.science-direct.com/science/article/pii/S1096717611000735> (accessed 17.06.15.).

[22] L. Pasteur, Animalcules infusoires vivant sans gaz oxygene libre et determinant des fermentations, Mallet-Bachelier 3 (5) (1861) 570–577.

[23] O.P. Ward, A. Singh, Bioethanol technology: development and perspectives, Adv. Appl. Microbiol. 51 (2002) 53–80.

[24] M.R. Ladisch, Fermentation-derived butanol and scenarios for its uses in energyrelated applications, Enzyme Microbiol. Technol. 13 (1991) 280–283.

[25] S.A. Khan, Rashmi, M.Z. Hussain, S. Prasad, U.C. Banerje, Prospects of biodiesel production from microalgae in India, Renew. Sustain. Energy Rev. 13 (2009) 2361–2372.

[26] Y.K. Lee, Commercial production of microalgae in the Asia-Pacific rim, J. Appl. Phycol. 9 (1997) 403–411.

[27] W. Xiong, X. Li, J. Xiang, Q. Wu, High-density fermentation of microalga Chlorella protothecoides in bioreactor for microbiodiesel production, Appl. Microbiol. Biotechnol. 78 (2008) 29–36.

[28] P.J.L.B. Williams, L.M.L. Laurens, Microalgae as biodiesel & biomass feedstocks: review & analysis of the biochemistry, energetics & economics, Energy Environ. Sci. 3 (2010) 554–590.

[29] M.A. Hassaan, A. Pantaleo, F. Santoro, M.R. Elkatory, G. De Mastro, A.E. Sikaily, et al., Techno-economic analysis of ZnO nanoparticles pretreatments for biogas production from barley straw, Energies 13 (19) (2020) 5001.

[30] M.A. Hassaan, A. Pantaleo, L. Tedone, M.R. Elkatory, R.M. Ali, A.E. Nemr, et al., Enhancement of biogas production via green ZnO nanoparticles: experimental results of selected herbaceous crops, Chem. Eng. Commun. (2019) 1–14.

[31] R. Amirante, G. Demastro, E. Distaso, M.A. Hassaan, A. Mormando, A.M. Pantaleo, et al., Effects of ultrasound and green synthesis ZnO nanoparticles on biogas production from Olive Pomace, Energy Procedia 148 (2018) 940–947.

[32] S. Sawayama, S. Inoue, Y. Dote, S.Y. Yokoyama, CO_2 fixation and oil production through microalga, Energy Convers. Manage 36 (1995) 729–731.

[33] E.N. Efremenko, A.B. Nikolskaya, I.V. Lyagin, O.V. Senko, T.A. Makhlis, N.A. Stepanov, et al., Production of biofuels from pretreated microalgae biomass by anaerobic fermentation with immobilized Clostridium acetobutylicum cells, Bioresour. Technol. 114 (2012) 342–348.

[34] A. Hirano, R. Ueda, S. Hirayama, Y. Ogushi, CO_2 fixation and ethanol production with microalgal photosynthesis and intracellular anaerobic fermentation, Energy. 242 (2–3) (1997) 137–142.

[35] A.M. Illman, A.H. Scragg, S.W. Shales, Increase in Chlorella strains calorific values when grown in low nitrogen medium, Enzyme Microb. Technol. 27 (8) (2000) 631–635.

[36] G. Dragone, B.D. Fernandes, A.P. Abreu, A.A. Vicente, J.A. Teixeira, Nutrient limitation as a strategy for increasing starch accumulation in microalgae, Appl. Energy 88 (10) (2011) 3331–3335.

[37] S.H. Ho, S.W. Huang, C.Y. Chen, T. Hasunuma, A. Kondo, J.S. Chang, Characterization and optimization of carbohydrate production from an indigenous microalga Chlorella vulgaris FSP-E, Bioresour. Technol. 135 (2013) 157–165.

[38] S.P. Choi, M.T. Nguyen, S.J. Sim, Enzymatic pretreatment of Chlamydomonas reinhardtii biomass for ethanol production, Bioresour. Technol. 101 (14) (2010) 5330–5336.

[39] M. Kim, J. Baek, Y. Yun, S. Junsim, S. Park, S. Kim, Hydrogen production from Chlamydomonas reinhardtii biomass using a two-step conversion process: anaerobic conversion and photosynthetic fermentation, Int. J. Hydrogen Energy 31 (6) (2006) 812–816.

References

[40] A. Ike, N. Toda, N. Tsuji, K. Hirata, K. Miyamoto, Hydrogen photoproduction from CO_2-fixing microalgal biomass: application of halotolerant photosynthetic bacteria, J. Ferment. Bioeng. 84 (6) (1997) 606–609.

[41] S. Rodjaroen, N. Juntawong, A. Mahakhant, K. Miyamoto, High biomass production and starch accumulation in native green algal strains and cyanobacterial strains of Thailand, Kasetsart J. (Nat. Sci.) 41 (2007) 570–575.

[42] S.H. Ho, C.Y. Chen, J.S. Chang, Effect of light intensity and nitrogen starvation on CO_2 fixation and lipid/carbohydrate production of an indigenous microalga Scenedesmus obliquus CNW-N, Bioresour. Technol. 113 (2012) 244–252.

[43] M.R. Brown, M.A. McCausland, K. Kowalski, The nutritional value of four Australian microalgal strains fed to Pacific oyster Crassostrea gigas spat, Aquaculture 165 (3–4) (1998) 281–293.

[44] L. Sánchez Rizza, et al., Bioprospecting for native microalgae as an alternative source of sugars for the production of bioethanol, Algal Res. 22 (2017) 140–147.

[45] A. Demirbas, Hydrogen from mosses and algae via pyrolysis and steam gasification, Energy Sources, A: Recover. Util. Environ. Eff. 32 (2010) 172–179.

[46] J. Trivedi, M. Aila, D.P. Bangwal, S. Kaul, M.O. Garg, Algae based biorefinery – how to make sense? Renew. Sustain. Energy Rev. 47 (2015) 295–307.

[47] G.W. Huber, S. Iborra, A. Corma, Synthesis of transportation fuels from biomass: chemistry, catalysts, and engineering, Chem. Rev. 106 (2006) 4044–4098.

[48] J.S. Chang, P.L. Show, T.C. Ling, C.Y. Chen, S.H. Ho, C.H. Tan, et al., Photobioreactors, in: C. Larroche, M. Sanroman, G. Du, A. Pandey (Eds.), Current Developments in Biotechnology and Bioengineering: Bioprocesses, Bioreactors and Controls, Elsevier, Atlanta, GA, 2017, pp. 313–352.

[49] R. Craggs, S. Heubeck, T. Lundquist, J. Benemann, Algal biofuels from wastewater treatment high rate algal ponds, Water Sci. Technol. 63 (4) (2011) 660–665.

[50] A. de la Jara, P. Assunção, E. Portillo, K. Freijanes, H. Mendoza, Evolution of microalgal biotechnology: a survey of the European Patent Office database, J. Appl. Phycol. 28 (5) (2016) 2727–2740.

[51] M.T. Nguyen, et al., Hydrothermal acid pretreatment of Chlamydomonas reinhardtii biomass for ethanol production, J. Microbiol. Biotechnol. 19 (2) (2009) 161–166.

[52] F.S. Eshaq, et al., Spirogyra biomass a renewable source for biofuel (bioethanol) production, Eng. Sci. Technol. 2 (12) (2010) 7045–7054.

[53] R. Harun, W.S.Y. Jason, T. Cherrington, M.K. Danquah, Exploring alkaline pre-treatment of microalgal biomass for bioethanol production, Appl. Energy 88 (10) (2011) 3464–3467.

[54] S. Lee, Y. Oh, D. Kim, D. Kwon, C. Lee, J. Lee, Converting carbohydrates extracted from marine algae into ethanol using various ethanolic Escherichia coli strains, Appl. Biochem. Biotechnol. 164 (6) (2011) 878–888.

[55] J.R. Miranda, P.C. Passarinho, L. Gouveia, Bioethanol production from Scenedesmus obliquus sugars: the influence of photobioreactors and culture conditions on biomass production, Appl. Microbiol. Biotechnol. 96 (2) (2012) 555–564.

[56] H. Guo, M. Daroch, L. Liu, G. Qiu, S. Geng, G. Wang, Biochemical features and bioethanol production of microalgae from coastal waters of Pearl River Delta, Bioresour. Technol. 127 (2013) 422–428.

[57] M.J. Scholz, M.R. Riley, J.L. Cuello, Acid hydrolysis and fermentation of microalgal starches to ethanol by the yeast Saccharomyces cerevisiae, Biomass Bioenergy 48 (2013) 59–65.

[58] K.H. Kim, I.S. Choi, H.M. Kim, S.G. Wi, H.J. Bae, Bioethanol production from the nutrient stress-induced microalga Chlorella vulgaris by enzymatic hydrolysis and immobilized yeast fermentation, Bioresour. Technol. 153 (2014) 47–54.

[59] V. Ashokkumar, Z. Salam, O.N. Tiwari, S. Chinnasamy, S. Mohammed, F.N. Ani, An integrated approach for biodiesel and bioethanol production from Scenedesmus bijugatus cultivated in a vertical tubular photobioreactor, Energy Convers. Manage. 101 (2015) 778–786.

[60] M.M. El-Dalatony, et al., Long-term production of bioethanol in repeatedbatch fermentation of microalgal biomass using immobilized Saccharomyces cerevisiae, Bioresour. Technol. 219 (2016) 98–105.

[61] C. Park, J.H. Lee, X. Yang, H.Y. Yoo, J.H. Lee, S.K. Lee, et al., Enhancement of hydrolysis of Chlorella vulgaris by hydrochloric acid, Bioprocess. Biosyst. Eng. 39 (6) (2016) 1015–1021.

[62] L.M. Chng, D.J. Chan, K.T. Lee, Sustainable production of bioethanol using lipid-extracted biomass from Scenedesmus dimorphus, J. Clean. Prod. 130 (2016) 68–73.

[63] L.M. Chng, K.T. Lee, D.J.C. Chan, Synergistic effect of pretreatment and fermentation process on carbohydrate-rich Scenedesmus dimorphus for bioethanol production, Energy Convers. Manage. 141 (2017) 410–419.

[64] F.S. Eshaq, M.N. Ali, M.K. Mohd, Production of bioethanol from next generation feed stock alga *Spirogyra* species, Int. J. Eng. Sci. Technol. 3 (2011) 1749–1755.
[65] P.M. Schenk, S.R. Thomas-Hall, E. Stephens, U.C. Marx, J.H. Mussgnug, C. Posten, et al., Second generation biofuels: high- efficiency microalgae for biodiesel production, Bioener. Res. 1 (2008) 20–43.
[66] B. Subhadra, M. Edwards, An integrated renewable energy park approach for algal biofuel production in United States, Energy Policy (2010). Available from: https://doi.org/10.1016/j.enpol.2010.04.036.
[67] K. Hon-Nami, A unique feature of hydrogen recovery in endogenous starch to alcohol fermentation of the marine microalga, *Chlamydomonas perigranulata*, Appl. Biochem. Biotechnol. 131 (2006) 808–828.
[68] K.W. Chew, J.Y. Yap, P.L. Show, N.H. Suan, J.C. Juan, T.C. Ling, et al., Microalgae biorefinery: high value products perspectives, Bioresour. Technol. 229 (2017) 53–62.
[69] S.Y. Chiu, C.Y. Kao, M.T. Tsai, S.C. Ong, C.H. Chen, C.S. Lin, Lipid accumulation and CO_2 utilization of Nanochloropsis oculata in response to CO_2 aeration, Bioresour. Technol. 100 (2) (2009) 833–838.
[70] J. Doucha, F. Straka, K. Lívanský, Utilization of flue gas for cultivation of microalgae (*Chlorella* sp.) in an outdoor open thin-layer photobioreactor, J. ApplPhycol 17 (5) (2005) 403–412.
[71] R. Halim, B. Gladman, M.K. Danquah, P.A. Webley, Oil extraction from microalgae for biodiesel production, Bioresour. Technol. 102 (1) (2011) 178–185.
[72] M. Nayak, A. Karemore, R. Sen, Sustainable valorization of flue gas CO_2 and wastewater for the production of microalgal biomass as a biofuel feedstock in closed and open reactor systems, RSC Adv. 6 (2016) 91111–91120.
[73] A.C. Apel, C.E. Pfaffingera, N. Basedahla, N. Mittwollena, J. Göbela, J. Sautera, et al., Open thin-layer cascade reactors for saline microalgae production evaluated in a physically simulated Mediterranean summer climate, Algal Res. 25 (2017) 381–390.
[74] G.A. De Andrade, M. Berenguel, J.L. Guzmán, D.J. Pagano, F.G. Acién, Optimization of biomass production in outdoor tubular photobioreactors, J. Process. Control. 37 (2016) 58–69.
[75] Y. Chisti, Large-scale production of algal biomass: raceway ponds, in: F. Bux, Y. Chisti (Eds.), Algae Biotechnology: Products and Processes, Springer, New York, 2016, pp. 21–40.
[76] D. Chiaramonti, M. Prussi, D. Casini, M.R. Tredici, L. Rodolfi, N. Bassi, et al., Review of energy balance in raceway ponds for microalgae cultivation: rethinking a traditional system is possible, Appl. Energy 102 (2013) 101–111.
[77] Y. Chisti, Biodiesel from microalgae, Biotechnol. Adv. 25 (3) (2007) 294–306.
[78] O. Jorquera, A. Kiperstok, E.A. Sales, M. Embiruçu, M.L. Ghirardi, Comparative energy life-cycle analyses of microalgal biomass production in open ponds and photobioreactors, Bioresour. Technol. 101 (2010) 1406–1413.
[79] N. Eriksen, Production of phycocyanin – a pigment with applications in biology, biotechnology, foods and medicine, Appl. Microbiol. Biotechnol. 80 (1) (2008) 1–14.
[80] E. Jacob-Lopes, L.Q. Zepka, L.G.R. Merida, M.M. Maroneze, C. Neves, Fotobiorreator híbrido industrial e bioprocesso realizado por meio do dito fotobiorreator, Patent PT2016041028, 2016.
[81] L. Rodolfi, G.C. Zittelli, N. Bassi, G. Padovani, N. Biondi, G. Bonini, et al., Microalgae for oil: strain selection, induction of lipid synthesis and outdoor mass cultivation in a low-cost photobioreactor, Biotechnol. Bioeng. 102 (1) (2008) 100–112.
[82] R.M. Ali, M.R. Elkatory, H.A. Hamad, Highly active and stable magnetically recyclable $CuFe_2O_4$ as a heterogenous catalyst for efficient conversion of waste frying oil to biodiesel, Fuel 268 (2020) 117297.
[83] M.R. Andrade, J.A.V. Costa, Mixotrophic cultivation of microalga Spirulina platensis using molasses as organic substrate, Aquaculture 264 (1–4) (2007) 130–134.
[84] C.Y. Chen, et al., Microalgae-based carbohydrates for biofuel production, Biochem. Eng. J. 78 (2013) 1–10.
[85] B. Wang, Y. Li, N. Wu, C. Lan, CO_2 bio-mitigation using microalgae, Appl. Microbiol. Biotechnol. 79 (5) (2008) 707–718.
[86] E. Molina-Grima, E.H. Belarbi, A.F.G. Fernandez, M.A. Robles, Y. Chisti, Recovery of microalgal biomass and metabolites: process options and economics, Biotechnol. Adv. 20 (2003) 491–515.
[87] T. Chatsungnoen, Y. Chisti, Harvesting microalgae by flocculation sedimentation, Algal Res. 13 (2016) 271–283.
[88] D. Vandamme, I. Foubert, K. Muylaert, Flocculation as a low-cost method for harvesting microalgae for bulk biomass production, Trends Biotechnol. 31 (4) (2013) 233–239.
[89] R. Divakaran, V.N.S. Pillai, Flocculation of algae using chitosan, J. Appl. Phycol. 14 (5) (2002) 419–422.

References

[90] T. Bruton, H. Lyons, Y. Lerat, M. Stanley, M.B. Rasmussen, A review of the potential of marine algae as a source of biofuel in Ireland, Sustain. Energy Ireland, Dublin (2009) 88.

[91] X. Wang, Y.H. Liu, D.X. Hu, S. Balamurugan, Y. Lu, W.D. Yang, et al., Identification of a putative patatin-like phospholipase domain-containing protein 3 (PNPLA3) ortholog involved in lipid metabolism in microalga Phaeodactylum tricornutum, Algal Res. 12 (2015) 274–279.

[92] C. Wan, M.A. Alam, X.Q. Zhao, X.Y. Zhang, S.L. Guo, S.H. Ho, et al., Current progress and future prospect of microalgal biomass harvest using various flocculation technologies, Bioresour. Technol. 184 (2015) 251–257.

[93] T. Willke, K.D. Vorlop, Industrial bioconversion of renewable resources as an alternative to conventional chemistry, Appl. Microbiol. Biotechnol. 66 (2) (2004) 131–142.

[94] Y. Sun, J. Cheng, Hydrolysis of lignocellulosic materials for ethanol production: a review, Bioresour. Technol. 83 (2002) 1–11.

[95] J. Xuan, M.K. Leung, D.Y. Leung, M. Ni, A review of biomass-derived fuel processors for fuel cell systems, Renew. Sustain. Energy Rev. 13 (6) (2009) 1301–1313.

[96] H.H. Cheng, L.M. Whang, K.C. Chan, M.C. Chung, S.H. Wu, C.P. Liu, et al., Biological butanol production from microalgae-based biodiesel residues by *Clostridium acetobutylicum*, Bioresour. Technol. 184 (2015) 379–385.

[97] Y. Wang, W. Guo, C.L. Cheng, S.H. Ho, J.S. Chang, N. Ren, Enhancing biobutanol production from biomass of Chlorella vulgaris JSC-6 with sequential alkali pretreatment and acid hydrolysis, Bioresour. Technol. 200 (2016) 557–564.

[98] M. Daroch, S. Geng, G. Wang, Recent advances in liquid biofuel production from algal feedstocks, Appl. Energy 102 (2013) 1371–1381.

[99] L. Brennan, P. Owende, Biofuels from microalgae—a review of technologies for production, processing, and extractions of biofuels and co-products, Renew. Sustain. Energy Rev. 14 (2010) 557–577.

[100] A.-M. Lakaniemi, C.J. Hulatt, D.N. Thomas, O.H. Tuovinen, J.A. Puhakka, Biogenic hydrogen and methane production from Chlorella vulgaris and Dunaliella tertiolecta biomass, Biotechnol. Biofuels 4 (1) (2011) 34.

[101] D.T. Jones, D. Woods, Acetone-butanol fermentation revisited, Microbiol. Rev. 50 (1986) 484–524.

[102] V. Hönig, M. Kotek, J. Mařík, Use of butanol as a fuel for internal combustion engines, Agron. Res. 12 (2014) 333–340.

[103] M. Pospíšil, J. Šiška, G. Šebor, BioButanol as Fuel in Transport, Biom [online], 2014. Available from: <http://www.biom.cz>.

[104] P.H. Pfromm, V. Amanor-Boadu, R. Nelson, et al., Bio-butanol versus bio-ethanol: a technical and economic assessment for corn and switchgrass fermented by yeast or *Clostridium acetobutylicum*, Biomass Bioenergy 34 (2010) 515–524.

[105] H. van der Wal, B.L. Sperber, B. Houweling-Tan, R.R. Bakker, W. Brandenburg, A.M. Lopez-Contreras, Production of acetone, butanol, and ethanol from biomass of the green seaweed *Ulva lactuca*, Bioresour. Technol. 128 (2013) 431–437.

[106] J.T. Ellis, N.N. Hengge, R.C. Sims, C.D. Miller, Acetone, butanol, and ethanol production from wastewater algae, Bioresour. Technol. 111 (2012) 491–495.

[107] A. Jernigan, M. May, T. Potts, B. Rodgers, J. Hestekin, P.I. May, et al., Effects of drying and storage on year-round production of butanol and biodiesel from algal carbohydrates and lipids using algae from water remediation, Environ. Prog. Sustain. Energy 32 (4) (2013) 1013–1022.

[108] T. Potts, J. Du, M. Paul, P. May, R. Beitle, J. Hestekin, The production of butanol from Jamaica bay macro algae, Environ. Prog. Sustain. Energy 31 (1) (2012) 29–36.

[109] T. Ezeji, N. Qureshi, H.P. Blaschek, Butanol production from agricultural residues: impact of degradation products on *Clostridium beijerinckii* growth and butanol fermentation, Biotechnol. Bioeng. 97 (6) (2007) 1460–1469.

[110] T.W. Jesse, T.C. Ezeji, N. Qureshi, H.P. Blaschek, Production of butanol from starchbased waste packing peanuts and agricultural waste, J. Indust. Microbiol. Biotechnol. 29 (2002) 117–123.

[111] A. Mahdy, L. Mendez, M. Ballesteros, C. González-Fernández, Enhanced methane production of Chlorella vulgaris and Chlamydomonas reinhardtii by hydrolytic enzymes addition, Energy Convers. Manage. 85 (2014) 551–557.

[112] M. Notoya, Production of biofuel by macroalgae with preservation of marine resources and environment, Seaweeds and Their Role in Globally Changing Environments, Springer, Dordrecht, 2010, pp. 217–228.

[113] K. Luning, S.J. Pang, Mass cultivation of seaweeds: current aspects and approaches, J. Appl. Phycol. 15 (2003) 115—119.

[114] C.X. Chan, C.L. Ho, S.M. Phang, Trends in seaweed research, Trends Plant. Sci. 11 (2006) 165—166.

[115] D.J. McHugh, A guide to the seaweed industry, FAO Fisheries Technical Paper No 441, 105, 2003.

[116] S. Kraan, Mass cultivation of carbohydrate rich macroalgae, a possible solution for sustainable biofuel production, Mitig. Adapt. Strateg. Glob Change (2010).

[117] C.S. Goh, K.T. Lee, A visionary and conceptual macroalgae-based third generation bioethanol (TGB) biorefinery in Sabah, Malaysia as an underlay for renewable and sustainable development, Renew. Sustain. Energy Rev. 14 (2010) 842—848.

[118] S.G. Wi, H.J. Kim, S.A. Mahadevan, D. Yang, H. Bae, The potential value of the seaweed Ceylon moss (*Gelidium amansii*) as an alternative bioenergy resource, Bioresour. Technol. 100 (2009) 6658—6660.

[119] R. Maceiras, M. Rodriguez, A. Cancela, S. Urrejola, A. Sanchez, Macroalgae: raw material for biodiesel production, Appl. Energy 88 (2011) 3318—3323.

[120] S.J. Horn, I.M. Aasen, K. Ostgaard, Ethanol production from seaweed extract, J. Ind. Microbiol. Biotechnol. 25 (2000) 249—254.

[121] A. Ross, J.M. Jones, M.L. Kubacki, T.G. Bridgeman, Classification of macroalgae as fuel and its thermochemical behavior, Bioresour. Technol. 99 (2008) 6494—6504.

[122] J.M. Adams, J.A. Gallagher, I.S. Donnison, Fermentation study on *Saccharina latissima* for bioethanol production considering variable pre treatments, J. Appl. Phycol. 21 (5) (2009) 569—574.

[123] Y.B. Seo, Y.W. Lee, C.H. Lee, H.C. You, Red algae and their use in paper making, Bioresour. Technol. 101 (2010) 2549—2553.

[124] M. Lahaye, B. Ray, Cell wall polysaccharides from the marine green alga *Ulva rigida* (Ulvales, Chlorophyta)-NMR analysis of ulvan oligosaccharides, Carbohydr. Res. 283 (1996) 161—173.

[125] M.D.N. Meinita, Y. Hong, G. Jeong, Comparison of sulfuric acid and hydrochloric acids as catalysts in hydrolysis of *Kappaphycus alvarezii* (cottonii), Bioprocess. Biosyst. Eng. 35 (1—2) (2011) 123—128.

[126] S. Karunakaran, R. Gurusamy, Bioethanol production as renewable biofuel from rhodophytes feedstock, Int. J. Biol. Biotechnol. 2 (2) (2011) 94—99.

[127] K.H. Mody, P.K. Ghosh, B. Sana, G. Gopalasamy, A.D. Shukla, K. Eswaran, et al., A process for integrated production of ethanol and seaweed sap from *Kappaphycus alvarezii*, Patent Filed Indian Application No. 1839/DEL/2009 dated 07/09/09; WO 2011/027360A1 dated 10.03.11, 2009.

[128] M. Benjamin, Methods and compositions for producing metabolic products for algae, United States Patent No. 5,270,175, 1993.

[129] M. Uchida, M. Murata, Isolation of a lactic acid bacterium and yeast consortium from a fermented material of *Ulva* spp. (Chlorophyta), J. Appl. Microbiol. 97 (2004) 1297—1310.

[130] R.P. John, G.S. Anisha, K.M. Nampoothiri, A. Pandey, Micro and macroalgal biomass: a renewable source for bioethanol, Bioresour. Technol. 102 (2011) 186—193.

[131] M. Yanagisawa, S. Kawai, K. Murata, Strategies for the production of high concentrations of bioethanol from seaweeds: production of high concentrations of bioethanol from seaweeds, Bioengineered 4 (2013) 224—235.

[132] M.G. Borines, M.P. McHenry, R.L. de Leon, Integrated macroalgae production for sustainable bioethanol, aquaculture and agriculture in Pacific island nations, BIOFPR 5 (6) (2011) 599—608.

[133] L. Ge, P. Wang, H. Mou, Study on saccharification techniques of seaweed wastes for the transformation of ethanol, Renew. Energy. 36 (2011) 84—89.

[134] J.-H. Yeon, S.-E. Lee, W.Y. Choi, D.H. Kang, H.-Y. Lee, K.H. Jung, Repeatedbatch operation of surface-aerated fermentor for bioethanol production from the hydrolysate of seaweed *Sargassum sagamianum*, J. Microbiol. Biotechnol. 21 (3) (2011) 323—331.

[135] R.G. Allen, L.S. Pereira, D. Raes, M. Smith, FAO Irrigation and drainage paper No. 56, Rome: Food and Agriculture Organization of the United Nations 56 (97) (1998) e156.

[136] D.P. Chynoweth, Review of biomethane from marine biomass, Review of history, results and conclusions of the "United States Marine Biomass Energy Program" (1968—1990), 194, 2002.

[137] Z. Yang, F. Kong, Z. Yang, M. Zhang, Y. Yu, S. Qian, Benefits and costs of the grazer induced colony formation in *Microcystis aeruginosa*, Ann. Limnol. Int. J. Limnol 45 (2009) 203—208. Available from: https://doi.org/10.1051/limn/2009020.

References

[138] M. Balat, H. Balat, C. Oz, Progress in bioethanol processing, Prog. Energy Combust. Sci. 34 (2008) 551−573.

[139] J.S. Ventura, E.C. Escobar, D. Jahng, Bio-butanol production by Clostridium saccharoperbutylacetonicum N1−4 using selected species of brown and red macroalgae, Iran. (Iran.) J. Energy Environ. 7 (4) (2016) 325−333.

[140] J. Gressel, Transgenics are imperative for biofuel crops, Plant. Sci. 174 (2008) 246−263.

[141] A. Golberg, E. Vitkin, G. Linshiz, S.A. Khan, N.J. Hillson, Z. Yakhini, et al., Proposed design of distributed macroalgal biorefineries: thermodynamics, bioconversion technology, and sustainability implications for developing economies, Biofuels Bioprod. Bioref. 8 (2014) 67−82.

[142] K.A. Jung, S.R. Lim, Y. Kim, J.M. Park, Potentials of macroalgae as feedstocks for biorefinery, Bioresour. Technol. 135 (2013) 182−190.

[143] X. Yuan, X. Shi, D. Zhang, Y. Qiu, R. Guo, L. Wang, Biogas production and microcystin biodegradation in anaerobic digestion of blue green algae, Energ. Environ. Sci. 4 (2011) 1511−1515.

[144] P. Morand, B. Carpentier, R.H. Charlier, J. Maze, M. Orlandini, B.A. Plunkett, et al., Bioconversion of seaweeds, in: M.D. Guiry, G. Blunden (Eds.), Seaweed Resources in Europe, Uses and Potential, Wiley, Chichester, 1991, p. 432.

[145] Y. Khambhaty, K. Mody, M.R. Gandhi, S. Thampy, P. Maiti, H. Brahmbhatt, *Kappaphycus alvarezii* as a source of bioethanol, Bioresour. Technol. 103 (2012) 180−185.

[146] M.H. Huesemann, L.J. Kuo, L. Urquhart, G.A. Gill, G. Roesijadi, Acetone-butanol fermentation of marine macroalgae, Bioresour. Technol. 108 (2012) 305−309.

[147] M.D.N. Meinita, B. Marhaeni, T. Winanto, G.T. Jeong, M.N.A. Khan, Y.K. Hong, Comparison of agarophytes (*Gelidium*, *Gracilaria*, and *Gracilariopsis*) as potential resources for bioethanol production, J. Appl. Phycol. 25 (2013) 1957−1961.

[148] M. Aizawa, K. Asaoka, M. Atsumi, T. Sakou, Seaweed bioethanol production in Japan—the ocean sunrise project, in: Proceedings of the OCEANS 2007, Vancouver, BC, Canada, 29 September−4 October 2007, pp. 1−5.

[149] A. Philippsen, Energy Input, Carbon Intensity, and Cost for Ethanol Produced from Brown Seaweed, University of Victoria, Victoria, 2013.

[150] M. Huesemann, G. Roesjadi, J. Benemann, F.B. Metting, Biofuels from microalgae and seaweeds, Biomass to Biofuels, Blackwell Publishing Ltd, Oxford, 2010, pp. 165−184.

[151] L. Nikolaison, J. Dahl, K.S. Bech, A. Bruhn, M.B. Rasmussen, A.B. Bjerre, et al., Energy production from macroalgae, in: Proceedings of the 20th European Biomass Conference, Milan, Italy, June 2012, pp. 18−22.

[152] J.R. Gapes, The economics of acetone-butanol fermentation—theoretical and market considerations, J. Mol. Microbiol. Biotechnol. 2 (1) (2000) 27−32.

[153] A. Scalbert, Antimicrobial properties of tannins, Photochemistry 30 (1991) 3875−3883.

13

Thermochemical conversion of algal biomass

Sabariswaran Kandasamy[1], Narayanamoorthy Bhuvanendran[1], Mathiyazhagan Narayanan[2] and Zhixia He[1,3]

[1]Institute for Energy Research, Jiangsu University, Zhenjiang, P.R. China [2]PG and Research Centre in Biotechnology, MGR College, Adhiyamaan Educational Research Institute, Krishnagiri, India [3]School of Energy and Power Engineering, Jiangsu University, Zhenjiang, P.R. China

13.1 Introduction

Since the industrial revolution in the 18th century, fossil fuels have been considered to be the primary source of energy. However, substantial environmental concerns, such as overexploitation of natural resources, air pollution, deteriorating environmental greenhouse effects, and global warming, have resulted from the extensive use of fossil fuels. In recent decades there has been a great deal of interest in the production of clean energies to alleviate these problems. The fourth-largest source of primary energy generation is renewables or bioenergy, which makes biomass a potential alternative to fossil fuels [1]. The carbon in biomass comes from plant organic matter and is released into the atmosphere when the plants are burned.

Biomass shall be treated as a pollution-neutral source regardless of whether the above cycle has a negative net carbon footprint. First-generation food crops and biofuels such as bioethanol were derived from the sucrose and starch consumed extensively due to huge demand. Biofuels from nonfood crops, such as lignocellulosic bioethanol, have been produced to avoid the food shortages. The third-generation biofuels are biodiesel, bioethanol, and biobutanol derived from macroalgae and microalgae [2]. Due to the fast growth rate and CO_2 remediation in algae biomass, they have been used to convert into as biofuel. Thermochemical transformations in carbohydrate polymers and proteins were mainly uncovered for microalgal-based biofuel technology [3]. Potential value-added products were formed during the direct combustion, torrefaction, gasification, liquefaction, and pyrolysis [4]. At present, tremendous interest has been paid to the industrialization of low-lipid microalgal organisms for biofuel and other value-added products. Over the last decade, most biodiesel algae conversions have been tracked and carried out for a long time through conventional methods. Recent techniques have used various conversion methods that are feasible and essential to large-scale operations in order to produce algae-based biofuel [5]. The thermochemical conversion has received huge attention and gained a lot of importance due to its broad range of applications and economic feasibility for biofuel production.

Biomass conversion is generally classified into biochemical (photofermentation, dark fermentation, generally microbial) and thermochemical processes (torrefaction, liquefaction, pyrolysis, gasification and direct combustion), which allow effective conversion into biofuels at different temperatures [6]. The low-lipid microalgae species lipid content was less than 15wt.%, whereas the higher biomass productivity and growth rates were obtained in the high lipid microalgae species and required stringent and controlled cultivation. The algae lipid portion is only utilized in the transesterification process. During the thermochemical conversion process, the entire algae are converted into biooil, including the proteins, carbohydrates, and lipids. The algae are also produced in highly diluted water systems, which are more suitable for wet (hydrothermal) treatment. The overwhelming majority of research is focused on the thermochemical conversion of algae into biofuel, through hydrothermal production, in particular, the hydrothermal liquefaction (HTL) process to biocrude products [7]. Due to the negative energy balance caused by the prerequisite drying of feedstock, pyrolysis is limited to laboratory research and causes a massive barrier to large-scale industrial production. The HTL process simply renders the thermochemical algae into scaled-up pilot plant operations from the laboratory [8]. Still, the process is not yet matured due to the extraction of the algae oil at low cost and limited algae in the biofuels market, making it difficult for large-scale industrial production. Thermochemical conversion reactors can be handled for the downstream processing of algae using a completely different method from the separation process [9,10]. This chapter focuses on the pyrolysis, HTL, gasification, torrefaction, and direct combustion for the production of biooil from algae.

13.2 Thermochemical conversion

Biodiesel is produced through the process of lipid extraction from algal biomass which will undergo a transesterification process to be converted into the methyl ester. The main components present in the microalgae cell are lipids, proteins, and carbohydrates. In comparison, the growth rate of the algae biomass is 5–10 times higher than conventional food crops. Due to lipid productivity, the algae have been chosen as an alternative feedstock and can be used to produce biodiesel. Furthermore, there is an ease of adaptation to the environment, the requirement for less arable land than terrestrial plants, and competition with cultivated food crops can be avoided. The high and low lipid microalgae can serve as a feedstock in the HTL reactor [11,12]. The typical thermochemical conversion of algae biomass is shown in Fig. 13.1.

This technique is appropriate for dry and wet biomass microalgae, but lignocellulosic materials from feedstocks of the first and second generation biofuels are much easier than microalgae biomass to process. Dewatering and then cost-intensive drying phases are

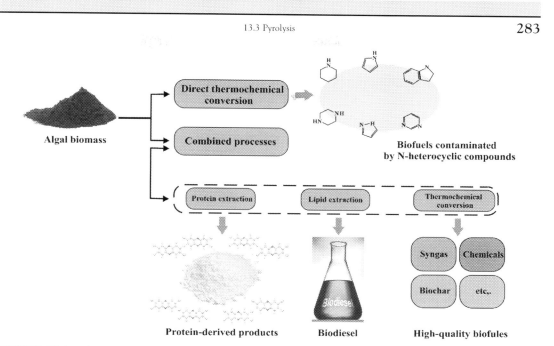

FIGURE 13.1 Thermochemical conversion of algae biomass. Source: *Adapted from H.V. Ly, S.-S. Kim, H.C. Woo, J.H. Choi, D.J. Suh, J. Kim, Fast pyrolysis of macroalga* Saccharina japonica *in a bubbling fluidized-bed reactor for bio-oil production. Energy 93 (2015) 1436–1446.*

needed, such as sedimentation, flocculation, and wet algal feedstock [14,15]. The pyrolytic mass char obtained is greater than 10.98 MJ/kg for a solid char from thermochemical liquefaction (TCL), with energy rates of 23.77–26.12 MJ/kg. In contrast, the primary microalgal biomass chemical strength has been turned in the TCL into functional liquids and gaseous products. At the same time, most is retained as an unconverted constant char in pyrolysis.

13.3 Pyrolysis

The pyrolysis process generally occurs at heating above 430°C, typically at ambient pressure with a total absence of oxygen (under inert atmosphere). Products of pyrolysis (char, crude oil, pyrogas) are mainly dependent on process conditions, such as the type of reactor, feedstock characteristics, and reaction temperature. At the temperature range of 450°C–550°C, the biooil yield is between 50% and 70% w/w in rapid and flash pyrolysis. The process of the pyrolysis unit is shown in Fig. 13.2.

Acidic and very complex liquid substances are found in the lignocellulosic biomass and can be separated into different fractions. Due to the intermediate energy/chemical carriers, pyrolysis oil could be ideal for biorefining in downstream plants. When compared to lignocellulosic biomass, the pyrolysis of algae feedstock has very distinct and peculiar characteristics. Moreover, in the pyrolysis step the critical products are oil and char. The exhaust gas contains up to 10% w/w of CO_2, which is needed for algae cultivation.

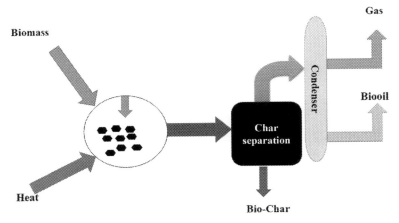

FIGURE 13.2   Pyrolysis experiment setup for algae biomass.

The hydrothermal liquefaction of microalgae produced such components as bio-oil (18–57.9 wt.%) product yield and water-soluble (15–30 wt.%, gases (10–60 wt.%), and biochar, respectively. These compositions were mostly due to reaction type conditions, diverse species of microalgae, and cultivation conditions. To this day, only lab-scale pyrolysis has been tested—the slow process of pyrolysis using thermogravimetry analysis (TGA) in the early 1990s. In most of the reports, a pyrolysis study was observed, although a few studies were collected only via rapid pyrolysis. The slow pyrolysis and fast heating rate of pyrolysis were obtained at less than 1°C/s and 10°C/s, respectively. The Py-GC/Ms (pyrolyzer combined with gas chromatography/mass spectrometry) was performed in some experiments due to a limited volume of sample (0.2–2 mg) fed into the preheated reactor, so this phase is called fast pyrolysis. The pyrolytic products from the feedstock have been identified only through Py-GC/Ms but without secondary reactions [16].

Pyrolysis is considered to be one of the potential biofuel production methods using the algae biomass from the selected particular microalgae. Macroalgae could be a possible candidate for environmental sustainability. An essential advancement is required for reaching the large-scale aquatic biomass fuels [17,18]. *Spirulina* and *Chlorella* pyrolysis for various applications has been reported in over 5000 articles. Still, a comprehensive analysis of green algal pyrolysis for different micro- and macroalgae has not yet been performed [19]. Aromatics, sugars, and several other high value-added chemical compounds are present in these biooils. For the production of biodiesel [20], different chemical transfer techniques can be employed. Algae are considered third generation biofuels. The incorporation of macroalgae into biorefineries was prioritized by research. The natural source of biooil is green microalgae [21,22]. The maximum biooil yields have been reported from *Spirulina* and *Chlorella* [23]. Extensive pyrolysis of algae species has been performed using different pyrolytic techniques and methods and these must be further studied extensively [24]. As mentioned above, the pyrolysis relates to the process through which, in the presence of inertia or even lower stoichiometric oxygen, the thermal oxidative effects of smaller molecules at high temperatures occur. A variety of thermal degradation products can be produced by pyrolysis [25]. The devolatilization of algae begins pyrolysis at 200°C−550°C, and the solid decomposition of algae occurs above 550°C [26]. Algae thermal decomposition can be separated into three main phases: protein and carbohydrate decomposed at temperatures below 400°C during algal pyrolysis, whereas lipids decompose mainly at temperatures above 550°C. Moreover, increased heating of biomass at higher temperatures (600°C) could induce secondary burning and reduce the higher-molecular-weight hydrocarbons into low-molecular weight hydrocarbons, thus limiting biooil use. The elimination of bound water algae at 150°C at a high decomposition indicates the instability of the elevated organic matter at temperatures from 200°C to 600°C (3) [27]. Furthermore, the carbohydrates and proteins decompose through pyrolysis of biomolecules from the algae at below 400°C. The lipid loss can be detected at 550°C, significantly higher than for proteins or carbohydrates. Table 13.1 and Table 13.2 show the pyrolysis of various micro- and macroalgae species.

Because the hydrocarbon chain is detached into simple linear chains, at a higher temperature above 600°C, less biooil and higher gas is obtained. The pyrolysis method breaks down a substance in three phases: spontaneous split, side-group separation, and monomer reverse. Monomeric inversion and random separation for biomass pyrolysis have usually

TABLE 13.1 Pyrolysis of various microalgae species.

S. no	Microalgae species	Temperature (°C)	Biooil yield (%)	References
1	*Microcystis aeruginosa*	500	32.5	[28,29]
2	*Desmodesmus* sp.	750	42.25	[30]
3	*Chlorella* sp.	550	62	[31]
4	*Polytrichum commune, Dicranum scoparium, Thuidium tamarascinum, Sphagnum palustre, Drepanocladus revolvens, Cladophora fracta,* and *Chlorella protothecoides*	775	4.3−55.3	[32]
5	*Spirulina platensis*	550	25	[33,34]
6	*Scenedesmus* sp.	900	20.8	[34]
7	Microalgae consortium	500	44	[35]
8	*Chlorella vulgaris, Schizochytrium limacinum, Arthrospira platensis,* and *Nannochloropsis oculate*	700	39, 48, 36, and 48	[33,35]
9	*Spirulina* sp.	550	30.00	[32]
10	*Chlorella pyrenoidosa*	500	39.4	[36]
11	*Nannochloropsis* sp.	550	22	[37]
12	*Chlorella* sp.	550 and 569	57 and 28.6	[38]
13	*Scenedesmus almeriensis*	400	23	[39]
14	*Green algae* and *Cyanobacteria*	750	22.5	[33]
15	*C. vulgaris*	550	44.42	[35]

13. Thermochemical conversion of algal biomass

TABLE 13.2 Pyrolysis of various macroalgae species.

S. no	Macroalgae species	Temperature (°C)	Biooil yield (wt. %)	Reference
1	*Porphyra tenera*	300−600	37.5 and 47	[40]
2.	*Laminaria digitata, Fucus serratus*	500	11 and 17	[41]
3.	*Saccharina japonica*	350	44.99	[13]
4.	*Enteromorpha clathrata*	550	44.78	[42]
5.	*E. clathrata*	550	52.6	[43]
6.	*E. clathrata*	550	74.24	[44]
7.	*Saccharina japonica*	430−530	48.4	[45]
8.	*Sargassum* spp.	350	18.4 ± 0.1 and 22.2 ± 0.1	[46]
9.	*S, japonica*	450	47	[47]
10.	*Gracilaria gracilis*	500	42	[48]
11.	*S. japonica*	400−500	39.70−45.36−	[49]
12.	*Ulva prolifera*	400	41.3	[50]
13.	*Oscillatoria*	550	32−36	[51]
14.	*Saccharina japonica*	400	88	[52]
15.	*Cystoseira barbat*	461−663	29.9%−34.8%	[53]

been observed [54]. Moreover, spontaneous heat damage in biomass dissolves the carbon chain, where a uniform cellulose-like structure is present. Instead of the main chain, a side partition is seen as the element bond is twisted on the carbon side. A monomer reversal [55] was seen as the full breakup of the polymer into component monomers. Any biomass composition consists of large compounds such as cellulose, hemicellulose, and lignin, such as algal biomass. The amounts of these substances differ depending on the biomass used, for example, timber, rice, chicken, or algae. Instead of laboratory concepts, pyrolysis focuses on experimental and non-substantial evidence that cellulose is a key component of green-algae biomass [56]. As a result, two mechanisms for breakdown must be recognized in opposition: fragmentation of the char reactions and light volatiles such as gases, aldehydes, and ketones [57]. When high temperatures and heating velocities are detected in slow pyrolysis, the first stage is typical. Volatiles are generally assumed to grow by free radical processes, particularly by separation. Simultaneous, noncompetitive responses are conceived for the generation of multiple products, and the reactions are of the first order. The ultimate process of algal pyrolysis involves relative proteins, lipids, and hydrocarbons [58]. The general method for extracting the response phase is thermogravimetric analysis. Furthermore, carbohydrate breakdown usually results in macroalgae mass loss. Protein pyrolysis is required for the final heating, with depolymerization being the most important part. The depolymerization of proteins contributes to a release of nitrogen in algal pyrolysis, producing ammonium and nitrous oxides. Lipid disintegration is usually found in the hemicellulose pathways [59]. The reaction criteria for biochar, biooil, and pyrogas processing vary. Slow pyrolysis, known as torrefaction, occurs with the biomass volatiles at 200°C–250°C extracted in an inert atmosphere [60].

13.4 Hydrothermal liquefaction

The HTL process is a promising solution to the cost-effective dewatering/drying method for the microalgae. In the 1940s and 1980s, the batch reactors were mentioned first. The thermo-chemistry of microalgae is classified into the solvent slurry at around mild temperatures of 280°C–623°C and pressure at 10–25 MPa for ensuring the solvent fluid status [61]. Furthermore, a drop in dielectric constant and nonpolarity is found during the hydrogen bonding heating in water molecules [62]. At this point, biooil is easily extracted by using the previously defined processes with organic solvents, including dichloro/trichloro-methane, tetrahydrofuran, and n-hexane. The concept of HTL of algae biomass is demonstrated in Fig. 13.3.

Direct conversion into lipid fatty acids consisting of triacylglycerides, the existence and lack of a water medium catalyst, which may also respond under essential temperatures, in TAGs (triacylglycerides), heterocyclic nitrogen and carbohydrates, such as polysaccharides, starch, and wet biomass phenols are significant. Several solvents, like water as a reaction medium, are absorbed during the HTL [63,64]. Moreover, Glycose, the main ingredient that rapidly becomes fructose isomerized and deteriorated, and is disintegrated to generate glycolaldehydes and glyceraldehydes are quickly transformed into monosaccharides. Furthermore, coke and other hazardous products, including H_2, CH_4, CO, etc., are processed. The relationship between the functional, structural hydrolyzes, NH_2 and COOH, in proteins, is causing amino acid that is subjected to DCO, HC, $-NH_2$, $-CHO$

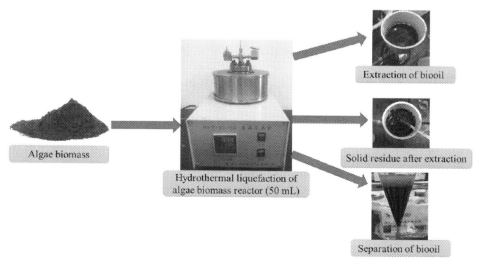

FIGURE 13.3 Concept of hydrothermal liquefaction of algae biomass.

and —COOH deformation and production [65]. The result of long-chain HCs is a partial degradation of FFAs. The carbohydrate hydrolysates associate with proteins to create heterocyclic nitrogen-based cells that adopt Maillard reactions. Generally, macro-molecules are hydrolyzed to structure molecular chips as fatty acids, amino acids and monosaccharides [66] to remove biooil from microalgae residues under HTL. The precise DCO-accompanied amino acids are found in HCs of amines, aldehydes and acids. Several researchers have studied the HTL mechanisms for further analysis [67]. Generation of biooil mechanisms, a combination of a broad spectrum of *Desmodesmus* sp. compounds and peptides. The macromolecular structure of HCs is found in the long sequence. The organic solvent portion of cells' microalgal constituents is retained at relatively low temperatures throughout the first phase, utilizing the microalgal lipids, algal, and some hydrophobic protein portions. Multiple routes are now feasible for the relation of protein-carbohydrate and lipid [68,69]. The second stage consists of producing the protein and cell constituents of diketopiperazine (DKP) at around 573K–648K. The amino acid and carbohydrate (furan) derivatives and similar structure to asphaltene are depolymerized by peptide. In contrast, biooil nitrogen content is degraded by the carbon components [70], where protein degradation increases the biooil yields. For thermal protein fragmentation, a mechanical approach was suggested to process microalgae biooil under HTL conditions [71,72]. To serve as a weak acidic reaction medium, the co-solvent is water sub/supercritical ethanol. To produce HTL catalyzed biooil under acidic conditions, *Tertiolecta species* suggested the mechanical pathway [73]. This process involves the technique of affecting lipids, proteins, and carbohydrates. Then protein becomes amino acids of long-chain by acidic hydrolysis; through rupture, condensation, DCO, and deamination, it becomes

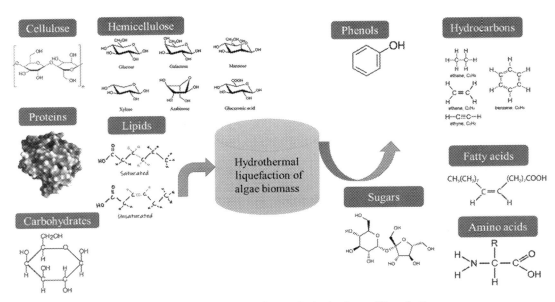

FIGURE 13.4 Potential algal biomass reaction pathways for hydrothermal liquefaction.

liquefied content. Various compounds produced the algae biomass during the HTL reaction (Fig. 13.4.)

Moreover, another segment deals with liquid protonation products [74]. The formed acid creates ester and amide in alcohol and ammonolysis, respectively. Acid-catalyzed protein NH_3 should be used for ammonolysis. The resulting amide undergoes ethanolysis in the presence of large ethanol concentrations to create esters. In the intoxication and ammonolysis directions, the formed acid generates ester and amide, respectively [75,76]. Acid-catalyzed protein-based NH_3 may be used for the ammonolysis process. The produced amide undergoes ethanolysis in the presence of a high ethanolic concentration to produce esters. The common. *S. occulta* and *Spirulina* contain more than their initial lipid components (5%−30% by weight). A low lipid source of microalgal *Spirulina* and *Desmodesmus* sp. is possible with an HHV of 22−36 MJ/kg. The highest yield for biooil was reached respectively by 43% and 66%. *Spirulina platensis* (20wt.%) achieved a biooils yield of about 40wt.% with 39.9 HHV MJ/kg at 623K for 60 min in a 1.8 L batch reactor [77,78]. Furthermore, at 623K for 3 min and moderate pressure of 20 MPa for *chlorella* and *Spirulina* for the varied *Chlorella* varieties, the HTL pilot-scale produced 41.7% biooil

TABLE 13.3 Hydrothermal liquefaction of various microalgae species.

S. no	Microalgae species	Temperature (°C)	Biooil yield (%)	References
1	*Chlorella Minutissima, Chlorella sorokiniana* UUIND6, *Chlorella singularis* UUIND5, and *Scenedesmus abundans*	270	20.10	[79]
2	Microalgae−bacteria consortium	299.7	21.7	[80]
3	*Chlorella vulgaris*	300	23	[81]
4	*Euglena* sp.	350	22.45−25.41	[82]
5	Pretreated *Spirulina*	225	8.6	[83]
6	*Coelastrum* sp.	320	25−32	[37]
7	*C. vulgaris, C. sorokiniana*, and *G. sulphuraria*	240	31	[84]
8	*Spirulina* sp.	225	6.5−34.3	[85]
9	*S. quadricauda*	300	18	[83]
10	*Sargassum* sp.	350	7.20	[86]
11	*Scenedesmus obliques*	300	24.57 and 39.6	[87]
12	*Scenedesmus* sp.	350	20−36	[88]
13	*Cyanobacteria* sp.	325	21.10	[88]
14	*Chlorella sorokiniana*	240	30	[89]
15	*Bacillariophyta* sp.	325	18.21	[88]
16	*Nannochloropsis* sp.	400 and 500	36−46 and 52	[90−92]

290   13. Thermochemical conversion of algal biomass

yields. The 66% of the maximum biomass oil yield at around *Nannochloropsis* sp. reaction temperatures. The HTL of various micro and macroalgae species biooil production is shown in Table 13.3 and Table 13.4, respectively.

The significant heteroatomic profile of biooil might lead to the upgrade criteria by the time it is launched. Therefore the overall biooil production depends on the protein profile and the cellular carbohydrates of the microalgal species. It also uses depolymerization kinetics and methods used for the test conditions of HTL [108]. Furthermore, the biooil output of different microalgae organisms relies on ideal conditions for responding, higher heating values (HHV), and product distribution. Moreover, HHV has an essential impact on both carbon and water quality. The improved biooil production can be accomplished by homogeneous catalysis or heterogeneous catalysis. Furthermore, the utilization and coliquidation of the microalgal HTL [109] are other methods. In acid catalysis, homogenous HTL catalyzed mechanisms are used to enhance the gasification and water−gas by inhibiting the biomass dehydration. The deoxygenation and dehydration of DCO in place of a single char and tar polymerization increases the biooil yield at pH [9,110]. The homogeneous Na_2CO_3 basis of the *Braunii* species HTL was catalyzed. The biooil yields of *Tertiolecta* have decreased by 28%−32% at 573K−613K, and the content of oxygen is reduced by 24.2%−19.7%. Moreover, *Tertiolecta* with 5% Na_2CO_3 and biooil production contributed to lipids > proteins > carbohydrates. In some cases, the yield of ethyl hexadecanoate is not improved by the FeS catalyst. The product of saponification for high-lipid

TABLE 13.4   Hydrothermal liquefaction of various macroalgae species.

S. no	Macroalgae species	Temperature (°C)	Biooil yield (wt. %)	Reference
1.	*Gracilaria corticata*	280	56.2	[93]
2.	*Ulva lactuca*	280	55.2	[94]
3.	*Ascophyllum nodosum*	345	29.9	[95]
4.	*Ulva prolifera*	300	50.6	[96]
5.	*Undaria pinnatifida*	240	69.22	[97]
6.	*Sargassum tenerrimum*	280	23.8	[98]
7.	*Sargassum tenerrimum*	280	33	[99]
8.	*Saccharina japonica*	300	20.26	[100]
9.	*Ulva fasciata*	300	44	[101]
10.	*Oedogonium*	300−350	25	[102]
11.	*Enteromorpha prolifera*	290	28.4	[103]
12.	*kelp Laminaria saccharina*	350	79	[104]
13.	*Sargassum tenerrimum*	280	16.33	[105]
14.	*Oedogonium*	350	26.2	[106]
15.	*Amphiroa fragilissima*	320	28.9	[107]

Handbook of Algal Biofuels

alkaline catalyst species. For *Chlorella* and *Spirulina* species as KOH and Na_2CO_3 and HCOOH and its mixture, it was performed and expanded as catalysts. As a result, the biooil production of $Na_2CO_3 > CH_3COOH > HCOOH$ by the order has occurred. Moreover, alkaline catalysts were unfeasible for high-carbohydrate HTL algal biomass, and strongly lipid microalgae acids were satisfactory [111–113]. The utilization and contrast of biooil production with that of Na_2CO_3 by alkaline earth catalyst [$Ca_3(PO_4)_2$] for microalgae and transition metal [114] (NiO) reveals that biooil production increased to 51.6% and 29.2% for a noncatalytic reaction. The catalyst for alkaline earth and transition metals increases the content of gas and reduces biooil content. In the sulfuric acid-glycol process as a catalyst, at 392K–467K, the biooil yield increased to 0.3%–2.4%. The FAMEs and carbon chain alkanes C17–C20 at 443K in 33 min are the biooil yield observed percentage. The synergetic effect of *tertiolecta* and carbon improves the biooil conversion (70.6 wt.%) and enhances the quality of biooil (40.3 wt.%) production, because of its high hydrogen content, where the microalgae involved as a hydrogen donor [115–117].

13.5 Gasification

The hydrothermal gasification shall occur in the supercritical water supply at a temperature above 374°C and pressure above 22 MPa for carbonate components to produce CH_4. The viscosity and diffusivity of water are similar to the gases under those experimental conditions. Hydrothermal gasification consists of an environmentally sustainable conversion into a gasifier [118] (about 4–6 MJ/kg) of microalgae biofuel with partial vapor oxidation at large microalgae biomass. The gas product is made of CH_4, CO_2, N_2 and can be directly burnt for the generation of diesel/gas turbines. The drying of biomass at high waters often does not need hydrothermal gasification. In the presence of a catalyst, it can be accomplished at varying temperatures and pressures. The gasification can be done at hydrothermal gasification (> 500°C) at a noncatalytic high temperature, in catalytic conditions (225°C–265°C and 2.9–5.6 MPa), and supercritical catalytic requirements (500°C or higher). Endothermic biomass (CHaOb) decomposition processes involve the synthesis of H_2 and CO and exothermic gas reaction; additional H_2 and CO_2 were produced [119]. The CH_4 begins this stage of exothermic carbon and CO_2 methanation to generate water. Supercritical Water Gasification (SCWG) is the solution for wet-biomass gasification without a catalyst under biofuel continuous/batch conditions. Moreover, traditional gasification, SCWG, and energy conservation is contrasted. The complete process of the gasification unit is shown in Fig. 13.5.

The gasification path is challenging to understand because of the dynamic mechanism of biomass [120]. Microalgal cell monomer degradation is formed under free-radical conditions in intermediate acids and furfurals at 300°C–400°C. Then smaller organic acid concentrations are decreased to produce gas products at 450°C–650°C by dehydration and decarboxylation. In subcritical environments, polymerization in automotive formation is caused by furfurals. The heating temperatures of 250°C–265°C by hydrolysis, amino acids in sub- and supercritical water are decomposed for gasification. Moreover, by releasing NH_3 gas, amino acids may be turned into amino compounds and intermediate acids. After

Handbook of Algal Biofuels

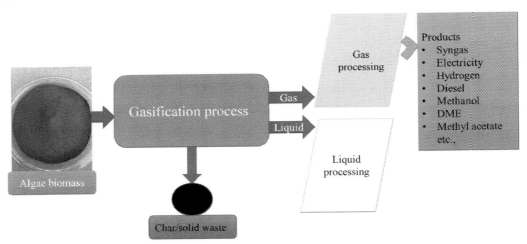

FIGURE 13.5 Typical gasification process.

acid intermediates have decomposed, smaller intermediates and gases are produced. A strong black substance production starts at 500°C relatives to fluids and gasses [121].

13.6 Torrefaction

Due to low carbon and water content characteristics, the microalgae biomass thermochemical conversion was a little expensive. Nevertheless, this drawback significantly lowers the HHV and hence the volumetric energy density of microalgal biomass. The high excess of water content contributes to increased pyrolysis, gasification, and direct combustion costs for dewatering or drying. Furthermore, due to microalgae high-water content, the whole conversion process has also been affected [122]. Torrefaction has recently been recognized as a viable method of transforming biomass into hydrophobic, high-energy-dense fuels that can thermochemically overcome the drawbacks mentioned above. Many torrefaction experiments and reports on lignocellulose have been reported, but few industries have used microalgae as feedstock or residues. However, the lignocellulosic biomass composition differentiates dramatically from microalgae biomass composition. It can expose different torrefaction behavior and various effects on biochar materials, an outline of the torrefaction mechanism is essential [123]. Torrefaction is an oxygen-free environment at a temperature range of 200°C–300°C, defined as a thermal biotreatment process. In the present scale, three distinct grades can be divided: moderate (200°C–235°C), medium (235°C–275°C), and severe (275°C–300°C). The HHV increased to 300°C for microalgae improved for 30 min between 20.46 and 25.92 MJ/kg and thus more adapted to partial industrial gas replacement by the modified biomass. The TGA for kinetic torrefaction has shown that the CNW-N activation force of *Scenedesmus obliquus* in the nonisothermal region of the torrefaction was 57.52 kJ/M and 40.14–88.41 MJ/kg [124]. A comparative analysis of isothermal torrefactions and nonisothermal torrefactions showed that both were exacerbated at the same average temperature as the previous. The significant phases of the microalgae exist: (1) depolymerization and (2) thermal deterioration of protein and carbohydrates (under 200°C); and (3) decarbonization and cracking. The latest transformative spectrum TGA and FTIR showed that carbohydrates with an enhanced potential to torrefaction were first collected combined with protein usage. Moreover, the temperature followed by resident time close to lignocellulosic biomass, is the most significant torrefaction parameter. In comparison, the major parameters for torrefaction are the variables of solid production and energy improvement (or energy densification), showing the mass proportion of the residual solid to its parent biomass and the energy densification of the solid, respectively [124,125]. Certain microalgae have large amounts of lipid and are essential oils for the production of biodiesel. As shown by the literature, some torrefaction experiments rely on microalgae residues to generate solid fuels. Thermochemical microalgae products improve fuel content by raising temperatures, increasing the reaction time, increasing the consumption of oil, and reducing energy production in intense environments. The collection of optimal fuel production and energy usage is also vital for methodological optimization [126].

13.7 Direct combustion

In the presence of oxygen at a temperature around 850°C, combustion burns microalgae. In the boiler or furnace, the combustion occurs where microalgal biomass can be loaded with >50% moisture content. In general, to facilitate the release of heat and the completion of reactions, the air pumped into a combustor is almost 10% in excess concerning the feedstock. During the combustion process, the chemical energy obtained using photosynthesis in the microalgae is converted into hot gases [127]. The heat generated by combustion cannot be stored and converted into a valuable product through electricity generation using a turbine. The cost of energy production using the microalgae is higher, mainly because of the effect of pretreatment phases. One of the potential alternatives for process intensification is Combined Heat and Power Generation. Due to the high moisture content, algal biomass combustion processes have been extensively explored [128,129]. However, technological advances in low-energy drying could potentially increase the adherence of microalgae to the operation of combustion. Co-firing of algae and either coal or natural gas are other options to mitigate the emissions of greenhouse gases from fossil sources only. Indeed, the process absorbs carbon dioxide and uses it for algal cultivation [130].

13.8 Economic feasibility

A few countries are undertaking biofuel projects, with first-generation biofuels on a commercial scale, second-generation biofuels on the market, and the third-generation biofuels only on a laboratory scale. Predictably, they will all change the direction at the global scale [127]. At this point, such theoretical methods, such as energy balance and GHG pollution, need to be well known in terms of the variations between biofuels, in addition

294

to capital demand, job opportunities, and the development of suitable technology, taking into account the study of the life cycle. Innovative processes on a laboratory scale involve commercializing biofuels to produce large-scale production close to potential without compromising the performance. The algae are grown in water, before pyrolysis, dewatering, and energy-intensive drying are required. Most of the cost-effective methods for water usage employed as a fraction in the pyrolysis phase as a hydrogen source to reforming pyrolytic products. Biooil refineries could be a close field to generate algal biomass and then transfer to biooil as a transport fuel, thus offering a trade-off between transport costs and economic scale to provide for higher pyrolysis [128]. Torrefied microalgal biomass has low moisture content and increased torrefied biofuels, leading to economical transport and quick packaging.

Even if several cost-effective torrefaction techniques are practically accessible these days, more study, particularly reaction kinetics, is mandatory for large-scale reactor design. In the case of the combustion of microalgae biomass, feasibility data are limited. Thus further research and development are required; however, total combustion efficiency can be improved if combusted with coal. Technically speaking, microalgal biomass can be an economically feasible solution for conserving chemical energy, with a good evolution towards technological conversion [129].

13.9 Conclusion

The microalgal feedstocks for thermochemical conversion techniques need to be worked on in order to make them viable. The advances in establishing and processing biofuel production from microalgae remains a challenge in terms of making it cost-effective. In contrast, flue gas can be a necessity for agriculture as a source of carbon. The drying method absorbs extreme energy throughout the assessment of other species. The sustainability, land area, and water criteria for microalgae biomass are the main factors in agriculture and harvest. Microalgae can consume carbon dioxide from coal-fired power plants and biosorb the heavy metals from industrial waste disposal. A possible cost-effective solution is to reduce microalgae and energy recovery development in the mixture of algal growth and the existing wastewater treatment methods. Furthermore, this chapter presents a comprehensive review of recent research on pyrolysis, liquefaction gasification, and torrefaction. At the present time, most of the scientific works have been undertaken in microalgae-based biofuels due to the feasible economic and energetic balance.

Moreover, the HTL technique was one of the best approaches compared to other thermochemical conversion methods. The biooil production by pyrolysis has fewer advantages than the HTL process due to the technological feasibility and product quality. Thus the critical aspects of pyrolysis and the HTL process have significant technical downstream pathways for microalgae. Moreover, the use of multicomponent catalysts could be a promising strategy to increase biooil efficiency and yield with the short duration of HTL reaction and resident time. A longer life span and lower costs dependent on catalytic mechanism discovery remain critical in preparing catalysts with higher operation and stability. Recent literature has promoted biofuel production through the thermochemical conversion pathway, which can help to preserve the climate and conserve the environment.

Acknowledgment

This research was supported by the National Natural Science Foundation of China (51876083, 51776088), a Project Funded by the Priority Academic Program Development of Jiangsu High Education Institutions and High-tech Research Key laboratory of Zhenjiang (SS2018002).

References

[1] S. Gold, S. Seuring, Supply chain and logistics issues of bio-energy production, J. Clean. Prod. 19 (2011) 32–42. Available from: https://doi.org/10.1016/j.jclepro.2010.08.009.

[2] M.V. Rodionova, R.S. Poudyal, I. Tiwari, R.A. Voloshin, S.K. Zharmukhamedov, H.G. Nam, et al., Biofuel production: challenges and opportunities, Int. J. Hydrog. Energy 42 (2017) 8450–8461. Available from: https://doi.org/10.1016/j.ijhydene.2016.11.125.

[3] U. Suparmaniam, M.K. Lam, Y. Uemura, J.W. Lim, K.T. Lee, S.H. Shuit, Insights into the microalgae cultivation technology and harvesting process for biofuel production: a review, Renew. Sustain. Energy Rev. 115 (2019) 109361. Available from: https://doi.org/10.1016/j.rser.2019.109361.

[4] B. de Caprariis, P. De Filippis, A. Petrullo, M. Scarsella, Hydrothermal liquefaction of biomass: influence of temperature and biomass composition on the bio-oil production, Fuel 208 (2017) 618–625. Available from: https://doi.org/10.1016/j.fuel.2017.07.054.

[5] Y. Chen, Y. Wu, D. Hua, C. Li, M.P. Harold, J. Wang, et al., Thermochemical conversion of low-lipid microalgae for the production of liquid fuels: challenges and opportunities, RSC Adv. 5 (2015) 18673–18701. Available from: https://doi.org/10.1039/C4RA13359E.

[6] B.B. Uzoejinwa, X. He, S. Wang, A.E.-F. Abomohra, Y. Hu, Q. Wang, Co-pyrolysis of biomass and waste plastics as a thermochemical conversion technology for high-grade biofuel production: recent progress and future directions elsewhere worldwide, Energy Convers. Manag. 163 (2018) 468–492. Available from: https://doi.org/10.1016/j.enconman.2018.02.004.

[7] R. Bhujade, M. Chidambaram, A. Kumar, A. Sapre, Algae to economically viable low-carbon-footprint oil, Annu. Rev. Chem. 8 (2017) 335–337. Available from: https://doi.org/10.1146/annurev-chembioeng-060816-101630.

[8] R.A. Voloshin, M.V. Rodionova, S.K. Zharmukhamedov, T.N. Veziroglu, S.I. Allakhverdiev, Biofuel production from plant and algal biomass, Int. J. Hydrog. Energy 41 (2016) 17257–17273. Available from: https://doi.org/10.1016/j.ijhydene.2016.07.084.

[9] M.M. El-Dalatony, E.-S. Salama, M.B. Kurade, K.-Y. Kim, S.P. Govindwar, J.R. Kim, et al., Whole conversion of microalgal biomass into biofuels through successive high-throughput fermentation, Chem. Eng. J. 360 (2019) 797–805. Available from: https://doi.org/10.1016/j.cej.2018.12.042.

[10] J. Allen, S. Unlu, Y. Demirel, P. Black, W. Riekhof, Integration of biology, ecology and engineering for sustainable algal-based biofuel and bioproduct biorefinery, Bioresour. Bioprocess. 5 (2018) 47. Available from: https://doi.org/10.1186/s40643-018-0233-5.

[11] M. Montingelli, S. Tedesco, A. Olabi, Biogas production from algal biomass: a review, Renew. Sustain. Energy Rev. 43 (2015) 961–972. Available from: https://doi.org/10.1016/j.rser.2014.11.052.

[12] Y.M. Isa, E.T. Ganda, Bio-oil as a potential source of petroleum range fuels, Renew. Sustain. Energy Rev. 81 (2018) 69–75. Available from: https://doi.org/10.1016/j.rser.2017.07.036.

[13] H.V. Ly, S.-S. Kim, H.C. Woo, J.H. Choi, D.J. Suh, J. Kim, Fast pyrolysis of macroalga Saccharina japonica in a bubbling fluidized-bed reactor for bio-oil production, Energy 93 (2015) 1436–1446. Available from: https://doi.org/10.1016/j.energy.2015.10.011.

[14] U. Jena, K. Das, Comparative evaluation of thermochemical liquefaction and pyrolysis for bio-oil production from microalgae, Energy Fuels 25 (2011) 5472–5482. Available from: https://doi.org/10.1021/ef201373m.

[15] H. Chen, D. Zhou, G. Luo, S. Zhang, J. Chen, Macroalgae for biofuels production: progress and perspectives, Renew. Sustain. Energy Rev. 47 (2015) 427–437. Available from: https://doi.org/10.1016/j.rser.2015.03.086.

[16] S. Charoensiddhi, M.A. Conlon, C.M. Franco, W. Zhang, The development of seaweed-derived bioactive compounds for use as prebiotics and nutraceuticals using enzyme technologies, Trends Food Sci. Technol. 70 (2017) 20–33. Available from: https://doi.org/10.1016/j.tifs.2017.10.002.

[17] B. Bharathiraja, M. Chakravarthy, R.R. Kumar, D. Yogendran, D. Yuvaraj, J. Jayamuthunagai, et al., Aquatic biomass (algae) as a future feed stock for bio-refineries: a review on cultivation, processing and products, Renew. Sustain. Energy Rev. 47 (2015) 634−653. Available from: https://doi.org/10.1016/j.rser.2015.03.047.

[18] F. Masarin, F.R.P. Cedeno, E.G.S. Chavez, L.E. De Oliveira, V.C. Gelli, R. Monti, Chemical analysis and biorefinery of red algae *Kappaphycus alvarezii* for efficient production of glucose from residue of carrageenan extraction process, Biotechnol. Biofuels 9 (2016) 122. Available from: https://doi.org/10.1186/s13068-016-0535-9.

[19] O.M. Adeniyi, U. Azimov, A. Burluka, Algae biofuel: current status and future applications, Renew. Sustain. Energy Rev. 90 (2018) 316−335. Available from: https://doi.org/10.1016/j.rser.2018.03.067.

[20] F. Alam, A. Date, R. Rasjidin, S. Mobin, H. Moria, A. Baqui, Biofuel from algae-is it a viable alternative? Procedia Eng. 49 (2012) 221−227. Available from: https://doi.org/10.1016/j.proeng.2012.10.131.

[21] M.L. Wells, V.L. Trainer, T.J. Smayda, B.S. Karlson, C.G. Trick, R.M. Kudela, et al., Harmful algal blooms and climate change: learning from the past and present to forecast the future, Harmful Algae 49 (2015) 68−93. Available from: https://doi.org/10.1016/j.hal.2015.07.009.

[22] N. Pragya, K.K. Pandey, P. Sahoo, A review on harvesting, oil extraction and biofuels production technologies from microalgae, Renew. Sustain. Energy Rev. 24 (2013) 159−171. Available from: https://doi.org/10.1016/j.rser.2013.03.034.

[23] S. Yaman, Pyrolysis of biomass to produce fuels and chemical feedstocks, Energy Convers. Manag. 45 (2004) 651−671. Available from: https://doi.org/10.1016/S0196-8904(03)00177-8.

[24] S. Pourkarimi, A. Hallajisani, A. Alizadehdakhel, A. Nouralishahi, Biofuel production through micro-and macroalgae pyrolysis−a review of pyrolysis methods and process parameters, J. Anal. Appl. Pyrolysis 142 (2019) 104599. Available from: https://doi.org/10.1016/j.jaap.2019.04.015.

[25] X. Zhuang, H. Zhan, Y. Song, X. Yin, C. Wu, Structure-reactivity relationships of biowaste-derived hydrochar on subsequent pyrolysis and gasification performance, Energy Convers. Manage. 199 (2019) 112014. Available from: https://doi.org/10.1016/j.enconman.2019.112014.

[26] T. Li, F. Guo, X. Li, Y. Liu, K. Peng, X. Jiang, et al., Characterization of herb residue and high ash-containing paper sludge blends from fixed bed pyrolysis, Waste Manage 76 (2018) 544−554. Available from: https://doi.org/10.1016/j.wasman.2018.04.002.

[27] N.H. Zainan, S.C. Srivatsa, F. Li, S. Bhattacharya, Quality of bio-oil from catalytic pyrolysis of microalgae *Chlorella vulgaris*, Fuel 223 (2018) 12−19. Available from: https://doi.org/10.1016/j.fuel.2018.02.166.

[28] Y. Shirazi, S. Viamajala, S. Varanasi, In situ and ex situ catalytic pyrolysis of microalgae and integration with pyrolytic fractionation, Front. Chem. 8 (2020) 786. Available from: https://doi.org/10.3389/fchem.2020.00786.

[29] K. Chaiwong, T. Kiatsiriroat, N. Vorayos, C. Thararax, Study of bio-oil and bio-char production from algae by slow pyrolysis, Biomass Bioenerg. 56 (2013) 600−606. Available from: https://doi.org/10.1016/j.biombioe.2013.05.035.

[30] N. Zhou, N.T. Dunford, Thermal degradation and microwave-assisted pyrolysis of green algae and cyanobacteria isolated from the great salt plains, T Asabe 60 (2017) 561−569. Available from: https://doi.org/10.13031/trans.12028.

[31] P. Bhoi, A. Ouedraogo, V. Soloiu, R. Quirino, Recent advances on catalysts for improving hydrocarbon compounds in bio-oil of biomass catalytic pyrolysis, Renew. Sustain. Energy Rev. 121 (2020) 109676. Available from: https://doi.org/10.1016/j.rser.2019.109676.

[32] F. Vargas e Silva, L.O. Monteggia, Pyrolysis of algal biomass obtained from high-rate algae ponds applied to wastewater treatment, Front. Energy Res. 3 (2015) 31. Available from: https://doi.org/10.3389/fenrg.2015.00031.

[33] M. Kröger, F. Müller-Langer, Review on possible algal-biofuel production processes, Biofuels 3 (2012) 333−349. Available from: https://doi.org/10.4155/bfs.12.14.

[34] Q. Xie, M. Addy, S. Liu, B. Zhang, Y. Cheng, Y. Wan, et al., Fast microwave-assisted catalytic co-pyrolysis of microalgae and scum for bio-oil production, Fuel 160 (2015) 577−582. Available from: https://doi.org/10.1016/j.fuel.2015.08.020.

[35] F.C. Borges, Q. Xie, M. Min, L.A.R. Muniz, M. Farenzena, J.O. Trierweiler, et al., Fast microwave-assisted pyrolysis of microalgae using microwave absorbent and HZSM-5 catalyst, Bioresour. Technol. 166 (2014) 518−526. Available from: https://doi.org/10.1016/j.biortech.2014.05.100.

[36] D. Beneroso, J. Bermúdez, A. Arenillas, J. Menéndez, Microwave pyrolysis of microalgae for high syngas production, Bioresour. Technol. 144 (2013) 240−246. Available from: https://doi.org/10.1016/j.biortech.2013.06.102.

References

[37] P.K. Kumar, S.V. Krishna, K. Verma, K. Pooja, D. Bhagawan, K. Srilatha, et al., Bio oil production from microalgae via hydrothermal liquefaction technology under subcritical water conditions, J. Microbiol. Methods. 153 (2018) 108–117. Available from: https://doi.org/10.1016/j.mimet.2018.09.014.

[38] B.B. Makut, G. Goswami, D. Das, Evaluation of bio-crude oil through hydrothermal liquefaction of microalgae-bacteria consortium grown in open pond using wastewater, Biomass Convers. Biorefin. (2020). Available from: https://doi.org/10.1007/s13399-020-00795-x.

[39] C. Jazrawi, P. Biller, Y. He, A. Montoya, A.B. Ross, T. Maschmeyer, et al., Two-stage hydrothermal liquefaction of a high-protein microalga, Algal Res. 8 (2015) 15–22. Available from: https://doi.org/10.1016/j.algal.2014.12.010.

[40] Y.J. Bae, C. Ryu, J.-K. Jeon, J. Park, D.J. Suh, Y.-W. Suh, et al., The characteristics of bio-oil produced from the pyrolysis of three marine macroalgae, Bioresour. Technol. 102 (2011) 3512–3520. Available from: https://doi.org/10.1016/j.biortech.2010.11.023.

[41] J. Yanik, R. Stahl, N. Troeger, A. Sinag, Pyrolysis of algal biomass, J. Anal. Appl. Pyrolysis 103 (2013) 134–141. Available from: https://doi.org/10.1016/j.jaap.2012.08.016.

[42] B. Cao, Y. Sun, J. Guo, S. Wang, J. Yuan, S. Esakkimuthu, et al., Synergistic effects of co-pyrolysis of macroalgae and polyvinyl chloride on bio-oil/bio-char properties and transferring regularity of chlorine, Fuel 246 (2019) 319–329. Available from: https://doi.org/10.1016/j.fuel.2019.02.037.

[43] B. Cao, S. Wang, Y. Hu, A.E.-F. Abomohra, L. Qian, Z. He, et al., Effect of washing with diluted acids on Enteromorpha clathrata pyrolysis products: towards enhanced bio-oil from seaweeds, Renew. Energy 138 (2019) 29–38. Available from: https://doi.org/10.1016/j.renene.2019.01.084.

[44] S. Wang, B. Cao, X. Liu, L. Xu, Y. Hu, S. Afonaa-Mensah, et al., A comparative study on the quality of bio-oil derived from green macroalga Enteromorpha clathrata over metal modified ZSM-5 catalysts, Bioresour. Technol. 256 (2018) 446–455. Available from: https://doi.org/10.1016/j.biortech.2018.01.134.

[45] J.H. Choi, S.-S. Kim, D.J. Suh, E.-J. Jang, K.-I. Min, H.C. Woo, Characterization of the bio-oil and bio-char produced by fixed bed pyrolysis of the brown alga *Saccharina japonica*, Korean J. Chem. Eng. 33 (2016) 2691–2698. Available from: https://doi.org/10.1007/s11814-016-0131-5.

[46] L.M. Díaz-Vázquez, A. Rojas-Pérez, M. Fuentes-Caraballo, I.V. Robles, U. Jena, K. Das, Demineralization of Sargassum spp. macroalgae biomass: selective hydrothermal liquefaction process for bio-oil production, Front. Energy Res. 3 (2015) 6. Available from: https://doi.org/10.3389/fenrg.2015.00006.

[47] J.H. Choia, H.C. Woob, D.J. Suha, Pyrolysis of seaweeds for bio-oil and bio-char production, Chem. Eng. J. 37 (2014) 121–126. Available from: https://doi.org/10.3303/CET1437021.

[48] O. Norouzi, A. Tavasoli, S. Jafarian, S. Esmailpour, Catalytic upgrading of bio-products derived from pyrolysis of red macroalgae Gracilaria gracilis with a promising novel micro/mesoporous catalyst, Bioresour. Technol. 243 (2017) 1–8. Available from: https://doi.org/10.1016/j.biortech.2017.06.072.

[49] H.V. Ly, J.H. Choi, H.C. Woo, S.-S. Kim, J. Kim, Upgrading bio-oil by catalytic fast pyrolysis of acid-washed Saccharina japonica alga in a fluidized-bed reactor, Renew. Energy. 133 (2019) 11–22. Available from: https://doi.org/10.1016/j.renene.2018.09.103.

[50] C. Ma, J. Geng, D. Zhang, X. Ning, Non-catalytic and catalytic pyrolysis of Ulva prolifera macroalgae for production of quality bio-oil, J. Energy Inst. 93 (2020) 303–311. Available from: https://doi.org/10.1016/j.joei.2019.03.001.

[51] H.D. Kawale, N. Kishore, Production of hydrocarbons from a green algae (Oscillatoria) with exploration of its fuel characteristics over different reaction atmospheres, Energy 178 (2019) 344–355. Available from: https://doi.org/10.1016/j.energy.2019.04.103.

[52] H. Zeb, J. Park, A. Riaz, C. Ryu, J. Kim, High-yield bio-oil production from macroalgae (*Saccharina japonica*) in supercritical ethanol and its combustion behavior, Chem. Eng. J. 327 (2017) 79–90. Available from: https://doi.org/10.1016/j.cej.2017.06.078.

[53] O.A. Fakayode, E.A.A. Aboagarib, C. Zhou, H. Ma, Co-pyrolysis of lignocellulosic and macroalgae biomasses for the production of biochar—a review, Bioresour. Technol. 297 (2020) 122408. Available from: https://doi.org/10.1016/j.biortech.2019.122408.

[54] J. Lee, E.E. Kwon, Y.-K. Park, Recent advances in the catalytic pyrolysis of microalgae, Catal. Today 355 (2019) 263–271. Available from: https://doi.org/10.1016/j.cattod.2019.03.010.

[55] K. Azizi, M.K. Moraveji, H.A. Najafabadi, Characteristics and kinetics study of simultaneous pyrolysis of microalgae *Chlorella vulgaris*, wood and polypropylene through TGA, Bioresour. Technol. 243 (2017) 481–491. Available from: https://doi.org/10.1016/j.biortech.2017.06.155.

[56] X. Wu, Y. Wu, K. Wu, Y. Chen, H. Hu, M. Yang, Study on pyrolytic kinetics and behavior: the co-pyrolysis of microalgae and polypropylene, Bioresour. Technol. 192 (2015) 522–528. Available from: https://doi.org/10.1016/j.biortech.2015.06.029.

[57] X. Wang, B. Zhao, X. Tang, X. Yang, Comparison of direct and indirect pyrolysis of micro-algae isochrysis, Bioresour. Technol. 179 (2015) 58–62. Available from: https://doi.org/10.1016/j.biortech.2014.11.015.

[58] J. Yu, K. Maliutina, A. Tahmasebi, A review on the production of nitrogen-containing compounds from microalgal biomass via pyrolysis, Bioresour. Technol. 270 (2018) 689–701. Available from: https://doi.org/10.1016/j.biortech.2018.08.127.

[59] L. Leng, J. Li, Z. Wen, W. Zhou, Use of microalgae to recycle nutrients in aqueous phase derived from hydrothermal liquefaction process, Bioresour. Technol. 256 (2018) 529–542. Available from: https://doi.org/10.1016/j.biortech.2018.01.121.

[60] J. Jiang, P.E. Savage, Metals and other elements in biocrude from fast and isothermal hydrothermal liquefaction of microalgae, Energy Fuels 32 (2017) 4118–4126. Available from: https://doi.org/10.1021/acs.energyfuels.7b03144.

[61] S. Leow, J.R. Witter, D.R. Vardon, B.K. Sharma, J.S. Guest, T.J. Strathmann, Prediction of microalgae hydrothermal liquefaction products from feedstock biochemical composition, Green Chem. 17 (2015) 3584–3599. Available from: https://doi.org/10.1039/C5GC00574D.

[62] E.A. Ramos-Tercero, A. Bertucco, D.W.F. Brilman, Process water recycle in hydrothermal liquefaction of microalgae to enhance bio-oil yield, Energy Fuels 29 (2015) 2422–2430. Available from: https://doi.org/10.1021/ef502773w.

[63] X. Tang, C. Zhang, Z. Li, X. Yang, Element and chemical compounds transfer in bio-crude from hydrothermal liquefaction of microalgae, Bioresour. Technol. 202 (2016) 8–14. Available from: https://doi.org/10.1016/j.biortech.2015.11.076.

[64] Z. He, B. Wang, B. Zhang, H. Feng, S. Kandasamy, H. Chen, Synergistic effect of hydrothermal co-liquefaction of *Spirulina platensis* and Lignin: optimization of operating parameters by response surface methodology, Energy (2020) 117550. Available from: https://doi.org/10.1016/j.energy.2020.117550.

[65] B. Zhang, J. Chen, S. Kandasamy, Z. He, Hydrothermal liquefaction of fresh lemon-peel and Spirulina platensis blending-operation parameter and biocrude chemistry investigation, Energy 193 (2020) 116645. Available from: https://doi.org/10.1016/j.energy.2019.116645.

[66] S. Kandasamy, B. Zhang, Z. He, H. Chen, H. Feng, Q. Wang, et al., Effect of low-temperature catalytic hydrothermal liquefaction of Spirulina platensis, Energy 190 (2020) 116236. Available from: https://doi.org/10.1016/j.energy.2019.116236.

[67] S. Kandasamy, B. Zhang, Z. He, H. Chen, H. Feng, Q. Wang, et al., Hydrothermal liquefaction of microalgae using Fe_3O_4 nanostructures as efficient catalyst for the production of bio-oil: optimization of reaction parameters by response surface methodology, Biomass Bioenerg. 131 (2019) 105417. Available from: https://doi.org/10.1016/j.biombioe.2019.105417.

[68] B. Zhang, J. Chen, Z. He, H. Chen, S. Kandasamy, Hydrothermal liquefaction of fresh lemon-peel: parameter optimisation and product chemistry, Renew. Energy. 143 (2019) 512–519. Available from: https://doi.org/10.1016/j.renene.2019.05.003.

[69] H. Chen, Z. He, B. Zhang, H. Feng, S. Kandasamy, B. Wang, Effects of the aqueous phase recycling on bio-oil yield in hydrothermal liquefaction of Spirulina Platensis, α-cellulose, and lignin, Energy 179 (2019) 1103–1113. Available from: https://doi.org/10.1016/j.energy.2019.04.184.

[70] H. Feng, Z. He, B. Zhang, H. Chen, Q. Wang, S. Kandasamy, Synergistic bio-oil production from hydrothermal co-liquefaction of Spirulina platensis and α-Cellulose, Energy 174 (2019) 1283–1291. Available from: https://doi.org/10.1016/j.energy.2019.02.079.

[71] B. Zhang, Z. He, H. Chen, S. Kandasamy, Z. Xu, X. Hu, et al., Effect of acidic, neutral and alkaline conditions on product distribution and biocrude oil chemistry from hydrothermal liquefaction of microalgae, Bioresour. Technol. 270 (2018) 129–137. Available from: https://doi.org/10.1016/j.biortech.2018.08.129.

[72] S.Z. Hossain, Biochemical conversion of microalgae biomass into biofuel, Chem. Eng. Technol. 42 (2019) 2594–2607. Available from: https://doi.org/10.1002/ceat.201800605.

References

[73] M. Parsa, H. Jalilzadeh, M. Pazoki, R. Ghasemzadeh, M. Abduli, Hydrothermal liquefaction of *Gracilaria gracilis* and *Cladophora glomerata* macro-algae for biocrude production, Bioresour. Technol. 250 (2018) 26–34. Available from: https://doi.org/10.1016/j.biortech.2017.10.059.

[74] J.D. Sheehan, P.E. Savage, Modeling the effects of microalga biochemical content on the kinetics and biocrude yields from hydrothermal liquefaction, Bioresour. Technol. 239 (2017) 144–150. Available from: https://doi.org/10.1016/j.biortech.2017.05.013.

[75] F. Obeid, T.C. Van, R. Brown, T. Rainey, Nitrogen and sulphur in algal biocrude: a review of the HTL process, upgrading, engine performance and emissions, Energy Convers. Manage. 181 (2019) 105–119. Available from: https://doi.org/10.1016/j.enconman.2018.11.054.

[76] S. Arvindnarayan, K.K.S. Prabhu, S. Shobana, G. Kumar, J. Dharmaraja, Upgrading of micro algal derived bio-fuels in thermochemical liquefaction path and its perspectives: a review, Int. Biodeter. Biodegr. 119 (2017) 260–272. Available from: https://doi.org/10.1016/j.ibiod.2016.08.011.

[77] K.P.R. Dandamudi, T. Muppaneni, J.S. Markovski, P. Lammers, S. Deng, Hydrothermal liquefaction of green microalga *Kirchneriella* sp. under sub-and super-critical water conditions, Biomass Bioenerg. 120 (2019) 224–228. Available from: https://doi.org/10.1016/j.biombioe.2018.11.021.

[78] H.A. Baloch, S. Nizamuddin, M. Siddiqui, S. Riaz, A.S. Jatoi, D.K. Dumbre, et al., Recent advances in production and upgrading of bio-oil from biomass: a critical overview, J. Environ. Chem. Eng. 6 (2018) 5101–5118. Available from: https://doi.org/10.1016/j.jece.2018.07.050.

[79] B. Zhang, Q. Lin, Q. Zhang, K. Wu, W. Pu, M. Yang, et al., Catalytic hydrothermal liquefaction of *Euglena* sp. microalgae over zeolite catalysts for the production of bio-oil, RSC Adv. 7 (2017) 8944–8951. Available from: https://doi.org/10.1039/C6RA28747F.

[80] W. Costanzo, U. Jena, R. Hilten, K. Das, J.R. Kastner, Low temperature hydrothermal pretreatment of algae to reduce nitrogen heteroatoms and generate nutrient recycle streams, Algal Res. 12 (2015) 377–387. Available from: https://doi.org/10.1016/j.algal.2015.09.019.

[81] K. Prapaiwatcharapan, S. Sunphorka, P. Kuchonthara, K. Kangvansaichol, N. Hinchiranan, Single-and two-step hydrothermal liquefaction of microalgae in a semi-continuous reactor: effect of the operating parameters, Bioresour. Technol. 191 (2015) 426–432. Available from: https://doi.org/10.1016/j.biortech.2015.04.027.

[82] J.S. Martinez-Fernandez, S. Chen, Sequential hydrothermal liquefaction characterization and nutrient recovery assessment, Algal Res. 25 (2017) 274–284. Available from: https://doi.org/10.1016/j.algal.2017.05.022.

[83] Z. Huang, A. Wufuer, Y. Wang, L. Dai, Hydrothermal liquefaction of pretreated low-lipid microalgae for the production of bio-oil with low heteroatom content, Process. Biochem. 69 (2018) 136–143. Available from: https://doi.org/10.1016/j.procbio.2018.03.018.

[84] D. Xu, G. Lin, S. Guo, S. Wang, Y. Guo, Z. Jing, Catalytic hydrothermal liquefaction of algae and upgrading of biocrude: a critical review, Renew. Sustain. Energy Rev. 97 (2018) 103–118. Available from: https://doi.org/10.1016/j.rser.2018.08.042.

[85] J. Arun, K. Gopinath, S. Shreekanth, R. Sahana, M. Raghavi, D. Gnanaprakash, Effects of process parameters on hydrothermal liquefaction of microalgae biomass grown in municipal wastewater, Pet. Chem. 59 (2019) 194–200. Available from: https://doi.org/10.1134/S0965544119020026.

[86] C. Miao, M. Chakraborty, S. Chen, Impact of reaction conditions on the simultaneous production of polysaccharides and bio-oil from heterotrophically grown *Chlorella sorokiniana* by a unique sequential hydrothermal liquefaction process, Bioresour. Technol. 110 (2012) 617–627. Available from: https://doi.org/10.1016/j.biortech.2012.01.047.

[87] P.J. Valdez, M.C. Nelson, H.Y. Wang, X.N. Lin, P.E. Savage, Hydrothermal liquefaction of *Nannochloropsis* sp.: systematic study of process variables and analysis of the product fractions, Biomass Bioenerg. 46 (2012) 317–331. Available from: https://doi.org/10.1016/j.biombioe.2012.08.009.

[88] J.L. Faeth, P.J. Valdez, P.E. Savage, Fast hydrothermal liquefaction of *Nannochloropsis* sp. to produce biocrude, Energy Fuels 27 (2013) 1391–1398. Available from: https://doi.org/10.1021/ef301925d.

[89] D.L. Barreiro, W. Prins, F. Ronsse, W. Brilman, Hydrothermal liquefaction (HTL) of microalgae for biofuel production: state of the art review and future prospects, Biomass Bioenerg. 53 (2013) 113–127. Available from: https://doi.org/10.1016/j.biombioe.2012.12.029.

[90] A.V. Bridgwater, Review of fast pyrolysis of biomass and product upgrading, Biomass Bioenerg. 38 (2012) 68–94. Available from: https://doi.org/10.1016/j.biombioe.2011.01.048.

13. Thermochemical conversion of algal biomass

[91] T.V. Choudhary, D. Goodman, CO-free fuel processing for fuel cell applications, Catal. Today 77 (2002) 65–78. Available from: https://doi.org/10.1016/S0920-5861(02)00233-X.

[92] L. Fan, H. Zhang, J. Li, Y. Wang, L. Leng, J. Li, et al., Algal biorefinery to value-added products by using combined processes based on thermochemical conversion: a review, Algal Res. 47 (2020) 101819. Available from: https://doi.org/10.1016/j.algal.2020.101819.

[93] Y. Li, C. Zhu, J. Jiang, Z. Yang, W. Feng, L. Li, et al., Catalytic hydrothermal liquefaction of *Gracilaria corticata* macroalgae: effects of process parameter on bio-oil up-gradation, Bioresour. Technol. 319 (2020) 124163. Available from: https://doi.org/10.1016/j.biortech.2020.124163.

[94] Y. Li, C. Zhu, J. Jiang, Z. Yang, W. Feng, L. Li, et al., Hydrothermal liquefaction of macroalgae with in-situ-hydrogen donor formic acid: effects of process parameters on products yield and characterizations, Ind. Crop. Prod. 153 (2020) 112513. Available from: https://doi.org/10.1016/j.indcrop.2020.112513.

[95] S. Raikova, C. Le, T.A. Beacham, R. Jenkins, M. Allen, C. Chuck, Towards a marine biorefinery through the hydrothermal liquefaction of macroalgae native to the United Kingdom, Biomass Bioenerg. 107 (2017) 244–253. Available from: https://doi.org/10.1016/j.biombioe.2017.10.010.

[96] J. Xu, X. Dong, Y. Wang, Hydrothermal liquefaction of macroalgae over various solids, basic or acidic oxides and metal salt catalyst: products distribution and characterization, Ind. Crop. Prod. 151 (2020) 112458. Available from: https://doi.org/10.1016/j.indcrop.2020.112458.

[97] Y. Chen, Q. Wei, X. Ren, The effect of hydrophilic amines on hydrothermal liquefaction of macroalgae residue, Bioresour. Technol. 243 (2017) 409–416. Available from: https://doi.org/10.1016/j.biortech.2017.06.148.

[98] B. Biswas, A.A. Kumar, Y. Bisht, R. Singh, J. Kumar, T. Bhaskar, Effects of temperature and solvent on hydrothermal liquefaction of *Sargassum tenerrimum* algae, Bioresour. Technol. 242 (2017) 344–350. Available from: https://doi.org/10.1016/j.biortech.2017.03.045.

[99] B. Biswas, A. Kumar, A.C. Fernandes, K. Saini, S. Negi, U.D. Muraleedharan, et al., Solid base catalytic hydrothermal liquefaction of macroalgae: effects of process parameter on product yield and characterization, Bioresour. Technol. (2020) 123232. Available from: https://doi.org/10.1016/j.biortech.2020.123232.

[100] H. Niaz, B. Brigljevic, Y.B. Park, H.-C. Woo, J.J. Liu, Comprehensive feasibility assessment of combined heat, hydrogen, and power production via hydrothermal liquefaction of *Saccharina japonica*. ACS Sustain, Chem. Eng. 8 (2020) 8305–8317. Available from: https://doi.org/10.1021/acssuschemeng.0c01951.

[101] R. Singh, T. Bhaskar, B. Balagurumurthy, Effect of solvent on the hydrothermal liquefaction of macro algae Ulva fasciata, Process. Saf. Environ. 93 (2015) 154–160. Available from: https://doi.org/10.1016/j.psep.2014.03.002.

[102] Y. He, X. Liang, C. Jazrawi, A. Montoya, A. Yuen, A.J. Cole, et al., Continuous hydrothermal liquefaction of macroalgae in the presence of organic co-solvents, Algal Res. 17 (2016) 185–195. Available from: https://doi.org/10.1016/j.algal.2016.05.010.

[103] W. Yang, X. Li, S. Liu, L. Feng, Direct hydrothermal liquefaction of undried macroalgae Enteromorpha prolifera using acid catalysts, Energy Convers. Manag. 87 (2014) 938–945. Available from: https://doi.org/10.1016/j.enconman.2014.08.004.

[104] Q.-V. Bach, M.V. Sillero, K.-Q. Tran, J. Skjermo, Fast hydrothermal liquefaction of a Norwegian macro-alga: screening tests, Algal Res. 6 (2014) 271–276. Available from: https://doi.org/10.1016/j.algal.2014.05.009.

[105] B. Biswas, A.C. Fernandes, J. Kumar, U.D. Muraleedharan, T. Bhaskar, Valorization of *Sargassum tenerrimum*: value addition using hydrothermal liquefaction, Fuel 222 (2018) 394–401. Available from: https://doi.org/10.1016/j.fuel.2018.02.153.

[106] N. Neveux, A. Yuen, C. Jazrawi, M. Magnusson, B. Haynes, A. Masters, et al., Biocrude yield and productivity from the hydrothermal liquefaction of marine and freshwater green macroalgae, Bioresour. Technol. 155 (2014) 334–341. Available from: https://doi.org/10.1016/j.biortech.2013.12.083.

[107] J. Arun, K.P. Gopinath, P. SundarRajan, R. Malolan, P. AjaySrinivaasan, Hydrothermal liquefaction and pyrolysis of *Amphiroa fragilissima* biomass: comparative study on oxygen content and storage stability parameters of bio-oil, Bioresour. Technol. Rep. 11 (2020) 100465. Available from: https://doi.org/10.1016/j.biteb.2020.100465.

[108] S.K. Maity, Opportunities, recent trends and challenges of integrated biorefinery: part II, Renew. Sustain. Energy Rev. 43 (2015) 1446–1466. Available from: https://doi.org/10.1016/j.rser.2014.08.075.

[109] R. Shakya, J. Whelen, S. Adhikari, R. Mahadevan, S. Neupane, Effect of temperature and Na_2CO_3 catalyst on hydrothermal liquefaction of algae, Algal Res. 12 (2015) 80–90. Available from: https://doi.org/10.1016/j.algal.2015.08.006.

[110] G. Yu, Y. Zhang, L. Schideman, T. Funk, Z. Wang, Distributions of carbon and nitrogen in the products from hydrothermal liquefaction of low-lipid microalgae, Energy Environ. Sci. 4 (2011) 4587–4595. Available from: https://doi.org/10.1039/C1EE01541A.

[111] U. Jena, K. Das, J. Kastner, Comparison of the effects of Na_2CO_3, $Ca_3(PO_4)_2$, and NiO catalysts on the thermochemical liquefaction of microalga *Spirulina platensis*, Appl. Energy. 98 (2012) 368–375. Available from: https://doi.org/10.1016/j.apenergy.2012.03.056.

[112] S. Nanda, J. Mohammad, S.N. Reddy, J.A. Kozinski, A.K. Dalai, Pathways of lignocellulosic biomass conversion to renewable fuels, Biomass Convers. Biorefin 4 (2014) 157–191. Available from: https://doi.org/10.1007/s13399-013-0097-z.

[113] C. Tian, B. Li, Z. Liu, Y. Zhang, H. Lu, Hydrothermal liquefaction for algal biorefinery: a critical review, Renew. Sustain. Energy Rev. 38 (2014) 933–950. Available from: https://doi.org/10.1016/j.rser.2014.07.030.

[114] S. Hongthong, S. Raikova, H.S. Leese, C.J. Chuck, Co-processing of common plastics with pistachio hulls via hydrothermal liquefaction, Waste Manage. 102 (2020) 351–361. Available from: https://doi.org/10.1016/j.wasman.2019.11.003.

[115] A.H. Zacher, M.V. Olarte, D.M. Santosa, D.C. Elliott, S.B. Jones, A review and perspective of recent bio-oil hydrotreating research, Green Chem. 16 (2014) 491–515. Available from: https://doi.org/10.1039/C3GC41382A.

[116] B.R. Pinkard, D.J. Gorman, K. Tiwari, J.C. Kramlich, P.G. Reinhall, I.V. Novosselov, Review of gasification of organic compounds in continuous-flow, supercritical water reactors, Ind. Eng. Chem. Res. 57 (2018) 3471–3481. Available from: https://doi.org/10.1021/acs.iecr.8b00068.

[117] M. Schubert, J.B. Müller, Fdr Vogel, Continuous hydrothermal gasification of glycerol mixtures: autothermal operation, simultaneous salt recovery, and the effect of K_3PO_4 on the catalytic gasification, Ind. Eng. Chem. Res. 53 (2014) 8404–8415. Available from: https://doi.org/10.1021/ie5005459.

[118] J.A. Okolie, R. Rana, S. Nanda, A.K. Dalai, J.A. Kozinski, Supercritical water gasification of biomass: a state-of-the-art review of process parameters, reaction mechanisms and catalysis, Sustain. Energy Fuels. 3 (2019) 578–598. Available from: https://doi.org/10.1039/C8SE00565F.

[119] W.Y. Cheah, T.C. Ling, P.L. Show, J.C. Juan, J.-S. Chang, D.-J. Lee, Cultivation in wastewaters for energy: a microalgae platform, Appl. Energy 179 (2016) 609–625. Available from: https://doi.org/10.1016/j.apenergy.2016.07.015.

[120] Y.Y. Gan, H.C. Ong, W.-H. Chen, H.-K. Sheen, J.-S. Chang, C.T. Chong, et al., Microwave-assisted wet torrefaction of microalgae under various acids for coproduction of biochar and sugar, J. Clean. Prod. 253 (2020) 119944. Available from: https://doi.org/10.1016/j.jclepro.2019.119944.

[121] K.-T. Wu, C.-J. Tsai, C.-S. Chen, H.-W. Chen, The characteristics of torrefied microalgae, Appl. Energy 100 (2012) 52–57. Available from: https://doi.org/10.1016/j.apenergy.2012.03.002.

[122] N. Phusunti, W. Phetwarotai, S. Tekasakul, Effects of torrefaction on physical properties, chemical composition and reactivity of microalgae, Korean J. Chem. Eng. 35 (2018) 503–510. Available from: https://doi.org/10.1007/s11814-017-0297-5.

[123] S.-H. Ho, C. Zhang, W.-H. Chen, Y. Shen, J.-S. Chang, Characterization of biomass waste torrefaction under conventional and microwave heating, Bioresour. Technol. 264 (2018) 7–16. Available from: https://doi.org/10.1016/j.biortech.2018.05.047.

[124] Q.-V. Bach, W.-H. Chen, S.-C. Lin, H.-K. Sheen, J.-S. Chang, Wet torrefaction of microalga *Chlorella vulgaris* ESP-31 with microwave-assisted heating, Energy Convers. Manage. 141 (2017) 163–170. Available from: https://doi.org/10.1016/j.enconman.2016.07.035.

[125] R.E. Sims, W. Mabee, J.N. Saddler, M. Taylor, An overview of second generation biofuel technologies, Bioresour. Technol. 101 (2010) 1570–1580. Available from: https://doi.org/10.1016/j.biortech.2009.11.046.

[126] M. Hajjari, M. Tabatabaei, M. Aghbashlo, H. Ghanavati, A review on the prospects of sustainable biodiesel production: a global scenario with an emphasis on waste-oil biodiesel utilization, Renew. Sustain. Energy Rev. 72 (2017) 445–464. Available from: https://doi.org/10.1016/j.rser.2017.01.034.

[127] S.S. Hassan, G.A. Williams, A.K. Jaiswal, Moving towards the second generation of lignocellulosic biorefineries in the EU: drivers, challenges, and opportunities, Renew. Sustain. Energy Rev. 101 (2019) 590–599. Available from: https://doi.org/10.1016/j.rser.2018.11.041.

[128] G. Kumar, S. Shobana, W.-H. Chen, Q.-V. Bach, S.-H. Kim, A. Atabani, et al., A review of thermochemical conversion of microalgal biomass for biofuels: chemistry and processes, Green. Chem. 19 (2017) 44–67. Available from: https://doi.org/10.1039/C6GC01937D.

[129] S. Ali-Ahmad, A. Karbassi, G. Ibrahim, K. Slim, Pyrolysis optimization of Mediterranean microalgae for bio-oil production purpose, Int. J. Environ. Sci. Technol. 17 (2020) 4281–4290. Available from: https://doi.org/10.1007/s13762-020-02735-8.

[130] L. Gang, X. Shunan, J. Fang, Z. Yuguang, H. Zhigang, Thermal cracking products and bio-oil production from microalgae *Desmodesmus* sp, Int. J. Agric. Biol. Eng. 10 (2017) 198–206. Available from: https://doi.org/10.25165/j.ijabe.20171004.3348.

CHAPTER 14

Direct biohydrogen production from algae

Eithar El-Mohsnawy[1], Ali Samy Abdelaal[2] and Mostafa El-Sheekh[3]

[1]Botany and Microbiology Department, Faulty of Science, Kafrelsheikh University, Kafr Al Sheikh, Egypt [2]Department of Genetics, Faculty of Agriculture, Damietta University, Damietta, Egypt [3]Department of Botany, Faculty of Science, Tanta University, Tanta, Egypt

14.1 Introduction

Hydrogen production capacity is starting to be an effective solution for solving the energy crisis, providing sustainable and renewable energy for industry. Applying hydrogen (H_2) energy takes place in four main phases: production, storage, security, and applications. Hydrogen production technology depends on the type of raw materials. Since hydrogen production methods and associated technologies are the main challenging issues nowadays, developing new techniques for production and storage is notable. Although hydrogen does not have any energy for emissions of carbon dioxide at the point of end-use, hydrogen production requires a production process with high cleanliness [1].

With the rapid development of human activities (industry, mechanical agriculture, transportation, etc.), the demand for energy is increasing, leading to major crises. Traditional energy resources depend mainly on carbon-rich fuels (fossil fuels), which decrease and then become depleted [1]. Fossil fuels and their production of carbon dioxide are the main cause of global warming [2]. Because of this, finding new, sustainable, and clean renewable energies is a major challenge for many scientific groups around the world [2]. In order to replace the energy source with 100% renewable energy, many technical adjustments are required, which vary depending on the energy source.

14.1.1 Hydrogen production and applications

Hydrogen is expected to be the primary energy carrier in several industrial applications. For example, many industrial processes require a high temperature for which hydrogen combustion can be used as an efficient method. Hydrogen is also an essential material in the production of many high-value materials, such as ammonia, methanol, and polymers. Also, it has beneficial uses in this purification of fuel from sulfur, the steel industry, and the reuse of carbon dioxide that has risen as a result of industrial activity [3].

However, technological progress is required to enable the hydrogen content to be increased, and accordingly, the increase in global hydrogen applications has led to the promotion of investment and the development of the production of renewable hydrogen and related technologies in many countries. The United States, Australia, Japan, and Germany are among the major players in the field of hydrogen production [3].

Based on 2016 reports, the source of hydrogen production for industry is basically fossil fuels (55 million tons per year) [4]. On the other hand, the thermochemical conversion of coal and steam reforming of methane are the most common methods of hydrogen production due to their cost-efficiency [5], where hydrogen is produced through the chemical reaction of purified methane with natural gas and an increase in temperature. It should be noted that the steam reforming of methane produces carbon dioxide. Due to its production from fossil raw materials, hydrogen is not considered renewable energy here, whereas hydrogen produced from the electrical splitting of water is considered renewable energy. Hydrogen production in this way currently represents only 4% of the total but it is expected to reach 22% by 2050. Scientists have developed many methods for producing renewable hydrogen, including photocatalytic hydrogen production from photoactive material and biological sources. Some of these technologies are expected to mature and enter commercial production by 2030. As a result, green hydrogen is becoming more available and widely used in power generation and powertrain systems [3].

14.1.2 Hydrogen as new vehicle energy source

Due to the strong pressure on companies that use fossil fuels to minimize the CO_2 emissions, many companies have developed fleets of machines and cars that produce less CO_2. Accordingly, several machine designs have been developed and approved for the progressive replacement of the internal combustion engine (ICE) drive technology with fuel cells (FC) and battery-powered electric vehicles [6]. Hence, hydrogen has been recognized due to its role in the elimination of carbon emissions (carbon dioxide, carbon monoxide, and soot) and its high energy efficiency [7]. The first successful commercial use of hydrogen as a vehicle energy source was in the 1930s. More than 1000 vehicles have been developed with a hybrid motor that can rely on either hydrogen or gasoline as a source of energy. The development of hydrogen-powered ICEs has now ceased, with scientific research continuing but practical applications being limited. After 1950, hydrogen-powered ICEs were primarily demonstration projects. More recently, efforts to decarbonize and tighten emissions standards have led to many developments in renewable hydrogen technologies, including production, transportation, storage, and use. These developments led to a revival of global interests regarding the use of hydrogen as a power source for engines [8].

The Hydrogen Council has predicted that by 2050 about a quarter of passenger cars and a fifth of nonelectrified trains will be powered by hydrogen, so the daily consumption of oil will be reduced to 20%, and then it can be consumed for other purposes [3].

14.1.3 Development of hydrogen-dependent energy storage systems

Hopes are pinned on developing the Hydrogen Energy Storage Systems (HydESS) to find the solution for large-scale hydrogen storage, transportation, and export [9]. The entire hydrogen-dependent economy remains controversial and inaccessible, although it has recently begun to show high potential. Hydrogen can be gained from different resources using various raw materials, methods, and technologies, including fossil fuels and renewable energy sources. The classic method is to crack or reform fossil fuels in order to produce hydrogen for industrial purposes in a cost-effective manner; in 2016 this was estimated (worldwide) to produce 85 million tons (over 600 billion Nm/year). Since hydrogen is an industrial raw material, the energy content of hydrogen produced by the Clean Energy Index is one of the most important indicators for calculating fuel quality. With the growing interest in reducing greenhouse gas emissions, hydrogen has gained attention as a clean source and as a carbon-free energy carrier [10]. Renewable hydrogen provides the link between renewable energy sources, the modernization of energy supply, transportation, and industry, and the export of renewable energies

14.2 Biohydrogen as efficient future fuels

Concern has been raised about the sustainability of the use of fossil fuels due to fuel depletion and the emission of greenhouse gases. Among the various renewable energy sources, biofuels, natural gas, and hydrogen are the three most important renewable energy sources. Since hydrogen gas does not contain carbon, it does not produce carbon dioxide, and therefore, it has no effect on raising the Earth's temperature and can be easily converted into electricity. Thus it can be considered an effective source of future alternative energy.

The use of hydrogen reduces environmental risks such as climate change and its impact on biological species, melting ice, and the rise in water levels [11]. This framework is in line with the advice given by environmental scientists in this regard, represented by the production of clean energy and the reduction of emissions of harmful gases [12]. Hence, these concerns regarding environmental safety and energy security have encouraged researchers to find, invite, and develop renewable, clean, sustainable, and economical energy sources [13]. For this reason, renewable energy plays a critical role nationally and globally in reducing concerns about energy needs and climate changes. So, working on converting energy supplies to hydrogen as an alternative to petroleum in the future could increase environmental security and create a source of income for farmers [13]. However, the conditions for producing sustainable renewable energy and replacing it with petroleum are a very difficult task and may take more time to implement [13].

Because most of the sources of commercial biofuel production come from food crops, this has led to the emergence of serious social and economic concerns due to the competition between energy and food. From here comes biorenewable hydrogen as a strong competitor with the possibility of using it as a fuel for various needs such as electricity, heating, and transportation [14]. Enhancing biomass production is considered an indirect way of effective biohydrogen production. Enhancing algal biomass leads to increasing hydrogen production enzymes as well as increasing the fermentation process that are discussed in depth later in this chapter.

14.2.1 Benefits of biohydrogen and future prospects

The biohydrogen industry has attracted global attention due to the environmental, economic, and social benefits it provides [15]. Since biohydrogen can be generated by solar energy with trace concentrations of nutrients or organic wastewater as a source of nutrients, this has environmental advantages and competitive production costs. This source of energy is not competing with food or animal feed and does not require fertile space for agriculture. Hydrogen gas is renewable and does not emit carbon dioxide, yet it emits a large burning energy per unit weight, making it easier to convert into electricity by FCs, and for this reason, biohydrogen is considered to be one of the strong competitors for future energy (http://www.oilgee.com, 2012). It should be noted here that there are some technological limitations currently for the industrial production of biohydrogen, but its multiple advantages enable it to be the most popular alternative in comparison to other renewable energy sources.

There are three successful strategies for biohydrogen production: biodegradation (using microalgae), dark fermentation, and light fermentation [14]. One of the main challenges that face the production of biohydrogen is that the production rate is slow, and hydrogen production is low. However, available results over the past two decades have indicated the availability of data on the production of biohydrogen. However, for its industrial use, which is essential for the economy, the current rate of biohydrogen production and yield must necessarily exceed current successes [16]. Theoretical data indicate the possibility of producing 80 kg of hydrogen per acre per day from photosynthesis by algae, but actually the production of biological hydrogen is half the theoretical data, and its production cost is about $ 2.80 per kg [17]. Because in the current scenario 10% of the algae's photosynthetic capacity to generate biohydrogen is used [16], biotechnology approaches are currently being researched to improve photosynthetic biohydrogen production in algae and these show promising results.

14.3 Direct cellular biohydrogen production

At the beginning of life on Earth, it is assumed by many scientists that hydrogen was the only/most important source of energy for primitive life. Up to now, there were no recorded oxygen-resistant hydrogenases [18]. According to theoretical estimations, biohydrogen production could be enhanced by up to 140% via coupling with photosynthesis

[19]. Nevertheless, the actual performance is close to 10% efficiency [20]. Based on photosynthesis, the strategy to increase hydrogen production deals with the flow of high-energy electrons from the optical systems (PSI, PSII) to the hydrogenase enzyme, which uses these high-energy electrons to reduce protons and then sublime the hydrogen. Combining one or two photosynthetic complexes (PSI and PSII) with hydrogenase by direct or indirect fusion enhances electron pumping to hydrogenase, and hydrogen production is promoted. The nature and structure of these complexes aids in the construction of the appropriate designs to achieve this goal. Here, we will highlight the structure of the main complexes involved in hydrogen production via photosynthesis as well as describe electron transport during the light reaction.

14.3.1 The photosynthetic electron transport chain in the natural system

By tracking the photosynthetic electron transport in oxygen-producing organisms, it is discovered that there are two fundamental chains of electron transport in photosynthesis, namely, cyclic and linear [21]. The light-dependent reaction is associated with a catalytic power separation reaction, which is carried out via water splitting by photosystem II (PSII) in algae and cyanobacteria. Electrons migrate through a range of intrinsic electron cofactors within PSII (fovitin and kinone) and PSI (phylloquinone and 4Fe4S groups) to attain the ultimate electron receptors for plastoquinone (PQ) and ferredoxin, respectively [21,22].

The solar energy absorbed through the chlorophyll system is efficaciously and shortly transferred to the PSII and PSI reaction centers, where energy is separated and the water split in PSII (Fig. 14.1). The first initial power switch to electrochemical potentials occurs in PSII with a thermodynamic efficiency of about 70%, which creates a $P680^{\bullet+}Pheo^{\bullet-}$ [21]. The yielded redox potential of $P680^{\bullet+}$ is strongly oxidizing and is estimated to be about +1.2 V, while the capacity of $Pheo^{\bullet-}$ is around 0.5 V [21]. $P680^{\bullet+}$ generated in PSII

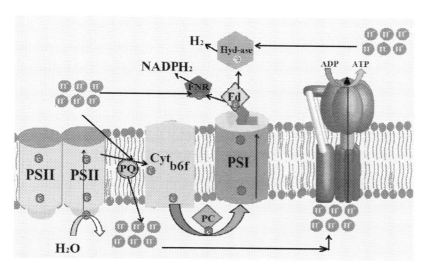

FIGURE 14.1 Linear photosynthetic electron transport chain in green algae. The final electron acceptor ferredoxin transfers electrons from photosystem I to either ferredoxin-NADP-reductase or H_2-ase according to cultivation condition.

recovers the liberated electrons by splitting water molecules in the oxygen evolving complex center. Two water molecules are split into two oxygen atoms, four protons, and four electrons [21]. In oxygen-containing photosynthetic organisms, this response continues as the QH2 molecules form. At the same time, the PSI antenna absorbs solar light, and the P_{700} reaction center is stimulated in order to elevate it to a decreasing voltage of about 1.0 V or more [21]. The launched electrons go through the external electron carriers reaching ferredoxin (Fd), the final electron acceptor of oxygen-producing organisms. Fd is reduced to ferredoxin—NADP reductase (FNR) allowing reduced nicotinamide adenine dinucleotide phosphate (NADPH) to be formed, whilst the reduced equivalent is transported alongside the electron transport chain to regain liberated electrons from PSI. For this reason, PSII is regarded to be a hydrophobic PQ oxidoreductase, which catalyzes the splitting of water into O_2, four protons, and electrons:

$$2H_2O \xrightarrow{P_{680}} 4H^+ + 4e^- + O_2$$

Then the PQ cyclizes electrons between PSII and cytochrome b6f (Cyt-b6f) within the lipid section in the thylakoid membrane (TM), which then redirects these electrons to the mobile electron carrier cytochrome c6 (Cyt-c6) or plastocyanin (Pc), or they are bound to the lumen side of Cyt-b6f. These electron carriers switch the electrons to the PSI. In addition, a difference in proton awareness throughout the TM is formed, leading to the electrochemical force to the TM, which derives the protons to be pumped throughout the ATP-synthase embedded within the TM. This action yields enough energy to produce ATP from ADP and Pi [2]. The created free chemical energy in the form of NADPH and ATP is used in light-independent reactions that are carried out in the cytosol of cyanobacteria and other oxygen-containing photosynthetic organisms [2]. PSII, Cyt-b6f, and PSI are three supercomplexes embedded in the TM that oversee in sequence a linear photosynthetic electron transport chain [2]. So this process is summarized as: solar energy is used to transfer electrons from water to $NADP^+$ and protons, O_2, ATP, and NADPH are produced.

Besides the linear photosynthetic electron transport chain, there is a cycle route (electronic cyclic transport). Under inactivation of PSII and/or a low availability of $NADP^+$ at the stromal side, the amount of electron flow to Fd becomes very limited, so Fd redirects the electrons toward the Cyt-b6f, and electrons are transferred back to plastocyanin or Cyt-c6 via the Cyt-b6f complex. Under this scheme, this system does not now require any input energy (electrons) from PSII, so PSII is not now incorporated in this chain, and hence there is O_2 produced [21].

The energetic system of photosynthesis in the thylakoid membrane acts to the direct manufacturing of biohydrogen by converting the yielded electrons generated throughout photosynthesis into a hydrogenase enzyme instead of FNR (ferredoxin reductase—NADP) [19,20,23]. There are some herbal structures in which some organisms can produce hydrogen whilst producing NADPH [23]. Three supercomplex enzymes are concerned with hydrogen production in algal cells: PSII (splitting H_2O and pumping electrons), PSI (pumping electrons), and hydrogenases (using electrons and protons to produce hydrogen). Other protein complexes concerned with electron transport are considered electron carriers [23].

14.3.2 Photosystem II

Photosystem II is one of the largest protein supercomplexes embedded in the TM of phototrophic oxygen-producing organisms. The main roles of PSII are splitting water to liberate electrons, pumping these electrons via internal electron carriers to PQ, pumping protons, and, of course, evolving molecular oxygen. The exact composition of PSII was solved with precision from 3.8 to 2.9 Å in two thermophilic cyanobacterial strains (*Thermosynechococcus elongatus* and *Thermosynechococcus vulcanus*) [24–26]. The estimated molecular weight of PSII in higher photosynthetic organisms is about 350 kDa and it consists of 17 subunits of embedded protein plus three peripheral protein subunits and a number of cofactors. The Mn4Ca group is present in the center of the oxygen-evolving complex where it is responsible for splitting water into electrons, protons, and molecular oxygen [26]. PSII is always current in binary shape, and each monomer includes 19 protein subunits, 35 chlorophyll molecules, 11 ß-carotenes, two phaeophytins, two PQ, heme, nonheme iron, four manganese atoms, and three or four atoms of calcium (one of which is in the Mn_4Ca group), three chlorine ions (two of which are close to the Mn_4Ca group), one bicarbonate ion, more than 20 fats, more than 15 detergents, and more than 1300 water molecules [25,26]. The electron density of the manganese and calcium atoms in the oxygen development complex (OEC) was well resolved by the use of X-ray measurements of the PSII crystal shape [26]. It used to be determined that the electron density of the four calcium atoms is much less than that of the manganese atoms, while the five oxygen atoms act as oxo bridges connecting the five metal atoms [26].

As shown in Fig. 14.2, the Mn_4CaO_5 network atoms build a cube-like structure in which the calcium and manganese atoms are located at four corners, while oxygen atoms occupy the other corners (Fig. 14.2). The bond distances between the oxygen atoms and the calcium atom in the mantle are usually in the range of 2.4–2.5 Å, and those between the other four oxygen atoms and manganese atoms are in the range of 1.8–2.1 Å [26]. The estimated distance of the oxygen atoms (O5) in the cup attitude and calcium atom is 2.7 Å, whilst those between O5 atom and Mn atoms range between 2.4 and 2.6 Å. Beside these five oxygen atoms, there are four water molecules (W1–W4) that coordinate with Mn4 with precise distances ranging from 2.1 to 2.2 Å. The water atom W3 and W4 particles are coordinated with Ca atom and reach about 2.4 Å [26]. Ultrafast kinetic measurements within PSII emphasize the presence of radical pairs earlier than electrons switching to quinone QA [27,28]. In spite of high PSII activity in native cells, most activity is normally lost in vitro due to the turnover of the D1 subunit that directly affects the stability of complex. For this reason, extensive work has been undertaken to solve this problem [28]. The combination of high light intensity and iron deficiency on the thermophilic cyanobacteria *T. elongatus* (30°C and 1000 mol/mgs photons) showed a complete inhibitory effect on the Psb27 mutant strain, whilst the wild cells survived to develop [28]. Previous investigations have reported the presence of two distinct complexes. The first one is observed in the PSII-Psb27 monospecies, which is believed to be involved in PSII synthesis, whilst the second is discovered in the 2D-PSII-Psb27 complex, which may also have a function in the PSII repair cycle [28].

The electron transport pathway inside PSII depends upon the redox potentials of the core of the reaction center P680$^+$/P680 (primary oxidizing agent) and other internal

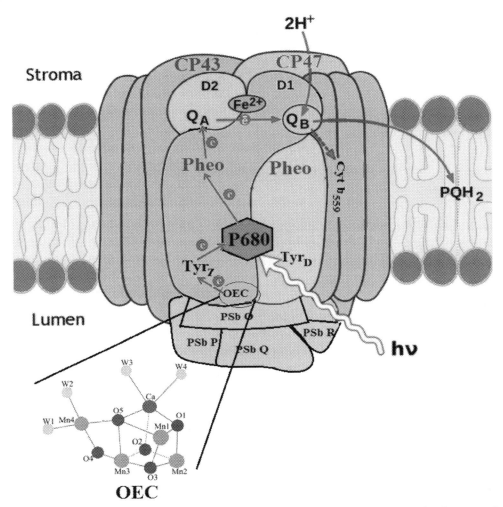

FIGURE 14.2 Diagrammatic illustration of photosystem II, electron transport via internal cofactors, and cubic structure of oxygen evolving complex.

electron carriers. The estimated redox attainable of the PSII reaction core (P680$^+$/P680) was +1.27 V [29], whilst the redox potentials recorded for the intermediate electron carriers have been 640 mV for Pheo a/Pheo a$^-$ [30], 30 mV for Q_{A-}/Q_A, +970 mV for TyrZ$^•$/TyrZ, +930 mV for O_2/H_2O, and +30 mV for Q_B/Q_{B-} [31]. Beside the importance of redox potentials of the PSII reaction center and internal electron carriers in electron flow within PSII, the actual distances between these cofactors (carriers) have great structural importance in controlling the electron flow. Based on the high resolution of the PSII crystal structure, the distances between P$_{680}$ and the other electron carriers involved in the electron transport chain have been recorded, and confirmed. Close distances between P$_{680}$ and different intermediate electron carriers facilitate rapid electron transport with the flow to Q_B as well as the recovery of electrons from the OEC (oxygen evolution complex) to P_{680}. The estimated distance between P_{680} and Pheo a is 14 Å, Pheo an and Q_A is 14 Å, TyrZ and P_{680} is 8.4 Å, OEC (Mn4Ca) and TyrZ is 7 Å, and Q_A and Q_B is 14.3 Å [32]. Kinetic measurements showed that the electron transfer from P680 to Pheo a occurs in 3 ps, while water oxidation takes place at around 1.4 ms [30,31,33].

14.3.3 Photosystem I

Photosystem I is a multiprotein complex embedded within the TM and acts to absorb solar energy for electron pumping to a final electron acceptor, ferredoxin (Fd), in higher plants, algae, and cyanobacteria [31]. In higher plants and eukaryotic algae, PSI improves light-harvesting capacity through carrying an antenna system. These antennas are composed of a protein—chlorophyll complex that absorbs the incident light energy and transfers it to the reaction center (P_{700}). So PSI acts to enable light-driven electron transfer from the lumen side by plastocyanin (Pc) or cytochrome C6 (Cyt C6) to the stromal side to ferredoxin [34—40].

The crystal structure of PSI from flowering plants and cyanobacteria show considerable similarity, with minor variations in the outer protein subunits [41]. The PSI subunits PsaH and PsaG are found only in eukaryotic algae and flowering plants, whereas they are missing in cyanobacteria. On the other hand, the small subunits PsaM and PsaX are observed only in cyanobacteria [22,42,43]. Subunit PsaL has a different specific structural position in eukaryotic photosynthetic organisms and cyanobacteria. While PsaL acts to stabilize trimeric configuration in cyanobacteria, it prevents oligomerization in eukaryotic algae and flowering plants [37,39,43]. Another fundamental structural difference between cyanobacteria and eukaryotic algae is that the eukaryotic algal PSI binds to four light-harvesting complexes, LHC, that enhance light absorption through large chlorophyll a/b binding to LHC complexes [36,38,44]. Intensive research has emphasized that PSI of thermophilic cyanobacteria are mainly in trimeric form, while those of mesophilic cyanobacteria contain monomeric and trimeric forms [44—46]. Trimeric PSI molecular weight is estimated to be 1,068,000 Da [22,34,35]. The X-ray crystal structure of *T. elongatus* shows nine transmembrane subunits: PsaA, PsaB, PsaF, PsaI, PsaJ, PsaK, PsaL, PsaM, and PsaX; and three luminal subunits: PsaC, PsaD, and PsaE. One hundred and twenty-seven cofactors, are noncovalently bound to protein subunits. PSI carries 96 chlorophyll molecules, 22 ß-carotenes, three (4Fe4S) clusters, two phylloquinones, and four lipid molecules [22,35]. PsaA and PsaB are the bulk of monomeric PSI, where the antenna chlorophyll network, reaction center, and most electron carriers are found [34,35,45—47]. It should be clarified that PsaC has a functional role in electron transport, where it is bound to F_A and F_B, the terminal electron acceptor. Extrinsic subunits, PsaC, PsaD, and PsaE facilitate the electron transport to the external electron acceptor ferredoxin/flavodoxin by saving docking site to them [22,35]. Eleven cofactors, six chlorophylls, two phylloquinones, and three 4Fe4S clusters share the electron transport in PSI. [22,34,35,37,44—46].

The catalyzing role of PSI could be summarized as light assembly, excitation of P_{700}, energy transfer, and electron transport. Although charge separation was widely thought to be initiated at P_{700} [48], Holzwarth and coworkers proposed that P_{700} is not oxidized at the second electron-transfer process [47]. As light energy is captured, energy is assembly at reaction center P_{700} causing it to oxidize, then the electron is transferred stepwise to A_0, A_1, and from there subsequently to the three 4Fe4S clusters, named F_X, F_A, and F_B [35,42].

After the docking of ferredoxin at PSI stromal side, the electron is transferred from the final electron carrier in PSI, F_B, to the 2Fe2S cluster of ferredoxin, which forwards these electrons p to the ferredoxin-NADP oxidoreductase FNR that reduces $NADP^+$ into NADPH [35,42]. To complete the cycle, P_{700}^+ is rereduced with the aid of soluble electron carrier proteins in the lumen side, plastocyanin or cytochrome. In some algae, the electrons switch to hydrogenase in competing with FNR, leading to the production of hydrogen. Because the hydrogen production rate is affected mainly by pumped electrons from PSI to hydrogenase, PSI—hydrogenase fusion could improve electron flow [2].

14.3.4 Hydrogenases

Hydrogenases are a group of enzymes that catalyze the reversible reaction $2H^+ + 2e^- \leftrightarrow H_2$ [49]. Although some classify hydrogenases into only two classes [49], others differentiate them into three divisions, NiFe, FeFe, and Hmd-Fe hydrogenases (Fe hydrogenases), based on the linker of bimetallic active site cofactors [50]. Generally, hydrogenases are detected in bacteria, cyanobacteria, and chlorophytes, and they are able to produce hydrogen and/or fix nitrogen [51]. There is some evidence that some cyanobacteria contain heterocysts, so they cannot fix nitrogen in the same cell, whereas some nonheterocysts cyanobacteria can fix nitrogen in the same cell. These organisms create a time interval between the oxygen evolution during photosynthesis and the production of hydrogen by hydrogenase/nitrogenase such as *Synechococcus* PCC 7942 and *Spirulina* sp. [52].

It should be noted that FeFe-hydrogenases are isolated from some eukaryotic algae (green algae) and a few prokaryotes [50]. FeFe-hydrogenases are very active, but are greatly affected by the presence of oxygen. On the other hand, some other organisms can produce hydrogenases that tolerate oxygen, such as *Aquifex aeolicus*, as they are able to produce hydrogen in the presence or absence of oxygen [53]. With regard to NiFe-hydrogenases, they are produced mainly in prokaryotes and are characterized by their tolerance to oxygen compared to FeFe-hydrogenases [52].

14.3.4.1 *NiFe-hydrogenases*

According to the analysis of X-ray crystal imaging and electron paramagnetic resonance (EPR), as well as by phylogenetic analysis, nickel—iron (NiFe)-hydrogenases have a spatial structure of the bimetallic active region linking Ni to the protein through four cysteines, two of which act as the final coordinator while the other two act as a bridge between the bimetallic centers (Fig. 14.3) [54,55]. NiFe-hydrogenases are composed of two protein subunits in which the active center resides in a large protein subunit, while the FeS centers are linked to a small protein subunit that connects to the active center for hydrogen production with external redox partners [54,55]. During the catalytic cycle, the center and oxidation state change with changes in the nature of the ligand at the fifth focal site of nickel

Handbook of Algal Biofuels

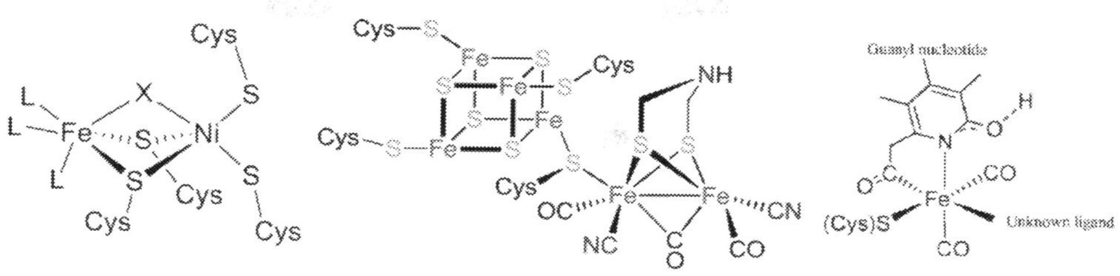

FIGURE 14.3   Structures of functional active site of different types of hydrogenases.

[54,55]. The triple state of Ni is inactive oxidation, whereas under reduction activation, hydrogen can bind to the Ni^{2+} state producing Ni^{3+}-μ-hydrido complex [54,55].

Presently, NiFeSe-hydrogenases are getting a lot of attention due to their ability to work in oxygen conditions [55]. Structurally, they are considered a subset of NiFe-hydrogenases, as they are only detected and isolated from microorganisms that are grown on a medium containing selenium [54].

14.3.4.2 FeFe-hydrogenases

Algae hydrogenases or FeFe-Hydrogenases are known by this name because they are mainly found in green algae [54]. Investigations of FeFe-hydrogenases active sites reveal the presence of six iron atoms (catalytic H group) of which four construct a (4Fe4S) group by binding to the S-Cys (FeFe) site while the separated iron atoms are classified as distant (Fig. 14.3) [54–58]. The hydrogen-cluster active site of the algal FeFe-hydrogenase is inserted into a single polypeptide chain. However, the structures of most bacterial FeFe-hydrogenases appear strongly similar to those of NiFe-hydrogenase species, with additional FeS groups, which bind electrons between the active site and the protein surface [54,55].

FTIR and EPR spectroscopic analysis of various FeFe-hydrogenases showed that the nature of the hydrogen-cluster is highly conserved, and all of these types of enzymes showed almost the same spectral properties, as well as the 4Fe4S and FeFe components of the hydrogen-cluster active site group [54–58]. It should be noted here that a highly potent enzyme FeFE-hydrogenase of *Clostridium pasteurianum* produces a hydrogen rate of 3400 μL/min mg (in 1 min/1 g of the enzyme produced 83 L of atmospheric hydrogen) [55]. But under aerobic conditions, the rate of hydrogen production is irreversibly inhibited due to the damaging effect of oxygen on mass (4Fe4S) [56]. There have been a lot of attempts by scientists to overcome the problem of O_2 inhibition of the enzyme, but there is no mechanism reported to explain how FeFe-hydrogenases can be reengineered to neutralize the inhibitory effect of oxygen and thus remain active in the presence of oxygen [54,55].

14.3.4.3 Fe-hydrogenases

In 1990 the Fe-hydrogenases were isolated and identified. Fe-hydrogenases are also named as methylenetetrahydromethanopterin or iron-sulfur-cluster free hydrogenases [50,54,59,60]. Cytochrome-free hydrogenotroph methanogens catalyze the reverse conversion of hydride to methylene tetrahydromethanopterin by hydrogen molecules [60]. This reaction takes place in one step while converting carbon dioxide to methane [54]. In addition, the single iron atom is the only mineral center in each protein subunit, which is present in the hydrogen-activating Fe guanylpyridinol catalytic region (Fig. 14.3) [60]. Two moles of iron per mole of 76 kDa homodimer were detected [54,60]. This class of hydrogenase is hypersensitive to O_2 [59].

14.3.4.4 Improving hydrogenases activity

The extensive study of the ultrastructure of hydrogenase enzymes, the synthesis of their active sites, and the investigation of hydrogen production, the mechanisms, and active site inhibitors paved the way for the synthesis of artificially mimic compounds similar to the active site of hydrogenase enzymes [55].

Since the presence of carbon dioxide limits the inhibitory effect of oxygen, it is considered a prerequisite for the activity of hydrogenases as it also coordinates the iron atom within the active site of hydrogenase as a bond in the active site [54,60]. As a result of iron stabilization in lower oxidation states with the help of a carbon dioxide acceptor bond, the iron atoms behave like Pt [61]. The practical investigations applied on the enzyme isotopes of both active FeFe-hydrogenases and NiFe-hydrogenases have demonstrated the importance of the enzyme construction mechanism for the active site in order to have the proton transport near the center of the hydrogen-activating metal in a hydrogen catalyst [54,60−62]. By transferring different active mimics of FeFe-hydrogenase active sites, which contain a dithiolate coated with carbon, oxygen, nitrogen, or ligand into a [4Fe4S] group containing protein hydrogenase, only the nitrogen-containing molecule produces the component catalytically active enzyme [54,60−62]. Also, in Ni-model compounds, a second coordination ball geometry that could act as a proton relay greatly enhanced and accelerated the rate of hydrogen evolution [54,61,62].

14.4 Photosynthetic hydrogen production—cyanobacteria

Indeed, the processes of converting solar energy directly into hydrogen fuel are some of the most important biological processes to be applied. In photosynthesis, the captured light energy by photoactive complexes is consumed on charge separation and the reduction of ferredoxin which consequently reduces hydrogenases and/or nitrogenases, leading to the evolving of hydrogen. In indirectly producing hydrogen via biodegradation, the chemical energy generated by photosynthesis is firstly stored in the form of carbohydrates before being used to reduce hydrogenases and/or nitrogenases; these two operations could be separated either in time or in space. Some unicellular and filamentous cyanobacteria are able to produce hydrogen. The main reason for producing hydrogen by cyanobacteria instead of green algae is their possibly easier potential to be genetically manipulated, paving the way for future enhancement of hydrogen production via genetic engineering [63]. Several reports have revealed the responsibility of nitrogenases for the majority of the observed cyanobacteria hydrogen production [64]. Although some hydrogen is produced, with dinitrogen reduction as the inevitable side reaction, hydrogen production could be promoted in limited N_2, while the enzyme continues to circulate. Because of the slow reaction rate of converting protons to hydrogen (6.4/s), the reaction needs high energy to be achieved [64]. Some *Nostocales* cyanobacteria can produce specialized cells for nitrogen fixation and hydrogen production [65,66]. Under conditions of nitrogen reduction, approximately 10% of cell in the filaments can differentiate into a heterocyst leading to high cysts numbers and consequently high N_2-fixation and high hydrogen production [67]. These cells provide an aerobic environmental condition allowing nitrogenases sensitive to oxygen to work efficiently. Of course gene expression in heterocysts is altered, and thus the metabolism of the heterocyst is structurally different to that of the vegetative cells. PSII is not expressed in heterocysts, and thus the water is not dissociated, and consequently O_2 does not evolve. Since heterocysts lack the Calvin–Benson–Basham cycle, they cannot fix carbon dioxide; therefore, metabolism in heterocysts depends on supplying organic compounds imported from the neighboring cyanobacterial vegetative cells [68]. For this reason, hydrogen production in heterocysts is considered to be an indirect way as it depends upon imported organic compounds, that is, sucrose, so it is the second step [68].

The metabolic pathways that support hydrogen production by nitrogenase within the cyanobacterial heterocyst show that CO_2 cannot be fixed because there is no Rubisco (ribulose-1,5-bisphosphate carboxylase/oxygenase) in heterocysts. Thus the electrons reduce the proton from the organic compounds (possibly sucrose) that are synthesized during photosynthesis by neighboring vegetative cells. Monosaccharides are passed to glucose-6-phosphate-1-dehydrogenase, a starting enzyme in the pentose pathway. PSI creates sufficient photoelectrons to promote the reduction of ferredoxin (FdxH). The energy required for nitrogenases activity, 4 ATP/H_2, is obtained from cyclic photophosphorylation of PSI. Following up, the mechanism of action for oxygen-sensitive nitrogenase reveals that these enzymes require an anaerobic environment for activity. For this reason, heterocysts have special plasma membranes that are able to limit the spread or diffusion of oxygen inside heterocysts. It is possible that the presence of H_2-ase uptake (HupLS) decreases net hydrogen production [69,70].

Some unicellular cyanobacteria that have nitrogenase do not have heterocysts; they can fix nitrogen can also produce hydrogen. While inhibitions occurred due to photosynthetic evolving O_2 are circumvented by daily transcription control, resulting in maximum photosynthesis during daylight and maximum nitrogenase function in the dark [71]. Investigation of hydrogen production by nitrogenase hydrogen in unicellular cyanobacteria showed a maximum hydrogen production at a low light intensity and the presence of glycerin, which can consume O_2 through respiration and avoid argon, which removes the photovoltaic O_2 [71].

These organisms can also evolve hydrogen through the indirect photolysis process, in which photosynthesis of carbon takes place in the first stage of light and then uses the carbohydrates stored in the second stage, anaerobically, to make hydrogen. Thus both spatially and temporally, such a process separates photosynthesis to produce O_2 from the proton reduction reaction, which is sensitive to O_2. This has been shown experimentally with *Plectonema boryanum*, which was cycled several times through a nitrogen-limited aerobic phase that allowed glycogen accumulation and a second phase for anaerobic hydrogen production. Dark fermentation was performed on hydrogen during the second phase, with a yield of 12% (i.e., 1.44 moles H_2/mole hexose) [72].

14.5 Enhancing hydrogen production in microalgae by gene technology

Several studies have been conducted to improve the biohydrogen production in microalgae using advanced genetic engineering techniques. Development in genome sequence analysis [73], genetic transformation [74], genome editing [75,76], and other molecular biology techniques have paved the way for genetic engineering development to enhance biohydrogen production.

Random mutagenesis was the most applied popular approach to generate new mutants in the nuclear genes of green algae [77], but it requires high-throughput screening and is time-consuming. In prokaryotic algae and chloroplasts of Chlorophytes, homologous recombination is applied exclusively for knockout genes [78]. Vitamin-mediated riboswitch has been used to inhibit the chloroplast genes expression [79]. Also, RNA interference has been used to knockdown genes for a long time [80].

Recently, targeted genome-editing tools such as CRISPR-Cas9 or Cpf1 system, zinc finger nucleases (ZFNs), and transcription activator-like effector nucleases (TALENs) have been developed to specifically edit genomes of microalgae [75,76,81].

Complete genome sequences of several microalgae, including *Thalassiosira pseudonana*, *Cyanidioschyzon merolae*, *Ostreococcus tauri*, *Chlamydomonas* sp., *Ostreococcus lucimarinus*, *Phaeodactylum tricornutum*, and *Micromonas pusilla*, etc. have significantly improved genome-editing efficiency [82].

Besides the nuclear genome, a sequencing of chloroplast and mitochondrial genomes has been performed [83,84]. The DNA of the chloroplast and nucleus of *Chlamydomonas* sp. has been widely manipulated by genetic engineering technology to enhance their potential hydrogen production since they are easily transformed [74,85,86].

The key challenges for the commercialization of hydrogen production are the oxygen sensitivity of hydrogen production, and competition between hydrogenases and NADPH-dependent CO_2 fixation. The molecular genetic techniques have been applied to increase biohydrogen production mainly by overcoming oxygen sensitivity of hydrogenase, elimination of competing pathways, etc. A comprehensive list of key studies in genetic engineering for microalgal biohydrogen production along with their outcomes has been presented in Table 14.1.

14.5.1 Overcoming O_2 sensitivity of hydrogenase

The main challenge for sustainable biohydrogen production from microalgae is the sensitivity of hydrogenase to oxygen [99]. There are different approaches proposed to overcome the O_2 sensitivity of the hydrogenases, for examples:

1. Bioprospecting for new natural O_2-tolerant enzymes. A diversity of [FeFe] hydrogenase (HydA) was identified using degenerate primers in a microbial habitat representing adaptation to the environmental conditions [100].
2. Applying random mutations followed by screening and identifying more O_2-tolerant enzymes. A number of efforts have been performed to produce O_2-tolerant hydrogenases using random mutagenesis in vivo [101] and in vitro [102]. Unsurprisingly, these attempts have shown minor improvements in the tolerance of O_2.

TABLE 14.1 Genetic manipulation of microalgae to enhance hydrogen production.

Challenges	Species	Algal strain	Target	Type of modification	Results	Reference
Oxygen sensitivity	*Chlamydomonas reinhardtii*	L1591-N230Y	D1 protein subunit	Site-directed mutagenesis	Lower chlorophyll *a* content, higher photosynthetic rate, prolonged hydrogen production period, and higher H_2 production	[87]
	Chlorella sp. DT	siRNA-PsbO	PsbO	RNAi	HydA induced and increased photobiological H_2 production 10 times	[88]
	C. reinhardtii	cy6Nac2.49	Photosystem II	Riboswitch and homologous recombination	Knockdown of PsbD, less oxygen production, and an increase in hydrogenase activity	[79]
	C. reinhardtii	aprl	P/R ratio	DNA insertional mutagenesis	Promoted anaerobiosis and the rate of hydrogen production enhanced to double	[89]
	C. reinhardtii	CC-849 codon optimized hemH-lbA	Leghemoglobin	Heterologous expression	Slightly inhibited growth, a lower oxygen content, and improved H_2 yield	[90]
	Chlamydomonas reinhardtii	ccPHC	Pyruvate oxidase	Heterologous expression	Consumed more oxygen and hydrogen is promoted three-times more	[91]
Competition for electron	*C. reinhardtii*	CC-2803	RuBisCO large subunit	DNA insertional mutagenesis	Suppressed Calvin cycle, and decreased O_2 evolution	[92]
	C. reinhardtii	Y67A	RuBisCO small subunit	Site-directed mutagenesis	Suppressed Calvin cycle, decreased O_2 evolution and low photosynthesis rate	[93]
	C. reinhardtii	FDX1/HydA1	Ferredoxin-hydrogenase fusion protein	Prokaryotic expression	Increased hydrogen production six times	[94]
	Synechocystis strain M55	ndhB	The respiratory complex I, NADPHdehydrogenase	Homologous recombination	Low amount of O_2 in the light, and continuous hydrogen production for several minutes	[95]
	Synechocystis sp. PCC 6803	ctaI/ctaII/cyd	Respiratory terminal oxidases	Homologous recombination	higher Hox-hydrogenase activity and hydrogen production in the light	[96]
	Synechocystis sp. PCC 6803	narB/nirA	Nitrate assimilation pathway	Homologous recombination	Improved hydrogen production	[97]
	Synechococcus sp. PCC 7002	ldhA	NADH-dependent reduction of pyruvate to d-lactate	Homologous recombination	Inhibition in lactate production and enhanced H_2 production up to fivefold	[98]

318　14. Direct biohydrogen production from algae

3. Molecular engineering of hydrogenases to decrease the binding of O_2 to their respective catalytic site [103]. Modification of the structure of the hydrogenases may be an effective solution to reduce their sensitivity to oxygen. The specific amino acids were substituted to alter the conformation of the gas channel in the hydrogenases preventing the passage of oxygen and protecting the active sites [104,105]. The hydrogenases of these mutants are not folded properly, which has directly led to lower enzyme activity.

Unfortunately, none of these approaches has yet generated O_2 insensitive species, which is attributed to the complexity of the O_2 tolerance being conferred by multiple and dynamic mechanisms.

4. Anaerobiosis induction either by decreasing the O_2 evolution rates through partial PSII inactivation or increasing uptake/sequestration within the cell.

14.5.1.1 Partial photosystem II inactivation

The PSII large protein subunits D1 and D2 have a specific structural and functional role, where they carry all cofactors involved electron transport within PSII. A mutant *Chlamydomonas reinhardtii* strain, sulfur-deprived mutant, carrying a double amino acid substitution (L159I-N230Y) in D1 protein was modified for inactivating PSII activity in order to improve hydrogen photoproduction [87]. Compared to its wild type, this mutant has a low chlorophyll *a* content, a high photosynthetic rate, and high H_2 production as a result of the prolonged hydrogen production period.

A decrease in the evolved O_2 was attained to reduce the oxygen concentration in the cell, which inhibits hydrogenase. Knockdown of PsbO, a PSII subunit involved evolution of oxygen, using short interference RNA in unicellular alga *Chlorella* sp. DT led to HydA induction and a 10-fold increase in photobiological H_2 production compared to the wild type [88].

To regulate PSII activity, the chloroplast gene expression system has been modified to control the oxygen level and electron flow within the cell in order to enhance hydrogen production. The nucleus-encoded Nac2 chloroplast protein required for the stable accumulation of the psbD RNA (which encodes PSII D2 protein) was fused to the copper-sensitive *Cyc6* inducible promoter in *C. reinhardtii nac2−26* mutant strain deficient in *Nac2* to control the expression of the D2 protein. In the presence of copper in the medium, the results showed a clear reduction in PSII to a level in which oxygen consumption by respiration exceeded the evolved oxygen by PSII, which of course, led to an increase in hydrogenase activity [79].

14.5.1.2 Increased O_2 consumption/sequestration

C. reinhardtii apr1 mutant showed a reduction of photosynthesis/respiration ratio (P/R ratio), continuously remained in anaerobiosis, and induced hydrogenase synthesis in the light. Hydrogen was produced under normal growth conditions after supplying exogenous glycolaldehyde (GA) to media, which involved the Calvin−Benson cycle by using inhibition of the activation of phosphorylation in order to keep electron dietary supplements for hydrogenases. As a result, the rate of hydrogen production of apr1 mutant in vivo was enhanced to double that of sulfur-deprived WT [89].

Another way to promote anaerobiosis is the introduction of O_2 sequesters inside the chloroplast. Both leghemoglobin (lba), and ferrochelatase (hemH) genes were transferred

Handbook of Algal Biofuels

and expressed into the chloroplast of *C. reinhardtii* to sequester oxygen. Consequently, O_2 was more rapidly consumed, and H_2 yield was improved in the mutant compared with the control strain [90].

An alternative approach to eliminate O_2 from the culture medium is a new pathway in *Chlamydomonas* for O_2 utilization. Decarboxylation of pyruvate is catalyzed by pyruvate oxidase (PoX), where it consumes O_2 and produces acetyl phosphate and CO_2. This enzyme was loaded into the chloroplast of *C. reinhardtii* to consume the created oxygen and hence hydrogen is promoted three-times more than in the wild type under low-light conditions [91].

14.5.2 Elimination of competing pathways

Another effective and promising approach for enhancing biohydrogen production is eliminating or redirecting the competitive electron pathway(s). The main competing pathway for hydrogenase is reducing the FNR that converts NADP into NADPH required for CO_2 fixation using the Calvin–Benson cycle. Under an aerobic atmosphere, liberated electrons are directed from FDX1 to FNR, while under anaerobic conditions that promote hydrogen production, hydrogenase competes with FNR to get electrons from ferredoxin [94,106,107].

A bioengineered ferredoxin–hydrogenase fusion was thought to be an effective route to direct the electrons to hydrogenase instead of FNR. This fusion increases hydrogen production sixfold in vitro compared to isolated hydrogenase with added ferredoxin [94].

The carbon dioxide is fixed by ribulose diphosphate using the enzyme RuBisCO. RuBisCO consists of two protein subunits, the large one is encoded by chloroplast DNA, and the smaller one is nuclear-encoded. Hence, inhibiting RuBisCO production through the inhibition of the expression of big and/or small subunit(s) is(are) an ideal way to increase the production of biohydrogen in *C. reinhardtii* mutants [93,108].

In *Synechocystis* sp. PCC 6803, inhibition of the respiratory complex I by deleting the large subunit of type 1 NADPH-dehydrogenase (NDH-1) resulted in a lower amount of O_2 produced in the light. The *ndhB* mutant M55 was able to continuously produce hydrogen for several minutes in the presence of glucose [95].

In addition, the interruption of terminal respiratory oxidases (ΔctaI, ΔctaII, and/or Δcyd) in *Synechocystis* PCC 6803 to redirect the electron flow for hydrogen production showed that the absence of the quinol oxidase (Cyd) resulted in higher Hox-hydrogenase activity and hydrogen production in the light than for the wild-type strain [96].

The nitrate assimilation pathway is a considerable competitive pathway that can reduce the electron flow for hydrogen biosynthesis. In *Synechocystis* PCC 6803 this pathway was disrupted with regard to nitrate reductase (ΔnarB), nitrite reductase (ΔnirA), or both genes (ΔnarB/ΔnirA). The deletion of both ΔnarB and ΔnirA has been reported to improve hydrogen production compared to the wild type [97].

In *Synechococcus* sp. strain PCC 7002, the deletion of *ldhA* gene, which encodes for NADH-dependent reduction of pyruvate to lactate, showed an inhibition in lactate production and enhanced H_2 production upto fivefold [98].

14.6 Biosystem and semiartificial system for photocurrent and biohydrogen productions

Chlamydomonas hydrogenase is considered to be one of the highest activity hydrogenases reaching 2000 H_2/s, although it shows high oxygen-sensitivity [109,110]. A successful coupling of hydrogen production with photosynthesis has been reported [109,110]. This coupling directly affected algal growth due to the inactivation of PSII under an anaerobic atmosphere and finally inhibited algal growth completely [111,112]. Although cyanobacterial hydrogenases are less active compared to that of *Chlamydomonas* sp., they are highly tolerant and able to prolong biohydrogen production. Throughout phototrophic nutrition, blue—green algae can utilize inorganic nutrients and light for mass production and biohydrogen production [110,112]. Several techniques have been applied to solve or limit the inhibitory effects of oxygen and promote the hydrogen production inside algal or cyanobacterial cell [112].

Galvanic cells based on electron transport of photosynthesis have been suggested since 1970 [113,114]. Throughout the Z-scheme, the charge separation processes of the natural system gave the basic backbone of semiartificial photosynthesis systems to convert light energy into chemical energy [114]. So, enhancing light harvesting improves the efficiency of absorbing and converting light energy into usable energy [113]. Some of these techniques are highlighted in the following sections.

14.6.1 Hydrogenase-ferredoxin fusion

Since ferredoxin-NADP-reductase (FNR) is located on the TM surface of cyanobacteria and eukaryotic algae, it strongly impedes the effective flow of electrons to ydA and hence hydrogen production is significantly affected [20]. The Fd-HydA construction can both structurally and functionally replace HydA (Fig. 14.4) [115]. This strategy relies on controlling the electron distribution between the FNR and the hydrogenase that has been successfully applied to many donors and acceptors of the electron. By adding dichlorophenyl dimethylurea to the culture, PSII is blocked and the link between bacterial FeFe-hydrogenase and Fd promotes hydrogen production from organic sources, that is, glucose, which is another effective source for hydrogen production [115]. The hydrogen production by photosynthesis requires low light intensity, since light is needed to synthesize the sugar molecules supplied to the cyanobacteria [94,116,117]. Also, this link reduced the electron flow to FNR and enhanced the electron flow on the ferredoxin-hydrogenase route and consequently promoted the rate of hydrogen production via photosynthesis (Fig. 14.4) [94]. These results reveal that the reorientation of photoelectronic transport improves biological hydrogen production [94,115—117]. The presence of the singular Fd is essential for cell survival while producing hydrogen [94].

14.6.2 Complex of photosystem I and NiFe-hydrogenases via PsaE

This complex has been reported by Ihara et al. (2006), who fused the stromal PSI subunit, PsaE, from *Synechocystis* sp. to tolerant NiFe-hydrogenase from *Ralstonia eutropha*

14.6 Biosystem and semiartificial system for photocurrent and biohydrogen productions 321

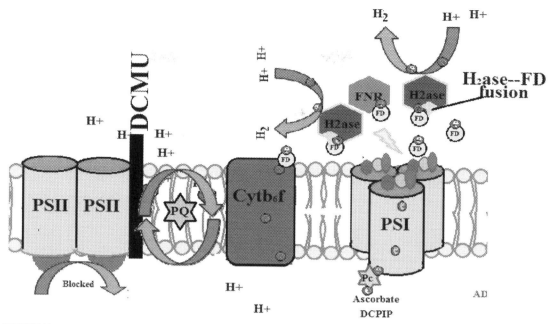

FIGURE 14.4 Electron distribution between ferredoxin-NADP-reductase and the Fd-HydA construction. The schematic shows that electrons flow to Fd and Fd-HydA, which supports hydrogen production.

through the C-terminus. This complex was achieved in vitro by incubating the pure PSIΔPsaE mutant from *Synechocystis* sp. PCC 6803 with the Hyd-PsaE complex [118]. The formed PSI-hydrogenase hybrid showed light-driven hydrogen production using dichlorophenol indophenol and sodium ascorbate as artificial electron donors (Fig. 14.5) [118]. The amount and the rate of produced hydrogen were too low to be quantified, which may be due to the partial binding of Hyd-PsaE with ΔPsaE. This issue was overcome by Schwarze and coworkers, who fabricated a new complex, the Hyd-PsaE, as well as PSI-ΔPsaE [115]. This new complex showed a pure fusion, where the performed protocol enables the removal of contaminated MBH forms and purification of homogenous complexes. Hyd-PsaE-PSI exhibited high biohydrogen production activity compared to that of the wild-type. PSI-His10-tag was fabricated on the luminal side N terminus of PsaF, which facilitated the orientation of PSI immobilization onto modified gold electrodes [116]. It should be clarified that PSI wild type and Hyd-PsaE-PSI showed the same activity of P_{700} (700 μmol O_2/mg Chl/h) [118].

14.6.3 Wiring photosystem I through nanoconstruction

The existing fusion depends upon connecting hydrogenase to PSI through nanowire. The PSI complex was driven from thermophilic cyanobacterium *T. elongatus* and tolerant (FeFe)-hydrogenase extracted from *C. pasteurianum*. This fusion was fabricated by binding both protein complexes using dithiol linkers (Fig. 14.6) [119–121].

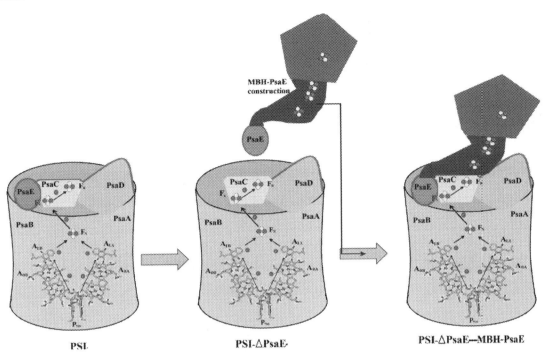

FIGURE 14.5 Models of the hydrogenase-PsaE-photosystem I(PSI) complex. Membrane-bound hydrogenase of *Ralstonia eutropha* H16. and PSIΔ PsaE of *Synechocystis* sp. PCC 6803 were mixed and incubated forming MBHPsaE and PSI-PsaE.

Lubner and coworkers designed and fabricated an organic nanowire that was able to bind F_B, the terminal 4Fe4S cluster within PSI from *T. elongatus* to FeFe-hydrogenase, at the distal 4Fe4S cluster of *Clostridium acetobutylicum* [119]. At sufficient light intensity, electrons flow from PSI to hydrogenase via the dithiol nanowire and evolve biohydrogen at a rate of 2200 ± 460 μmol/mg per chlorophyll per h, which is equivalent to 105 ± 22 $e^- PSI^{-1} s^{-1}$ (Fig. 14.6) [119]. At PSI to PSII ratio of 1.8, cyanobacteria can produce oxygen at a rate of 400 μmol/mg per chlorophyll per h, which is equivalent to 47 $e^- PSI^{-1} s^{-1}$ [119]. Lubner and coworkers reported that the hybrid biological/organic nanocomplex saves double the electron flow compared to in vivo [119].

As dithiol linkers, several lengths of dithiol derivatives, that is, decanedithiol, octanedithiol, and hexanedithiol, have been fabricated and tested for establishing complexes between FeFe-Hydrogenase and PSI [120]. PSI and FeFe-hydrogenase fusion was confirmed via crystal structures of both complexes. Moreover, the movement of the protein through square root fluctuations resulted in binding via the shortest hexandithiol bond which improved the complexes atomic fluctuations, hydrogenase, and PSI, within these complexes [120]. While the internal distances in these complexes involved electron flow, they were structurally changed in the enzyme FeFe hydrogenase due to binding to PSI via the shortest hexanedithiol link, affecting electron flow to the hydrogenase. This may clarify why the wire-length octanethiol bond produced the

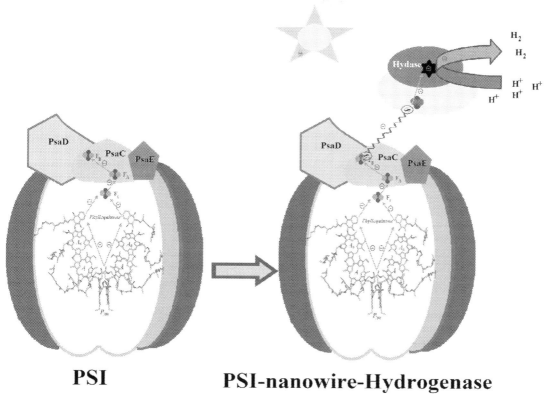

FIGURE 14.6 Schematic shows the nanocomplexes, dithiol linkers, and electron flow during photosynthesis to FeFe-hydrogenase.

highest hydrogen evolving rate [120]. The observed different distances between the final electron acceptor of PSI, FB, of PSI and active site of hydrogenase is considered another main result of this fusion. Because of creating various distances between the coupling complexes due to dithiol linkers, Applegate and coworkers fabricated and optimized a new nanocomplex, PSI—hydrogenase fusion, where they incubated cytochrome c6 with PSI, FeFe-hydrogenase, and the nanowire, octanedithiol [121]. Good results were obtained through this complex, where cytochrome c6 promoted the fabrication of PSIC13G−1,8-octanedithiol-[FeFe]-hydrogenaseC97G (PSI—hydrogenase) nanocomplex [121]. Although the theoretical hydrogen production throughout this complex is 0.5 H_2 molecules/photon, the routine practical applications revealed that the actual rate was 0.10−0.15 H_2 molecules/photon in visible light [121]. These differences might be due to the formation of various conjugates such as Hyd-Hyd conjugates and PSI—PSI conjugates as well as nonproductive PSIC13G−1,8-octanedithiol−PSIC13G, which could also absorb the light but without hydrogen evolution [121].

14.6.4 Photosystem I-hydrogenase complex via nanowire from phylloquinone

Photosystem IΔ menB mutants of *Synechocystis* sp. PCC 6803 were wired to nanoparticles via a molecular wire consisting of 15-(3-methyl-1,4-naphthoquinone-2-yl) pentadecyl sulfide (Fig. 14.7).

The Δ menB mutant strain creates an interruption in the biosynthetic pathway at phylloquinone, leading to the presence of a displaceable PQ-9 in the A1A/A1B sites. The artificial quinone has a typical headgroup as the native phylloquinone with a long tail of 15-carbons ended with a thiol. The thiol tail is bound to an [FeFe]-hydrogenase variant from *C. acetobutylicum* that contains an iron on the distal [4Fe-4S] cluster afforded by mutating the surface exposed Cys97 residue to Gly [122].

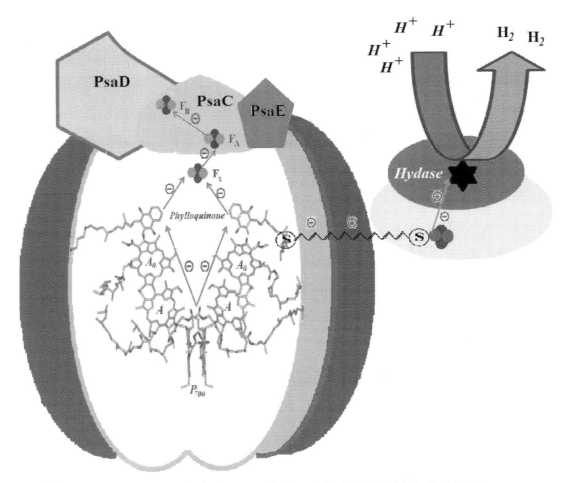

FIGURE 14.7 Proposed scheme for hydrogen production via PS I-NQ (CH 2) 15 S- [FeFe] hydrogenase.

The thiol molecular wire coordinates the iron atom at the corner of the iron–sulfur cluster. Under illumination and in the presence of an electron donor and an electron carrier, the PS I-NQ(CH2)15S-[FeFe] hydrogenase nanocomplex produced molecular hydrogen at a rate of 50.3 ± 9.96 μmol of H_2 mg/Chl/h at pH 8.3. The successful results obtained via in vitro experiments pave the way for constructing a PSI-NQ(CH2)15S-[FeFe]H2ase nanocomplex in vivo in the menB mutants of *Synechocystis* sp. PCC 6803 [122].

14.6.5 Fabrication of PsaD-hoxYH complex

Hydrogen production during photosynthesis through PSI–PsaD-H2ase fusion within a cell is highlighted. The NiFe-H2ase HoxYH was obtained from *Synechocstis* sp. PCC 6803 constructed with a PSI-PsaD subunit in order to maintain close proximity to the 4Fe4S cluster of F_B, so that it facilitates electron flow to ferredoxin (Fig. 14.8). The mutant strain PsaD-hoxYH grew photoautotrophically and had the ability to produce hydrogen of 500 μM under illumination and aerobic condition [123].

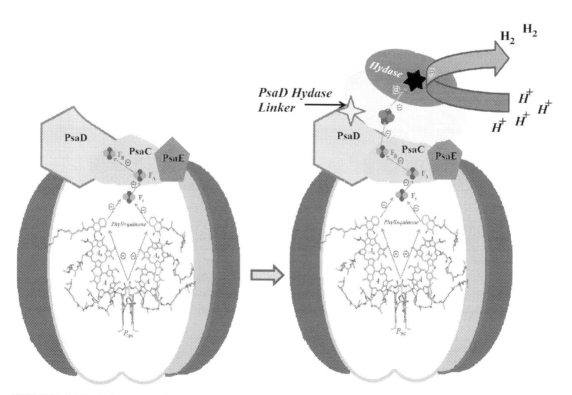

FIGURE 14.8 Fabrication of a complex between photosystem I-PsaD-NiFe-hydrogenase HoxYH. The F_b is close to the hydrogenase active site, enabling this complex to enhance hydrogen production.

14.7 Conclusion

Due to high energy demand, the exploration and development of new energy sources are the only way for continued urban development and political stability. Here, the applied progress of renewable energy resources are introduced, and new and advanced successfully strategies for hydrogen production are summarized.

There are two effective routes for biohydrogen production. The first way is biomass fermentation, so enhancing biomass production necessary. The second way is biohydrogen production via electrons produced during light reaction and this is considered the most advanced research currently. For this reason, induced mutation is an effective tool to reach a high photocurrent that consequently leads to high biohydrogen production either in vivo or in vitro. These induced mutations include subunit(s) deletion, reorientation of electrons, fabrication of different nanocomplex linkers, some cofactor(s), activation or inactivation of some pathways, and the combination of native and artificial cofactors, polymers, and/or electron carriers. With the ability of nitrogenase to produce biohydrogen, some heterocystous cyanobacteria are considered to show an alternative route for biohydrogen production.

References

[1] F. Dawood, M. Anda, G.M. Shafiullah, Hydrogen production for energy: an overview, Int. J. Hydrog. Energy 45 (7) (2020) 3847–3869.

[2] M.E. El-Khouly, E. El-Mohsnawy, S. Fukuzumi, Solar energy conversion: from natural to artificial photosynthesis, J. Photochem. Photobiol. C 31 (2017) 36–83.

[3] H.L. Yip, et al., A review of hydrogen direct injection for internal combustion engines: towards carbon-free combustion, Appl. Sci. 9 (22) (2019).

[4] S. Banerjee, M.N. Musa, A.B. Jaafar, Economic assessment and prospect of hydrogen generated by OTEC as future fuel, Int. J. Hydrog. Energy 42 (1) (2017) 26–37.

[5] G. Maggio, A. Nicita, G. Squadrito, How the hydrogen production from RES could change energy and fuel markets: a review of recent literature, Int. J. Hydrog. Energy 44 (23) (2019) 11371–11384.

[6] X. Dong, et al., *CO2 Emission of electric and gasoline vehicles under various road conditions for China, Japan, Europe and world average—prediction through year 2040*, Appl. Sci. 9 (11) (2019).

[7] G.D. Berry, et al., Hydrogen as a future transportation fuel, Energy 21 (4) (1996) 289–303.

[8] A.M. Abdalla, et al., Hydrogen production, storage, transportation and key challenges with applications: a review, Energy Convers. Manage. 165 (2018) 602–627.

[9] J.O. Abe, et al., Hydrogen energy, economy and storage: review and recommendation, Int. J. Hydrog. Energy 44 (29) (2019) 15072–15086.

[10] S.Y. Gómez, D. Hotza, Current developments in reversible solid oxide fuel cells, Renew. Sustain. Energy Rev. 61 (2016) 155–174.

[11] R.E. Gullison, et al., Environment. Tropical forests and climate policy, Science 316 (5827) (2007) 985–986.

[12] B. Subhadra, M. Edwards, An integrated renewable energy park approach for algal biofuel production in United States, Energy Policy 38 (9) (2010) 4897–4902.

[13] A. Singh, S.I. Olsen, Key issues in life cycle assessment of biofuels, in: Sustainable Bioenergy and Bioproducts, 2012. pp. 213–228.

[14] D. Rathore, A. Singh, Biohydrogen production from microalgae, in: Biofuel Technologies, 2013, pp. 317–333.

[15] P.T. Sekoai, M.O. Daramola, Biohydrogen production as a potential energy fuel in South Africa, Biofuel Res. J. 2 (2) (2015) 223–226.

[16] K.Y. Show, et al., Biohydrogen production: current perspectives and the way forward, Int. J. Hydrog. Energy 37 (20) (2012) 15616–15631.

References

[17] A. Melis, T. Happe, Hydrogen production. Green algae as a source of energy, Plant Physiol. 127 (3) (2001) 740–748.

[18] U. Eberle, B. Müller, R. von Helmolt, Fuel cell electric vehicles and hydrogen infrastructure: status 2012, Energy Environ. Sci. 5 (10) (2012).

[19] R.C. Prince, H.S. Kheshgi, The photobiological production of hydrogen: potential efficiency and effectiveness as a renewable fuel, Crit. Rev. Microbiol. 31 (1) (2008) 19–31.

[20] B. Esper, A. Badura, M. Rogner, Photosynthesis as a power supply for (bio-)hydrogen production, Trends Plant Sci. 11 (11) (2006) 543–549.

[21] J. Barber, P.D. Tran, From natural to artificial photosynthesis, J. R. Soc. Interface 10 (81) (2013).

[22] I. Grotjohann, P. Fromme, Structure of cyanobacterial Photosystem I, Photosyn. Res. 85 (1) (2005) 51–72.

[23] D.S. Horner, et al., Iron hydrogenases – ancient enzymes in modern eukaryotes, Trends Biochem. Sci. 27 (3) (2002) 148–153.

[24] N. Kamiya, J.R. Shen, Crystal structure of oxygen-evolving photosystem II from Thermosynechococcus vulcanus at 3.7-A resolution, Proc. Natl. Acad. Sci. 100 (1) (2002) 98–103.

[25] K.N. Ferreira, Architecture of the photosynthetic oxygen-evolving center, Science 303 (5665) (2004) 1831–1838.

[26] Y. Umena, et al., Crystal structure of oxygen-evolving photosystem II at a resolution of 1.9 Å, Nature 473 (7345) (2011) 55–60.

[27] A.R. Holzwarth, et al., Kinetics and mechanism of electron transfer in intact photosystem II and in the isolated reaction center: pheophytin is the primary electron acceptor, Proc. Natl. Acad. Sci. 103 (18) (2006) 6895–6900.

[28] Y. Miloslavina, et al., Charge separation kinetics in intact photosystem II core particles is trap-limited. A picosecond fluorescence study†, Biochemistry 45 (7) (2006) 2436–2442.

[29] J. Nield, K. Redding, M. Hippler, Remodeling of light-harvesting protein complexes in chlamydomonas in response to environmental changes, Eukaryot. Cell 3 (6) (2004) 1370–1380.

[30] B.A. Diner, F. Rappaport, Structure, dynamics, andenergetics of theprimaryphotochemistry ofphotosystemiiofoxygenicphotosynthesis, Annu. Rev. Plant Biol. 53 (1) (2002) 551–580.

[31] B.A. Diner, G.T. Babcock, Structure, dynamics, and energy conversion efficiency in photosystem II, in: Oxygenic Photosynthesis: The Light Reactions, 2004. pp. 213–247.

[32] C.C. Moser, C.C. Page, P. Leslie Dutton, Tunneling in PSII, Photochem. Photobiol. Sci. 4 (12) (2005) 933–939.

[33] M. Szczepaniak, et al., Charge separation, stabilization, and protein relaxation in photosystem II core particles with closed reaction center, Biophys. J. 96 (2) (2009) 621–631.

[34] P. Fromme, P. Jordan, N. Krauß, Structure of photosystem I, Biochim. Biophys. Acta 1507 (1–3) (2001) 5–31.

[35] P. Jordan, et al., Three-dimensional structure of cyanobacterial photosystem I at 2.5 Å resolution, Nature 411 (6840) (2001) 909–917.

[36] P.R. Chitnis, PHOTOSYSTEMI: function and physiology, Annu. Rev. Plant Physiol. Plant Mol. Biol. 52 (1) (2001) 593–626.

[37] P. Fromme, et al., Structure and function of photosystem I: interaction with its soluble electron carriers and external antenna systems, FEBS Lett. 555 (1) (2003) 40–44.

[38] M.K. Şener, et al., Excitation migration in trimeric cyanobacterial photosystem I, J. Chem. Phys. 120 (23) (2004) 11183–11195.

[39] M.K. Şener, et al., Comparison of the light-harvesting networks of plant and cyanobacterial photosystem I, Biophys. J. 89 (3) (2005) 1630–1642.

[40] E. El-Mohsnawy, et al., Structure and function of intact photosystem 1 monomers from the cyanobacteriumThermosynechococcus elongatus, Biochemistry 49 (23) (2010) 4740–4751.

[41] M. Sarcina, M.J. Tobin, C.W. Mullineaux, Diffusion of phycobilisomes on the thylakoid membranes of the Cyanobacteriumsynechococcus7942, J. Biol. Chem. 276 (50) (2001) 46830–46834.

[42] A. Ben-Shem, F. Frolow, N. Nelson, Crystal structure of plant photosystem I, Nature 426 (6967) (2003) 630–635.

[43] N. Nelson, A. Ben-Shem, The complex architecture of oxygenic photosynthesis, Nat. Rev. Mol. Cell Biol. 5 (12) (2004) 971–982.

[44] F. Klimmek, et al., *Structure of the higher plant light harvesting complex I: in vivo characterization and structural interdependence of the Lhca proteins†*, Biochemistry 44 (8) (2005) 3065–3073.

328   14. Direct biohydrogen production from algae

[45] E.J. Boekema, et al., A giant chlorophyll–protein complex induced by iron deficiency in cyanobacteria, Nature 412 (6848) (2001) 745–748.

[46] R. Kouřil, et al., Photosystem I: a search for green plant trimers, Photochem. Photobiol. Sci. 4 (12) (2005).

[47] A.R. Holzwarth, et al., Charge recombination fluorescence in photosystem I reaction centers from Chlamydomonas reinhardtii, J. Phys. Chem. B 109 (12) (2005) 5903–5911.

[48] C.W. Mullineaux, J.F. Allen, State 1-State 2 transitions in the cyanobacterium Synechococcus 6301 are controlled by the redox state of electron carriers between photosystems I and II, Photosynth. Res. 23 (3) (1990) 297–311.

[49] D.C. Ducat, G. Sachdeva, P.A. Silver, Rewiring hydrogenase-dependent redox circuits in cyanobacteria, Proc. Natl. Acad. Sci. 108 (10) (2011) 3941–3946.

[50] J. Appel, R. Schulz, Hydrogen metabolism in organisms with oxygenic photosynthesis: hydrogenases as important regulatory devices for a proper redox poising? J. Photochem. Photobiol. B 47 (1) (1998) 1–11.

[51] K. Schütz, et al., *Cyanobacterial H_2 production? A comparative analysis*, Planta 218 (3) (2004) 350–359.

[52] H. Bothe, et al., Nitrogen fixation and hydrogen metabolism in cyanobacteria, Microbiol. Mol. Biol. Rev. 74 (4) (2010) 529–551.

[53] A. Volbeda, et al., *[NiFe]-hydrogenases revisited: nickel–carboxamido bond formation in a variant with accrued O_2-tolerance and a tentative re-interpretation of Ni-SI states*, Metallomics 7 (4) (2015) 710–718.

[54] A. Parkin, Understanding and harnessing hydrogenases, biological dihydrogen catalysts, in: The Metal-Driven Biogeochemistry of Gaseous Compounds in the Environment, 2014, pp. 99–124.

[55] W. Lubitz, et al., Hydrogenases, Chem. Rev. 114 (8) (2014) 4081–4148.

[56] S.T. Stripp, et al., How oxygen attacks [FeFe] hydrogenases from photosynthetic organisms, Proc. Natl. Acad. Sci. 106 (41) (2009) 17331–17336.

[57] C.S.A. Baltazar, et al., Nickel-iron-selenium hydrogenases - an overview, Eur. J. Inorg. Chem. 2011 (7) (2011) 948–962.

[58] T. Sakai, D. Mersch, E. Reisner, Photocatalytic hydrogen evolution with a hydrogenase in a mediator-free system under high levels of oxygen, Angew. Chem. Int. (Ed.) 52 (47) (2013) 12313–12316.

[59] R.K. Thauer, et al., *Hydrogenases from methanogenic archaea, nickel, a novel cofactor, and H_2storage*, Annu. Rev. Biochem. 79 (1) (2010) 507–536.

[60] H. Tamura, et al., *Crystal structures of [Fe]-hydrogenase in complex with inhibitory isocyanides: implications for the H_2-activation site*, Angew. Chem. Int. (Ed.) 52 (37) (2013) 9656–9659.

[61] Y.A. Small, et al., Proton management as a design principle for hydrogenase-inspired catalysts, Energy Environ. Sci. 4 (8) (2011).

[62] R.M. Bullock, A.M. Appel, M.L. Helm, Production of hydrogen by electrocatalysis: making the H–H bond by combining protons and hydrides, Chem. Commun. 50 (24) (2014) 3125–3143.

[63] T.J. Johnson, et al., Molecular genetic improvements of cyanobacteria to enhance the industrial potential of the microbe: a review, Biotechnol. Prog. 32 (6) (2016) 1357–1371.

[64] N. Khanna, P. Lindblad, Cyanobacterial hydrogenases and hydrogen metabolism revisited: recent progress and future prospects, Int. J. Mol. Sci. 16 (12) (2015) 10537–10561.

[65] E. Flores, A. Herrero, Compartmentalized function through cell differentiation in filamentous cyanobacteria, Nat. Rev. Microbiol. 8 (1) (2009) 39–50.

[66] A. Herrero, et al., The multicellular nature of filamentous heterocyst-forming cyanobacteria, FEMS Microbiol. Rev. 40 (6) (2016) 831–854.

[67] S. Joshua, C.W. Mullineaux, Phycobilisome diffusion is required for light-state transitions in cyanobacteria, Plant. Physiol. 135 (4) (2004) 2112–2119.

[68] H. Masukawa, et al., Increased heterocyst frequency by patN disruption in Anabaena leads to enhanced photobiological hydrogen production at high light intensity and high cell density, Appl. Microbiol. Biotechnol. 101 (5) (2017) 2177–2188.

[69] Ro López-Igual, E. Flores, A. Herrero, *Inactivation of a heterocyst-specific invertase indicates a principal role of sucrose catabolism in heterocysts of Anabaena* sp, J. Bacteriol. 192 (20) (2010) 5526–5533.

[70] S.F. Salleh, et al., Investigation of the links between heterocyst and biohydrogen production by diazotrophic cyanobacterium A. variabilis ATCC 29413, Arch. Microbiol. 198 (2) (2015) 101–113.

[71] N.J. Skizim, et al., Metabolic pathways for photobiological hydrogen production by nitrogenase- and hydrogenase-containing unicellular Cyanobacteria cyanothece, J. Biol. Chem. 287 (4) (2012) 2777–2786.

Handbook of Algal Biofuels

References

[72] P. Dechatiwongse, G. Maitland, K. Hellgardt, Demonstration of a two-stage aerobic/anaerobic chemostat for the enhanced production of hydrogen and biomass from unicellular nitrogen-fixing cyanobacterium, Algal Res. 10 (2015) 189−201.

[73] S.S. Merchant, et al., The Chlamydomonas genome reveals the evolution of key animal and plant functions, Science 318 (5848) (2007) 245−250.

[74] P.T. Pratheesh, M. Vineetha, G.M. Kurup, An efficient protocol for the Agrobacterium-mediated genetic transformation of microalga Chlamydomonas reinhardtii, Mol. Biotechnol. 56 (6) (2014) 507−515.

[75] S.E. Shin, et al., CRISPR/Cas9-induced knockout and knock-in mutations in Chlamydomonas reinhardtii, Sci. Rep. 6 (2016) 27810.

[76] A. Ferenczi, et al., Efficient targeted DNA editing and replacement in Chlamydomonas reinhardtii using Cpf1 ribonucleoproteins and single-stranded DNA, Proc. Natl. Acad. Sci. U S A 114 (51) (2017) 13567−13572.

[77] X. Li, et al., A genome-wide algal mutant library and functional screen identifies genes required for eukaryotic photosynthesis, Nat. Genet. 51 (4) (2019) 627−635.

[78] N.J. Gumpel, J.D. Rochaix, S. Purton, Studies on homologous recombination in the green alga Chlamydomonas reinhardtii, Curr. Genet. 26 (5−6) (1994) 438−442.

[79] R. Surzycki, et al., Potential for hydrogen production with inducible chloroplast gene expression in Chlamydomonas, Proc. Natl. Acad. Sci. U S A 104 (44) (2007) 17548−17553.

[80] T. Zhao, et al., Gene silencing by artificial microRNAs in Chlamydomonas, Plant. J. 58 (1) (2009) 157−164.

[81] I. Sizova, et al., Nuclear gene targeting in Chlamydomonas using engineered zinc-finger nucleases, Plant. J. 73 (5) (2013) 873−882.

[82] E.V. Armbrust, et al., The genome of the diatom Thalassiosira pseudonana: ecology, evolution, and metabolism, Science 306 (5693) (2004) 79−86.

[83] C. Vahrenholz, et al., Mitochondrial DNA of Chlamydomonas reinhardtii: the structure of the ends of the linear 15.8-kb genome suggests mechanisms for DNA replication, Curr. Genet. 24 (3) (1993) 241−247.

[84] J.M. Archibald, et al., Lateral gene transfer and the evolution of plastid-targeted proteins in the secondary plastid-containing alga Bigelowiella natans, Proc. Natl. Acad. Sci. U S A 100 (13) (2003) 7678−7683.

[85] K. Shimogawara, et al., High-efficiency transformation of Chlamydomonas reinhardtii by electroporation, Genetics 148 (4) (1998) 1821−1828.

[86] C. Economou, et al., A simple, low-cost method for chloroplast transformation of the green alga Chlamydomonas reinhardtii, Methods Mol. Biol. 1132 (2014) 401−411.

[87] G. Torzillo, et al., Increased hydrogen photoproduction by means of a sulfur-deprived Chlamydomonas reinhardtii D1 protein mutant, Int. J. Hydrog. Energy 34 (10) (2009) 4529−4536.

[88] H.-D. Lin, et al., *Knockdown of PsbO leads to induction of HydA and production of photobiological H_2 in the green alga Chlorella sp. DT*, Bioresour. Technol. 143 (2013) 154−162.

[89] T. Rühle, et al., A novel screening protocol for the isolation of hydrogen producing Chlamydomonas reinhardtii strains, BMC Plant Biol. 8 (2008) 1.

[90] S. Wu, et al., Improved hydrogen production with expression of hemH and lba genes in chloroplast of Chlamydomonas reinhardtii, J. Biotechnol. 146 (3) (2010) 120−125.

[91] F.-Q. Xu, W.-M. Ma, X.-G. Zhu, Introducing pyruvate oxidase into the chloroplast of Chlamydomonas reinhardtii increases oxygen consumption and promotes hydrogen production, Int. J. Hydrog. Energy 36 (17) (2011) 10648−10654.

[92] A. Hemschemeier, et al., Hydrogen production by Chlamydomonas reinhardtii: an elaborate interplay of electron sources and sinks, Planta 227 (2) (2007) 397−407.

[93] T.S. Pinto, et al., Rubisco mutants of Chlamydomonas reinhardtii enhance photosynthetic hydrogen production, Appl. Microbiol. Biotechnol. 97 (12) (2013) 5635−5643.

[94] I. Yacoby, et al., *Photosynthetic electron partitioning between [FeFe]-hydrogenase and ferredoxin:NADP + -oxidoreductase (FNR) enzymes* in vitro, Proc. Natl. Acad. Sci. 108 (23) (2011) 9396−9401.

[95] L. Cournac, et al., *Sustained photoevolution of molecular hydrogen in a mutant of Synechocystis sp. strain PCC 6803 deficient in the type I NADPH-dehydrogenase complex*, J. Bacteriol. 186 (6) (2004) 1737−1746.

[96] F. Gutthann, et al., *Inhibition of respiration and nitrate assimilation enhances photohydrogen evolution under low oxygen concentrations in Synechocystis sp. PCC 6803*, Biochim. Biophys. Acta 1767 (2) (2007) 161−169.

[97] W. Baebprasert, et al., *Increased H_2 production in the cyanobacterium Synechocystis* sp. *strain PCC 6803 by redirecting the electron supply via genetic engineering of the nitrate assimilation pathway*, Metab. Eng. 13 (5) (2011) 610—616.

[98] K. McNeely, et al., Redirecting reductant flux into hydrogen production via metabolic engineering of fermentative carbon metabolism in a cyanobacterium, Appl. Environ. Microbiol. 76 (15) (2010) 5032—5038.

[99] M.L. Ghirardi, *Implementation of photobiological H_2 production: the O_2 sensitivity of hydrogenases*, Photosynth. Res. 125 (3) (2015) 383—393.

[100] E.S. Boyd, J.R. Spear, J.W. Peters, [FeFe] hydrogenase genetic diversity provides insight into molecular adaptation in a saline microbial mat community, Appl. Environ. Microbiol. 75 (13) (2009) 4620—4623.

[101] M.L. Ghirardi, R.K. Togasaki, M. Seibert, *Oxygen sensitivity of algal H_2- production*, Appl. Biochem. Biotechnol. 63—65 (1997) 141—151.

[102] J.A. Stapleton, J.R. Swartz, *Development of an* in vitro *compartmentalization screen for high-throughput directed evolution of [FeFe] hydrogenases*, PLoS One 5 (12) (2010) e15275.

[103] A. Melis, M. Seibert, M.L. Ghirardi, Hydrogen fuel production by transgenic microalgae, Adv. Exp. Med. Biol. 616 (2007) 110—121.

[104] C.H. Chang, et al., Atomic resolution modeling of the ferredoxin:[FeFe] hydrogenase complex from Chlamydomonas reinhardtii, Biophys. J. 93 (9) (2007) 3034—3045.

[105] D.W. Mulder, et al., Stepwise [FeFe]-hydrogenase H-cluster assembly revealed in the structure of HydA (DeltaEFG), Nature 465 (7295) (2010) 248—251.

[106] A. Dubini, M.L. Ghirardi, Engineering photosynthetic organisms for the production of biohydrogen, Photosynth. Res. 123 (3) (2015) 241—253.

[107] A.M. Terauchi, et al., Pattern of expression and substrate specificity of chloroplast ferredoxins from Chlamydomonas reinhardtii, J. Biol. Chem. 284 (38) (2009) 25867—25878.

[108] A. Hemschemeier, et al., Hydrogen production by Chlamydomonas reinhardtii: an elaborate interplay of electron sources and sinks, Planta 227 (2) (2008) 397—407.

[109] A. Melis, Green alga hydrogen production: progress, challenges and prospects, Int. J. Hydrog. Energy 27 (11—12) (2002) 1217—1228.

[110] G. Berggren, et al., Biomimetic assembly and activation of [FeFe]-hydrogenases, Nature 499 (7456) (2013) 66—69.

[111] T. Happe, A. Kaminski, Differential regulation of the Fe-hydrogenase during anaerobic adaptation in the green alga Chlamydomonas reinhardtii, Eur. J. Biochem. 269 (3) (2002) 1022—1032.

[112] L. Girbal, et al., Homologous and heterologous overexpression in Clostridium acetobutylicum and characterization of purified clostridial and algal Fe-only hydrogenases with high specific activities, Appl. Environ. Microbiol. 71 (5) (2005) 2777—2781.

[113] T. Kothe, et al., Combination of a photosystem1-based photocathode and a photosystem2-based photoanode to a Z-scheme mimic for biophotovoltaic applications, Angew. Chem. Int. (Ed.) 52 (52) (2013) 14233—14236.

114] W. Haehnel, H.J. Hochheimer, 303 - On the current generated by a galvanic cell driven by photosynthetic electron transport, Bioelectrochem. Bioenerg. 6 (4) (1979) 563—574.

[115] A. Schwarze, et al., Requirements for construction of a functional hybrid complex of photosystem I and [NiFe]-hydrogenase, Appl. Environ. Microbiol. 76 (8) (2010) 2641—2651.

[116] J.R. Benemann, Hydrogen production by microalgae, J. Appl. Phycol. 12 (3/5) (2000) 291—300.

[117] A. Melis, M. Seibert, T. Happe, Genomics of green algal hydrogen research, Photosynth. Res. 82 (3) (2004) 277—288.

[118] M. Ihara, et al., Light-driven hydrogen production by a hybrid complex of a [NiFe]-hydrogenase and the cyanobacterial photosystem I, Photochem. Photobiol. 82 (2006) 3.

[119] C.E. Lubner, et al., Solar hydrogen-producing bionanodevice outperforms natural photosynthesis, Proc. Natl. Acad. Sci. 108 (52) (2011) 20988—20991.

[120] B.J. Harris, X. Cheng, P. Frymier, Structure and function of photosystem I—[FeFe] hydrogenase protein fusions: an all-atom molecular dynamics study, J. Phys. Chem. B 120 (4) (2016) 599—609.

[121] A.M. Applegate, et al., *Quantum yield measurements of light-induced H_2 generation in a photosystem I—[FeFe]-H2ase nanoconstruct*, Photosynth. Res. 127 (1) (2014) 5—11.

[122] M. Gorka, J.H. Golbeck, Generating dihydrogen by tethering an [FeFe]hydrogenase via a molecular wire to the A1A/A1B sites of photosystem I, Photosynth. Res. 143 (2) (2019) 155—163.

[123] J. Appel, et al., *Cyanobacterial in vivo solar hydrogen production using a photosystem I—hydrogenase (PsaD-HoxYH) fusion complex*, Nat. Energy 5 (6) (2020) 458—467.

CHAPTER 15

Biojet fuels production from algae: conversion technologies, characteristics, performance, and process simulation

Medhat Elkelawy[1], Hagar Alm-Eldin Bastawissi[1], Ahmed Mohamed Radwan[1], Mohamed Taha Ismail[1] and Mostafa El-Sheekh[2]

[1]Mechanical Power Engineering Department, Faculty of Engineering, Tanta University, Tanta, Egypt [2]Department of Botany, Faculty of Science, Tanta University, Tanta, Egypt

Nomenclature and abbreviations

ABE	acetone−butanol−ethanol
APR	aqueous phase reforming
ASTM	American society for testing and materials
ATJ	alcohol to jet
BTL	biomass to liquid
CH	catalytic hydrothermolysis
CTL	coal to liquid
DCN	cetane number (derived)
DMF	2,5-dimethylfuran
DSH	direct sugar to hydrocarbons
DXP	deoxyxylulose-5- phosphate
FAA	federal aviation administration
FFA	free fatty acid
FT	Fisher Tropsch
GGE	gallon of gasoline equivalent
GHG	greenhouse gases
TTW	tank to wake
GTJ	gas to jet
HDCJ	hydrotreated depolymerized cellulosic jet

Handbook of Algal Biofuels
DOI: https://doi.org/10.1016/B978-0-12-823764-9.00003-0

© 2022 Elsevier Inc. All rights reserved.

HEFA	fatty acids and hydroprocessed esters
HMF	5-hydroxymethylfurfural
HRJ	hydroprocessed renewable jet
HTFT	high temperature Fischer–Tropsch
HyChem	hybrid chemistry
IATA	international air transport association
IPP	isopentenyl pyrophosphate
LCA	life cycle assessment
LTFT	low temperature Fischer–Tropsch
MIL	military
MVA	mevalonic acid pathway
OEM	original equipment manufacturer
OTJ	oil to jet
PM	particulate matter
SPK	synthetic paraffinic kerosene
STJ	sugar to jet

15.1 Introduction

There are various technologies to convert biomass materials into biojet fuels. Some technological pathways are either at the commercial or precommercial stage, whereas other conversion technologies are still at the stage of research and development [1]. The type of feedstock utilized is the main parameter to determine the conversion pathway [2]. Hydroprocessing technologies, which include isomerization/hydrocracking, hydrotreating, and deoxygenation, are considered as the main conversion pathways to convert oil-based feedstock into biobased jet fuels [3]. Also, hydro–thermolysis (CH) processes development is made to treat oils based on triglyceride [4]. Gasification process is used to convert the solid-based feedstock into biomass-derived intermediates, where it is converted into alcohols by using thermochemical or biochemical processes [5]. The derived alcohols through biochemical processes are converted into sugars, where pyrolysis processes are used to obtain biooils. There are various processes of catalytic, synthesis, or fermentation used to upgrade Syngas, biooils, alcohols, and sugars into biobased jet fuels [6]. Hydroprocessing of oil and Fischer–Tropsch (FT) synthesis are conversion technologies of biobased jet fuel that met ASTM international method D7566, where 50% of biobased jet fuels can be blended into the jet [3]. The conversion technologies of hydroprocessing, which use refined and wasted oils as feedstock, are considered as the only ready and available conversion pathways for the Implementation of a large scale [7]. The optimum processes which can produce economically and utilize sustainable feedstocks are currently being developed by the industrial sector.

The production cost plays a vital rule in the commercial feasibility of biobased jet fuels [8]. The biobased jet fuels production cost varies depending on various parameters, where the process design, conversion efficiency, feedstock price, and economic return of coproducts have the most significant effect [9]. Also, the continuous improvement in the conversion technology could decrease the production cost of biojet fuels and replace the annual fuel consumption of airlines, predicted to be by 30% through using biojet fuel by 2030 [10]. The petroleum-derived jet fuel increases environmental pollution, and the contribution of commercial aviation is approximately 2%–6% of the global carbon emissions [11]. Therefore global environmental awareness is growing rapidly to improve air quality and reduce the global greenhouse gas emissions (GHG).

Researchers found that biojet fuels have great potential as an alternative energy source in the reduction of aviation GHG emissions [12]. The GHG emissions comparison between the conventional jet fuel and biojet fuel is performed by considering the overall production chain emissions. Life cycle assessment (LCA), which normally is named as the analysis of well-to-wake (WTW), can provide a detailed study of the emissions that begin with the field of harvested feedstock down to the wake behind the aircraft [13]. WTW analysis begins with a well-to-tank (WTT) analysis, where the second part of the analysis is tank-to-wake (TTW) [3]. Feedstock production, conversion process, and land-use change are the main parameters of the WTT analysis that have a significant contribution to GHG emissions [14]. The TTW analysis considers the emissions from burning jet fuels. GHG emissions highly depend on the conversion technology applied due to the wide range of available feedstock and the reaction process [3,15]. The optimum platform for biojet fuel production can be determined by making a comprehensive review of the GHG emissions data of the various conversion pathways [11]. The academic research fields, coupled with the industrial and agricultural field, are seeking the development of commercial biobased jet fuels, which depend mainly on feedstock availability, conversion technology development, GHG emissions reduction, and policy. A comprehensive review is important to understand conversion pathways for the future development of biojet fuel.

Algae are one of the most important sources of biodiesel nowadays. Algae are mainly aquatic microorganisms whose growth depends on photosynthesis [16]. Biojet fuels are an alternative, sustainable, near- and long-range solution for airline application in the industry and military field, where they can provide a reduction in environmental pollution as compared to petroleum jet fuels [9]. A comprehensive and detailed review is given in this chapter to describe the various conversion pathways of bioderived jet fuel, whether in research, demonstration, or commercialization phases.

15.2 Biomass jet fuel conversion pathways

The biomass conversion pathways are classified based on the type of feedstocks and the conversion process into four basic categories: alcohol-to-jet (ATJ), oil-to-jet (OTJ), gas to jet (GTJ), and sugar to jet (STJ). Biochemical and thermochemical conversion pathways are the main technical conversion approaches for biofuel production [17]. There are various pathways to produce biojet fuel, depending on the type of feedstock [5]. Hydrogenated esters and fatty acids (HEFA) and the catalytic hydrothermolysis (CH) are the common conversion pathways for biofuel production [3].

15.2.1 Alcohol-to-jet conversion pathways

The oligomerization of alcohol or ATJ conversion technology is a method used to convert long-chain fatty alcohols, methanol, butanol, and ethanol into biojet fuel [18]. Currently, 10%−15% of ethanol blend is the maximum percentage used to power the gasoline vehicles, where there are difficulties in making further penetration of ethanol into the market as an alternative blend for gasoline [9]. Therefore converting ethanol into biojet fuel is an alternative pathway for further achievements in the biojet fuel market.

15.2.1.1 Process description

The main concept of the ATJ conversion pathway is to minimize the physiochemical characteristics between alcohols and petroleum jet fuels [5]. The percentage of purity required from anhydrous ethanol is 99.5%–99.9% to avoid separation in the gasoline/ethanol blend, as stated in the United States [3]. A typical series of reactions, is used to convert alcohols into biojet fuels: dehydration, oligomerization, hydrogenation, and distillation [3]. The dehydration of alcohols is used to generate olefins, then a middle distillate is produced by alcohol oligomerization in the presence of catalyst, and a biojet fuel-ranged hydrocarbon is produced by hydrogenate of the middle distillate, where distillation is the final step in the process [9]. Several companies are continuously developing alcohols-converting technologies for biojet fuels [5]. The production pathway can be categorized as the production of biojet fuel with ethylene as an intermediate, through a propylene intermediate, a higher alcohol intermediate, or through a carbonyl intermediate [5]. The dehydration is a well-developed process, where the most common dehydration catalysts are alumina (Al_2O_3), transition metal oxides, heteropolyacid catalysts, and H-ZSM-5 zeolite catalyst [3]. Direct oligomerization of ethylene to biojet fuel is the most straightforward approach. Homogeneous catalysts are the most commonly used in the catalytic oligomerization process to produce long-chain alpha-olefins [9]. These catalysts have been long used to produce valuable olefins such as octene and hexene rather than the production of a wide range of products. All reactants, catalysts, and products of the homogeneous catalyzed reaction are in the same phase and dissolved in the same solvent. A high turnover frequency is an important parameter to commercialize the catalysts in the industrial process [19].

Heterogeneous catalysis is more acceptable and preferred in the process due to the separation problems arising with the homogeneous catalysts, which make the process expensive [20]. Polyphosphoric acid is one of the earliest heterogeneous catalysts used for olefin oligomerization. However, it has undesirable characteristics, such as low productivity, poor flexibility, corrosion issues, and poor thermal stability [21]. Recent research has focused on using zeolites and acidic mesoporous catalysts in the oligomerization of ethylene, where the carbocation intermediates in the acidic catalyst are utilized in the olefins oligomerization. A highly branched hydrocarbon is generated through tertiary carbocations that have greater stability, where the product is improved, and the melting point is lowered. The oligomerization of ethylene using ZSM-5 zeolite catalyst was also investigated to form higher hydrocarbons [22]. The incorporation of active metals, for example, nickel into zeolites, and amorphous or mesoporous silica–alumina catalysts is a method used to extend the length of the hydrocarbon chain and improve the yield of the jet-fuel. The high partial pressure of hydrogen is used to hydrogenate alpha olefins into the saturation after the oligomerization step. Modest reaction condition levels of temperature, hydrogen, and pressure are required for these olefins to obtain nearly 100% conversion to paraffin. Although, direct ethylene to jet is the simplest available process, it requires significant recycling, and the economics of these processes is not favorable [9]. Therefore further development beyond the laboratory scale has not been recorded. The oligomerization of intermediate olefins is an effective alternative way to overcome the difficulties of ethylene direct oligomerization. The process consists of two steps, the first step is converting ethylene to an intermediate olefin of a mixture of C4-C10 as these intermediates of higher olefins are more easily oligomerized into biojet fuel in the second step. This process is also considered with butene and hexene intermediates [5]. Propylene intermediate is another solution to convert ethanol into biojet fuel, as ethanol is converted to propylene rather than ethylene.

Biojet fuel can be obtained through the conversion of alcohols via a carbonyl intermediate through hydroformylation or oxosynthesis. The aromatic content is about 25 wt.% in the conventional jet fuel, where paraffinic hydrocarbons can't meet all the specifications of jet fuel but still can be blended with conventional jet fuel [5]. ATJ conversion technologies provide the production of both aromatics and paraffin via the direct methanol approach or olefins intermediate approach. Blending aromatics with conventional jet fuels is not favorable, as poor combustion characteristics are obtained using aromatics. The renewable aromatics/conventional fuels blend has an aromatic content that can exceed the limits [23]. As can be seen from the above discussion, each approach has its advantages and disadvantages. Low jet fuel yield is obtained from simple processes, where more complex and additional feedstock or catalysts are required for improved processes. In most cases, advanced technologies are required to obtain a cost-competitive biojet fuel.

15.2.1.2 Economic perspective

This section considers biobased jet fuel production from renewable alcohols with low-cost rather than direct alcohol selling as chemical material. The conventional petroleum price varies; an increase of 262% was recorded between 2001 and 2012 [24]. Now, jet fuel contributes nearly 40% of the total operating costs of airline, and the fluctuations in the jet fuel price inhibit the growth of the airline industry [9]. Therefore biojet fuels play a vital role as sustainable, competitive, and readily available fuels in decreasing the risk of the fluctuating airline market and helping in airline industry growth [6].

The production cost of alcohols, such as ethanol and butanol, needs to be determined to evaluate the production cost of biojet fuels from alcohols. Also, an extensive study was performed for the biochemical and thermochemical processes in the field of technoeconomic analysis [24]. The lowest selling price of the produced biobased jet fuel is calculated on the bases of some economic assumptions. The lowest ethanol selling price from lignocellulosic biomass based on the biochemical conversion process was about 2.15$/gal or 3.27$/gal of gasoline equivalent (GGE) in 2007 and 2.76$/gal or 4.18$/gal of GGE in 2011. The thermochemical pathway produced ethanol with a minimum selling price of 2.05$/gal or 3.11$/gal of GGE in 2007 and 2.5$/gal or 3.8$/gal of GGE in 2011. The acetone−butanol−ethanol (ABE) process produces butanol with a selling price of 0.34$/kg or 1.04$/gal based on the cost of corn feedstock (79.23$/ton), and the price of butanol was reported to be 3.7$/gal or 4.1$/gal of GGE in 2011 [3].

The fluctuation in the price of agricultural feedstock has attracted attention for the utilization of feedstock that is cheaper, renewable, and noncompetitive with food such as algal biomass and agricultural waste as compared to traditional feedstocks. There are other important factors that have a significant effect in the economic evaluation of alcohols production, such as the substrate cost, the technology of solvent recovery, credit for byproducts, and product molar ratio [25]. The overall evaluation of ATJ pathway and the estimation of commercial feasibility, the economics related to upgrading conversion processes that include dehydration, oligomerization, hydrogenation, and distillation, were also considered. Research efforts are made continuously in the field of developing these processes to complete the target.

15.2.1.3 Assessment of life cycle

The environmental concerns about climate changes are increasing due to the rapid consumption of conventional jet fuel, which has a negative impact on the environment [6]. Therefore the airline industry is looking to reduce emissions per passenger mile and to increase engine efficiency [26]. Boeing's 787 jets lowered the emissions per passenger by 20% compared to other similar-sized planes. However, the industry is now seeking alternative biojet fuels for further reduction of CO_2 emissions. Bio aviation jet fuels can reduce around 80% of CO_2 life cycle emissions compared to conventional jet fuels [5]. The standardized environmental LCA methodology can assist the environmental profile of the biojet fuel produced from alcohols. LanzaTech and collaborators have gained a scientific perception of the value chain of the ATJ process, as the value chain involves feedstock preparation. In each LCA stage the energy and input materials must be tracked, and the LCA database and software modeling packages use transparent and efficient impact assessment methods to qualify the environmental impacts related to these inputs [5]. Also, the GHG emissions ultimate quantification is very important, as the carbon moves in different forms in and out of the life cycle of the fuel in different locations; it can be quantified by performing a clear carbon accounting.

The researchers of Michigan Technological University in cooperation with LanzaTech and PNNL conducted initial LCA analysis of a jet fuel produced from methanol [5]. The LCA of the ATJ pathway focused mainly on ethanol, butanol, and isobutanol production. Feedstock or change in land usage, enzyme production on-site, biorefinery process, and the credits of biorefinery coproducts are the main areas that were included in the LCA for alcohol production as a fuel [27]. The performance of the biochemical process is slightly different compared with the thermochemical process considering GHG emissions and the consumption of fuel and water if various pathways are considered. The studies of the conversion of butanol and isobutanol are mainly focused on emissions, the use of consumptive water, the global warming problem, and fossil fuel depletion [9]. Direct emissions are extensive, and include CO_2, NO_2, and SO_2 which are abundately released from the production process of butanol compared with the production process of isobutanol [28]. However, more water is consumed in the butanol refining process compared with isobutanol biorefining. Furthermore, the toxic air emissions were reduced as well as the reduction in the CO_2 life cycle emissions by utilizing biojet fuels, as they contain aromatics and sulfur at negligible levels [3]. Future research and studies are required for the LCA of the ATJ conversion technologies, especially for the upgrading processes.

15.2.2 Oil-to-jet conversion pathways

The OTJ conversion pathways are classified into three processes: hydroprocessing of renewable jet (HRJ), also named as HEFA; CH that is also known as hydrothermal liquefaction; and hydrotreated depolymerized cellulosic jet (HDCJ), which is termed as fast pyrolysis with jet fuel upgrading [3]. The biobased jet fuel produced from the HRJ process is the only product that has met the ASTM specification and been approved to be blended with conventional jet fuels [10]. Triglyceride is the main feedstock used for the HRJ and CH processes, but the production of free fatty acids (FFAs) is different in both processes [29]. Glycerides are cleaved by using propane to make FFAs in the HRJ process, whereas the formation of FFAs in the CH process occurs by thermal hydrolysis. The pyrolysis of biomass feedstock is the way to produce biooil in the process of HDCJ [9]. The hydrotreating process for HRJ, CH, and HDCJ are similar, and a detailed discussion and review of the three processes are shown below.

15.2.2.1 Hydrogenated esters and fatty acids

HEFA or hydroprocessing of renewable jet is considered as the most important process with a high level of maturity, as it is available commercially and is utilized to produce biobased jet fuel for military aircraft [30]. The production of biobased jet fuel from the process of HRJ had superior characteristics such as higher cetane number, lower GHG emissions, lower aromatic content, and lower sulfur content compared with conventional jet fuel [31]. HEFA is the process of hydrotreating triglycerides, animal fats, saturated or unsaturated fatty acids, and used cooking oils to produce biojet fuels. A wide range of catalytic hydrogenation, hydroisomerization, hydrocracking, and deoxygenation have been successfully improved and commercialized over the past 60 years [9]. The HEFA conversion process is divided into two steps. The first step is the conversion of triglycerides and unsaturated fatty acids into saturated ones by utilizing catalytic hydrogenation method. A β-hydrogen elimination reaction acts on the triglycerides, and a fatty acid is the yield of the process. Propane cleavage is an intermediate process used to produce three moles of FFAs. FFAs and glycerol are obtained through processing oil and fats with water under high temperature ($250-260°C$) and high pressure [3]. An energy-intensive process is required for glycerol purification, which would raise the overall cost of the process. However, the selling value of glycerol might compensate for the additional cost of the purification process [9]. Hydrodeoxygenation and decarboxylation are two processes used to produce C15-C18 straight-chain alkanes by converting the obtained saturated fatty acid. Propane, CO_2, H_2O, and CO are the byproducts of the process [3]. Noble metals are supported with zeolites or oxides used as early developed catalysts, where transition metals such as Mo, Ni, and Co or their supported bimetallic composites are the latest successfully developed catalysts [32].

The second step of the HEFA process is the hydrocracking and isomerization reactions, where the reactions are either sequential or concurrent. The hydrocracking and hydroisomerization reaction is used to convert the deoxygenated normal paraffins or straight-chain alkanes to a biojet fuel known as synthetic paraffinic kerosene (SPK) that has a carbon chain ranging from C9 to C15 [31]. Researchers have found that the isomerization reaction occurs first, followed by cracking. The isomerization process is associated with a hydrocracking process that produces various yields from isomerized species [3]. The exothermic property of the hydrocracking reactions results in gas and lighter liquid products.

A low yield of jet-fuel-range alkanes will result from overcracking, and there is a high yield of light products ranging from C1 to C4 and naphtha with a carbon chain ranging from C5 to C8. The products from overcracking are out of the jet fuel range with low economic value compared with conventional diesel or jet fuel. The cracking process will be varied by varying the catalyst type at the paraffin molecule end, resulting in jet-fuel-range yield adjustment [33].

15.2.2.2 Process of catalytic hydrothermolysis

The process of CH is another pathway to convert biomass feedstocks into biojet fuel, which is also known as hydrothermal liquefaction (HTL) [34]. This a well-developed process patented by the Associates of Applied Research, Inc., for renewable, aromatic, and drop-in fuel production, which is also named ReadiJet or ReadiDiesel from algal or plant oils [3]. The CH conversion pathway has a series of reactions, which include cracking, isomerization, decarboxylation, hydrolysis, and cyclization. Triglycerides are converted through the reaction series to a mixture of hydrocarbons chains (straight, cyclic, and branched). The conversion process of CH occurs at a temperature range of $450°C-475°C$ and a pressure of 210 bar in the presence of water, with or without a catalyst [9]. The mild reaction conditions allow us to have high energy efficiency and use wet feedstock [35]. Decarboxylation and hydrotreating processes are used to saturate and remove oxygen from the resulting products that include unsaturated molecules, carboxylic acids, and oxygenated species [36]. The treated products have a carbon chain ranging from C6 to C28, and include aromatics, n-alkanes, cycloalkanes, and isoalkanes. The resulting products require a fractionation process to separate the mixture of naphtha, diesel fuel, and jet fuel [9]. The biojet fuel produced by the CH conversion pathway meets the ASTM and the military specifications, having excellent stability, quality of combustion, and cold flow characteristics [37]. Recent studies have found that biojet fuel can be produced through the CH conversion process from various triglyceride feedstocks that include tung, soybean, camelina, and jatropha oils [9].

15.2.2.3 Hydrotreated depolymerized cellulosic jet

The HDCJ is a recent oil-upgrading conversion technology. The lignocellulose biomass produced by the hydrothermal process or biooils produced by pyrolysis are converted through the HDCJ pathway [9]. Lignocellulosic biomass has attracted attention due to its remarkable advantages such as ready availability, low cost, and not representing competition to food supplies [3,9]. There are undesirable characteristics of biomass pyrolysis oil, such as high corrosivity, poor thermal instability, and low energy density due to the high oxygen content. A series of hydrotreating processes are used to convert biooils produced through pyrolysis to jet-fuel-range products [38]. Hydrotreating and fractionation processes are applied to biooil produced through pyrolysis to produce jet blend fuels if there is no further upgrading using a catalyst. The biooil upgrading process is divided into two main steps of hydroprocessing. The first step is to hydrotreat the biooil in the presence of a catalyst under mild reaction conditions. Hydrodeoxygenation of biooils could be promoted using organics to overcome the formation of coke problem. The second step is to obtain the hydrocarbon fuel by using a catalyst and a conventional hydrogenation process under high temperature [3,9]. Recent studies have focused on using an integrated biorefinery system, which combines commercial Rapid Thermal Processing and catalytic hydroconversion. The jet fuel produced from the HDCJ pathway has high aromatic content, few impurities, and low oxygen content. However, further research and development are required for the HDCJ conversion process, as this process requires high deoxygenation and hydrogen consumption, and that leads to a considerable expense. Furthermore, the humble yields of hydrocarbon and the short lifetime of the catalyst can be considered as challenges for the utilization in the aviation sector.

15.2.2.4 Economic perspective

The economic perspective of the fuel produced from OTJ conversion process is discussed in different studies [6]. The fuel price produced through the HEFA conversion pathway was about 4.1$/gal or 3.6$/GGE in 2011 for a plant with a capacity of 98.28 MM gal/yr, and for a plant with a capacity of 30.16 MM gal/yr the price of the produced fuel was about 4.8$/gal or 4.2$/GGE in 2011. The production of the maximum yield of jet fuel requires the addition of 0.27–0.31$/gal as hydrogen usage is increased and yields of jet fuel are decreased. Recent studies have focused on economic analysis of biojet fuel produced from Pongamia oils and microalgae, where the estimation of jet fuel selling prices were 8.9 or 7.9$/GGE and 31.98 or 28.3$/gal, respectively, in 2011. Technology and market development will reduce the prices to be 6.07 or 5.4$/GGE and 9.2 or 8.1$/GGE, respectively [3].

A hydrotreatment process is required for the production of biobased jet fuel through the HRJ conversion pathway; therefore the capital cost is suggested to be 20% higher than the production of biodiesel. However, the HRJ byproducts, including liquefied petroleum gas (LNG), HRJ-naphtha, diesel, and propane, have higher credits than the byproducts of the transesterification process that include the glycerol product [24]. The biomass feedstocks represent a significant contribution to the overall production cost. The edible and nonedible oils have remarkable characteristics such as being renewable, environment-friendly, and produced locally. These make these feedstocks a promising alternative solution. The choice of feedstock affects the operating cost as oil feedstocks with high yield will result in decreasing the operating cost [35]. The development of the HRJ coproducts will increase the overall yield or allow the sale of these byproducts with high-value-added. Coproducts depend upon the conversion technology used, where 4% of the propane is created in the fractionation step, and its current price is about 2.48$/gal. LNG formed in the separation step has a current price of 4.58$/1000 cubic feet, where the crackdown of diesel to a jet fuel-range product will increase LNG to 6%. Naphtha, which is created in the distillation step, has a current selling price of 2.03$/gal, whereas the current biodiesel price is about 3.64$/gal [3,9].

15.2.2.5 Assessment of life cycle

Numerous researchers have analyzed the GHG emissions on a life cycle basis. Soybean oil releases about 40%–80% GHG emissions of conventional jet fuel, which releases about 89 g CO_2 e/MJ [3]. Soybean oil yield requires hydrogen for the hydrotreatment process, produce liming emissions, and N_2O emissions from fertilizer. A considerable increase is noted by changing the land use; therefore 800% more emissions have been indicated from low-yield soybean oil obtained from the tropical rainforest compared to petroleum jet fuel. The process of converting palm oil to biobased jet fuel has emissions that are 30%–40% of the petroleum jet fuel production process [3,15]. It was found that by accounting for the change in land use, there was 40%−800% increase in the GHG emissions of conventional jet fuel. The emissions from rapeseed oil were about 45%−87% of conventional jet fuel and 87%−147% with the consideration of the land-use change, whereas emissions from jatropha oil were 36%−52% of conventional jet fuel and 20% of the total emissions are represented by N_2O emissions [9,18]. Also, another study illustrates that the emissions from the OTJ conversion pathway using jatropha were 40 g CO_2 e/MJ of fuel produced, which represents about 45% that of conventional jet fuel. The range of emissions for algal oil was about 16%−220% of those from petroleum jet fuel. The GHG emissions are lowered by 45% relative to petroleum jet fuels by utilizing the hydrogen generated from biochar and natural gas to support the energy process. The reforming of biochar and pyrolysis oil generate hydrogen that is used as fertilizer, as a result a 103% reduction in GHG emissions relative to conventional jet fuels is indicated.

15.2.3 Process of gas-to-jet fuel

The GTJ conversion pathway discussed here explains how different feedstocks such as biogas or natural gas, or syngas can be converted to biojet fuel. The most common conversion methods are FT and the fermentation of gas process, which are categorized and reviewed below.

15.2.3.1 *Process illustration*

15.2.3.1.1 Process of Fisher Tropsch biomass to liquid

The process that converts syngas to liquid hydrocarbon fuel is knowing as *Fisher Tropsch* (FT) [39]. Lower emissions are formed from fuels produced by FT when used as a fuel in the jet engines as a result of being free from sulfur and lower aromatics content compared to diesel and gasoline [40]. Some studies have indicated that using the FT method of converting biomass into artificial fuel can produce carbon-neutral alternatives to fossil fuels like gasoline, diesel, and kerosene [41]. Fig. 15.1 shows a FT-BTL process diagram that begins with biomass feedstock drying to reduce the size of particles during the pretreatment process [42].

Biomass has been converted to syngas through many gasification processes. Firstly, the gasification process is performed on dried and pressurized biomass at a temperature of around 1300°C. The gasification process is conducted in the presence of steam and pure oxygen. The biomass drying process is provided due to the presence of a combustor that also provides heat. Syngas is then cooled by a direct water quenching method that also helps to eliminate tar and ash. After quenching, the H_2 to Co ratio is adjusted using water gas shift system [42]. Several researchers also investigated indirect gasification on the FT process. In this process, hot olivine circulation indirectly heats the gasification process. Also, steam fluidizes the material in the gasifier, and the gasification process is conducted at 880°C and atmospheric conditions. In a fluid catalytic cracker, the conditioning process is performed on syngas to convert light hydrocarbon, residual tar, and methane to syngas. Zinc oxide is used to polish the produced syngas. An absorbent for activated carbon is also used and then the syngas is pressurized to FT process pressure, which equals 25 bar

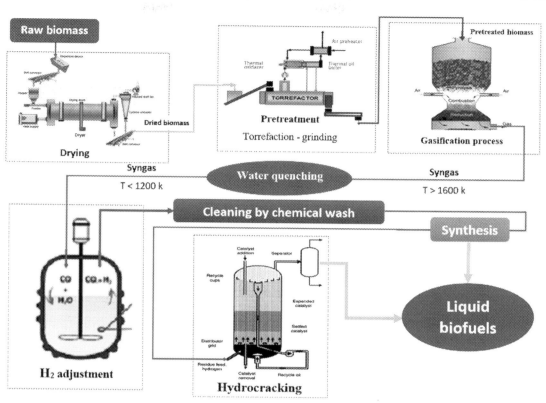

FIGURE 15.1 The process of converting raw biomass into liquid biofuels through Fisher Tropsch-catalytic hydrothermolysis block diagram.

in some conditions. Treated syngas is then converted to liquid fuel directly through the FT process.

The process of FT is classified according to its operating temperatures, such as high-temperature Fischer–Tropsch (HTFT) and low-temperature Fischer–Tropsch (LTFT) [43]. High-temperature FT process is performed at a temperature range of 300°C–350°C using an iron-based catalyst, which is used mainly to produce gasoline [44]. HTFT is characterized by the production of oil containing a high percentage of short-chain hydrocarbons with a carbon atom count below 10. The low-temperature FT process is performed at a temperature range of 200°C–240°C using an iron or cobalt catalyst. LTFT is used to produce higher-molecular-weight hydrocarbons, and the products range from methane to long-chain hydrocarbons. One of the most important things to consider in the process of FT is the need to remove the heat used in the reaction, since it is highly exothermic, to avoid deactivation of the catalyst and overheating [45]. There are many catalysts that can be used in the FT process, including iron, nickel, cobalt, and ruthenium, which have been studied in the literature [46]. Syngas that isn't converted into liquid fuel is returned to the FT reactor again and some of it is sent to the system that removes acid gas [41].

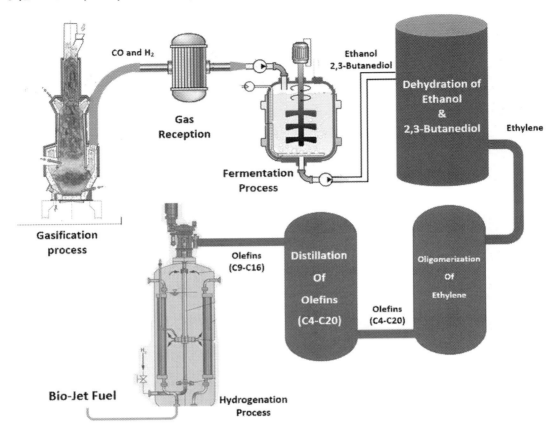

FIGURE 15.2 The process of converting syngas into liquid biojet fuel through gas fermentation pathway.

15.2.3.1.2 Process of gas fermentation

The process of fermentation into a liquid jet biofuel transforms the syngas in the sequence shown in Fig. 15.2. Acetogenic bacteria are able to ferment CO and H_2 into ethanol, and 2,3-butanediol [47]. Ethanol and butanediol are directly converted to biojet fuel using ATJ conversion technology that includes processes of hydrogenation, distillation, oligomerization, and dehydration [48]. The fermentation process has the advantage of being able to produce several products other than those produced from other conversion processes [49]. Production costs are low because they do not need a high-temperature or high-pressure operation [49]. The overall energy efficiency in this system is high, at around 57%, compared to a decreased overall efficiency for FT process of 45% [50].

15.2.3.2 Economic perspective

Biomass to liquid (BTL) fuel production cost depends on the price of the feedstock. High-cost agricultural crops increase the cost of the production process by 70% more than high-cost forest wastes and residues [51]. Also, improved crop yield and reduced conversion cost reduce the overall production costs. Small plants with lower energy efficiencies and a high-cost gasification process would lead to BTL production costs higher than CTL production costs [52]. The gasification temperature also had a significant effect on the BTL FT process cost analysis [53]. The gasifier with low-temperature has many drawbacks that include the restriction of reaction kinetics and the formation/flow of slag [54]. For the same capacity, the capital cost is about UD\$ 610 million for low-temperature (870°C), direct fluidized bed, the process of nonslagging gasification; while it is US\$ 516 million for the high-temperature (1300°C), indirect gasification process [55]. Capital cost for the gasification process is higher than for biochemical processes and pyrolysis [56]. The FT liquid cost of the output varies from 1.86 to 3.3 \$/gal, and 75% of the capital cost of production is spent on pretreatment, gas cleaning, and gasification processes.

15.2.3.3 Assessment of life cycle

The greenhouse gases (GHG) produced from fuel synthesis processes, and syngas production are methane, CO, CO_2, H_2, and N_2O. CO_2 emissions are produced from gasification process and FT synthesis. The N_2O emission comes from biomass cultivation and fuel combustion. In addition, methane emissions are produced from incomplete combustion and methane released from coal bed. The process of FT BTL had GHG emissions that were 92%−95% lower than GHG produced from petroleum jet fuel [51]. GHG emissions life cycle when using biomass such as Switchgrass, corn stover, and forest residue are recorded to be −2, 9, 12.2 CO_2 e/MJ of those produced when using petroleum jet fuel [57]. The GHG and other emissions of FT process can be reduced by identifying technologies and strategies such as coproduction of fuels and power, coal and biomass coprocessing, carbon sequestration, and improved vehicle technology [58]. Carbon capture and storage has a major impact on the climate impacts of peat FT fuel. The climatic impacts of peat FT fuel without the capture and storage of carbon are 30%−40% less than for conventional fuel. The climate impacts of peat FT fuel with the capture and storage of carbon are 50%−84% less than for petroleum fuel [59].

15.2.4 Process of sugar to jet fuel

Biochemical conversion methods of liquid fuels and chemicals can be provided by the conversion of sugar to hydrocarbons using biological and catalytic routes [60,61]. Catalytic upgrading includes chemical and biochemical processes that are performed to separate sugars from biomass. The hydrocarbons are produced by fermenting biomass sugars. The hydrocarbons produced are then purified, recovered, and upgraded to decay hydrocarbon fuels.

15.2.4.1 Sugar to jet processes

Conversion of STJ fuel includes two main processes: sugars to hydrocarbons by catalytic upgrading; and fermentation of sugars to hydrocarbons. In the following section, the two methods will be introduced and discussed in detail.

15.2.4.1.1 Process of sugars to hydrocarbons catalytic upgrading

Pretreatment and enzymatic hydrolysis processes are performed on biomass feedstock to produce lignocellulosic material. The APR process is performed to catalytically convert sugars to hydrocarbons and hydrogen, but this process needs a certain level of concentration and purification of biomass sugars. The fractionation process is performed to hydrocarbon molecules to separate them into jet, diesel, gasoline fuels, and other bioderived chemicals as coproducts [62].

An example of converting plant sugar to hydrocarbon mixtures is the bioreforming platform from Virent [61]. After pretreatment and fractionation processes, cellulose, hemicellulose, and lignin are derived from lignocellulosic biomass. The lignin produced is used in the combustor to provide heat to the process. Fractionated and hemicellulose are converted to five and six-carbon sugars using acid or enzymatic hydrolysis. The APR process is the primary step in bioforming in which water reacts with a stream of product from hydrotreating in the presence of the heterogeneous catalyst and at a temperature range of 450°C−575°C and pressure of 10−90 bar. Oxygenates from the APR process can be converted into jet fuel hydrocarbons through three potential routes. Acid condensation is the first route in which oxygenates turn into alkanes, aromatics, and isoalkanes with the presence of ZSM-5 catalyst. This route includes the dehydration of oxygenates to alkanes and then alkanes are oligomerized to heavier alkanes, which are cracked, cyclized, and dehydrogenated to aromatics [63]. Jet fuel is produced from the distillation and blending of the heavier species produced. The second route includes several direct condensation reactions through multifunctional solid-base catalysts [64,65]. This route produces products that are in the jet fuel range. The third route includes hydrogenation−dehydration, and dehydration reactions are carried out for the conversion of oxygenates produced from APR into alkanes and alkenes. Kerosene is produced from alkenes through the oligomerization process with the presence of zeolite or solid phosphoric catalysts [66]. ABE fermentation and chemical catalysis were integrated into a process to convert cane sugars and lignocellulosic biomass into jet fuel, gasoline, and diesel by scientists at Berkeley [67].

15.2.4.1.2 Direct sugar to hydrocarbons

In this process, the hemicellulosic sugars are released due to the breaking down of cell walls of biomass. Solid materials have to be removed by enzymatic hydrolysis and then 50% of water is removed, which results in the concentration of liquid sugars. An anaerobic fermentation process is performed using fed-batch or continuous fermentation to produce hydrocarbon intermediates. Hydrocarbon fuels are to be recovered through the phase separation stage of the resulting products from the previous stage.

15.2.4.2 Economic prospects

Sugar and sugar intermediates separation plays an important role in the analysis of STJ process economics. Separation of biomass contributes about 50%−70% of the operation and capital cost [60]. Separation process efficiency at high temperatures can be increased by flux increases, resulting in avoiding the formation of precipitants and decreasing viscosities [60]. Production cost is affected by four major process steps: acid-catalyzed dehydration, APR process, hydrogenation, and pretreatment [68]. Based on the technoeconomic analysis for the production of 5-hydroxymethylfurfural (HMF) and 2,5-dimethylfuran (DMF) for a 20-year operating period and 300-ton/day capacity of fructose, the minimum estimated selling prices for DMF and HMF were 7.63 $/gal and 5.03 $/gal [69]. The utilized catalyst in the operating process plays a vital role in the economic analysis of the DMF process production cost. Technoeconomic analysis for the direct sugar to hydrocarbons pathway has been conducted and shown that the sugarcane biobased jet fuel selling price was about 7.16 $/gal.

15.2.4.3 Assessment of life cycle

The analysis of GHG emissions life cycle of sugarcane sugars biobased jet fuel was conducted by the International Trade Negotiations institute based on the parameters of the Amryis process [70]. GHG emissions of the life cycle were found to be about 15 g CO_2 e/MJ, which shows a reduction of 82% compared to conventional jet fuel [70]. Farm input and emissions of N_2O from the soil are dominating the GHG emissions from sugarcane production and transport. N_2O emissions from soil and farm input are about 45 and 32 g CO_2e/MJ, respectively [71].

15.3 Algae biojet fuel

Algae are one of the most important sources of biojet fuels, but they are not used in their primary form. They are converted into four different feedstock types, which are lipids. Algae are converted into lipids through the extraction process and can also be converted into biooil through the liquefaction process. They can also be converted into syngas through the gasification process, and a pretreatment process can also be used to convert algae into sugars. After converting algae to one of the four images, one of the conversion methods used in the production of biojet fuel is used, which is suitable for each type of different feedstocks, as shown in Fig. 15.3.

The biojet conversion method is then chosen according to the type of algae feedstock. Whereas, when algae are converted into lipids, they are transformed into biojet by hydrocracking (isomerization) to produce HEFA-SPK. When the algae were converted into biooil

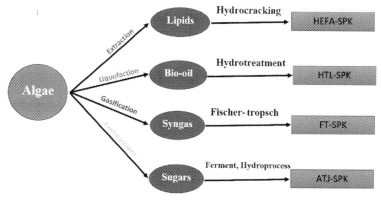

FIGURE 15.3 The conversion pathways of algae feedstocks to biojet fuel through different pathways.

by liquefaction, this biooil is then converted to biojet using a hydrotreatment process to produce HTL-SPK. When the algae are converted into syngas by the gasification process, syngas is then converted to biojet fuel using the Fischer–Tropsch method that produces FT-SPK. When the algae are converted into sugars through the pretreatment process, these sugars are converted to biojet through the fermentation process to produce ATJ-SPK.

15.4 Biojet fuel performance characteristics

Standard specifications are required to be met for all biojet fuels characteristics before the approval of any commercial activities in the aviation field. The renewable aviation fuels or biobased jet fuels refer to the mixture of petroleum jet and synthesized hydrocarbons instead of 100% renewable jet fuels. The specifications and the target compositions of biobased jet fuels are determined in terms of the performance requirements [72]. The biojet fuel must achieve a specific requirement such as acceptable minimum density of energy by mass, maximum permissible deposits in the standard heating tests, and maximum permissible freeze point temperature, viscosity, sulfur, and aromatics content, the concentration of acidity and mercaptan. The American Society for Testing and Materials (ASTM) and the British Ministry of Defense (MOD) generate various standards for jet fuels. ASTM (D1655) is referred to as the International Air Transport Association Guidance Material (Kerosene Type), and DEF STAN 91–91 is the standard of jet fuel governed by the British MOD, where these two standards are the most used for jet fuels [72,73]. Therefore numerous efforts have been made by ASTM and British MOD to make a comprehensive evaluation of the possibility to blend biobased synthetic hydrocarbon with fossil jet fuels in conventional aviation turbines. ASTM (D4054) issued in 2009 is the standard practice for the Qualification and Approval of New Aviation Turbine Fuels and Fuel Additives, a very strict process.

The approval process for a new biobased jet fuel comprises test programs, the specification change, and internal reviews. Also, to ensure that the new biojet fuel performance is acceptable for the engine through all the environmental conditions of the aircraft, the properties of fit-for-purpose described in the Handbook of Aviation Fuel Properties were also assessed [72]. The fit-for-purpose properties, which include flammability limits versus altitude, thermal coefficient of expansion, thermal conductivity, and surface tension, are not specified in the ASTM (D1655) standard. The United States Federal Aviation Administration and the original equipment manufacturer are responsible for the test results to be reviewed. ASTM (D7566) is a new standard coordinated by the UK DEF STAN and the United States-based ASTM organization issued in 2009 and referred to as the Standard Specification for Aviation Turbine Fuel Containing Synthesized Hydrocarbons. The major properties of jet fuels are similar for different standards, where the most important property is to have a high flash point in consideration of fire hazard. Also, good cold flow properties are important to ensure the flow of jet fuel at high altitude [73]. Numerous investigations on alternative renewable jet fuels have been performed, and ASTM (D7566–18) illustrates the various types of synthesized hydrocarbons, which have been certified as blending ingredients with petroleum jet, such as ATJ-SPK (2018), SPK/A by FT (2015), SPK by HEFA (2011), and

SIP-SPK (2015). The comprehensive evaluation of the biojet fuel performance characteristics coupled with a detailed comparison with the specified ASTM (D7566−18) is very important for jet fuel to be safe, compatible, and reliable with aeroengine parts. The performance characteristics are very important in clarifying the feasibility of alternative biojet fuels. Therefore the aim is to make a comprehensive evaluation of the performance characteristics of biojet fuels coupled with a complete understanding of the physicochemical characteristics and biobased jet fuels chemical composition. Hereinafter, biobased jet fuel performance characteristics can be categorized into thermal oxidation stability, the compatibility of fuel with the current system of aircraft, aircraft range, and metering of fuel, low-temperature fluidity, combustion property, and fuel volatility [3,74]. The evaluation of performance properties was made on the basis of their corresponding characteristics.

15.4.1 Stability of thermal oxidation

Oxidation and stability characteristics of fuel at operating temperature of the aircraft are considered one of the most important performance requirements, where the desirable requirement is to have a high stability of thermal oxidation. The thermal oxidation stability can be categorized into two various perspectives, including the oxidation stability and thermal stability.

15.4.1.1 Thermal stability

Thermal stability is measured relative to deposit amount in the aeroengine fueling system at different operating conditions. Jet Fuel Thermal-Oxidation-Stability Test (JFTOT) is included in the ASTM (D3241) standard and can be used for biojet fuels thermal stability. Two quantitative metrics are provided by the JFTOT to measure the tendency of deposit forming, which includes the deposits on surface of the testing tube and drop in pressure subsequent to trapping polymerization/degradation products of biobased jet fuel. The two metrics were also specified in the ASTM (D7566−18) to ensure that the biobased jet fuels have proper thermal stability [75]. The scale of surface deposits ranges from 0 to 4 as a heavy deposit, where the proper surface deposit should be less than 3. The pressure drops after 2.5-h JFTOT test at 325°C as minimum temperature should be rated less than 25 mm Hg. Generally, studies have found that biobased jet fuels had even better thermal stability than petroleum jet fuels, and fully synthesized jet fuel (FSJF) at a high operating temperature of 360°C has excellent thermal stability. Also, Shell SPK by FT, castor HEFA, Sasol SPK by FT, camelina HEFA, Rentech SPK by FT, and HEFA from tallow show high thermal stability, as the HEFA jet fuel has a surface deposit rating of 1 and almost no drop in pressure at standard operating condition of 325°C after 2.5-h [72]. Petroleum jet fuel has low thermal stability, which may be due to the presence of heteroatom components, benzothiophenes (C_8H_6S), and other compounds containing cyclic sulfur structures, where the FSJF has high thermal stability as compared to petroleum jet fuels, as it is free of (benzothiophenes). Biojet fuels are distinguished from conventional jet fuels as they consist of n-paraffins, iso-paraffins, and cycloparaffins and are commonly free of aromatics [72,75]. The tendency to form deposits at high temperatures for paraffinic compounds is weaker compared to aromatic compounds. Thus biobased jet fuels have proper thermal stability, despite the limited availability of literature about the thermal stability of biobased jet fuels.

15.4.1.2 Oxidation stability

The jet fuel oxidation stability is a measure of the fuel's ability to resist oxidation in the presence of oxygen at a moderate temperature of $100°C-160°C$. The induction period (IP) can be used to quantitatively measure the oxidation of the jet fuel. ASTM (D7566−18) doesn't include the oxidation stability of biojet fuel, even though a few scientific researchers have studied biojet fuels oxidation stability [73]. The IP of HEFA was about 1 h at $140°C$ test temperature and 7 bar oxygen pressure, where jet fuel (A-1) has an IP of nearly 2.3 h. Jet A-1 has a high IP due to the aromatic content in the jet fuel. It was suggested to improve the oxidation stability of HEFA by mixing the biojet fuel with jet fuel (A-1) or/and compounds of the aromatic model. HEFA/jet fuel (A-1) blend has an IP of about 3 h by adding 25% as a volumetric percentage of jet fuel (A-1) to HEFA. Many attempts have been made to investigate the effect of the molecular structure of the model aromatic compounds on the IP, where the attempts lead to further evaluation of the influence of the addition of model compounds into HEFA and presenting the IP values of the mixtures. The effect of diaromatics, monoaromatics, and cyclic alkane on the IP of HEFA was investigated, where a significant improvement in the IP of HEFA was indicated by adding 5% volumetric percentage of 1-MN (diaromatics) as IP of HEFA increased from 1 to 8 h. The oxidation stability of HEFA exhibits a moderate improvement by adding monoaromatics compounds, where there was almost no improvement in the oxidation stability of HEFA by adding the cyclic alkane [72].

15.4.2 Combustion characteristics

The combustion characteristics of biobased jet fuels are one of the most important properties, especially with regard to the growth in global awareness concerning GHG emissions and climate change. The combustion mechanism of biobased jet fuels is continuous vaporization and ignition with rapid flux of high temperature air in the aeroengine of the aircraft. The incomplete combustion will form visible unburned hydrocarbons and particulates such as smoke and/or soot in the case of high smoke concentration. Parameters that can help to fully describe the combustion characteristics are smoke point, particulate matter (PM) emissions, carbon monoxide (CO), carbon dioxide (CO_2), and the derived cetane number (DCN).

15.4.2.1 Smoke point

The smoke point is one of the parameters used for the evaluation of jet fuels combustion characteristics. It requires a flame without smoking with a minimum height of 25 mm. The jet fuel with a low tendency to produce smoke has a high smoke point. ASTM (D7566) does not specify the limit of synthesized hydrocarbons smoke point; however, researchers evaluated biojet fuels smoke point [73]. FT-SPK and HEFA are biojet fuels, that are characterized by high smoke points (larger than 40 mm) compared with petroleum jet fuels such as JP-8, which has a smoke point of 25 mm. Therefore the combustion performance of biojet fuels is outstanding, and this was attributed to the presence of aromatics in conventional jet fuels, which have a high tendency to form soot as compared to paraffin compounds [12]. Also, the tendency for soot formation for petroleum and biobased jet fuels is evaluated using the sooting threshold index, where it can be treated as one of the emulated jet fuel characteristics through the jet fuel surrogate improvement [72].

15.4.2.2 Particulate matter emissions

Haze and smog are formed due to particulate emissions and can be harmful to humans. The aromatic compounds are one of the main factors that are responsible for PM emissions formation [73]. Biobased jet fuels have a low aromatic content that is attributed to the lower PM emissions as compared to conventional jet fuels, for example, a 52% and 62% reduction in PM (number and mass) was accomplished by using 100% of FT-SPK. Also, the blended jet fuels such as 50% FT-SPK, which has 9.25vol.% of aromatic content, and camelina HEFA/Jet (50:50vol.%) used in NASA DC-8 aircraft can reduce the PM number and PM mass emissions to a large extent [72].

15.4.2.3 Gaseous emissions

There is a strong relationship between gaseous emissions produced by jet fuels and the operating conditions of the aircraft, including take-off, ground idle, top of the climb, low power, and cruise. Gaseous emissions refer to the CO, unburned hydrocarbon (UHC), NO_x, and CO_2 emissions, and biojet fuels emissions have been extensively investigated [3,73]. Studies have found that FT-SPK and blend fuel (50:50) reduce the NO_x emissions by 10% and 5%, respectively. Other studies stated that FT-SPK and HEFA produce NO_x and CO_2 emissions similar to those of JP-8, where the emissions of CO and UHC were 10%−25% lower than that of JP-8. Also, a comprehensive numerically simulated investigation was performed on the biobased jet fuel combustion characteristics in a turbofan aeroengine, where the results indicated a reduction in CO and NO_x emissions of 6.8% and 8%−18%, respectively [73]. However, UHC emissions increased by 5%−10%. The fuel/air ratio has a significant effect on CO and UHC emissions, which depend on the type of the aeroengine and also on the operating conditions. Generally, the reduction in PM emissions is higher than gaseous emissions from the combustion of biojet fuels, where about 72% of the reduction occurring in the pollution emissions is from the PM emissions reduction [12,72].

15.4.2.4 Derived cetane number

The ignition characteristic of the fuel is represented by the DCN, where the ASTM (D7566−18) standard does not specify DCN. Complete combustion with high power, better performance, and lower harmful emissions is observed with fuels that have high DCN, which leads to shorter ignition delay [73]. Biojet fuels have DCN > 60, which is higher than that of conventional jet fuels, for example, Jet A has a DCN of 49.35, with the exception of Sasol FT-SPK which has a DCN of 33.46. Sasol FT-SPK has lower DCN due to the low content of n-paraffins and high content of iso-paraffins (>80 wt.%), as the reactivity of iso-paraffins is lower than n-paraffins. The aromatic compounds have a significant effect on the ignition characteristics of jet fuels; the aromatic compounds are stable as the benzene ring has remarkable bonding. Therefore the conventional jet fuels have low DCN compared with biobased jet fuels that are free of aromatics and have high DCN [72].

15.5 Fuel compatibility with current fueling system of aircraft

Jet fuels have to be compatible with the current aircraft fueling system. This is one of the most important concerns in the aviation industry, and it can be categorized into two important characteristics: lubricity and volume-swell of seal material. The aromatic content in aircraft fuels has a significant effect on the compatibility of aviation fuels with elastomers in the current aircraft fueling system. Petroleum jet fuels have aromatic content about 10%–20%, which can properly soften the swell O-ring seal. On the contrary, biojet fuels are free of aromatics, and therefore biojet fuels tend to shrink and harden O-ring seals, resulting in the leakage of fuel [72]. A comprehensive discussion about lubricity and volume-swell of seal material is made in this section to illustrate the compatibility of biobased jet fuels with the current fueling system of aircraft.

15.5.1 Volume-swells of seal material

The volume-swell of seal material in the aircraft system is mainly dependent on the interaction strength between seal materials and the strength of the interaction between jet fuels. There is a strong relationship between the interaction strength and the chemical compounds structure and composition in biobased jet fuels [73]. The volume-swell of seal material percentage is not specified in the ASTM (D7566–18) standard for either blend fuels or synthesized hydrocarbons. However, to ensure a desirable volume swell of elastomers, the blend fuel is required to have 8 wt.% of aromatics.

A few attempts were made to investigate the effect of biobased jet fuels on the volume-swell of seal material, and it was found that the molecular structure compositions and distributions of FT-SPK and HEFA have significant effects on the penetration ability of these fuels through O-ring seals (polymer-based) [72,75]. The addition of aromatics was feasible for biojet fuels with low aromatic content; however, a small impact was observed on the change of elastomer compared with conventional jet fuels. In addition, the different materials of O-rings have significant effects on the compatibility of biobased jet fuels with elastomers [72,74]. Even though the aromatics addition can improve alternative jet fuels compatibility with the existing fueling system of aircraft, the soot formation tendency might increase, resulting in an increase in the PM emissions, as mentioned before. In conclusion, there is a low capability to obtain a satisfactory percentage of volume swell with the aromatics-free alternative jet fuels compared to petroleum jet fuels [72]. The aromatics addition can improve biobased jet fuels compatibility with the current fueling system of the aircraft. However, the dependence between the soot formation tendency and the volume-swell needs to be considered.

15.5.2 Lubricity

The definition of lubricity is the ability of the fuel to minimize the wear of engine components. ASTM (D7566−18) standard does not specify the lubricity limit of synthesized hydrocarbons; however, aircraft turbine fuel/synthesized hydrocarbon blend jet fuel has a wear scar diameter of about 0.85 mm, as specified by (D5001) test method. Also, lubricity and wear scar diameter data of synthesized hydrocarbons are not specified in the literature, but it was reported that there is a relation between the biobased jet fuels lubricity and the presence of compounds with a polar nature [72,73]. The compounds of conventional jet fuels that occur naturally and are responsible for the desirable lubricity such as oxygen, nitrogen, and sulfur can be removed during the processes of severe hydrotreatment from synthesized hydrocarbons, result in biojet fuels with poor lubricity [72,75].

The utilization or blending of additives at low levels (as low as 10 ppm) with petroleum jet fuels, which contain sulfur of 700 ppm, are two possible solutions to eliminate the weak lubricity of biobased jet fuels. Fatty acid methyl ester can be used as an additive to enhance the lubricity of HEFA fuel [72,76]. In conclusion, biojet fuels have less satisfactory lubricity compared to conventional jet fuels due to the low presence of polar-nature compounds. In addition, numerous research efforts are required for the investigation and improvement of the lubricity of biobased jet fuels to achieve the fully (drop-in) biobased jet fuels, instead of mixing with conventional jet fuels.

15.5.3 Low-temperature-fluidity

Biobased jet fuel fluidity at low temperatures is one of the most important characteristics. Kinematic viscosity and freezing point can characterize the biobased jet fuels' low-temperature fluidity. To ensure good fluidity of the biojet fuel in the engine, the kinematic viscosity and freezing point must be low to overcome the low fuel tank temperature at high altitudes.

15.5.3.1 Freezing point

Biofuel freezing-point is one of the most important determinants of low-temperature properties. SIP has recorded a maximum freezing point of −60°C and SPK by FT, HEFA, SPK/A by FT, and SPK by ATJ all have a highest freezing point of −40°C. Table 15.1 presents the freezing-point and kinematic-viscosity of selected types of biobased jet fuels. The freezing point of biobased jet fuel is influenced mainly by three factors: bioparaffins' carbon chain length, the content of alkylated aromatics, and isoparaffin content.

The first and most decisive factor in determining the biojet fuels' freezing point is the length of the carbon chain of the bioparaffins [95]. The lower the number of carbon atoms, the lower the freezing point, for biofuels of C10 content, the freezing point is −97°C, while it is −40°C for C15 farnesane [96]. High freezing points of −15°C and −10°C were reported for palm kernel biokerosene and coconut biokerosene resulting from the (catalytic) distillation of triglyceride-based oil [85]. The absence of proper hydrocracking and long carbon chains were the possible causes of the high freezing points of biokerosene. Blending biokerosene with conventional jet fuel reduces the freezing point of the mixture.

352   15. Biojet fuels production from algae: conversion technologies, characteristics, performance, and process simulation

TABLE 15.1   The freezing-point and kinematic-viscosity of some of the biojet fuels.

Jet fuel	Freezing point of jet fuel ($^\circ$C)	Kinematic viscosity of jet fuel (mm^2/s)	Density at 15°C (kg/m^3)	Net heat of combustion (MJ/kg)	Reference
SIP-AMJ-300-A	-97.71	3.821 @ -20.00°C	768	43.33	[77]
fuel SIP	<-80.00	3.011 @ -20.00°C	773.1	–	[78]
FT-SPK-Shell	-55.00	2.610 @ -20.00°C	737	44.1	[79]
SIP-UQI-1-A	-40.41	7.721 @ -20.00°C	778	43.93	[80]
HEFA-1B coconut	9.51	6.461 @ -20.00°C	759	42.48	[81]
Fuel branched decalin	<-51.00	22.000 @ -40.00°C	880	42	[82]
Acetone derived fuel of geranyl	-72	–	800	45	[83]
Triglyceride-based oil distillate	-37.00	1.821 @ 20.00°C	–	–	[84]
Coconut biokerosene	-10.00	8.261 @ -20.00°C	867	35.06	[85]
Palm biodiesel-jet A-1 5% distillate	-27.00	2.951	820	–	[80]
FT-SPK 100%	-50.00	4.710 @ -20.00°C	761.2		[86]
HEFA castor	-62.00	5.311 @ -20.00°C	758		[87]
S-8 (FT-SPK) syntroleum	-59.00	4.612 @ -20.00°C	757	44.1	[88]
100% synjet (FT-SPK)	-59.00	4.611 @ -20.00°C	–	–	[89]
Dicyclohexylmethane + perhydrofluorene	-40	2138.000 @ -10.0°C	930	–	[90]
Perhydrofluorene	-15	1752.000 @ -20.0°C	959	43.12	[91]
Fraction of green jet fuel	–	2.851 @ 40.00°C	792	–	[92]
Hydrolyzed olein oil HEFA	-30.00	–	–	–	[93]
Camelina oil HEFA	–	5.014 @ -20.00°C	751	–	[76]
Babassu biodiesel distillate	-10.31	3.910 @ 20.00°C	875	38.1	[94]

The table is modified from the original work of J. Yang, Z. Xin, K. Corscadden, H.J.F. Niu, An overview on performance characteristics of biojet fuels, Fuel, 237 (2019) 916–936.

The second factor influencing biojet fuels' freezing points is content of alkylated aromatics. The addition of propyl-benzene has been studied by Hong et al. [81] to decrease the freezing point of HEFA, and the extent of the decrease was proportional to the amount of propyl-benzene added. Biojet fuel produced from the pyrolyzing process of waste cooking oil, rubber seed oil, and soybean oil with 5wt.% base catalyst has a freezing point of -37°C [84].

Corporal et al. [79] have demonstrated the effect of the isoparaffin on the freezing temperature, as the large proportion of branched paraffins in Sasol FT-SPK resulted in a poor freezing point ($<-77^\circ$C). Branched alkanes (farnesan) are considered as the main

Handbook of Algal Biofuels

component of SIP fuel, which has −90°C freezing point [69]. Han et al. also reported that the branched cyclohexane fuel has a poor freezing point (−80°C) [77]. Branched decalin fuel that is directly produced from monocyclic alkanes and alcohols through H_2SO_4 one-pot synthesis route has a freezing-point of (<− 51°C). A low freezing-point (−92°C) for the branched amyl ether resulted in satisfying freezing points by mixing it in various ratios in carbon jet fuel (QAV-1). Biojet fuel derived from branched geranyl-acetone (C13H22O) recorded a poor freezing point of −72°C [83].

15.5.3.2 Kinematic viscosity at −20°C

One of the most critical considerations evaluating aircraft fuel fluidity at low temperatures is the kinematic viscosity (at −20°C). Although ASTM (D7566) did not specify limits for kinematic viscosity, the biojet fuel kinematic viscosity and its blends should not be less than 8 mm^2/s at −20°C. If the kinematic viscosity of jet fuel increases, it may cause incomplete combustion, weak atomization, and pumping problems, and may, in the worst conditions, lead to a blockage in the fuel injectors [97]. Table 15.1 presenting different biojet fuels kinematic viscosity and its blends. It is common for the kinematic viscosity of biojet fuels to be determined at −20°C, but some studies have been conducted at higher or lower temperatures making the obtained values less comparable [82]. The kinematic viscosity was generally satisfactory for the majority of biobased jet fuel (<8 mm^2/s at 20°C), despite the fact that some biokerosene that was distilled from triglyceride oil, which has relatively high viscosity [81,85]. The viscosity of 5.3 mm^2/s was reported HEFA castor biojet fuel and 3.3 mm^2/s for its blend fuel with Jet fuel (A-1) (50/50) at −20°C were reported that demonstrates satisfactory fluidity at low temperatures [87]. FT-SPK viscosity was 4.65 mm^2/s and for its blended fuel with Jet fuel A-1 (50/50) was 4.4 mm^2/s at −20°C. ATJ-SPK fuel's kinematic viscosity at −20°C has been reported by Scheuermann et al. [78] to be 4.795 mm^2/s.

Some researchers investigated the relationship between biofuels chemical composition and their kinematic viscosity. Kinematic viscosity was evaluated by Chuck and Donnelly [97] for nine different biofuels at temperatures ranging between −30°C and 40°C. The chosen biofuels were ethyl cyclohexane, farnesane, methyl linolenate, limonene, ethyl octanoate, butyl butyrate, butyl levulinate, n-hexanol, and n-butanol also. The viscosity evaluation for 20% and 50% (volumetric percentage) biofuel/Jet fuel (A-1) was made as well. The viscosity was found to increase at decreased temperatures approximately according to the ideal fluid behavior. The viscosity at −20°C was high due to the existence of hydrogen bonds through the groups of alcoholics such as 12.84 mm^2/s of n-butanol and 36.21 mm^2/s of n-hexanol. The viscosity was less than 8 mm^2/s at −20°C for Butyl-butyrate (C8), ethyl-octanoate (C10), and their blends, which indicates favorable fluidity at low temperature. The higher viscosity for methyl linolenate (C18) was about 20.68 mm^2/s at −20°C, while the viscosity was 12 mm^2/s at −20°C for its 50vol.% blends.

15.5.3.3 Fuel volatility

The biobased jet fuel volatility can be expressed as the fuel's ability to evaporate, which is specified by its flash point and distillation characteristics. The flash point of biobased jet fuel is known as the lowest temperature, where the fuel with air can form a flammable

354   15. Biojet fuels production from algae: conversion technologies, characteristics, performance, and process simulation

mixture. Biojet fuel distillation characteristic can be defined at various temperatures by the percentage of recovery fraction.

15.5.3.4 Flash point

A typical indication for fuel volatility is the flashpoint. It is an important factor in dealing with fuel because it determines the risks due to fuel flammability during storage and shipping. ASTM (D7566−18) has set a minimum flashpoint of 38°C for SPK by FT, HEFA, SPK by FT/A, and SPK by ATJ. Whereas the minimum flash point for SIP fuel was 100°C because this fuel contains long carbon chains of farnesane (C15), which in turn makes the flash level naturally high. The specifications of ASTM (D6751) set a flashpoint of 93°C for biodiesel with a long carbon chain length [42]. For HEFA derived from used cooking oil, a flashpoint of 42°C (>38°C) has been reported [98]. Also, flashpoints of 46°C and 47.5°C were reported for FT-SPK and ATJ SPK, respectively [78]. The only results available for the effect of the chemical composition of biofuels on the flashpoint are studies by Scheuermann et al. [78]. They showed that the flashpoint of biojet fuel rises with higher fuel content of high boiling-point aromatic compounds and the flashpoint decreases with higher fuel content of lower boiling point aliphatic compounds.

15.5.4 Distillation property

The biojet fuel distillation property is very important during the production process, as it is critical for the controllability of process, product optimization, and energy integration [99]. ASTM (D7566−18) specifications determined that a recovery rate for SPK by FT, HEFA, SPK by FT/A, and SPK by ATJ was about 10%, where 205°C (T10) is the maximum temperature to ensure adequate ignition starting. In addition, to exclude heavy compound fractions a temperature of 300°C was set to be the maximum temperature limitation for the final boiling point (FBP). Nevertheless, for SIP fuel, it has a relatively narrow distillation range, as it achieves a recovery rate of 10% and maximum FBP at temperatures of 250 and 255°C due to the high unity of SIP fuel that contains about 97 wt.% of isoparaffins. FT-SPK (50% vol.%) and JP-8 blended fuel has a FBP of 268°C [89]. The distillation property of different carbon chain lengths fuels was evaluated, and it was found that the boiling point and, consequently, distillation property is greatly affected by carbon chain length [97]. The longer the length of the carbon chain of the fuel the higher the boiling point and vice versa. In the end, it turns out that biobased jet fuels had a satisfying distillation property and flashpoint. The presence of high-boiling aromatic compounds and low-boiling aliphatic compounds affected the flashpoint, while the length of the carbon chain mainly affected the distillation property.

15.5.5 Fuel-metering and aircraft-range

Chemical energy conversion in biobased jet fuels into useful heat energy is one of the most important factors controlling aircraft operation. The lower heat energy resulting from biofuel combustion means more fuel consumption, which will increase operating costs. Aircraft range performance and fuel metering have two characterizing parameters: net

Handbook of Algal Biofuels

heat of combustion and fuel density. The importance of the two previous factors stems from the fact that the thermal energy (MJ) produced from jet fuel combustion is proportional to the net combustion heat (MJ/kg), the fuel density (kg/m^3), and the volume of the fuel tank (m^3). Therefore with the constant volume of the fuel tank, the resulting total thermal energy increases with increasing density and net combustion heat of fuel, and thus allows for greater flight time and greater payload. A detailed summary and discussion of the fuel density and combustion net heat will be presented in the following.

15.5.6 Fuel density

ASTM (D7566−18) specifies the density range at 15°C to be 730−770 kg/m^3. The density for SPK by FT was about 737 kg/m^3 and for HEFA it was 751 kg/m^3 [79]. ASTM (D7566−18) specifies the density range for SIP fuel to be 765−780 kg/m^3; SPK by ATJ density was 757.1 kg/m^3 [78]. The density of SIP fuel blends was not available, although Amyris produced SIP fuel that recorded a density of 773 kg/m^3. Chuck and Donnelly [97] reported the density of pure farnesene to be 795 kg/m^3, 20vol.% farnesene/jet fuel (A-1) has a density of 789 kg/m^3, and 50vol.% farnesene/jet A-1 is 785 kg/m^3.

The density for FT-SPK/A is allowed to be up to 798 kg/m^3 by ASTM (D7566−18) specifications. It also allows a density range between 774 and 835 kg/m^3 for any type of aviation fuel, whether biological, fossil, or their blends, as long as it contains high aromatics content. The density of conventional jet A-1 fuel was about 803 kg/m^3 and it was 799 kg/m^3 for conventional JP-8 fuel [79]. The density of ATJ-SPK/A containing 15.8vol.% of aromatics was 785.9 kg/m^3 and it was 805.2 kg/m^3 for HEFA/A containing 19.7vol.% [78]. Several studies have been conducted to demonstrate that cycloalkanes containing short carbon chains have a high density [77,82]. The density of biobased jet fuel is influenced significantly by aromatic content, and the high content of aromatics will lead to an increased density for biobased jet fuel. Aromatics content can significantly affect the net heat of combustion and the density of bio-jet fuel due to the aromatics' heavy mass and low H/C ratio. Despite the high density of aromatic biofuel, which was reported to be 865 kg/m^3 compared to the density of cyclic-alkane biofuel (817 kg/m^3), the net combustion heat for aromatic biofuel (42.5 MJ/kg) was lower than that of cyclic alkane biofuel (45.9 MJ/kg).

15.6 Process simulation

Simulation is considered as one of the most critical and important strategies in the scientific field. Researchers are seeking to apply the simulation process in almost every scientific steep, as it can be used to discover the appropriate process conditions to obtain the best possible result. Simulation can be categorized based upon the process itself, as there are many processes in the field of biojet fuels such as production, transportation, utilization, and LCA. Multivariable statistical techniques for analytical optimization through different methods such as response surface methodology (RSM) is one of the most important ways to improve the process performance and decrease the experimental time and

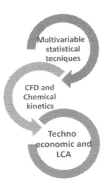

FIGURE 15.4   Different strategies for process simulation and optimization.

operation cost. The behavior of the data set can be described through RSM by fitting the polynomial equation to the experimental data, as RSM is considered as a collection of statistical and mathematical techniques. RSM aims to simultaneously optimize the process variables to improve system performance. Another way to simulate the process is through computational fluid dynamics (CFD), as it represents a cost-effective method to simulate real flows through the numerical solution of the governing equations.

Computational fluid dynamics can be categorized into one, two, and three-dimensional flow models either steady or unsteady flow, where the turbulent model is gaining great attention recently. Today, numerical methods are widely used in many research fields such as biomass gasification, combustion process, enhanced oil recovery problems, chemical reactions, nanofluid applications, and molecular dynamics. Chemical kinetics coupled with combustion models, referring to as CFD, is utilized to investigate the reaction rates, reaction mechanisms, and to improve the combustion process results in GHG reduction. The production of renewable hydrocarbons from microalgae through process simulation has gained attention as a promising raw renewable material. The hydrotreating process of microalgae can be modeled and simulated using different simulators, such as Aspen Plus processes simulator, based upon the experimental data obtained or recently reported [100]. Recently, a Hybrid Chemistry (HyChem) approach was developed to understand the multicomponent liquid fuels combustion chemistry and model real fuels, allowing it to overcome the approach of conventional surrogate fuel. Also, the development was extended to model the combustion behavior of bio-based jet fuels and its blending with petroleum jet fuel [101]. Fig. 15.4 shows a schematic diagram of the different strategies for process simulation and optimization.

15.7 Conclusions

Alternative biojet fuels are gaining great attention due to the environmental and economic concerns of using conventional petroleum-derived jet fuels. Microalgae species represent a promising renewable and cost-effective raw material for biojet fuel production. This chapter includes a comprehensive review of different pathways and methods for biojet fuel production and conversion technologies with technoeconomic and LCA analyses. In addition, the chapter provides a wide review of the field including the different performance and combustion characteristics of biojet fuels and their blends. This chapter summarizes an overview of the different methodologies and strategies for process simulation and optimization.

References

[1] I.J.A.R., IATA, International Air Transport Association, 2017.

[2] C. Gutiérrez-Antonio, F. Gómez-Castro, J. de Lira-Flores, S.J.R. Hernández, A review on the production processes of renewable jet fuel, Renew. Sustain. Energy Rev. 79 (2017) 709–729.

[3] W.-C. Wang, L.J.R. Tao, Bio-jet fuel conversion technologies, Renew. Sustain. Energy Rev. 53 (2016) 801–822.

[4] D.J. Luning Prak, M. Romanczyk, K.E. Wehde, S. Ye, M. McLaughlin, P.J. Luning Prak, et al., Analysis of catalytic hydrothermal conversion jet fuel and surrogate mixture formulation: components, properties, and combustion, Energy Fuels 31 (2017) 13802–13814.

[5] C. Chuck, Biofuels for Aviation: Feedstocks, Technology and Implementation, Academic Press, 2016.

[6] N.R. Baral, O. Kavvada, D. Mendez-Perez, A. Mukhopadhyay, T.S. Lee, B.A. Simmons, et al., Techno-economic analysis and life-cycle greenhouse gas mitigation cost of five routes to bio-jet fuel blendstocks, Energy Environ. Sci 12 (2019) 807–824.

[7] M.Y. Kim, J.-K. Kim, M.-E. Lee, S. Lee, M.J.Ac Choi, Maximizing biojet fuel production from triglyceride: importance of the hydrocracking catalyst and separate deoxygenation/hydrocracking steps, ACS Catal. 7 (2017) 6256–6267.

[8] G.W. Diederichs, M.A. Mandegari, S. Farzad, J.F.J.Bt Görgens, Techno-economic comparison of biojet fuel production from lignocellulose, vegetable oil and sugar cane juice, Bioresour. Technol. 216 (2016) 331–339.

[9] H. Wei, W. Liu, X. Chen, Q. Yang, J. Li, H.J.F. Chen, Renewable bio-jet fuel production for aviation: a review, Fuel 254 (2019) 115599.

[10] A. Fivga, L.G. Speranza, C.M. Branco, M. Ouadi, A.J.A.E. Hornung, A review on the current state of the art for the production of advanced liquid biofuels, AIMS Energy 7 (2019) 46.

[11] A. O'Connell, M. Kousoulidou, L. Lonza, W.J.R. Weindorf, Considerations on GHG emissions and energy balances of promising aviation biofuel pathways, Renew. Sustain. Energy Rev. 101 (2019) 504–515.

[12] R.H. Sundararaj, R.D. Kumar, A.K. Raut, T.C. Sekar, V. Pandey, A. Kushari, et al., Combustion and emission characteristics from biojet fuel blends in a gas turbine combustor, Energy 182 (2019) 689–705.

[13] A.T. Ubando, D.R.T. Rivera, W.-H. Chen, A.B.J.Bt Culaba, A comprehensive review of life cycle assessment (LCA) of microalgal and lignocellulosic bioenergy products from thermochemical processes, Bioresour. Technol. 291 (2019) 121837.

[14] J. Han, L. Tao, M.J.Bfb Wang, Well-to-wake analysis of ethanol-to-jet and sugar-to-jet pathways, Biotechnol. Biofuels 10 (2017) 21.

[15] N. Dangol, D.S. Shrestha, J.A.J.B. Duffield, Life-cycle energy, GHG and cost comparison of camelina-based biodiesel and biojet fuel, Biofuels 11 (2020) 399–407.

[16] T. Suganya, M. Varman, H. Masjuki, S. Renganathan, Macroalgae and microalgae as a potential source for commercial applications along with biofuels production: a biorefinery approach, Renew. Sustain. Energy Rev. 55 (2016) 909–941.

[17] G. Kumar, J. Dharmaraja, S. Arvindnarayan, S. Shoban, P. Bakonyi, G.D. Saratale, et al., A comprehensive review on thermochemical, biological, biochemical and hybrid conversion methods of bio-derived lignocellulosic molecules into renewable fuels, Fuel 251 (2019) 352–367.

[18] S. De Jong, K. Antonissen, R. Hoefnagels, L. Lonza, M. Wang, A. Faaij, et al., Life-cycle analysis of greenhouse gas emissions from renewable jet fuel production, Biotechnol. Biofuels 10 (2017) 64.

[19] K. Brooks, L. Snowden-Swan, S. Jones, M. Butcher, G.-S. Lee, D. Anderson, et al., Low-carbon aviation fuel through the alcohol to jet pathway, Biofuels for Aviation, Elsevier, 2016, pp. 109–150.

[20] D. Kennes, H.N. Abubackar, M. Diaz, M.C. Veiga, C.J.Jo.C.T. Kennes, Bioethanol production from biomass: carbohydrate versus syngas fermentation, J. Chem. Technol. Biotechol. 91 (2016) 304–317.

[21] M. Carter, Direct catalytic conversion of sugars to ethanol, Google Patents, 2017.

[22] H. Wang, B. Yang, Q. Zhang, W.J.R. Zhu, Catalytic routes for the conversion of lignocellulosic biomass to aviation fuel range hydrocarbons, Renew. Sustain. Energy Rev. 120 (2020) 109612.

[23] M.C. Vasquez, E.E. Silva, E.F.J.B. Castillo, Hydrotreatment of vegetable oils: a review of the technologies and its developments for jet biofuel production, Biomass Bioenergy 105 (2017) 197–206.

Handbook of Algal Biofuels

358 15. Biojet fuels production from algae: conversion technologies, characteristics, performance, and process simulation

[24] U. Neuling, M.J.F.P.T. Kaltschmitt, Techno-economic and environmental analysis of aviation biofuels, Fuel Process. Technol. 171 (2018) 54−69.

[25] G. Yao, M.D. Staples, R. Malina, W.E.J.Bfb Tyner, Stochastic techno-economic analysis of alcohol-to-jet fuel production, Biotechnol. Biofuels 10 (2017) 1−13.

[26] J.R. Hannon, L.R. Lynd, O. Andrade, P.T. Benavides, G.T. Beckham, M.J. Biddy, et al., Technoeconomic and life-cycle analysis of single-step catalytic conversion of wet ethanol into fungible fuel blendstocks, Proc. Natl. Acad. Sci. USA 117 (2020) 12576−12583.

[27] B.C. Klein, M.F. Chagas, T.L. Junqueira, M.C.A.F. Rezende, T. de Fátima Cardoso, O. Cavalett, et al., Techno-economic and environmental assessment of renewable jet fuel production in integrated Brazilian sugarcane biorefineries, Appl. Energy 209 (2018) 290−305.

[28] D.S. Braz, A.P.J.Bt Mariano, Jet fuel production in eucalyptus pulp mills: economics and carbon footprint of ethanol versus butanol pathway, Bioresour. Technol. 268 (2018) 9−19.

[29] S. Hansen, A. Mirkouei, L.A.J.R. Diaz, A comprehensive state-oFTechnology review for upgrading bio-oil to renewable or blended hydrocarbon fuels, Renew. Sustain. Energy Rev. 118 (2020) 109548.

[30] L. Tao, A. Milbrandt, Y. Zhang, W.-C.J.Bfb Wang, Techno-economic and resource analysis of hydroprocessed renewable jet fuel, Biotechnol. Biofuels 10 (2017) 261.

[31] X. Du, D. Li, H. Xin, W. Zhou, R. Yang, K. Zhou, et al., The conversion of jatropha oil into jet fuel on NiMo/Al-MCM-41 catalyst: intrinsic synergic effects between Ni and Mo, Energy Technol. 7 (2019) 1800809.

[32] V. Itthibenchapong, A. Srifa, R. Kaewmeesri, P. Kidkhunthod, K.J.Ec Faungnawakij, Deoxygenation of palm kernel oil to jet fuel-like hydrocarbons using $Ni-MoS_{2/-,}-Al_2O_3$ catalysts, Energy Convers. Manage. 134 (2017) 188−196.

[33] M. Shahinuzzaman, Z. Yaakob, Y.J.R. Ahmed, Non-sulphide zeolite catalyst for bio-jet-fuel conversion, Renew. Sustain. Energy Rev. 77 (2017) 1375−1384.

[34] P. Biller, A. Roth, Hydrothermal liquefaction: a promising pathway towards renewable jet fuel, Biokerosene, Springer, 2018, pp. 607−635.

[35] Y. Nie, X.T.J.E. Bi, Techno-economic assessment of transportation biofuels from hydrothermal liquefaction of forest residues in British Columbia, Energy 153 (2018) 464−475.

[36] K.F. Tzanetis, J.A. Posada, A.J.Re Ramirez, Analysis of biomass hydrothermal liquefaction and biocrude-oil upgrading for renewable jet fuel production: the impact of reaction conditions on production costs and GHG emissions performance, Renew. Energy 113 (2017) 1388−1398.

[37] C. Miao, O. Marin-Flores, S.D. Davidson, T. Li, T. Dong, D. Gao, et al., Hydrothermal catalytic deoxygenation of palmitic acid over nickel catalyst, Fuel 166 (2016) 302−308.

[38] T.J. Morgan, A. Youkhana, S.Q. Turn, R. Ogoshi, M.J.E. Garcia-Pérez, Review of biomass resources and conversion technologies for alternative jet fuel production in Hawai'i and tropical regions, Energy Fuels 33 (2019) 2699−2762.

[39] M.E.J.Ct Dry, The Fischer−Tropsch process: 1950−2000, Catal. Today 71 (2002) 227−241.

[40] A. Bauen, J. Howes, L. Bertuccioli, C.J.F.r.p.f.t.C.o.C.C.E.t.S. Chudziak, Review of the potential for biofuels in aviation, 2009.

[41] T.G. Kreutz, E.D. Larson, G. Liu, R.H. Williams, Fischer-Tropsch fuels from coal and biomass, in: Proceedings of the Twenty-Fifth Annual International Pittsburgh Coal Conference, International Pittsburgh Coal Conference Pittsburgh, Pennsylvania, 2008, p. 2.10.

[42] F. You, B.J.I. Wang, E.C. Research, Life cycle optimization of biomass-to-liquid supply chains with distributed−centralized processing networks, Ind. Eng. Chem. Res. 50 (2011) 10102−10127.

[43] M. Gupta, J.J. Spivey, New and future developments in catalysis, in: Chapter 5Chapters Catalytic Processes for the Production of Clean Fuels, Elsevier Inc, 2013.

[44] L. Hagey, H.I.J.I.Jo.C.R.E. de Lasa, Conversion of synthesis gas into light hydrocarbons. Modelling of the catalytic reaction network, Int. J. Chem. React. Eng. 2 (2004).

[45] S. Sie, R.J.A.C.A.G. Krishna, Fundamentals and selection of advanced Fischer−Tropsch reactors, Appl. Catal. A Gen. 186 (1999) 55−70.

[46] J.P. den Breejen, J.R. Sietsma, H. Friedrich, J.H. Bitter, K.P.J.Jo.C. de Jong, Design of supported cobalt catalysts with maximum activity for the Fischer−Tropsch synthesis, J. Catal. 270 (2010) 146−152.

[47] J. Daniell, M. Köpke, S.D.J.E. Simpson, Commercial biomass syngas fermentation, Energies 5 (2012) 5372−5417.

Handbook of Algal Biofuels

References

[48] T.R. Pray, J. Pramanik, D. McPhee, J. Galazzo, K. Reiling, J. Cherry, et al., Amyris' integrated biorefinery operations for domestic renewable diesel and chemical production, in: Proceedings of the Thirty-Second Symposium on Biotechnology for Fuels and Chemicals, 2010.

[49] Advanced Biofuels USA. Syngas fermentation: the third pathway for cellulosic ethanol; 2011.

[50] D.W. Griffin, M.A.J.Ep Schultz, Fuel and chemical products from biomass syngas: a comparison of gas fermentation to thermochemical conversion routes, Environ. Prog. Sustain. Energy 31 (2012) 219–224.

[51] Bauen A., Howes J., Bertuccioli L., Chudziak C., Review of the potential for biofuels inaviation. E4tech; 2009.

[52] J.I. Hileman, D.S. Ortiz, J.T. Bartis, H.M. Wong, P.E. Donohoo, M.A. Weiss, et al., Near-Term Feasibility of Alternative Jet Fuels, Rand Corporation, 2009.

[53] R.M. Swanson, A. Platon, J. Satrio, R. Brown, D.D. Hsu, Techno-Economic Analysis of Biofuels Production Based on Gasification, National Renewable Energy Lab. (NREL), Golden, CO, 2010.

[54] A. Dutta, R.L. Bain, M.J.J.E.P. Biddy, Techno-economics of the production of mixed alcohols from lignocellulosic biomass via high-temperature gasification, Environ. Prog. Sustain. Energy 29 (2010) 163–174.

[55] A. Dutta, M. Talmadge, J. Hensley, M. Worley, D. Dudgeon, D. Barton, et al., Process Design and Economics for Conversion of Lignocellulosic Biomass to Ethanol: Thermochemical Pathway by Indirect Gasification and Mixed Alcohol Synthesis, National Renewable Energy Lab.(NREL), Golden, CO, 2011.

[56] R.P. Anex, A. Aden, F.K. Kazi, J. Fortman, R.M. Swanson, M.M. Wright, et al., Techno-economic comparison of biomass-to-transportation fuels via pyrolysis, gasification, and biochemical pathways, Fuel 89 (2010) S29–S35.

[57] R.W. Stratton, H.M. Wong, J.I.J.P.p. Hileman, Life cycle greenhouse gas emissions from alternative jet fuels, 28 (2010) 133.

[58] C. Taylor, J. Baltrus, D.J.N.E.T.L. Driscoll, Golden, CO, USA, Fischer-Tropsch Fuels, 2011.

[59] K. Holmgren, L. Hagberg, Life cycle assessment of climate impact of Fischer-Tropsch diesel based on peat and biomass, 2009.

[60] D. In, Conversion technologies for advanced biofuels: preliminary roadmap & workshop report, Arlington, VA, Arlington, VA, 2011.

[61] P.G. Blommel, G.R. Keenan, R.T. Rozmiarek, R.D.J.Isj Cortright, Catalytic conversion of sugar into conventional gasoline, diesel, jet fuel, and other hydrocarbons, Int. Sugar J. 110 (2008) 672–679.

[62] R.D. Cortright, Catalytic conversion of sugars to conventional liquid fuels, Abstracts of papers of the American Chemical Society, Amer Chemical Soc 1155 16th st, NW, Washington, DC, 2009.

[63] P. Blommer, R.J.I. Cortright, Madison, W.I., Online Whitepaper, production of conventional liquid fuels from sugars, Virent Energy Systems, (2008).

[64] J.N. Chheda, J.A.J.C.T. Dumesic, An overview of dehydration, aldol-condensation and hydrogenation processes for production of liquid alkanes from biomass-derived carbohydrates, Catal. Today 123 (2007) 59–70.

[65] F. King, G. Kelly, E. Stitt, 98 Improved base catalysts for industrial condensation reactions, Studies in Surface Science and Catalysis, Elsevier, 2003, pp. 443–446.

[66] E.I. Gürbüz, J.A. Dumesic, Catalytic strategies and chemistries involved in the conversion of sugars to liquid transportation fuels, 2013.

[67] P. Anbarasan, Z.C. Baer, S. Sreekumar, E. Gross, J.B. Binder, H.W. Blanch, et al., Integration of chemical catalysis with extractive fermentation to produce fuels, 2012.

[68] N.A.B. Consortium, Catalysis of lignocellulosic sugars process strategy, 2011.

[69] F.K. Kazi, A.D. Patel, J.C. Serrano-Ruiz, J.A. Dumesic, R.P.J.C.E.J. Anex, Techno-economic analysis of dimethylfuran (DMF) and hydroxymethylfurfural (HMF) production from pure fructose in catalytic processes, Chem. Eng. J. 169 (2011) 329–338.

[70] G. Granço, A. Nassar, J. Seabra, L. Harfuch, M. Moreira, P. Moura, Sustainability of sugarcane-derived renewable jet fuel: lifecycle GHG emissions and benchmark of major sustainability standards. ICONE, 2012.

[71] A. Total, Breaking the barriers with break through jet fuel solutions. Berlin Airshow, Berlin, 2012.

[72] J. Yang, Z. Xin, K. Corscadden, H.J.F. Niu, An overview on performance characteristics of bio-jet fuels, Fuel 237 (2019) 916–936.

[73] N. Yilmaz, A.J.E. Atmanli, Sustainable alternative fuels in aviation, Energy 140 (2017) 1378–1386.

[74] E. Corporan, J.T. Edwards, S. Stouffer, M. DeWitt, Z. West, C. Klingshirn, et al., Impacts of fuel properties on combustor performance, operability and emissions characteristics, in: Proceedings of the Fifty-Fifth AIAA Aerospace Sciences Meeting, 2017, p. 0380.

Handbook of Algal Biofuels

[75] A.P. Pires, Y. Han, J. Kramlich, M.J.B. Garcia-Perez, Chemical composition and fuel properties of alternative jet fuels, BioResources 13 (2018) 2632–2657.

[76] B. Gawron, T.J.I.Jo.E.S. Białecki, Technology, Impact of a Jet A-1/HEFA blend on the performance and emission characteristics of a miniature turbojet engine, Int. J. Environ. Sci. Technol. 15 (2018) 1501–1508.

[77] P. Han, G. Nie, J. Xie, E. Xiu-tian-feng, L. Pan, X. Zhang, et al., Synthesis of high-density biofuel with excellent low-temperature properties from lignocellulose-derived feedstock, Fuel Process. Technol. 163 (2017) 45–50.

[78] S. Scheuermann, S. Forster, S.J.E. Eibl, In-depth interpretation of mid-infrared spectra of various synthetic fuels for the chemometric prediction of aviation fuel blend properties, Energy Fuels 31 (2017) 2934–2943.

[79] E. Corporan, T. Edwards, L. Shafer, M.J. DeWitt, C. Klingshirn, S. Zabarnick, et al., Chemical, thermal stability, seal swell, and emissions studies of alternative jet fuels, Energy Fuels 25 (2011) 955–966.

[80] M. ElGalad, K. El-Khatib, E. Abdelkader, R. El-Araby, G. ElDiwani, S.J.Joar Hawash, Empirical equations and economical study for blending biofuel with petroleum jet fuel, J. Adv. Res. 9 (2018) 43–50.

[81] T.D. Hong, T.H. Soerawidjaja, I.K. Reksowardojo, O. Fujita, Z. Duniani, M.X.J.C.E. Pham, et al., A study on developing aviation biofuel for the tropics: production process—experimental and theoretical evaluation of their blends with fossil kerosene, Chem. Eng. Process. 74 (2013) 124–130.

[82] G. Nie, X. Zhang, L. Pan, M. Wang, J.-J.J.C.E.S. Zou, One-pot production of branched decalins as high-density jet fuel from monocyclic alkanes and alcohols, Chem. Eng. Sci. 180 (2018) 64–69.

[83] C. Ju, M. Wang, Y. Huang, Y. Fang, T.J.A.S.C. Tan, Engineering, high-quality jet fuel blend production by oxygen-containing terpenoids hydroprocessing, ACS Sustain. Chem. Eng. 6 (2018) 4871–4879.

[84] F. Li, J. Jiang, P. Liu, Q. Zhai, F. Wang, C.-y Hse, et al., Catalytic cracking of triglycerides with a base catalyst and modification of pyrolytic oils for production of aviation fuels, Sustain. Energy Fuels 2 (2018) 1206–1215.

[85] A. Llamas, M.J. García-Martínez, A.-M. Al-Lal, L. Canoira, M.J.F. Lapuerta, Biokerosene from coconut and palm kernel oils: production and properties of their blends with fossil kerosene, Fuel 102 (2012) 483–490.

[86] M.T. Timko, S.C. Herndon, E. De La Rosa Blanco, E.C. Wood, Z. Yu, R.C. Miake-Lye, et al., Combustion products of petroleum jet fuel, a Fischer–Tropsch synthetic fuel, and a biomass fatty acid methyl ester fuel for a gas turbine engine, Combust. Sci. Technol. 183 (2011) 1039–1068.

[87] Q. Zhu, S. Liu, Q. Guan, L. He, W. Li, Bio-aviation fuel production from hydroprocessing castor oil promoted by the nickel-based bifunctional catalysts, Bioresour. Technol. 183 (2015).

[88] K. Kumar, X. Hui, C.-J. Sung, T. Edwards, D. Gardner, Experimental studies on the combustion characteristics of alternative jet fuels, Fuel (2012).

[89] E. Corporan, M.J. DeWitt, V. Belovich, R. Pawlik, A.C. Lynch, J.R. Gord, et al., Emissions characteristics of a turbine engine and research combustor burning a Fischer – Tropsch jet fuel, Energy Fuels 21 (2007) 2615–2626.

[90] G. Nie, X. Zhang, P. Han, J. Xie, L. Pan, L. Wang, et al., Lignin-derived multi-cyclic high density biofuel by alkylation and hydrogenated intramolecular cyclization, Chem. Eng. Sci. 158 (2017) 64–69.

[91] G. Nie, X. Zhang, L. Pan, P. Han, J. Xie, Z. Li, et al., Hydrogenated intramolecular cyclization of diphenylmethane derivatives for synthesizing high-density biofuel, Chem. Eng. Sci. 173 (2017) 91–97.

[92] A. Mancio, S. da Mota, C. Ferreira, T. Carvalho, O. Neto, J. Zamian, et al., Separation and characterization of biofuels in the jet fuel and diesel fuel ranges by fractional distillation of organic liquid products, Fuel 215 (2018) 212–225.

[93] F.P. Sousa, L.N. Silva, D.B. de Rezende, L.C.A. de Oliveira, V.M.J.F. Pasa, Simultaneous deoxygenation, cracking and isomerization of palm kernel oil and palm olein over beta zeolite to produce biogasoline, green diesel and biojet-fuel, Fuel 223 (2018) 149–156.

[94] V.F. de Oliveira, E.J. Parente Jr, C.L. Cavalcante Jr, F.M.T.J.T.C.Jo.C.E. Luna, Short-chain esters enriched biofuel obtained from vegetable oil using molecular distillation, Can. J. Chem. Eng. 96 (2018) 1071–1078.

[95] J.C. Alger, H.J. Robota, L. Shafer, Converting algal triglycerides to diesel and HEFA jet fuel fractions, Energy Fuels (2013).

[96] T.C. Brennan, C.D. Turner, J.O. Krömer, L.K.J.B. Nielsen, Alleviating monoterpene toxicity using a two-phase extractive fermentation for the bioproduction of jet fuel mixtures in Saccharomyces cerevisiae, Biotechnol. Bioeng. 109 (2012) 2513–2522.

[97] J. Donnelly, C.J. Chuck, The compatibility of potential bioderived fuels with Jet A-1aviation kerosene, Appl. Energy (2014).

[98] M. Buffi, A. Valera-Medina, R. Marsh, D. Pugh, A. Giles, J. Runyon, et al., Emissions characterization tests for hydrotreated renewable jet fuel from used cooking oil and its blends, Appl. Energy 201 (2017) 84–93.

Handbook of Algal Biofuels

References

[99] C. Gutiérrez-Antonio, A.G. Romero-Izquierdo, F. Israel Gómez-Castro, S.J.I. Hernández, Energy integration of a hydrotreatment process for sustainable biojet fuel production, Ind. Eng. Chem. Res. 55 (2016) 8165–8175.

[100] C. Gutiérrez-Antonio, A. Gómez-De la Cruz, A.G. Romero-Izquierdo, F.I. Gómez-Castro, S.J.C.T. Hernández, Modeling, simulation and intensification of hydroprocessing of micro-algae oil to produce renewable aviation fuel, Clean. Technol. Environ. Policy 20 (2018) 1589–1598.

[101] K. Wang, R. Xu, T. Parise, J. Shao, A. Movaghar, D.J. Lee, et al., A physics-based approach to modeling real-fuel combustion chemistry–IV. HyChem modeling of combustion kinetics of a bio-derived jet fuel and its blends with a conventional Jet A, Combust. Flame 198 (2018) 477–489.

CHAPTER 16

Photosynthetic microalgal microbial fuel cells and its future upscaling aspects

Mohd Jahir Khan[1], Vishal Janardan Suryavanshi[2], Khashti Ballabh Joshi[2], Praveena Gangadharan[3] and Vandana Vinayak[1]

[1]Diatom Nanoengineering and Metabolism Laboratory (DNM), School of Applied Sciences, Dr. Harisingh Gour Central University Sagar, Sagar, India [2]School of Chemical Science and Technology, Department of Chemistry, Dr. Harisingh Gour Central University Sagar, Sagar, India [3]Department of Civil Engineering, Indian Institute of Technology, Palakkad, India

16.1 Introduction

With an increase in globalization and industrialization, recent studies show that electricity production by solar panels, wind power, and hydropower stations faces challenges to fulfill the current demand for electricity by electric vehicles (EVs) [1]. However, lithium ion (Li$^+$) batteries, which are an alternative solution to conventional power storage for EVs are not feasible in many countries where neither Li$^+$ reserves nor Li$^+$ manufacturing industries exist [2–4]. Hence there is need for a stable and renewable energy source for the production of electricity [5]. Even though EVs run by electricity and are superior to gasoline in having zero carbon emissions [6], a discontinuity in electricity supply may switch them back to gasoline vehicles, hence not only a continuous and reasonable electricity supply but also gasoline and biofuel with a zero-carbon footprint is the need of an hour [7]. A lot of research has been done on alternative sources to generate electricity and gasoline reservoirs to run hybrid electric vehicles which operate by both electricity and gasoline. Such renewable resources should have the following attributes: readily available, abundant, feasible, economical, and have zero carbon emissions [8,9]. A report from the International Energy Agency shows that biofuels from renewable energy resources produce 10% of the world's total primary energy compared to 2.4% from water and 1.1% from sun and wind [10]. Biofuel from algae are supposed to have zero carbon emissions [11]. They are promising candidates as they are readily available and abundantly found compared to nonrenewable energy resources.

Of the different approaches employed to harvest energy, the microbial fuel cell (MFC) is one such technique that not only produces bioelectricity but also acts as a synergistic approach to harvest metabolites from microalgae by simultaneously recycling wastewater. MFC are thus very advantageous bioelectrochemical devices which have their anode and cathode chamber separated by a proton exchange membrane. They produce electricity and degrade microbes at the anode while reducing oxygen at the cathode [12,13]. This method of producing electricity is not only cost-effective but also completely carbon-neutral [1]. The growth of microorganisms in a MFC depends mainly on the anolyte and type of substrate characteristics [14,15]. Generally, energy produced by mixed microbial cultures is higher than by one pure culture but exceptions also exist for example, *Rhodopseudomonas palustris* DX-1 generates higher power in pure culture than mixed and this increases with increasing light intensity, whereas *R. palustris* ATCC 17001 does not produce power [16]. The highest power densities obtained are with *Shewanella oneidensis, Escherichia coli* (acclimated, nonmediated), *E. coli K12* at 3000, 600, and 760 mWm^{-2}, respectively [17]. Another very interesting aspect of MFC is the use of wastewater rich in microbes as anolyte. This not only helps in treating wastewater for its recycling on agricultural crops but is also beneficial because of no energy requirement and high output [18,19]. Further it has been confirmed that MFC using microalgae have higher power densities compared to MFC employing other substrates [20]. Such redox reactions that are carried out by algae or cyanobacteria as substrates are known as algal microbial fuel cells (AMFC) [21].

In AMFC, microalgae can be used either in living form or as dry powder at the anode with wastewater as anolyte to produce power [22]. At the anode, microalgal biomass (dead algae) serves as a substrate for bacteria which degrades biochemical metabolites like carbohydrates and protein to produce electrons [23–26]. In a single-chamber MFC, *Chlorella vulgaris* while utilizing dead microalgae powder produced a power density of 980 mWm^{-2}, whereas with *Ulva lactuca* it produced 760 mWm^{-2} [26]. The biomass dry powder of other microalgae, such as *Scenedesmus*, also served as a feedstock for MFC. This feedstock results in instant electricity generation due to the excessive amount of fatty acids. The cell voltage enhancement from 258 ± 74 to 584 ± 18 mV was observed with a chemical oxygen demand (COD) of 147 ± 18 and 353 ± 15 mg/L, respectively, using *Scenedesmus* powder as a substrate. However, coulombic efficiency was in the range of 3.5%–6.3% for the substrate of 400–2500 mg COD L^{-1} of *Scenedesmus* powder [23]. This is much less than that produced using *C. vulgaris* (about 10%–28%) and macroalgae *U. lactuca* powder as MFC feedstock [26].

However, the major drawback of AMFC is that since it uses dried algal powder as a bacterial substrate, the algae needs pretreatment such as thermal treatment and ultrasonication etc. [27]. This is necessary since wastewater sludge rich in electricity producing bacteria can't feed directly on algal powder due to the thick cell wall of algal cells. Therefore pretreatment is a must for the algal cells to be digested by microbes. Also the power density of nontreated algal biomass is less compared to pretreated algae, as seen in *Laminaria* used as substrate for bacteria at the anode. A maximum power density of 218 Wm^{-2} was produced with autoclaved pretreated *Laminaria* compared to 118 Wm^{-2} microwave pretreated and 86 Wm^{-2} without any pretreatment [28]. Hence, pretreatment is not only a time-consuming but also an expensive process [25]. Both MFC and AMFC techniques produce electricity and CO_2 but need a constant supply of oxygen from the cathode for the electrochemical reactions to produce an electric current [29].

If the alga is in its live state at the cathode and capable of carrying out photosynthesis it is known as a photosynthetic microalgal microbial fuel cell (PMMFC) [30]. To overcome the production of CO_2, photosynthetic microalgae microbial fuel cells (PMMFC) are most suited, as the photosynthetic algae take CO_2 for photosynthesis and liberate O_2 [31,32]. Hence, microalgae in return reduce the capital cost of producing metabolites in MFC by generating the free nascent O_2. The functioning of such PMMFC depends upon several factors like type of microalgae, pH, nutrients, temperature, and most importantly intensity of light. In both MFC and PMMFC, electrons are transported extracellularly to the anode and protons are exchanged via plasma exchange membrane. However, the major difference is that PMMFC operates without the addition of any externally substrates, whereas MFC operates in the presence of added substrates to be degraded by microbes to produce bioelectricity [33]. However, microalgae if present at the anode serve as a substrate for exoelectrogenic bacteria [34]. Even though electron transfer in bacteria and algae is extracellular with electron shuttles either inside the organism or externally added, due to their similar electron shuttling mechanism there is no reason why algae can't be used at the anode in a MFC as an electron donor [21]. The electron transfer system in a conventional bacterial MFC and PMMFC are shown in Fig. 16.1A and B, respectively.

Here in this chapter the factors responsible for an efficient PMMFC are discussed. The photosynthetic algae, besides acting as biobatteries, also produce value-added metabolites, mainly carotenoids, lipids, and biofuel, while cleaning the wastewater that is acting as the anolyte. In addition, future approaches, like nanopore DNA technology, to increase the efficiency of MFC by selecting the right microbes/microalgae are discussed.

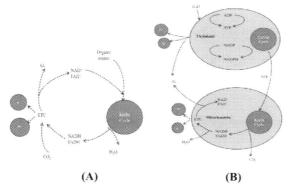

FIGURE 16.1 Schematic representation of electron shuttling in (A) bacteria and (B) algae. [21]. *Source: Reproduced with permissions number 4923820751929 from M. Shukla, S. Kumar, Algal growth in photosynthetic algal microbial fuel cell and its subsequent utilization for biofuels, Renew. Sustain. Energy Rev., 82 (2018) 402–414.*

16.2 What are photosynthetic microalgal microbial fuel cell

During the 1980–90s, Tanaka et al. established aerobic photosynthetic microbial fuel cells using cyanobacterial cell cultures [35–37]. They recorded an instant increase in electricity production on light illuminating the MFC containing photosynthetic microalgae at the cathode. On the offset, live microalgal cells are advantageous as they utilize CO_2 for photosynthesis and accept electrons simultaneously generated from the degradation of substrates in the anode by the microbes. Oxygen is on the other hand released at the cathode during photosynthesis and acts as a terminal electron acceptor [38]. Oxygen and organic substrates produced in the light phase via photosynthesis are utilized in the dark phase, for example, *Spirulina platensis* produces 10 times more power density during a dark phase when used at the anode at both light and dark phase [39]. Live microalgae at cathode produced enough in situ oxygen to increase the power efficiency by much greater than that achieved by mechanical aeration in a AMFC. The algae at the cathode could be further harvested for biomass biofuel and carotenoids production at regular intervals [24,40]. Fig. 16.2A shows microalgae at the cathode in a double chambered cell with wastewater at the anode, whereas Fig. 16.2B shows microalgae at the anode in a single cell electrolytic chamber containing wastewater microbes [20]. Yadav et al. have shown that blue–green algae at the cathode in PMMFC attained a power density of $78.12 \, mWm^{-2}$ and removed COD from wastewater by 89.23% [41]. The advantage of a photosynthetic or PMMFC over MFC is its facile way to extract energy from algae for sustainable electricity production and the much cleaner way to clean wastewater than MFC.

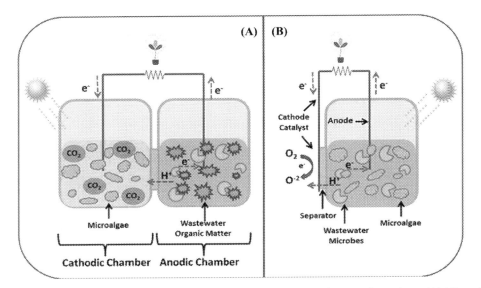

FIGURE 16.2 Schematic illustration of microbial fuel cell with microalgae configurations. (A) Microalgae in the cathodic chamber consume CO_2 and (B) microalgae as a carbon source for microorganisms in the anodic chamber. [20]. *Source: Reproduced with permissions number 4923820979043 from Z. Baicha, et al., A critical review on microalgae as an alternative source for bioenergy production: a promising low cost substrate for microbial fuel cells. Fuel Process. Technol., 154 (2016) 104–116.*

The reaction of light photosynthesis that occurs at the cathode utilizing CO_2 and producing biomass proceeds as an overall redox reaction

At anode:

$$Organic\ degradation(Wastewater + Microbes) + O_2 \rightarrow CO_2 + H_2O + External\ Power$$

At cathode:

$$CO_2 + H_2O + Light\ for\ photosynthesis \rightarrow Living\ Algae(Biomass + Metabolites) + O_2$$

The degradation of organic substrate at anode, growth of algae at cathode, and O_2 generated in electrochemical reaction determine the amount of electricity produced. In PMMFC, live microalgae can also be employed in open water systems along with self-sustaining catalyst bacteria and cost-effective electrodes like graphite with the subsequent generation of free oxygen. Thus photosynthetic MFC seek to convert light into electricity by exploiting the photosynthetic activity of live, phototrophic microorganisms. The live microalgae in PMMFCs by the mechanism of photosynthesis have the advantage over MFC as they reduce the requirement for oxygen, which is required as a substrate for bacteria, from other sources [26,42,43].

Among several structural designs of photosynthetic MFCs, a few like tubular, single chambered, sediment, and dual-chamber PMMFC are shown in Fig. 16.3. Photosynthetic bacteria at the anode along with artificial mediators to shuttle electrons from photosynthetic microorganisms to the anode are shown in Fig. 16.3A, while Fig. 16.3B shows hydrogen generating photosynthetic bacteria at the anode which helps to liberate in situ H_2. However, this MFC limits its usage due to easy inactivation and the cost of the electrocatalyst. Fig. 16.3C—E shows photosynthesis coupled with mixed heterotrophic bacteria at the anode and Fig. 16.3F shows direct electron transfer between photosynthetic bacteria and electrodes. Fig. 16.3G shows photosynthesis at the cathode and liberated nascent O_2. Among these PMMFCs, the dual chamber with algae at the cathode, as shown in Fig. 16.3G, is the best as it not only harvests bioelectricity but also recycles wastewater, reduces CO_2, and produces O_2, important pigments, and lipids/biofuel.

The live microalgae and electrochemically active bacteria in PMMFC are economical as they reduce the cost of the mechanical aeration that is required in other MFCs. They also have an advantage as they don't require redox mediators, which are the compounds needed for gaining or losing an electron in a redox reaction. Strik et al. worked on PMMFCs and generated a maximum current density of $539\ mAm^{-2}$ with naturally selected algae *Chlorella* [22]. They worked on comparing the power efficiency of PMMFCs and MFCs and found that maximum power efficiency in a PMMFC was $110\ mWm^{-2}$ compared to $67\ mWm^{-2}$ in a MFC. Though ohmic losses are a common factor in MFCs, they are not a limiting factor in PMMFCs which can produce a maximum current density of $110\ mWm^{-2}$ for more than a 100 days due to the continuous production of oxygen along with oxidation of organic substrates at the anode to generate bioelectricity [21]. This is mainly because of its nonexhaustive biomass due to the live algal cells at the cathode which continuously produce the oxygen required for the redox reaction to produce bioelectricity. Table 16.1 lists the different microalgae which are used in MFCs, AMFCs, and PMMFCs to produce electricity and other by-products.

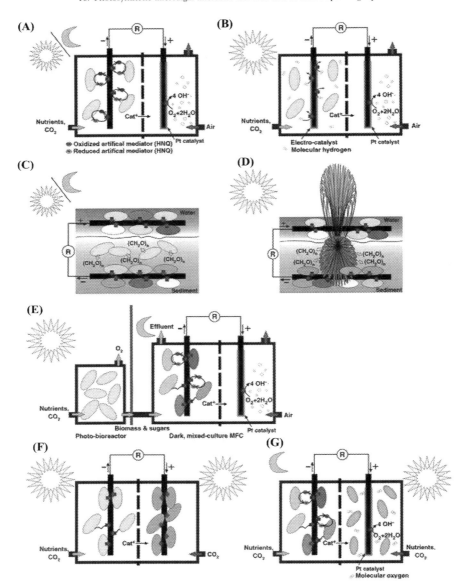

FIGURE 16.3 Seven different configurations of photosynthetic microbial fuel cells (A)–(G) corresponds to separate sections of the review. The sun/moon symbols represent reactions that are dependent on the day/night cycle, respectively. [30]. *Source: Reproduced with permissions number 4924030989746 from M. Rosenbaum, Z. He, L.T. Angenent, Light energy to bioelectricity: photosynthetic microbial fuel cells. Curr. Opin. Biotechnol., 21 (3) (2010) 259–264.*

16.2 What are photosynthetic microalgal microbial fuel cell

369

TABLE 16.1 Different algae used in microbial fuel cell as algal microbial fuel cell and photosynthetic microalgal microbial fuel cells with their power efficiencies.

S. No.	Algae acting as substrate/cathode/anode	Cell type	Power density (mWm^{-2})	References
1	*Dunaliella tertiolecta*	Double chamber	5.3	[44]
2	*Spirulina platensis*	Double chamber	6.5	[45]
3	*Chlamydomonas reinhardtii* transformation F5	Double chamber	12.94	[46]
4	*Microcystis aeruginosa, Chlorella vulgaris*	Dual chamber	13	[47,48]
5	*C. vulgaris*	Double chamber	13.5 187	[49,50]
6	*C. vulgaris*	Dual chamber	15	[44]
7	*Arthrospira axima (Spirulina maxima)*	Double chamber	20.5	[51]
8	*C. vulgaris*	Modified MFC + tubular photobioreactor	27.5	[52]
9	*Scenedesmus obliquus*	Double chamber	30	[53]
11	*Dunaliella tertiolecta*	Single chamber + sediment	38	[44]
12	*Desmodesmus* sp. AS	Double chamber	64.2	[52]
13	Cyanobacteria	Single chamber	72	[54]
14	*C. reinhardtii*	Single chamber	75	[55]
15	Mixed algae	single	76	[43]
16	*C. reinhardtii*	Single chamber	78	[55]
17	*Desmodesmus* sp. A8	Double chamber	99.09	[56]
18	*S. obliquus*	Double chamber	102	[53]
19	Cyanobacteria	Single	114	[57]
20	Mixed algae	Double chamber	207	[58]
21	*Laminaria saccharina*	Dual chamber	250	[28]
22	*C. vulgaris*	Single chamber	980	[26]
23	*Arthrospira maxima*	Dual chamber	—	[51]
24	*Ulva lactuca*	Single chamber	—	[26]
25	*C. vulgaris*	Three chamber	—	Kokabian and Gude [59]
26	*Synechococcus leopoliensis*	Stack of 9 MFC	—	Walter et al. [60]

Handbook of Algal Biofuels

16.3 Effect of light on the performance of photosynthetic microalgal microbial fuel cells

Photosynthetic pigments in all plants including algae absorb light at different wavelengths, for example, chlorophyll a (*chl a*) absorbs at 410, 435, and 665 nm and chlorophyll b(*chl b*) absorbs at 460, 600, and 645 nm [61]. The electricity produced in a PMMFC is dependent on the source and intensity of the light being utilized by live algae [23,62]. A PMMFC seeks to convert light into electricity by exploiting the photosynthetic activity of living, phototrophic microorganisms. A single-chambered PMMFC retaining biofilm in freshwater displays small positive light responses with an increase in cell voltage when illuminated with light. The voltage rate and amplitude of the light response, however, improved significantly when PMMFC anodes were coated with electrically conductive polymers [63]. Thus PMMFCs can be distinguished into two types: (1) Type 1; and (2) Type 2. Type 1 uses light for photosynthesis and is referred to as a microbial solar cell since its fuel is solar energy. Type 2 uses light to facilitate the generation of electric current from the anaerobic microbial oxidation of organic matter. The major metabolic path used in this type of fuel cell is anaerobic fermentation by photosynthetic bacteria, also known as photofermentation [64]. Both types of PMMFC have fundamentally different operating principles from each other, leading to different cell designs and fields of application [65]. It was seen that there was fall in the coulombic efficiency by 53.4% on the increase in algal biomass from 3.6 g/L (5000 Lux) to 4.0 g/L (10,000 Lux) in a PMMFC. However, the coulombic efficiency increased by 24% when the biomass of algae increased slightly, that is, from 3.25 to 3.6 g/L [66]. This is probably due to the fact that when biomass reaches its maximum threshold concentration the dissolved oxygen (DO) accumulated due to photosynthesis also reaches its maximum value. The DO is required by microbes at the anode for organic degradation but when it is increased it leads to the backdiffusion of oxygen to the cathode thus reducing the power density of the PMMFC. Therefore for a PMMFC to work efficiently not only the intensity of the light but also the microalgal biomass growth needs to be optimized, which can be done by regular harvesting for value metabolites.

16.3.1 Light-emitting diodes and photosynthetic microalgal microbial fuel cells

Sunlight is the natural source of light for microalgal photosynthesis. Just like terrestrial plants they photosynthesize and produce food in the form of lipids and pigments which may vary for different microalgae. However, changing seasonal and weather conditions lead to their uncontrolled growth, especially if some metabolites are required for commercial production. In such a scenario artificial light is well-suited for their controlled growth and monitoring. Therefore microalgae are grown in manmade photobioreactors under artificial light at fixed intervals and in the same conditions without interruptions to avoid the changes in weather conditions or day and night.

The source of artificial light is very important as it is directly related to the photosynthetic pigments that absorb light at different wavelength intensities. The use of light emitting diodes (LED) artificial light as is seen to appreciably boost the growth of microalgae. Thus the performance of PMMFCs are likely to show crucial results related to their power density if

16.3 Effect of light on the performance of photosynthetic microalgal microbial fuel cells 371

illuminated by LEDs compared to natural sunlight [67]. Furthermore, the selected source should give enough light of the desired wavelength for the microalgae to photosynthesize without raising the temperature of the surroundings. Therefore LEDs and fluorescence lights are more reasonable and economical since they dissipate less heat and have narrow/single wavelengths, longer life, and consume less power [68]. Also LEDs have more light penetration in a photobioreactor (PBR) than florescent lights which increase the microalgal growth, as proven by increased growth rate and the total metabolites (carotenoids and lipids). LEDs have an important impact on different types of PBR. It has been noted that the transmission and refraction of light is different in suspended and biofilm PBR [38,69]. Plant pigments absorb light at different wavelengths. The effects of four different wavelengths 460, 540, 650, and 400−700 nm which belong to blue, red, green, and white light are very crucial for the growth of microalgae. Lights of these wavelengths directly influence the power density of a PMMFC.

It was seen that power density increased with an increase in light intensity from 100−900 Lux under the influence of blue and red light [46]. However red light leads to an amplification in the current due to the high absorption of light by *Chlamydomonas reinhardtii* leading to a maximum power density of 12.947 mWm^{-2}. In contrast, blue light affects the lower absorption of light energy hence leading to lower power generation (4.741 mWm^{-2}) [46,67].

In yet another case *Scenedesmus obliquus* cells in a PMMFC were grown on wastewaters in plastic poly bags illuminated to different monochromatic light sources so as to see their impact on growth and metabolites [68]. It was found that the light of a particular wavelength (red, blue, and green) illuminated at night not only affects the growth of microalgae in a different manner but also the composition of metabolites produced as a result of photosynthesis. The exponential phase of microalgal growth with all these color lights LED ended after 14 days. However, blue light showed the highest algal cell number which declined at around the 18th day. Average cellular chlorophyll (ACC) content was highest for red light, being 19.5% and 30.2% higher than blue and control, respectively. Though, both blue and red light incubated cultures increased the photosynthetic activity, red light stimulated pigment formation and blue light enhanced the efficiency of photosystem II working at the wavelength of 680 nm in the process of photosynthesis [70]. Many researchers found coherence with this aspect in many algae. It was seen that *Nannochloropsis* showed maximum growth in blue light [71], *C. vulgaris* under red light showed an enhancement in its pigmentation by 1.5-fold [72]. However, blue light results in photoinhibition in some microalgae, like *S. platensis*, and in a recent work by Abomohra et al. [68] it was seen that blue light photoinhibited cell growth of *S. obliquus* on the 16−18th days, as seen in Fig. 16.4. On the other hand green light showed no significant differences compared to the control.

The intensity of light also plays a very important role in increasing the power density as it directly affects photosynthesis [73,74]. The effects of different light intensities (3500, 5000, 7000, and 10,000 lx) and light/dark regimes (24/00, 12/12, 16/8 h) were studied on PMMFC using *C. vulgaris* and the best result was observed at a higher light intensity (10,000 Lux) [74,75]. He et al. studied the different spectra of LED light on the growth of microalgae and the efficiency of photosynthetic MFC in terms of electric current, algal biomass, lipids, and pigments. Water collected from a lake was used as the source of photosynthetic planktons [69]. Generally blue light is best for enhanced microalgal growth and its metabolites when compared to white, red, and green light.

Handbook of Algal Biofuels

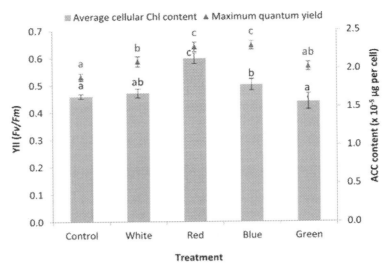

FIGURE 16.4 Average cellular chlorophyll content and maximum quantum yield of *Scenedesmus obliquus* grown under different light regimes. Cells were harvested at the end of the exponential phase. Columns of the same series with the same letter showed insignificant differences (at $P \leq .05$). [68]. *Source: Reproduced with permission number 501604468 from A.E.-F. Abomohra, et al., Night illumination using monochromatic light-emitting diodes for enhanced microalgal growth and biodiesel production. Bioresour. Technol., 288 (2019) 121514.*

They showed that a photosynthetic MFC (air chamber and two-chamber MFC) operated for about 7 months with graphite as an anode and phototropic lake water at the cathode and when operated in different LED spectra showed significant results. During the first month of PMFC operation, the current was high during the day and less during night. Even after several hours when the dark mode was discontinued, a peak current of 0.041 ± 0.002 mA was obtained which was delayed until the end of dark period. Initially upon using dazzling light on PMFC, the current decreased and increased slightly as shown in Fig. 16.5A. The lowest peak current reached a negative value when the light was switched on. However, after 5 months, this pattern followed a reverse pattern, as can be seen in Fig. 16.5B: a rise of 0.054 ± 0.002 mA and fall of 0.045 ± 0.003 mA in current when the light was shut or turned on, respectively. On resetting the light on and off, the current generating profile was restored in about 15 days, as seen in Fig. 16.5B. He et al. also exposed sediment type PMFCs to red (620–750 nm), and blue filters (450–495 nm) and compared them with control or white light [69]. When the red light was off, the current increased to 0.060 ± 0.002 mA in about 4 h. However, with blue light the current was irregular and when both red and blue lights were on together, the current level was fluctuating but never below zero. The current was always hiked up when the lights were shut on in both sediment and air cathode type PMMFCs, which is mainly due to the fact that during light conditions microalgae photosynthesize organic compounds providing food for heterotrophic bacteria in the sediment of the cathode cell. Further, these compounds are oxidized during the dark and current is generated in the electrochemical reaction process. Alternatively,

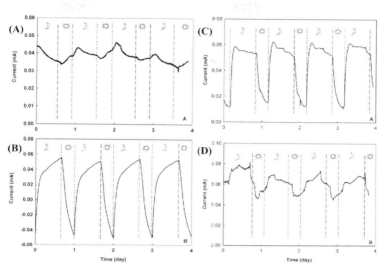

FIGURE 16.5 Electric current productions from the sediment of phototrophic microbial fuel cell under the full-spectrum light after 1 month (A) and 5 months (B); under red light (C) and blue light (D); the symbols of moon and sun represent dark and light conditions, respectively [69].

when the lights were switched on, the oxygen synthesized at the cathode via photosynthesis is utilized by the microbes at the anode of the PMFC.

Since a dark period is essential for the health and growth rate of photosynthetic microorganisms, a decrease in period of light/dark cycles results in lower electricity and biomass production. The best performance of PMFC has been reported to be 16/8 h of light/dark. Furthermore, besides light, performance of PMFC depends upon wastewater treatment and biomass production which are of paramount importance in future work. A PMMFC with *C. vulgaris* was optimized to operate under controlled light intensity of 3500, 5000, 7000, and 10,000 Lux for wastewater treatment and biomass production along with electricity production [74]. An anode with activated sludge and a cathode with *C. vulgaris* with an optical density of 0.8 at 680 nm were tested for different light intensities. The light intensity of 5000 and 6000 Lux was identified as the optimum intensity for maximum power generation in the *C. vulgaris* PMMFC. The light had a simultaneous effect on coulombic efficiency, COD removal, and accumulation of biomass. In the presence of light, the maximum drift in COD was 1.3%, however it was 30% in coulombic efficiency. Maximum coulombic efficiencies of about 78% and 48% were obtained at 5000 and 10,000 Lux, respectively [74].

16.4 DNA sequencing of microbial genomes

The in situ diversity of microorganisms in a PMMFC if analyzed can assist in determining the microbial compositions giving the most efficient power density of

PMMFC. The community of microbes in the anode and in the cathode determines the efficiency of a PMMFC. Besides microscopic identification, which sometimes becomes difficult in mixed cultures, DNA sequencing of microbial communities is very propitious since wastewater used in the anode has a mixed bacterial, fungal, and other microbial community. Researchers have tried to unravel the genomes of microbes in wastewater for many applications. But if the genome of microbes in the anode and the plankton water in the cathode of a PMMFC is explored it will help in standardizing the microbial community bringing the highest efficiency. Khanaum studied the genome of microbial communities in a MFC and observed obligatory anaerobes and rod-shaped bacteroidetes which produce the majority of electricity in a MFC [76].

In a PMMFC, an interdependent relationship between microbes and heterotrophic bacteria may occur. The denaturing gradient gel electrophoresis (DGGE) demonstrated the presence of photosynthetic cyanobacteria and microalagae in sediment MFC [69]. No photosynthetic microbes were present in the anodic community. On molecular sequencing of sediments in cathode, photosynthetic microorganisms were predominant, which included cyanobacteria and microalgae, as seen in Fig. 16.6 [69].

Apart from this, nanopore DNA sequencing is a technique that helps to study the metagenomic profile not only in a pure or mixed culture but also microbial communities in wastewater serving as anolyte. This technology gives real-time analysis of the dominant genomes which play an important role in electricity generation in an MFC. The technique was first demonstrated by Oxford Nanopore Technologies' MinION, which produces long-read DNA sequences using a mobile DNA sequencer in a very short time [77]. The MinION unraveled the genome of a multiplexed genome assembly of bacterial samples combined with existing Illumina short-read data. Generally, wastewater contamination is assessed by *E. coli* count which can be easily determined by economical Oxford Nanopore MinION shotgun metagenomic sequencing method. The performance of a MiniON for a sample containing bacteria, *E. coli* (strain K12, MG1655), lambda bacteriophage, and sample containing mice *Mus musculus* in a terrestrial laboratory and on the International Space research station were compared [78]. The nine sequencing runs performed both on the ISS and in the terrestrial lab showed almost similar pairwise consensus identity results. This can be very useful not only for astronauts to do real-time screening of microbes for wet lab research, but also shows the possibility of using PMMFC to recycle human excreta and recover metabolites which could be used as a source of food in spaceships. The economic burden of analyzing the presence of genome diversity can be compared using old approaches versus nanopore DNA technology, as indicated in Fig. 16.7, which shows a comparison of nanopore DNA technology for metagenome sequencing with conventional DNA technologies. The demarcation by three bars in the sketch illustrates the possible time required in the nanopore-based method. Consequently, the time required for sequencing and interpretation of data was determined on the basis of computing power. The pointed arrow below MiSeq further illustrates the sampling of five samples for processing. In the Coliret-18 test the time required for sequencing does not alters with the change in sample size. However. the time required in the DNA nanopore has been labeled as "B" which can be continued even while the sequence is ongoing its amplification and data mining process. The fourth and fifth

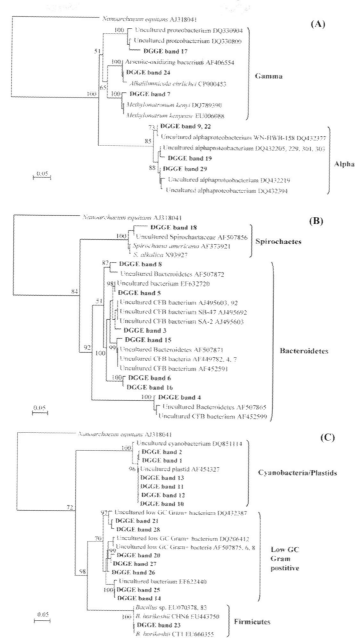

FIGURE 16.6 Phylogenetic analysis of denaturing gradient gel electrophoresis band sequences obtained in the current study and most closely related representatives from GenBank: (A) Proteobacteria; (B) Spirochates and Bacteroidetes; (C) other bacterial groups including Cyanobacteria/plastids, low GC Gram-positive and Firmicutes. Scale bars indicate the number of substitutions per site. Bootstrap values were based on 1000 resampling data sets. Only bootstrap values relevant to the interpretation of groupings are shown [69].

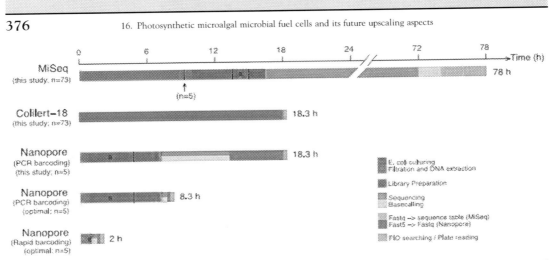

FIGURE 16.7 Expenditures of time for assessing levels of human fecal contamination in water samples with different approaches [79].

bars on the nanopore timeline show the possible time required for data mining of five processed samples using different barcoding methods to the extent of 1000 reads per sample [79].

16.5 Integrated approaches of photosynthetic microalgal microbial fuel cells

16.5.1 Bioelectricity production

Zou et al. tested PMMFC performance using cyanobacteria *Synecho cystis* and natural fresh water biofilm as the anode and open air cathode without organic buffers or electron shuttles [63]. The PMMFC was operated in an open circuit with external resistance of 100–10 kOhm and showed a sharp fall in voltage on illumination of light mainly due to the negative impact of the evolved photosynthetic oxygen. They observed that the rate of voltage in such a single-chamber PMFC declined as a function of the current in a carbon paint-coated anode with an internal resistance of 12.1–0.4 kOhm and power density of 0.35 W/m². However, in polymer-coated anodes, the fall in voltage was less with internal resistance of 3.9–0.01 kOhm with a power density of 0.95 Wm^{-2} with a coating of polyaniline (polyA) anode and 1.3 Wm^{-2} with polypyrrole (poly P) anode, respectively. These coatings are believed to enhance the growth of the biofilm responsible for a high photosynthesis rate. The biofilm-based PMMFC showed immediate response to light which increased cell voltage despite increasing DO and pH. The depth of biofilm may help in reducing the DO [63]. A high exchange current density for oxygen reduction with effective biofilm formation of *S. obliquus* on platinized cathode is observed because of its direct oxygen transfer by algal biofilms. A simple scraping method was used for the recovery of algal biomass from the biofilm. So, algal biofilm production is important and attractive for harvesting the biomass. However, a detailed study on algal biofilm formation in order to improve O_2 transfer, obtain better MFC performance, and enable high algal biomass production and its effective harvesting from the cathode compartment is required [66,80].

16.5.2 Recycling wastewater

Wastewaters from industrial and domestic sources have large amounts of organic matters and some toxic chemicals [66]. The major obstacles in conventional wastewater treatments are the high energy consumption and operational costs. Several activated sludge plant units are required for treating the wastewater, thus getting the maximum efficiency from each unit remains an important issue [74,81]. These hurdles can be overcome by culturing microalgae in which nutrients from highly polluted wastewater can be reused, which helps in treating high-nutrient-strength wastewater. Using microalgae for the treatment of wastewater removes nutrients and numerous pollutants like heavy metals, nitrogen compounds, and harmful chemicals. The main advantage of using PMMFC in the treatment of wastewater is the low energy requirement with low operation and maintenance costs [82,83].

16.5.3 Production of value-added chemicals

Several steps are required for production of value-added products from PMMFC via the harvesting and separation of algal biomass from cathode chamber. The economical methods for separation of biomass are biofilm formation, membrane separation, flocculation, and cell immobilization. The next step is the recovery of value-added product from microalgal biomass, which may require pretreatment. The use of photosynthetic microalgae in PMMFC produces various pigments, such as chlorophyll, β-carotene, canthaxanthin, lutein, and zeaxanthin, as well as biofuels. Besides pigments and biofuels the harvested biomass can be used as a source of methane gas (biogas) production, and CO_2 as a by-product which can be used as an inorganic carbon source for algae. After oil/lipid extraction, the biomass residue further can be served as animal feed [24]. Therefore besides generating bioelectricity, PMMFC generates several high- and low-value metabolites which are alternative energy sources [84–86].

16.6 Future of photosynthetic microalgal microbial fuel cells using diatoms

Among microalgae, not much work has been done on PMMFCs utilizing diatoms for bioelectricity and biofuel. Also diatoms are the best candidates for biofuel production, since geologists claim that about 30% of the crude oil comes from them [87] and the chemical analysis of their lipid shows that they have almost zero carbon footprint [11]. They are unicellular microscopic golden-brown algae with silica shells which play a major role in oil production. In our earlier work, we developed different techniques to harvest Diafuel (biofuel from diatoms) [87]. Among the biofuels from algae, Diafuel could help the transition/shift to EV in the near future with almost zero carbon emissions [11,87]. It is noteworthy that the dependency on imported oil and hike in oil prices has resulted in the upcoming era of EVs [88–90]. However, Diafuel would be a boon to the oil industry to run vehicles if Diafuel is harvested in an economical resonance PBR which runs economically using solar energy [91,92]. This transition to Diafuel not only reduces the dependence on fossil fuel but also brings down global warming as diatoms are primary producers which can fix approximately 25% of atmospheric carbon dioxide [11,93]. Diatoms are also a rich source of fatty acids like myristic acid, palmitic acid, palmitoleic acid, silica, and many amino acids which are used in medicine and material science. However, due to their thick silica wall oil does not come out easily from diatoms. In earlier studies various attaempts were made to help diatoms ooze oil. The beautiful nanoarrays of diatom frustules were doped with specifically designed biotinylated peptides that acted as carriers of gold nanoparticles. It was observed that the thermoplasmonic heat of gold nanoparticles opens the pores of diatom allowing the oil to ooze out [94]. In another work, fabricated titanium engineered diatoms were used in dye-sensitized solar cells (DSSC) that generated power more efficiently than the traditional DSSC [95]. More facile DSSC were made with diatoms doped with silver nanoparticles to produce bioelectricity [96].

However, there is a possibility of designing a diatom-based PMMFC to not only produce bioelectricity but also high-value metabolites from diatoms. Furthermore, if these diatom photosynthetic microbial fuel cell (DPMFC) are stacked together, they produce amplified electricity compared with a single DPMFC [93,97,98]. The concept of stacked DPMFC is hypothesized not only to produce bioelectricity but also to produce Diafuel which may have future benefits in hybrid electric vehicles [4,87]. Hence, DPMFC could be used to generate both electricity and Diafuel for possible use in hybrid electric vehicles. No doubt, further studies are needed to scale up the technology of such DPMFCs to run motor-based machinery or vehicles dependent on electricity and biofuel [99,100]. Here, we hypothesize that DPMFC could be tethered to charging stations not only for electricity production but also for Diafuel to run hybrid EVs. The schematic comparison of EVs run with solar, wind, and coal energy shift versus hybrid EVs run via DPMFC is shown in Fig. 16.8. Diatoms, like other living algae in PMMFC, require CO_2 for growth which they accept from the cathode during the oxidation–reduction process, thus releasing free oxygen.

These DPMFC could be stacked three-dimensionally at a commercial scale to produce electricity to run EVs and also to generate Diafuel to run gasoline or combustion engines, or together could be used to run hybrid EVs. Our recent critical report suggests that the era of EVs alone would not result in zero emissions of carbon dioxide [4]. Diatoms that have 10 times more oil than our common seed crops could help in preserving the technology of gasoline vehicles without completely switching to EVs [4]. Even though EVs are going to be emission-free, the electricity from renewable energy sources is certainly a concern in the future of EVs [101]. Besides this, the increasing population and urbanization has brought a huge rise in industrialization which has increased CO_2 emissions to much above the threshold value [102].

Handbook of Algal Biofuels

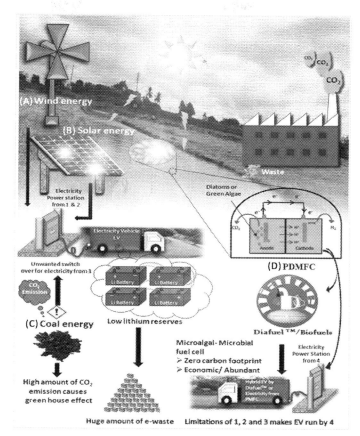

FIGURE 16.8 Schematic representation of electric vehicles being driven by electricity from (A) solar and (B) wind energies showing the accumulation of a huge amount of e-waste, limited Li$^+$ reserves, and an unavoidable switch over to coal energy (C) for electricity generation which would be an alarming situation for electric vehicles. (D) Diatom photosynthetic microbial fuel cells acting as biobatteries and source of Diafuel with almost zero carbon footprint and abundant in nature.

16.7 Conclusions

Despite generating electricity, microbial fuel cells have many limitations. The power generated by abiotic anodes of MFC seems to be equivalent to algal-based anodes but it is unstable for long-term use. Therefore, biotic algal-based systems provide sustained operation without replacement. The microalgal photosynthetic biofilm acts as a biocatalyst that is economic and effective for the real-world application of PMMFC for sustainable energy generation. PMMFC technology needs numerous modulations to make it a sustainable alternative source of energy. The plausible theoretical values of cell potential are very much less for single PMMFCs hence the stacked PMMFC would be an effective way to achieve feasible power. Various research groups are trying to overcome the limitations of this technology and make it an innovative and promising source of alternate energy to the society. Stacked PMMFC technology is advantageous as it not only produces enough bioelectricity to drive larger electric devices and vehicles but also produces biofuel which could help in the switch over to running hybrid EVs with zero carbon emissions. However, PMMFCs enriched with diatom, that is, DPMFC, give the possibility to not only produce bioelectricity but also to recover Diafuel. They harvest solar energy to reduce biomass in wastewater for bioelectricity and other important products. Owing to the current energy shortage and grave concerns over global warming due to accelerated carbon emission from fossil fuels, the DPMFC technology in its various forms represents a promising green technology for renewable and sustainable energy production, especially in a distributed energy system. The prospect of DPMFC technology is enabling energy production with carbon fixation from intractable waste materials and diatom biomass production. Besides, the stacking of DPMFCs can be used to produce the desired electricity and thus the charging stations can be connected to stacked DPMFCs to charge hybrid EVs and simultaneously the chemical metabolites, specially Diafuel, can run the gasoline and electricity-driven hybrid EVs. Our future research efforts in these directions will scale up the DPMFC technique as a pilot project scheme.

Acknowledgments

MJK thank to DST-Nanomission, Government of India and VJS to CSIR for postdoctoral fellowship. Authors acknowledge Dr. Anitha V efforts for critic comments in the chapter. This work is supported by DST-Nanomission, Government of India Project No. SR/NM/NT-1090/2014(G) to VV (Principal Investigator) and KBJ.

Conflict of Interest

The authors declare no conflict of interest.

References

[1] N.A. Sepulveda, et al., The role of firm low-carbon electricity resources in deep decarbonization of power generation, Joule 2 (11) (2018) 2403–2420.
[2] S. Tian, et al., Lithium isotopic evidence for subduction of the Indian lower crust beneath southern Tibet, Gondwana Res. 77 (2020) 168–183.
[3] Mahajan, S., *Lithium Ion Battery Market: Industry Analysis and Forecast Till 2026*, 2020.
[4] V. Vinayak, K.B. Joshi, P.M. Sarma, Diafuel™(diatom biofuel) versus electric vehicles, a basic comparison: a high potential renewable energy source to make india energy independent, Diatoms (2019) 537–582.
[5] H. Lund, W. Kempton, Integration of renewable energy into the transport and electricity sectors through V2G, Energy policy 36 (9) (2008) 3578–3587.
[6] C.C. Chan, The state of the art of electric, hybrid, and fuel cell vehicles, Proc. IEEE 95 (4) (2007) 704–718.
[7] W. Beckerman, Economic growth and the environment: whose growth? Whose environment? World Dev. 20 (4) (1992) 481–496.
[8] F.H. Abanda, Renewable energy sources in Cameroon: potentials, benefits and enabling environment, Renew. Sustain. Energy Rev. 16 (7) (2012) 4557–4562.
[9] B.K. Sovacool, R.F. Hirsh, Beyond batteries: an examination of the benefits and barriers to plug-in hybrid electric vehicles (PHEVs) and a vehicle-to-grid (V2G) transition, Energy Policy 37 (3) (2009) 1095–1103.

References

[10] Hoeven, M.V.D., World energy outlook special report, in: International Energy Agency, Paris, 2015, pp. 1–200.

[11] T. Ramachandra, D.M. Mahapatra, R. Gordon, Milking diatoms for sustainable energy: biochemical engineering versus gasoline-secreting diatom solar panels, Ind. Eng. Chem. Res. 48 (19) (2009) 8769–8788.

[12] C. Santoro, et al., Microbial fuel cells: from fundamentals to applications. A review, J. Power Sources 356 (2017) 225–244.

[13] H. Liu, R. Ramnarayanan, B.E. Logan, Production of electricity during wastewater treatment using a single chamber microbial fuel cell, Environ. Sci. Technol. 38 (7) (2004) 2281–2285.

[14] B.E. Logan, Exoelectrogenic bacteria that power microbial fuel cells, Nat. Rev. Microbiol. 7 (5) (2009) 375.

[15] Y. Sun, et al., Microbial community analysis in biocathode microbial fuel cells packed with different materials, AMB. Express 2 (1) (2012) 21.

[16] D. Xing, et al., Electricity generation by Rhodopseudomonas palustris DX-1, Environ. Sci. Technol. 42 (11) (2008) 4146–4151.

[17] V.J. Watson, B.E. Logan, Power production in MFCs inoculated with Shewanella oneidensis MR-1 or mixed cultures, Biotechnol. Bioeng. 105 (3) (2010) 489–498.

[18] A. Rhoads, H. Beyenal, Z. Lewandowski, Microbial fuel cell using anaerobic respiration as an anodic reaction and biomineralized manganese as a cathodic reactant, Environ. Sci. Technol. 39 (12) (2005) 4666–4671.

[19] M. Zhou, et al., Recent advances in microbial fuel cells (MFCs) and microbial electrolysis cells (MECs) for wastewater treatment, bioenergy and bioproducts, J. Chem. Technol. Biotechnol. 88 (4) (2013) 508–518.

[20] Z. Baicha, et al., A critical review on microalgae as an alternative source for bioenergy production: a promising low cost substrate for microbial fuel cells, Fuel Process. Technol. 154 (2016) 104–116.

[21] M. Shukla, S. Kumar, Algal growth in photosynthetic algal microbial fuel cell and its subsequent utilization for biofuels, Renew. Sustain. Energy Rev. 82 (2018) 402–414.

[22] D.P. Strik, et al., Renewable sustainable biocatalyzed electricity production in a photosynthetic algal microbial fuel cell (PAMFC), Appl. Microbiol. Biotechnol. 81 (4) (2008) 659–668.

[23] Y. Cui, et al., Electricity generation and microalgae cultivation in microbial fuel cell using microalgae-enriched anode and bio-cathode, Energy Convers. Manag. 79 (2014) 674–680.

[24] L. Gouveia, et al., Effect of light on the production of bioelectricity and added-value microalgae biomass in a photosynthetic alga microbial fuel cell, Bioresour. Technol. 154 (2014) 171–177.

[25] N. Rashid, et al., Enhanced electricity generation by using algae biomass and activated sludge in microbial fuel cell, Sci. Total. Environ. 456 (2013) 91–94.

[26] S.B. Velasquez-Orta, T.P. Curtis, B.E. Logan, Energy from algae using microbial fuel cells, Biotechnol. Bioeng. 103 (6) (2009) 1068–1076.

[27] P. Rajesh, D. Jadhav, M. Ghangrekar, Improving performance of microbial fuel cell while controlling methanogenesis by Chaetoceros pretreatment of anodic inoculum, Bioresour. Technol. 180 (2015) 66–71.

[28] V. Gadhamshetty, et al., Evaluation of Laminaria-based microbial fuel cells (LbMs) for electricity production, Bioresour. Technol. 127 (2013) 378–385.

[29] V. Krey, et al., Getting from here to there – energy technology transformation pathways in the EMF27 scenarios, Climatic Change 123 (3) (2014) 369–382.

[30] M. Rosenbaum, Z. He, L.T. Angenent, Light energy to bioelectricity: photosynthetic microbial fuel cells, Curr. Opin. Biotechnol. 21 (3) (2010) 259–264.

[31] A. ElMekawy, et al., Techno-productive potential of photosynthetic microbial fuel cells through different configurations, Renew. Sustain. Energy Rev. 39 (2014) 617–627.

[32] I. Gajda, et al., Photosynthetic cathodes for Microbial Fuel Cells, Int. J. Hydrog. Energy 38 (26) (2013) 11559–11564.

[33] L. Semenec, et al., Electron transfer between bacteria and electrodes, Functional Electrodes For Enzymatic and Microbial Electrochemical Systems (2017) 93.

[34] V.G. Gude, Wastewater treatment in microbial fuel cells—an overview, J. Clean. Prod. 122 (2016) 287–307.

[35] K. Tanaka, R. Tamamushi, T. Ogawa, Bioelectrochemical fuel-cells operated by the cyanobacterium, Anabaena variabilis, J. Chem. Technol. Biotechnol. 35 (3) (1985) 191–197.

[36] K. Tanaka, N. Kashiwagi, T. Ogawa, Effects of light on the electrical output of bioelectrochemical fuel-cells containing Anabaena variabilis M-2: mechanism of the post-illumination burst, J. Chem. Technol. Biotechnol. 42 (3) (1988) 235–240.

Handbook of Algal Biofuels

[37] T. Yagishita, T. Horigome, K. Tanaka, Effects of light, CO_2 and inhibitors on the current output of biofuel cells containing the photosynthetic organism Synechococcus sp, J. Chem. Technol. Biotechnol. 56 (4) (1993) 393−399.

[38] Z. He, L.T. Angenent, Application of bacterial biocathodes in microbial fuel cells, Electroanalysis 18 (19−20) (2006) 2009−2015.

[39] C.-C. Fu, et al., Effects of biomass weight and light intensity on the performance of photosynthetic microbial fuel cells with Spirulina platensis, Bioresour. Technol. 100 (18) (2009) 4183−4186.

[40] J.-K. Wang, M. Seibert, Prospects for commercial production of diatoms, Biotechnol. Biofuels 10 (1) (2017) 16.

[41] A.K. Yadav, et al., Performance assessment of innovative constructed wetland-microbial fuel cell for electricity production and dye removal, Ecol. Eng. 47 (2012) 126−131.

[42] S.V. Mohan, et al., Waste biorefinery models towards sustainable circular bioeconomy: critical review and future perspectives, Bioresour. Technol. 215 (2016) 2−12.

[43] R. Chandra, G.V. Subhash, S.V. Mohan, Mixotrophic operation of photo-bioelectrocatalytic fuel cell under anoxygenic microenvironment enhances the light dependent bioelectrogenic activity, Bioresour. Technol. 109 (2012) 46−56.

[44] A.-M. Lakaniemi, O.H. Tuovinen, J.A. Puhakka, Production of electricity and butanol from microalgal biomass in microbial fuel cells, BioEnergy Res. 5 (2) (2012) 481−491.

[45] C.-C. Fu, et al., Current and voltage responses in instant photosynthetic microbial cells with Spirulina platensis, Biochemical Eng. J. 52 (2−3) (2010) 175−180.

[46] J.C.-W. Lan, et al., The impact of monochromatic blue and red LED light upon performance of photo microbial fuel cells (PMFCs) using Chlamydomonas reinhardtii transformation F5 as biocatalyst, Biochem. Eng. J. 78 (2013) 39−43.

[47] Y.-P. Wang, et al., A microbial fuel cell−membrane bioreactor integrated system for cost-effective wastewater treatment, Appl. Energy 98 (2012) 230−235.

[48] X. Wang, et al., Use of carbon mesh anodes and the effect of different pretreatment methods on power production in microbial fuel cells, Environ. Sci. Technol. 43 (17) (2009) 6870−6874.

[49] A.G. del Campo, et al., Microbial fuel cell with an algae-assisted cathode: a preliminary assessment, J. Power Sources 242 (2013) 638−645.

[50] T. Liu, et al., Bioelectricity generation in a microbial fuel cell with a self-sustainable photocathode, Sci. World J. 2015 (2015) 8.

[51] A.E. Inglesby, A.C. Fisher, Downstream application of a microbial fuel cell for energy recovery from an Arthrospira maxima fed anaerobic digester effluent, RSC Adv. 3 (38) (2013) 17387−17394.

[52] Y.-c Wu, et al., Light intensity affects the performance of photo microbial fuel cells with Desmodesmus sp. A8 as cathodic microorganism, Appl. Energy 116 (2014) 86−90.

[53] S. Kondaveeti, et al., Microalgae Scenedesmus obliquus as renewable biomass feedstock for electricity generation in microbial fuel cells (MFCs), Front. Environ. Sci. Eng. 8 (5) (2014) 784−791.

[54] J. Zhao, et al., Electricity generation from Taihu Lake cyanobacteria by sediment microbial fuel cells, J. Chem. Technol. Biotechnol. 87 (11) (2012) 1567−1573.

[55] K. Nishio, K. Hashimoto, K. Watanabe, Digestion of algal biomass for electricity generation in microbial fuel cells, Biosci. Biotechnol. Biochem. 77 (3) (2013) 670−672.

[56] D.-J. Lee, J.-S. Chang, J.-Y. Lai, Microalgae−microbial fuel cell: a mini review, Bioresour. Technol. 198 (2015) 891−895.

[57] Y. Yuan, et al., Bioelectricity generation and microcystins removal in a blue-green algae powered microbial fuel cell, J. Hazard. Mater. 187 (1) (2011) 591−595.

[58] P. Singhvi, M. Chhabra, Simultaneous chromium removal and power generation using algal biomass in a dual chambered salt bridge microbial fuel cell, J. Bioremed Biodeg 4 (5) (2013).

[59] B. Kokabian, V.G. Gude, Photosynthetic microbial desalination cells (PMDCs) for clean energy, water and biomass production, Environ. Sci. 15 (12) (2013) 2178−2185.

[60] X.A. Walter, et al., Microbial fuel cells continuously fuelled by untreated fresh algal biomass, Algal Res. 11 (2015) 103−107.

[61] F. Zhang, et al., Applying and comparing two chemometric methods in absorption spectral analysis of photopigments from Arctic microalgae, J. Microbiol. Methods 83 (2) (2010) 120−126.

References

[62] C. Xu, et al., Using live algae at the anode of a microbial fuel cell to generate electricity, Environ. Sci. Pollut. Res. 22 (20) (2015) 15621–15635.

[63] Y. Zou, et al., Photosynthetic microbial fuel cells with positive light response, Biotechnol. Bioeng. 104 (5) (2009) 939–946.

[64] P.C. Hallenbeck, J.R. Benemann, Biological hydrogen production; fundamentals and limiting processes, Int. J. Hydrog. Energy 27 (11–12) (2002) 1185–1193.

[65] M. Rosenbaum, U. Schröder, Photomicrobial solar and fuel cells, Electroanalysis 22 (7–8) (2010) 844–855.

[66] S. Arun, et al., Algae based microbial fuel cells for wastewater treatment and recovery of value-added products, Renew. Sustain. Energy Rev. 132 (2020) 110041.

[67] H.-B. Chen, et al., Modeling on chlorophyll a and phycocyanin production by Spirulina platensis under various light-emitting diodes, Biochem. Eng. J. 53 (1) (2010) 52–56.

[68] A.E.-F. Abomohra, et al., Night illumination using monochromatic light-emitting diodes for enhanced microalgal growth and biodiesel production, Bioresour. Technol. 288 (2019) 121514.

[69] Z. He, et al., Self-sustained phototrophic microbial fuel cells based on the synergistic cooperation between photosynthetic microorganisms and heterotrophic bacteria, Environ. Sci. Technol. 43 (5) (2009) 1648–1654.

[70] K. Tagawa, H. Tsujimoto, D.I. Arnon, Role of chloroplast ferredoxin in the energy conversion process of photosynthesis, Proc. Natl Acad. Sci. U S Am. 49 (4) (1963) 567.

[71] P. Das, et al., Enhanced algae growth in both phototrophic and mixotrophic culture under blue light, Bioresour. Technol. 102 (4) (2011) 3883–3887.

[72] D.G. Kim, et al., Manipulation of light wavelength at appropriate growth stage to enhance biomass productivity and fatty acid methyl ester yield using Chlorella vulgaris, Bioresour. Technol. 159 (2014) 240–248.

[73] P. Heydarizadeh, et al., Plastids of marine phytoplankton produce bioactive pigments and lipids, Mar. Drugs 11 (9) (2013) 3425–3471.

[74] E. Bazdar, et al., The effect of different light intensities and light/dark regimes on the performance of photosynthetic microalgae microbial fuel cell, Bioresour. Technol. 261 (2018) 350–360.

[75] E.E. Powell, et al., Growth kinetics of Chlorella vulgaris and its use as a cathodic half cell, Bioresour. Technol. 100 (1) (2009) 269–274.

[76] Khanaum, M.M., et al., Wastewater treatment and electricity generation from sugarbeet processing wastewater (SBWW) using microbial fuel cell (MFC), in: 2020 ASABE Annual International Virtual Meeting, American Society of Agricultural and Biological Engineers, 2020.

[77] S.C. Bayliss, et al., The use of Oxford nanopore native barcoding for complete genome assembly, Gigascience 6 (3) (2017) gix001.

[78] S.L. Castro-Wallace, et al., Nanopore DNA sequencing and genome assembly on the international space station, Sci. Rep. 7 (1) (2017) 1–12.

[79] Y.O. Hu, et al., Stationary and portable sequencing-based approaches for tracing wastewater contamination in urban stormwater systems, Sci. Rep. 8 (1) (2018) 1–13.

[80] I. Gajda, et al., Self-sustainable electricity production from algae grown in a microbial fuel cell system, Biomass Bioenergy 82 (2015) 87–93.

[81] Z. Du, H. Li, T. Gu, A state of the art review on microbial fuel cells: a promising technology for wastewater treatment and bioenergy, Biotechnol. Adv. 25 (5) (2007) 464–482.

[82] A.F. Mohd Udaiyappan, et al., A review of the potentials, challenges and current status of microalgae biomass applications in industrial wastewater treatment, J. Water Process. Eng. 20 (2017) 8–21.

[83] J. Greenman, I. Gajda, I. Ieropoulos, Microbial fuel cells (MFC) and microalgae; photo microbial fuel cell (PMFC) as complete recycling machines, Sustain. Energy Fuels 3 (10) (2019) 2546–2560.

[84] A.P. Faaij, Bio-energy in Europe: changing technology choices, Energy Policy 34 (3) (2006) 322–342.

[85] O. Edenhofer, et al., Renewable Energy Sources and Climate Change Mitigation: Special Report of the Intergovernmental Panel on Climate Change, Cambridge University Press, 2011.

[86] Konur, O., Current state of research on algal bioelectricity and algal microbial fuel cells, Marine Bioenergy: Trends and Developments, 2015, pp. 527–556.

[87] Vinayak, V., et al., "Diafuel." 2018 Trademark application no 3778882, Trade Marks Journal No: 1846, 23/04/2018 Class 4, 2018.

[88] T. Lane, M. Vanderschuren, Potential transport energy demand and oil dependency mitigation measures, WIT Trans. Ecol. Environ. 121 (2009).

[89] M. Vanderschuren, R. Jobanputra, T. Lane, Potential transportation measures to reduce South Africa's dependency on crude oil, J. Energy South. Afr. 19 (3) (2008) 20–29.

[90] M. Vanderschuren, T. Lane, J. Wakeford, Can the South African transport system surmount reduced crude oil availability? Energy Policy 38 (10) (2010) 6092–6100.

[91] Vinayak, V., et al., Fabrication of resonating microfluidic chamber for biofuel production in diatoms (Resonating device for biofuel production), in: 2016 3rd International Conference on Emerging Electronics (ICEE), IEEE, 2016.

[92] M.J. Khan, et al., Cultivation of diatoms for lipid production: a comparison of methods for harvesting the lipid from the cells, Bioresour. Technol. (2020) 124129.

[93] V. Vinayak, et al., Diatom milking: a review and new approaches, Mar. Drugs 13 (5) (2015) 2629–2665.

[94] V. Kumar, et al., Biomimetic fabrication of biotinylated peptide nanostructures upon diatom scaffold; a plausible model for sustainable energy, RSC Adv. 6 (77) (2016) 73692–73698.

[95] S. Gautam, et al., Metabolic engineering of TiO_2 nanoparticles in Nitzschia palea to form diatom nanotubes: an ingredient for solar cells to produce electricity and biofuel, RSC Adv. 6 (99) (2016) 97276–97284.

[96] S. Gupta, et al., Peptide mediated facile fabrication of silver nanoparticles over living diatom surface and its application, J. Mol. Liq. 249 (2018) 600–608.

[97] A. Dekker, et al., Analysis and improvement of a scaled-up and stacked microbial fuel cell, Environ. Sci. Technol. 43 (23) (2009) 9038–9042.

[98] M. Hildebrand, et al., The place of diatoms in the biofuels industry, Biofuels 3 (2) (2012) 221–240.

[99] V. Mimouni, et al., The potential of microalgae for the production of bioactive molecules of pharmaceutical interest, Curr. Pharm. Biotechnol. 13 (15) (2012) 2733–2750.

[100] Z. Yi, et al., Exploring valuable lipids in diatoms, Front. Mar. Sci. 4 (2017) 17.

[101] S.F. Tie, C.W. Tan, A review of energy sources and energy management system in electric vehicles, Renew. Sustain. energy Rev. 20 (2013) 82–102.

[102] J. Rogelj, et al., Energy system transformations for limiting end-of-century warming to below 1.5 C, Nat. Clim. Change 5 (6) (2015) 519.

Sequential algal biofuel production through whole biomass conversion

Mahdy Elsayed[1] and Abd El-Fatah Abomohra[2,3]

[1]Department of Agricultural Engineering, Faculty of Agriculture, Cairo University, Giza, Egypt
[2]Department of Environmental Engineering, School of Architecture and Civil Engineering, Chengdu University, Chengdu, P.R. China [3]Botany and Microbiology Department, Faculty of Science, Tanta University, Tanta, Egypt

17.1 Introduction

In an era of accelerated changes, the imperative of curbing climate changes and achieving a sustainable growth is driving the global energy transition. The elevated energy security around the world together with the negative environmental impacts of the use of fossil fuels has led to an urgent need to explore new biofuel feedstocks [1,2]. Therefore more research is needed to achieve energy security and reduce the environmental pollution. Among the different biofuel feedstocks, algal biomass has showed more economically attractive routes. Although the first- and second-generation candidates for biofuel production, including lignocellulosic wastes such as crop straw, manure, and urban wastes, have been widely discussed due to their abundance, the elimination/degradation of lignin to improve cellulose and hemicellulose accessibility by destroying the recalcitrant structures represent the main challenges in their utilization [3,4]. Therefore the third-generation biofuel production from the wild-type algae have recently gained considerable interest [5–7], while the genetically engineered microalgae are currently being tailored as novel feedstocks for fourth-generation biofuel [8].

Algae have been recently recognized as a competitive forum for biorefineries and the production of biofuels. Since microalgae grown in wastewater will not create competition for food fuels, algal biofuels are more economically attractive [9]. In addition, compared to terrestrial biofuel crops such as maize, soybeans, and rapeseed, the algal biomass production results in a far smaller land/water footprint due to higher algal oil productivity [10]. Algae can also be grown on arid lands that are worthless for plant crops and have poor conservation value, and can use salt water from aquifers that is not valuable for farming or drinking. In addition, algae can grow in bags or floating screens on the ocean floor [10]. Thus seaweeds and microalgae may provide a renewable energy source with no effect on the provision of sufficient food and water or biodiversity conservation. Moreover, algal biofuel provides a chance to solve the ethical problems raised from renewable energy production from edible human food. It also would create green cities, create more jobs, enable faster economic growth, and improve the overall welfare. In economic terms, it will reduce the environmental pollution and thus reduce the costs of remedying the environment. However, the fast response of the world today is very important to create an industrial sustainable bioenergy system.

As discussed in previous chapters, algal biomass utilization represents one of the most promising future solutions for the global energy crisis, particularly the problems associated with conventional fuel utilization. In that context, algal growth has only a few requirements including CO_2, sunlight, inorganic nutrients, and water. Algal cells can grow in nonarable land using wastewater or seawater [11]. Algae have a short generation time, and some microalgal species can double their biomass multiple times a day, resulting in a high growth rate with higher macromolecules content, which might reach 30 times higher production of oil per acre than oil crops [9,10]. Many species have shown a wide range of oil contents (20%−60%dw), which can reach to higher levels under stress conditions [12,13]. In that regard, the production of 1 ton of dry microalgal biomass needs 1.83 tons of CO_2 [14], which was reported to be 10−50 times more efficient than higher plants in terms of CO_2 biofixation. Therefore the *"algae for green fuel"* concept has recently gained increasing interest in terms of the circular bioeconomy. In addition, the ability to grow in seawater/wastewater and under severe stress conditions, such as high salinity, reduces the dependence on freshwater, which provides a promising technique for large-scale microalgal biomass production, minimizing the competition on the available resources of water and agricultural land. In that context, recent studies confirmed the efficiency of marine microalgae that can grow optimally at elevated levels of salinity, that is, "halophilic microalgae," for the biodesalination of seawater as well as biodiesel production from the generated algal biomass [11,15].

A previous study evaluated the capability of CO_2 from flue gases to cultivate microalgae, and confirmed overall reduction in CO_2 emissions by 90% [16], suggesting microalgae coupled with biological CO_2 sequestration to be a hopeful approach for biofuel production. However, their efficient implementations in that field are still restricted to some pilot trials, which require further R&D to achieve commercialization. Hence, recent studies have focused on developing new integrated routes of bioconversion and biorefinery technologies that convert the whole algal biomass into different forms of biofuel, such as biodiesel, bioethanol, and biomethane. However, there is a gap in the literature concerning the sequential biofuel production from microalgae, especially those coupled with wastewater treatment. This chapter aims to provide a timely outlook and evaluation of sequential biofuel production from algal biomass and related challenges. The common biofuels through various conversion routes of algal biomass are briefly discussed. In addition, the energy recovery potential from algal biomass through different routes of biofuels production (e.g., biogas, biodiesel, bioethanol, biobutanol, and bioelectricity) as well as suggested integrated routes are evaluated. This work provides a summary of the full utilization of algal constituents through complete conversion of algal biomass into valuable bioproducts and sustainable biofuels with a *"zero-waste"* approach.

Handbook of Algal Biofuels

17.2 Different processes of algal biofuel production

As discussed in previous chapters, different types of biofuels can be produced from algal biomass, depending on the process and the part of the cells used. As the most energetic component, lipids can be extracted and converted to biodiesel through a process similar to that used for any other plant oil. Subsequently, the lipid-free residues can be used for further energy production or other refining processes. Therefore a proper process conversion implemented should be economically viable, with minimum chemical usage, low energy consumption, low toxic by-product effluents, and be easy to be applied at the large scale. To produce biofuel from algal biomass, many processes have been suggested, which include dark fermentation, anaerobic digestion (AD), transesterification of oil, alcoholic fermentation, and algal fuel cells (AFCs) (Fig. 17.1), which are summarized briefly in this chapter.

17.2.1 Biohydrogen

Biohydrogen is an auspicious renewable clean energy source, since it can be applied as a biofuel producing water as the only exhaust by-product without any pollutants or undesired emissions (such as SO_x and NO_x). In addition, it has many advantages as a biofuel, including the possibility to be converted into electricity and high energy density fuel (with HHV of ≈ 142 MJ/kg) [17]. Practically, hydrogen is generated through dark fermentation of organic substrates using hydrogenic bacteria that convert the complex organic compounds into a mixture of hydrogen and organic acids with low molecular weight [18]. Under nonoxygen conditions in a closed culture environment, microalgae can directly use water and sunlight for biohydrogen generation [19]. During the photosynthesis, cells can detach two water molecules throughout photolysis to create four hydrogen ions and one oxygen molecule, where the former can be transformed further to produce two hydrogen molecules [20]. Some microalgae, such as *Scenedesmus obliquus* and *Chlamydomonas reinhardtii*, showed hydrogenase activity for biohydrogen production [21,22]. The oxygen produced has a strong inhibitory effect on hydrogenases, which can be recovered by cultivation of microalgae under sulfur deficiency for 2—3 days, leading to anaerobic conditions for the

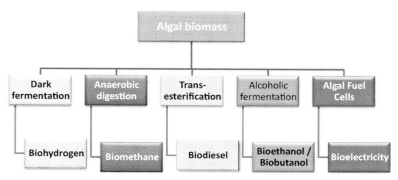

FIGURE 17.1 Different processes of algal biomass conversion into different forms of bioenergy.

production of biohydrogen [23]. Winkler et al. [23] designed a novel bioreactor to cultivate *C. reinhardtii* in the dark for efficient biohydrogen production, and Chatzitakis et al. [24] suggested a photoelectrocatalytic enzymatic hybrid biosystem for concurrent biohydrogen generation and reduction of organic pollutants.

Lee et al. [25] evaluated the dark fermentation of the marine brown seaweed *Laminaria japonica* and achieved a total biohydrogen yield of 444 mL from 10 g/L of dry algae from 100 mL culture grown for 62 h. However, practically, reaching the theoretical maximum yield of 20 g H_2 m^{-2} day^{-1} is not possible, which makes the current large-scale algal bio-hydrogen production unfeasible [26]. Thus one of the main factors limiting the large-scale biohydrogen production is the lack of effective microalgal species to yield the theoretical biohydrogen production in full scale. Therefore it needs further screening studies to select the promising candidate, as well as to improve the cultivation and operating conditions. In addition, genetic engineering to maximize the use of solar energy, CO_2 sequestration, and enhance the hydrogenase activity might help to capitalize on the viable and sustainable hydrogen output. Overall, algal hydrogen production is still some distance from commercial feasibility, but its eventual promise will be testified by continuing success.

17.2.2 Biomethane

Biomethane production through AD is a fascinating technology for simultaneous energy and resource recovery from various organic biowaste streams. AD of biowastes is the process of breaking down the organic compounds by anaerobic microorganisms through four main stages (hydrolysis, acetogenesis, acidogenesis, and methanogenesis) which results in the production of gaseous product, mainly biomethane and CO_2 [27]. AD technology is an attractive algae management strategy for renewable energy production, mainly methane-rich biogas. In addition, biomethane fuel (with HHV of ≈ 55 MJ/kg) is now commonly used in stationary applications, but it is also still suitable for use in road vehicles, often buses, such as liquid natural gas [28]. Potential methane yields of 1.01, 0.42, and 0.50, L g^{-1} VS, respectively, can be produced from lipids, carbohydrates, and proteins [29]. Assuming the biochemical composition of various algal species based on the study of Mussgnug et al. [30], the theoretical biomethane yield from different biomass components is presented in Fig. 17.2. From this figure, it can be determined that proteins play a significant role as a main biogas feedstock in microalgae. Practically, the AD efficiency of algae biomass is strongly correlated with the algal species, pretreatment, and nutrients balance (specially the C/N ratio). A previous study examined six algal species for mesophilic AD and recorded a wide range of actual biomethane yields (from 161 to 435 L CH_4 kg^{-1} volatile solids), confirming also a wide range of conversion efficiency based on the biomass heating value (from 26% to 79%) [31].

Biomethane generation through AD of algal biomass can play a significant role in the restoration of the aquatic ecosystem, through elimination and storage of "hazardous algal blooms" such as cyanobacteria and diatoms in lakes, wetlands, rivers, reservoirs, or seas. This route of biomass utilization may minimize the possible threats, such as water oxygen depletion, blocking of sunlight from reaching fish, and production of toxic metabolites that kill fish, mammals, and birds [28]. However, algal biomass conversion into

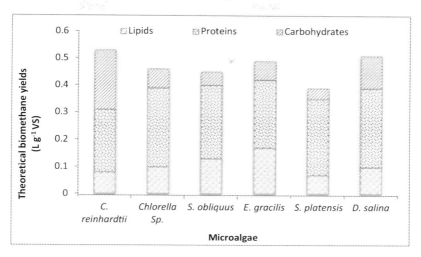

FIGURE 17.2 The theoretical biomethane yield from lipids, proteins, and carbohydrates of various microalgal species as calculated by Angelidaki and Sanders [29]. The data of biochemical composition were extracted from Mussgnug et al. [30].

biomethane is still in the pretrading stages at the moment, due to the high energy consumption for digesters heating and the higher surface areas required to produce 1 MJ of biomethane-based biofuel compared with 1 MJ of algal biodiesel.

In addition, the average hydrolytic retention time (HRT) required to reach a high digestibility during AD is 20–30 days, which increases the costs of biogas production at the large scale [32]. Conversely, from the energy consumption aspect, biodiesel production from algal biomass requires high energy for algal dewatering and oil extraction steps which are not required for biomethane production, making algal biomethane production highly competitive compared to biodiesel [32,33]. On the other hand, algal cells require cell disruption (pretreatment) before injection to AD to decrease the cell wall resistance and augment the solubilization of biomass. Pretreatments such as hydrothermal, mechanical, and chemical treatments, can solve problems, increasing the digestion speed and the total biogas yields from algal biomass [34]. Therefore two main points should be taken into consideration to select the suitable pretreatment method for algal biomethane production. First, low energy consumption in order to avoid the possible negative energy balance of biogas production. Second, the preferred pretreatment can not only improve the solubility of algae biomass, but also improve the biodigestibility [35].

Alternatively, to increase the AD efficiency of algae biomass, the C/N needs to be set in the appropriate range for high biogas yield, because C/N ratio is directly influenced by the supply of macronutrients, ammonia concentration, and AD alkalinity [36]. Zhong et al. [37] proposed 20/1 as an optimum C/N ratio for algae codigestion with maize straw. Park and Li [38] achieved biomethane productivity of 0.54 L CH_4 g^{-1} VS day^{-1} from algal biomass residue codigested with lipid-rich fat, oil, and grease (FOG) waste. In addition, it was observed that codigestion of *Chlorella* sp. with waste sludge increases the biogas yields by 73%–79%, and also improves the dewaterability of digestate effluents [39].

Solé-Bundó et al. [34] examined the effect of thermal pretreatment of microalgae on the codigestion with primary sludge or FOG. The biomethane yield increased by 25% and 42% by adding FOG in a ratio of 10% and 20% VS, respectively. Therefore pretreatment of microalgae and codigestion with other wastes is a promising approach to improve biomethane production from algae.

17.2.3 Biodiesel

Biodiesel is currently the major biofuel in the global market and the most commonly used liquid biofuel [1]. It is a mixture of fatty acid methyl esters or fatty acid ethyl esters produced by the transesterification/esterification of lipids with methanol or ethanol, respectively. In a typical transesterification reaction, oil reacts with alcohol, such as methanol, in the presence of a catalyst, such as KOH or NaOH. Although 3:1 is the alcohol:oil theoretical molar ratio, a higher molar ratio of 6:1 is suggested to accurately complete the reaction [40]. The used lipid feedstocks usually contain 90%−98% (w/w) triacylglycerols (TAGs) and small proportions of free fatty acids, monoacylglycerols (MAGs), diacylglycerols (DAGs), and traces of water, sulfur compounds, phospholipids, and tocopherols [9]. The transesterification is a complex process that takes place in three main stages; TAGs are decomposed to form DAGs, then the latter are decomposed to form MAGs, which are finally converted to biodiesel and glycerol as a by-product. Compared to other renewable biofuels, biodiesel is advantageous because it is eco-friendlier, cleaner burning, and can be used directly in the present engines' infrastructures. In addition, it has a low oxygen content of about 10% with no aroma nor sulfur, resulting in full clean burning. Moreover, its combustion dramatically eliminates the emissions of unburned hydrocarbons, carbon monoxide, and the particulate substances including smut. Furthermore, its lubrication is a superior characteristic over other liquid biofuels, which is attributed to the small glycerin residues. Like diesel fuel and due to its high oxidative stability, biodiesel can be stored safely using the present infrastructures, but has high ignition quality due to the higher cetane number [6].

The use of edible oil plants (mainly rapeseed and soybeans) to produce biodiesel has a domino effect, as it has doubled the prices of oilseeds during the last decade [41]. This causes the food chain to fluctuate, the foodstuffs prices can double quickly, and few oilseeds will be available for emergency food assistance. Stephenson et al. [42] studied the biodiesel production from rapeseed and reported 1.4 tons ha^{-1} y^{-1} oil productivity. Hill et al. [43] reported an oil productivity from soybean by 0.48 tons ha^{-1} y^{-1}, while Upham et al. [44] estimated oil productivity from *Jatropha* to be 2.4 tons ha^{-1} y^{-1}. In a recent study, Abomohra et al. [10] cultivated *S. obliquus* in wastewater at a pilot scale to compare the biodiesel production with oilseeds. The study reported that the annual biodiesel productivity from *S. obliquus* could be nearly five times as much than that of *Jatropha*. This finding supports microalgae to be at the top of biodiesel feedstocks that can effectively substitute the fossil diesel without competition with human food [9,45]. In addition, several screening experiments have shown that green microalgae are the largest group from a taxonomic aspect from which the most promising biodiesel candidates have been identified [46−48]. It might be attributed to the fact that many green algae can be isolated easily from diverse habitats and grow faster than other species in varied conditions. Thus both lipid content and growth for biomass production are the main parameters that simultaneously affect the lipid productivity of a certain algal species [13,46,49].

Despite the advantages of microalgae, the major hindrance for commercial biodiesel production is the elevated costs due to harvest and lipid extraction. In addition, a recent life cycle assessment study [50] pinpointed the main challenges for commercial large-scale production of algal biodiesel. According to the study, the major bottleneck is the high energy required for different stages of biomass production including cultivation, harvest, and drying. Compared to raceway ponds, all the normalized photobioreactors models have higher biomass productivity but consume more energy. In tubular photobioreactors, the energy used to only circulate the algal culture was estimated to be 86%−92%, compared to 22%−79% in raceway ponds [51]. In addition, the system construction and nutritious costs are among the other parameters that increase the final cost of the biodiesel [51]. Therefore the reduction of the overall cost through new approaches without biomass drying, such as in situ hydrothermal liquefaction [2,50], or through development of efficient dryers [52], is receiving great attention. In addition, microalgal biomass can be handled directly as a whole culture (wet processing) for the production of different bioliquids, reducing the cost required for harvest and dewatering stages [53,54]. Hence, there is still an urgent need to enhance the technoeconomic performance of algal biodiesel production in order to make the process less energy intensive and more commercially viable.

17.2.4 Bioethanol/biobutanol (bioalcohols)

Alcoholic biofuel represents one of the main renewable clean fuels in the global market [55]. Ethanol is recognized as an octane booster for gasoline, which decreases the gasoline consumption by 3.0%−4.4% when blending up to 40% ethanol, thus increasing the performance of the internal combustion engine and minimizing the annual CO_2 emissions by 20−30 tons [56]. Bioethanol can be generated from carbohydrate-rich algal biomass through three steps: hydrolysis of algal biomass to produce simple monomers including glucose and xylose, which are fermented by yeast; then the bioethanol can be obtained by a distillation process, producing a liquor effluent called stillage as a by-product [57,58]. Due to the high efficiency of the fermentative yeast at mild conditions, the algal bioethanol production is very competitive which results in high bioethanol output at low energy consumption. On the other hand, the key limitations of this process are the high cost and its complexity due to the use of expensive hydrolytic enzymes for biomass hydrolysis [59]. Harun et al. [60] studied bioethanol production from *Chlorococcum infusionum* using *Saccharomyces cerevisiae* and recorded the highest bioethanol yield of 0.26 g/g at 120°C for 30 min after alkaline pretreatment using 0.75% (*w/v*) NaOH. Since lipid-based microalgal biodiesel production has been broadly applied, the lipid-free microalgal biomass residues should be taken into consideration to enhance the overall bioenergy recovery.

Biobutanol is a four-carbon alcohol ($C_4H_{10}O$) that has been proposed as a renewable liquid fuel produced from a special fermentation route of biomass. The production of biobutanol by biological means was first discovered in 1861 by Louis Pasteur. However, the process was further modified to coproduce acetone, ethanol, and butanol simultaneously. Therefore biological

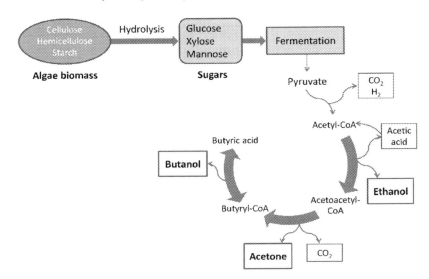

FIGURE 17.3 The simplified metabolic pathways of clostridial acetone–butanol–ethanol fermentation [63].

production of these fuels is commonly called "Acetone–Butanol–Ethanol" (ABE) fermentation, with around 6:3:1 ratio [61,62]. The fermentation of ABE (Fig. 17.3) can be conducted using saccharolytic butyric acid-producing bacteria, such as *Clostridium acetobutylicum*, which is also called the "Weizmann organism" as it was developed by the chemist Chaim Weizmann. Unlike yeasts, clostridia can generate alcohols from a wide variety of carbohydrates, primarily using hexoses or pentoses as the organic carbon source [64]. Because butanol is a totally nonpolar hydrocarbon, it has a similar chemical composition to gasoline, making it suitable for operating in gasoline engines without any mechanism alterations or engine modification [65]. Biobutanol has higher vaporization temperature, low volatility, high miscibility, high energy content (33.07–36.1 MJ/kg), and a high density of 810 kg/m^3 [65]. Hence, a butanol–gasoline blend didn't show cold start problems, and pure biobutanol can be used instead of gasoline [66].

The ABE fermentation process is bacterial anaerobic metabolism of the normal butyric-type fermentation that can be divided into two main phases: acidogenesis and solventogenesis. In acidogenesis, clostridial bacteria utilize monosaccharides for their growth, producing butyric and acetic acids along with hydrogen (H$_2$) and carbon dioxide (CO$_2$) gases. This process is accompanied by the solventogenic phase, with organic acid transformation into acetone, butanol, and ethanol (Fig. 17.4) in a typical ratio of 3:6:1, as by-products [63]. In the early 1980s, extensive studies focused on this fermentation with different aims to make butanol fermentation commercially feasible. The problems associated with this kind of fermentation include the elevated butanol recovery cost using distillation technology, the low concentration of the produced butanol (<20 g/L), and the low production rate (<0.50 g/Lh) due to the inhibition of ABE fermentation by the products [61]. The main limiting factor is the toxicity and inhibition of butanol production in the ABE fermentation system. Various strategies have been

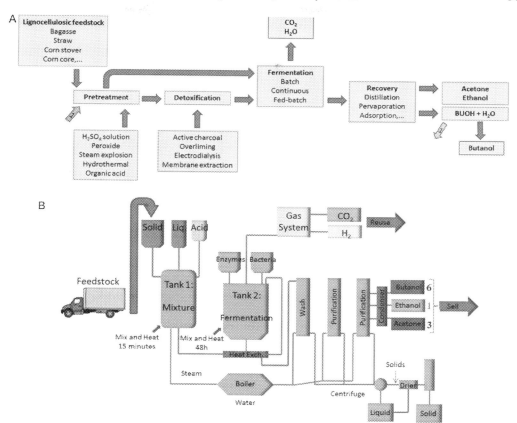

FIGURE 17.4 A flowchart of the biobutanol production process (A) and a schematic of a suggested biobutanol production unit (B). [66] *Source: Modified from W.R. da Silva Trindade, R.G. dos Santos, Review on the characteristics of butanol, its production and use as fuel in internal combustion engines. Renew. Sustain. Energy Rev.69 (2017) 642–651 after the required copyright permission (No. 4923561252157).*

developed to achieve in situ biobutanol recovery to simultaneously alleviate butanol toxicity and improve the efficacy of ABE fermentation. Such strategies include solvent extraction, pervaporation, adsorption, reverse osmosis, and gas stripping [67]. Among these, gas stripping is an effective, low-energy technique that is simple to incorporate within the ABE fermentation process for in situ recovery of butanol from the fermentation medium to relieve its toxicity and prevent the inhibitory effect [68].

Most of algae, including cyanobacteria, store carbohydrates in the form of starch. *Scenedesmus* sp., *Dunaliella* sp., *Chlorella* sp., and *Chlamydomonas* sp. have been reported as promising green microalgae for industrial applications, with more than 50%dw digestible starch, glycogen, and cellulose [69]. In contrast to lignocellulosic biomass that are the main second-generation biofuel feedstocks, pentosaccharides such as mannose, xylose, and galactose are found in small amounts in algal cells [6]. However, direct conversion of the intact algal cells reduces the bioconversion rate and produces poor biobutanol yield [70]. Thus a suitable pretreatment step is essential to reduce the inhibitors release and enhance the subsequent hydrolysis and fermentation, increasing the yield of the target product [6].

So far, few studies were done on biobutanol production from microalgae. Different methods can be used for the pretreatment including heat treatment under pressure, hydrolysis/saccharification, and nourishment [71]. However, saccharification was recommended as the most efficient and commonly used method for complex carbohydrates conversion into simple fermentable sugars. Therefore commercial production of biobutanol can be achieved through three main stages: (1) biomass pretreatment, (2) ABE fermentation, and (3) product separation (Fig. 17.4). Ellis et al. [72] studied biobutanol production from algal biomass pretreated with 10% (w/v) of diluted acid and base via ABE fermentation using *Clostridium saccharoperbutylacetonicum* N1–4. The pretreatment resulted in ABE yield of 2.74 g/L, while pretreatment using enzymatic hydrolysis (cellulases and xylanases) showed 9.74 g/L. Thus enzymatic pretreatment remains superior from the yield aspect, while the high cost limits its commercialization. Generally, the pretreatment cost for enhanced biobutanol production varies influentially, representing about 40%–70% of the product selling prices [73]. Thus selection of a proper method for algal pretreatment and saccharification requires to take the final cost into consideration. There has been a continued improvement in biobutanol technologies in terms of production cost and return of investment. Green [74] expected that biobutanol had the potential to replace both bioethanol and biodiesel in the approximate $247 billion market by 2020. In addition, the future global production of biobutanol, to be marketed as a chemical and part of the fuel industry, was expected to significantly rise exponentially by 2020 (from 25 million gal in 2008) [75].

17.2.5 Algal fuel cells

Algal fuel cells are bioelectric instruments that transform light and chemical energy into bioelectrical energy using photosynthetic microalgae [76]. A typical AFC is composed of a cathode and an anode bound by an external electrical circuit and internally divided by a membrane in which algal growth is evaluated. Growing algae in AFC (Fig. 17.5) can create bioelectricity where, in the presence of sunlight, the microalgae behave as a biocatholyte, assimilate CO_2, receive electrons, and produce O_2 during the normal photosynthetic process.

Microalgae are eco-friendly microorganisms that have relatively higher photosynthetic rate, rapid growth, and also provide a variety of biofuels due to their unique cellular composition. Therefore novel designs have been developed and the cell operation has been moved from microbial fuel cells (MFCs) toward AFCs for generating bioelectricity and biomass through photosynthesis [76]. Many studies recommended the MFCs using the right microalgal strain to maximize the bioelectricity production. Innovative designs were suggested with different electrode materials to enable the microalgae in a MFC for enhanced electricity generation. A recent study developed a cost-effective photosynthetic AFC prototype with highly reproducible electrochemical properties that can be used for photosynthetic electrogenic activity. Among the

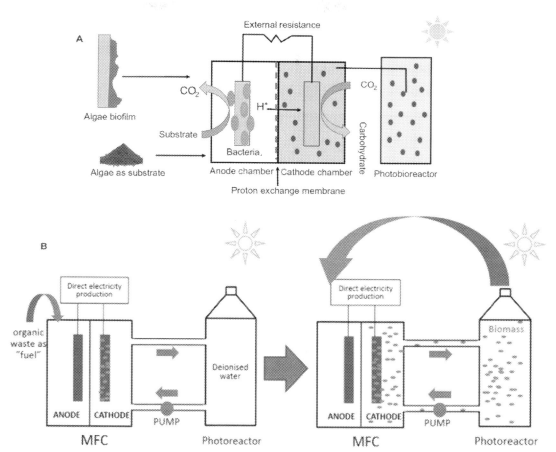

FIGURE 17.5 Configuration of algal fuel cells (A). [77, 76] Source: *Adopted from B. Saba, A.D. Christy, Bioelectricity generation in algal microbial fuel cells. Handb. Algal Sci. Technol. Med., Academic Press, 2020 after the required permission (No. 4923590602980), and microbial Fuel Cell (MFC) with microalgal biomass harvesting photobioreactor adopted from the Open Access I. Gajda, J. Greenman, C. Melhuish, I. Ieropoulos, Self-sustainable electricity production from algae grown in a microbial fuel cell system. Biomass Bioenergy 82 (2015) 87−93.*

different screened microalgal species, *Paulschulzia pseudovolvox* (Chlorophyceae) was recognized as a strong electrogenic species [78].

In AFCs, microalgae and bacteria can be grown together in a single chamber, which doesn't require a membrane, wherein a microalgal biofilm is formed on the anode and some are usually configured with an air cathode [79]. Compared to other configurations, the single AFC chamber is cost-effective, easy to manage, to operate, and to scale up for commercial usability. In general, the energy analysis of AFCs is expressed in normalized energy recovery as "kWh/m^3" based on the size of the used bioreactor. Due to easy degradation, simple anode substrates are recommended to generate more bioelectricity than complex substrates. For instance, the strength of acetate is significantly greater than that of glucose (<0.03), sucrose (<0.01) and complex wastewater (<0.01). Similarly, for acetate, glucose, and wastewater, the total normalized energy recovery is 0.25, 0.18 and 0.04 kWh/m^3, respectively [80].

Kakarla and Min [81] cultivated *S. obliquus* in a cathode chamber to provide oxygen for the sustainable operation of MFC. The MFC with algal aeration contained in the polarization test exhibited a maximum power density of 153 mW/m^2, which was 32% higher than the value with mechanical aeration (116 mW/m^2). Gajda et al. [76] evaluated a fully biotic biosystem that is capable of simultaneously producing bioelectricity with the production of algal biomass in the cathodic photoreactor (Fig. 17.5B). The algae operations assisted the in situ oxygen reduction through providing active oxygenation for the reaction of oxygen reduction with simultaneous CO_2 capture. In addition, the MFC process provided microalgal growth cations in the cathode of the MFC. The study concluded that the development of the MFC cathode with algae for wastewater treatment has a considerable potential.

Saar et al. [82] developed a recombinant blue—green alga (*Synechocystis* sp. PCC 6803) that have been genetically engineered to increase the characteristics of microscale flow-based design and electron export in order to allow an independent optimization of charging and power production processes. In addition, it allows membrane-free operation through the separation of both anolyte and catholyte streams using laminar flow. This, along with a novel design, made it possible to create a biophotovoltaic cell with a power density of 0.5 W/m^2, five times that of the previously described biophotovoltaic devices. Although this is still just around one-tenth of the power density given by traditional solar fuel cells, these modern biophotovoltaic devices have some attractive features. AFCs involve efficient nutrient cycling techniques and light capturing coupled with electricity generation by the system, which is relatively more cost-effective than photobioreactors for large-scale use. Nevertheless, it still requires a high-cost infrastructure and energy input to harvest the biomass and scale up. In addition, the pH membrane gradient, which decreases the power output and cell voltage, is another issue associated with MFCs. Another obstacle is the need for enhanced innovation of AFC's downstream algal biofuel processing for renewable energy production. Therefore integrated AFCs with other energy bioprocesses such as biogas, biohydrogen, and algal biorefinery will increase the overall energy recovery and the economic value. In addition, process modification for simultaneous nutrient recovery, seawater desalination, and valuable chemicals production will also increase the AFCs benefits. Thus application of a single biofuel route from algal biomass is not economically feasible and recent trends in this field focus on the sequential biofuel production.

17.3 Recent trends in sequential algal biofuel production

In recent years, there have been multiple techniques for wastewater treatment/remediation through algae cultivation with simultaneous production of biofuels that have been presented and discussed. In addition, sequential biofuel production has been employed to enhance the efficiency of the microalgal biofuel industry. A modern approach to full energy recovery from microalgal biomass is biomass conversion into biofuels at lower waste production rate. The selection of a suitable algal biomass feedstock depends on its biochemical composition, mainly carbohydrates and lipids as the energy products, which

17.3 Recent trends in sequential algal biofuel production

are varied based on the algal species and growth conditions. During screening studies, the algal biomass with high lipid productivity might be not the same with the high carbohydrate yield. Therefore selection should be done based on both carbohydrates and lipids content, which was calculated for the first time as Relative Increase in Energy Compounds (RIEC, %) by Osman et al. [83] using Eq. 17.1;

$$\text{RIEC} = \left[\left(\frac{P_H - P_L}{P_L} \right)_{\text{Carb}} + \left(\frac{P_H - P_L}{P_L} \right)_{\text{Lip}} \right] \times 100 \tag{17.1}$$

where P_L represents the lowest recorded productivity of lipids (Lip) and carbohydrates (Carb) in all species, while P_H represents the productivities of the studied organism.

Some studies have been done on the dual utilization of microalgal biomass for biodiesel production followed by carbohydrates fermentation of the microalgal residue to other biofuels such as ethanol, biobutanol, and biogas [6,83]. For instance, Xu et al. [6] evaluated the bioenergy recovery from the whole biomass of S. obliquus through production of biodiesel followed by bioethanol through sequential lipid extraction for biodiesel production followed by fermentation for bioethanol production, with concurrent recycling of the waste glycerol (WG) as a nutrient. Seven different routes were studied and the maximum energy output of 21.4 GJ ton^{-1} dry microalgae was recorded from the integrated route, where microalga was cultivated in WG-enriched growth medium with sequential biodiesel and bioethanol production. Thus the fermentation of lipid-free algal biomass improved the rate of ethanol production and the overall energy output. El-Dalatony et al. [57] implemented a novel optimized approach to achieve remarkable biomass conversion from *Chlamydomonas mexicana* through sequential phases. The first phase involved carbohydrate fermentation for bioethanol production, while the second phase concentrated on the production of higher alcohols through protein fermentation. The third phase included the transesterification of the remaining lipid-rich biomass for biodiesel production (Fig. 17.6). The study showed that 0.5 g of ethanol g^{-1} sugar was produced through the first phase (carbohydrate yeast fermentation), 0.37 g of higher alcohols g^{-1} amino acids as produced from the protein portion through the second phase (fermentation of biomass residue from the first phase), and 0.5 g of biodiesel g^{-1} fatty acids was produced by transesterification of the lipid portion. In addition, sequential fermentations resulted in unparalleled microalgal biomass conversion efficiencies (89%) to useful biofuels with the highest energy recovery of 46% of the microalgal biomass.

In addition, Elshobary et al. [84] also evaluated eight seaweeds species collected from the coastal area of Marsa Matruh, Egypt for sequential biofuel production. They studied three routes for direct production of biodiesel or bioethanol from the whole macroalgal biomass (R1 and R2, respectively), and sequential biodiesel and bioethanol (R3). The study concluded that R3 showed the highest energy output of 9.96 MJ/kg, which represented 28.3% efficiency, and six times higher than R2 and R1. In the recent study conducted by Osman et al. [83], 22 seaweeds were collected from Abu Qir Bay, Alexandria, Egypt in different seasons and screened for sequential biofuel production. *Ulva intestinalis* was selected as a promising candidate because it showed the highest RIEC value among the studied species. Further, it showed the highest gross energy output of 3.44 GJ ton^{-1} dw through a sequential route of biodiesel and bioethanol production.

Handbook of Algal Biofuels

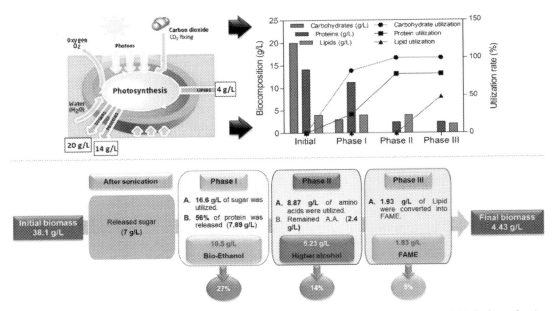

FIGURE 17.6 Mass balance of the *Chlamydomonas mexicana* biomass during sequential biofuel production through three phases. [57] Source: M.M. El-Dalatony, E.-S. Salama, M.B. Kurade, K.-Y. Kim, S.P. Govindwar, J.R. Kim, et al. *Whole conversion of microalgal biomass into biofuels through successive high-throughput fermentation*. Chem. Eng. J. 360 (2019) 797–805 after the required permission (No. 4924080138146).

Chalima et al. [85] examined the ability of a *Crypthecodinium cohnii* to grow on dark fermentation effluent for omega-3 fatty acids production. They found that *C. cohnii* after 60 h of fed-batch cultivation, fully consumed the organic acids from the ultrafiltrated dark fermentation effluent. In addition, docosahexaenoic acid represented as high as 29.8% of the total fatty acids, thereby offering a bioprocess capable not only of reducing the environmental toxins, but also providing a solution for renewable energy combined with valuable compounds production.

In a recent study, Abomohra and Almutairi [5] suggested a novel biorefinery approach to enhance the energy recovery from microalgae and seaweeds through integration of biodiesel production from microalgae with agar- and biogas-production from seaweeds. The study concluded an energy output of 3969.9 and 5098.9 MJ ton^{-1} of dry *S. obliquus* grown on wastewater (WW) and supplementation of the WW with 20% (v/v) of anaerobic digestate, respectively. In addition, extraction of agar from seaweeds before AD followed by anaerobic digestate recirculation for microalgal growth enhanced the overall energy recovery from the biomass.

Fradler et al. [86] investigated an integrated anaerobic biorefinery route including dark fermentation, AD, and MFC for converting municipal biowaste to useful bioenergy carriers. The study found that MFC could reduce the chemical oxygen demand (COD) and recover further bioenergy from the effluent by-product of a two-stage (biomethane and biohydrogen) system. The power yield of 3.1 W/m^{-3} was generated at organic loading rates of 0.572 g COD L^{-1} d^{-1}, which resulted in the highest coulombic efficiency (60%)

and energy conversion efficiency (81%). Accordingly, integrated AFCs with other anaerobic bioprocesses such as biomethane, biohydrogen, and algal biorefinery will increase the overall energy recovery and the economic value.

Li et al. [87] applied an integrated dark fermentation and/or digestion of biowastes with microalgae cultivation, thus being a very promising technology relative to the traditional methods of biofuel development. The usage of effective innovation techniques for the assessment of large-scale crops to further improve algal bioenergy applications capacity as well as to eliminate toxins from sludge digestion was also addressed to make the integrated framework more practical [87]. The economic viability of biochar generated from residues of sludge/microalgae for integrated application in the supercapacitors, adsorber, and catalyst has also been studied. In addition, the reaction mechanisms were also addressed in these systems (Fig. 17.7). In this system, the composition of anaerobic digestate, including TN, TP, COD, and trace elements, significantly affects the microalgal growth under favorable environmental conditions and CO_2 concentrations. The microalgae will utilize these pollutants to produce the biomass, which can be sequentially converted to bioenergy through the different aforementioned routes. It is of note that this analysis will encourage the production and productivity of the reuse of waste in the production of an economically sustainable optimized energy system. Overall, biofuel production using an integrated route is very promising, however, the research on that topic is still enormously limited. In order to increase the applicability of the integrated system, further study is required of the

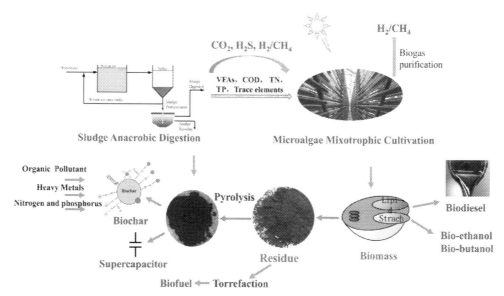

FIGURE 17.7 Integrated microalgae biorefinery concept with dark fermentation and/or anaerobic digestion of biowastes. [87] Source: S. Li, S.-H. Ho, C. Wang, Y.-C. Lin, D. Nagarajan, J.-S. Chang, et al., Integration of sludge digestion and microalgae cultivation for enhancing bioenergy and biorefinery. Renew. Sustain. Energy Rev., 96 (2018) 76−90 after the required permission (No. 4924100229372).

utilization of suitable environmental engineering strategies and life cycle assessments of pilot- and large-scale operations to further enhance the microalgal potential to produce bioenergy accompanied with pollutant removal.

17.4 Conclusion

The recent emerging strategy for enhanced biofuel production from algal biomass is to develop new biorefinery routes and biotransformation technologies to convert the whole algal biomass into different forms of biofuel like biodiesel, bioethanol, biobutanol, biomethane, and bioelectricity. Although sequential conversion improved the gross energy output, studying the environmental impact and life cycle assessment of the different suggested conversion routes are of great importance for future application. In addition, the engineering challenges related to sequential biofuel production from algal biomass need to be extensively studied in order to achieve a *"zero-waste"* approach.

References

[1] A. Abomohra, M. Elsayed, S. Esakkimuthu, M. El-Sheekh, D. Hanelt, Potential of fat, oil and grease (FOG) for biodiesel production: A critical review on the recent progress and future perspectives, Prog. Energy Combust. Sci. 81 (2020) 100868.

[2] A.E.-F. Abomohra, X. Zheng, Q. Wang, J. Huang, R. Ebaid, Enhancement of biodiesel yield and characteristics through in-situ solvo-thermal co-transesterification of wet microalgae with spent coffee grounds, Bioresour. Technol. (2020) 124640.

[3] R. Ebaid, H. Wang, C. Sha, A.E.-F. Abomohra, W. Shao, Recent trends in hyperthermophilic enzymes production and future perspectives for biofuel industry: a critical review, J. Clean. Prod. 238 (2019) 117925.

[4] M. Elsayed, Y. Ran, P. Ai, M. Azab, A. Mansour, K. Jin, et al., Innovative integrated approach of biofuel production from agricultural wastes by anaerobic digestion and black soldier fly larvae, J. Clean. Prod. 263 (2020) 121495. Available from: https://doi.org/10.1016/j.jclepro.2020.121495.

[5] A.E.-F. Abomohra, A.W. Almutairi, A close-loop integrated approach for microalgae cultivation and efficient utilization of agar-free seaweed residues for enhanced biofuel recovery, Bioresour. Technol. (2020) 124027.

[6] S. Xu, M. Elsayed, G.A. Ismail, C. Li, S. Wang, A.E.F. Abomohra, Evaluation of bioethanol and biodiesel production from Scenedesmus obliquus grown in biodiesel waste glycerol: a sequential integrated route for enhanced energy recovery, Energy Convers. Manag. 197 (2019). Available from: https://doi.org/10.1016/j.enconman.2019.111907.

[7] S. Wang, S. Zhao, B.B. Uzoejinwa, A. Zheng, Q. Wang, J. Huang, et al., A state-of-the-art review on dual purpose seaweeds utilization for wastewater treatment and crude bio-oil production, Energy Convers. Manag. 222 (2020) 113253.

[8] J. Lü, C. Sheahan, P. Fu, Metabolic engineering of algae for fourth generation biofuels production, Energy Env. Sci. 4 (2011) 2451−2466.

[9] A.E.-F. Abomohra, W. Jin, R. Tu, S.-F. Han, M. Eid, H. Eladel, Microalgal biomass production as a sustainable feedstock for biodiesel: current status and perspectives, Renew. Sustain. Energy Rev. 64 (2016) 596−606.

[10] A.E.F. Abomohra, M. El-Sheekh, D. Hanelt, Pilot cultivation of the chlorophyte microalga Scenedesmus obliquus as a promising feedstock for biofuel, Biomass Bioenergy 64 (2014). Available from: https://doi.org/10.1016/j.biombioe.2014.03.049.

[11] A.E.-F. Abomohra, A.H. El-Naggar, S.O. Alaswad, M. Elsayed, M. Li, W. Li, Enhancement of biodiesel yield from a halophilic green microalga isolated under extreme hypersaline conditions through stepwise salinity adaptation strategy, Bioresour. Technol. (2020) 123462.

Handbook of Algal Biofuels

References

[12] M. El-Sheekh, A.E.-F. Abomohra, D. Hanelt, Optimization of biomass and fatty acid productivity of Scenedesmus obliquus as a promising microalga for biodiesel production, World J. Microbiol Biotechnol. 29 (2013). Available from: https://doi.org/10.1007/s11274-012-1248-2.

[13] J.Q.M. Almarashi, S.E. El-Zohary, M.A. Ellabban, A.E.F. Abomohra, Enhancement of lipid production and energy recovery from the green microalga Chlorella vulgaris by inoculum pretreatment with low-dose cold atmospheric pressure plasma (CAPP), Energy Convers. Manag. 204 (2020). Available from: https://doi.org/10.1016/j.enconman.2019.112314.

[14] Y. Chisti, Biodiesel from microalgae beats bioethanol, Trends Biotechnol. 26 (2008) 126−131.

[15] E. Sahle-Demessie, A.A. Hassan, A. El Badawy, Bio-desalination of brackish and seawater using halophytic algae, Desalination 465 (2019) 104−113.

[16] J.N. Rosenberg, A. Mathias, K. Korth, M.J. Betenbaugh, G.A. Oyler, Microalgal biomass production and carbon dioxide sequestration from an integrated ethanol biorefinery in Iowa: a technical appraisal and economic feasibility evaluation, Biomass Bioenergy 35 (2011) 3865−3876. Available from: https://doi.org/10.1016/j.biombioe.2011.05.014.

[17] M.E. Nissilä, Y.-C. Li, S.-Y. Wu, C.-Y. Lin, J.A. Puhakka, Hydrogenic and methanogenic fermentation of birch and conifer pulps, Appl. Energy 100 (2012) 58−65.

[18] R. Moscoviz, E. Trably, N. Bernet, H. Carrère, The environmental biorefinery: state-of-the-art on the production of hydrogen and value-added biomolecules in mixed-culture fermentation, Green. Chem. 20 (2018) 3159−3179.

[19] A. Sharma, S.K. Arya, Hydrogen from algal biomass: a review of production process, Biotechnol. Rep. 15 (2017) 63−69.

[20] H. Sakurai, H. Masukawa, M. Kitashima, K. Inoue, Photobiological hydrogen production: bioenergetics and challenges for its practical application, J. Photochem. Photobiol. C. Photochem Rev. 17 (2013) 1−25.

[21] J.E. Meuser, G. Ananyev, L.E. Wittig, S. Kosourov, M.L. Ghirardi, M. Seibert, et al., Phenotypic diversity of hydrogen production in chlorophycean algae reflects distinct anaerobic metabolisms, J. Biotechnol. 142 (2009) 21−30.

[22] L. Florin, A. Tsokoglou, T. Happe, A novel type of iron hydrogenase in the green algascenedesmus obliquus is linked to the photosynthetic electron transport chain, J. Biol. Chem. 276 (2001) 6125−6132.

[23] M. Winkler, A. Hemschemeier, C. Gotor, A. Melis, T. Happe, [Fe]-hydrogenases in green algae: photofermentation and hydrogen evolution under sulfur deprivation, Int. J. Hydrog. Energy 27 (2002) 1431−1439.

[24] A. Chatzitakis, E. Nikolakaki, S. Sotiropoulos, I. Poulios, Hydrogen production using a photoelectrocatalytic−enzymatic hybrid system, Catal. Today 209 (2013) 60−65.

[25] J.-H. Lee, D.-G. Lee, J.-I. Park, J.-Y. Kim, Bio-hydrogen production from a marine brown algae and its bacterial diversity, Korean J. Chem. Eng. 27 (2010) 187−192.

[26] van Iersel S., Gamba L., Rossi A., Alberici S., Dehue B., Van de Staaij J., et al. Algae-based biofuels: a review of challenges and opportunities for developing countries, Italy Food Agric Organ Unites Nations, 2009.

[27] M. Elsayed, A.E.-F. Abomohra, P. Ai, K. Jin, Q. Fan, Y. Zhang, Acetogenesis and methanogenesis liquid digestates for pretreatment of rice straw: a holistic approach for efficient biomethane production and nutrient recycling, Energy Convers. Manag. 195 (2019) 447−456.

[28] L.D. Zhu, E. Hiltunen, E. Antila, J.J. Zhong, Z.H. Yuan, Z.M. Wang, Microalgal biofuels: flexible bioenergies for sustainable development, Renew. Sustain. Energy Rev. 30 (2014) 1035−1046.

[29] I. Angelidaki, W. Sanders, Assessment of the anaerobic biodegradability of macropollutants, Rev. Env. Sci. Biotechnol. 3 (2004) 117−129.

[30] J.H. Mussgnug, V. Klassen, A. Schlüter, O. Kruse, Microalgae as substrates for fermentative biogas production in a combined biorefinery concept, J. Biotechnol. 150 (2010) 51−56.

[31] K.P. Roberts, S. Heaven, C.J. Banks, Comparative testing of energy yields from micro-algal biomass cultures processed via anaerobic digestion, Renew. Energy 87 (2016) 744−753.

[32] P. Collet, A. Hélias, L. Lardon, M. Ras, R.-A. Goy, J.-P. Steyer, Life-cycle assessment of microalgae culture coupled to biogas production, Bioresour. Technol. 102 (2011) 207−214.

[33] P. Collet, A. Hélias, L. Lardon, J.-P. Steyer, O. Bernard, Recommendations for life cycle assessment of algal fuels, Appl. Energy 154 (2015) 1089−1102.

[34] M. Solé-Bundó, M. Garfí, I. Ferrer, Pretreatment and co-digestion of microalgae, sludge and fat oil and grease (FOG) from microalgae-based wastewater treatment plants, Bioresour. Technol. 298 (2020) 122563.

Handbook of Algal Biofuels

[35] H. Tijani, N. Abdullah, A. Yuzir, Integration of microalgae biomass in biomethanation systems, Renew. Sustain. Energy Rev. 52 (2015) 1610−1622.

[36] M. Solé-Bundó, F. Passos, M.S. Romero-Güiza, I. Ferrer, S. Astals, Co-digestion strategies to enhance microalgae anaerobic digestion: a review, Renew. Sustain. Energy Rev. 112 (2019) 471−482.

[37] W. Zhong, Z. Zhang, Y. Luo, W. Qiao, M. Xiao, M. Zhang, Biogas productivity by co-digesting Taihu blue algae with corn straw as an external carbon source, Bioresour. Technol. 114 (2012) 281−286.

[38] S. Park, Y. Li, Evaluation of methane production and macronutrient degradation in the anaerobic co-digestion of algae biomass residue and lipid waste, Bioresour. Technol. 111 (2012) 42−48.

[39] M. Wang, A.K. Sahu, B. Rusten, C. Park, Anaerobic co-digestion of microalgae Chlorella sp. and waste activated sludge, Bioresour. Technol. 142 (2013) 585−590.

[40] M. Canakci, J. Van Gerpen, Biodiesel production from oils and fats with high free fatty acids, Trans. ASAE 44 (2001) 1429.

[41] J. Gressel, Transgenics are imperative for biofuel crops, Plant. Sci. 174 (2008) 246−263.

[42] A.L. Stephenson, J.S. Dennis, S.A. Scott, Improving the sustainability of the production of biodiesel from oilseed rape in the UK, Process. Saf. Env. Prot. 86 (2008) 427−440.

[43] J. Hill, E. Nelson, D. Tilman, S. Polasky, D. Tiffany, Environmental, economic, and energetic costs and benefits of biodiesel and ethanol biofuels, Proc. Natl Acad. Sci. 103 (2006) 11206−11210.

[44] P. Upham, P. Thornley, J. Tomei, P. Boucher, Substitutable biodiesel feedstocks for the UK: a review of sustainability issues with reference to the UK RTFO, J. Clean. Prod. 17 (2009) S37−S45.

[45] M. Ashour, M.E. Elshobary, R. El-Shenody, A.W. Kamil, A.E.F. Abomohra, Evaluation of a native oleaginous marine microalga Nannochloropsis oceanica for dual use in biodiesel production and aquaculture feed, Biomass Bioenergy 120 (2019) 439−447. Available from: https://doi.org/10.1016/j.biombioe.2018.12.009.

[46] A.E.-F. Abomohra, M. Wagner, M. El-Sheekh, D. Hanelt, Lipid and total fatty acid productivity in photoautotrophic fresh water microalgae: screening studies towards biodiesel production, J. Appl. Phycol. 25 (2013). Available from: https://doi.org/10.1007/s10811-012-9917-y.

[47] M. El-Sheekh, A.E.F. Abomohra, H. Eladel, M. Battah, S. Mohammed, Screening of different species of Scenedesmus isolated from Egyptian freshwater habitats for biodiesel production, Renew. Energy 129 (2018) 114−120. Available from: https://doi.org/10.1016/j.renene.2018.05.099.

[48] Y. Li, W. Zhou, B. Hu, M. Min, P. Chen, R.R. Ruan, Integration of algae cultivation as biodiesel production feedstock with municipal wastewater treatment: strains screening and significance evaluation of environmental factors, Bioresour. Technol. 102 (2011) 10861−10867.

[49] Y. Zhang, M. He, S. Zou, C. Fei, Y. Yan, S. Zheng, et al., Breeding of high biomass and lipid producing Desmodesmus sp. by Ethylmethane sulfonate-induced mutation, Bioresour. Technol. 207 (2016) 268−275. Available from: https://doi.org/10.1016/j.biortech.2016.01.120.

[50] N. Misra, P.K. Panda, B.K. Parida, B.K. Mishra, Way forward to achieve sustainable and cost-effective biofuel production from microalgae: a review, Int. J. Env. Sci. Technol. 13 (2016) 2735−2756.

[51] R. Slade, A. Bauen, Micro-algae cultivation for biofuels: cost, energy balance, environmental impacts and future prospects, Biomass Bioenergy 53 (2013) 29−38.

[52] H.S. El-Mesery, A.E.-F. Abomohra, C.-U. Kang, J.-K. Cheon, B. Basak, B.-H. Jeon, Evaluation of infrared radiation combined with hot air convection for energy-efficient drying of biomass, Energies 12 (2019) 2818.

[53] R.R. Soomro, T. Ndikubwimana, X. Zeng, Y. Lu, L. Lin, M.K. Danquah, Development of a two-stage microalgae dewatering process—a life cycle assessment approach, Front. Plant. Sci. 7 (2016) 113.

[54] M. Lakshmikandan, A.G. Murugesan, S. Wang, A.E.F. Abomohra, P.A. Jovita, S. Kiruthiga, Sustainable biomass production under CO_2 conditions and effective wet microalgae lipid extraction for biodiesel production, J. Clean. Prod. (2020) 247. Available from: https://doi.org/10.1016/j.jclepro.2019.119398.

[55] S.R. Chia, H.C. Ong, K.W. Chew, P.L. Show, S.-M. Phang, T.C. Ling, et al., Sustainable approaches for algae utilisation in bioenergy production, Renew. Energy 129 (2018) 838−852.

[56] M. Amine, Y. Barakat, Properties of gasoline-ethanol-methanol ternary fuel blend compared with ethanol-gasoline and methanol-gasoline fuel blends, Egypt. J. Pet. 28 (2019) 371−376.

[57] M.M. El-Dalatony, E.-S. Salama, M.B. Kurade, K.-Y. Kim, S.P. Govindwar, J.R. Kim, et al., Whole conversion of microalgal biomass into biofuels through successive high-throughput fermentation, Chem. Eng. J. 360 (2019) 797−805.

References

[58] M. Elsayed, A.E.-F. Abomohra, P. Ai, D. Wang, H. El-Mashad, Y. Zhang, Biorefining of rice straw by sequential fermentation and anaerobic digestion for bioethanol and/or biomethane production: comparison of structural properties and energy output, Bioresour. Technol. 268 (2018) 183–189.

[59] C.E. de Farias Silva, A. Bertucco, Bioethanol from microalgae and cyanobacteria: a review and technological outlook, Process. Biochem. 51 (2016) 1833–1842.

[60] R. Harun, W.S.Y. Jason, T. Cherrington, M.K. Danquah, Exploring alkaline pre-treatment of microalgal biomass for bioethanol production, Appl. Energy 88 (2011) 3464–3467.

[61] N. Qureshi, T.C. Ezeji, Butanol, 'a superior biofuel' production from agricultural residues (renewable biomass): recent progress in technology, Biofuels, Bioprod. Bioref. Innov. a Sustain. Econ. 2 (2008) 319–330.

[62] A.E.-F. Abomohra, M. Elshobary, Biodiesel, Bioethanol, and Biobutanol Production from Microalgae. Microalgae Biotechnol. Dev. Biofuel Wastewater Treat, Springer, 2019, pp. 293–321.

[63] N. Qureshi, X. Li, S. Hughes, B.C., Saha, M.A. Cotta, Butanol production from corn fiber xylan using Clostridium acetobutylicum, Biotechnol. Prog. 22 (2006) 673–680.

[64] T. Yoshida, Y. Tashiro, K. Sonomoto, Novel high butanol production from lactic acid and pentose by Clostridium saccharoperbutylacetonicum, J. Biosci. Bioeng. 114 (2012) 526–530.

[65] V. Hönig, M. Kotek, J. Mařík, Use of butanol as a fuel for internal combustion engines, Agron. Res. 12 (2014) 333–340.

[66] W.R. da Silva Trindade, R.G. dos Santos, Review on the characteristics of butanol, its production and use as fuel in internal combustion engines, Renew. Sustain. Energy Rev. 69 (2017) 642–651.

[67] C. Lu, J. Dong, S.-T. Yang, Butanol production from wood pulping hydrolysate in an integrated fermentation–gas stripping process, Bioresour. Technol. 143 (2013) 467–475.

[68] D. Cai, Z. Chang, L. Gao, C. Chen, Y. Niu, P. Qin, et al., Acetone–butanol–ethanol (ABE) fermentation integrated with simplified gas stripping using sweet sorghum bagasse as immobilized carrier, Chem. Eng. J. 277 (2015) 176–185.

[69] H.M. El-Mashad, Biomethane and ethanol production potential of Spirulina platensis algae and enzymatically saccharified switchgrass, Biochem. Eng. J. 93 (2015) 119–127.

[70] Y. Wang, W. Guo, C.-L. Cheng, S.-H. Ho, J.-S. Chang, N. Ren, Enhancing bio-butanol production from biomass of Chlorella vulgaris JSC-6 with sequential alkali pretreatment and acid hydrolysis, Bioresour. Technol. 200 (2016) 557–564.

[71] Y. Li, W. Tang, Y. Chen, J. Liu, F.L. Chia-fon, Potential of acetone-butanol-ethanol (ABE) as a biofuel, Fuel 242 (2019) 673–686.

[72] J.T. Ellis, N.N. Hengge, R.C. Sims, C.D. Miller, Acetone, butanol, and ethanol production from wastewater algae, Bioresour. Technol. 111 (2012) 491–495.

[73] V. García, J. Päkkilä, H. Ojamo, E. Muurinen, R.L. Keiski, Challenges in biobutanol production: how to improve the efficiency? Renew. Sustain. Energy Rev. 15 (2011) 964–980.

[74] E.M. Green, Fermentative production of butanol—the industrial perspective, Curr. Opin. Biotechnol. 22 (2011) 337–343.

[75] R. Cascone, Biobutanol-A replacement for bioethanol? Chem. Eng. Prog. 104 (2008). S-4.

[76] I. Gajda, J. Greenman, C. Melhuish, I. Ieropoulos, Self-sustainable electricity production from algae grown in a microbial fuel cell system, Biomass Bioenergy 82 (2015) 87–93.

[77] B. Saba, A.D. Christy, Bioelectricity generation in algal microbial fuel cells, Handbook of Algal Science, Technology and Medicine, Academic Press, 2020.

[78] V.M. Luimstra, S.-J. Kennedy, J. Güttler, S.A. Wood, D.E. Williams, M.A. Packer, A cost-effective microbial fuel cell to detect and select for photosynthetic electrogenic activity in algae and cyanobacteria, J. Appl. Phycol. 26 (2014) 15–23.

[79] H. Jiang, Combination of microbial fuel cells with microalgae cultivation for bioelectricity generation and domestic wastewater treatment, Env. Eng. Sci. 34 (2017) 489–495.

[80] Z. Ge, J. Li, L. Xiao, Y. Tong, Z. He, Recovery of electrical energy in microbial fuel cells: brief review, Env. Sci. Technol. Lett. 1 (2014) 137–141.

[81] R. Kakarla, B. Min, Photoautotrophic microalgae Scenedesmus obliquus attached on a cathode as oxygen producers for microbial fuel cell (MFC) operation, Int. J. Hydrog. Energy 39 (2014) 10275–10283.

[82] K.L. Saar, P. Bombelli, D.J. Lea-Smith, T. Call, E.-M. Aro, T. Müller, et al., Enhancing power density of biophotovoltaics by decoupling storage and power delivery, Nat. Energy 3 (2018) 75–81.

Handbook of Algal Biofuels

404   17. Sequential algal biofuel production through whole biomass conversion

[83] M.E.H. Osman, A.M. Abo-Shady, M.E. Elshobary, M.O. Abd El-Ghafar, A.E.-F. Abomohra, Screening of seaweeds for sustainable biofuel recovery through sequential biodiesel and bioethanol production, 2020. https://doi.org/10.1007/s11356-020-09534-1.

[84] M.E. Elshobary, R. El-Shenody, A.E. Abomohra, Sequential biofuel production from seaweeds enhances the energy recovery: a case study for biodiesel and bioethanol production, Int. J. Energy Res. (2020). Available from: https://doi.org/10.1002/er.6181.

[85] A. Chalima, A. Hatzidaki, A. Karnaouri, E. Topakas, Integration of a dark fermentation effluent in a microalgal-based biorefinery for the production of high-added value omega-3 fatty acids, Appl. Energy 241 (2019) 130–138.

[86] K.R. Fradler, J.R. Kim, G. Shipley, J. Massanet-Nicolau, R.M. Dinsdale, A.J. Guwy, et al., Operation of a bioelectrochemical system as a polishing stage for the effluent from a two-stage biohydrogen and biomethane production process, Biochem. Eng. J. 85 (2014) 125–131.

[87] S. Li, S.-H. Ho, C. Wang, Y.-C. Lin, D. Nagarajan, J.-S. Chang, et al., Integration of sludge digestion and microalgae cultivation for enhancing bioenergy and biorefinery, Renew. Sustain. Energy Rev. 96 (2018) 76–90.

18

By-products recycling of algal biofuel toward bioeconomy

Hanan M. Khairy[1], Heba S. El-Sayed[1], Gihan M. El-Khodary[2] and Salwa A. El-Saidy[2]

[1]National Institute of Oceanography and Fisheries (NIOF), Cairo, Egypt [2]Department of Zoology, Faculty of Science, Damanhour University, Damanhour, Egypt

18.1 Introduction

Increasing world urbanization and industrialization is increasing the interest in finding different renewable energy sources to meet the demands to replace petroleum-derived fuels and products. Global environmental awareness is focused on investigations and research into sustainable biodiesel by utilizing biological wastes and biomasses. Renewable biofuels are crucial energy resources due to helping to protect the environment from terrible carbon emissions. Biomass is the organic material that makes up the Earth's living organisms and stores chemical energy from sunlight. Biomass by-products provide an inexpensive renewable source of biofuels [1]. About 12.2% of global primary energy consumption comes from biomass, and it is considered to be one of the largest energy sources, making up approximately 73.1% of the world's renewable energy [2]. Barampouti et al. [3] presented various technologies of the organic part derived from municipal wastes for biofuels generation. Awasthi et al. [4] reviewed the biorefinery of organic manure to generate sustainable bioproducts. As shown in Fig. 18.1, the biorefinery approach integrated biology, engineering, and ecology leading to useful technology based on algae to produce numerous bioproducts as well as biofuels [5]. The most promising biomass sources are animal, agricultural, and forestry residues, sewage, aquatic crops, and algae. High-value biofuel products are efficiently obtained from biomass, food wastes, microbial-treated wastes, and algal biomass [6]. Biomass-conversion processes are used to generate several bioproducts like biofuels (biobutanol), bioenergy and biomaterials. Energy from biomass can be generated from chemical or thermochemical (pyrolysis and gasification) and biological (fermentation) processes [7]. Catalytic processes are interestingly applied in the development

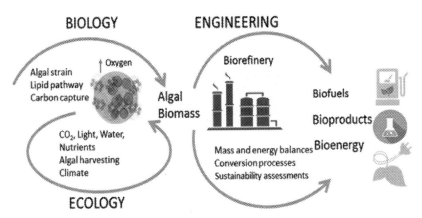

FIGURE 18.1 Algal-based technology is integrated with biological, ecological, and engineering approaches to produce biofuels and bioproducts [5].

TABLE 18.1 Pharmacological effects of algal carbohydrates.

Microalgae species	Carbohydrate type	Pharmacological effect	References
Chlorella spp.	Polysaccharides	Antiinflammatory, immunomodulating, and antioxidant	[14,15]
Gyrodinium impudicum KG-03	Sulfated polysaccharides	Antiinfluenza virus	[16]
Haematococcus lacustris	Water-soluble polysaccharides	Immune stimulating activities	[17]
Phaeodactylum tricornutum	Polysaccharides	Antiinflammatory and immunomodulating activities	[14]
Rhodella reticulate	Extracellular polysaccharides	Antioxidant effect	[18]
Scenedesmus quadricuada	Crude polysaccharides	Antioxidant effect	[15]

of biofuel, including nanocatalysts [8]. Recent enzymatic technologies have been applied in advanced biofuels generation [9]. Ferreira et al. [10] reviewed valuable strategies of numerous by-products from bioethanol waste through the microbial conversion approach. Recent breakthroughs were reported by Delbecq et al. [11] who advanced conversion pathways for the sustainable production of furfural from carbohydrates feedstocks.

Carbohydrates provided by microalgae undergo the fermentation process for bioethanol biosynthesis. These carbohydrates are considered to be an alternative source of carbon instead of lingocellulose biomass [12]. Algal carbohydrates, such as glucose and starch, are mainly utilized for biofuels generation including biohydrogen and bioethanol by fermentation [13].

Algae polysaccharides are potential emulsifiers, stabilizers, textiles, clinical drugs, cosmetics, thickening agents, and water-soluble lubricants. As shown in Table 18.1, polysaccharides exhibit several biomedical applications due to their pharmaceutical properties, encompassing antiinflammatory, antioxidant, and antitumor activities [19]. It has been reported that algal polysaccharides could trigger nitric oxide synthesis, induce reactive oxidative species, cytokines, and the function of macrophages, therefore leading to immune system modulation [20]. Matsui et al. [21] demonstrated that sulfated polysaccharide derived from red *Porphyridium* microalgae showed potential antiinflammatory activities. Furthermore, Tannin-Spitz et al. [22] found that sulfated polysaccharides provided from the cell wall of *Porphyridium* could protect against oxidative stresses. Sulfated polysaccharides from algae exert antiviral activity through their interactions with several molecules in the cellular compartment of the virus [16].

Obtaining biodiesels from different algal species has been reviewed and discussed thoroughly in the literature [23,24]. There are four main groups of microalgae species: *Cyanophyceae*, *Bacillariophyceae*, *Chlorophyceae*, and the golden algae *Chrysophyceae* [25]. The most common microalgal strains that are well-known in the field of biodiesel production are *Isochrysis*, *Chaetoceros*, *Chlorella*, *Spirulina*, and *Dunaliella*.

Microalgae could fix approximately 40% of globally produced carbon, leading to the conversion of inorganic carbons into valuable biochemical organic compounds. Using these organisms gives a great advantage for their capability to convert sunlight, water, and carbon dioxide into different metabolites and chemicals that end up in algal biomass. Microalgae could influence the biofixation of carbon dioxide waste, thus decreasing greenhouse gas emissions [26]. Algae have been extensively reported to produce various biofuels, for instance, biodiesel, bioethanol, biogas, biokerosene, biohydrogen, and bio-oil [27].

The remaining algal biomass after biofuel extraction is mainly made of proteins and carbohydrates. These by-products can be isolated and collected post extraction of biodiesel sequentially or simultaneously. These compounds can be recycled for many applications, including anaerobic digestion or production of methane gas and biohydrogen gas to power energy systems, or production of slurry rich in nutrients, which can then be returned back to algal ponds as biofertilizer. Maurya et al. [28] have reported that the lipid biomass derived from *Lyngbya majuscule* and *Chlorella vanabilis* could decrease the use of chemical fertilizers. Also, these by-products can be used in producing industrial enzymes, cosmetics, pharmaceuticals, and nutraceuticals or animal feedstocks [13,29] (Figs. 18.2 and 18.3).

High carbohydrate and lipid contents in microalgae cells efficiently influence the biofuels yield. Their carbohydrate biomasses, including starch, sugar, and cellulose serve as substrates

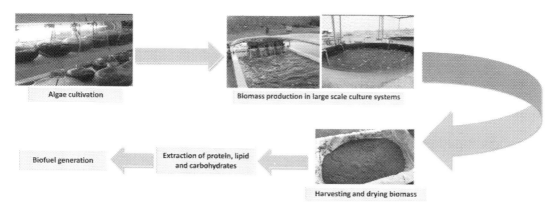

FIGURE 18.2 The generation of biofuel and valuable by-products from microalgae.

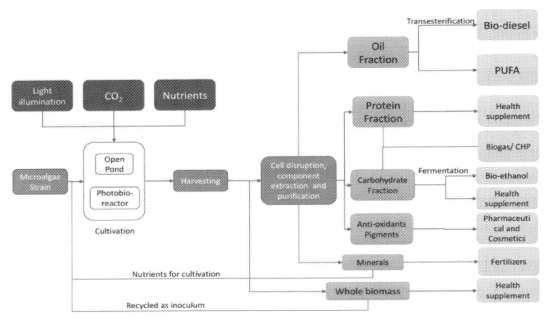

FIGURE 18.3 Diagram showing the biorefinery products from microalgae [13].

for environmental-friendly and commercially viable bioethanol fuels production by using fermentation [30]. Furthermore, microalgal lipid as well as protein constituents produce higher alcohols and biodiesel via transesterification with glycerol as a by-product and the fermentative Ehrlich pathway, respectively. Additionally, microalgae provide other applications in different fields, such as wastewater treatment and pharmaceuticals. Therefore microalgae have become one of the most promising organisms for human beings. Biodiesels are produced by carbohydrates and proteins fermentation, and lipids trans-esterification from microalgal strains, such as *Pseudochlorella* sp. and *Chlamydomonas mexicana* [31]. Ashour et al. [32] have characterized a microalgal strain named *Nannochloropsis oceanica* NIOF15/001, and evaluated its application for using in biodiesel generation and aquaculture feed.

A biological pretreatment may be considered as an eco-friendly approach, including microbial fermentation application [30]. The microalgae potentially generate commercially valuable by-products. Around 30% of their harvested biomass are lipids, and this biomass has potential to be applied in energy-related products, animal feed, biogas, and hydrogen fuel. Generally, by-products of algae are also good sources of food after extraction of their oil contents. Microalgal pigments were reported to be by-products during the generation of biodiesel [33]. Interestingly, the commercial viability of microalgal biofuel depends on the suitable commercial use of their by-products [34].

Economic activity that converts waste products to valuable products, such as bioproducts, bioenergy, and food, is known as bioeconomy in waste management. The utilization of biomaterials in production cycles from the biological cycle provides opportunities for remanufacturing that can be potentially applied in different fields [35]. The major aim of this review is to discuss different by-products that result from biomass conversion, the current status of by-products' applications, the environmental impact of by-products' recycling, the challenges faced to further expand energy derived from biomass, the future perspectives, and the economic feasibility of recycling.

18.2 Applications of algae by-products

Annually, a number of algae are cultivated to produce compounds with beneficial values to humans and animals. The development of numerous bioproducts is an important proposal in the algal field. Regarding the important applications provided by algal by-products, it is essential to focus on their use in economic industries, like food nutrition, pharmaceuticals, and biofuel [36]. Bioproducts of algae show promising prospects due to the valuable metabolites of algae (Fig. 18.4). However, their production costs should be taken into account, mainly regarding cultivation systems, large-scale production systems and the recycling of the end by-products. Various essential compounds can be produced from microalgae, for example, *Nannochloropsis*, *Spirulina*, *Stichococcus*, *Tetraselmis*, and *Pavlova pinguis* contain high levels of vitamins; *Ecklonia radiate* microalgae are good sources of antioxidants; *Phormidium autumnale* produces different pigments; proteins and fatty acids can be obtained from *Chlorella* sp., *Chlorella minutíssima*, *Chlorella saccharophila*, and *Chlorella vulgaris*; and various bioproducts are produced by *Kirchneriella*. All these microalgal species have several commercial applications, and can be potentially applied in biofuel production, as well as the production of other essential compounds and human and animal food [37].

Proteins from microalgae can be utilized in many applications, including as nutraceuticals, probiotics, enzymes, and food additives. Protein recovery from microalgae is gaining particular attention for feedstuffs for in livestock rearing and aquaculture, especially for

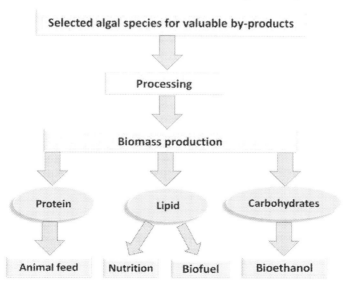

FIGURE 18.4 Different by-products and their applications from microalgal biomass.

fish, shrimp, and mollusks. In the aquaculture field, approximately 40%−70% of the costs required for fish production are the costs of protein in aquaculture feeds [13]. The cyanobacteria *Arthrospira* (*Spirulina*) have a high percentage of protein content with potential as a food supplement for various organisms [38]. Protein waste generated from *Chlorella vulgaris* has been reported to show antioxidant activity against peroxyl radicals and in anticancer activity [39].

Li et al. [40] demonstrated that algae represent an excellent source of natural antioxidants at levels higher than those of some plants due to their enormous biodiversity. The antioxidant status of some microalgae that have been naturally isolated has been identified and characterized. Different compounds can be used as antioxidants, such as sulfated polysaccharides, long-chain fatty acids, phenolics, and pigments. Natural pigments show several biological impacts, such as antiinflammatory, neuroprotective, antiobesity, antioxidants, and anticancer activity [41]. Algae carotenoids can be used in different applications like animal and human food, food colorants, additives, pharmaceuticals, as well as in biomaterials production. In addition, several vitamins can be found in microalgae, including carotene, alpha-tocopherol, thiamin, pyridoxine, biotin, and folic acid [12].

Moreover, microalgal biomass contains various pigments like astaxanthin, chlorophyll, and carotenoids. These pigments can be employed in the pharmaceutical and cosmetic industries [19]. Chlorophyll of microalgae shows potential health benefits due to its antioxidant, antiinflammatory, and antimicrobial activities [42]. β-Carotene is an orange-colored pigment that is extracted from the marine algae *Dunaliella salina* (upto 14% dry weight [43]), and is used as a natural food colorant. Also, it is a precursor of vitamin A that has antioxidant effects [44]. Astaxanthin is a red pigment produced by the freshwater algae, *Haematococcus pluvialis*, which can accumulate upto 4% of its dry weight [45]. It is used primarily as a feed additive. Also, it can be used as a food supplement for humans, and has an extraordinary antioxidant activity [46]. In addition, phycobiliproteins are minor commercial products from microalgae (*Arthrospira* and *Porphyridium*), and are used as food dyes [47]. Also, their extracts from *Nannochloropsis* and *Dunaliella* are used for cosmetics [48]. These pigments exhibit antiobesity, neuroprotective, antioxidative, antiangiogenic, antiinflammatory, and anticarcinogenic properties [49].

Using algae in nutrition offers the advantage of the availability of many essential compounds simultaneously. Barka and Blecker [36] showed that correctly processing some microalgal strains like *Dunaliella, Scenedesmus, Spirulina*, and *Chlorella*, led to an amazing flavor that can be used in different food types. In addition to the richness of protein content in microalgae, microalgae also contain significant amounts of pigments, vitamins, fatty acids, and essential amino acids, thus making these microalgae useful for application in food industries [50]. Bioactive substances derived from microalgae, such as omega-3, have essential roles as secondary metabolites, hence they can be used in several biomedical fields and in the development of pharmaceutical products. Microalgae can be potentially applied for carbon dioxide capture and in the field of bioenergy production. These microalgae can be considered to be promising sources for renewable biofuels production, that include biodiesel resulted from microalgae oil, methane through anaerobic digestion of algal biomass, and biohydrogen by photo bioproduction [51].

Methane and biohydrogen are generated from microalgae biomass via fermentation. The model proposed by Yang et al. [52] showed the two-stage hydrogen generation with methane and hydrogen yields were 393 and 46 mL/g-VS, respectively. Miranda et al. [53] demonstrated that biofilms of microalgae are potentially applied for bioenergy generation and wastewater treatment. Furthermore, Scherer et al. [54] showed that the production of microalgal biomass and cattle manure effluent bioremediation occurred via microalgae cultivation. Moreover, the photosynthetic microalgae have potential for bioremediation and biodegradation applications [55] and can be used as biofertilizers [56].

18.3 Microalgal by-products of biomasses conversion processes

To isolate intracellular metabolites from algae requires extensive processing of upstream and downstream procedures beginning with algal cultivation, harvesting, or biomass recovery, followed by other processing, for instance, drying, dewatering, cell disruption, extraction, and finally with the purification of the end products [57]. Biomasses are nondepleted sources of carbon, showing multiple benefits, from the sequestration of carbon to the production of bioenergy. Biomass utilization as an energy source depends on the maximum conversion yields. Diverse parameters are determined in the choice of the biomass conversion processes, including the required renewable end-product, the quantity and quality of biomass, and the cost of these processes [58]. The use of the biomass is augmented by several conversion pathways and technologies, including physicochemical, biochemical, thermochemical, and microorganism growth platforms, which convert it to sustainable bioproduct streams like bioenergy, biochemicals, and other valuable bioproducts [10]. Selecting a switchable solvent for microalgae oil extraction reduces the production cost, and allows reuse and easy separation from the biomass. Downstream processes of microalgae biodiesel depend on the method and type of catalyst. Effective downstream processing involving the major product purification, biodiesel, glycerol, excess alcohol, by-products, and waste products is important to recycle energy from the whole process. The main challenge in the microalgal application achievement is the evaluation of the method that permits downstream processing procedures, and which is adaptable for biofuel and other by-products production [59]. There are different by-products generated from microalgae biomass during the production of biodiesel, such as the following:

18.4 By-products from ethanol production

Filtration and/or centrifugation are the two main processes used for separation of the nonfermentable products, residual catalyst, and the solvent of the fermentation process. The nonfermentable residual slurry consists of organic acids, proteins, and lipids. These by-products can be used for the production of methane hydrocarbons, which lead to a reduction in the waste by upto 10%. Furthermore, it can be used for cell rupture for the enzymes or proteins release as a by-product [60]. The difficulty of deoxyribonucleic acid (DNA) deliberation is considered as the main obstacle in the cells rupture process because it can cause increase in the suspension viscosity, which in turn causes problem in the pumping system. Therefore using a distillation process is the preferred procedure in ethanol purification with a high percentage of purification (95.6%) without losing the main product [61]. The evaporation method can be used instead of the distillation process, and the evaporated water can be recycled to the system to improve ethanol production without the loss of by-products [62]. Park et al. [63] reported that using spent microalgal biomass post bioethanol residue in methane production led to the recovery of the energy by methane generation increasing by 2.24 times more than the energy generated from bioethanol generation as the main product.

18.5 Glycerol by-products of biodiesel productions

The methyl ester produced as biodiesel from microalgae has a lower density than its glycerol by-products. Therefore it is easily separated by using settling tanks and membrane technology at applied pressure [64]. Gomes et al. [65] showed a stable permeate flux of $78.4 \, kg/m^2$ contains around 99.4% of glycerol retention. The isolated and collected methyl ester from glycerol can be refined in a distillation unit where a high content of the excess alcohol is eliminated, leading to 99.6% pure methyl ester. The washed methyl ester is neutralized and washed, these neutralized methyl esters then undergo a process of drying and are finally ready to be marketed. During the washing process, the residual alcohol and catalysts were eliminated and recycled to the reactor by using a low-velocity water spray injection. The glycerol by-product that is about 10% of the whole products is either purified and sold as crude glycerol or recycled into the system [66]. However, the crude glycerol without a purification process is considered to be a carbon source for microalgal cultivation [67].

18.6 By-products from bio-oil fuel production

Tar generation by the pyrolysis processes influences the produced bio-oil quality that may be related to the polymerization process of fine tar in the reactor. By using filtration processes, this tar can be retained and eliminated. However, the alkaline material concentrated in the tar will dissolve in bio-oil. According to the previous study conducted by Devi et al. [68], the most effective method in the formation of tar is the removal of the tars inside the gasifier. The main challenges faced by the pyrolysis and gasification process is the generation of ash, char, and tar, which can be reduced by using additives like char, dolomite, and olivine inside the gasifier or by filtering through the gas scrubber [69]. During hydrogenation, the fatty acids that are hydrogenated in the biomass cause an increase in the melting point and decrease in the iodine value of the oil. Therefore the biodiesel resulting from hydrogenated oil shows a low degree of polymerization and is more easily oxidized than the nonhydrogenated oil. This is the reason that the hydrogenation process is less common when compared with trans-esterification for biodiesel generation from microalgae. The determination of catalysts that can be adaptable for use in the reaction was the main interest in the trans-esterification process. There are several aqueous by-products generated by the liquefaction process, which are recycled for feeding microalgae in aquacultures [70].

18.7 Microalgal-based protein by-products

Microalgae offer sustainable value-adding processes for biomolecules development for different applications. Proteins account for a high fraction of the growing microalgae biomass, so microalgae have been identified as sources of proteins that can be potentially used in functional foods and nutraceuticals [71]. Proteins are expressed in the chloroplast, mitochondrial, and nuclear compartments isolated from some species of microalgae. Wijffels and Barbosa [72] reported that microalgal biomass production generates 0.4 billion m^3 of biodiesel containing approximately 40% of proteins as by-products. Barreiro et al. [73] reported that protein purification from *Scenedesmus almeriensis* and *Nannochloropsis gaditana* led to an improvement in the yield and quality of biocrude as the major product. Also, Sheih et al. [39] reported the therapeutic effects of anticancer peptides in microalgae protein by-products generated as waste from *C. vulgaris*. Lipid-derived microalgae biomass contains other by-product components, including protein, carbohydrate, and the residual biomass content. These compounds can be generally recovered for methane production and biohydrogen gases to power energy systems [29]. Recyclable surfactants and ionic liquids are preferred for isolating the protein by-product of interest and keeping their functionality. It has been demonstrated that efficient protein extraction to produce lipid-rich biofuel intermediates from *Scenedesmus* occurred through hydrolysis [74].

18.8 Environmental impact of biodiesel and by-products

Biodiesel has arisen as an alternative source of petroleum products due to its nontoxic, renewable, and biodegradable properties and its nonaccumulation of greenhouse gases [75]. Biodiesel has a low content of aromatic compounds and a higher cetane number than petroleum diesel [76]. Furthermore, other benefits of biodiesel include lower harmful exhaust gas emissions and a high lubricity level when compared to the petroleum diesel [77]. Comparing biodiesel products with petroleum diesel fuels, Schumacher et al. [78] reported that using biodiesel led to a decrease in the emissions of hydrocarbon by 45%, the emissions of carbon monoxide by 47%, and a reduction in particulate matter (PM) emission by 66%. PM is a mixture of carbonaceous soot with other solid and liquid material. Demirbas [75] also reported a reduction in carbon and PM emissions by 42% and 55%, respectively. These effects are related to the higher oxygen content and cetane number of biodiesel fuel. Although the impacts of biodiesel use on the environment are very positive, biodiesel increases the emissions of nitrogen oxides. However, Hess et al. [79] reported that modifications in combustion temperatures and injection time resulted in reductions in nitrogen oxides emissions.

Interestingly, microalgae have been of interest to many researchers, industries, and governments due to their economically valuable cellular components. They have potential for several applications in biotechnology, for instance, pharmaceutical products, biofuel, the capture of carbon dioxide, and nutritional value, and for the development of green processes [80]. Biofuels showed significant promise due to the sustained ability of their environment-friendly products. Abundant wastewater liberated from several industries is becoming a fabulous resources for biodiesel, and microalgae are used for toxin bioremediation. In addition, microalgae are potentially used to fix wastewater for biomass production, thus decreasing the costs [81].

Coupling biological waste treatments to bioenergy production can significantly increase the positive impact of microalgae application on the environment. A previous study reported that microalgae have potential regulation in the self-sanitization of organic waters purification and natural wastewater treatment [82]. The microalgae showed applications in the treatment of many industrial and domestic wastewaters [81]. The use of municipal wastewater and dairy manure wastewater has potential for the sustainable production of algal biofuel [83].

18.9 Economic feasibility of microalgae biodiesel

The Environment Committee of the European Parliament reported that the advanced biofuels sourced from seaweed or certain types of waste should account for at least 1.25% of energy consumption in transport by 2020. Microalgae are becoming more relevant in the biofuel industry. The feasibility of using microalgae for biodiesel production, as an alternative to fossil fuels, is being paid important interest, because macroalgae have high lipids content. The major advantages related to the generation of biofuels from microalgae are their high photosynthetic efficiency and high productivity. Production cost is the main point for algae-based biopower feedstock as a high initial investment is required for its generation [84]. Indigenous species have characterized and contributed to higher biomass and lipid production. Having optimal conditions improves the growth of microalgae with a high yield of their biochemical constituents. During the cultivation process, assessing the concentrations and growth rates of microalgae is critical for evaluating the economic feasibility of microalgal yield [85]. Several parameters have to be assessed for macroalgae-based production, including cultivation procedures, time and method of harvesting, seaweed species, and yield per hectare [86]. Microalgae could synthesize very promising, economically important metabolites and by-products that have potential uses in the agricultural, food, biotech, and pharmaceutical industries [87]. Microalgae are better organisms for generating biodiesel due to their rich lipid content, which is significantly higher than other biomasses used for this purpose [88]. However, the study demonstrated by Norsker et al. [89] stated that some technologies must be optimized to decrease the operating costs of using microalgae for biofuel generation. Advances in the biological carbon capture field via microalgae for its application for biodiesel production shed light on the significance of particular key characters and parameters, for instance, metabolic pathways of microalgae, strain selection, identification, production of microalgal biomass, cultivation systems, and also the requirements of the international biodiesel specifications and properties [90].

There is a high variation in the algal biomass, costs and methods that used in the oil extraction for biodiesel production. Conventionally, microalgae species, for instance, *Chlorella* (*C. emersonii* and *C. vulgaris*), *Nannochloropsis*, *Phaeodactylum tricornutum*, and *Tetraselmis suecica*, that were cultivated using glucose as the carbon source, have been reported for biodiesel production [91]. Based on acidic trans-esterification, Miao and Wu

Handbook of Algal Biofuels

[92] reported that biodiesel can be generated from *Chlorella protothecoides*, which is a heterotrophic cultivated microalga, by 100% sulfuric acid. One of the most important economic impacts to decrease biodiesel generation cost, in situ trans-esterification, which is the direct conversion of the biomass oil to biodiesel, has been obtained. Because of the use of a stirring reactor that promoted the mixing and decreased the resistances of mass transfer, about 91% conversion yield was obtained post 8 h at 60°C from *Chlorella* and a conversion yield upto 82% was achieved from *Chaetoceros gracilis* at 80°C [93]. Approximately 63% of the optimum conversion yield of biodiesel was produced with a methanol:oil molar ratio of 56:1. Carrero et al. [94] overcame the disadvantages of homogeneous catalysts by investigating the use of hierarchical zeolites as heterogeneous catalysts.

Norsker et al. [89] estimated the costs needed for biomass production for three production systems, and they reported that the cost of biomass production, including the dewatering process was €15 per kilogram. Optimizing the influential factors, including media constituents, mixing and irradiation resulted in a reduction in the production cost to €0.68 per kilogram. These make microalgae become an important feedstock for biopower generation [95]. Furthermore, the pretreatment also improved the solubilization of lipids and removed microalgal residues that enhanced volatile fatty acids and biomethane generation, which has the ability to reduce the algal biodiesel cost [96]. Several proposals have been presented for biofuels production from microalgae at larger scales with a low cost. In 2017 the Bioenergy Technologies Office presented a model that has ability to supply a million metric tons of ash-free dry weight cultivated algal biomass. In 2020 the Japanese Government introduced biojet fuel for flights, and the volume was determined to be 100,000 to 1 million liters [97]. By 2022, about 20 million metric tons of the sustainable models will be supplied, and in 2030, 5 billion gallons of algal biofuels will be produced per year [98].

18.10 Future research focus and perspectives

Algae growing in large cultures are harvested. Lipids extracted from algal biomass could be converted to biodiesel fuels. A pertinent inquiry is whether biofuels derived from microalgae can be made economically and at a sufficient scale to substitute for petroleum fuel. Various technical challenges will be required to achieve this target. Significant support should be given to the applied as well as the basic research on the engineering of sustainable microalgal systems for biofuels production. The technoeconomic analysis has illustrated that the productivity of algal biomass is the main determinant of the production cost, so more efforts are required to focus on different aspects of algal biology that provide the most important influence on the rate of algal growth and lipid biosynthesis. Furthermore, complete detailed environmental impact analysis is necessary to ensure a smooth commercialization path for large-scale deployment. The net energy recovery from the algal fuel compared with the input of energy supply from fossil fuels to produce renewable fuels should be evaluated and addressed. According to the previous reports on the energy calculations, microalgal biofuels have the ability to be generated sustainably; the energy ratios range from 3.3 to 7.5 depending on various parameters, including algal biomass productivity, biogas yield, algal oil yields, harvesting and wastewater treatment, extraction processes, and recycling [23].

Intensive research is necessary for the scale-up applications of microalgae biomass as biodiesel sources as well as their by-products generated via conversion processes. Based on the current trends and status, areas of interest are proposed to be the following: optimizing the conditions for fresh and marine algae separately; optimizing current pretreatment methods; optimizing the sludge and biomass concentration for codigestion; development of innovative methods of biomass pretreatment to decrease the cost vested on drying; developing a technoeconomic analysis; investigating the effect of upstream and downstream processes; and performing a comprehensive study of energy recovery from all possible routes and valuable by-products. One of them might be using recombinant DNA technology through metabolic engineering toward genetic modifications of the lipid/triglyceride biosynthetic pathways, leading to the production of oil-rich microalgal strains [84].

18.11 Conclusion

Microalgae provide interesting characteristics that qualify them to be potential substitute feedstocks for various environmental and industrial applications. Algae can be considered as the only renewable biodiesel that is able to displace liquid petroleum fuels. Nevertheless, many efforts are needed to assess the challenges of finding high-efficiency biofuels with low cost and recycling of the by-products generated during the biodiesel production. The microalgae are feasible as a biofuel feedstock by considering the cultivation technique. Using different solvents for lipid extraction from the microalgae led to a significant reduction in the production cost due to the ease of separate purification and reuse. The use of algal biomass in the integrated biorefinery concept should also be considered, taking advantage of most compounds found in the biomass and avoiding the production of waste streams. Improving the economics of algal biodiesel generation is necessary to be competitive with petroleum fuels. Reaping energy independence and the security benefits of algal feedstocks will need critical innovation for algal mass culture and fundamental algal physiology. Overall system engineering is required to ensure that technical and economic feasibility can be reached. In this regard, this chapter shows a clear outlook on the different by-products of algae biomass conversion, recycling, applications, the challenges facing biomass-derived energy, future perspectives, and its economic feasibility.

References

[1] V.R. Lebaka, Potential bioresources as future sources of biofuels production: an overview, in: V. Gupta, M.G. Tuohy (Eds.), Biofuel Technologies, Recent Developments, Springer, Berlin, 2013, pp. 223–258.

[2] REN 21, Renewable energy policy network for 21st century, Renewables Global Status Report, 2012.

[3] E.M. Barampouti, S. Mai, D. Malamis, K. Moustakas, M. Loizidou, Liquid biofuels from the organic fraction of municipal solid waste: a review, Renew. Sustain. Ene. Rev. 110 (2019) 298–314.

[4] M.K. Awasthi, S. Sarsaiya, S. Wainaina, K. Rajendran, S. Kumar, W. Quan, et al., A critical review of organic manure biorefinery models toward sustainable circular bioeconomy: technological challenges, advancements, innovations, and future perspectives, Renew. Sustain. En. Rev. 111 (2019) 115–131.

References

[5] J. Allen, S. Unlu, Y. Demirel, P. Black, W. Riekhof, Integration of biology, ecology and engineering for sustainable algal-based biofuel and bioproduct biorefinery, Bioresour. Bioprocess. 5 (1) (2018) 47.

[6] G. De Bhowmick, A.K. Sarmah, R. Sen, Lignocellulosic biorefinery as a model for sustainable development of biofuels and value-added products, Bioresour. Technol. 247 (2018) 1144−1154.

[7] B.R. Arias, C.G. Pevida, J.D. Fermoso, M.G. Plaza, F.G. Rubiera, J.J. Pis, Martinez, Influence of torrefaction on the grindability and reactivity of woody biomass, Fuel. Proce. Technol. 89 (2) (2008) 169−175.

[8] M. Akia, F. Yazdani, E. Motaee, D. Han, H. Arandiyan, A review on conversion of biomass to biofuel by nanocatalysts, Biofuel Res. J. 1 (1) (2014) 16−25.

[9] A. Singh, R.M. Rodríguez Jasso, K.D. Gonzalez-Gloria, M. Rosales, R.B. Cerda, C.N. Aguilar, et al., The enzyme biorefinery platform for advanced biofuels production, Bioresour. Technol. Rep. 7 (2019) 100257.

[10] J.A. Ferreira, P. Brancoli, S. Agnihotri, K. Bolton, M.J. Taherzadeh, A review of integration strategies of lignocelluloses and other wastes in 1st generation bioethanol processes, Process. Biochem. 75 (2018) 173−186.

[11] F. Delbecq, Y. Wang, A. Muralidhara, K. El Ouardi, G. Marlair, C. Len, Hydrolysis of hemicellulose and derivatives: a review of recent advances in the production of furfural, Front. Chem. 6 (2018) 146.

[12] K.W. Chew, J.Y. Yap, P.L. Show, N.H. Suan, J.C. Juan, T.C. Ling, et al., Microalgae biorefinery: high value products perspectives, Bioresour. Technol. 229 (2017) 53−62.

[13] A.K. Koyande, P.L. Show, R. Guo, B. Tang, C. Ogino, J.S. Chang, Bio-processing of algal bio-refinery: a review on current advances and future perspectives, Bioengineered 10 (1) (2019) 574−592.

[14] S. Guzmán, A. Gato, M. Lamela, M. Freire-Garabal, J.M. Calleja, Anti-inflammatory and immunomodulatory activities of polysaccharide from *Chlorella stigmatophora* and *Phaeodactylum tricornutum*, Phyther Res. 17 (6) (2003) 665−670.

[15] Z.A. Mohamed, Polysaccharides as a protective response against microcystin-induced oxidative stress in *Chlorella vulgaris* and *Scenedesmus quadricauda* and their possible significance in the aquatic ecosystem, Ecotoxicology 17 (6) (2008) 504.

[16] M. Kim, J.H. Yim, S.Y. Kim, H.S. Kim, W.G. Lee, S.J. Kim, et al., In vitro inhibition of influenza A virus infection by marine microalga-derived sulfated polysaccharide p-KG03, Antivir. Res. 93 (2) (2012) 253−259.

[17] J.K. Park, Z.H. Kim, C.G. Lee, A. Synytsya, H.S. Jo, S.O. Kim, et al., Characterization and immunostimulating activity of a water-soluble polysaccharide isolated from *Haematococcus lacustris*, Biotechnol. Bioproc. E 16 (6) (2011) 1090−1098.

[18] B. Chen, W. You, J. Huang, Y. Yu, W. Chen, Isolation and antioxidant property of the extracellular polysaccharide from *Rhodella reticulata*, World J. Microbiol. Biotechnol. 26 (5) (2010) 833−840.

[19] H.W. Yen, I.C. Hu, C.Y. Chen, S.H. Ho, D.J. Lee, J.S. Chang, Microalgae-based biorefinery from biofuels to natural products, Bioresour. Technol. 135 (2013) 166−174.

[20] I.A. Schepetkin, M.T. Quinn, Botanical polysaccharides: macrophage immunomodulation and therapeutic potential, Int. Immunopharmacol. 6 (3) (2006) 317−333.

[21] M.S. Matsui, N. Muizzuddin, S. Arad, K. Marenus, Sulfated polysaccharides from red microalgae have anti-inflammatory properties in vitro and in vivo, Appl. Biochem. Biotechnol. 104 (1) (2003) 13−22.

[22] T. Tannin-Spitz, M. Bergman, D. van-Moppes, S. Grossman, S.M. Arad, Antioxidant activity of the polysaccharide of the red microalga *Porphyridium* sp, J. Appl. Phycol. 17 (3) (2005) 215−222.

[23] Y. Chisti, Biodiesel from microalgae beats ethanol, Trends Biotechnol. 26 (3) (2008) 126−130.

[24] M. El-Sheekh, H.M. Eladel, A. Abomohra, M.G. Battah, S.A. Mohamed, Optimization of biomass and fatty acid productivity of *Desmodesmus intermedius* as a promising microalga for biodiesel production, Energ. Sources Part A: Recovery Util. Environ. Eff. 1 (2019) 1−14.

[25] S.A. Khan, M.Z. Hussain, S. Prasad, U.C. Banerje, Prospects of biodiesel production from microalgae in India, Renew. Sust. Energy Rev. 13 (9) (2009) 2361−2372.

[26] M. Hannon, J. Gimpel, M. Tran, B. Rasala, S. Mayfield, Biofuels from algae: challenges and potential, Biofuels 1 (5) (2010) 763−784.

[27] Z. Reyimu, D. Özçimen, Batch cultivation of marine microalgae *Nannochloropsis oculata* and *Tetraselmis suecica* in treated municipal wastewater toward bioethanol production, J. Clean. Prod. 150 (2017) 40−46.

[28] R. Maurya, K. Chokshi, T. Ghosh, K. Trivedi, I. Pancha, D. Kubavat, et al., Lipid extracted microalgal biomass residue as a fertilizer substitute for *Zea mays* L, Front. Plant. Sci. 6 (2016) 1266.

[29] D. Hernández, M. Solana, B. Riaňo, M.C. García-González, A. Bertucco, Biofuels from microalgae: lipid extraction and methane production from the residual biomass in a biorefinery approach, Bioresour. Technol. 170 (2014) 370−378.

Handbook of Algal Biofuels

[30] M. El-Dalatony, E.S. Salama, M. Kurade, S. Hassan, S.E. Oh, S. Kim, et al., Utilization of microalgal biofractions for bioethanol, higher alcohols, and biodiesel production: a review, Energies 10 (12) (2017) 2110–2128.

[31] B.H.H. Goh, H.C. Ong, M.Y. Cheah, W.H. Chen, K.L. Yu, T.M.I. Mahlia, Sustainability of direct biodiesel synthesis from microalgae biomass: a critical review, Renew. Sustain. Ene. Rev. 107 (2019) 59–74.

[32] M. Ashour, M.E. Elshobary, R. El-Shenody, A. Kamil, A. Abomohra, Evaluation of a native oleaginous marine microalga *Nannochloropsis oceanica* for dual use in biodiesel production and aquaculture feed, Biom. Bioen. 120 (2019) 439–447.

[33] M. Bai, C. Cheng, H. Wan, Y. Lin, Microalgal pigments potential as by-products in lipid production, J. Taiwan. Instit. Chem. Engin. 42 (5) (2011) 783–786.

[34] F. Alam, A. Date, R. Rasjidin, S. Mobin, H. Moria, A. Baqui, Biofuel from algae—is it a viable alternative? Procedia. Eng. 49 (2012) 221–227.

[35] A.T. Ubando, C.B. Felix, W.H. Chen, Biorefineries in circular bioeconomy: a comprehensive review, Bioresour. Technol. 299 (2020) 122585.

[36] A. Barka, C. Blecker, Microalgae as a potential source of single-cell proteins: a review, Biotechnol. Agron. Soc. Environ. 20 (3) (2016) 427–436.

[37] D.M. Frampton, R.H. Gurney, G.A. Dunstan, L.A. Clementson, M.C. Toifl, C.B. Pollard, et al., Evaluation of growth, nutrient utilization and production of bioproducts by a wastewater-isolated microalga, Bioresour. Technol. 130 (2013) 261–268.

[38] W. Becker, Microalgae in human and animal nutrition, in: A. Richmond (Ed.), Handbook of Microalgal Culture: Biotechnology and Applied Phycology, Blackwell Science, Oxford, UK, 2003, pp. 312–351.

[39] I.C. Sheih, T.K. Wu, T.J. Fang, Antioxidant properties of a new antioxidative peptide from algae protein waste hydrolysate in different oxidation systems, Bioresour. Technol. 100 (13) (2009) 3419–3425.

[40] H.B. Li, K.W. Cheng, C.C. Wong, K.W. Fan, F. Chen, Y. Jiang, Evaluation of antioxidant capacity and total phenolic content of different fractions of selected microalgae, Food Chem. 102 (3) (2007) 771–776.

[41] E.B. D'Alessandro, N.R. Antoniosi, Filho, Concepts and studies on lipid and pigments of microalgae: a review, Renew. Sustain. Ene. Rev. 58 (2016) 832–841.

[42] F.V. da Silva, C. Sant'Anna, Impact of culture conditions on the chlorophyll content of microalgae for biotechnological applications, World J. Microbiol. Biotechnol. 33 (1) (2017) 20.

[43] Z.W. Ye, J.G. Jiang, G.H. Wu, Biosynthesis and regulation of carotenoids in *Dunaliella*: progresses and prospects, Biotechnol. Adv. 26 (4) (2008) 352–360.

[44] R. Raja, S. Hemaiswarya, R. Rengasamy, Exploitation of *Dunaliella* for 3-carotene production, Appl. Microbiol. Biotechnol. 74 (3) (2007) 517–523.

[45] J. Eonseon, C.G. Lee, J.E.W. Polle, Secondary carotenoid accumulation in *Haematococcus* (Chlorophyceae): biosynthesis, regulation, and biotechnology, J. Microbiol. Biotechnol. 16 (6) (2006) 821–831.

[46] J. Li, D. Zhu, J. Niu, S. Shen, G. Wang, An economic assessment of astaxanthin production by large scale cultivation of *Haematococcus pluvialis*, Biotechnol. Adv. 29 (6) (2011) 568–574.

[47] S. Sekar, M. Chandramohan, Phycobiliproteins as a commodity: trends in applied research, patents and commercialization, J. Appl. Phycol. 20 (2) (2008) 113–136.

[48] P. Stolz, B. Obermayer, Manufacturing microalgae for skin care, Cosmetics Toiletries 120 (3) (2005) 99–106.

[49] S.P. Cuellar-Bermudez, I. Aguilar-Hernandez, D.L. Cardenas-Chavez, N. Ornelas-Soto, M.A. Romero-Ogawa, R. Parra-Saldivar, Extraction and purification of high-value metabolites from microalgae: essential lipids, astaxanthin and phycobiliproteins, Microb. Biotechnol. 8 (2) (2015) 190–209.

[50] A.M. Ibekwe, S.E. Murinda, M.A. Murry, G. Schwartz, T. Lundquist, Microbial community structures in high rate algae ponds for bioconversion of agricultural wastes from livestock industry for feed production, Sci. Total. Environ. 580 (2017) 1185–1196.

[51] P. Spolaore, C. Joannis-Cassan, E. Duran, A. Isambert, Commercial applications of microalgae, J. Biosci. Bioeng. 101 (2) (2006) 87–96.

[52] Z. Yang, G. Rongbo, X. Xiaohui, F. Xiaolei, L. Shengjun, Hydrogen and methane production from lipid-extracted microalgal biomass residues, Int. J. Hydrog. Energ. 36 (5) (2011) 3465–3470.

[53] A.F. Miranda, N. Ramkumar, C. Andriotis, T. Höltkemeier, A. Yasmin, S. Rochfort, et al., Applications of microalgal biofilms for wastewater treatment and bioenergy production, Biotechnol. Biofuels. 10 (1) (2017) 120.

[54] M.D. Scherer, A.C. de Oliveira, F.J.C. Magalhães Filho, C.M.L. Ugaya, A.B. Mariano, J.V.C. Vargas, Environmental study of producing microalgal biomass and bioremediation of cattle manure effluents by microalgae cultivation, Clean. Technol. Envir. 19 (6) (2017) 1745–1759.

[55] R. Munoz, B. Guieysse, Algal-bacterial processes for the treatment of hazardous contaminants: a review, Water Res. 40 (15) (2006) 2799–2815.

[56] A. Vaishampayan, R.P. Sinha, D.P. Hader, T. Dey, A.K. Gupta, U. Bhan, Cyanobacterial biofertilizers in rice agriculture, Bot. Rev. 67 (4) (2001) 453–516.

[57] K.Y. Show, D.J. Lee, J.H. Tay, T.M. Lee, J.S. Chang, Microalgal drying and cell disruption: recent advances, Bioresour. Technol. 184 (2015) 258–266.

[58] F. Dalena, A. Senatore, A. Tursi, A. Basile, Bioenergy production from second- and third-generation feed-stocks, in: F. Dalena, A. Basile, C. Rossi (Eds.), Bioenergy Systems for the Future: Prospects for Biofuels and Biohydrogen, Elsevier Publishing, London, 2017, pp. 559–599.

[59] L. Christenson, R. Sims, Production and harvesting of microalgae for wastewater treatment, biofuels, and bioproducts, Biotechnol. Adv. 29 (6) (2011) 686–702.

[60] R. Ueda, S. Hirayama, K. Sugata, H. Nakayama, Process for the production of ethanol from microalgae, Patent and Trademark Office, Washington, DC, United States Patent No. 5, 578, 1996, pp. 472.

[61] M.J. Waites, N.L. Morgon, J.S. Rockey, G. Higton, Industrial Microbiology: An Introduction, John Wiley and Sons, Blackwell Science Ltd, 2001, p. 288.

[62] C.A.C. Alzate, O.J.S. Toro, Energy consumption analysis of integrated flow sheets for production of fuel ethanol from lignocellulosic biomass, Energy 31 (13) (2006) 2447–2459.

[63] J.H. Park, J.Y. Jeong, D. Hee, J.L. Dong, Anaerobic digestibility of algal bioethanol residue, Bioresour. Technol. 113 (2012) 78–82.

[64] J. Saleh, A.Y. Tremblay, M.A. Dube, Glycerol removal from biodiesel using membrane separation technology, Fuel 89 (9) (2010) 2260–2266.

[65] M.C.S. Gomes, N.C. Pereira, S.T.D. de Barros, Separation of biodiesel and glycerol using ceramic membranes, J. Membr. Sci. 352 (1–2) (2010) 271–276.

[66] F. Sun, H. Chen, Organosolv pretreatment by crude glycerol from oleochemicals industry for enzymatic hydrolysis of wheat straw, Bioresour. Technol. 99 (13) (2008) 5474–5479.

[67] S.C. D'Angelo, A. Dall'Ara, C. Mondelli, J. Pérez-Ramírez, S. Papadokonstantakis, Techno-economic analysis of a glycerol biorefinery, ACS Sustain. Chem. Eng. 6 (12) (2018) 16563–16572.

[68] L. Devi, K.J. Ptasisnski, F.J.G. Janssen, A review of the primary measures for tar elimination in biomass gasification processes, Biomass Bioenerg. 24 (2) (2003) 125–140.

[69] J. Corella, M.P. Aznar, J. Gil, M.A. Caballero, Biomass gasification in fluidized bed: where to locate the dolomite to improve gasification? Energy Fuels 13 (6) (1999) 1122–1127.

[70] M.K. Lam, K.T. Lee, A.R. Mohamed, Homogeneous, heterogeneous and enzymatic catalysis for transesterification of high free fatty acid oil (waste cooking oil) to biodiesel: a review, Biotechnol. Adv. 28 (4) (2010) 500–518.

[71] M.A. Borowitzka, High-value products from microalgae their development and commercialization, J. Appl. Phycol. 25 (3) (2013) 743–756.

[72] R.H. Wijffels, M. Barbosa, An outlook on microalgal biofuels, Sci 329 (5993) (2010) 796–799.

[73] D.L. Barreiro, C. Samorì, G. Terranella, U. Hornung, A. Kruse, W. Prins, Assessing microalgae biorefinery routes for the production of biofuels via hydrothermal liquefaction, Bioresour. Technol. 174 (2014) 256–265.

[74] J.L. Garcia-Moscoso, W. Obeid, S. Kumar, P.G. Hatcher, Flash hydrolysis of microalgae (Scenedesmus sp.) for protein extraction and production of biofuels intermediates, J. Supercrit. Fluids 82 (2013) 183–190.

[75] A. Demirbas, Progress and recent trends in biodiesel fuels, Energ. Convers. Manage. 50 (1) (2009) 14–34.

[76] S. Al-Zuhair, Production of biodiesel: possibilities and challenges, Biofuel. Bioprod. Bior.: Innov. A Sustain. Econ. 1 (1) (2007) 57–66.

[77] A. Demirbas, Importance of biodiesel as transportation fuel, Energ. Policy 35 (9) (2007) 4661–4670.

[78] L.G. Schumacher, W. Marshall, J. Krahl, W.B. Wetherell, M.S. Grabowski, Biodiesel emissions data from series 60 DDC engines, Trans. ASAE 44 (6) (2001) 1465–1468.

[79] M.A. Hess, M.J. Haas, T.A. Foglia, W.N. Marmer, Effect of antioxidant addition on NOx emissions from biodiesel, Energ. Fuel. 19 (4) (2005) 1749–1754.

[80] A.A. Nesamma, K.M. Shaikh, P.P. Jutur, Genetic engineering of microalgae for production of value-added ingredients, in: S.K. Kim (Ed.), Handbook of Marine Microalgae: Biotechnology Advances, Academic Press, Elsevier Science, Boston MA, 2015, pp. 405–414.

[81] H. Kamyab, M.F.M. Din, S.E. Hosseini, S.K. Ghoshal, V. Ashokkumar, A. Keyvanfar, et al., Optimum lipid production using agro-industrial wastewater treated microalgae as biofuel substrate, Clean. Technol. Environ. 18 (8) (2016) 2513–2523.

[82] J. Venkatesan, P. Manivasagan, S.K. Kim, Marine microalgae biotechnology: present trends and future advances, in: S.K. Kim (Ed.), Handbook of Marine Microalgae: Biotechnology Advances, Academic Press, Elsevier Science, Boston MA, 2015, pp. 1–9.

[83] J.K. Pittman, A.P. Dean, O. Osundeko, The potential of sustainable algal biofuel production using wastewater resources, Bioresour. Technol. 102 (1) (2011) 17–25.

[84] N. Mallick, S.K. Bagchi, S. Koley, A.K. Singh, Progress and challenges in microalgal biodiesel production, Front. Microbiol. 7 (2016) 1019.

[85] H. Kamyab, S. Chelliapan, M.F.M. Din, C.T. Lee, S. Rezania, T. Khademi, et al., Isolate new microalgal strain for biodiesel production and using FTIR spectroscopy for assessment of pollutant removal from palm oil mill effluent (POME), Chem. Engin. Transac. 63 (2018) 91–96.

[86] L.M. Laurens, M. Chen-Glasser, J.D. McMillan, A perspective on renewable bioenergy from photosynthetic algae as feedstock for biofuels and bioproducts, Algal Res. 24 (2017) 261–264.

[87] K.H.M. Cardozo, T. Guaratini, M.P. Barros, V.R. Falcão, A.P. Tonon, N.P. Lopes, et al., Metabolites from algae with economical impact, Comp. Biochem. Phys. C: Toxicol. Pharmacol. 146 (1–2) (2007) 60–78.

[88] Z. Kang, B.H. Kim, R. Ramanan, J.E. Choi, J.W. Yang, H.M. Oh, et al., A cost analysis of microalgal biomass and biodiesel production in open raceways treating municipal wastewater and under optimum light wavelength, J. Microbiol. Biotechnol. 25 (1) (2015) 109–118.

[89] N.H. Norsker, M.J. Barbosa, M.H. Vermu, R.H. Wijffels, Microalgal production—a close look at the economics, Biotechnol. Adv. 29 (1) (2011) 24–27.

[90] M. Mondal, S. Goswami, A. Ghosh, G. Oinam, O.N. Tiwari, P. Das, et al., Production of biodiesel from microalgae through biological carbon capture: a review, Biotech. 7 (2) (2017) 99.

[91] S. Rocca, A. Agostini, J. Giuntoli, L. Marelli, Biofuels from algae: technology options, energy balance and GHG emissions, in; Sci. Tech. Res. Rep. Off. Eur. Union, Luxembourg, 2015.

[92] X. Miao, Q. Wu, Biodiesel production from heterotrophic microalgal oil, Bioresour. Technol. 97 (6) (2006) 841–846.

[93] B.D. Wahlen, R.M. Willis, L.C. Seefeldt, Biodiesel production by simultaneous extraction and conversion of total lipids from microalgae, cyanobacteria, and wild mixed-cultures, Bioresour. Technol. 102 (3) (2011) 2724–2730.

[94] A. Carrero, G. Vicente, R. Rodríguez, M. Linares, G.L. Del, Peso, Hierarchical zeolites as catalysts for biodiesel production from *Nannochloropsis* microalga oil, Catal. Today 167 (1) (2011) 148–153.

[95] S.R. Medipally, F.M. Yusoff, S. Banerjee, M. Shariff, Microalgae as sustainable renewable energy feedstock for biofuel production, Bio. Med. Res. Int. (2015) 13.

[96] A. Suresh, C. Seo, H.N. Chang, Y. Kim, Improved volatile fatty acid and biomethane production from lipid removed microalgal residue (LRIAR) through pretreatment, Bioresour. Technol. 149 (2013) 590–594.

[97] M. Iijima, J. Paulson, Japan biofuels annual 2017, Global Agricultural Information Network, USDA Foreign Agricultural Service, Tokyo, Japan, 2017, pp. 1–22.

[98] R. Efroymson, Sustainable development of algae for biofuel, DOE bioenergy tevhnologies office (BETO), Project Peer Review-Oak Ridge National Laboratory, 2017.

CHAPTER 19

Harnessing solar radiation for potential algal biomass production

Imran Ahmad[1], Norhayati Abdullah[2], Mohd Danish Ahmad[3], Iwamoto Koji[1] and Ali Yuzir[4]

[1]Algae and Biomass, Research Laboratory, Malaysia-Japan International Institute of Technology (MJIIT), Universiti Teknologi Malaysia, Jalan Sultan Yahya Petra, Kuala Lumpur, Malaysia
[2]UTM International, Level 8, Menara Razak, Universiti Teknologi Malaysia, Jalan Sultan Yahya Petra, Kuala Lumpur, Malaysia [3]Department of Post-Harvest Engineering and Technology, Aligarh Muslim University, Aligarh, India [4]Department of Chemical and Environmental Engineering (ChEE), Malaysia-Japan International Institute of Technology (MJIIT), Universiti Teknologi Malaysia, Jalan Sultan Yahya Petra, Kuala Lumpur, Malaysia

19.1 Introduction

The global increase in population, urbanization, economic expansion, and higher living standards are increasing energy consumption leading to the depletion of nonrenewable fossil reserves and the release of harmful greenhouse gases that cause global warming. This acute condition of dwindling resources and global climate change are stimulating research in the development of feedstocks for the production of biofuels and other value-added products, which can be retrieved, processed, and made viable in an environmentally friendly and sustainable manner [1].

Presently, the fossil reserves are exploited mainly for oil, coal, and natural gas accounting for ~80% (>400 EJ/year) of use [2,3]. As predicted by environmental engineers in accordance with the sustainable development goals if the initiatives are efficiently implemented then 60% of the electricity and 30% of the fuel use globally can be met by renewable energy alternatives by the year 2025 [4].

In recent years, algae as third-generation biofuel feedstocks are acting as a source for the production of renewable bioenergy against the odds of depleting fossil fuels, limited land and water, and environmental stress [5]. Algae are potential biotechnological resources which are driven by solar energy to produce environmentally sustainable products by the valorization of algal biomass [6]. The autotrophic capability of algae gives flexibility and an edge over other conventional/nonrenewable fuels.

Algae undergo photosynthesis and convert the captured solar energy into chemical energy by the fixation of CO_2. The bulk cultivation of algae has proven to be most widely accepted in the production of sustainable/carbon neutral products such as (1) biofuels, further classified as biocrude oil, biodiesel, bioethanol, biogas, and biohydrogen [7]; (2) value-added products for pharmaceutical and nutraceutical uses [8]; and (3) the production of biomaterials and cosmetics, as discussed in the previous chapters. Various products obtained from algae, along with their applications are shown in Fig. 19.1. The expedience and dominance of algae over other feedstocks for the production of biomass and subsequently other useful entities can be understood by its advantages: (1) higher lipid yield (60 m^3/ha as compared to 2 m^3/ha for jatropha and 0.2 m^3/ha for corn) [9]; (2) fast growth rate (doubling in few hours) and competence in photosynthetic conversion (10–50 times

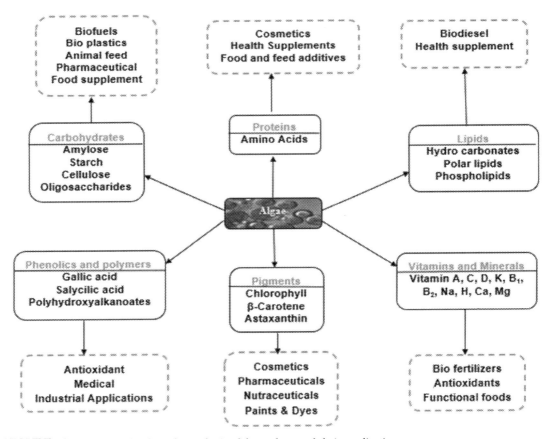

FIGURE 19.1 Value-added products obtained from algae and their applications.

greater than C4 plants) [10]; (3) high potential for CO_2 sequestration (1 kg of microalgal biomass can fix about 1.8 kg CO_2) [11]; (4) high value agricultural land is not required (solving the food vs. fuel feud); and (5) flexibility in the growth environment as it can be grown in sea water, domestic wastewater, and industrial/agricultural wastewater, thus overcoming the problem of freshwater dependence [12]. The mass production of algae is carried out using scrubbers, open ponds, raceway ponds, tanks, and closed photobioreactors (PBRs). Closed PBRs have an upper hand over open systems because of the following advantages: optimum growth conditions (light, temperature, and nutrients); control and minimization of contamination (bacteria, protozoa, and unwanted algae species); reduction in water and CO_2 losses; and attainment of higher algae densities and higher volumetric productivity of algal biomass [13,14]. Therefore PBRs are said to be the most effective in terms of improving the yield of algal biomass [15]. Biomass productivity is of prime importance as it is the governing factor for the value-added products obtained from algae, as the quality and quantity of products obtained are directly proportional to the biomass produced.

This chapter will provide an insight into various strategies to optimize the utilization of solar energy radiation for efficient conversion to algal biomass, which can be further harvested for different useful metabolites. The subsequent sections of the chapter will focus on different types of solar cells, solar panels, and their applications. Coming to the main theme of the chapter is the utilization of solar energy for algal biomass and the efficacy of a solar tracker coupled with a PBR. Thermochemical chemical processes involved in the conversion of algal biomass to useful products and lastly the technoeconomic aspect of microalgal cultivation and harvesting are also discussed.

19.2 Solar cells

A French physicist Alexandre-Edmond Becquerel performed an experiment with an electrolytic cell and recognized the photovoltaic (PV) effect in 1839 [16]. Afterwards, Russel Ohl patented the modern PV cell which was silicon based in 1946 [17]. PV solar cells are made with silicon wafers that convert sunlight radiation into electrical energy. Currently commercial photovoltaic cells are made by using the semiconductor material that produces free carriers and absorbs light in the visible region. In this type of cell, there are two different types of semiconductors—p-type and n-type—and when the sunlight strikes them the electrons move from one layer to another, thus producing a potential difference at the terminals of the solar cell. The outcome is a direct current (DC) of electric power [18]. Generally, PV cells are grouped into four categories termed as generations (Fig. 19.2). If *photovoltaic solar panels* are made up of individual photovoltaic cells connected, then the solar photovoltaic array, also known simply as a solar array, is a system made up of a group of solar panels connected.

A *photovoltaic array* is therefore multiple solar panels electrically wired together to form a much larger PV installation (PV system) called an array, and in general the larger the total surface area of the array, the more solar electricity will be produced (Fig. 19.3).

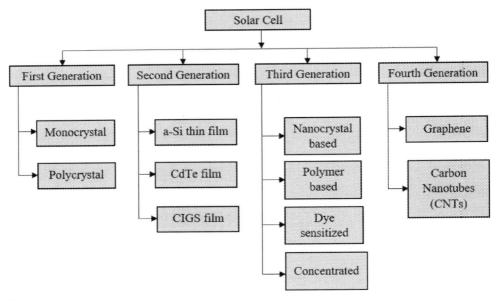

FIGURE 19.2 Different generations of solar cells.

19.2.1 First-generation solar cell

The first generation is the oldest and most popular photovoltaic technology comprising solar cells that are based on silicon wafers, which not only leads to high efficiency, but also to inflated cost. The silicon wafer is further classified into two subgroups, namely [19], monocrystal solar cell and polycrystal solar cell.

19.2.2 Second-generation (thin film) solar cells

Second-generation PV cells are thin film solar cells, like amorphous silicon, cadmium telluride, and copper indium gallium di-selenide. They are more economical compared to the first-generation cells because although these types of solar cells have lower efficiency, they are much cheaper to produce and the cost per watt is lower. The thickness of light absorbing layers in silicon wafer cell is 350 µm, whereas the thin-film solar cells have layer thickness of about 1 µm [20].

19.2.3 Third-generation solar cells

Third-generation solar cells are a newly developed technology but most of the technologies in the third generation are not yet used at commercial scale. However, research is going on in this area with the aim of making PV solar cells cheap and efficient [21]. The third generation of PV cells are grouped into four subcategories: (1) nanocrystal-based solar cells; (2) polymer-based solar cells; (3) dye sensitized solar cells (DSSC); and (4) concentrated solar cells (CSC). All of these are recently developed technologies that are

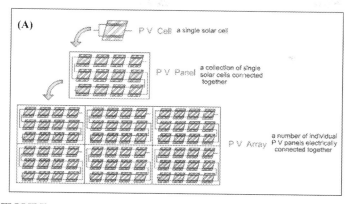 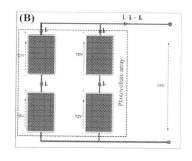

FIGURE 19.3 (A) Photovoltaic solar array and (B) solar array connections.

effective but are not used at commercial scale. The two highly growing solar cell types in this generation are DSSC and CSC.

Dye sensitized solar cells are capable of converting any visible light into potential difference and are based on dye molecules between electrodes. Electron—hole pairs exist in dye molecules and are transferred by titanium dioxide (TiO_2) nanoparticles. Its manufacturing is easy as compared with other alternative technologies. Furthermore, CSC is also an efficient technology. It works on the principle of convergence in which a high quantity of solar radiation is concentrated into a minor region or place where the PV cell is situated with the help of lenses and curved mirrors, which allows a reduction in the quantity of expensive semiconductor material. The degree of concentration ranges from 10 sol to 1000 sol and thus lowers the total cost compared to other conventional systems [22].

19.2.4 Fourth-generation solar cells

Fourth-generation cells are also known as *"inorganics-in-organics."* They are the combination of economic flexibility of thin polymer films with the stability of inorganic nanostructures like metal oxides and metal nanoparticles or organic-based nanomaterials like carbon nanotubes, graphene, and its derivatives. They are not only less costlier to manufacture but also improve the charge separation and transport inside the solar cells with the use of nanomaterials, especially graphene (G), which will become the optimal future nanomaterial in scientific and technological field. Recently, researchers observed graphene to be the elementary unit of graphite [23]. In 1991, Japanese physicist Sumio Iijima discovered carbon nanotubes (CNTs), which are allotropes of carbon. A tube-shaped structure is prepared by using curling graphene sheets [24]. These are divided into two main categories: single-wall carbon nanotubes (SWCNTs), and multiwall carbon nanotubes (MWCNTs). The SWCNTs are composed of a single graphene tube, whereas MWCNTs are made with some concentric tubes of graphene [25].

Arc discharge, laser ablation, high-pressure carbon monoxide disproportionation, and chemical vapor deposition (CVD) processes are various techniques for the synthesis of

426   19. Harnessing solar radiation for potential algal biomass production

CNTs [26]. CNTs are very effective materials as they have various applications due to their excellent electronic, chemical, and mechanical properties [27]. They have a very high length-to-diameter ratio, ranging from 10^3 to 10^5, and are among the strongest and stiffest materials known, with an elastic modulus close to 1 TPa [28]. CNTs have very superior heat conduction along the tube, display ballistic conduction at room temperatures, and also have very good electrical conductivity. Metallic nanotubes are capable of transporting an electric current density of 4×10^9 ampere/cm^2, which is 1000 times higher than the metals such as copper [29].

19.3 Solar panel

An individual solar cell can only generate power at a very small level, such as 2–3 watts only, which can be used in various small-scale applications. In order to generate adequate power, many solar cells can be connected in series to form a solar panel by combining 36, 54, 72, or 96 cells together and for larger capacity these panel should be connected in both series and parallel for high voltage and current, respectively. The cells in a solar panel are covered by a tough and weather-resistant box which is composed of glass or plastic cover, an antireflective coating; front contact semiconductor material (n-type) is used to allow negative charge (electrons) to enter into the electrical circuit, the (p-type) semiconductor layers; a back contact is normally made of aluminum sheet to complete the electrical circuit, and a substrate for avoiding moisture [30]. The layers are shown in Fig. 19.4.

19.4 Different applications of solar radiation

19.4.1 Agrophotovoltaic

In 1982 Goetzberger and Zastrow introduced the concept of agrophotovoltaics (APV), and this method is also known as the agrivoltaic system [31]. The agricultural lands are used for both food and photovoltaic electricity production, creating the opportunity to increase photovoltaic capacity significantly, and at the same time preserving land that is suitable for agriculture. Recently the developments in this technology have shown dynamic growth in all regions around the world. A 5 MW$_p$ capacity was installed in 2012 and by 2019 it was already increased to 2.1 GW$_p$ [32,33].

In 2019 the Fraunhofer Institute for solar energy systems gave an idea to merge artificial intelligence and robotics for field analysis and turned it into APV system. This is based on the concept of the Farm Bot, but the only difference is that the Farm Bot is for domestic applications at a small-scale and without photovoltaic power generation [34].

19.4.2 Aquavoltaic

Water bodies are considered as an essential element for eco-systems, the existence of life, and for almost all human activities. The production of seafood from aquaculture farms

Handbook of Algal Biofuels

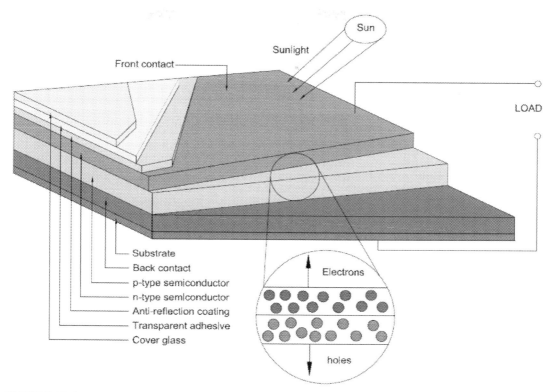

FIGURE 19.4 Microscopic view of layers in a solar panel.

is reported to be about **40%** [35]. The floatovoltaic (FV) is a kind of photovoltaic system that floats on different water bodies, thereby reducing evaporation by up to 70%–85%. The integration of FV in aquaculture is termed as aquavoltaics It benefits food production and energy generation by efficient utilization of water. In relation to this development, farmers benefit from this technology by using a PV power generation technique and efficient use of water stored in natural water bodies (lakes & ponds) or in water storage facilities (canals & reservoirs). These systems may be installed for both commercial level of farms and also for minor villages in remote areas [36,37].

19.4.3 Solar tractors

In current years vehicles used for agricultural purposes are mostly run using internal combustion engines that rely on conventional fuels. Alternatives to these are electric engines which can be used in farm vehicles and electricity for these engines can be provided from renewable sources like solar, hydro, and wind [38].

Internal combustion engines are the most economical but they have low efficiency and also require more maintenance if compared with electric motors. Their maximum

19.4.4 Solar photovoltaic (PV) systems

Sustainable agricultural activities can be run by using PV technology. It was reported that this technology has become widely accepted and dominant among all other technologies because of its significant merits [40]. Solar photovoltaic systems have been installed to provide electricity or both electricity and heat for various activities that require energy in agriculture [41]. In small farms, these systems are favorable and may be used for distributed electricity production. Farm photovoltaic systems can be helpful to overcome the problems of peak load, lower supply from grid, and transmission losses by the onsite production of electricity. The cost-effectiveness depends on the solar radiation availability in that region [42]. Recently, the integration of PV solar systems shows rapid progress in the agricultural sector, but still requires more studies and practical investigation of both technical and economic aspects to make this integration viable to a greater extent. PV solar systems are promising techniques that are used as alternative sources for power supply requirements for various activities in smart and precision farming [43].

19.4.5 Solar water pumping

Pumping water using solar energy is very promising application for farming purposes because of the unavailability of grid electricity and high cost of conventional fuels in remote areas. PV systems are like traditional systems but the only difference is that they utilize solar radiation for the electricity generation, which is eco-friendly and requires minimal maintenance, as well as saving the costs of fuels. The integration of PV panels and solar water pumps was first introduced in the late 1970s with some limitations in initial designs, for example, in performance parameters due to designing errors [44].

Centrifugal pumps were used in the first generation of solar PV-powered water pumping, as they can be operated with the help of DC/AC motors and they have hydraulic efficiency in the range of 25%–35%. Positive displacement pumps, progressing cavity pumps, or diaphragm pumps are used in the second generation. Generally, they are operated at low solar power input and their hydraulic efficiencies arealso high (upto 70%) [45].

19.4.6 Solar dryers

The mechanism of removing moisture from the product is called as drying and it includes both the transfer of heat and mass. Sun-drying is a traditional technique mainly practised in rural areas, which is widely used to preserve agricultural crops. This method of drying also has some limitations like spoiling of crops due to rain, moisture present in wind, and affect of dust. The reasons for deterioration are susceptibility to fungal, bacterial, and insect attacks etc. [46]. PV solar drying is seen as one of the most efficient methods of utilizing solar radiation compared with different applications of solar energy systems. It is an alternative technique to dry various agricultural goods in clean and aseptic surroundings which helps in minimizing losses in crops, energy utilization, and the time of drying required to improve the final quality of the products [47].

19.4.7 Solar distillation

Potable or drinking water is a primary and most essential requirement of human beings for their survival. Drinking water scarcity is a critically grave issue in the developing and underdeveloped nations which is due to insufficient or polluted water resources available. Most of the underdeveloped and developing nations have one of their main aim of supplying potable water to all of their citizens [48].

To meet this demand various techniques are utilized to make water potable from different sources like river, sea etc. because only about 1% of freshwater is available in nature. PV solar distillation technique utilizes solar energy which is abundantly available to obtain potable water using distillation from brackish water [49]. The traditional method of distillation uses fossils fuels for the distillation process. In solar distillation, brackish water is heated at the operating pressure to reach the boiling point and produce steam, and the steam is condensed to freshwater in a condenser. Solar distillation technology is very simple and environment-friendly as it uses solar energy (renewable source) for its input. This technique requires little maintenance and as a result there is no need for skilled labor for its operation and can be installed at the desired locations [50].

19.4.8 Biomass to electricity conversion using solar radiation

Photobioreactors are special type of bioreactors which are extensively used to culture microorganisms that photosynthesize. They convert solar radiation into chemical energy by photosynthesis utilizing carbon dioxide (CO_2) to generate biomass [51]. The efficiency of photosynthesis relies on the surface-to-volume ratio. The main microorganisms that are cultivated by this technique are purple bacteria, mosses, macro- and microalgae, and cyanobacteria.

The applications are different according to the type of microorganisms used like the production of proteins, lipids, and carbohydrates etc. or the other value-added products [52]. The research on the photobiorectors gained momentum after it was studied that they can be utilized for the production of algal biomass. Algal biomass can later be exploited for obtaining biofuels and other value added products. Thus solving the issue of crisis for agricultural land.

19.4.9 Solar cooker

Utilization of solar radiation for cooking is not a new idea, Ehrenfried Walther Von Tschirnhaus, a German physicist, was the first scientist who gave the concept of cooking using solar energy. He performed an experiment using a large lens to focus the sun rays to boil water present in a clay pot. A highly reflective surface or mirror is used to converge sunlight to a specific cooking point, but it also depends on the shape of the cooking device. Solar cookers consist of a cooking pan on which sunlight is usually concentrated.

430   19. Harnessing solar radiation for potential algal biomass production

When sunlight comes into contact with the material to be cooked it is converted to heat. The pots or pan that are used in cooking must be black in color and a glass lid on the top of pot provides insulation to reduce convective losses during the heating process in solar cooker [53].

19.4.10 Solar water heater

Solar water heaters are classified according to their dependency on the heating capacity and fluid applied for heat transfer (water or antifreeze fluid) through the collector. Solar water heaters are usually classified into two groups, namely, direct or indirect, which can be further subdivided into active or passive systems [54]. In direct systems water is heated inside the collectors, whereas indirect systems require a heat exchanger between the collector and a hot water storage tank. Active systems are electrically driven while passive systems do not require electrical pumps. They are based on convective heat transfer to allow the flow of hot water between the collector and tank. A thermosyphon is an example of a passive system because it requires neither pump nor controller [55].

19.5 Conversion of solar radiation to algal biomass

Light is the elemental and vital source of energy for algae cultivation. Light exclusively from electromagnetic radiation is one of the most crucial and decisive growth limiting factors that administer the microalgal proliferation and the efficiency of PBRs. Photosynthesis taking place in algae involves light and dark reaction cycles. The function of the light reaction cycle, mainly taking place in thylakoids, is the conversion of light energy to chemical energy. While in the dark reaction cycles the chemical energy gets converted into stable chemical energy [56]. The light and dark cycles are illustrated in Fig. 19.5. The overall photosynthetic reaction fixes CO_2 and produces glucose by synthesizing proteins, carbohydrates, and lipids (Eq. 19.1). 1.57 g of CO_2 is required to produce 1 g of glucose.

$$6CO_2 + 12H_2O + \text{light} + \text{chlorophyl} \rightarrow C_6H_{12}O_6 + 6O_2 + 6H_2O \tag{19.1}$$

There is always a difficulty for the light to reach uniformly and adequately to all the portions of the PBR, and this is always an issue of concern in PBR carrying high-density cultures [58]. Moreover, a significant amount of light reaching PBR (IR and UV wavelength) does not contribute to the photosynthetic solar conversion. Exclusively visible light (400−700 nm), termed as photosynthetically active radiation (PAR), can be harnessed and converted to chemical energy in biomass, accomplished by the algal light harnessing pigments [59]. UV radiation having wavelength <400 nm and categorized as high radiation causes an ionizing effect and cell damage, while the IR radiation that has a wavelength >750 nm and categorized as low energy radiation causes excessive heating of the cultures in the PBRs.

The resulting effect of both UV and IR radiation may cause a considerable decrease in the microalgal bioproductivity. Overheating is fatal for the growth of algae and the control

Handbook of Algal Biofuels

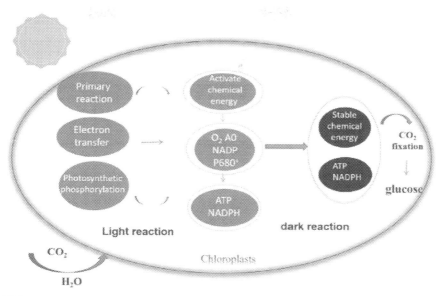

FIGURE 19.5 The light/dark cycles illustrated in algal chloroplast [57].

of the temperature in the optimum range and cooling of PBRs is energy intensive, costly, and environmentally unsustainable [60].

Therefore manipulation of incident irradiation insolation should be achieved to get the maximum photons from the desirable PAR wavelength, which will in turn enhance the efficiency of PBRs and ultimately the productivity of biomass.

19.5.1 The mechanism of light absorption in algae

The energy which is absorbed by algae depends greatly on the chemical characteristics of the pigments present in it, with specific colors absorbing specific wavelengths in the band of PAR. The main pigments responsible for photosynthetic light capture and processing are chlorophylls, carotenoids, and phycobilins. The major group of pigments is chlorophyll and chlorophyll-a is the most important, while the carotenoids and phycobilins are termed accessory pigments [61].

During the photosynthesis firstly the photons are captured by chlorophyll-a, assisted by other light harvesting antenna/complex, and the energy is efficiently transferred to photochemical reaction centers consisting of protein–pigment complexes, where it can be utilized for the formation of biomass (photochemistry), fluorescence, or dissipated as heat [62]. The size of the antennae depends upon the prevailing light conditions in which the species of algae is found. To match the photosynthetic requirements the algal species found in the environments having low light generally have a larger number of antennae. As the condition of saturating irradiance prevails large antennae pigments are not efficient in absorbing the light leading to photoinhibition and photosaturation [63]. This situation has adverse effects on the photosynthesis and algal productivity. Those algal species found

432
19. Harnessing solar radiation for potential algal biomass production

in the environment of high light have a lower number of pigments to tackle the problem of photoinhibition [64]. The performance of algae species can be improved by tailoring of the pigments which harvest the light, thereby enhancing the spectral sensitivity of the algae and their resilience to photoinhibition and varying light conditions [65].

19.5.2 Strategic management of harnessing solar radiation for algal cultivation

The production of adequate quantities of biomass is essential to match the large-scale demand and economic viability, which greatly depends upon efficient utilization of solar irradiation during microalgal cultivation. Therefore tailoring and customizing the irradiance transmitted to algal cultures could help to enhance the photosynthetic efficiency, thereby increasing the productivity of algal biomass [66]. The various strategies being studied and experimented/trialed include increasing PAR quality and quantity, appropriate orientation, and location of PBRs, shift in wavelength, plasmonic scattering, use of light etc. These strategies are briefly elaborated in Table 19.1.

19.5.3 Constraints effecting the growth of algae

Algal biomass production is achieved using open ponds and closed PBRs. As discussed in the previous section closed PBRs have advantages over open ponds but at the same time open ponds have lower costs of production and operation [76]. It is unanimously accepted worldwide that raceway ponds are the most economic technology for the bulk production of algae. Problems encountered in open ponds are inadequate mixing, temperature fluctuation, and risk of contamination. The temperature of culture influences the intensity of solar irradiation needed by algae for its optimum growth and productivity [77].

In the optimum range of temperature ($20-25°C$) algae are tolerant to high light intensity. While overheating reduces the tolerance of microalgae, the impact of overheating is mostly encountered in closed photobioreactors (PBRs) as the open ponds have a self-evaporating cooling mechanism [78]. Fluctuations in temperature and seasonal variation can badly affect the culture conditions, thus affecting the microalgal photosynthesis and biomass productivity. Therefore to tackle the problem of overheating freshwater sprays or heat exchangers are used to keep the temperature within optimum range. This increases the energy demand and the water footprint thereby negatively affecting the environment [79].

Photosynthetic efficiency is a function of the intensity of irradiation and the productivity of PBR. In outdoor cultures photosynthetic efficiency varies from $8\%-12\%$, while at the industrial level of microalgal production photosynthetic efficiency is $1.5\%-2\%$ even after utilizing the optimum culture conditions. Table 19.2 provides a detailed account of different types of PBRs cultivating distinct species of algae with the photosynthetic efficiency, volumetric productivity, and other solar irradiance components for different combinations of algae cultivation. About $0.1\%-10\%$ of light becomes the net photosynthetic energy conversion efficiency in fixing CO_2 [91].

Handbook of Algal Biofuels

TABLE 19.1 Strategic management of harnessing solar radiation for algal cultivation.

Parameter/criterion	Details/effects/consequences	References
Photobioreactor location	The solar irradiance and trajectory vary with the change of latitude with maximum irradiance is near the equator. Weather also plays a significant role to the irradiance available for the installation.	[67]
Positioning and direction, solar tracking	More irradiance is intercepted with East–West vertical facing surfaces but at the same time they cast shadows which is problematic for large-area installations.	[68]
Spectrum of light	If PAR is increased by changing the spectral distribution the growth of algae is positively influenced.	[69]
Shift in wavelength	With the conversion of light from low to high photosynthetic utility the energy available for photosynthesis can be increased.	[70]
Artificial irradiance/light	Light emitting diode provides better spatial and spectral control of light hence their usage can be justified to produce high value products, but they are energy and cost incentive.	[71]
Nearfield confinement	Also known as plasmonic scattering, it can be utilized to direct/ confine suitable wavelength of light into the photobioreactor (PBR).	[72]
Density of microalgal culture and light path length	Controlling the microalgal cell density, path of light and mixing can lead to optimum areal efficiency and productivity.	[73]
Dilution of light for PBR	Light having high density can be distributed over a large surface area by employing light guiding elements or by using curved surfaces to dilute the light below the saturation threshold of algae.	[74]
Genetic engineering	This is achieved by genetic modification of the size of light harvesting antennae or by the usage of quantum dots/nanotubes to boost the energy transduction.	[75]

The efficient and sufficient supply of CO_2, not only contributes to replenish the deficiency caused by the light losses but also helps to enhance the biomass productivity. In normal atmospheric CO_2 high light intensity negatively effects the photosynthetic performance of algae. In terms of environmental sustainability algae is a better option for carbon capture and utilization [92]. The captured CO_2 based inorganic carbon is incorporated into algal cells as valuable biochemicals. Therefore algal cultivation can substantially contribute to reduce the carbon footprint [93].

19.6 Solar tracking system

In the last couple of years, static solar systems have been employed to extract solar energy but recently new techniques are being used for this purpose. As they can enhance the efficiency of solar systems by the application of single axis solar tracking (SATs) or dual axis solar tracking systems (DATs) that change the orientation of solar panels according to the sun's movements. The variation is due to the different seasons and time during

TABLE 19.2 Algal biomass productivity and solar irradiance components.

Algae species	Type of photobioreactor	Location, light system	Volume (L)	Photosynthetic efficiency (%)	Light path (cm)	Volumetric productivity (g/L/d)	Illuminated surface area (m²)	Light intensity μmol photons/ m²/s	References
Tetraselmis suecica	Green wall panel	Outdoor, Sunlight	315,000	–	4.5	0.02	14,000	915	[80]
Tetraselmis suecica	Annular column	Outdoor, sunlight	120	9.3	40	0.42	5.3	900	[81]
Nannochloropsis sp.	Horizontal tubular	Outdoor, sunlight	560	1.2–1.8	4.6	0.3–0.85	27		[82]
Muriellopsis sp.	Raceway pond	Outdoor, sunlight	100	0.69–0.97	30	0.04	–	1449	[83]
Chlorella sorokianiana	Rotating annular	Indoor, Artificial light	3.4	7	1.2	7.3	0.24	1500	[84]
Nannochloropsis salina	Algae raceway integrated design	Outdoor, sunlight	7500	1	15	0.02	50	490	[85]
Chlamydomonas reinhardtii	cylindrical	Indoor, artificial light	2	7	5	0.7	0.1	200	[86]
Chlorella spp.	Thin layer inclined cascades	Outdoor, sunlight	2200	4	0.6	1.9	224	540	[87]
Stichococcus bacillaris	Inclined bubble column	Indoor, artificial light	1.7	7	4	0.3	0.002	300	[88]
Chlorella vulgaris	Light emitting diode (LED) Photobioreactor	Indoor, LED	0.5	8	2	2.1	0.08	300	[89]
Chlorella sp.	Tubular with static mixers	Outdoor, sunlight	883	3	7.5	0.2	15	400	[90]

the day [94]. Eldin et al. [95] found that the increment of about 39% by use of these tracking systems was achieved in electrical output even in cold regions.

In hot cities the maximum electrical output is only 8%, because the elevated temperature overheats the solar panels thus reducing the efficiency of the system. However, the sun tracking system is not economical in countries with a hot climate, if the energy required to run the tracking system is 5%–10% of the electrical energy generated by them [95].

19.6.1 Classification of solar trackers

There are various types of solar tracking systems. They are broadly categorized by their control, drivers applied, tracking strategies, and their degree of freedom, i.e., axis of rotation as shown in Fig. 19.6.

19.6.1.1 Classification based on their control

There are two types of control systems: closed loop tracking systems and open loop tracking systems. In a closed loop tracking system sensors are used to identify the suitable orientation of the sun and give a signal to the system. Then, the error can be detected by using a comparator or microprocessor in the system that gives instructions through signals to motors to rectify the errors, and therefore this principle on which system is working is termed as a feedback control system [96].

In the open loop tracking control system driving signal to the motor is managed by a controller which solely works on the principle of current data inputs. It has no feature of observing and evaluating the output data regarding the desired output. They are simple and economical to install compared to closed loop sun tracking systems [94].

19.6.1.2 Classification based on driving systems

Driving systems can broadly be classified into passive and active solar tracking systems. A passive solar tracking system does not consist of a mechanical device to orient the solar panel in the direction of the sun. Instead it uses some low boiling point compressed gas fluid or shape memory alloys as actuators which on receiving unbalanced illumination, forces the panel to undergo some angular movement so as to re-establish equilibrium of

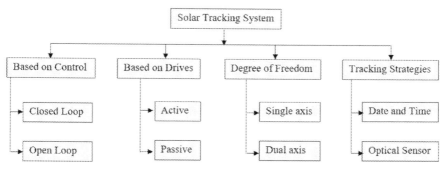

FIGURE 19.6 Classification of solar tracking systems.

irradiance by inducing thermal expansion in expansible gases or in on-shape memory alloys. When a portion of gas encounters more heat energy due to elevated temperature then gas will expand on other side and thus the tracker is moved toward the sun [97]. Active solar tracking systems use electrical drives and mechanical gear trains to orient the panels normal to the sun's radiation. They use sensors, motors, and microprocessors for the tracking and are more accurate and efficient than the passive solar trackers, but they are required to be powered and consume energy. This type of tracker system was used in studies to improve the efficiency and it was revealed that an overall gain of 40% in stored thermal energy is obtained in comparison to the fixed systems [98].

19.6.1.3 Classification based on degree of freedom

Degree of freedom signifies the number of directions in which independent movement can take place. Based on this tracking systems have been classified into single axis and double axis solar tracking systems.

Single axis solar tracking system (SATs) have only one axis about which the rotation is possible to align the panel normal to the sun's horizontal with tilted modules (HTSAT), polar aligned (PSAT), and tilted (TSAT).While a double axis tracking system (DATs) consists of two axes on which rotations are generally normal to each other. SATs are less costly but less efficient too. Furthermore, they are also less complex as compared with DATs. Sun et al. [99] studied the double axis tracking strategies and found that they increased the efficiency by about 15%−17% compared with the single axis tracking system [99].

19.6.2 Classification based on tracking strategy

Date and time-based tracking systems work solely on predefined algorithms which are based on mathematical models related to the sun's movement. A feedback control system or any type of sensor is not employed. Thus the algorithm is solely responsible for the system to work efficiently [100]. Microprocessors and electrooptical sensors based tracking systems require sensors for the detection of the sun's trajectory and send feedback to the microprocessor that gives instructions to motors to change the position of the PV panel. Kalogirou [97] constructed an accurate SATs with the application of three light-dependent resistors as sensors, a DC motor with reduction gears, and an electronic circuit. In this system sensors give feedback to the electronic circuit which then gives commands to the motor [101].

19.6.3 Efficacy of solar tracker in harnessing solar energy for algal cultivation

It was reported in 2015 that solar power contributed about 1% of the total electricity demand in the world. While in some European countries like Germany, Greece, and Italy about 7% electricity demand was met by solar photovoltaic technology [102].

The algal biomass productivity varies because of the following parameters: composition of culture medium, configuration of cultivation system, different light conditions, and temperature regimes (seasonal and geographical variations). The application of photobioreactors continuously encounters some problems, like overheating of the culture medium

(PBRs having high surface to volume ratios). During the diurnal cycles high solar irradiance takes place, which is above the light saturation of photosynthesis. This causes the problem of photoinhibition and thereby decreases the photosynthetic energy utilization leading to the reduced production of algal biomass [103]. The stress condition caused by prolonged periods of intense light may cause photobleaching of the cells and ultimately leading to the collapse of microalgal culture [58].

The technology of solar tracking can be executed to overcome the problems associated with the mass cultivation of algae for the efficient production of biomass. The definition, advantages, and applications for the solar tracker are discussed in the above sections. Solar trackers are turnable devices that helps to direct the photobioreactors to always face the direction of the sun. The orientation of the solar tracking devices is continuously angled to follow the trajectory of the sun, all through the day. The sun is tracked vertically as well as horizontally by solar tracking devices, therefore the maximum capture and collection of solar energy is attained by PBRs, thus enhancing the production of algal biomass [1].

Various advantages of solar tracked PBRs include (1) the use of a solar tracked photobioreactor reduces the chance of photoinhibition of photosynthesis taking place in microalgal cultures of low density by reducing the solar irradiance; (2) increasing the irradiance above 100% of the horizontal irradiance in the cultures of high density by exposing the reactor perpendicularly to the sunlight [104]; (3) the capability to regulate the temperature by providing adjustment of irradiance/cooling to minimize or avoid the development of heat stresses in the microalgal cultures.

19.6.4 Different modes of operation of a solar tracker coupled with photobioreactor

The incident irradiance is controlled by a solar tracking device by working in two different orientations. To attain maximum irradiance the photobioreactors are kept perpendicular to the light and it is termed as standard orientation. Moreover, minimum irradiance is attained by turning the PBRs at 90 degrees angle and this is termed as offset orientation. Since the photobioreactors are operated at these two orientations, therefore the microalgal cultures can automatically run at three different modes, that is, offset mode, standard mode, and temperate mode. Offset mode is a permanent offset orientation for minimum light, standard mode is a permanent standard orientation for maximum light, while in temperate mode the temperature of the culture controls the angle of the PBRs to the sun. Which means that when the temperature of the culture medium increases above the set threshold, the solar tracker turns away from the sun by attaining an offset orientation, and if the temperature falls below the set value the PBRs are positioned back to the standard orientation.

Hindersin et al. [105] experimentally studied solar tracked flat panel photobioreactors and found that they intercepted up to 79 mol $\cdot$ photons $\cdot$ m^{-2} $\cdot$ d^{-1} as compared to photobioreactors without solar tracking that intercepted about 55 mol $\cdot$ photons $\cdot$ m^{-2} $\cdot$ d^{-1}. The study concluded that photobioreactors with solar tracking technology help to overcome the limitation of light, reduce the irradiance thus controlling photoinhibition, resolve the problem of overheating by changing the orientation, and cumulatively increased the productivity of biomass in the photobioreactor [105].

The flat panel photobioreactors mounted on a solar tracking device and a functional line diagram including both outdoor and indoor facilities are shown in Fig. 19.7. The experiments were conducted at an outdoor pilot plant in Northern Germany which utilized sunlight and flue gas from a power plant as the energy and carbon sources, respectively. *Chlorella vulgaris* and *Scenedesmus obliquus* were used in the study as they can tolerate flue gas and can adapt in high irradiance [104].

19.7 Solar to heat for thermochemical conversion of algal biomass

Biomass is termed as a renewable source of energy and over its entire life cycle it is considered nearly carbon neutral. Algal biomass is considered to be a promising source to produce biofuels like bioethanol, biodiesel, biohydrogen, and biomethane, as they are composed of a mixture of carbohydrates, proteins, and lipids. The process of conversion of algal biomass to biofuels can be broadly classified into biochemical, chemical, and thermochemical processes. Thermochemical processes can be further categorized into five

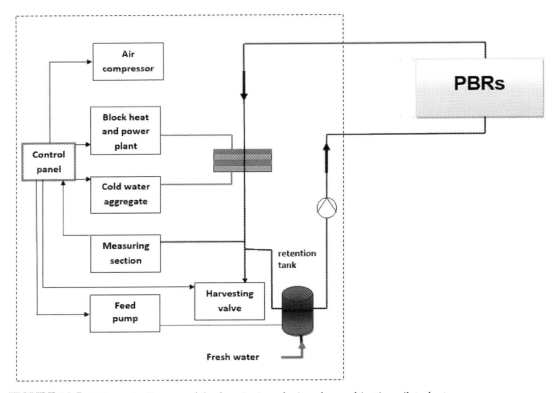

FIGURE 19.7 Schematic diagram of the functioning of microalgae cultivation pilot plant.

subprocesses that are hydrothermal liquefaction, pyrolysis, gasification, torrefaction, and combustion [106].

The selection process of thermochemical conversion is based on various factors like species of algae utilized, the desired form of bioenergy, environmental and economic aspects of the installation/setup, and the end products obtained [76]. The processes and the products obtained are illustrated in Fig. 19.8.

19.7.1 Classification of thermochemical processes

The thermochemical process is utilized to convert the algal biomass into different types of biofuels, as shown in Fig. 19.8. The thermochemical methods generally employed to convert algal biomass into biofuels are gasification, combustion, pyrolysis, liquefaction, and torrefaction [107].

(1) Hydrothermal liquefaction: It takes place at high pressure (5–20 MPa) and temperature (250°C–370°C) in an aqueous medium. The above-stated conditions are subcritical under which the complex biomass is decomposed into smaller molecules having high energy density by the help of repolymerization and hydrolysis reactions [108]. Microalgae are fed as slurry into the HTL reactor with the remaining medium as water. The product obtained is termed as biocrude or bio-oil which can be utilized for multipurposes after refining [109].

When the temperature comes below 250°C the product thus obtained is called hydrochar, which is like coal, and the process is termed as hydrothermal carbonization. From

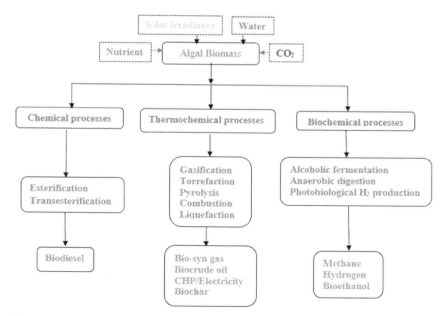

FIGURE 19.8 Conversion processes and the products obtained.

the microalgal perspective hydrochar is obtained mainly from the carbohydrate and protein fractions. When the temperature exceeds 380°C the process has the reactions of the nature of gasification. The product obtained is called syngas and the process is called hydrothermal gasification. The hydrothermal conversion process is an efficient method for the conversion of algae biomass under wet conditions as it does not require pretreatment (drying). Furthermore the small size of microalgae particles help to hasten the transfer, thereby providing advantage in accomplishing the process [110]. The productivity of microalgae is found by the yield of biocrude, ranging from 38% to 64% by weight on a dry basis which corresponds to the efficiency of 60%−78% [111].

(2) Pyrolysis: The thermal decomposition of algal biomass occurs at high temperature (400°C−800°C) in an environment which is inert in terms of atmospheric pressure. The slow and fast pyrolysis can be classified on the basis of biomass residence time and the rate of heating. If the biomass residence time is 60 min and the rate of heating is 5°C/min−10°C/min, then the pyrolysis is called slow pyrolysis. When the biomass residence time is only a few seconds and the rate of heating is 600°C/min, it is called fast pyrolysis. The products obtained from pyrolysis are bio-oil, solid char, and syngas. Microalgal biomass has an advantage over other lingocellulosic feedstock and other algae too as the bio-oil obtained from microalgae is more stable [112].

(3) Torrefaction: It is an exclusive process of mild pyrolysis which is used for augmenting the characteristics of algal biomass and it is achieved by subjecting the algal biomass to a temperature of 200°C−250°C under inert atmospheric conditions to remove the volatile matter present in it [113].

(4) Gasification: In the gasification process controlled and lesser amount of oxygen at high temperature range (700−1000°C) in association with a catalyst decomposes various molecules. The gasification process for microalgae can be grouped into two categories namely, conventional gasification and supercritical gasification. The conventional process of gasification involves partial oxidation of algal biomass by providing a gasification medium of air or steam. The gasification process can further be divided into four subprocesses: (1) drying of the remaining moisture; (2) pyrolysis to break the biomass into simpler molecules; (3) oxidation to burn the incoming biomass to generate heat for further endothermic reaction; and (4) gasification to convert the pyrolysis products into smaller molecules of high energy. The products obtained are syngas, char, tar, and waste ashes. *Scenedesmus* species has been selected for the solar-driven thermochemical processes. Pyrolysis and gasification were carried by irradiating the sample of *Scenedesmus* in 7 kW$_e$ high flux solar simulation; the gases released (H_2, CO_2, CO, and CH_4) and the temperature were continuously monitored (950°C−1050°C). The results showed that the syngas produced by algae in gasification has better quality than the gas produced during pyrolysis [114]. A schematic view is shown in Fig. 19.9.

(5) Combustion: The process of burning algae at 850°C in the presence of oxygen is called combustion. During combustion, the photosynthetically stored chemical energy in the microalgal biomass is converted to hot gases which are utilized immediately for the generation of electricity in the turbine. The algal biomass fed into the combustion chamber should have a moisture content less than 50% [115]. Moreover, the option for the utilization of microalgal biomass is combined heat and power generation.

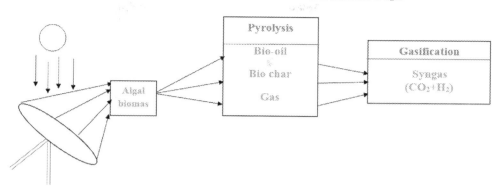

FIGURE 19.9 Illustration showing pyrolysis and gasification of solar irradiated reactors.

Furthermore the sustainable option for the combustion of algal biomass is coal–algae cofiring by which a reduction of CO_2 emissions is achieved by recycling CO_2 from the combustion to the cultivation of algae [116]. The suitable thermochemical process opted for different genera of microalgae, the products, and the operating conditions are shown in Table 19.3.

19.8 Technoeconomic considerations for different routes of conversion of algae

There are two modes for the cultivation of algae, that is, open ponds and closed photobioreactors. The cultivation of microalgae in closed PBRs results in high biomass yield and low risk of contamination but they have very high CAPEX and OPEX.

Open ponds like raceway ponds (paddle wheel driven) are having lesser CAPEX and OPEX but also have low productivity [125]. Raceway ponds are the most accepted and practiced method to produce different types of algal species like *Haematococcus*, *Arthrospira*, *Chlorella*, and *Dunaliella*. The main challenge for algae cultivation is the economic factor, therefore the researchers are promoting the concepts of biorefinery for the utilization of biofuels and other value-added products which are obtained from algal biomass. The productivity of different species of algae cultivated in raceway ponds are compiled in Table 19.4 [132].

19.8.1 Study based on four different scenarios

Four different scenarios of microalgae bioenergy routes were studied for economic aspects in terms of energy output and the abatement of carbon dioxide. Scenario 1, scenario 2, scenario 3, and scenario 4 were summarized as biodiesel production, scenario 1 integrated with anaerobic digestion, biogas production, and supercritical gasification, respectively.

The expenses for all the scenarios were greater than 650,000 $/year for the production of 1000 tons/year of dry microalgae (*C. vulgaris*).

442   19. Harnessing solar radiation for potential algal biomass production

TABLE 19.3 Conversion of distinct species of algae by the appropriate thermochemical process with its operating conditions.

Type of algae	Thermochemical conversion process	Catalyst	Temperature (°C)	Pressure (MPa)	Residence time (min)	Product	Yield (%w)	Biofuel HHV (MJ/kg)	References
Chlorella	FP	No	400–700	0.1	–	Bio-oil	28–72	27.9	[117]
Desmodesmus	HTL	No	300–350	–	60	Bio-oil	41	36	[118]
Dunaliella tertiolecta	SP	No	500	0.1	20	Bio-oil	24	–	[119]
Nannochloropsis	G	No	850	0.1	–	Syn gas		38.3	[120]
Botryococcus braunii	HTL	No	300	–	–	Bio-oil	64	45.9	[121]
Scenedesmus	FP	No	440	0.1	–	Bio-oil	22	29.6	[122]
Spirulina	SP	No	350	0.1	60	Bio-oil	23	29	[123]
Spirulina	SG	Yes	400	30	–	Syn gas	–	–	[124]

FP, Fast Pyrolysis; *HTL*, hydrothermal liquefaction; *SP*, short pyrolysis; *G*, gasification (conventional); *SG*, supercritical gasification.

TABLE 19.4 Estimated cost of distinct species of microalgae grown in raceway ponds.

Type of algae	Culture area (ha)	Productivity (g/m²/d)	Estimated cost (US$/kg)	References
Scenedesmus	4	20	7.56	[126]
Chlorella	10	25–30	12.42	[127]
Spirulina	2	12	12.57	[128]
Dunaliella salina	2	4	12.75	[129]
Nannochloropsis	0.2	8 (Winter) 16 (summer)	54.99	[130]
Spirulina	5	3.2	20.20	[131]

Scenario 4 surpassed the others in terms of net energy production (1282.42 kWh/ton algae) with a CO_2 abatement potential of 1.32-ton CO_2/ton algae. Scenario 2 had the lowest net cost, and the lowest energy was obtained through Scenario 1, while Scenario 3 was the costliest. The expected return on investment (ROI) for the bioenergy system is expected to be 20 years [133].

19.8.2 Study based on 3 different biorefinery routes

The integration of biofuel production with value added products reduces the capital and operating cost, thereby making it economically viable and sustainable [134]. Different routes studied for the production of biofuels and other value-added products are classified into routes 1, 2, and 3, respectively, and are elaborated accordingly, inclusive of their processes [135].

Handbook of Algal Biofuels

The biorefinery route 1 is inclusive of cultivation, harvesting, lipid, and biodiesel production from microalgae. The algae production cost falls in the range of 150−6000 $/tons depending upon the techniques adopted. In closed photobioreactors and open ponds a 10% return was obtained with a selling price of 8.52 $/gallon of triglycerides [136]. Bubble column PBRs have a high cost (81.17% of the total cost) and low productivity, while the tubular photobioreactors have higher biomass productivity and low cost, therefore decreasing the required number of reactors [137]. For the bulk production of microalgae, open ponds are less expensive but the cost increases in terms of the large requirements of water and productivity is also less compared to closed PBRs. The cost of cultivation of microalgae is about 45.73% of the total cost, therefore they are more profitable than closed PBRs as the investment of PBRs is four times higher than the open ponds.

The harvesting of microalgae accounts for 20%−30% of the production cost. Centrifugation is the common method to harvest microalgae, but it requires high energy and capital cost, while flocculation is better in terms of these two aspects. The cost of flocculation and filtration are 2000 and 9884 $/hm^2. The cost generated by biofuels is about 2.8 $/L, which is greater than 1.1 $/L for petroleum. With the introduction of innovative technologies, capable algal strains with high lipid content and high growth rates can increase the productivity of algae. The biodiesel cost for open ponds and PBR are 9.8 and 20.5 $/gallon, respectively [138].

The lipid production can be increased from 30% to 50% by using *Dunaliella* sp. and *Chlorella* sp. The cost of the production of biodiesel ranges from 9.8−20.5 $/gallon with the selling price of 5−22 $/gallon [139]. The system can be made more economical by cultivating *Dunaliella salina* in open ponds for β-carotene and *Haematococcus pluvialis* in PBR for astaxanthin, as they are rich in these pigments that have high market; thereby the system can be made economically viable [140].

In the biorefinery scenario 2, biomethane and biohydrogen production from microalgae is studied. The heat energy and electricity consumed in the process of anaerobic digestion are 2.5 and 0.39 MJ/kg$_{DM}$, respectively and the methane production varies from 0.16 to 0.8 CH$_4$/kg [141]. For the process of two-stage fermentation the cost of hydrogen production is about 12 to 15 times more than the cost of biogas produced for the same energy consumption. In the two-stage fermentation the production of methane is about 22 times higher than for single-stage fermentation [142].

In the biorefinery scenario 3 the production of pigments from *Spirogyra* species was 0.12 g/100 g of algal biomass, the energy utilized was 15 MJ/g$_{(microalgae)}$, and the process emitted about 13 gCO$_2$eq/g$_{(microalgae)}$ [143]. The pigments obtained were astaxanthin (56%), β-carotene (16%), lutein (5%), and canthaxanthin (5%). Hydrogen was produced at a cost of about 0.3 $/kg which is four times greater than the natural gas [144].

19.9 Conclusion

The chapter provides extensive details about different applications of solar photovoltaic systems and solar cells. The cultivation of algae is solely dependent on light, therefore different options and configurations that are utilized to harness solar energy for their cultivation and algal biomass production are elaborated. The harvesting of microalgal biomass is carried out to obtain biofuels and different value-added products by different processes like biochemical, chemical, and thermochemical processes. Thermochemical processes are classified with the perspective of obtaining different useful products like bio-oil and syngas. Lastly the economic aspect of different routes of biorefinery is discussed to show that the cultivation and harvesting of algae can be accomplished to obtain different value products. This chapter will provide an insight into the harnessing solar radiation for the cultivation of algal biomass.

Acknowledgment

The authors are thankful to Malaysia-Japan International Institute of Technology, Universiti Teknologi Malaysia for their support.

Conflict of interest

The authors declare no conflict of interest.

References

[1] E.G. Nwoba, et al., Light management technologies for increasing algal photobioreactor efficiency, Algal Res. 39 (2019) 101433.
[2] M.F. Hossain, Solar energy integration into advanced building design for meeting energy demand and environment problem, Int. J. Energy Res. 40 (9) (2016) 1293−1300.
[3] I.B. Fridleifsson, Status of geothermal energy amongst the world's energy sources, Geothermics 32 (4−6) (2003) 379−388.
[4] D. Kardaš, D. Bogdan, Biomass and waste water as sustainable energy sources, Industry 4.0 2 (1) (2017) 29−33.
[5] M. Stockenreiter, et al., Nutrient uptake and lipid yield in diverse microalgal communities grown in wastewater, Algal Res. 15 (2016) 77−82.
[6] E.G. Nwoba, et al., Energy efficiency analysis of outdoor standalone photovoltaic-powered photobioreactors coproducing lipid-rich algal biomass and electricity, Appl. Energy 275 (2020) 115403.
[7] Y. Chisti, Biodiesel from microalgae beats bioethanol, Trends Biotechnol. 26 (3) (2008) 126−131.
[8] A.P. Batista, et al., *Microalgae biomass as an alternative ingredient in cookies: sensory, physical and chemical properties, antioxidant activity and* in vitro *digestibility*, Algal Res. 26 (2017) 161−171.
[9] Y. Chisti, Biodiesel from microalgae, Biotechnol. Adv. 25 (3) (2007) 294−306.
[10] T. Ishika, N.R. Moheimani, P.A. Bahri, Sustainable saline microalgae co-cultivation for biofuel production: a critical review, Renew. Sustain. Energy Rev. 78 (2017) 356−368.
[11] R.A. Voloshin, et al., Biofuel production from plant and algal biomass, Международный научный журнал Альтернативная энергетика и экология 7−9 (2019) 12−31.
[12] F. Naaz, et al., Investigations on energy efficiency of biomethane/biocrude production from pilot scale wastewater grown algal biomass, Appl. Energy 254 (2019) 113656.
[13] M.M. Addy, et al., Co-cultivation of microalgae in aquaponic systems, Bioresour. Technol. 245 (2017) 27−34.
[14] C. Ugwu, H. Aoyagi, H. Uchiyama, Photobioreactors for mass cultivation of algae, Bioresour. Technol. 99 (10) (2008) 4021−4028.
[15] P. Maza-Márquez, et al., Full-scale photobioreactor for biotreatment of olive washing water: structure and diversity of the microalgae-bacteria consortium, Bioresour. Technol. 238 (2017) 389−398.
[16] A. Yadav, P. Kumar, M. RPSGOI, Enhancement in efficiency of PV cell through P&O algorithm, Int. J. Technol. Res. Eng. 2 (2015) 2642−2644.

References

445

[17] R.N. Castellano, Solar panel processing, Archives contemporaines, 2010.

[18] B. Srinivas, et al., Review on present and advance materials for solar cells, Int. J. Eng. Research-Online 3 (2015) 178—182.

[19] A.M. Bagher, M.M.A. Vahid, M. Mohsen, Types of solar cells and application, Am. J. Opt. Photonics 3 (5) (2015) 94—113.

[20] K. Chopra, P. Paulson, V. Dutta, Thin-film solar cells: an overview, Prog. Photovol.: Res. Appl. 12 (2-3) (2004) 69—92.

[21] P. Choubey, A. Oudhia, R. Dewangan, A review: solar cell current scenario and future trends, Recent. Res. Sci. Technol. 4 (2012) 8.

[22] M.T. Kibria, et al., A review: comparative studies on different generation solar cells technology, in: Proceeding of 5th International Conference on Environmental Aspects of Bangladesh, 2014.

[23] A.K. Geim, K.S. Novoselov, The rise of graphene, in: Nanoscience and Technology: A Collection of Reviews From Nature Journals, World Scientific, 2010, pp. 11—19.

[24] S. Iijima, Helical microtubules of graphitic carbon, Nature 354 (6348) (1991) 56—58.

[25] A.M. Díez-Pascual, et al., High performance PEEK/carbon nanotube composites compatibilized with polysulfones-I. Structure and thermal properties, Carbon 48 (12) (2010) 3485—3499.

[26] A.M. Díez-Pascual, et al., Development and characterization of PEEK/carbon nanotube composites, Carbon 47 (13) (2009) 3079—3090.

[27] L. Lavagna, et al., Grafting carbon nanotubes onto carbon fibres doubles their effective strength and the toughness of the composite, Compos. Sci. Technol. 166 (2018) 140—149.

[28] M.-F. Yu, et al., Strength and breaking mechanism of multiwalled carbon nanotubes under tensile load, Science 287 (5453) (2000) 637—640.

[29] T. Dürkop, et al., Extraordinary mobility in semiconducting carbon nanotubes, Nano Lett. 4 (1) (2004) 35—39.

[30] Y. Baghzouz, Basic photovoltaic theory, Handbook of clean energy systems, 2015, pp. 1—13.

[31] A. Goetzberger, A. Zastrow, On the coexistence of solar-energy conversion and plant cultivation, Int. J. Sol. Energy 1 (1) (1982) 55—69.

[32] A. Weselek, et al., Agrophotovoltaic systems: applications, challenges and opportunities. A review, Agron. Sustain. Dev. 39 (4) (2019) 35.

[33] M. Trommsdorff, An economic analysis of agrophotovoltaics: opportunities, risks and strategies towards a more efficient land use, The Constitutional Economics Network Working Papers, 2016.

[34] B.L. Ward, et al., Creating a circular economy with urban farming, 2019.

[35] R.L. Naylor, et al., 2000. Effect of aquaculture on world fish supplies. Nature, 405 (6790) (2000) 1017—1024.

[36] F.C. Prinsloo, Development of a GIS-based decision support tool for environmental impact assessment and due-diligence analyses of planned agricultural floating solar systems, 2019.

[37] J. Xue, Photovoltaic agriculture-new opportunity for photovoltaic applications in China, Renew. Sustain. Energy Rev. 73 (2017) 1—9.

[38] E. Ertrac, SmartGrids, European Roadmap to electrification of road transport, 2012.

[39] D.S. Khatawkar, P.S. James, D. Dhalin, Modern trends in farm machinery-electric drives: a review, Int. J. Curr. Microbiol. Appl. Sci. 8 (1) (2019) 83—98.

[40] REN21, R.E.P.N., Renewables 2017: global status report, vol 72, no. 2016, Oct.

[41] F. Caballero, E. Sauma, F. Yanine, Business optimal design of a grid-connected hybrid PV (photovoltaic)-wind energy system without energy storage for an Easter Island's block, Energy 61 (2013) 248—261.

[42] A. Soroudi, M. Ehsan, A possibilistic—probabilistic tool for evaluating the impact of stochastic renewable and controllable power generation on energy losses in distribution networks—a case study, Renew. Sustain. Energy Rev. 15 (1) (2011) 794—800.

[43] S. Gorjian, et al., Applications of solar PV systems in agricultural automation and robotics, Photovoltaic Solar Energy Conversion, Elsevier, 2020, pp. 191—235.

[44] Z. Gao, et al., Progress on solar photovoltaic pumping irrigation technology, Irrig. Drain. 67 (1) (2018) 89—96.

[45] S. Chandel, M.N. Naik, R. Chandel, Review of solar photovoltaic water pumping system technology for irrigation and community drinking water supplies, Renew. Sustain. Energy Rev. 49 (2015) 1084—1099.

[46] M. Fadhel, et al., Review on advanced of solar assisted chemical heat pump dryer for agriculture produce, Renew. Sustain. Energy Rev. 15 (2) (2011) 1152—1168.

[47] S. Tiwari, G. Tiwari, I. Al-Helal, Development and recent trends in greenhouse dryer: a review, Renew. Sustain. Energy Rev. 65 (2016) 1048—1064.

Handbook of Algal Biofuels

[48] D. González, J. Amigo, F. Suárez, Membrane distillation: perspectives for sustainable and improved desalination, Renew. Sustain. Energy Rev. 80 (2017) 238–259.

[49] S.E. Moore, et al., Process modeling for economic optimization of a solar driven sweeping gas membrane distillation desalination system, Desalination 437 (2018) 108–120.

[50] Y. Zhang, et al., Application of solar energy in water treatment processes: a review, Desalination 428 (2018) 116–145.

[51] M. Borowitzka, Phycology. In eLS. Volume, John Wiley & Sons, Ltd, 2012.

[52] N. Gardner, et al., Blue-emitting InGaN–GaN double-heterostructure light-emitting diodes reaching maximum quantum efficiency above 200 A/cm 2, Appl. Phys. Lett. 91 (24) (2007) 243506.

[53] H. Ronge, V. Niture, M.D. Ghodake, A Review Paper on Utilization of Solar Energy for Cooking, Imperial International Journal of Eco-Friendly Technologies (IIJET), 2016, p. 34.

[54] T.S. Vinubhai, R. Vishal, K. Thakkar. A review: solar water heating systems, in: The National Conference on Epmerging Vista of Technology in the 21st Century, Limda, Vadodara, 2014.

[55] J.A. Duffie, W.A. Beckman, N. Blair, Solar Engineering of Thermal Processes, Photovoltaics and Wind, John Wiley & Sons, 2020.

[56] V. Okoro, et al., Microalgae cultivation and harvesting: growth performance and use of flocculants-a review, Renew. Sustain. Energy Rev. 115 (2019) 109364.

[57] Z. Yin, et al., A comprehensive review on cultivation and harvesting of microalgae for biodiesel production: environmental pollution control and future directions, Bioresour. Technol. 301 (2020) 122804.

[58] A.P. Carvalho, et al., Light requirements in microalgal photobioreactors: an overview of biophotonic aspects, Appl. Microbiol. Biotechnol. 89 (5) (2011) 1275–1288.

[59] A. Vadiveloo, et al., *Effect of different light spectra on the growth and productivity of acclimated Nannochloropsis* sp. *(Eustigmatophyceae)*, Algal Res. 8 (2015) 121–127.

[60] E.G. Nwoba, et al., Growth comparison of microalgae in tubular photobioreactor and open pond for treating anaerobic digestion piggery effluent, Algal Res. 17 (2016) 268–276.

[61] P. Kuczynska, M. Jemiola-Rzeminska, K. Strzalka, Photosynthetic pigments in diatoms, Mar. Drugs 13 (9) (2015) 5847–5881.

[62] R. Croce, H. van Amerongen, Light-harvesting and structural organization of photosystem II: from individual complexes to thylakoid membrane, J. Photochem. Photobiol. B: Biol. 104 (1–2) (2011) 142–153.

[63] Z. Perrine, S. Negi, R.T. Sayre, Optimization of photosynthetic light energy utilization by microalgae, Algal Res. 1 (2) (2012) 134–142.

[64] R.E. Blankenship, et al., Comparing the efficiency of photosynthesis with photovoltaic devices and recognizing opportunities for improvement, Science 332 (2011). ANL/CSE/JA-69467.

[65] A. Vadiveloo, et al., *Photosynthetic performance of two Nannochloropsis* spp. *under different filtered light spectra*, Algal Res. 19 (2016) 168–177.

[66] M.A. Borowitzka, High-value products from microalgae—their development and commercialisation, J. Appl. Phycol. 25 (3) (2013) 743–756.

[67] J.C. Quinn, R. Davis, The potentials and challenges of algae based biofuels: a review of the techno-economic, life cycle, and resource assessment modeling, Bioresour. Technol. 184 (2015) 444–452.

[68] R. Bosma, et al., Design and construction of the microalgal pilot facility AlgaePARC, Algal Res. 6 (2014) 160–169.

[69] T. de Mooij, et al., Impact of light color on photobioreactor productivity, Algal Res. 15 (2016) 32–42.

[70] H.D. Amrei, et al., Using fluorescent material for enhancing microalgae growth rate in photobioreactors, J. Appl. Phycol. 27 (1) (2015) 67–74.

[71] E. Darko, et al., Photosynthesis under artificial light: the shift in primary and secondary metabolism, Philos. Trans. R. Soc. B: Biol. Sci. 369 (1640) (2014) 20130243.

[72] M.D. Ooms, Y. Jeyaram, D. Sinton, Wavelength-selective plasmonics for enhanced cultivation of microalgae, Appl. Phys. Lett. 106 (6) (2015) 063902.

[73] J. Wang, J. Liu, T. Liu, The difference in effective light penetration may explain the superiority in photosynthetic efficiency of attached cultivation over the conventional open pond for microalgae, Biotechnol. Biofuels 8 (1) (2015) 1–12.

[74] Y. Sun, et al., Enhancement of microalgae production by embedding hollow light guides to a flat-plate photobioreactor, Bioresour. Technol. 207 (2016) 31–38.

[75] K.K. Sakimoto, A.B. Wong, P. Yang, Self-photosensitization of nonphotosynthetic bacteria for solar-to-chemical production, Science 351 (6268) (2016) 74–77.

References

447

[76] L. Brennan, P. Owende, Biofuels from microalgae—a review of technologies for production, processing, and extractions of biofuels and co-products, Renew. Sustain. Energy Rev. 14 (2) (2010) 557–577.

[77] Ahmad et al., Evolution of photobioreactors: a review based on microalgal perspective. In: IOP Conference Series: Materials Science and Engineering (Vol. 1142, No. 1, p. 012004). IOP Publishing, 2021.

[78] Q. Béchet, A. Shilton, B. Guieysse, Modeling the effects of light and temperature on algae growth: state of the art and critical assessment for productivity prediction during outdoor cultivation, Biotechnol. Adv. 31 (8) (2013) 1648–1663.

[79] J. Pruvost, et al., Microalgae culture in building-integrated photobioreactors: biomass production modelling and energetic analysis, Chem. Eng. J. 284 (2016) 850–861.

[80] M.R. Tredici, et al., Techno-economic analysis of microalgal biomass production in a 1-ha Green Wall Panel (GWP®) plant, Algal Res. 19 (2016) 253–263.

[81] G.C. Zittelli, et al., Productivity and photosynthetic efficiency of outdoor cultures of Tetraselmis suecica in annular columns, Aquaculture 261 (3) (2006) 932–943.

[82] J.H. De Vree, et al., Comparison of four outdoor pilot-scale photobioreactors, Biotechnol. Biofuels 8 (1) (2015) 215.

[83] A.M. Blanco, et al., *Outdoor cultivation of lutein-rich cells of Muriellopsis* sp. *in open ponds*, Appl. Microbiol. Biotechnol. 73 (6) (2007) 1259–1266.

[84] A.M. Kliphuis, et al., Photosynthetic efficiency of Chlorella sorokiniana in a turbulently mixed short light-path photobioreactor, Biotechnol. Prog. 26 (3) (2010) 687–696.

[85] B. Crowe, et al., A comparison of Nannochloropsis salina growth performance in two outdoor pond designs: conventional raceways versus the ARID pond with superior temperature management, Int. J. Chem. Eng. 2012 (2012).

[86] A. Jacobi, et al., The application of transparent glass sponge for improvement of light distribution in photo-bioreactors, J. Bioprocess. Biotech. 2 (01) (2012) 1–8.

[87] J. Masojídek, et al., Productivity correlated to photobiochemical performance of Chlorella mass cultures grown outdoors in thin-layer cascades, J. Ind. Microbiol. Biotechnol. 38 (2) (2011) 307–317.

[88] G. Olivieri, et al., Effects of photobioreactors design and operating conditions on Stichococcus bacillaris biomass and biodiesel production, Biochem. Eng. J. 74 (2013) 8–14.

[89] W. Fu, et al., Maximizing biomass productivity and cell density of Chlorella vulgaris by using light-emitting diode-based photobioreactor, J. Biotechnol. 161 (3) (2012) 242–249.

[90] Q. Zhang, et al., Study of hydrodynamic characteristics in tubular photobioreactors, Bioprocess. Biosyst. Eng. 36 (2) (2013) 143–150.

[91] M.D. Ooms, et al., Photon management for augmented photosynthesis, Nat. Commun. 7 (1) (2016) 1–13.

[92] N. Moheimani, *Tetraselmis suecica culture for CO$_2$ bioremediation of untreated flue gas from a coal-fired power station*, J. Appl. Phycol. 28 (4) (2016) 2139–2146.

[93] M.J. Raeesossadati, et al., *CO$_2$ Environmental bioremediation by microalgae*, Biomass and Biofuels from Microalgae, Springer, 2015, pp. 117–136.

[94] A. Awasthi, et al., Review on sun tracking technology in solar PV system, Energy Rep. 6 (2020) 392–405.

[95] S.S. Eldin, M. Abd-Elhady, H. Kandil, Feasibility of solar tracking systems for PV panels in hot and cold regions, Renew. Energy 85 (2016) 228–233.

[96] I. Stamatescu, et al., Design and implementation of a solar-tracking algorithm, Procedia Eng. 69 (2014) 500–507.

[97] A. Hafez, et al., A comprehensive review for solar tracking systems design in Photovoltaic cell, module, panel, array, and systems applications, in: 2018 IEEE 7th World Conference on Photovoltaic Energy Conversion (WCPEC)(A Joint Conference of 45th IEEE PVSC, 28th PVSEC & 34th EU PVSEC), IEEE, 2018.

[98] M. Abdelghani-Idrissi, et al., Solar tracker for enhancement of the thermal efficiency of solar water heating system, Renew. Energy 119 (2018) 79–94.

[99] J. Sun, et al., An optimized tracking strategy for small-scale double-axis parabolic trough collector, Appl. Therm. Eng. 112 (2017) 1408–1420.

[100] R. Nuwayhid, F. Mrad, R. Abu-Said, The realization of a simple solar tracking concentrator for university research applications, Renew. Energy 24 (2) (2001) 207–222.

[101] S.A. Kalogirou, Design and construction of a one-axis sun-tracking system, Sol. Energy 57 (6) (1996) 465–469.

Handbook of Algal Biofuels

[102] R.C. Allil, et al., Solar tracker development based on a POF bundle and Fresnel lens applied to environment illumination and microalgae cultivation, Sol. Energy 174 (2018) 648–659.

[103] I.S. Suh, C.-G. Lee, Photobioreactor engineering: design and performance, Biotechnol. bioprocess. Eng. 8 (6) (2003) 313.

[104] C.J. Hulatt, D.N. Thomas, Energy efficiency of an outdoor microalgal photobioreactor sited at mid-temperate latitude, Bioresour. Technol. 102 (12) (2011) 6687–6695.

[105] S. Hindersin, et al., Irradiance optimization of outdoor microalgal cultures using solar tracked photobioreactors, Bioprocess. Biosyst. Eng. 36 (3) (2013) 345–355.

[106] C. Silva, et al., A comparison between microalgae virtual biorefinery arrangements for bio-oil production based on lab-scale results, J. Clean. Prod. 130 (2016) 58–67.

[107] S. Pourkarimi, et al., Biofuel production through micro-and macroalgae pyrolysis–a review of pyrolysis methods and process parameters, J. Anal. Appl. Pyrol. 142 (2019) 104599.

[108] D.L. Barreiro, et al., Hydrothermal liquefaction (HTL) of microalgae for biofuel production: state of the art review and future prospects, Biomass Bioenergy 53 (2013) 113–127.

[109] J.K. Mwangi, et al., Microalgae oil: algae cultivation and harvest, algae residue torrefaction and diesel engine emissions tests, Aerosol Air Qual. Res. 15 (1) (2014) 81–98.

[110] S.M. Heilmann, et al., Hydrothermal carbonization of microalgae II. Fatty acid, char, and algal nutrient products, Appl. Energy 88 (10) (2011) 3286–3290.

[111] D.C. Elliott, et al., Process development for hydrothermal liquefaction of algae feedstocks in a continuous-flow reactor, Algal Res. 2 (4) (2013) 445–454.

[112] A. Baldinelli, et al., Addressing the energy sustainability of biowaste-derived hard carbon materials for battery electrodes, Green. Chem. 20 (7) (2018) 1527–1537.

[113] K. Heimann, R. Huerlimann, Microalgal classification: major classes and genera of commercial microalgal species, Handbook of Marine Microalgae, Elsevier, 2015, pp. 25–41.

[114] L. Arribas, et al., Solar-driven pyrolysis and gasification of low-grade carbonaceous materials, Int. J. Hydrog. Energy 42 (19) (2017) 13598–13606.

[115] A. Raheem, et al., Thermochemical conversion of microalgal biomass for biofuel production, Renew. Sustain. Energy Rev. 49 (2015) 990–999.

[116] D.L. Medeiros, E.A. Sales, A. Kiperstok, Energy production from microalgae biomass: carbon footprint and energy balance, J. Clean. Prod. 96 (2015) 493–500.

[117] G. Belotti, et al., Effect of Chlorella vulgaris growing conditions on bio-oil production via fast pyrolysis, Biomass Bioenergy 61 (2014) 187–195.

[118] L. Garcia Alba, et al., Hydrothermal treatment (HTT) of microalgae: evaluation of the process as conversion method in an algae biorefinery concept, Energy Fuels 26 (1) (2012) 642–657.

[119] S. Grierson, et al., Thermal characterisation of microalgae under slow pyrolysis conditions, J. Anal. Appl. Pyrol. 85 (1–2) (2009) 118–123.

[120] H. Khoo, et al., Bioenergy co-products derived from microalgae biomass via thermochemical conversion–life cycle energy balances and CO_2 emissions, Bioresour. Technol. 143 (2013) 298–307.

[121] Y. Dote, et al., Recovery of liquid fuel from hydrocarbon-rich microalgae by thermochemical liquefaction, Fuel 73 (12) (1994) 1855–1857.

[122] S.W. Kim, B.S. Koo, D.H. Lee, A comparative study of bio-oils from pyrolysis of microalgae and oil seed waste in a fluidized bed, Bioresour. Technol. 162 (2014) 96–102.

[123] U. Jena, K. Das, Comparative evaluation of thermochemical liquefaction and pyrolysis for bio-oil production from microalgae, Energy Fuels 25 (11) (2011) 5472–5482.

[124] S. Stucki, et al., Catalytic gasification of algae in supercritical water for biofuel production and carbon capture, Energy Environ. Sci. 2 (5) (2009) 535–541.

[125] G.C. Zittelli, et al., Photobioreactors for mass production of microalgae, in: Handbook of Microalgal Culture: Applied Phycology and Biotechnology second ed., 2013, pp. 225–266.

[126] E. Becker, L. Venkataraman, Production and processing of algae in pilot plant scale experiences of the Indo-German project, in: G. Shelef, C.J. Soeder (Eds.), Algae Biomass: Production and Use/[Sponsored by the National Council for Research and Development, Israel and the Gesellschaft fur Strahlen-und Umweltforschung (GSF), Munich, Germany], 1980.

[127] K. Kawaguchi, Microalgae production systems in Asia, in: G. Shelef, C.J. Soeder (Eds.), Algae Biomass: Production and Use/[sponsored by the National Council for Research and Development, Israel and the Gesellschaft fur Strahlen-und Umweltforschung (GSF), Munich, Germany], 1980.

[128] M. Rebeller, Techniques de culture et de récolte des algues spirulines, Doc. IFP 14 (1982) 109−144.

[129] F.H. Mohn, O.C. Contreras, Harvesting of the alga Dunaliella: some considerations concerning its cultivation and impact on the production costs of b-carotene, Research Center. Jüllich, Antofagasta, Chile (1990).

[130] O. Zmora, A. Richmond, 20 Microalgae for aquaculture, in: Handbook of Microalgal Culture: Biotechnology and Applied Phycology, 2004, p. 365.

[131] A. Jassby, Spirulina: a model for microalgae as human food, Algae Hum. Aff. (1988) 149−179.

[132] M. Raeisossadati, N.R. Moheimani, D. Parlevliet, Luminescent solar concentrator panels for increasing the efficiency of mass microalgal production, Renew. Sustain. Energy Rev. 101 (2019) 47−59.

[133] J.-R.S. Ventura, et al., *Life cycle analyses of CO_2, energy, and cost for four different routes of microalgal bioenergy conversion*, Bioresour. Technol. 137 (2013) 302−310.

[134] R. Kothari, et al., Microalgal cultivation for value-added products: a critical enviro-economical assessment, 3 Biotech. 7 (4) (2017) 243.

[135] J.R. Banu, et al., Microalgae based biorefinery promoting circular bioeconomy-techno economic and life-cycle analysis, Bioresour. Technol. 302 (2020) 122822.

[136] J. Hoffman, et al., Techno-economic assessment of open microalgae production systems, Algal Res. 23 (2017) 51−57.

[137] M. Faried, et al., Biodiesel production from microalgae: processes, technologies and recent advancements, Renew. Sustain. Energy Rev. 79 (2017) 893−913.

[138] R. Katiyar, et al., *Utilization of de-oiled algal biomass for enhancing vehicular quality biodiesel production from Chlorella sp. in mixotrophic cultivation systems*, Renew. Energy 122 (2018) 80−88.

[139] A. Kouzuma, K. Watanabe, Exploring the potential of algae/bacteria interactions, Curr. Opin. Biotechnol. 33 (2015) 125−129.

[140] G. Thomassen, et al., A techno-economic assessment of an algal-based biorefinery, Clean. Technol. Environ. Policy 18 (6) (2016) 1849−1862.

[141] D. Mu, et al., Life cycle assessment and nutrient analysis of various processing pathways in algal biofuel production, Bioresour. Technol. 230 (2017) 33−42.

[142] S.C. Togarcheti, et al., Life cycle assessment of microalgae based biodiesel production to evaluate the impact of biomass productivity and energy source, Resour. Conserv. Recycl. 122 (2017) 286−294.

[143] R. Pacheco, et al., *The production of pigments & hydrogen through a Spirogyra sp. biorefinery*, Energy Convers. Manag. 89 (2015) 789−797.

[144] M. Kumar, A.O. Oyedun, A. Kumar, A comparative analysis of hydrogen production from the thermochemical conversion of algal biomass, Int. J. Hydrog. Energy 44 (21) (2019) 10384−10397.

Further reading

A. Energy, Alternative Energy Tutorials, 2017.

Ahmad, I., Abdullah, N., Koji, I., Yuzir, A. and Mohamad, S.E., 2021. Potential of Microalgae in Bioremediation of Wastewater. Bulletin of Chemical Reaction Engineering & Catalysis, 16(2), pp. 413−429.

Ahmad, I., Yuzir, A., Mohamad, S.E., Iwamoto, K. and Abdullah, N., 2021, February. Role of Microalgae in Sustainable Energy and Environment. In IOP Conference Series: Materials Science and Engineering (Vol. 1051, No. 1, p. 012059). IOP Publishing.

Ahmad, I., Abdullah, N., Yuzir, A., Koji, I. and Mohamad, S.E., 2020. Efficacy of microalgae as a nutraceutical and sustainable food supplement. In 3rd ICA Research Symposium (ICARS) 2020 (p. 6).

Physical stress for enhanced biofuel production from microalgae

Sivakumar Esakkimuthu[1], Shuang Wang[1] and Abd EL-Fatah Abomohra[2,3]

[1]New Energy Department, School of Energy and Power Engineering, Jiangsu University, Jiangsu, P. R. China [2]Department of Environmental Engineering, School of Architecture and Civil Engineering, Chengdu University, Chengdu, P. R. China [3]Botany and Microbiology Department, Faculty of Science, Tanta University, Tanta, Egypt

20.1 Introduction

Microalgae require a favorable environment for their growth and living. Hence, altering the growth conditions affects microalgal physiology and metabolism. The term stress refers to the unfavorable growth conditions faced by microalgae either naturally or artificially. The factors responsible for such stress may be either physical (light, temperature, etc.) or chemical (nutrients) [1]. Microalgae require several nutrients, light, and water to perform photosynthesis, in which light energy can be converted into chemical energy. Hence, the trivial changes of these vital requirements could significantly affect the process of photosynthesis which in turn affects the cascade of microalgal metabolism [2]. Like plants, photosynthesis is the energy-acquiring process which helps in the survival of plants. Photosynthesis requires carbon dioxide, light, water, and nutrients to convert inorganic carbon compounds into an energy reserve, such as starch [3]. Hence, the alteration of these nutrients or physical parameters could potentially initiate a cascade of reactions in microalgae [4]. The structural component and energy reserves of microalgae are mainly carbohydrates, proteins, and lipids. The ultimate reflection of such stress conditions would be the fluctuation of biochemical proportions such as carbohydrates, lipids, and proteins. These biochemical proportions are the raw materials for the production of biofuels. For instance, lipids are the feedstock for biodiesel production through transesterification, whereas carbohydrates are utilized for alcohol production through fermentation [5–7]. Also, all the three macromolecules proportions could

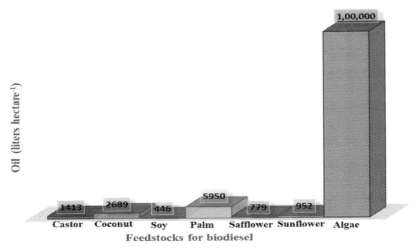

FIGURE 20.1 Oil productivity of algae compared to other biodiesel feedstocks [11].

be utilized for bio-oil production [8,9]. Hence, the strategy of altering these proportions for enhanced biofuel production potentials are practiced worldwide. For instance, immediate technological development is needed to increase the organic carbon by around 6 billion tons for effective biofuel production [10].

At present, it is clear that microalgae could potentially overtake the productivity of terrestrial crops and be established as promising feedstock for biofuel production (Fig. 20.1). Though there are novel raw materials (such as FOG and waste coffee grounds) that have been suggested for biodiesel production [12,13], microalgae still hold potential. However, the present productivity of microalgae and some technological bottlenecks are the prevailing challenges for microalgal-based biofuel production [14,15]. In such terms, tools, such as induction of stress, are effectively used to increase the productivity of microalgae [16]. This increase in the productivity results in an increase in the economy of microalgal-biofuel production [17]. Exploitation of such stress responding behavior of microalgae for an increase of biofuel productivity is of interest worldwide and has been practiced for more than a decade. However, there are challenges with regard to feasibility, failures of targeted compound productivity, the cost of the techniques, and energy expense in stress-mediated biofuel production [18]. On the other hand, there are numerous strategies that have been effectively devised by researchers worldwide and are successfully applied. This chapter deals with the common types of stress induction, cellular response, degree of productivity increases, and the challenges associated with the stress induction strategies.

20.2 Nutrient stress

As mentioned earlier, there are stringent nutrient requirements for microalgae, such as macronutrients (nitrogen, phosphorus, sulfur, magnesium, iron, and silicon) and micronutrients (zinc, molybdenum, copper, cobalt, and manganese), to carry out their essential processes such as photosynthesis, respiration, cell division, and protein synthesis [19]. Among these nutrients, nitrogen, and phosphorus are required in large quantities and other nutritional requirements slightly vary depending upon species. Since nitrogen is a primary element, "nitrogen mediated stress," especially nitrogen starvation, significantly influences lipid production. Nitrogen depletion significantly reduces cell division and radiates stress signals within the cell which in turn reflects the accumulation of energy-rich compounds, such as lipids. Precisely, the stress signals lead to reduced structural lipid synthesis and synthesis of more neutral lipids. Also, there is inhibition of citrate synthase, thereby preventing the acetyl CoA from entering into the tricarboxylic acid (TCA) cycle. Hence, the abundant acetyl CoA activates acetyl CoA carboxylase, which performs the first committed step of lipid biosynthesis [20]. Through nitrogen starvation, the lipid accumulation can be increased up to and above 50% of dry cell weight. For instance, nitrogen starvation promoted lipid content up to 40.7% and 54% in *Chlorococcum nivale* and *Scenedesmus destricola*, respectively [21]. Under nitrogen limitations, the lipid content was raised to 46% in *Chaetoceros muelleri* [22]. It is proven that varying nitrogen concentrations and types of nitrogen sources plays a major role in lipid increment. The nitrogen concentration variation also showed profound influences in lipid accumulation [21]. Hence, a nitrogen variation strategy is found to be the most familiar strategy among the nutrient stresses for improving lipid accumulation and hence act as the most promising approach for improving the biodiesel potential of microalgae. As lipids are potentially enriched and enhanced under nitrogen stress conditions, thus improving the biodiesel recovery of microalgae, nitrogen stress also improves the carbohydrate production and aids in increasing bioethanol production. For instance, a nitrogen stress strategy was employed in *Chlorella vulgaris* which resulted in 22% of carbohydrate, which finally led to the maximum ethanol yield of 88%, along with saccharification optimization [23].

Next to nitrogen, phosphorus is another important nutrient which can be adjusted to create metabolic stress in microalgae. Phosphorus is critical in various cellular processes, such as photosynthesis, energy transfer, and cellular signaling processes [2]. By using phosphorus limitation as a lipid enhancing tool, 53% of lipid content was achieved in *Scenedesmus* sp. [24]. Actually, under sufficient phosphorus conditions, microalgae exhibit high uptake of phosphorus and store it in the form polyphosphate which can be utilized in deficient conditions [25]. There are numerous studies worldwide involving microalgal species for maximizing lipid production using nitrogen and phosphorus limitation strategies, however, the results are selective and not universal. Secondly, the major bottleneck of these nutrient limitation strategies is lower biomass production. As these two nutrients, nitrogen and phosphorus, are required in large quantity, starvation obviously affects the growth of the microalgae and it ends up with lower productivity. Hence, other nutrients are also explored for a similar purpose. Iron and magnesium are the other nutrients which are showing some excellent promise for lipid production by microalgae. For example, magnesium ions act as cofactor for vital enzymes and hence the variation regime of magnesium could potentially influence the lipid accumulation property of microalgae. Hence magnesium are found to be crucial in influencing lipid accumulation as it acts as cofactor for the lipid biosynthetic enzyme acetyl-CoA carboxylase. On looking into such rare nutrient-mediated stress, calcium starvation was proposed to increase the lipid content of

Handbook of Algal Biofuels

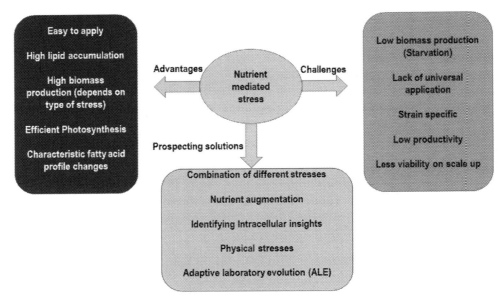

FIGURE 20.2 Advantages, challenges, and prospective solutions of nutrient-mediated stress.

microalgae without much reduction in growth [26]. As mentioned before, starvation of nutrient stress leads to a retardation of growth which was considered to be one of the greater limitations of this strategy. Hence augmenting nutrients was suggested to be effective strategy to circumvent this growth retardation issue. Magnesium augmentation (300 mg/L) promoted the lipid content (54.6%) in the microalgae *Scenedesmus* sp. Nutrient-mediated alteration of biochemical compositions of microalgae is widely of interest because it is easily devisable and applicable. Also, varying nutrient concentrations are apparently reflected in biochemical compositions which in turn can be exploited for enhanced biofuel production. Nitrogen, phosphorus, calcium, magnesium, iron, and a few micronutrients are exploited for this nutrient-mediated biochemical alteration. Although various counter strategies have been devised and tried widely, growth retardation during lipid elicitation was the major challenge of this strategy (Fig. 20.2). Nutrient/chemical stress is increasingly reaching benchmark status with many deep insights and yet there is need for a potential breakthrough to achieve a prompt response universally across microalgal species. On the other hand, physical stress for enhanced biofuel production is also mature enough with various innovative attempts having been reported.

20.3 Physical stress

Apart from nutrient stress, physical stresses created by altering the physical growth conditions, such as pH, temperature, pressure, and light, are important in altering the chemical composition of microalgae. The vital physical factors and the strategy of using it for manipulating the biochemical composition for benefitting the biofuel production are

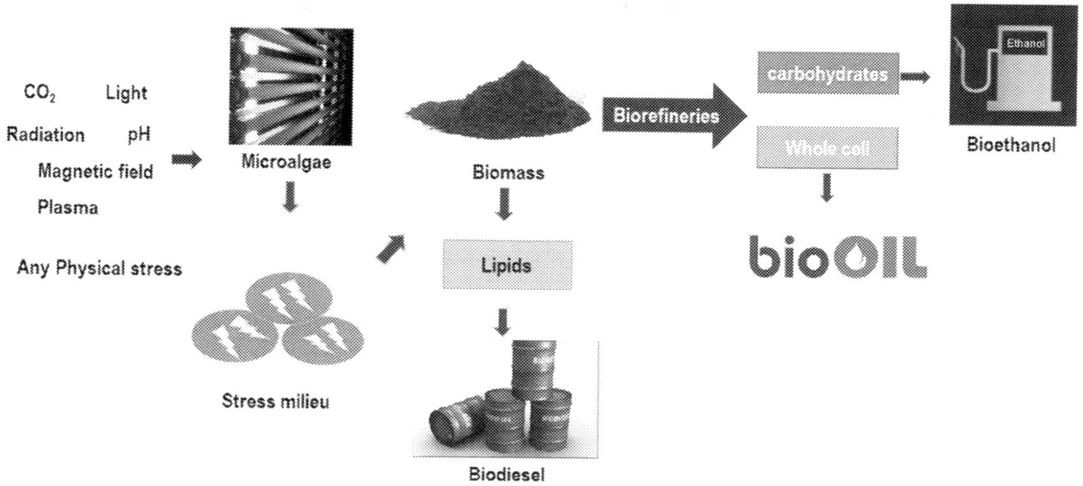

FIGURE 20.3 Influence of physical stress on biofuel potential of microalga.

elaborated in this chapter. Under physical stress, microalgal cells respond by alterations of biochemical composition which eventually affect the quantity and quality of the energy product (Fig. 20.3). Therefore the appropriate physical stress should be chosen based on the targeted energy product (biodiesel, bioethanol, crude bio-oil, etc.).

20.3.1 Temperature

Like other microorganisms, temperature is very critical for microalgal growth and the preferable temperature for most microalgal species varies between 15°C and 25°C; above or below these ranges potentially affects growth, thus leading to temperature-mediated stress [20]. Temperature variation is mostly reflected in the lipid production of microalgae. At 35°C of cultivation temperature, lipid content was increased to 22% with more than half of its proportion as neutral lipids [27]. In a wastewater treatment study, neutral lipids were increased by five times in microalga contaminants at a temperature of 30°C [28]. This shows that temperature not only influences the lipid quantity but also quality. For instance, lowering the temperature from 25°C to 10°C increased the content of docosahex-aenoic acid (DHA) in *Isochyris galbana* [29]. Similarly, about 50% increase in eicosapentae-noic acid (EPA) was achieved in *Pavlovalutheri* when the temperature was reduced to 15°C. Similarly, EPA was increased in *Phaeodactylum tricornutum* when the temperature was reduced to 10°C for a brief exposure of about 12 h. This exemplifies the potential of temperature stress as a fast technique to adjust the lipid content [30]. In *Nannochloropsis* sp., a more than threefold increase in EPA was observed under low temperature [31]. The major mechanism behind such a response by microalgal cell is maintaining the fluidity of the cellular membrane, which is mainly attributed to the presence of polyunsaturated fatty acid content (PUFA) [32]. Thus an increase in polyunsaturated fatty acids was the acclimatization response of microalgae.

It is quite frequently observed that temperature changes influence the polar and nonpolar lipids accumulation [33]. As mentioned before, very specific long-chain fatty acids are also elicited under lowered cultivation temperature. Its advantages from a biofuel perspective, especially biodiesel, depends upon the quality of the lipid used, and neutral lipids are most preferred for biodiesel conversion because of the higher conversion efficiency with good fuel properties. Actually, shifting from synthesis of membrane lipids to neutral lipids occurs when there is a shift in normal growth phase to the stressed phase. Under normal conditions, the fatty acids synthesized are primarily allocated for membrane lipid synthesis and are glycerol based, whereas under stress conditions, the major proportions of fatty acid synthesis are allocated for neutral lipid synthesis for storage as an energy reserve [34]. However, the degree of increasing or decreasing and predicted increase or decrease in this temperature-mediated stress is still elusive. There are contrasting responses of microalgae with variation of temperature. In general, lipid content was significantly enhanced under lower temperature, whereas it was decreased under higher temperature. This can be exemplified by the fact that lipid content was considerably reduced when the temperature was increased to 30°C in *C. vulgaris* [20]. In contrast to this, a significant increase in lipid quantity was observed when the cultivation temperature was increased to 25°C in *Nannochloropsis oculate* [20]. This report clearly stated that temperature-mediated stress can bring different responses in lipid accumulation fashion (increase/decrease) based on different species. Hence, on comparing with nutrient stress like nitrogen stress, this temperature mediated stress yielded highly species-specific responses. Reactive oxygen species (ROS) is one important parameter that is noted under temperature-mediated stress studies [34,35]. Increased ROS was found at lower temperature ranges. Similarly, membrane fluidity was the main reason for such varying lipid allocations, as saturated fatty acids are produced more under increased temperature and polyunsaturated fatty acids under lower temperature, thus facilitating the adjustment of membrane fluidity to withstand such temperature variations [36]. However, growth retardation was also a consequence here as it occurs on raising temperature above 35°C and reducing temperature below 15°C. This is again a matter of concern as growth retardation decreases the overall lipid productivity. Apart from such inherent disadvantages, a temperature variation strategy was effectively used as a technique for enhanced biofuel production, especially biodiesel production. *Acutodesmus dimorphus* was studied by exposing it to higher temperatures upto 35−38°C. High temperature (38°C) supported a fair amount of carbohydrate content (35.08%) in addition to lipid content (32.3%) [27]. Such potential of temperature stress in creating a fair amount of both carbohydrate and lipid content would be useful in producing dual biofuels such as biodiesel (from lipids) and bioethanol (from carbohydrate). Hence, temperature stress-related studies are not only important for improving the desirable biochemical composition for particular biofuel production, but also essential for large-scale production as this requires robust withstanding microalgal species.

20.3.2 Carbon dioxide

CO_2 is important for plants and algae who convert it via photosynthesis into the energy required for their survival. It is obtained from a variety of sources, such as primarily from atmosphere and from other sources like industrial exhaust carbonates [37]. Actually, such

CO$_2$ sequestration helps greatly in maintaining the carbon dioxide levels in the atmosphere and contributes to mitigating greenhouse effects and thereby global warming [38]. Urbanization and CO$_2$ emissions from a variety of sources have been effectively used by microalgae. Such a strategy of utilizing waste flue gases for microalgal biomass and biofuel production are attractive for contemporary research. Like other physical parameters, carbon dioxide is of great importance in photosynthesis and hence the alterations could potentially influence the microalgal metabolism. Hence, to accelerate the biosynthesis of lipids, CO$_2$ supply in the phototrophic culture was varied and widely investigated. CO$_2$ variation significantly influenced the lipid production. For instance, a maximum lipid content was obtained at 2% of CO$_2$ and the lowest lipid content was obtained at excessive (20%) and minimal CO$_2$ concentrations (0.03%) in *Tribonema minus* [39]. At lower CO$_2$ concentrations not many energy-rich compounds can be accumulated, whereas growth can be accelerated with minimal CO$_2$. This could be understood from the investigations made on varying CO$_2$ concentrations in which microalgae at a minimal concentration triggered growth and at a higher concentration promoted fatty acid synthesis [40]. Indeed, these CO$_2$ variations could impact the fatty acid profile of microalgae greatly. This is because fatty acid synthesis primarily depends upon long-chain fatty acids such as C16 and C18 which are then elongated and desaturated for microalgal growth. As these CO$_2$-mediated stress conditions affect the desaturation process, they can potentially influence the profile by forming polyunsaturated fatty acids at times [41]. This can be exemplified by an investigation where long-chain fatty acids are increased under CO$_2$ aeration of 15% [42].

The process of CO$_2$ stress induction is also effectively coupled with contaminants removal which can combine multiple purposes, such as wastewater treatment and bioenergy production along with CO$_2$ sequestration. Microalgae are efficient CO$_2$ sequestering organisms, which is mainly attributed to their high rate of photosynthesis [43]. The response to different CO$_2$ supplementation varies based on species. For example, *Scenedesmus obliquus* SJTU 3 and *Chlorella pyrenoidosa* SJTU 2 showed higher biomass production under 40% of CO$_2$ [44], whereas *Auxenchlorella prothecoides* have not shown any variation for different CO$_2$ treatment [44]. In a recent study, stepwise removal of CO$_2$ concentration variation along with nutrient removal was tested among eight newly isolated microalgal strains [45]. About 2% of CO$_2$ increased the growth of nine strains, whereas not all the strains survived at increased CO$_2$ concentration of 20%. However, they were adapted with a stepwise increase from 2% to 20%. The nutrient removal efficiency was high at 2% of CO$_2$ supplementation and it was predicted that it was directly proportional to the growth of microalgae. However, a well-adapted strain at 20% of CO$_2$ showed considerable removal of nitrogen and phosphorus in simulated wastewater. Hence, this strategy can serve multiple purposes, including flue gases reduction, wastewater removal, and bioenergy production. Utilizing carbon for such bioenergy production was termed as bioenergy with carbon capture and sequestration (BECCS) and is considered to be the best way of satisfying the growing demand for energy [46]. Such carbon capturing and bioenergy production from microalgae was one of the cost-effective strategies which has reduced capital and operational cost [47]. The actual mechanism of CO$_2$ assimilation involves mainly two enzymes: RuBisCo (intracellular) (ribulose-1,5-bisphosphate carboxylase/oxygenase) and carbonic anhydrase (extracellular). In fact, supplying external CO$_2$ or fuel gas as a carbon source would promote the growth and biomass production of

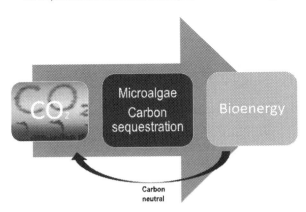

FIGURE 20.4 Bioenergy with carbon capture and sequestration strategy and carbon neutrality.

microalgae. In addition, fixed carbon acts as an endogenous source for acetyl COA which can be eventually converted to lipids [48]. Though there have been several fruitful attempts at using CO_2-mediated stress for biofuel production, multiple factor optimization is significantly needed to improve the strategy. For instance, 20% of CO_2 was primarily used as a good stress induction to influence metabolic responses and acclimatization was further needed to improve the concentrations. The need for multiple factor optimization is critical because microalgal growth crucially depends on other factors too like light, pH, nutrients, etc. Apart from that, there are several advantages of BECCS (Fig. 20.4), such as flue gases can be used, thereby avoiding the additional expense for a CO_2 supply. Importantly, flue gases could be utilized without any pretreatment requirements, which is very attractive and economical. As anthropogenic emissions are utilized, this BECCS is carbon neutral. The strategy is increasing the renewability with a lesser negative impact on the environment. The strategy also offers an advantage in that the development of relevant technology could be an added improvement for a conventional cultivation setup providing atmospheric CO_2 [49]. Considering these potential advantages, CO_2-mediated stress and subsequent bioenergy production could have many advantages over the conventional technique.

20.3.3 pH

pH is the one important physical factor, the variation of which could have a considerable impact on microalgal metabolism and act as a tool for enhancing biofuel potential. The favorable pH for microalgal growth is mildly acidic and near neutral (6—7). Lowering of pH is usually achieved by increasing the carbonic acids concentrations of the growth medium [39,50]. However, lowering of pH leads to decreased carbon flux for lipid synthesis due to lesser bicarbonate concentration [50]. Similarly, high pH also affects the cellular division and induces the utilization of storage lipids [51]. pH of the growth medium basically influences the microalgal metabolism by controlling the influx and efflux of anions and cations across the cells. In *Pavlova lutheri*, the optimum pH range was found to be between pH 7 and 9, which supported fair growth and lipid production (36%) [52]. Indeed, the optimum pH range varies for different microalgal species based on its geographical location [53]. For example, microalgae grown in a high salinity region can adapt better to higher pH than to low salinity levels. pH variation interferes with the transport of ions across the cellular membrane, and pH combined with other parameters or growth conditions is found to be effective. Sakarika and Kornaros [54] stated that there are no strict ranges for pH-mediated impact on microalgal growth or lipid production. Hence, they combined other growth conditions such as heterotrophic and sulfur deprivation. The optimal pH was found to be 7.5 under sulfur limitation, which promoted the lipid content to about 53.43% along with good growth in *C. vulgaris*. Similarly, light intensity variation ($27-94.5\ \mu\mathrm{mol\ m^{-2}s^{-1}}$) along with pH variation (5—9) was performed in *S. obliquus*. It was observed that light intensity variation significantly promoted the lipid content whereas pH has a profound influence on growth and biomass production [55]. The actual influence of pH on microalgal growth has been studied for a long time and it was stated that a variation in extracellular pH potentially influences the intracellular pH. Hence, such changes potentially cause variation of vital enzymes inside the cell as the optimum range for each enzyme may vary [56]. Microalgal cells are so sensitive and are unable to maintain the constant intracellular pH when the extracellular pH was varied. In *Skeletonema costatum*, pH variation showed profound influence on carbohydrate production and the amino acid profile [57]. However, the dominant amino acids such as glutamine, asparagine, glutathione, alanine, and arginine were the amino acids that are not changed with pH variation. The carbohydrate production was enhanced at a pH of 9.4. Although pH variation alone plays little role in enhancing the biofuel potential of microalgae, combining it with other factors and the ability to influence the microalgal metabolism with regard to lipids, carbohydrates, and proteins suggest that pH is a potential tool for enhancing biofuel potential.

20.3.4 Light

Light acts as a primary resource for energy, illuminating the world and also igniting photosynthesis in plants and algae [58]. Chlorophyll of plants and microalgae is the receiving pigment and the chloroplast acts as a site of lipid biosynthesis. Hence the light intensity, type of light (natural or artificial), and duration of light (photoperiod) will influence the microalgal cellular processes effectively. The light source and its effect on growth and microalgal metabolism varies within laboratory (indoor) and outdoor cultivation [59]. Natural light from the primary reservoir, the sun, is the most economical route at a large scale. However, the inexorable potential of light and uncontrolled seasonal fluctuations are the major challenges at the large-scale level. Microalgae possess good growth rate at natural light conditions, but can't be used for biofuel potential. To address such challenges, controlled light, varying light intensity, and different light sources are used as stress-inducing tools for microalgal-based biofuel production. Even at an outdoor cultivation setup, LED lights are used for larger photobioreactors to provide a controlled light intensity [59]. Light actually helps in initiating the photooxidation process and aids in microalgal growth. Increasing the light intensity to a particular level might induce the growth, whereas it can diminish the growth once it reaches the saturation point [60]. The

20. Physical stress for enhanced biofuel production from microalgae

TABLE 20.1 Light-mediated stress combined with other stress conditions in enhancing lipid and carbohydrate production (biofuel potential).

Microalgal species	Light intensity ranges	Accompanying stress	Biomass	Lipid	FA	Carbohydrate	References
Botryococcus spp	82.5 lE m^{-2} s^{-1}	Nitrogen starvation + high level iron	NR	35.9%	NR	NS	[62]
Parietochloris incisa	35–400 μmol photons m^{-2} s^{-1}	Nitrogen starvation	NR	0.25 g/L	C20	NS	[63]
Scenedesmus obliquus CNW-N	420 μmolm^{-2}s^{-1}	Nitrogen starvation	0.44 g/L	22.4%	C16; C18	46.65%	[64]
Scenedesmus sp	400 μmol photons m^{-2} s^{-1}	N starvation	3.88 g/L	41.1%	C18:1	NS	[65]
Nannochloropsis sp	100 μmol m^{-2} s^{-1}	Photoperiod 18 L:6D dark	NR	31.3%	NS	NS	[66]
Ankistrodesmus falcatus	60 μmol m^{-2} s^{-1}	Media composition and photoperiod (12 L:12D)	0.224 g/L	67.2%	NS	7.77%	[67]
Chlorella minutissima	33.75 μmol m^{-2} s^{-1}	Pentose	1.34 g/L	NS	NS	53.8%	[68]
Tetradesmus obliquus	150 and 650 μmol photons m^{-2} s^{-1}	Nitrogen limitation	0.80 g/L	NS	NS	45%	[69]
Chlorella kessleri.	90μE m^{-2}μs^{-1}	Glucose; Photoperiod 20L:4D	1.03 g/L	17.3%	NS	27.8%	[70]

FA, fatty acid; *NR*, not reported as direct values; *NS*, not studied; % expressed to the total biomass; values expressed are the maximum under the treatment of the study.

optimum ranges of light intensity vary between 15 and 150 μmol of photons m^{-2} s^{-1} and hence exceeding or being below this range significantly affects the microalgal growth and leads to stress induction [61]. Hence, light intensity variation was found to be greater in producing profound changes in microalgal lipid production and microalgal physiology. There are ample reports dealing with the light-mediated stress for lipid production and other parameters like nutrient stress are also combined with light-mediated stress for enhancing the biofuel potential of microalgae Table 20.1.

Higher light intensity primarily promoted the neutral lipids of microalgae along with suppressing the polar lipids proportions. However, this is not the case with every microalgal species and response varies from species to species. Low light intensity tends to promote higher polar lipids accumulation which is not favorable for biodiesel conversion [71]. However, light intensity above a certain level will cause greater damage and reduce the growth completely. For example, an increase of light intensity upto 300 μmol of photons m^{-2} s^{-1} simultaneously increased the biomass and lipid content, whereas above that it reduced the growth of *P. tricornutum* completely [60].

Handbook of Algal Biofuels

Next to light intensity, the photoperiod of light is found to be an important factor for its growth and necessary metabolic process. A photosynthetic organism requires two distinct phases, a light-dependent phase (photoperiod) and a light-independent phase (dark period), and such photo- and dark periods are 12 h each in a day. Hence, increasing or decreasing the photoperiod brings potential changes in microalgal metabolism. Wahidin et al. [66] combined the light intensity variation and photoperiod changes to observe changes in growth and lipid accumulation of *Nannochloropsis* sp. In the study, the authors gave three different sets of light intensities (50, 100, and 200 μmol of photons m^{-2} s^{-1}) and photoperiod regimes (24 h dark:0 h light; 12 h dark:12 h light; and 18 h light:06 h dark) in which lipid content reached upto 31% with a higher growth rate under the light intensity of 100 μmol of photons m^{-2} s^{-1} given for a photoperiod of 18:6. The authors observed that both high light intensity and a longer photoperiod affects the growth. For example, higher light intensity of about 200 μmol of photons m^{-2} s^{-1} showed slower growth and the highest photoperiod of 24:0 reduced the growth of microalgae. During light reactions, the energy from light is converted to energy molecules, whereas in dark period the energy molecules are used to synthesize sugars. Thus both periods are essential and need to be long enough. Usual preferred period durations are 12 h light:12 h dark or 18 h light:6 h dark for indoor microalgal cultivation. Lipid production was enhanced along with biomass production in *Nannochlorpsis* sp. under 100 μmol of photons m^{-2} s^{-1} of light under an 18:6 photoperiod.

Apart from light intensity and photoperiod, the spectral quality of light is also a vital physical factor in influencing the photosynthesis of microalgae. The specialized pigments such as chlorophyll, carotenoids, and phycobilin are responsible for absorbing the spectra for inducing photosynthesis. The preferred spectral ranges by microalgal chlorophyll are 430−475 (blue) and 630−675 nm (red) [72]. The optimum wavelength for microalgal pigments is greatly reflected in the growth and lipid production of microalgae. For instance, blue light was the preferred wavelength of the different wavelengths tested for lipid accumulation in *Isochrysis galbana, Tetraselmis* sp. etc. [72,73]. Subjecting microalgae to a wavelength of 430−510 nm, promoted the production of enzymes such as carbonic anhydrase and ribulose bisphosphate carboxylase, which in turn was reflected in neutral lipid productivity [74]. Apart from the quantities of lipids, spectral quality also influences the fatty acid profile of microalgae. For instance, subjecting microalga *C. vulgaris* to a green wavelength promoted C16:3 and C18:3 fatty acids, and thus adjusting the wavelength could produce fatty acids that are significant for biodiesel. There have been a lot of innovations with regard to altering the wavelength for microalgal biomass production and a few of them are listed in Table 20.2. Hence, the application of light can influence the quantity of lipids as well as lipid quality in order to produce a higher amount of biodiesel with good fuel properties.

20.3.5 Radiation

20.3.5.1 *Ultraviolet radiation*

The major advantage of giving UV radiation as a stress to microalgae is its effectiveness and a controlled dosage can be delivered. The solar radiation comprises sections of

462   20. Physical stress for enhanced biofuel production from microalgae

TABLE 20.2 Innovations on adjusting the wavelength of irradiance toward microalgal culture and resulting biofuel potential.

Microalgal species	Strategy	Biomass	Lipid	FA	Carbohydrate	References
Acutodesmus obliquus	Different LEDs (red, yellow, green, and white)	0.039 g/L	0.02 g/L	C16, C18:2; C18:3	NS	[75]
Scenedesmus obliquus	Different LEDs (white, blue, green, and red)	0.198 g/L	24.7%	C18:1; C18:2; C18:3	NS	[76]
Chlorella vulgaris	Organic dyes (rhodamine101 and 9,10-diphenylanthracene)	1.5 g/L	30%	C18; C18:1	NS	[77]
Chlorella sp.	Fluorescent paint	1.7 g/L	30%	NS	NS	[78]
Chlorella sp.	Fluorescent material (fluorescent dye Uvitex OB) coated on reactor	NR	NS	NS	NS	[79]

FA, fatty acid; *NR*, not reported as direct values; *NS*, not studied; % expressed to the total biomass; values expressed are the maximum under the treatment of the study.

ultraviolet radiations (UV-A, UV-B, and UV-C). Each section of radiations is characterized by its wavelength, for example, 400−315 nm is UV-A, 315−280 nm is UV-B, and 280−100 nm is UV-C. Among these three sections, UV-C possesses more energy, that is, 4.43 to 12.4 eV per photon, and creates greater stress in microalgal growth [80]. An interesting study was carried out by Sharma et al. [80] by combining UV-C treatment and nutrient starvation treatment. Different dosages of UV-C radiation were given to nitrogen-starved cells and it was observed that it increased the cellular size and lipid accumulation. The dosage of 250 mJ/cm^2 of UV-C radiation was given to nitrogen-starved *Tetraselmis* sp. The same condition also induced the characteristic changes in fatty acid profile of microalgae, that is, unsaturated fatty acids were increased along with C16 and C18 fatty acids in saturated fatty acid proportions. It was also observed that lipid accumulation was faster under UV-C radiation when compared to other UV radiation such as UV-A and UV-B. But interestingly, most of the reports have dealt with UV-A and UV-B irradiance, and even a combined UV-A and UV-B treatment was given to *Chaetoceros mualleri* and resulted in an increased proportion of monounsaturated fatty acids [81]. Hence, beyond total lipid enhancement, UV radiation is likely to alter the fatty acid chains greatly as a stress response. For instance, UV irradiation treatment to *P. lutheri* elevated the total lipid content along with increased unsaturated fatty acid content such as EPA and DHA [82]. Similarly, UV irradiation increased the lipid content of EPA to around 20% in *P. tricornutum*. This is because of the fact that under UV stress, microalgae tend to promote a greater amount of ROS and eventually lipid peroxidation which attributed to a greater amount of PUFA content. Liu et al. [83] created a mutant of *Chlorella* sp. by the application of UV radiation at the wavelength of about 253.7 nm and achieved a considerable increase of about 7.6% in growth and lipid production (28.1%). Therefore there are considerable reports that dealt with

Handbook of Algal Biofuels

UV-mediated stress induction in microalgae, which is reflected in lipid enhancement for efficient biodiesel production. However, there are fewer reports on UV-mediated stress induction for enhancing the bioethanol potential of microalgae or carbohydrate enhancement.

20.3.5.2 Gamma radiation

Gamma radiation is the shorter wavelength possessing electromagnetic radiation which can be potentially used as stressors for enhancing the microalgal lipids. Gamma radiation tends to induce oxidative stress and thereby induce lipid production in micro-alga. Gamma radiation dosages of about 100−1100 Gy of Co60 were administered to *Chlorella sorokiniana* and investigated for lipid increase and observed that 500 Gy dosage was found to be effective in triggering the lipid content upto 48%, which was a 100% increase compared to the control. However, the effect of lipid accumulation was found to be dose dependent, and above 500 Gy dosage it failed to accumulate lipids, which are confirmed by the fact that ROS formation was observed upto 500 Gy as an indication of stress response [84]. Similarly, diatoms cells of *Nitzschia* sp. were irradiated with gamma radiation at 900 Gy and mutated which reflected in the accumulation of lipids upto 28.1% in mutants along with salinity stress conditions of upto 30%. The study demonstrated the gamma radiation-induced mutation as a way of increasing the lipid content and also helping in toleration of the lipid salinity condition, which altogether improved the lipid production [85]. In addition, the specified mutant also performed well with nitrogen and silicon starvation in accumulating lipids (51.2%). Hence gamma radiation was considered to be a potential tool for creating mutants with effective tolerance and lipid accumulation potential.

20.3.5.3 Ion beam radiation

Ion beams are another widely used radiation for various applications which are densely ionizing with high linear energy transfer. Tu et al. (2015) mutated microalga *C. pyrenoidosa* using a nitrogen ion beam implantation technique which is familiarly used for inducing mutations for agricultural crops [86]. The dosage of 1.5×1015 ions/cm^{2}s of N increased the biofuel potential of microalgae by increasing the lipid content by 32% along with growth improvement, which altogether increased the lipid productivity upto 35%. Unfortunately, ion beam treatment reduced the saturated fatty acid content and increased the polyunsaturated fatty acid. Similarly, nitrogen ion beam implantation produced mutants in *S. obliquus* with increased lipid content by 24% than the wild strain [87]. In addition, this biofuel potential mutant also showed excellent 80% removal of total nitrogen, total phosphate, and ammonia. The mutant strain was proven to be very successful in different media and its stability was confirmed by random amplified polymorphic DNA analysis. The reported studies proved that the ion beam irradiation was found to be promising in inducing lipid accumulation but the results and degree vary among species. For instance, heavy ion beam radiation induced the lipid production increase by 28% in *Nannochloropsis* sp., whereas the fatty acid profile remains unchanged under ion beam radiation [88] (Table 20.3).

464   20. Physical stress for enhanced biofuel production from microalgae

TABLE 20.3   Influence of irradiation on biofuel potential of microalgae.

Microalgal species	Irradiance	Biomass	Lipid	FA	Carbohydrates	References
Scenedesmus dimorphus	Gamma radiation (800 Gy)	NR	0.648 g/L	NS	NS	[89]
Chlorella sp.	UV radiation	NR	21.7%	NS	52.02 mg/L	[83]
Parachlorella kessleri	UV radiation	5.8 g/L	66%	NS	3.41 g/L	[90]
Chlorella pyrenoidosa	Low energy ion implantation	0.738 g/L	44%	C16:2; C18:2	NS	[86]
C. pyrenoidosa	^{60}Co-γ irradiation	2.6 g/L	32%	C16	NS	[91]
Pavlova lutheri	UV radiation	NR	0.015 g/L	C20:5	NS	[82]
Nannochloropsis oculata	UV radiation	1.4 g/L	59%	C18:2	18.18%	[92]
Scenedesmus obliquus	UV radiation	0.31 g/L	24%	C18:1; C18:3	NS	[93]
Chlorella sorokiniana	UV radiation	0.3 g/L	40%	C18:3	NS	[93]
C. pyrenoidosa	UV radiation	NR	25%	C18:3 increment	Starchless	[94]
Scenedesmus sp.	UV radiation	NR	60%	C18:1	NS	[95]

FA, fatty acid; *NR*, not reported as direct values; *NS*, not studied; % expressed to the total biomass; values expressed are the maximum under the treatment of the study.

20.3.6 Magnetic field

Application of a magnetic field to microalgal cultures can attain relevant improvements in microalgal growth characteristics. For instance, periodic exposure of *Chlorella kessleri* to a magnetic field increased biomass production along with a considerable increase in protein, zinc, calcium, and pigments [96]. In *Chlorella fusca*, a magnetic field was applied to enhance the CO_2 biofixation rate by 50% along with enhanced protein content and biomass accumulation [97]. A static magnetic field (SMF) was applied to *Spirulina platensis* for about 6 h per day which improved the cadmium removal efficiency by 90% [98]. A SMF is achieved by keeping magnets on both sides of a microalgal culture which are kept in a static state in a tank. Han et al. [99] utilized this magnetic field in a culture to increase the efficiency of lipid accumulation and wastewater treatment of microalga *C. pyrenoidosa* in which the lipid productivity was enhanced by 10% [99]. Chu et al. suggested the optimum magnetic field strength for inducing the growth and lipid productivity along with appropriate nitrate concentrations in *Nannochloropsis occulata*. 20 MT of magnetic field strength was proven to be effective in increasing the lipid productivity by 65% compared with the control [100]. Also, as the magnetic strength increased to 30 MT and 40 MT, the growth

Handbook of Algal Biofuels

rate was affected which in turn affected the lipid productivity. Although nitrate concentration was an important parameter for growth and lipid production, the combination of magnetic field and nitrate has not achieved any significant influences in the lipid productivity. The actual mechanism behind the magnetic field application is that it induces the membrane permeability and enhances the transportation of substances toward cytoplasm, thus interfering with the microalgal metabolism [101].

Feng et al. [102] demonstrated the bottom and bypass modes of magnetic application to microalgae in wastewater. Magnetic field-treated wastewater and microalgal culture were studied and both modes greatly promoted the biomass and lipid accumulation, whereas the bypass mode of applying magnetic field was found to be better by operation and economy. Like other physical stress, a magnetic field above a certain strength reduces the growth of microalgae, which is mainly due to the production of free radicals at higher magnetic field strength [96]. Apart from the utilization of magnetic fields as a physical stress, magnetic fields are used for harvesting microalgal cells from the liquid media [103]. Interestingly, a magnetic field applied to *Spirulina* sp. enhanced te biomass production along with an increase in protein and carbohydrate content. An increase of 133% of carbohydrate content was achieved by 30 mT magnetic field application. This feature of eliciting the carbohydrate content by magnetic field would be attractive as it helps in improving the yield of bioethanol conversion [104]. The actual reason for carbohydrate elevation was suggested to be that the application of magnetic field acts as a stimulus for the Calvin—Benson cycle, resulting in more absorption of CO_2 and is reflected in carbohydrate production [104]. However, the selection of targeted lipid induction or an increase in carbohydrate production has not yet been studied. Hence, the selective application of magnetic field either based on field strength or exposure in order to target lipid and carbohydrate induction would be a great breakthrough in microalgal-based biofuels.

20.3.7 Atmospheric room temperature plasma

Atmospheric room temperature plasma (ARTP) is a novel mutagenic system that was developed just few years ago by the chemical engineering department of Tsinghua University, China. Plasma is the process of creating free ions and free electron by causing ionization with rapid electrons in gas. In ARTP, radio frequency power is used to generate plasma at atmospheric pressure and it can be controlled at room temperature which is essential for mutating microorganisms [105]. The chemical species produced in ARTP enter the cellular membrane and cause damage to DNA called mutations (Fig. 20.5) [107]. Though the exact phenomenon of mutation under this ARTP is unclear, the chemical species are found to be cause mutations. One suggested possible way is the indirect reaction between the chemical species and intracellular molecules, which results in the alteration of DNA bases [108]. It is also stated that mutation also depends upon the gases used, for example, noble gases may not produce adequate UV radiation and eventually free radicals. Hence the addition of oxygen and nitrogen to these noble gases would generate many numbers of free radicals [108]. Although these are improvements for attaining effective mutagenesis, the higher level of DNA damage would diminish the growth.

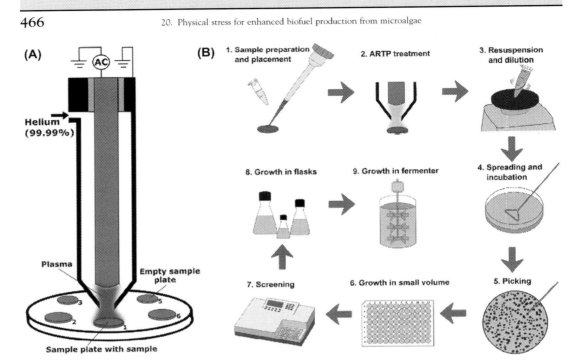

FIGURE 20.5 (A) Radio-frequency atmospheric-pressure glow discharge (RF APGD) plasma jet generator [main component of Atmospheric room temperature plasma (ARTP)] with microbial sample. (B) Procedure for ARTP mediated mutagenesis and mutant identification. [106]. *Source: The open access C. Ottenheim, M. Nawrath, J.C. Wu, Microbial mutagenesis by atmospheric and room-temperature plasma (ARTP): the latest development. Bioresour. Bioprocess. 5 (2018) 12.*

Atmospheric room temperature plasma-mediated mutagenesis was carried out in *S. platensis* and a significant increase in growth and carbohydrate content was observed [105]. It was observed that carbohydrate content showed a maximum 70% increase and 10.5% growth increase in the mutants produced. In addition, the produced mutants showed good stability with successive subculturing. The study suggested a vital point that dosage is directly dependent upon the radiation exposure time and it needs to be less than 60 s to prevent cell death. Almarashi et al. (2019) conducted a comprehensive study on cold atmospheric pressure plasma treatment (CAPP) for *C. vulgaris* and observed biodiesel potential [109]. CAPP is another kind of plasma technique which is mainly performed with ROS along with UV radiation, charged particles, and shockwaves [110]. The CAPP treatment influenced the growth as well as increased lipids by 7.5% and 6.9% compared with the control at 30 s and 60 s exposure time, respectively. Among such exposure time, 30 s was found to increase both the lipid content and growth resulting in contributing to a higher lipid productivity (17%). The maximum fatty acid methyl ester recovery was obtained at a 60 s treatment which is 43% higher than the control. Oxidation reduction potential was suggested to be the major factor creating damage within the cell in this CAAP technique. ARTP was even effective in reducing the chlorophyll size, which in turn was reflected in increased photosynthesis and photolysis which helped in hydrogen production. Exposure time varied between 20 s and

60 s which resulted in the mutants exhibiting 5.2% more hydrogen production than the wild strains after successive subculturing. The major inference from the study was that genes related to photosystem I, II, light-harvesting complex (LHC), and cytochrome have shown increased expression, resulting in improved photosynthesis efficiency. In *C. pyrenoidosa*, ARTP mutagenesis produced mutants that showed a 16% increase in lipid productivity [111]. In *Desmodesmus* sp. a more than twofold increase was observed by ARTP treatment [112]. Hence, ARTP mutagenesis is an upcoming effective tool for improving the biofuel characteristics of microalgae. Indeed, ARTP is the only tool which has been so effective in eliciting the different biofuel potential of microalgae, such as carbohydrates, hydrogen, and lipids for bioethanol, biohydrogen, and biodiesel production, respectively. However, still it is not clear which condition of ARTP is associated with a targeted biofuel potential increase. Hence, more studies and optimization of conditions for targeted enhancement of lipids/carbohydrates/hydrogen need to be addressed. With regard to artificially inducing stress conditions, ARTP will definitely occupy a special position in the near future.

20.4 Challenges and future directions of physical stress

The primary challenge of physical stress induction is the response yields are mostly species and strain specific. This poses an extra challenge in that microalgal lipid enhancement for each strain requires an initial optimization rather than finding one tool that can be effective and can be applied for all the species [113]. In such cases, chemical stress, like a N-limitation strategy, is found to be mostly similar in many reported microalgal species. On the other hand, stress conditions, whether physical or chemical, will definitely impact the metabolic flux and trigger lipid accumulation and hence biomass production is not always unaffected. Hence identifying the stress condition at the optimum state that triggers the lipid/carbohydrate accumulation but that does not affect the growth needs to be assessed. This assessment would be a greater challenge for each species but it ensures a potential breakthrough in the growth—lipid trade-off. As mentioned earlier, the choice of physical stress or its dosage/strength/range for targeted compound increases is not clarified yet. For instance, ARTP can stimulate carbohydrate in one species and lipids in another species. This provides a greater area of research for identifying and optimizing the relevant conditions for a particular compound enhancement. In another way, most of the studies except ARTP or physical mutagenesis, genetic expression in response to particular physical stress is not yet explored. Such insights might give good control over the physical stress strategy toward targeted compound enhancement. Also, it is noteworthy that microalgal biofuel production is not only confined to enhance the biofuel potential, whereas it has many other sections of improvement such as cultivation optimization, pretreatment techniques, biofuel conversion techniques, etc. [5]. However, optimizing the yield enhancement could help greatly for the overall production cost. Therefore to address the overall cost of production and to enhance the economic feasibility, the integration of different biofuels, such as biodiesel and bio-oil production together and biodiesel—bioethanol production together, has been proposed in recent days [113].

Handbook of Algal Biofuels

The operational challenges at a larger scale for applying this physical stress induction need to be considered. For instance, the application of a magnetic field to a larger cultivation area, either a raceway pond or larger photobioreactor, will be highly challenging. Hence, considerable attention should be given to the scale-up of all these potential physical stress agents. In such a case, ARTP mutagenesis is a better option where you can induce mutant within the ARTP system and mutants can be further used for scale-up. Though the reports showed that ARTP mutants are stable for successive subculturing, holding the enhanced biofuel potential on a large scale needs to be investigated. As whole stress induction strategies aim to improve the yield of targeted compounds for reducing the overall cost, open or closed ponds are preferred as being cheaper than large photobioreactors [114]. In such case, if light is used as physical stress, controlling the light intensity would be an unanswered question. On the other hand, the closed systems like photobioreactors (flat panel/tubular) could perform well for producing axenic biomass production but the energy and economic expense again affects the overall production cost. Even with photobioreactors, providing illumination that is able to be calibrated in order to induce light stress would be highly challenging. Also, in larger photobioreactors, controlling the required intensity to every cell at each depth would be laborious. Hence, considerable innovation in setting illumination for PBRs that can be calibrated according to desired illumination intensity needs to be devised. On considering temperature as stress-inducing tool, controlling temperature at a large scale is again challenging. Hence, these are the operational challenges in executing the physical stress which are optimized at lab scale to large scale [1]. To address those large-scale challenges and as an alternative to genetic engineering strategies, adaptive laboratory evolution (ALE) was proposed as a promising technique in which microalgae are allowed to adapt and grow in laboratory conditions defined for developing certain desired characters (lipids or carbohydrates) or phenotype [115]. Such a technique could be the better option to overcome the scale-up shortcomings as the instant stress is given and expected to attain a desired response. For instance, ALE experiments were performed with *Chlorococcum littorale* on continuous nitrogen starvation for about 75 days and the results showed not much difference in the growth and biomass production [116]. However, ALE for microalgal-based biofuels is in its infancy and wider studies and relevant insights are needed. From an economical point of view, providing physical stress like a magnetic field, or radiation-mediated stress conditions at the large scale is not cheap, hence the strategy of combining multiple stress conditions would be the better options [2]. Combining a chemical stress with a physical stress might reduce the energy expenditure for generating physical stress to a certain proportion of that if it was given alone. Also, a combination of different stresses might nullify the undesirable effect (e.g., growth reduction) exerted by the other stress [117]. Also, the possibility of synergistic effect might also help in better enhancement of the desired trait. However, a negative impact on microalgal cells also occurs when it is stressed by more than one factor at a time. Hence, combining physical stress requires significant optimization too. Therefore the economic, operational, and productivity challenges associated with physical stresses could be well addressed by practicing a combination of suitable physical/chemical stresses, ALE technique, and biorefinery approach.

20.5 Conclusion

Physical stress induction in microalgae has been comprehensively focused for more than a decade on enhancing the biofuel potential of microalgae. However, most of the physical stresses have dealt with lipid induction in microalgae. It is a quite obvious biological phenomenon that whenever microalgae are subjected to a stress stimulus, they accumulate lipids. However, carbohydrates and a good macromolecular composition are very essential for bioethanol and bio-oil production, respectively. Hence, more stress-related studies beyond the biodiesel perspective need to be the future directions. Stress conditions optimized for inducing finer macromolecular proportions for optimized bio-oil production could be an interesting direction. A few studies were reported giving importance to biochemical composition in bio-oil production. For instance, high lipid and low protein-containing microalgal species (*Nannochloropsis* sp.) and low lipid and high protein-containing microalgal species (*Chlorella* sp.) were studied for bio-oil production and low lipids provided the better yield [118]. Similarly, Chen et al. [119] proposed a review emphasizing the importance of microalgal biochemical composition in bio-oil production. However, that too was not focused on stress related studies for microalgal bio-oil production. Therefore for microalgal bio-oil production, there have been fewer stress-related studies, thus offering a greater space for exploration. Apart from that, prevailing stressors are promising for microalgal biodiesel and bioethanol production to an extent. ARTP, magnetic field, irradiation, and combined stress conditions are promising for enhancing microalgal biofuel production.

References

[1] X.-M. Sun, L.-J. Ren, Q.-Y. Zhao, X.-J. Ji, H. Huang, Microalgae for the production of lipid and carotenoids: a review with focus on stress regulation and adaptation, Biotechnol. Biofuels 11 (2018) 272.

[2] P. Singh, S. Kumari, A. Guldhe, R. Misra, I. Rawat, F. Bux, Trends and novel strategies for enhancing lipid accumulation and quality in microalgae, Renew. Sustain. Energy Rev. 55 (2016) 1−16.

[3] J. Masojídek, M. Koblížek, G. Torzillo, Photosynthesis in microalgae, Handbook of Microalgal Culture: Biotechnology and Applied Phycology, 2004, p. 20.

[4] Q. Hu, M. Sommerfeld, E. Jarvis, M. Ghirardi, M. Posewitz, M. Seibert, et al., Microalgal triacylglycerols as feedstocks for biofuel production: perspectives and advances, Plant. J. 54 (2008) 621−639.

[5] K.W. Chew, J.Y. Yap, P.L. Show, N.H. Suan, J.C. Juan, T.C. Ling, et al., Microalgae biorefinery: high value products perspectives, Bioresour. Technol. 229 (2017) 53−62. Available from: https://doi.org/10.1016/j.biortech.2017.01.006.

[6] S. Xu, M. Elsayed, G.A. Ismail, C. Li, S. Wang, A.E.F. Abomohra, Evaluation of bioethanol and biodiesel production from Scenedesmus obliquus grown in biodiesel waste glycerol: a sequential integrated route for enhanced energy recovery, Energy Convers. Manag. 197 (2019). Available from: https://doi.org/10.1016/j.enconman.2019.111907.

[7] M.E.H. Osman, A.M. Abo-Shady, M.E. Elshobary, M.O. Abd El-Ghafar, A.E.-F. Abomohra, Screening of seaweeds for sustainable biofuel recovery through sequential biodiesel and bioethanol production, Environ. Sci. Pollut. Res. Int. (2020). Available from: https://doi.org/10.1007/s11356-020-09534-1.

[8] S. Wang, S. Zhao, B.B. Uzoejinwa, A. Zheng, Q. Wang, J. Huang, et al., A state-of-the-art review on dual purpose seaweeds utilization for wastewater treatment and crude bio-oil production, Energy Convers. Manag. 222 (2020) 113253.

[9] S. Wang, B.B. Uzoejinwa, A.E.-F. Abomohra, Q. Wang, Z. He, Y. Feng, et al., Characterization and pyrolysis behavior of the green microalga Micractinium conductrix grown in lab-scale tubular photobioreactor using Py-GC/MS and TGA/MS, J. Anal. Appl. Pyrol. 135 (2018) 340–349.

[10] Y. Chisti, Biodiesel from microalgae beats bioethanol, Trends Biotechnol. 26 (2008) 126–131.

[11] A. Demirbas, M.F. Demirbas, Importance of algae oil as a source of biodiesel, Energy Convers. Manag. 52 (2011) 163–170.

[12] A. Abomohra, M. Elsayed, S. Esakkimuthu, M. El-Sheekh, D. Hanelt, Potential of fat, oil and grease (FOG) for biodiesel production: a critical review on the recent progress and future perspectives, Prog. Energy Combust. Sci. 81 (2020) 100868.

[13] A.E.-F. Abomohra, X. Zheng, Q. Wang, J. Huang, R. Ebaid, Enhancement of biodiesel yield and characteristics through in-situ solvo-thermal co-transesterification of wet microalgae with spent coffee grounds, Bioresour. Technol. (2020) 124640.

[14] S. Esakkimuthu, V. Krishnamurthy, S. Wang, A.E.-F. Abomohra, S. Shanmugam, S.G. Ramakrishnan, et al., Simultaneous induction of biomass and lipid production in Tetradesmus obliquus BPL16 through polysorbate supplementation, Renew. Energy (2019).

[15] A.E.-F. Abomohra, A.W. Almutairi, A close-loop integrated approach for microalgae cultivation and efficient utilization of agar-free seaweed residues for enhanced biofuel recovery, Bioresour. Technol. (2020) 124027.

[16] J.P. Maity, J. Bundschuh, C.-Y. Chen, P. Bhattacharya, Microalgae for third generation biofuel production, mitigation of greenhouse gas emissions and wastewater treatment: present and future perspectives—a mini review, Energy 78 (2014) 104–113.

[17] S. Esakkimuthu, V. Krishnamurthy, S. Wang, X. Hu, K. Swaminathan, A.E.-F. Abomohra, Application of p-coumaric acid for extraordinary lipid production in Tetradesmus obliquus: a sustainable approach towards enhanced biodiesel production, Renew. Energy 157 (2020) 368–376.

[18] A.E.-F. Abomohra, W. Jin, R. Tu, S.-F. Han, M. Eid, H. Eladel, Microalgal biomass production as a sustainable feedstock for biodiesel: current status and perspectives, Renew. Sustain. Energy Rev. 64 (2016) 596–606.

[19] A.J. Klok, D.E. Martens, R.H. Wijffels, P.P. Lamers, Simultaneous growth and neutral lipid accumulation in microalgae, Bioresour. Technol. 134 (2013) 233–243.

[20] A. Converti, A.A. Casazza, E.Y. Ortiz, P. Perego, M. Del Borghi, Effect of temperature and nitrogen concentration on the growth and lipid content of Nannochloropsis oculata and Chlorella vulgaris for biodiesel production, Chem. Eng. Process. Process Intensif. 48 (2009) 1146–1151.

[21] T. Li, L. Wan, A. Li, C. Zhang, Responses in growth, lipid accumulation, and fatty acid composition of four oleaginous microalgae to different nitrogen sources and concentrations, Chin. J. Oceanol. Limnol. 31 (2013) 1306–1314.

[22] Y. Gao, M. Yang, C. Wang, Nutrient deprivation enhances lipid content in marine microalgae, Bioresour. Technol. 147 (2013) 484–491.

[23] K.H. Kim, I.S. Choi, H.M. Kim, S.G. Wi, H.-J. Bae, Bioethanol production from the nutrient stress-induced microalga Chlorella vulgaris by enzymatic hydrolysis and immobilized yeast fermentation, Bioresour. Technol. 153 (2014) 47–54.

[24] L. Xin, H. Hong-Ying, G. Ke, S. Ying-Xue, Effects of different nitrogen and phosphorus concentrations on the growth, nutrient uptake, and lipid accumulation of a freshwater microalga Scenedesmus sp, Bioresour. Technol. 101 (2010) 5494–5500.

[25] H. Qi, J. Wang, Z. Wang, A comparative study of maximal quantum yield of photosystem II to determine nitrogen and phosphorus limitation on two marine algae, J. Sea Res. 80 (2013) 1–11.

[26] S. Esakkimuthu, V. Krishnamurthy, R. Govindarajan, K. Swaminathan, Augmentation and starvation of calcium, magnesium, phosphate on lipid production of Scenedesmus obliquus, Biomass Bioenergy 88 (2016) 126–134.

[27] K. Chokshi, I. Pancha, K. Trivedi, B. George, R. Maurya, A. Ghosh, et al., Biofuel potential of the newly isolated microalgae Acutodesmus dimorphus under temperature induced oxidative stress conditions, Bioresour. Technol. 180 (2015) 162–171.

[28] G.V. Subhash, M.V. Rohit, M.P. Devi, Y.V. Swamy, S.V. Mohan, Temperature induced stress influence on biodiesel productivity during mixotrophic microalgae cultivation with wastewater, Bioresour. Technol. 169 (2014) 789–793.

[29] C.J. Zhu, Y.K. Lee, T.M. Chao, Effects of temperature and growth phase on lipid and biochemical composition of Isochrysis galbana TK1, J. Appl. Phycol. 9 (1997) 451–457.

References

[30] M. Mitra, S.K. Patidar, S. Mishra, Integrated process of two stage cultivation of Nannochloropsis sp. for nutraceutically valuable eicosapentaenoic acid along with biodiesel, Bioresour. Technol. 193 (2015) 363–369.

[31] T. Menegol, A.B. Diprat, E. Rodrigues, R. Rech, Effect of temperature and nitrogen concentration on biomass composition of Heterochlorella luteoviridis, Food Sci. Technol. 37 (2017) 28–37.

[32] I. Nishida, N. Murata, Chilling sensitivity in plants and cyanobacteria: the crucial contribution of membrane lipids, Annu. Rev. Plant. Biol. 47 (1996) 541–568.

[33] S.M. Renaud, L.-V. Thinh, G. Lambrinidis, D.L. Parry, Effect of temperature on growth, chemical composition and fatty acid composition of tropical Australian microalgae grown in batch cultures, Aquaculture 211 (2002) 195–214.

[34] M.Y. Roleda, S.P. Slocombe, R.J.G. Leakey, J.G. Day, E.M. Bell, M.S. Stanley, Effects of temperature and nutrient regimes on biomass and lipid production by six oleaginous microalgae in batch culture employing a two-phase cultivation strategy, Bioresour. Technol. 129 (2013) 439–449.

[35] L. Xin, H. Hong-Ying, Z. Yu-Ping, Growth and lipid accumulation properties of a freshwater microalga Scenedesmus sp. under different cultivation temperature, Bioresour. Technol. 102 (2011) 3098–3102.

[36] V. Ananthi, K. Brindhadevi, A. Pugazhendhi, A. Arun, Impact of abiotic factors on biodiesel production by microalgae, Fuel 284 (2021) 118962. Available from: https://doi.org/10.1016/j.fuel.2020.118962.

[37] D.A. Los, N. Murata, Membrane fluidity and its roles in the perception of environmental signals, Biochim. Biophys. Acta 1666 (2004) 142–157.

[38] L. Moraes, G.M. da Rosa, B.B. Cardias, L.O. dos Santos, J.A.V. Costa, Microalgal biotechnology for greenhouse gas control: carbon dioxide fixation by Spirulina sp. at different diffusers, Ecol. Eng. 91 (2016) 426–431.

[39] L.D. Zhu, Z.H. Li, E. Hiltunen, Strategies for lipid production improvement in microalgae as a biodiesel feedstock, Biomed. Res. Int. (2016) 2016.

[40] D. Tang, W. Han, P. Li, X. Miao, J. Zhong, CO_2 biofixation and fatty acid composition of Scenedesmus obliquus and Chlorella pyrenoidosa in response to different CO_2 levels, Bioresour. Technol. 102 (2011) 3071–3076.

[41] J. Cheng, H. Lu, Y. Huang, K. Li, R. Huang, J. Zhou, et al., Enhancing growth rate and lipid yield of Chlorella with nuclear irradiation under high salt and CO_2 stress, Bioresour. Technol. 203 (2016) 220–227.

[42] A. Roncarati, A. Meluzzi, S. Acciarri, N. Tallarico, P. Meloti, Fatty acid composition of different microalgae strains (Nannochloropsis sp., Nannochloropsis oculata (Droop) Hibberd, Nannochloris atomus Butcher and Isochrysis sp.) according to the culture phase and the carbon dioxide concentration, J. World Aquac. Soc. 35 (2004) 401–411.

[43] B. Hu, M. Min, W. Zhou, Y. Li, M. Mohr, Y. Cheng, et al., Influence of exogenous CO_2 on biomass and lipid accumulation of microalgae Auxenochlorella protothecoides cultivated in concentrated municipal wastewater, Appl. Biochem. Biotechnol. 166 (2012) 1661–1673.

[44] M. Min, B. Hu, W. Zhou, Y. Li, P. Chen, R. Ruan, Mutual influence of light and CO_2 on carbon sequestration via cultivating mixotrophic alga Auxenochlorella protothecoides UMN280 in an organic carbon-rich wastewater, J. Appl. Phycol. 24 (2012) 1099–1105.

[45] F. Hussain, S.Z. Shah, W. Zhou, M. Iqbal, Microalgae screening under CO_2 stress: growth and micronutrients removal efficiency, J. Photochem. Photobiol. B Biol. 170 (2017) 91–98.

[46] D.L. Sanchez, D.S. Callaway, Optimal scale of carbon-negative energy facilities, Appl. Energy 170 (2016) 437–444.

[47] S. Gollakota, S. McDonald, CO_2 capture from ethanol production and storage into the Mt Simon Sandstone, Greenh. Gas Sci. Technol. 2 (2012) 346–351.

[48] G. De Bhowmick, L. Koduru, R. Sen, Metabolic pathway engineering towards enhancing microalgal lipid biosynthesis for biofuel application—a review, Renew. Sustain. Energy Rev. 50 (2015) 1239–1253.

[49] Y.Y. Choi, A.K. Patel, M.E. Hong, W.S. Chang, S.J. Sim, Microalgae bioenergy with carbon capture and storage (BECCS): an emerging sustainable bioprocess for reduced CO_2 emission and biofuel production, Bioresour. Technol. Rep. 7 (2019) 100270. Available from: https://doi.org/10.1016/j.biteb.2019.100270.

[50] Y.Y. Choi, J.M. Joun, J. Lee, M.E. Hong, H.-M. Pham, W.S. Chang, et al., Development of large-scale and economic pH control system for outdoor cultivation of microalgae Haematococcus pluvialis using industrial flue gas, Bioresour. Technol. 244 (2017) 1235–1244.

[51] L. Peng, C.Q. Lan, Z. Zhang, C. Sarch, M. Laporte, Control of protozoa contamination and lipid accumulation in Neochloris oleoabundans culture: effects of pH and dissolved inorganic carbon, Bioresour. Technol. 197 (2015) 143–151.

Handbook of Algal Biofuels

[52] S.M.U. Shah, C.C. Radziah, S. Ibrahim, F. Latiff, M.F. Othman, M.A. Abdullah, Effects of photoperiod, salinity and pH on cell growth and lipid content of Pavlova lutheri, Ann. Microbiol. 64 (2014) 157—164.

[53] E. Cucchiari, F. Guerrini, A. Penna, C. Totti, R. Pistocchi, Effect of salinity, temperature, organic and inorganic nutrients on growth of cultured Fibrocapsa japonica (Raphidophyceae) from the northern Adriatic Sea, Harmful Algae 7 (2008) 405—414.

[54] M. Sakarika, M. Kornaros, Effect of pH on growth and lipid accumulation kinetics of the microalga Chlorella vulgaris grown heterotrophically under sulfur limitation, Bioresour. Technol. 219 (2016) 694—701.

[55] A. Difusa, J. Talukdar, M.C. Kalita, K. Mohanty, V.V. Goud, Effect of light intensity and pH condition on the growth, biomass and lipid content of microalgae Scenedesmus species, Biofuels 6 (2015) 37—44.

[56] J.R. Coleman, B. Colman, Inorganic carbon accumulation and photosynthesis in a blue-green alga as a function of external pH, Plant. Physiol. 67 (1981) 917—921.

[57] M. Taraldsvik, S. Myklestad, The effect of pH on growth rate, biochemical composition and extracellular carbohydrate production of the marine diatom Skeletonema costatum, Eur. J. Phycol. 35 (2000) 189—194.

[58] T.M. Mata, A.A.A. Martins, N.S. Caetano, Microalgae for biodiesel production and other applications: a review, Renew. Sustain. Energy Rev. 14 (2010) 217—232. Available from: https://doi.org/10.1016/j.rser.2009.07.020.

[59] C.-Y. Chen, G.D. Saratale, C.-M. Lee, P.-C. Chen, J.-S. Chang, Phototrophic hydrogen production in photobioreactors coupled with solar-energy-excited optical fibers, Int. J. Hydrog. Energy 33 (2008) 6886—6895.

[60] D.P.K. Nogueira, A.F. Silva, O.Q.F. Araújo, R.M. Chaloub, Impact of temperature and light intensity on triacylglycerol accumulation in marine microalgae, Biomass Bioenergy 72 (2015) 280—287.

[61] D. Simionato, S. Basso, G.M. Giacometti, T. Morosinotto, Optimization of light use efficiency for biofuel production in algae, Biophys. Chem. 182 (2013) 71—78.

[62] C. Yeesang, B. Cheirsilp, Effect of nitrogen, salt, and iron content in the growth medium and light intensity on lipid production by microalgae isolated from freshwater sources in Thailand, Bioresour. Technol. 102 (2011) 3034—3040.

[63] A.E. Solovchenko, I. Khozin-Goldberg, S. Didi-Cohen, Z. Cohen, M.N. Merzlyak, Effects of light intensity and nitrogen starvation on growth, total fatty acids and arachidonic acid in the green microalga Parietochloris incisa, J. Appl. Phycol. 20 (2008) 245—251.

[64] S.-H. Ho, C.-Y. Chen, J.-S. Chang, Effect of light intensity and nitrogen starvation on CO_2 fixation and lipid/carbohydrate production of an indigenous microalga Scenedesmus obliquus CNW-N, Bioresour. Technol. 113 (2012) 244—252.

[65] J. Liu, C. Yuan, G. Hu, F. Li, Effects of light intensity on the growth and lipid accumulation of microalga Scenedesmus sp. 11—1 under nitrogen limitation, Appl. Biochem. Biotechnol. 166 (2012) 2127—2137.

[66] S. Wahidin, A. Idris, S.R.M. Shaleh, The influence of light intensity and photoperiod on the growth and lipid content of microalgae Nannochloropsis sp, Bioresour. Technol. 129 (2013) 7—11.

[67] B. George, I. Pancha, C. Desai, K. Chokshi, C. Paliwal, T. Ghosh, et al., Effects of different media composition, light intensity and photoperiod on morphology and physiology of freshwater microalgae Ankistrodesmus falcatus—a potential strain for bio-fuel production, Bioresour. Technol. 171 (2014) 367—374.

[68] B.C.B. Freitas, A.P.A. Cassuriaga, M.G. Morais, J.A.V. Costa, Pentoses and light intensity increase the growth and carbohydrate production and alter the protein profile of Chlorella minutissima, Bioresour. Technol. 238 (2017) 248—253.

[69] C.E. de Farias Silva, E. Sforza, A. Bertucco, Stability of carbohydrate production in continuous microalgal cultivation under nitrogen limitation: effect of irradiation regime and intensity on Tetradesmus obliquus, J. Appl. Phycol. 30 (2018) 261—270.

[70] X. Deng, B. Chen, C. Xue, D. Li, X. Hu, K. Gao, Biomass production and biochemical profiles of a freshwater microalga Chlorella kessleri in mixotrophic culture: effects of light intensity and photoperiodicity, Bioresour. Technol. 273 (2019) 358—367.

[71] Q. He, H. Yang, L. Wu, C. Hu, Effect of light intensity on physiological changes, carbon allocation and neutral lipid accumulation in oleaginous microalgae, Bioresour. Technol. 191 (2015) 219—228.

[72] C.L. Teo, M. Atta, A. Bukhari, M. Taisir, A.M. Yusuf, A. Idris, Enhancing growth and lipid production of marine microalgae for biodiesel production via the use of different LED wavelengths, Bioresour. Technol. 162 (2014) 38—44.

[73] M. Yoshioka, T. Yago, Y. Yoshie-Stark, H. Arakawa, T. Morinaga, Effect of high frequency of intermittent light on the growth and fatty acid profile of Isochrysis galbana, Aquaculture 338 (2012) 111—117.

References

[74] E. Roscher, K. Zetsche, The effects of light quality and intensity on the synthesis of ribulose-1, 5-bisphosphate carboxylase and its mRNAs in the green alga Chlorogonium elongatum, Planta 167 (1986) 582–586.

[75] D.X. Hurtado, C.L. Garzón-Castro, J. Cortés-Romero, E. Tello, Using different wavelengths and irradiance on the microalgae Acutodesmus obliquus batch culture, J. Chem. Technol. Biotechnol. 94 (2019) 2141–2147.

[76] A.E.-F. Abomohra, H. Shang, M. El-Sheekh, H. Eladel, R. Ebaid, S. Wang, et al., Night illumination using monochromatic light-emitting diodes for enhanced microalgal growth and biodiesel production, Bioresour. Technol. (2019) 121514.

[77] Y.H. Seo, Y. Lee, D.Y. Jeon, J.-I. Han, Enhancing the light utilization efficiency of microalgae using organic dyes, Bioresour. Technol. 181 (2015) 355–359.

[78] Y.H. Seo, C. Cho, J.-Y. Lee, J.-I. Han, Enhancement of growth and lipid production from microalgae using fluorescent paint under the solar radiation, Bioresour. Technol. 173 (2014) 193–197.

[79] H.D. Amrei, R. Ranjbar, S. Rastegar, B. Nasernejad, A. Nejadebrahim, Using fluorescent material for enhancing microalgae growth rate in photobioreactors, J. Appl. Phycol. 27 (2015) 67–74.

[80] K. Sharma, Y. Li, P.M. Schenk, UV-C-mediated lipid induction and settling, a step change towards economical microalgal biodiesel production, Green. Chem. 16 (2014) 3539–3548.

[81] Y. Liang, J. Beardall, P. Heraud, Effect of UV radiation on growth, chlorophyll fluorescence and fatty acid composition of Phaeodactylum tricornutum and Chaetoceros muelleri (Bacillariophyceae), Phycologia 45 (2006) 605–615.

[82] F. Guihéneuf, M. Fouqueray, V. Mimouni, L. Ulmann, B. Jacquette, G. Tremblin, Effect of UV stress on the fatty acid and lipid class composition in two marine microalgae Pavlova lutheri (Pavlovophyceae) and Odontella aurita (Bacillariophyceae), J. Appl. Phycol. 22 (2010) 629–638.

[83] S. Liu, Y. Zhao, L. Liu, X. Ao, L. Ma, M. Wu, et al., Improving Cell Growth and Lipid Accumulation in Green Microalgae Chlorella sp. via UV Irradiation, Appl. Biochem. Biotechnol. 175 (2015) 3507–3518. Available from: https://doi.org/10.1007/s12010-015-1521-6.

[84] M.P. Tale, R. devi Singh, B.P. Kapadnis, S.B. Ghosh, Effect of gamma irradiation on lipid accumulation and expression of regulatory genes involved in lipid biosynthesis in Chlorella sp, J. Appl. Phycol. 30 (2018) 277–286.

[85] J. Cheng, J. Feng, J. Sun, Y. Huang, J. Zhou, K. Cen, Enhancing the lipid content of the diatom Nitzschia sp. by 60Co-γ irradiation mutation and high-salinity domestication, Energy 78 (2014) 9–15.

[86] R. Tu, W. Jin, M. Wang, S. Han, A.E.-F. Abomohra, W.-M. Wu, Improving of lipid productivity of the biodiesel promising green microalga Chlorella pyrenoidosa via low-energy ion implantation, J. Appl. Phycol. 28 (2016) 2159–2166.

[87] F. Qu, W. Jin, X. Zhou, M. Wang, C. Chen, R. Tu, et al., Nitrogen ion beam implantation for enhanced lipid accumulation of Scenedesmus obliquus in municipal wastewater, Biomass Bioenergy 134 (2020) 105483.

[88] Y. Ma, Z. Wang, M. Zhu, C. Yu, Y. Cao, D. Zhang, et al., Increased lipid productivity and TAG content in Nannochloropsis by heavy-ion irradiation mutagenesis, Bioresour. Technol. 136 (2013) 360–367.

[89] J. Choi, M. Yoon, M. Joe, H. Park, S.G. Lee, S.J. Han, et al., Development of microalga Scenedesmus dimorphus mutant with higher lipid content by radiation breeding, Bioprocess. Biosyst. Eng. 37 (2014) 2437–2444.

[90] T. Takeshita, I.N. Ivanov, K. Oshima, K. Ishii, H. Kawamoto, S. Ota, et al., Comparison of lipid productivity of Parachlorella kessleri heavy-ion beam irradiation mutant PK4 in laboratory and 150-L mass bioreactor, identification and characterization of its genetic variation, Algal Res. 35 (2018) 416–426.

[91] W. Wang, T. Wei, J. Fan, J. Yi, Y. Li, M. Wan, et al., Repeated mutagenic effects of 60 Co-γ irradiation coupled with high-throughput screening improves lipid accumulation in mutant strains of the microalgae Chlorella pyrenoidosa as a feedstock for bioenergy, Algal Res. 33 (2018) 71–77. Available from: https://doi.org/10.1016/j.algal.2018.04.022.

[92] J.D. Moha-León, I.A. Pérez-Legaspi, L.A. Ortega-Clemente, I. Rubio-Franchini, E. Ríos-Leal, Improving the lipid content of Nannochloropsis oculata by a mutation-selection program using UV radiation and quizalofop, J. Appl. Phycol. 31 (2019) 191–199. Available from: https://doi.org/10.1007/s10811-018-1568-1.

[93] H. Vigeolas, F. Duby, E. Kaymak, G. Niessen, P. Motte, F. Franck, et al., Isolation and partial characterization of mutants with elevated lipid content in Chlorella sorokiniana and Scenedesmus obliquus, J. Biotechnol. 162 (2012) 3–12.

[94] A. Ramazanov, Z. Ramazanov, Isolation and characterization of a starchless mutant of Chlorella pyrenoidosa STL-PI with a high growth rate, and high protein and polyunsaturated fatty acid content, Phycol. Res. 54 (2006) 255–259.

[95] R. Sivaramakrishnan, A. Incharoensakdi, Enhancement of lipid production in Scenedesmus sp. by UV mutagenesis and hydrogen peroxide treatment, Bioresour. Technol. 235 (2017) 366–370. Available from: https://doi.org/10.1016/j.biortech.2017.03.102.

[96] D.P. Small, N.P.A. Hüner, W. Wan, Effect of static magnetic fields on the growth, photosynthesis and ultrastructure of Chlorella kessleri microalgae, Bioelectromagnetics 33 (2012) 298–308.

[97] K.M. Deamici, L.O. Santos, J.A.V. Costa, Use of static magnetic fields to increase CO_2 biofixation by the microalga Chlorella fusca, Bioresour. Technol. 276 (2019) 103–109.

[98] W. Shao, R. Ebaid, A.E. Abomohra, M. Shahen, Enhancement of Spirulina biomass production and cadmium biosorption using combined static magnetic field, Bioresour. Technol. (2018). Available from: https://doi.org/10.1016/j.biortech.2018.06.009.

[99] S. Han, W. Jin, Y. Chen, R. Tu, A.E.-F. Abomohra, Enhancement of lipid production of chlorella pyrenoidosa cultivated in municipal wastewater by magnetic treatment, Appl. Biochem. Biotechnol. 180 (2016) 1043–1055. Available from: https://doi.org/10.1007/s12010-016-2151-3.

[100] F.-J. Chu, T.-J. Wan, T.-Y. Pai, H.-W. Lin, S.-H. Liu, C.-F. Huang, Use of magnetic fields and nitrate concentration to optimize the growth and lipid yield of Nannochloropsis oculata, J. Env. Manage 253 (2020) 109680.

[101] H. Nezammahalleh, F. Ghanati, T.A. Adams II, M. Nosrati, S.A. Shojaosadati, Effect of moderate static electric field on the growth and metabolism of Chlorella vulgaris, Bioresour. Technol. 218 (2016) 700–711.

[102] X. Feng, Y. Chen, J. Lv, S. Han, R. Tu, X. Zhou, et al., Enhanced lipid production by Chlorella pyrenoidosa through magnetic field pretreatment of wastewater and treatment of microalgae-wastewater culture solution: magnetic field treatment modes and conditions, Bioresour. Technol. (2020) 123102.

[103] L. Xu, C. Guo, F. Wang, S. Zheng, C.-Z. Liu, A simple and rapid harvesting method for microalgae by in situ magnetic separation, Bioresour. Technol. 102 (2011) 10047–10051.

[104] K.M. Deamici, J.A.V. Costa, L.O. Santos, Magnetic fields as triggers of microalga growth: evaluation of its effect on Spirulina sp, Bioresour. Technol. 220 (2016) 62–67.

[105] M. Fang, L. Jin, C. Zhang, Y. Tan, P. Jiang, N. Ge, et al., Rapid mutation of Spirulina platensis by a new mutagenesis system of atmospheric and room temperature plasmas (ARTP) and generation of a mutant library with diverse phenotypes, PLoS One 8 (2013) e77046.

[106] C. Ottenheim, M. Nawrath, J.C. Wu, Microbial mutagenesis by atmospheric and room-temperature plasma (ARTP): the latest development, Bioresour. Bioprocess. 5 (2018) 12.

[107] L. Wang, Z. Huang, G. Li, H. Zhao, X. Xing, W. Sun, et al., Novel mutation breeding method for Streptomyces avermitilis using an atmospheric pressure glow discharge plasma, J. Appl. Microbiol. 108 (2010) 851–858.

[108] J.-W. Lackmann, J.E. Bandow, Inactivation of microbes and macromolecules by atmospheric-pressure plasma jets, Appl. Microbiol. Biotechnol. 98 (2014) 6205–6213.

[109] J.Q.M. Almarashi, S.E. El-Zohary, M.A. Ellabban, A.E.F. Abomohra, Enhancement of lipid production and energy recovery from the green microalga Chlorella vulgaris by inoculum pretreatment with low-dose cold atmospheric pressure plasma (CAPP), Energy Convers. Manag. 204 (2020). Available from: https://doi.org/10.1016/j.enconman.2019.112314.

[110] S. Bekeschus, K. Masur, J. Kolata, K. Wende, A. Schmidt, L. Bundscherer, et al., Human mononuclear cell survival and proliferation is modulated by cold atmospheric plasma jet, Plasma Process. Polym. 10 (2013) 706–713.

[111] S. Cao, X. Zhou, W. Jin, F. Wang, R. Tu, S. Han, et al., Improving of lipid productivity of the oleaginous microalgae Chlorella pyrenoidosa via atmospheric and room temperature plasma (ARTP), Bioresour. Technol. 244 (2017) 1400–1406.

[112] X. Sun, P. Li, X. Liu, X. Wang, Y. Liu, A. Turaib, et al., Strategies for enhanced lipid production of Desmodesmus sp. mutated by atmospheric and room temperature plasma with a new efficient screening method, J. Clean. Prod. 250 (2020) 119509.

[113] B. Chen, C. Wan, M.A. Mehmood, J.-S.S. Chang, F. Bai, X. Zhao, Manipulating environmental stresses and stress tolerance of microalgae for enhanced production of lipids and value-added products—a review, Bioresour. Technol. 244 (2017). Available from: https://doi.org/10.1016/j.biortech.2017.05.170.

[114] M.A. Borowitzka, High-value products from microalgae—their development and commercialisation, J. Appl. Phycol. 25 (2013) 743–756.

References

[115] H.J. Kim, H. Jeong, S. Hwang, M.-S. Lee, Y.-J. Lee, D.-W. Lee, et al., Short-term differential adaptation to anaerobic stress via genomic mutations by Escherichia coli strains K-12 and B lacking alcohol dehydrogenase, Front. Microbiol. 5 (2014) 476.

[116] I.T.D. Cabanelas, D.M.M. Kleinegris, R.H. Wijffels, M.J. Barbosa, Repeated nitrogen starvation doesn't affect lipid productivity of Chlorococcum littorale, Bioresour. Technol. 219 (2016) 576–582.

[117] P. Singh, A. Guldhe, S. Kumari, I. Rawat, F. Bux, Investigation of combined effect of nitrogen, phosphorus and iron on lipid productivity of microalgae Ankistrodesmus falcatus KJ671624 using response surface methodology, Biochem. Eng. J. 94 (2015) 22–29.

[118] H. Li, Z. Liu, Y. Zhang, B. Li, H. Lu, N. Duan, et al., Conversion efficiency and oil quality of low-lipid high-protein and high-lipid low-protein microalgae via hydrothermal liquefaction, Bioresour. Technol. 154 (2014) 322–329.

[119] Y. Chen, Y. Wu, D. Hua, C. Li, M.P. Harold, J. Wang, et al., Thermochemical conversion of low-lipid microalgae for the production of liquid fuels: challenges and opportunities, RSC Adv. 5 (2015) 18673–18701.

CHAPTER 21

Microalgal—bacterial consortia for biomass production and wastewater treatment

Muhammad Usman Khan[1,2], Nalok Dutta[1], Abid Sarwar[3], Muhammad Ahmad[4], Maryam Yousaf[4,5], Yassine Kadmi[6] and Mohammad Ali Shariati[7]

[1]Bioproducts Science and Engineering Laboratory, Washington State University, Richland, WA, United States [2]Department of Energy Systems Engineering, Faculty of Agricultural Engineering and Technology, University of Agriculture, Faisalabad, Pakistan [3]Department of Irrigation and Drainage, Faculty of Agricultural Engineering and Technology, University of Agriculture, Faisalabad, Pakistan [4]School of Chemistry and Chemical Engineering, Beijing Institute of Technology, Beijing, P.R. China [5]Department of Chemistry, University of Agriculture Faisalabad, Faisalabad, Pakistan [6]LASIRE CNRS UMR 8516, Sciences and Technologies, University of Lille, Lille, France [7]Kazakh Research Institute of Processing and Food Industry (Semey Branch), Semey, Kazakhstan

21.1 Introduction

Over the years, fossil fuels have been persistently utilized worldwide leading to the depletion of the existing fuel resources [1,2]. Moreover, the continuous use of fossil fuel as a source of energy has caused environmental pollution [3]. In this regard, biobased fuels derived from renewable feedstock provide an ecofriendly solution in contrast to hazardous chemical processes [4]. Among these alternative processes, biologically derived energy conversion mediated by microalgae has gained substantial popularity, owing to its unique ability to convert solar-based energy to bioenergy. The advantages of the energy conversion process mediated by microalgae are reflected by increased content of lipids, reduced generation time, and its adaptive nature in various environments [5,6]. Moreover, the utility of the microalgae-derived processes has been extended to applications ranging from pharmaceutical industries, food-processing companies, greenhouse gas mitigation, and power generation pertaining to heat/electricity.

The advancement in microalgal research has led to its application in wastewater treatment methodologies, making use of the microalgal ability to mitigate contamination by sequestering organic nutrients with simultaneous conversion to value-added products [7–11]. The limitation of this method of wastewater bioremediation is the high cost incurred in the overall process, which stems from the fact that the algae is present in suspension, which causes a bottleneck and requires troubleshoot in harvesting. Moreover, the process optimization also gets slowed due to the ineffectiveness of the extractive methodologies applied to the biomass and biofuel produced in the course of the process. To solve this problem, bacterial–microalgal processes have been practiced in tandem owing to a symbiotic interrelationship between these microbial communities that promotes the algal growth rate coupled with increased contaminant mitigation in wastewater and microalgal flocculation [12,13]. Of late, microalgae–bacterial consortia (MBC) have become popular due to their impact on the net efficiency in biomass harvest mediated by algal aggregation and nutrient removal [5,6]. The optimal functioning of the MBC is based on the altruistic interactions between microalgae and bacteria, which is centered around the exchange of nutrients [14]. The oxygen produced by microalgae is taken up by the bacterial species as an electron acceptor for organic valorization apart from dissolved carbon species. Again, the carbon dioxide produced by the bacterial species during mineralization is taken up by the algae coupled with vitamin B, N_2, and siderophores required for microalgal functional processes [9]. The optimal functioning of the overall process is dependent on the reciprocal nutrition procurement of the interspecific microbial communities, which are mediated by a cumulative response to chemicals like indole-3-acetic acid and N-acyl-homoserine lactones [12]. The efficacy of this behavior is reflected in nutrient screening, ecological manifestation, and the currency of reproduction. The interrelationship between these two communities is extended to production of nutrients in the form of polysaccharides and proteins by the bacteria which aids in algal flocculation, thereby facilitating downstream recovery and enhanced harvest. Moreover, the phycotoxins produced by the bacteria mitigates microalgal bloom, thereby keeping the community niche in check.

Lipases [triacylglycerol hydrolases (E.C. 3.1.1.3)] have been utilized for their versatility in the processes of lipid biotechnology owing to their selective properties in chemical reactions. This unique property of lipase has been exercised across numerous industries ranging from emulsifiers and pharmaceuticals to lipid-enriched wastewater pretreatment for effective bioremediation of oils and biodiesel synthesis. Lipases catalyze transesterification reactions, which make them an ideal candidate for environmental bioremediation, contributing to lipid mitigation in wastewater and biofuel synthesis [15]. One of the factors that contribute to the wide use of lipase stems from the fact that microbial organisms secreting lipase grow rapidly on low-cost media. Of late, microbial treatment of wastewater with biocatalysts is being implemented in the water treatment plants to degrade dissolved lipids and fats, which is an ecofriendly process compared to the chemical ones. The combined use of chemical coagulants and flocculants remove a small fraction of dissolved lipids that remains as emulsified droplets in the wastewater [16]. These droplets are carried over passively to secondary biological treatment systems where the final effluent volume is adjusted before clearing [17].

Biological systems pertaining to fabricated wetlands cannot degrade high concentrations of dissolved lipids and fats. Concentrations of these substances above the permissible levels negatively impact the environment. According to the literature review and case studies, ~90% of these wetlands are sensitive to dissolved lipids, biological oxygen demand, and chemical oxygen demand (COD) concentrations. Microorganism-based techniques have been successfully executed in complementary treatment systems to process wastewater with a high lipid content. These practices are manifested using pure enzymatic extracts obtained from pure cultures through solid-state fermentation and/or active consortia grown in the wastewater to be treated [18,19].

Biological treatment processes are economically feasible, effective, repeatable, and environmentally compatible with the prevailing wastewater treatment methods. These processes degrade the chemicals present in municipal wastes to nontoxic forms producing stable reaction intermediates and value-added compounds ranging from biopolymers to biofuels. Different forms of waste in water sources comprised of organic loads, biostatic compounds, and low pH values coupled with the unbalanced composition of nutrients are the determining factors that affect the overall biological treatment process. The choice of microorganisms as alternate fuel sources requires a nutrient medium, which needs to be provided in the domestic wastewater streams. However, the bulk of these wastewaters is overloaded with high concentrations of nutrients, namely, total N and total P concentration, along with the toxic metals, thus nullifying the requirement for harsh chemical-based treatments [20]. A thorough assessment of the individual components of wastewater is essential in the design and operation of treatment processes. The concentration of total N and P in these water sources ranges between 10−90 mg/L in municipal wastewater and more than 1000 mg/L in agricultural effluents.

With reference to the above context, we want to present an in-depth analysis of MBC and its systemic analysis pertaining to wastewater treatment and production of value-added products. We would investigate the recent advancements of the implication of lipase as a biocatalyst in its role of removing contaminants in wastewater. The essence of this review is to gain a cumulative outlook in depicting the microalgal−bacterial interaction. This will provide a thorough background required for the assessment of biotechnological advancement in the frontiers of this domain to affect scaled-up processes and applications mediated by biocatalysts and MBC.

21.2 Interchange of substrates, intercellular communication, and horizontal gene transfer in microalgal−bacterial consortia

The MBC functions in an optimal condition when there is a reciprocal nutrient exchange between microalgae and bacteria, forming the basis of this interspecific relationship. The primary mode of relationship is initiated by a simple gaseous exchange, where the microalgae is facilitated by bacterial CO_2 and inorganic compounds and the bacterial consortia benefit from the organic substances and O_2 provided by the algae [21]. This mutualistic nature of coexistence reduces operational costs in bioreactors and sludge beds where a significant cost is incurred with regard to oxygenation. To add to this, there is also the mitigation effect on the greenhouse gases (GHGs) emanating from the water treatment plants [22]. Micronutrients in the form of vitamin B_{12}, siderophores, and phytohormones secreted by bacteria have a positive impact on the overall microalgal growth and metabolism [12]. Apart from micronutrients, nitrogenous exchange takes place between algae and heterotrophic bacteria in nutrient-limiting environments. A study showed an instance of such mutualistic behavior where *Rhizobium* sp, a bacterial species promoted N_2 assimilation by *Chlorella* (algae) from wastewater in return for the supply of organic matter from the algae [23]. A communalistic behavior was reflected in the interaction between nitrogenous bacteria *Azotobacter* and microalgae *Neochloris* sp. where the latter received N_2 from the former [24]. This study is effective in the sense that in the process involved large-scale production of microalgae, the cost can be curtailed by using the aforementioned source of N_2. Moreover, as cited in the literature there have been several instances where certain bacterial species can also supply the microalgae with inorganic phosphate. Transfer of phosphate to microalgae *Microcystis* by *Pseudomonas* sp. bacteria was reported in a study [25]. Another study showcased the metabolites produced by *Alteromonas* sp. which induced death of microalgae by lysis and a change in the calcium-mediated signal transduction, whereas *Stenotrophomonas* sp. reduced the growth of cyanobacteria mediated by siderophores by a lowered availability of iron [26,27]. In context with this, a study depicted a secretion of antibacterial compounds by microalgal genera of *Prasinophyceae* and *Bacillariophyceae*, preventing the growth of the bacterial counterparts in a consortium [27,28].

Intercellular communication across MBC is mediated by cross-talk signals generated through quorum sensing. Indole acetic acid secreted by *Sulfitobacter* species facilitates cell division and promotes communication with profitable microbial consortia, whereas a certain threshold of *Pseudoalteromonas* sp. population generates algicidal activity triggering microalgal cell lysis [29,30]. In another research, it was shown that the *Chlorophyta* sp. formed bioaggregation in the presence of N-acyl-homoserine lactones, whereas it remained in the suspension in absence of N-acyl-homoserine lactones. Discharge of spores in *Acrochaetium* sp. was induced upon generation of molecular signal by *Gracilaria* sp. [31]. Some studies also put forth antagonistic nature of microalgae consortia where the microalgae, for example, *Chlamydomonas reinhardtii* (Chlorophyta), produced a vitamin B_2 derivative called lumichrome, mimicking molecular signals equivalent to the N-acyl-homoserine lactone analog to engage the binding sites of *Pseudomonas aeruginosa* [32,33]. Another modus operandi of the microalgae to save themselves from detrimental bacteria is by quorum quenching, wherein the N-acyl-homoserine lactone is deactivated by these specifically secreted enzyme machinery constituted by amylases and oxidases (e.g., vanadium haloperoxidases) [33—35]. This inactivation of N-acyl-homoserine lactone is brought about by hydrolyzing the homoserine lactone ring and the cleavage of the amide bond which prevents the interaction between quorum sensing components and their reciprocating counterparts [34]. There are several instances which showcase the interspecific communication between microalgae and bacteria. The currency of this interrelationship is based on the mode of cross-talk signaling mediated by several factors, namely distribution, chemotactic movement, mobility of the spores, cell lysis, nutrient procurement, biofilm generation, ecological niche, adaptation, virulence factor, and reproducibility [36,37].

Horizontal gene transfer is a concerted process of bacterial gene incorporation in the algal genome in engineered MBCs. This renders an advantage to microalgal species to overcome the environmental stresses. In the ornithine cycle, gene transfer occurs from bacteria to algae (diatoms) which revives the diatoms from conditions pertaining to nitrogen mitigation as reflected in metabolomics studies [38]. In another study, genetic analysis of *G. sulphuraria* genome showed that it can survive in harsh and toxic environments owing to the transfer of archeal and bacterial genes [39]. The above studies provide valuable insights on gene transfer across MBC. To cite an example in this regard, genes coding for actin and the operationally connected profilin have been found next to each other in the genome of the cyanobacterium *M. aeruginosa*. The lateral positioning of actin and profilin genes in its structure bears testimony to gene transfer [39]. The transfer of genetic materials from algae to bacteria has a slow onset owing to the structural complexity of the bacterial species (Fig. 21.1).

A number of factors account for the optimum functioning of MBC ranging from the thickness and tenacity of the algal cell wall to cultivation conditions and the hydraulic retention time [40]. Physiological and environmental factors (biotic and abiotic) ranging from pH, light, temperature, nutrient procurement, and intermixing intensity can influence the interactions within MBC [26].

21.3 Distribution and role of microalgal—bacterial consortia in the wastewater treatment

Belonging to the category of photosynthetic eukaryotes, the microalgae population has been classified into two major categories, namely, multicellular and unicellular macroalgae. The microalgae population is found in close proximity to heterotrophic bacteria in naturally occurring aquatic environments. Being autotrophs, microalgae synthesize organic food, which is used up by the bacteria in the course of decomposition of the organic carbon molecule. In return, heterotrophic bacteria boost algal growth by spore germination [41]. However this mutually exclusive relationship can turn antagonistic with regard to the availability of phosphate and carbon [26]. MBC recently have been widely studied and implemented in wastewater plants with the aim of devising feasible and economic strategies for effecting contaminant remediation [42,43]. MBC interactions are beneficial in formulating wastewater remediation strategies since MBC interactions (1) promote nutrient sharing to overcome metabolite depletion, (2) maintain solidarity amongst consortium resisting the influx of foreign invasive species, and (3) promote adaptation to adverse environments with an increased uptake of nutrients [44].

21.3.1 Role of microalgal—bacterial consortia in mitigation of carbon and nutrients

Open and closed bioreactor systems with MBC suspensions have been implanted for the removal of carbon and nutrients from wastewater [45]. Natural and artificial water systems constitute open-end bioreactors which also include high-rate algal ponds, in vogue nowadays for nutrient removal from water by using MBC suspension, thus promoting GHG mitigation [46]. Since the open bioreactors have a direct exposure to the

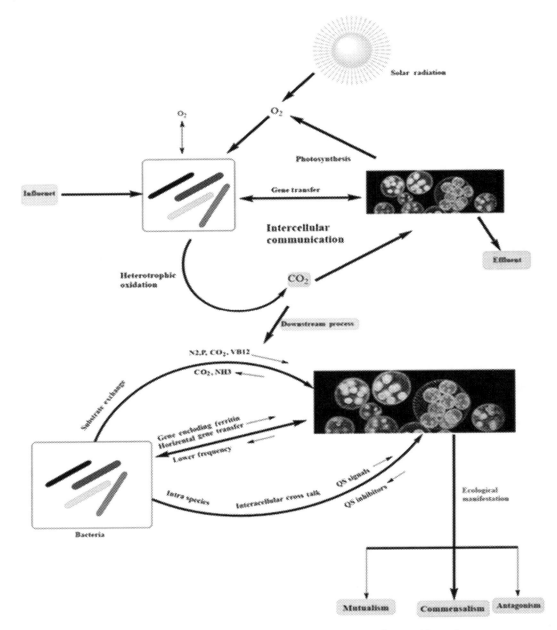

FIGURE 21.1 Schematic diagram of Interactional patterns between microalgae and bacteria.

surrounding environment, biotic and abiotic factors in the form of species invasion, temperature, pH, and evaporation emerge as pertinent factors which limit their efficacy [7]. Closed reactors provide the solution to address these problems as these kinds of bioreactors avoid possible contamination with strictly controlled parameters for cell growth. The most popular and widely used closed bioreactor systems are constituted by photobioreactors (PBRs) for seamless cultivation of MBC. Among the different types of closed bioreactors, tubular, column, and closed configurations of flat plate bioreactors are being used comprehensively. The pros and cons of the different types of PBRs range in terms of their efficacy of nutrient removal, space, operative complexity, and energy requirement [45]. A detailed pictorial representation of different PBRs shown in Fig. 21.2.

The effect of photoexposure on the removal of COD, nitrogen, and phosphorus of a MBC was studied by a group of researchers who revealed heightened efficiency following a biphased photoexposure of 12 h light/60 h darkness with 12 h night/day following the previous exposure [47]. The optimum removal efficacies were reflected in COD, nitrogen, and phosphorus with 93%, 96%, and 98% mitigation. In spite of the widespread use of MBC and its proven effectiveness, the operational procedure at times is difficult, particularly pertaining to the separation of the suspended biomass to ensure a healthy consortium. Table 21.1 describes the mitigation of carbon, nitrogen, and phosphorus levels in diverse wastewater sources by MBC with percentile efficiency of removal.

To design an optimally functioning attached MBC, an appropriate bioreactor setup is required. The maximal growth of MBC is seen on solid matrices, namely, wood, bamboo,

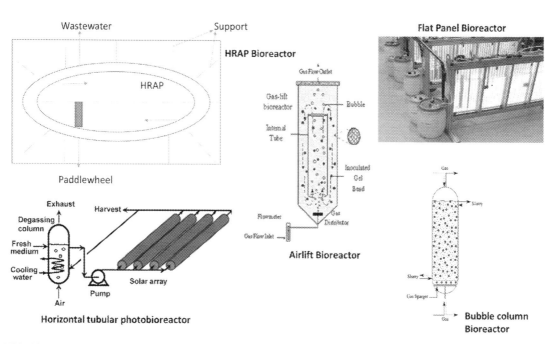

FIGURE 21.2 Different types of photobioreactor setups used for mitigation of carbon and nutrients.

484
21. Microalgal—bacterial consortia for biomass production and wastewater treatment

TABLE 21.1 Mitigation of carbon, nitrogen, and phosphorus levels in different wastewater sources by microalgal—bacterial consortia with percentile efficiency of removal.

MBC	Wastewater (WW) source	Type of bioreactor	Treatment time (in days)	Carbon removal efficiency (%)	Nitrogen removal efficiency (%)	Phosphorus removal efficiency (%)	Reference
Scenedesmus sp. + nitrifying sludge bacteria	Synthetic WW	PBR	95	NA	100	NA	[48]
Chlorella sp. + activated sludge	Piggery WW	PBR	13	47	21	54	[49]
Euglena viridis + activated sludge	Piggery WW	PBR	13	51	34	53	[49]
Clamydomonous sp. + *Microspora* sp. + *Chlorella* sp. + *Achnanthes* sp.	Piggery WW	HRAP	245	58.7—76.5	78—90	NA	[50]
Chlorella sp. + *Scenedesmus* sp. + *Stigeoclonium* sp.	Municipal WW	HRAP	28	85	92	83	[23]
Chlorella sp. + *Acutodesmus obliquus* + *Oscillatoria* sp.	Piggery WW	PBR	180	86	82—85	92	[51]
Scenedesmus sp. + indigenous bacteria	Municipal WW	PBR	14	92	96	98	[47]
Chlorella vulgaris + *Bacillus licheniforms in ratio of* 1:3	Synthetic WW	PBR	10	86	89	80	[52]
C. vulgaris + *Proteobacteria*	Synthetic WW	PBR	180	86—90	68—79	NA	[53]
C. vulgaris + native bacteria	Synthetic WW	Closed suspended system	8—20	NA	31—100	65—98	[54]
Chlorella sp. + wastewater bacteria	Process WW	Pilot-scale bioreactor	84	85.8	69	83	[21]

HRAP, high-rate algal ponds; *MBC*, microalgal—bacterial consortia; *PBR*, photobioreactor.

plastic, nylon, and PVC platforms [55]. PVC platforms are the chosen matrices for designing a properly functioning PBR owing to its withstanding texture and extensive availability [5]. MBC are bestowed with ability of self-aggregation and such aggregates are formed without the need of solid supports with a size range of 0.1—0.5 mm. Therefore optimized utilization of the MBC aggregates eventually curtails the incurred cost by 30% with

Handbook of Algal Biofuels

efficient CO_2-O_2 mass transport as suggested by the Luedeking–Piret model [55–57]. Generation of sequential batch culture in operative mode is imperative for creating MBC aggregates. Bacteria facilitates microalgae aggregation by evading surface predators and promoting synergistic growth of MBCc [58]. Studies have shown that microalgal cell lysis provides nutrients for the bacterial population promoting aggregation [59]. Aggregation of MBC is moderated by certain factors which range from ion concentration, photointensity, operative method, carbon concentration, and the microalgae/bacteria ratio [60–62]. The attached form of the MBC has been studied for the treatment of domestic wastewater with *Bacteroidetes* and *Proteobacteria* being the predominant population.

21.3.2 Heavy metal removal and sequestering of hazardous waste by microalgal–bacterial consortia

Improper modes of heavy metal disposal remain major environmental concerns owing to their toxicity, recalcitrant nature, and their eventual deposition in the biosphere pertaining to soil and water, thereby jeopardizing the normal features of the ecosystem [63]. MBC has been successfully implemented for the removal of the heavy metals as depicted in previous literature. *Ulothrix* sp. (a genre of microalgae) was utilized for measuring its ability in an MBC to mitigate acid mine drainage inundated with heavy metal pollution [64]. These results clearly indicate the efficacy of MBC in water body treatments polluted with heavy metals. MBC can successfully remove heavy metal contamination in the descending order of copper, nickel, manganese, zinc, selenium, and cobalt. Wastewater contaminated with copper, chromium, and nickel ranging from 50–100 ppm was effectively mitigated by *C. vulgaris* and *Exiguobacterium* sp. [65] with microalgal to bacterial ration in the MBC at 3:1. One interesting fact concerning heavy metal mitigation by MBC is that both the alive and the dead biomass is bestowed with the ability to sequester metal ions from wastewater as reflected in MBC consortium, efficient in the removal of copper and cadmium within ~5 min of contact time [66]. The underlying mechanism toward the removal of heavy metal pollutants by MBC is termed biosorption, a two-step process in which the first step is involved in physical modes of adsorption of metals to the ligand groups and the second step entails the sequestering of the metals by the cells by an active metabolism [63]. The anionic groups and the acidic moieties in microalgae and bacteria, respectively, make them suitable sequestering agents for heavy metal cations. The biosorption process is an economically feasible, efficient process, and has high-throughput implementation across a broad range of pH and temperature, followed by bioaccumulation and precipitation in the bacterial and microalgae organelles [64]. Physical processes like ion exchange, redox reactions, and cellular biotransformations further contribute to the metal removal [44]. Therefore an appropriate selection of MBC is crucial in devising an optimum delivery system pertaining to heavy metal removal, effecting seamless bioremediation. *Bacillus* sp, *Pseudomonas* sp, *Streptomyces* sp., and *Zoogloea* sp., have been routinely scrutinized to decipher a sustainable biosystem for sequestrating heavy metals in commercial applications [67]. Microalgal populations pertaining to *C. vulgaris* and *Tetraselmis suecica* have been proved efficient in the optimum removal of lead and cadmium in water systems [68,69]. MBC has been implemented by several groups of researchers for effective degradation of

21. Microalgal–bacterial consortia for biomass production and wastewater treatment

hazardous compounds in various wastewater systems in the form of complex hydrocarbons, atrazine, insecticides, thiocyanate, etc. [70–72].

21.4 Biofuel and bioproducts generation by microalgal–bacterial consortia

MBC possess the ability to produce copious amounts of renewable fuel making them potential candidates for alternative fuel generation within a minimalistic setup following filtration, flocculation, sedimentation, and subsequent thermochemical and biochemical biofuel conversion [73,74]. Bacteria act as a growth promoter in MBC, wherein they accelerate the manifestation of microalgal flocculation, as seen in HW001T, a bacterial strain associated with the family *Pseudomonadaceae*, aggregating the microalgae *Nannochloropsis oceanica* as well as other cyanobacteria and diatoms [59]. The efficacy of this method was reflected in the ease of separating the microalgae from the bottom of the culture facilitated by the MBC aggregation that synchronizes the bacterial Extracellular Polymeric Substances (EPS) and the electrostatic moiety of the microalgal cell surface [14]. For efficient conversion to biodiesel and bioethanol, the rupture of the microalgal intractable cell wall is an important step for the release of the lipid and carbohydrates toward methane and hydrogen production by the anaerobes or yeasts. This phenomenon is instrumental in dictating the economy of the eventual process [75]. Microalgae are instrumental in the generation of a large number of valuable product intermediates and industrially important biofuels and chemicals ranging from omega fatty acids, collagen, (Polyunsaturated fatty acids) PUFA, and biofertilizers [76]. Polyhydroxyalkanoate (PHA), a biopolymer and a green alternative to the plastics, is one of the prime examples of valuable bioproducts produced by the MBC [77]. For example, the MBC consisting of alpha and gamma proteobacteria were able to synthesize PHA (amounting to 20%) in a photosynthetic bioreactor utilizing only sunlight, thereby keeping the product costs to a bare minimum. Table 21.2 depicts the facets of thermochemical

TABLE 21.2 Thermochemical and biochemical conversions with percentile efficiency of the respective biofuel.

Biofuel conversion type	Process	Conversion conditions	Percentile efficiency (%)	Biofuel type
Thermochemical conversion	Pyrolysis	300°C–700°C	50	Gas mixture
	Liquefication	300°C–350°C	65–85	Bio-oil
	Combustion	800°C	35–45	Syngas + charcoal
	Gasification	900°C–1000°C	40–55	Steam + hot gases
Biochemical conversion	Anaerobic digestion	Anoxic	30–50	MeOH, CO_2
	Fermentation(alcoholic)	Anoxic, lignocellulose biomass	50–90	EtOH
	Biological hydrogen production	Anaerobic conditions	30–50	Hydrogen

Handbook of Algal Biofuels

and biochemical conversions with a focus on percentile efficiency of the respective biofuel as products.

21.5 Reduction in CO_2 emission and electricity generation

CO_2 is considered as a foremost GHG emitted as an outcome of fossil fuel's combustion, contributing to global warming, reaching the level of 400 ppm as per World Meteorological Organization [78]. Green microalgae have emerged as one of the most sought-after options for mitigating CO_2 levels in water owing to the increased tolerance of MBC toward CO_2 and the ability to convert CO_2 to chemical energy through a photochemical reaction by CO_2 fixation at rates 30–50 times higher compared to terrestrial plants [79]. The cooperative exchanges between microalgae and fungi in MBC enhance the effectiveness of CO_2 fixation [44]. An MBC consisting of *C. vulgaris* and *Pseudomonas* sp. was found to be effective in the treatment of polluted mixtures comprising phenol and pyridine in a continuously stirred tank bioreactor as reflected in the MBC productivity and illumination period (CSTR) [80].

Photosynthetic algae-mediated microbial fuel cells (PAMFC) are one of the latest biotechnological advancements that support the functionalities of photosynthetic microalgae and electrochemically active bacteria for biological conversion of photoenergy to alternative electricity [81]. In comparison to bacteria-based microbial fuel cells, PAMFC promote the generation of metabolic oxygen, resulting in the mitigation of oxygen bubbling [81]. PAMFC facilitate symbiotic relationships between photosynthetic and heterotrophic counterparts of the niche and utilize a wide range of wavelengths for power generation. Research studies have shown that MBC-assisted batteries generated ~1.5 times higher electricity than a bare cathode system [82,83]. PAMFC consisting of chlorophytes, firmicutes and proteobacterium were able to generate unlimited power supply even with a dearth of externally added nutrients/organics [84]. Power generation is a manifestation of the rates of cathode oxygen production, circuit resistance, and thickness of the biofilms on the cathode surface.

21.5.1 Implementation of genomic approaches for wastewater treatment

Omics-related methodologies have been implemented by various research studies of late for elucidation of the MBC regulation precursors, flux (influent and effluent), and integration of interspecies network. Data analytics derived from omics studies have paved the way for improved understanding of the MBC facilitating efficient consortia following genetic modifications. Functional genomics was applied to investigate the removal of nutrients from wastewater by a study [85]. Several research studies coupled PCR reaction with denaturing gradient gel electrophoresis to analyze MBC in wastewater by pyrosequencing [86]. NGS platform-mediated multiomics data analysis have put forth a novel approach toward qualitative and quantitative representation of consortium analysis pertaining to wastewater treatment. Increased biomass production by *Auxenochlorella* sp. in MBC was investigated by a metabolomics study through ultraparticulate liquid chromatography and quadrupole time of flight which depicted the dependency of the consortia

21. Microalgal–bacterial consortia for biomass production and wastewater treatment

TABLE 21.3 Summary of recent omics approaches in microalgal–bacterial consortia studies.

Wastewater source	Type of omics approach	Core area of study	MBC type	Reference
Municipal wastewater	16 s rRNA sequencing	Nutrient removal	*Chlorella vulgaris + B. lichenoformis*	[52]
Municipal wastewater + activated sludge	PCR reaction with denaturing gradient gel electrophoresis	Carbon, nitrogen, and phosphorus	Microalgae based	[91]
Municipal wastewater	qPCR, 16 s rRNA sequencing, T-RFLP	Nitrogen and COD removal	*C. vulgaris + S. gracile + Pseudomonas* sp. + *Acinetobacter* sp.	[47]
Municipal wastewater	Proteomics + isotope fractionation	Ammonia sequestration	*C. vulgaris*	[26]
Domestic wastewater	16 s rRNA sequencing	Biomass production	*Pseudanabena chlamydomonas + Nitrospira nitrosomonas*	[92]
Municipal wastewater + activated sludge	qRT-PCR + mRNA sequencing	Acylhomoserine lactone production	*Cholorophya* sp	[93]

MBC, microalgal–bacterial consortia.

on the optimal dosage of the cofactors TPP and amino pyrimidine. This was reflected in the parameters of enhanced lipid uptake, glucose sequestration, and the cumulative wastewater treatment [87]. Metagenomics and metabolomics platforms were implemented toward investigating biohydrogen production in MBC [88–90]. Table 21.3 summarizes several such approaches, which have been implemented successfully via metabolomics, metagenomics, and transcriptomics.

21.6 Role of lipase in wastewater treatment

The concomitant release of lipid-rich waste in running water systems without pretreatment imposes serious hazards to the ecosystem. Therefore it is imperative to remove the lipid contaminants from water for the overall impact on the environment. Microorganisms pertaining to wastewater treatment in various management plants elucidate the bioremediation process of carbon-derived waste materials and xenobiotic compounds. However, the persistence rate and optimal activity of microorganisms in wastewater treatment processes is suppressed by high concentrations of lipid, oil, and grease. The incumbent layers of oils/grease on the water surface restrict the penetrability of air in water, thereby affecting the progression of aerobic microbes in the contaminated water [94,95]. In this respect, studies have shown that microbial cultures producing free or bound lipases are very useful in this regard owing to their expansive substrate specificity, enzyme stability, and reusability.

Handbook of Algal Biofuels

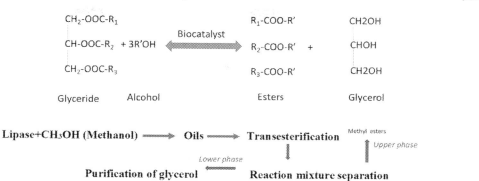

FIGURE 21.3 Lipase mediated esterification of free fatty acids to alkyl esters.

The chemical composition of the individual constituents in the culture medium play a pivotal role in sustained production of lipase [96]. A variety of different substrates involving waste materials have been tried out to optimize the production of lipase. These experimental analyses provide mitigation of environmental pollution related to waste disposal with concomitant conservation of valuable resources. Lipases, upon induction, promote the utilization of residual lipid constituents in waste streams as substrates for enzyme generation. The catalytic process involving the action of the lipase enzyme is provided in Fig. 21.3

Biocatalyst-mediated esterification is a feasible method for monitoring the level of oil content in water systems which can eventually be converted to alkyl esters. Enzyme catalysis mediated through immobilized lipase utilizes whole cell biocatalysts and pure lipase enzyme. Immobilized lipases are preferred to free lipases as the immobilization of enzymes increased stability with concomitant potential for continuous usage without the need of physical separation.

The process of alcoholysis generates glycerol, which is oil-insoluble and is adsorbed to the immobilized lipase surface with a reduction in enzyme activity. Removal of glycerol is a complicated process which poses a hindrance for larger-scale operations. Whole cell biocatalysts on the other hand exploit intracellular lipases and it is an unscaled process, which reduces the production costs by eliminating the requirements of enzyme characterization and immobilization. Cell-based immobilization is carried out in the course of cultivation of the organism and is optimized over the period of time for upscaled use as a biocatalyst. However, whole cell biocatalysis requires suitable pretreatment procedures to overcome the mass transfer resistance. Innovative research is being conducted across various research labs to produce engineered whole cell biocatalysts directed toward overproduction of intracellular lipase for biodiesel. The overall costs incurred by implementing this technology is economically feasible for the production of industrial biodiesel.

21.6.1 Microbial lipase-mediated biocatalysis

Lipase-facilitated biocatalysis is conducted in a liquid culture system which promotes the recovery of the enzyme in due time. Metal oxide nanoparticles with acidic or basic properties have been implemented as catalysts for the esterification of triacylglycerol from soybean oil with upto 90% conversion efficiency. This exclusive ability vis-à-vis their large surface areas make them heterogeneous catalysts of choice for carrying out alcoholysis of long-chained fatty acids. These metal oxide nanoparticles have less toxicity compared to conventional catalysts though substantial research needs to be conducted to gauge its impact on the production of biodiesel [97]. Another strategy to mitigate lipid pollutants is reusing the wastes for the production of value-added products by using the waste as a potential carbon source to formulate microbial lipases. In one of the studies, *Staphylococcus pasteuri* was cultured in a medium with coconut oil mill waste as a substrate to produce lipase and other industrially important isozymes [98].

In another study, two different strains of purified lipases obtained from *Yarrowia lipolytica* were used to hydrolyze diverse kind of hydrocarbons and oils to produce citric acid and other stable intermediates [99]. This capability of lipase enzyme to utilize agroindustrial or hydrocarbon wastes in producing value-added products is a beacon which can be henceforth applied to bioremediation plants for significant reduction of costs. Again, wastewater-mediated microbial growth reduces the overall cost associated with cultivation medium preparation and contaminant removal processes.

Waste grease was utilized as a substrate by *Penicillium chrysogenum* for fatty acid production in submerged culture conditions [100]. In light of the previous studies it was found that grease waste pretreatment with the sequential addition of lipases enhanced the degradation index and the rates of conversion [101]. Therefore lipase-mediated pretreatment serves as a quintessential step for the degradation of lipid/oil waste in wastewater [102]. In another study involving solid-state fermentation, lipase isolated from *Penicillium* sp. was investigated to analyze the degradation rates of oily substrates from dairy wastewater [103]. The lipid-containing wastes generated by meat and/or slaughterhouse-related industries are approximately equal to 50% weight of animals used for slaughter which can be in the future utilized [104].

Pseudomonas sp. generated a type of lipase which was used for the treatment of lipid-contaminated wastewater. A highly proficient strain of *Aspergillus awamori* producing lipase was reported in this study, which degraded ∼90% of fats in waste oil obtained from mills [105]. A similar research indicated that *Serratia*, *Kosakonia*, and *Mucor* genera possessed lipase facilitated transesterification functionality [106]. To control the decolorization and biodegradation of olive mill wastewater, transesterification activities of lipase obtained from *Geotrichum candidum* were studied [107]. Moreover, effective approaches were put in place to explore the immobilized lipases for wastewater detoxification and degradation. Lipases immobilized on large exposed surfaces in contact with wastewater can bring about efficient degradation of lipid components. Another group of researchers explored the ability of immobilized lipase to degrade FOG and pollutant removal from wastewater in cafeteria grease [106]. Lipase obtained from *B. subtilis* was found to be the most potent enzyme for wastewater remediation compared to bacterial lipases obtained from *S. epidermidis* and *P. aeruginosa*. *P. aeruginosa* lipase brought about a reduction in the thickness of the fat layer after prolonged use [106]. One of the recent studies showed the ability of lipase obtained from *B. pumilus* to hydrolyze palm oil wastewater in an anaerobic reactor [108].

A host of pollutants belonging to the ester category exist in aquatic environments and engineered aquatic environments in the form of parabens, phthalates, carbamates, and organophosphate pesticides [80]. In this study, the presence of methanol or ethanol in the sludge as an external carbon source brought about transesterification reaction by lipases which promoted the degradation of parabens. These findings emphasize the fact that the usefulness of the microbial lipases is not restricted to hydrolysis of oils or complex lipids but their controlled usage can be also extended to the biotransformation of parabens.

Pretreatment of wastewater systems to effect substantial lipid hydrolysis via conventional biological treatment and enzymes emerge as promising solutions which have been extensively reviewed in several studies [89,109–111].

21.6.2 Potential microbial strains used for the processing of complex oil and lipid

Numerable biological systems comprising lipase-producing microbial strain consortia have been successfully utilized to optimize lipid bioremediation. Bacterial strains of *Pseudomonas* spp. and *P. aeruginosa* have emerged as chosen candidates for effecting continuous bioremediation. Statistically derived formulations have been implemented for maximizing the lipase production process and simultaneous oil hydrolysis by *Pseudomonas* sp. [112]. *B. stearothermophilus* and a specific *Bacillus* sp. obtained from tannery and petroleum contaminated soil, respectively, were investigated thoroughly for their efficacy [113,114]. Edible oil decomposition under mildly acidic conditions were brought about by *Burkholderia* sp. and *Raoultella* sp. [115,116]. Another group of researchers investigated the efficacy of a bacterial consortium consisting of *P. aeruginosa.*, *Acinetobacter caloaceticus*, *Bacillus* sp., *Serratia marsescens*, *P. aeruginosa*, and *Staphylococcus aureus* for the degradation of wastewater contaminants [117]. Static fungal cultures of *G. candidum* and *P. chrysogenum* bearing lipase activity have been implemented to carry out effective bioremediation strategies of wastewater pollutant mitigation and degradation of cooking oil, respectively [105] (Table 21.2).

Yeast culture consortia bearing *Y. lipolytica* strains have been successfully utilized for wastewater bioremediation pertaining to oil mills [46]. A psychrophilic strain of *Mrakia blollopsis* (yeast strain derived from Antarctica) brought about transformation of milk fat at presumably low temperatures [118]. A symbiotic microbial yeast–bacteria consortium belonging to *Burkholderia arboris* and *C. cylindracea* was instrumental in bringing about substantial lipid biodegradation [115].

21.6.3 Analysis of the cumulative effect of enzymatic prehydrolysis on anaerobic digestion of various industrial wastewaters

For bioremediation of lipid rich wastewater, bioreactor systems in the form of anaerobic sludge blanket reactor (ASB), packed bed reactor, membrane bioreactor, and sequencing batch reactor are being used extensively [119,120]. For carrying out optimized lipase-catalyzed hydrolysis, an anaerobic setup is preferred to an aerobic one as studies have shown that lipase-mediated pretreatment promotes biodegradation via the anaerobic pathway [121]. Different types of microorganisms used for obtaining lipase are shown in Table 21.4.

Handbook of Algal Biofuels

21. Microalgal−bacterial consortia for biomass production and wastewater treatment

TABLE 21.4 Application of lipase obtained from microorganisms in lipid and oil bioremediation.

Bacteria	References
Pseudomonas aeruginosa	Mobarak et al. [122]
Bacillus sp.	Bayoumi et al. [113]
Bacillus stearothermophilus	Granzotto et al. [114]
Burkholderia sp.	Matsumiya et al. [115]
Microthrix parvicella	Nielsen et al. [123]
Fungi	
Geotrichum candidum	Asses et al. [105]
Penicillium chrysogenum	Kumar et al. [124]
Penicillium restrictum	Valladão et al. [125,126]
Rhizopus oryzae	Efremenko et al. [127]
Yeast	
Candida rugosa	Chakraborty et al. [128]
Yarrowia lipolytica	Goncalves et al. [129]
Lipomyces starkey	Yousuf et al. [130]
Mrakia blollopsis	Tsuji et al. [118]
Consortia	
P. aeruginosa, Bacillus sp., and *Acinetobacter caloaceticus*	Mongkolthanaruk et al. [117]
Bacillus sp., *Serratia marsescens, P. aeruginosa*, and *Staphylococcus aureus*	Prasad et al. [131]
Burkholderia arboris and *Caulerpa cylindracea*	Matsuoka et al. [132]

Microbiological and omics domain analysis of bacteria and archaea domains exposed noteworthy changes in the genomic profiles of the organisms which needs to be taken into consideration for building experimental setups for undergoing/not undergoing prehydrolytic steps [133]. Lipase derived from *P. aeruginosa* enhanced the bioremediation index of contaminated dairy wastewater with simultaneously increased COD eradication and production of biogas [122].

A lipase-rich enzymatic mixture derived from *Penicillium restrictum* was used in bioremediation of poultry wastewater in an ASB reactor [125,126]. Commercially obtained lipase enzyme from novozymes was used to treat wastewater from a piggery [134].

Another research study associated with *Candida rugosa* lipase; approx. fourfold increase in biogas production was achieved through the pretreated effluent compared to the untreated one. A biocatalytic mixture of lipase−protease−carboxyhydrase in the ratio of 1:2:1 effected increased solubilization and volatile fatty acid production in the course of food waste pretreatment [135]. Other methods of enzymatic hydrolysis in bioremediation of oil/grease contaminants of wastewater have been investigated thoroughly [26].

Handbook of Algal Biofuels

21.6.4 Role of lipase enzymes in activated sludge systems

Lipids constitute approximately ~40% of wastewater (COD) in activated sludge systems, thereby facilitating filamentous growth. In such setups lipase activates adsorption/breakdown of triglycerides on the activated sludge surface along with hydrolysis of the unbreakable/hard fats [119,120]. Implementation of this setup was reflected in uninterrupted COD removal for approximately 270 days in a bioreactor [136]. The effective bioremediation strategy mediated by an enzyme mixture obtained from *P. restrictum* was helpful in bioremediation of dairy wastewater with high lipid content. COD eradication levels of upto 15% and higher coupled with 50% lowered accumulation of lipids and up to two times higher biomass concentration with 1.5 times advanced oxygen uptake rate was encountered in this respect [137]. A bacterial consortium belonging to the genera of *Microthrix parvicella* was instrumental in sequestering fatty acids anaerobically in the presence of nitrate or oxygen as electron acceptors [136].

21.6.5 Immobilized enzyme and whole-cell biocatalysts in lipid bioremediation

Calcium alginate beads were used to immobilize lipase obtained from *C. rugosa* which was henceforth used in the hydrolysis of wastewater generated from the pet food industry [119,120]. Matrix-based system of capture was found to be effective when the nondigestable lipid concentration surpassed 5000 mg/L, which resulted in increased lipid and COD mitigation [138]. Such a kind of system was designed with a polyvinyl cryogel matrix which had immobilized *Rhizopus oryzae* cells for bioremediation purposed in food industries [127].

21.6.6 Conversion of lipid contaminated wastewater into value-added products

Lipid fractions obtained from the wastewater can be reconstituted by physical processes, that is, centrifugation and filtration. These fractions can then be utilized for downstream processing of biofuel and biodiesel through transesterification reactions mediated by lipase. Free fatty acid content of the obtained lipid was enhanced upto 15% through the lipase biocatalyzed esterification process in one research study [139], whereas *Lipomyces starkey* effected viable transformation of wastewater lipids to biofuels [19,43,130]. Another research highlighted the increased phosphorus content in wastewater derived from olive mill by an acidogenic *Aspergillus* sp [140]. Lipases were instrumental in stabilization of the compost waste of oil-bleached soil residues to mediate effective bioremediation [141]. A lipid-rich wastewater biodegradation was accelerated by a combinatorial treatment of microwave irradiation and lipase pretreatment [142].

21.7 Conclusions

This chapter focused on the microalgal—bacterial consortia-centered practices which pave the way for renewable and sustainable technology. This is in accordance with the environmental standards of low-carbon emission guidelines facilitating environmental protection and resource management. This review will serve as a beacon for the researchers who are trying to find scientific process optimizations based on symbiotic interactions between microalgae, bacteria, and the multiple factors that govern the process in the form of substrate exchange, cell-to-cell signaling, and horizontal gene transfer. Due to the rewarding cooperative work flow processes mediated between microalgae and bacteria, MBC are efficient in the removal of pollutants from wastewater. The unexploited potential of the attached MBC are encouraging researchers to explore their various applications. The microbial populations expressing lipases constitute the important biomachinery of catalysts, which provide the pathway for approaching environmental bioremediation, as discussed in this review. The diversity of the enzymes obtained across different genera of organisms can through research be studied in order to extract an optimum output. However, the recent advancement in lipase-mediated bioremediation processing has given a beacon to the scientific world to explore even further through cutting-edge research. Optimized usage of agroindustrial residues alongside effluents as potential substrates can help mitigate the operational cost and the approach is an ecofriendly one. Immobilization methods on novel matrices and solid supports permitting continuous use of the enzyme can be investigated with regard to facilitating increased and effective bioremediation. Biomolecular engineering can help us to develop robust lipases which can sustain conditions of adverse temperature and pH. Furthermore, biosynthetic lipases can be developed through metagenomics and bioinformatics studies based on mutagenic studies and cloning followed by expression in suitable hosts. These research objectives will aid in the fight against environmental pollution by means of lipolytic enzymes playing a crucial role in waste mitigation and transformation into value-added products.

Acknowledgments

This work was supported by the University of Agriculture, Faisalabad, Pakistan to Muhammad Usman Khan for research on Alternative Energy Sources.

References

[1] M.K. Enamala, S. Enamala, M. Chavali, J. Donepudi, R. Yadavalli, B. Kolapalli, et al., Production of biofuels from microalgae—a review on cultivation, harvesting, lipid extraction, and numerous applications of microalgae, Renew. Sustain. Energy Rev. 94 (2018) 49–68.

[2] S.P. Cuellar-Bermudez, G.S. Aleman-Nava, R. Chandra, J.S. Garcia-Perez, J.R. Contreras-Angulo, G. Markou, et al., Nutrients utilization and contaminants removal. A review of two approaches of algae and cyanobacteria in wastewater, Algal Research. 24 (2017) 438–449.

[3] M.I. Khan, B. Ahring, Anaerobic digestion of biorefinery lignin: effect of different wet explosion pretreatment conditions, Bioresour. Technol. 298 (2020) 122537. Available from: https://doi.org/10.1016/j.biortech.2019.122537.

[4] M.U. Khan, B.K. Ahring, Lignin degradation under anaerobic digestion: influence of lignin modifications—a review, Biomass Bioenergy 128 (2019) 105325. Available from: https://doi.org/10.1016/j.biombioe.2019.105325.

[5] L.-L. Zhuang, D. Yu, J. Zhang, F. Liu, Y.-H. Wu, T.-Y. Zhang, et al., The characteristics and influencing factors of the attached microalgae cultivation: a review, Renew. Sustain. Energy Rev. 94 (2018) 1110–1119. Available from: https://doi.org/10.1016/j.rser.2018.06.006.

[6] B. Abdullah, S.A.F.S. Muhammad, Z. Shokravi, S. Ismail, K.A. Kassim, A.N. Mahmood, et al., Fourth generation biofuel: a review on risks and mitigation strategies, Renew. Sustain. Energy Rev. 107 (2019) 37–50.

[7] E.S. Shuba, D. Kifle, Microalgae to biofuels: 'promising' alternative and renewable energy, review, Renew. Sustain. Energy Rev. 81 (2018) 743–755. Available from: https://doi.org/10.1016/j.rser.2017.08.042.

Handbook of Algal Biofuels

References

[8] A. Mehrabadi, R. Craggs, M.M. Farid, Wastewater treatment high rate algal ponds (WWT HRAP) for low-cost biofuel production, Bioresour. Technol. 184 (2015) 202–214. Available from: https://doi.org/10.1016/j.biortech.2014.11.004.

[9] R. Xiao, Y. Zheng, Overview of microalgal extracellular polymeric substances (EPS) and their applications, Biotechnol. Adv. 34 (2016) 1225–1244. Available from: https://doi.org/10.1016/j.biotechadv.2016.08.004.

[10] E. Posadas, R. Muñoz, B. Guieysse, Integrating nutrient removal and solid management restricts the feasibility of algal biofuel generation via wastewater treatment, Algal Res. 22 (2017) 39–46. Available from: https://doi.org/10.1016/j.algal.2016.11.019.

[11] S. Van Den Hende, E. Carré, E. Cocaud, V. Beelen, N. Boon, H. Vervaeren, Treatment of industrial wastewaters by microalgal bacterial flocs in sequencing batch reactors, Bioresour. Technol. 161 (2014) 245–254. Available from: https://doi.org/10.1016/j.biortech.2014.03.057.

[12] R. Ramanan, B.-H. Kim, D.-H. Cho, H.-M. Oh, H.-S. Kim, Algae–bacteria interactions: evolution, ecology and emerging applications, Biotechnol. Adv. 34 (2016) 14–29. Available from: https://doi.org/10.1016/j.biotechadv.2015.12.003.

[13] M.M. El-Sheekh, E.A. Alwaleed, A. Ibrahim, H. Saber, Detrimental effect of UV-B radiation on growth, photosynthetic pigments, metabolites and ultrastructure of some cyanobacteria and freshwater chlorophyta, Int. J. Radiat. Biol. 0 (2020) 1–15. Available from: https://doi.org/10.1080/09553002.2021.1851060.

[14] J.L. Fuentes, I. Garbayo, M. Cuaresma, Z. Montero, M. González-del-Valle, C. Vílchez, Impact of microalgae-bacteria interactions on the production of algal biomass and associated compounds, Mar. Drugs 14 (2016) 100. Available from: https://doi.org/10.3390/md14050100.

[15] A. Escobar-Niño, C. Luna, D. Luna, A.T. Marcos, D. Cánovas, E. Mellado, Selection and characterization of biofuel-producing environmental bacteria isolated from vegetable oil-rich wastes, PLoS One 9 (2014) e104063. Available from: https://doi.org/10.1371/journal.pone.0104063.

[16] S. Facchin, P.D.D. Alves, F. de, F. Siqueira, T.M. Barroca, J.M.N. Victória, et al., Biodiversity and secretion of enzymes with potential utility in wastewater treatment, Open. J. Ecol. 3 (2013) 34–37. Available from: https://doi.org/10.4236/oje.2013.31005.

[17] B. Primasari, S. Ibrahim, M.S.M. Annuar, L.X.I. Remmie, Aerobic treatment of oily wastewater: effect of aeration and sludge concentration to pollutant reduction and PHB accumulation, World Acad. Sci. Eng. Technol. 54 (2011) 6–20.

[18] A. Fadile, F.Z.E. Hassani, H. Aissam, M. Merzouki, M. Benlemlih, Aerobic treatment of lipid-rich wastewater by a bacterial consortium, AJMR 5 (2011) 5333–5342. Available from: https://doi.org/10.5897/AJMR11.700.

[19] M. El-Sheekh, H. Morsi, L. Hassan, Growth enhancement of *Spirulina platensis* through optimization of media and nitrogen sources, Egypt. J. Bot 0 (2020). Available from: https://doi.org/10.21608/ejbo.2020.27927.1487.

[20] J. Gasperi, S. Garnaud, V. Rocher, R. Moilleron, Priority pollutants in wastewater and combined sewer overflow, Sci. Total Environ. 407 (2008) 263–272. Available from: https://doi.org/10.1016/j.scitotenv.2008.08.015.

[21] H. Liu, Q. Lu, Q. Wang, W. Liu, Q. Wei, H. Ren, et al., Isolation of a bacterial strain, Acinetobacter sp. from centrate wastewater and study of its cooperation with algae in nutrients removal, Bioresour. Technol. 235 (2017) 59–69. Available from: https://doi.org/10.1016/j.biortech.2017.03.111.

[22] F.G. Acién, C. Gómez-Serrano, M. del, M. Morales-Amaral, J.M. Fernández-Sevilla, E. Molina-Grima, Wastewater treatment using microalgae: how realistic a contribution might it be to significant urban wastewater treatment? Appl. Microbiol. Biotechnol. 100 (2016) 9013–9022.

[23] B.-H. Kim, R. Ramanan, D.-H. Cho, H.-M. Oh, H.-S. Kim, Role of Rhizobium, a plant growth promoting bacterium, in enhancing algal biomass through mutualistic interaction, Biomass Bioenergy 69 (2014) 95–105. Available from: https://doi.org/10.1016/j.biombioe.2014.07.015.

[24] J.A. Villa, E.E. Ray, B.M. Barney, Azotobacter vinelandii siderophore can provide nitrogen to support the culture of the green algae *Neochloris oleoabundans* and *Scenedesmus* sp. BA032, FEMS Microbiol. Lett. 351 (2014) 70–77. Available from: https://doi.org/10.1111/1574-6968.12347.

[25] L. Jiang, L. Yang, L. Xiao, X. Shi, G. Gao, B. Qin, Quantitative studies on phosphorus transference occuring between *Microcystis aeruginosa* and its attached bacterium (Pseudomonas sp.), in: B. Qin, Z. Liu, K. Havens (Eds.), Eutrophication of Shallow Lakes with Special Reference to Lake Taihu, China, Springer, Netherlands, Dordrecht, 2007, pp. 161–165. Available from: https://doi.org/10.1007/978-1-4020-6158-5_18.

Handbook of Algal Biofuels

496   21. Microalgal—bacterial consortia for biomass production and wastewater treatment

[26] H. Liu, Y. Zhou, W. Xiao, L. Ji, X. Cao, C. Song, Shifting nutrient-mediated interactions between algae and bacteria in a microcosm: evidence from alkaline phosphatase assay, Microbiol. Res. 167 (2012) 292–298. Available from: https://doi.org/10.1016/j.micres.2011.10.005.

[27] M.R. Seyedsayamdost, R. Wang, R. Kolter, J. Clardy, Hybrid biosynthesis of roseobacticides from algal and bacterial precursor molecules, J. Am. Chem. Soc. 136 (2014) 15150–15153. Available from: https://doi.org/10.1021/ja508782y.

[28] A.E.-F. Abomohra, M. Elsayed, S. Esakkimuthu, M. El-Sheekh, D. Hanelt, Potential of fat, oil and grease (FOG) for biodiesel production: a critical review on the recent progress and future perspectives, Prog. Energy Combust. Sci. 81 (2020) 100868. Available from: https://doi.org/10.1016/j.pecs.2020.100868.

[29] S.A. Amin, L.R. Hmelo, H.M. Van Tol, B.P. Durham, L.T. Carlson, K.R. Heal, et al., Interaction and signalling between a cosmopolitan phytoplankton and associated bacteria, Nature 522 (2015) 98–101.

[30] A. Mitsutani, I. Yamasaki, H. Kitaguchi, J. Kato, S. Ueno, Y. Ishida, Analysis of algicidal proteins of a diatom-lytic marine bacterium Pseudoalteromonas sp. strain A25 by two-dimensional electrophoresis, Phycologia 40 (2001) 286–291. Available from: https://doi.org/10.2216/i0031-8884-40-3-286.1.

[31] F. Weinberger, J. Beltran, J.A. Correa, U. Lion, G. Pohnert, N. Kumar, et al., Spore release in Acrochaetium Sp. (rhodophyta) is bacterially controlled1, J. Phycol 43 (2007) 235–241. Available from: https://doi.org/10.1111/j.1529-8817.2007.00329.x.

[32] M. Manefield, S.L. Turner, Quorum sensing in context: out of molecular biology and into microbial ecology, Microbiology 148 (2002) 3762–3764. Available from: https://doi.org/10.1099/00221287-148-12-3762.

[33] S. Rajamani, W.D. Bauer, J.B. Robinson, J.M. Farrow, E.C. Pesci, M. Teplitski, et al., The vitamin riboflavin and its derivative lumichrome activate the LasR bacterial quorum-sensing receptor, MPMI 21 (2008) 1184–1192. Available from: https://doi.org/10.1094/MPMI-21-9-1184.

[34] Y.-H. Dong, L.-H. Zhang, Quorum sensing and quorum-quenching enzymes, J. Microbiol. 43 (2005) 101–109.

[35] A. Butler, M. Sandy, Mechanistic considerations of halogenating enzymes, Nature 460 (2009) 848–854.

[36] G.L. Wheeler, K. Tait, A. Taylor, C. Brownlee, I. Joint, Acyl-homoserine lactones modulate the settlement rate of zoospores of the marine alga *Ulva intestinalis* via a novel chemokinetic mechanism, Plant Cell Environ. 29 (2006) 608–618. Available from: https://doi.org/10.1111/j.1365-3040.2005.01440.x.

[37] J.S. Dickschat, Quorum sensing and bacterial biofilms, Nat. Product. Rep. 27 (2010) 343–369.

[38] A.E. Allen, C.L. Dupont, M. Oborník, A. Horák, A. Nunes-Nesi, J.P. McCrow, et al., Evolution and metabolic significance of the urea cycle in photosynthetic diatoms, Nature 473 (2011) 203–207.

[39] G. Schönknecht, W.-H. Chen, C.M. Ternes, G.G. Barbier, R.P. Shrestha, M. Stanke, et al., Gene transfer from bacteria and archaea facilitated evolution of an extremophilic eukaryote, Science 339 (2013) 1207–1210. Available from: https://doi.org/10.1126/science.1231707.

[40] M. Demuez, C. González-Fernández, M. Ballesteros, Algicidal microorganisms and secreted algicides: new tools to induce microalgal cell disruption, Biotechnol. Adv. 33 (2015) 1615–1625.

[41] A. Kouzuma, K. Watanabe, Exploring the potential of algae/bacteria interactions, Curr. Opin. Biotechnol. 33 (2015) 125–129. Available from: https://doi.org/10.1016/j.copbio.2015.02.007.

[42] M. El-Sheekh, M. Abu-Faddan, A. Abo-Shady, M.Z.A. Nassar, W. Labib, Molecular identification, biomass, and biochemical composition of the marine chlorophyte Chlorella sp. MF1 isolated from Suez Bay, J. Genet. Eng. Biotechnol. 18 (2020) 27. Available from: https://doi.org/10.1186/s43141-020-00044-8.

[43] M.M. El-Sheekh, E.A. Alwaleed, W.M.A. Kassem, H. Saber, Antialgal and antiproliferative activities of the algal silver nanoparticles against the toxic cyanobacterium Microcystis aeruginosa and human tumor colon cell line, Environ. Nanotechnol. Monit. Manage. 14 (2020) 100352. Available from: https://doi.org/10.1016/j.enmm.2020.100352.

[44] S.R. Subashchandrabose, B. Ramakrishnan, M. Megharaj, K. Venkateswarlu, R. Naidu, Consortia of cyano-bacteria/microalgae and bacteria: biotechnological potential, Biotechnol. Adv. 29 (2011) 896–907. Available from: https://doi.org/10.1016/j.biotechadv.2011.07.009.

[45] H.N.P. Vo, H.H. Ngo, W. Guo, T.M.H. Nguyen, Y. Liu, Y. Liu, et al., A critical review on designs and applications of microalgae-based photobioreactors for pollutants treatment, Sci. Total. Environ. 651 (2019) 1549–1568. Available from: https://doi.org/10.1016/j.scitotenv.2018.09.282.

[46] A.L. Gonçalves, J.C.M. Pires, M. Simões, A review on the use of microalgal consortia for wastewater treatment, Algal Res. 24 (2017) 403–415. Available from: https://doi.org/10.1016/j.algal.2016.11.008.

[47] C.S. Lee, H.-S. Oh, H.-M. Oh, H.-S. Kim, C.-Y. Ahn, Two-phase photoperiodic cultivation of algal–bacterial consortia for high biomass production and efficient nutrient removal from municipal

Handbook of Algal Biofuels

References 497

wastewater, Bioresour. Technol. 200 (2016) 867–875. Available from: https://doi.org/10.1016/j.biortech.2015.11.007.

[48] N.G.A.I. Karya, N.P. van der Steen, P.N.L. Lens, Photo-oxygenation to support nitrification in an algal–bacterial consortium treating artificial wastewater, Bioresour. Technol. 134 (2013) 244–250. Available from: https://doi.org/10.1016/j.biortech.2013.02.005.

[49] I. de Godos, V.A. Vargas, S. Blanco, M.C.G. González, R. Soto, P.A. García-Encina, et al., A comparative evaluation of microalgae for the degradation of piggery wastewater under photosynthetic oxygenation, Bioresour. Technol. 101 (2010) 5150–5158. Available from: https://doi.org/10.1016/j.biortech.2010.02.010.

[50] I. de Godos, S. Blanco, P.A. García-Encina, E. Becares, R. Muñoz, Long-term operation of high rate algal ponds for the bioremediation of piggery wastewaters at high loading rates, Bioresour. Technol. 100 (2009) 4332–4339. Available from: https://doi.org/10.1016/j.biortech.2009.04.016.

[51] D. García, E. Posadas, S. Blanco, G. Acién, P. García-Encina, S. Bolado, et al., Evaluation of the dynamics of microalgae population structure and process performance during piggery wastewater treatment in algal-bacterial photobioreactors, Bioresour. Technol. 248 (2018) 120–126. Available from: https://doi.org/10.1016/j.biortech.2017.06.079.

[52] X. Ji, M. Jiang, J. Zhang, X. Jiang, Z. Zheng, The interactions of algae-bacteria symbiotic system and its effects on nutrients removal from synthetic wastewater, Bioresour. Technol. 247 (2018) 44–50. Available from: https://doi.org/10.1016/j.biortech.2017.09.074.

[53] C. Alcántara, J.M. Domínguez, D. García, S. Blanco, R. Pérez, P.A. García-Encina, et al., Evaluation of waste-water treatment in a novel anoxic–aerobic algal–bacterial photobioreactor with biomass recycling through carbon and nitrogen mass balances, Bioresour. Technol. 191 (2015) 173–186.

[54] P.J. He, B. Mao, F. Lü, L.M. Shao, D.J. Lee, J.S. Chang, The combined effect of bacteria and Chlorella vulgaris on the treatment of municipal wastewaters, Bioresour. Technol. 146 (2013) 562–568. Available from: https://doi.org/10.1016/j.biortech.2013.07.111.

[55] G. Quijano, J.S. Arcila, G. Buitrón, Microalgal-bacterial aggregates: applications and perspectives for waste-water treatment, Biotechnol. Adv. 35 (2017) 772–781. Available from: https://doi.org/10.1016/j.biotechadv.2017.07.003.

[56] L. Christenson, R. Sims, Production and harvesting of microalgae for wastewater treatment, biofuels, and bioproducts, Biotechnol. Adv. 29 (2011) 686–702.

[57] L.B. Christenson, R.C. Sims, Rotating algal biofilm reactor and spool harvester for wastewater treatment with biofuels by-products, Biotechnol. Bioeng. 109 (2012) 1674–1684.

[58] V. Montemezzani, I.C. Duggan, I.D. Hogg, R.J. Craggs, A review of potential methods for zooplankton control in wastewater treatment high rate algal ponds and algal production raceways, Algal Res. 11 (2015) 211–226. Available from: https://doi.org/10.1016/j.algal.2015.06.024.

[59] H. Wang, R.T. Hill, T. Zheng, X. Hu, B. Wang, Effects of bacterial communities on biofuel-producing microal-gae: stimulation, inhibition and harvesting, Crit. Rev. Biotechnol. 36 (2016) 341–352. Available from: https://doi.org/10.3109/07388551.2014.961402.

[60] R.J. Powell, R.T. Hill, Mechanism of algal aggregation by Bacillus sp. strain RP1137, Appl. Environ. Microbiol. 80 (2014) 4042–4050. Available from: https://doi.org/10.1128/AEM.00887-14.

[61] O. Tiron, C. Bumbac, I.V. Patroescu, V.R. Badescu, C. Postolache, Granular activated algae for wastewater treatment, Water Sci. Technol. 71 (2015) 832–839. Available from: https://doi.org/10.2166/wst.2015.010.

[62] F. Meng, L. Xi, D. Liu, W. Huang, Z. Lei, Z. Zhang, et al., Effects of light intensity on oxygen distribution, lipid production and biological community of algal-bacterial granules in photo-sequencing batch reactors, Bioresour. Technol. 272 (2019) 473–481. Available from: https://doi.org/10.1016/j.biortech.2018.10.059.

[63] E. Priyadarshini, S.S. Priyadarshini, N. Pradhan, Heavy metal resistance in algae and its application for metal nanoparticle synthesis, Appl. Microbiol. Biotechnol. 103 (2019) 3297–3316. Available from: https://doi.org/10.1007/s00253-019-09685-3.

[64] S. Orandi, D.M. Lewis, N.R. Moheimani, Biofilm establishment and heavy metal removal capacity of an indigenous mining algal-microbial consortium in a photo-rotating biological contactor, J. Ind. Microbiol. Biotechnol. 39 (2012) 1321–1331. Available from: https://doi.org/10.1007/s10295-012-1142-9.

[65] S. Batool, A. Hussain, M.A. Iqbal, A. Javid, W. Ali, S.M. Bukhari, et al., Implication of highly metal-resistant microalgal-bacterial co-cultures for the treatment of simulated metal-loaded wastewaters, Int. Microbiol. 22 (2019) 41–48.

Handbook of Algal Biofuels

[66] S. Loutseti, D.B. Danielidis, A. Economou-Amilli, C. Katsaros, R. Santas, P. Santas, The application of a micro-algal/bacterial biofilter for the detoxification of copper and cadmium metal wastes, Bioresour. Technol. 100 (2009) 2099–2105. Available from: https://doi.org/10.1016/j.biortech.2008.11.019.

[67] S.S. Ahluwalia, D. Goyal, Microbial and plant derived biomass for removal of heavy metals from wastewater, Bioresour. Technol. 98 (2007) 2243–2257.

[68] N. Yoshida, R. Ikeda, T. Okuno, Identification and characterization of heavy metal-resistant unicellular alga isolated from soil and its potential for phytoremediation, Bioresour. Technol. 97 (2006) 1843–1849. Available from: https://doi.org/10.1016/j.biortech.2005.08.021.

[69] M. Pérez-Rama, J. Abalde Alonso, C. Herrero López, E. Torres Vaamonde, Cadmium removal by living cells of the marine microalga *Tetraselmis suecica*, Bioresour. Technol. 84 (2002) 265–270. Available from: https://doi.org/10.1016/S0960-8524(02)00045-7.

[70] H. Mahdavi, V. Prasad, Y. Liu, A.C. Ulrich, In situ biodegradation of naphthenic acids in oil sands tailings pond water using indigenous algae–bacteria consortium, Bioresour. Technol. 187 (2015) 97–105. Available from: https://doi.org/10.1016/j.biortech.2015.03.091.

[71] V. Matamoros, R. Gutiérrez, I. Ferrer, J. García, J.M. Bayona, Capability of microalgae-based wastewater treatment systems to remove emerging organic contaminants: a pilot-scale study, J. Hazard. Mater. 288 (2015) 34–42. Available from: https://doi.org/10.1016/j.jhazmat.2015.02.002.

[72] B.-G. Ryu, W. Kim, K. Nam, S. Kim, B. Lee, M.S. Park, et al., A comprehensive study on algal–bacterial communities shift during thiocyanate degradation in a microalga-mediated process, Bioresour. Technol. 191 (2015) 496–504. Available from: https://doi.org/10.1016/j.biortech.2015.03.136.

[73] M.L. Menegazzo, G.G. Fonseca, Biomass recovery and lipid extraction processes for microalgae biofuels production: a review, Renew. Sustain. Energy Rev. 107 (2019) 87–107. Available from: https://doi.org/10.1016/j.rser.2019.01.064.

[74] T. Suganya, M. Varman, H.H. Masjuki, S. Renganathan, Macroalgae and microalgae as a potential source for commercial applications along with biofuels production: a biorefinery approach, Renew. Sustain. Energy Rev. 55 (2016) 909–941. Available from: https://doi.org/10.1016/j.rser.2015.11.026.

[75] B.H.H. Goh, H.C. Ong, M.Y. Cheah, W.-H. Chen, K.L. Yu, T.M.I. Mahlia, Sustainability of direct biodiesel synthesis from microalgae biomass: a critical review, Renew. Sustain. Energy Rev. 107 (2019) 59–74. Available from: https://doi.org/10.1016/j.rser.2019.02.012.

[76] M. Rizwan, G. Mujtaba, S.A. Memon, K. Lee, N. Rashid, Exploring the potential of microalgae for new biotechnology applications and beyond: a review, Renew. Sustain. Energy Rev. 92 (2018) 394–404. Available from: https://doi.org/10.1016/j.rser.2018.04.034.

[77] R. a J. Verlinden, D.J. Hill, M.A. Kenward, C.D. Williams, I. Radecka, Bacterial synthesis of biodegradable polyhydroxyalkanoates, J. Appl. Microbiol. 102 (2007) 1437–1449. Available from: https://doi.org/10.1111/j.1365-2672.2007.03335.x.

[78] B. Yang, J. Liu, X. Ma, B. Guo, B. Liu, T. Wu, et al., Genetic engineering of the Calvin cycle toward enhanced photosynthetic CO_2 fixation in microalgae, Biotechnol. Biofuels 10 (2017) 229. Available from: https://doi.org/10.1186/s13068-017-0916-8.

[79] X. Xu, X. Gu, Z. Wang, W. Shatner, Z. Wang, Progress, challenges and solutions of research on photosynthetic carbon sequestration efficiency of microalgae, Renew. Sustain. Energy Rev. 110 (2019) 65–82. Available from: https://doi.org/10.1016/j.rser.2019.04.050.

[80] B. Wang, Y. Li, N. Wu, C.Q. Lan, CO_2 bio-mitigation using microalgae, Appl. Microbiol. Biotechnol. 79 (2008) 707–718. Available from: https://doi.org/10.1007/s00253-008-1518-y.

[81] M. Shukla, S. Kumar, Algal growth in photosynthetic algal microbial fuel cell and its subsequent utilization for biofuels, Renew. Sustain. Energy Rev. 82 (2018) 402–414. Available from: https://doi.org/10.1016/j.rser.2017.09.067.

[82] I. Gajda, J. Greenman, C. Melhuish, I. Ieropoulos, Self-sustainable electricity production from algae grown in a microbial fuel cell system, Biomass Bioenergy 82 (2015) 87–93. Available from: https://doi.org/10.1016/j.biombioe.2015.05.017.

[83] D.-B. Wang, T.-S. Song, T. Guo, Q. Zeng, J. Xie, Electricity generation from sediment microbial fuel cells with algae-assisted cathodes, Int. J. Hydrog. Energy 39 (2014) 13224–13230. Available from: https://doi.org/10.1016/j.ijhydene.2014.06.141.

References

[84] Z. He, J. Kan, F. Mansfeld, L.T. Angenent, K.H. Nealson, Self-sustained phototrophic microbial fuel cells based on the synergistic cooperation between photosynthetic microorganisms and heterotrophic bacteria, Environ. Sci. Technol. 43 (2009) 1648–1654. Available from: https://doi.org/10.1021/es803084a.

[85] I.A. Perera, S. Abinandan, S.R. Subashchandrabose, K. Venkateswarlu, R. Naidu, M. Megharaj, Advances in the technologies for studying consortia of bacteria and cyanobacteria/microalgae in wastewaters, Crit. Rev. Biotechnol. 39 (2019) 709–731. Available from: https://doi.org/10.1080/07388551.2019.1597828.

[86] D.-H. Cho, R. Ramanan, J. Heo, J. Lee, B.-H. Kim, H.-M. Oh, et al., Enhancing microalgal biomass productivity by engineering a microalgal–bacterial community, Bioresour. Technol. 175 (2015) 578–585.

[87] B.T. Higgins, I. Gennity, S. Samra, T. Kind, O. Fiehn, J.S. VanderGheynst, Cofactor symbiosis for enhanced algal growth, biofuel production, and wastewater treatment, Algal Res. 17 (2016) 308–315. Available from: https://doi.org/10.1016/j.algal.2016.05.024.

[88] P. Shetty, I.Z. Boboescu, B. Pap, R. Wirth, K.L. Kovács, T. Bíró, et al., Exploitation of algal-bacterial consortia in combined biohydrogen generation and wastewater treatment, Front. Energy Res. 7 (2019). Available from: https://doi.org/10.3389/fenrg.2019.00052.

[89] M.M. El-Sheekh, E. El-Mohsnawy, M.E.M. Mabrouk, W.F. Zoheir, Enhancement of biodiesel production from the green microalga *Micractinium reisseri* via optimization of cultivation regimes, J. Taibah Univ. Sci. 14 (2020) 437–444. Available from: https://doi.org/10.1080/16583655.2020.1745505.

[90] M.M. Ismail, B.S. Alotaibi, M.M. EL-Sheekh, Therapeutic uses of red macroalgae, Molecules. 25 (2020) 4411. Available from: https://doi.org/10.3390/molecules25194411.

[91] L. Sun, W. Zuo, Y. Tian, J. Zhang, J. Liu, N. Sun, et al., Performance and microbial community analysis of an algal-activated sludge symbiotic system: effect of activated sludge concentration, J. Environ. Sci. 76 (2019) 121–132. Available from: https://doi.org/10.1016/j.jes.2018.04.010.

[92] I.M. Bradley, M.C. Sevillano-Rivera, A.J. Pinto, J.S. Guest, Impact of solids residence time on community structure and nutrient dynamics of mixed phototrophic wastewater treatment systems, Water Res. 150 (2019) 271–282.

[93] D. Zhou, C. Zhang, L. Fu, L. Xu, X. Cui, Q. Li, et al., Responses of the microalga Chlorophyta sp. to bacterial quorum sensing molecules (N-acylhomoserine lactones): aromatic protein-induced self-aggregation, Environ. Sci. Technol. 51 (2017) 3490–3498. Available from: https://doi.org/10.1021/acs.est.7b00355.

[94] E.-H. Belarbi, E. Molina, Y. Chisti, Retracted: A Process for High Yield and Scaleable Recovery of High Purity Eicosapentaenoic Acid Esters from Microalgae and Fish Oil, Elsevier, 2000.

[95] S.P. Singh, D. Singh, Biodiesel production through the use of different sources and characterization of oils and their esters as the substitute of diesel: a review, Renew. Sustain. Energy Rev. 14 (2010) 200–216. Available from: https://doi.org/10.1016/j.rser.2009.07.017.

[96] N. Dutta, M.K. Saha, Chapter nine - immobilization of a mesophilic lipase on graphene oxide: stability, activity, and reusability insights, in: C.V. Kumar (Ed.), Methods in Enzymology, Academic Press, 2018, pp. 247–272. Available from: https://doi.org/10.1016/bs.mie.2018.05.009.

[97] N.H. Tran, J.R. Bartlett, G.S.K. Kannangara, A.S. Milev, H. Volk, M.A. Wilson, Catalytic upgrading of biorefinery oil from micro-algae, Fuel 89 (2010) 265–274. Available from: https://doi.org/10.1016/j.fuel.2009.08.015.

[98] P. Kanmani, K. Kumaresan, J. Aravind, Utilization of coconut oil mill waste as a substrate for optimized lipase production, oil biodegradation and enzyme purification studies in Staphylococcus pasteuri, Electron. J. Biotechnol. 18 (2015) 20–28. Available from: https://doi.org/10.1016/j.ejbt.2014.11.003.

[99] L. Mafakher, M. Mirbagheri, F. Darvishi, I. Nahvi, H. Zarkesh-Esfahani, G. Emtiazi, Isolation of lipase and citric acid producing yeasts from agro-industrial wastewater, N. Biotechnol. 27 (2010) 337–340. Available from: https://doi.org/10.1016/j.nbt.2010.04.006.

[100] S. Kumar, N. Katiyar, P. Ingle, S. Negi, Use of evolutionary operation (EVOP) factorial design technique to develop a bioprocess using grease waste as a substrate for lipase production, Bioresour. Technol. 102 (2011) 4909–4912. Available from: https://doi.org/10.1016/j.biortech.2010.12.114.

[101] T.J. Pilusa, E. Muzenda, M. Shukla, Thermo-chemical extraction of fuel oil from waste lubricating grease, Waste Manage. 33 (2013) 1509–1515. Available from: https://doi.org/10.1016/j.wasman.2013.02.014.

[102] D.R. Rosa, I.C.S. Duarte, N. Katia Saavedra, M.B. Varesche, M. Zaiat, M.C. Cammarota, et al., Performance and molecular evaluation of an anaerobic system with suspended biomass for treating wastewater with high fat content after enzymatic hydrolysis, Bioresour. Technol. 100 (2009) 6170–6176. Available from: https://doi.org/10.1016/j.biortech.2009.06.089.

[103] F. Cuadros, F. López-Rodríguez, A. Ruiz-Celma, F. Rubiales, A. González-González, Recycling, reuse and energetic valuation of meat industry wastes in Extremadura (Spain), Resour. Conserv. Recy 55 (2011) 393–399.

[104] S.M. Basheer, S. Chellappan, P.S. Beena, R.K. Sukumaran, K.K. Elyas, M. Chandrasekaran, Lipase from marine *Aspergillus awamori* BTMFW032: production, partial purification and application in oil effluent treatment, N. Biotechnol. 28 (2011) 627–638.

[105] N. Asses, L. Ayed, H. Bouallagui, I.B. Rejeb, M. Gargouri, M. Hamdi, Use of *Geotrichum candidum* for olive mill wastewater treatment in submerged and static culture, Bioresour. Technol. 100 (2009) 2182–2188.

[106] P. Lauprasert, J. Chansirirattana, J. Paengjan, Effect of selected bacteria as bioremediation on the degradation of fats oils and greases in wastewater from cafeteria grease traps, EJSD 6 (2017) 181. Available from: https://doi.org/10.14207/ejsd.2017.v6n2p181.

[107] P. Saranya, P.K. Selvi, G. Sekaran, Integrated thermophilic enzyme-immobilized reactor and high-rate biological reactors for treatment of palm oil-containing wastewater without sludge production, Bioprocess. Biosyst. Eng. 42 (2019) 1053–1064. Available from: https://doi.org/10.1007/s00449-019-02104-x.

[108] A. Kumar, R. Gudiukaite, A. Gricajeva, M. Sadauskas, V. Malunavicius, H. Kamyab, et al., Microbial lipolytic enzymes – promising energy-efficient biocatalysts in bioremediation, Energy 192 (2020) 116674. Available from: https://doi.org/10.1016/j.energy.2019.116674.

[109] A.A. Saber, M. El-Sheekh, A.Y. Nikulin, M. Cantonati, H. Saber, Taxonomic and ecological observations on some algal and cyanobacterial morphospecies new for or rarely recorded in either Egypt or Africa, Egypt. J. Bot. 0 (2020). Available from: https://doi.org/10.21608/ejbo.2020.50683.1587.

[110] E.A. Alwaleed, M. El-Sheekh, M.M. Abdel-Daim, H. Saber, Effects of Spirulina platensis and *Amphora coffeaeformis* as dietary supplements on blood biochemical parameters, intestinal microbial population, and productive performance in broiler chickens, Env. Sci. Pollut. Res. (2020). Available from: https://doi.org/10.1007/s11356-020-10597-3.

[111] S.S. Ali, R. Al-Tohamy, R. Xie, M.M. El-Sheekh, J. Sun, Construction of a new lipase- and xylanase-producing oleaginous yeast consortium capable of reactive azo dye degradation and detoxification, Bioresour. Technol. 313 (2020) 123631. Available from: https://doi.org/10.1016/j.biortech.2020.123631.

[112] R. Gaur, S.K. Khare, Statistical optimization of palm oil hydrolysis by *Pseudomonas aeruginosa* PseA lipase, Asia-Pacific J. Chem. Eng. 6 (2011) 147–153. Available from: https://doi.org/10.1002/apj.510.

[113] R.A. Bayoumi, H.M. Atta, M.H. El-Sehrawy, Bioremediation of khormah slaughter house wastes by production of thermoalkalistable lipase for application in leather industries, Life Sci. Journal. 9 (2012) 1324–1335.

[114] G. Granzotto, P.R.F. Marcelino, A. de, M. Barbosa, E.P. Rodrigues, M.I. Rezende, et al., Culturable bacterial pool from aged petroleum-contaminated soil: identification of oil-eating Bacillus strains, Ann. Microbiol. 62 (2012) 1681–1690. Available from: https://doi.org/10.1007/s13213-012-0425-8.

[115] Y. Matsumiya, D. Wakita, A. Kimura, S. Sanpa, M. Kubo, Isolation and characterization of a lipid-degrading bacterium and its application to lipid-containing wastewater treatment, J. Biosci. Bioeng. 103 (2007) 325–330. Available from: https://doi.org/10.1263/jbb.103.325.

[116] D. Sugimori, M. Watanabe, T. Utsue, Isolation and lipid degradation profile of Raoultella planticola strain 232–2 capable of efficiently catabolizing edible oils under acidic conditions, Appl. Microbiol. Biotechnol. 97 (2013) 871–880. Available from: https://doi.org/10.1007/s00253-012-3982-7.

[117] W. Mongkolthanaruk, S. Dharmsthiti, Biodegradation of lipid-rich wastewater by a mixed bacterial consortium, Int. Biodeterior. Biodegrad. 50 (2002) 101–105. Available from: https://doi.org/10.1016/S0964-8305(02)00057-4.

[118] M. Tsuji, Y. Yokota, K. Shimohara, S. Kudoh, T. Hoshino, An application of wastewater treatment in a cold environment and stable lipase production of antarctic basidiomycetous yeast mrakia blollopis, PLoS One 8 (2013) e59376. Available from: https://doi.org/10.1371/journal.pone.0059376.

[119] J. Jeganathan, A. Bassi, G. Nakhla, Pre-treatment of high oil and grease pet food industrial wastewaters using immobilized lipase hydrolyzation, J. Hazard. Mater. 137 (2006) 121–128. Available from: https://doi.org/10.1016/j.jhazmat.2005.11.106.

[120] J. Jeganathan, G. Nakhla, A. Bassi, Hydrolytic pretreatment of oily wastewater by immobilized lipase, J. Hazard. Mater. 145 (2007) 127–135. Available from: https://doi.org/10.1016/j.jhazmat.2006.11.004.

[121] M.C.M.R. Leal, D.M.G. Freire, M.C. Cammarota, G.L. Sant'Anna, Effect of enzymatic hydrolysis on anaerobic treatment of dairy wastewater, Process. Biochem. 41 (2006) 1173–1178. Available from: https://doi.org/10.1016/j.procbio.2005.12.014.

[122] Q.E. Mobarak, K.K. R, N. M, A. T, Enzymatic pre-hydrolysis of high fat content dairy wastewater as a pretreatment for anaerobic digestion, Int. J. Environ. Res. 6 (2012) 475–480.

References

[123] P.H. Nielsen, P. Roslev, T.E. Dueholm, J.L. Nielsen, Microthrix parvicella, a specialized lipid consumer in anaerobic–aerobic activated sludge plants, Water Sci. Technol. 46 (1–2) (2002) 73–80.

[124] S. Kumar, A. Mathur, V. Singh, S. Nandy, S.K. Khare, S. Negi, Bioremediation of waste cooking oil using a novel lipase produced by Penicillium chrysogenum SNP5 grown in solid medium containing waste grease, Bioresour. Technol. 120 (2012) 300–304.

[125] A.B.G. Valladão, P.E. Sartore, D.M.G. Freire, M.C. Cammarota, Evaluation of different pre-hydrolysis times and enzyme pool concentrations on the biodegradability of poultry slaughterhouse wastewater with a high fat content, Water Sci. Technol. 60 (2009) 243–249. Available from: https://doi.org/10.2166/wst.2009.341.

[126] A.B.G. Valladão, M.C. Cammarota, D.M.G. Freire, Performance of an anaerobic reactor treating poultry abattoir wastewater with high fat content after enzymatic hydrolysis, Environ. Eng. Sci. 28 (2011) 299–307. Available from: https://doi.org/10.1089/ees.2010.0271.

[127] E. Efremenko, O. Senko, D. Zubaerova, E. Podorozhko, V. Lozinsky, New biocatalyst with multiple enzymatic activities for treatment of complex food wastewaters, Food Technol. Biotechnol. 46 (2008) 208–212.

[128] S. Chakraborty, E. Drioli, L. Giorno, Development of a two separate phase submerged biocatalytic membrane reactor for the production of fatty acids and glycerol from residual vegetable oil streams, Biomass Bioenergy 46 (2012) 574–583.

[129] C. Gonçalves, M. Lopes, J.P. Ferreira, I. Belo, Biological treatment of olive mill wastewater by non-conventional yeasts, Bioresour. Technol. 100 (15) (2009) 3759–3763.

[130] A. Yousuf, F. Sannino, V. Addorisio, D. Pirozzi, Microbial conversion of olive oil mill wastewaters into lipids suitable for biodiesel production, J. Agric. Food Chem. 58 (2010) 8630–8635. Available from: https://doi.org/10.1021/jf101282t.

[131] M.P. Prasad, K. Manjunath, Comparative study on biodegradation of lipid-rich wastewater using lipase producing bacterial species, (2011).

[132] H. Matsuoka, A. Miura, K. Hori, Symbiotic effects of a lipase-secreting bacterium, Burkholderia arboris SL1B1, and a glycerol-assimilating yeast, Candida cylindracea SL1B2, on triacylglycerol degradation, J. Biosci. Bioeng. 107 (4) (2009) 401–408.

[133] M.C. Cammarota, D.M.G. Freire, A review on hydrolytic enzymes in the treatment of wastewater with high oil and grease content, Bioresour. Technol. 97 (2006) 2195–2210.

[134] E. Rigo, R.E. Rigoni, P. Lodea, D. De Oliveira, D.M.G. Freire, H. Treichel, et al., Comparison of two lipases in the hydrolysis of oil and grease in wastewater of the swine meat industry, Ind. Eng. Chem. Res. 47 (2008) 1760–1765. Available from: https://doi.org/10.1021/ie0708834.

[135] H.J. Kim, Y.G. Choi, G.D. Kim, S.H. Kim, T.H. Chung, Effect of enzymatic pretreatment on solubilization and volatile fatty acid production in fermentation of food waste, Water Sci. Technol. 52 (2005) 51–59. Available from: https://doi.org/10.2166/wst.2005.0678.

[136] F.R. Damasceno, D.M. Freire, M.C. Cammarota, Impact of the addition of an enzyme pool on an activated sludge system treating dairy wastewater under fat shock loads, J. Chem. Technol. Biotechnol. 83 (2008) 730–738.

[137] D.R. Rosa, M.C. Cammarota, D.M.G. Freire, Production and utilization of a novel solid enzymatic preparation produced by penicillium restrictum in activated sludge systems treating wastewater with high levels of oil and grease, Environ. Eng. Sci. 23 (2006) 814–823. Available from: https://doi.org/10.1089/ees.2006.23.814.

[138] G.M. Nisola, E.S. Cho, H.K. Shon, D. Tian, D.J. Chun, E.M. Gwon, et al., Cell immobilized FOG-trap system for fat, oil, and grease removal from restaurant wastewater, J. Environ. Eng. 135 (2009) 876–884. Available from: https://doi.org/10.1061/(ASCE)0733-9372(2009)135:9(876).

[139] M.J. Montefrio, T. Xinwen, J.P. Obbard, Recovery and pre-treatment of fats, oil and grease from grease interceptors for biodiesel production, Appl. Energy 87 (2010) 3155–3161. Available from: https://doi.org/10.1016/j.apenergy.2010.04.011.

[140] S. Crognale, A. D'Annibale, F. Federici, M. Fenice, D. Quaratino, M. Petruccioli, Olive oil mill wastewater valorisation by fungi, J. Chem. Technol. Biotechnol. 81 (2006) 1547–1555.

[141] A. Piotrowska-Cyplik, Ł. Chrzanowski, P. Cyplik, J. Dach, A. Olejnik, J. Staninska, et al., Composting of oiled bleaching earth: fatty acids degradation, phytotoxicity and mutagenicity changes, Int. Biodeterior. Biodegrad. 78 (2013) 49–57. Available from: https://doi.org/10.1016/j.ibiod.2012.12.007.

[142] N. Saifuddin, K.H. Chua, Biodegradation of lipid-rich wastewater by combination of microwave irradiation and lipase immobilized on chitosan, Biotechnology 5 (2006) 315–323.

CHAPTER 22

Process intensification for sustainable algal fuels production

Hector De la Hoz Siegler

Department of Chemical and Petroleum Engineering, University of Calgary, Calgary, AB, Canada

22.1 Introduction

Algae are widely recognized for their great potential as a source of biofuels and chemicals and ability to abate carbon dioxide emissions. The term *algae* encompasses aquatic photosynthetic organisms that lack vascular tissue, including multicellular seaweeds and unicellular microalgae. In the context of biofuel production, the term *algae* is also used when referring to cyanobacteria or "blue—green" algae. Algae have attracted significant interest as a source of biofuels and chemicals due to their ability to grow on relatively simple media, using carbon dioxide and light as their primary source of carbon and energy, respectively, and with areal productivities that are several times higher than that of terrestrial crops. Consequently, algae are viewed as key for producing carbon-neutral fuels and chemicals.

Notwithstanding their great potential, production of biofuels from microalgae is hindered by economic constraints. Fossil fuels remain a relatively cheap commodity and, in spite of mounting evidence of anthropogenic climate change, carbon pollution is not yet priced at the values required to incentivize a wider adoption of biofuels. In order for algal fuels to compete with fossil fuels, as well as to ensure their sustainability, it is necessary to reduce their overall production cost and to increase the efficiency of algal cultivation and conversion processes. The necessary gains in efficiency and cost reduction can be realized by the application of process intensification principles to the design of algal biofuel production processes. This chapter provides a review of recent developments in the area of process intensification applied to the cultivation of algal biomass and its conversion into fuels.

22.1.1 Principles and domains of process intensification

Process intensification is one of the major areas of growth in chemical engineering research and design [1]. Process intensification entails the application of scientific and engineering principles for the design of novel equipment and methods that results in "substantially smaller, cleaner, less costly, safer, and more energy efficient processes" [2]. Although there is not a unique definition of process intensification [3], there is a shared vision of process intensification being a dramatic improvement in efficiency, rather than an incremental change [4], that goes beyond simple parameter optimization [1].

The operating framework for process intensification has been formalized through the formulation of the fundamental principles that underpin process intensification and the domains on which these principles are applied [5,6]. Van Gerven and Stankiewicz [5] posited that intensification of any given process must be based on the following four guiding principles:

1. maximize the effectiveness inter- and intramolecular events (i.e., provide perfect reaction conditions);
2. ensure all molecules undergo the same processing history (e.g., maintain narrow residence time distribution and avoid gradients);
3. optimize driving forces by reducing fluid dynamics, heat transfer, and mass transfer limitations (e.g., increase concentration difference across interfaces); and
4. maximize synergistic effects from partial process steps (i.e., combine operations and steps).

These generic principles are then realized over four elementary domains: spatial, thermodynamic, functional, and temporal. The spatial domain refers to the implementation of changes in structure (of equipment or catalyst) to remove randomness. Removing randomness in the spatial domain improves predictability and controllability. The thermodynamic domain refers to the optimal transfer of energy and involves both selecting the best form of energy and the energy transfer mechanism. In the functional domain, the goal is to exploit the synergies between processing steps by combining multiple functions into one single device or processing step. Finally, the temporal domain refers to the exploitation of process dynamics to improve performance [3,6].

The application of process intensification to processes with a biological component adds an additional degree of freedom due to the possibility of directly manipulating the biocatalyst [1]. In the case of algal processes, modifying algal strains through metabolic or genetic engineering can be considered part of process intensification if these efforts are ultimately aimed at improving overall productivity. In this chapter, however, we will exclude direct manipulation of the algal genome in our study of process intensification of algal biofuels.

22.1.2 Algal to biofuels pathways

Production of biofuels from algae involves two distinct stages: upstream processing and downstream processing. A schematic of the steps required for algal biofuel production is shown in Fig. 22.1.

Upstream operations include all those steps necessary for producing algal biomass. Growing microalgae can be quite energy intensive and the cultivation efficiency is affected by the genetics of the particular algal strain, the source (type and concentration) of the carbon dioxide supply and other nutrients, the light source, environmental conditions, and

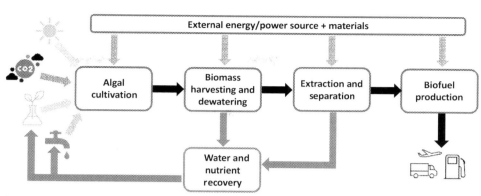

FIGURE 22.1 Simplified block flow diagram for the production of algal biofuels.

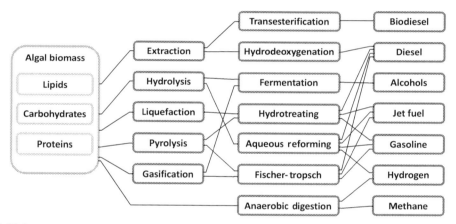

FIGURE 22.2 Downstream process alternatives for converting algal biomass into biofuels.

the type of cultivation system. Following cultivation, algal biomass must be harvested and after that undergo one or more of several alternative downstream processes. Each alternative process will result in a different biofuel, as shown in Fig. 22.2. Multiple downstream processes can be used in one biorefinery producing several different final products, which may improve overall economic feasibility.

Large-scale production of algal biofuels is at present hindered by low culture cell densities, inefficient and energy intensive harvesting methods, and suboptimal downstream processes. Process intensification at each step of the algal biofuels production chain will reduce the overall production cost of algal biofuels, while improving their sustainability and reducing environmental impacts. In the following sections, recent developments in the intensification of algal processes at each step of the algal biofuel production chain are presented.

22.2 Intensification of photobioreactors

In a typical algal culture, cells are suspended in a liquid medium that provides all required nutrients to support algal growth, while carbon dioxide must be transferred from the gas phase through the liquid medium and into the cells. Oxygenic photosynthesis performed by the cells results in the production of oxygen, while many algal species are known to experience inhibition at concentrations higher than 120% of the air saturation level [7]; consequently, oxygen must be transferred from the liquid medium into the gas phase. In addition, as light is needed to fuel algal metabolism, the culturing equipment must facilitate the penetration and distribution of light.

Large-scale cultivation of algae has traditionally been performed in open uncontrolled systems, such as lagoons and ponds. The main advantage of this type of system is that capital and some operating costs are kept to a minimum. Because of their uncontrolled nature, however, biomass quality and productivity are adversely affected. Furthermore, there is a high risk of contamination with unproductive algal strains or predators that feed on the algal biomass. To overcome some of these disadvantages, several culturing systems have been proposed and implemented. Closed photobioreactors enable the cultivation of algae under varying environmental conditions, reduce or eliminate contamination issues, prevent water loss through evaporation, and result in higher areal productivity [7]. The most commonly used closed photobioreactor geometries are flat panel, bubble column, and tubular. Other designs are typically variations of these basic geometries. The design and operation and performance of common photobioreactors have been reviewed by multiple authors [7–12].

Capital and operating costs of closed photobioreactors, however, are still significantly higher than open systems and further gains in productivity are required before these closed systems can compete with open ponds. Posten [7] identified a series of aspects affecting bioreactor performance:

1. Light saturation and light dilution: cultures should be operated in the nonlimited region of the light-saturation curve and the incident light should be spread over a large reactor surface area.
2. Light attenuation and light path length reduction: the length of the path that light travels inside the cultivation media should be as small as possible to ensure light gradients are minimized.
3. Light fluctuation and mixing: avoid persisting light/dark cycles by ensuring sufficiently low mixing times.
4. CO_2-supply and aeration: maintain a relatively high CO_2 partial pressure in the liquid phase (0.1–0.2 kPa) to avoid carbon limitation.
5. Mixing and auxiliary energy: minimize energy input for mixing while ensuring sufficient mass transfer, avoid cell settling, and avoid cell stress or damage due to excessive mechanical energy.

Acting on these aspects will generate significant gains in productivity, energy, and photosynthetic efficiency, achieving the goals of process intensification. These aspects identified by Posten [7] are used here as a guide to analyze current trends in the intensification of algal photobioreactors.

22.2.1 Light saturation and light dilution

The first guiding principle of process intensification states the need for maximizing the effectiveness of intermolecular and intramolecular events. When applied to an algal culture this principle implies providing the perfect lighting conditions to maximize biomass growth. Light provision must be adequate in terms of intensity, spectral quality, frequency, and duration [13]. Ideally, photobioreactors must be designed to avoid light inhibition; consequently, light intensity should be kept between the critical and the saturation intensities [14]. The irradiance incident over the bioreactor surface can be modulated by varying reactor geometry and orientation, achieving a light dilution effect [15]. Furthermore, only a fraction of the full spectral bandwidth can be converted to chemical energy by photosynthesis, while a substantial portion of the incident photons have deleterious effects on the cultures due to photodamage caused by UV radiation and thermal effects caused by IR photons [16,17]. Consequently, efforts to improve spectral quality by the use of spectral filtration or spectral shifting have also been pursued, as reviewed by Nwoba et al. [18].

22.2.1.1 Solar tracking systems

Solar tracking devices continually follow the trajectory of the sun throughout the day [18]. They can be used to maximize the light intensity reaching the culture or to achieve optimal light dilution. Solar trackers can be used to directly control the tilt of the bioreactor or in combination with solar collectors and optical fibers or guides to drive the light deep into the reactor.

Zijffers et al. [19] used Fresnel lenses mounted on a dual-axis rotating support. Correct dual-axis positioning of the Fresnel lenses was shown to enable the full focusing of sunlight on rectangular light guides all year around. Hindersin et al. [20] used a solar tracker to control the irradiance incident onto an outdoor solar flat panel photobioreactor. In their work, Hindersin et al. [20] concluded that the use of solar trackers resulted in decreased photoinhibition at low culture densities, enhancement of the incident irradiance beyond 100% of the horizontal irradiance in high cell density cultures, and effective temperature control.

Castrillo et al. [14] evaluated the gains in productivity that could be achieved in deep tank cultures when a solar tracker was used to control the orientation of cone-shaped light guides (see Fig. 22.3) to deliver the light deep into the culture medium. In this case, the cone-shaped structures "guide the light downward since they receive the incident solar light on their base and spread it out through its lateral surface" [14]. It was determined that a 2.7-fold increase in areal productivity could be achieved in comparison with open ponds bioreactors under identical irradiance conditions.

22.2.1.2 Spectral filtration and shifting

Improving the light spectral quality reaching the algal cells is an effective way to limit photodamage and inhibition, and to enhance overall productivity [18,21,22]. Improvements in spectral quality can be realized by filtering out the nonphotosynthetically active radiation or by applying wavelength shifters or photoluminescent spectral converters to convert photons of little or no photosynthetic value into photons of high photosynthetic potential [16,23,24].

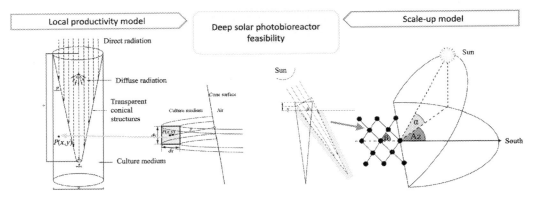

FIGURE 22.3 Scheme of the cone-shape light guides. Source: *Reproduced with permission from M. Castrillo, R. Díez-Montero, I. Tejero, Model-based feasibility assessment of a deep solar photobioreactor for microalgae culturing, Algal Res. 29 (2018) 304–318.*

Sharp et al. [25] used thin sheets of regenerated cellulose containing different dyes to selectively filtering out different portions of the light spectrum. It was demonstrated that cultures receiving only the red portion of the light were on average 53% more productive than the cultures receiving the full spectrum light. Organic solar cells have been proposed as a way to filtering the incoming solar radiation, letting the photosynthetically active radiation pass through the solar cell and converting the high-energy photons in the blue and UV spectra into electricity [26,27]. Organic solar cells are semitransparent, flexible and lightweight, and can be easily shaped to be integrated into an algal photobioreactor. They can be manufactured by solution processing using standard printing/coating techniques with minimal infrastructure [28,29]. These devices are flexible and lightweight and can be rolled into a scroll for easy transportation and storage. In addition, their low manufacturing cost will facilitate easy replacement and recycling.

Fluorescent or phosphorescent materials are able to convert photons from one wavelength into another [24]. Organic and inorganic dyes, phosphors, and quantum dots have all been shown to exhibit excellent properties to attain the wavelength shifting of light [23,30]. Wondraczek et al. [16] used phosphorescent dyes to shift the wavelength of green photons into red photons and improve the productivity of *Haematococcus pluvialis* under broadband irradiation. Biomass productivity increased by 36% due to the spectral shifting.

Luminescent solar concentrators combine spectrum shifting properties with spatial dilution to deliver high-quality light into the culture. As the concentrators absorb both direct and diffuse light, solar tracking systems are not required. Luminescent solar concentrators contain uniformly dispersed luminescent particles (e.g., organic dyes, quantum dots, or semiconducting polymers) [31] (Fig. 22.4).

Incident light is absorbed by luminescent dyes trapped inside the concentrator and are re-emitted at longer wavelength to the edges by total reflection. Raeisossadati et al. [32] reported a 18.5% increase in biomass productivity in *Scenedesmus* sp. cultures when a red luminescence solar concentrator was used in an outdoor raceway pond.

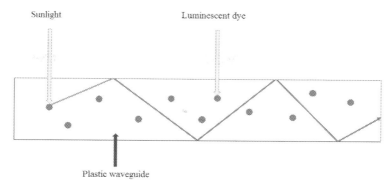

FIGURE 22.4 Scheme of the luminescent solar concentrator. Source: *Reproduced with permission from M. Raeisossadati, N.R. Moheimani, D. Parlevliet, Luminescent solar concentrator panels for increasing the efficiency of mass microalgal production, Renew. Sustain. Energy Rev. 101 (2019) 47–59.*

22.2.2 Light attenuation and light path length reduction

Large-scale efficiency of photobioreactors is a challenge as light is attenuated as it passes through the culture medium [33,34], due to mutual shading of the cells and scattering by the cells [7]. The decrease in light intensity follows an exponential decay as it travels inside the reactor [35]. At some point, light intensity is sufficiently low such that a fraction of the total reactor volume can be considered dark. This dark volume does not contribute to biomass growth but increases overall energy requirements. Furthermore, higher biomass concentration can be achieved if the dark volume is minimized [7,36].

Reduction in the light path can be achieved by increasing the surface to volume ratio (SVR). As the SVR increases, however, there may be a potential increase in overall system footprint, impacting land requirements and increasing environmental effects associated with land-use changes. SVR can be increased by narrowing the liquid path (e.g., narrow channels or panels, shallow ponds) or by internal illumination of the reactor contents. The latter approach achieves an increase in the internal (illuminated) surface area, while maintaining the overall footprint essentially unchanged.

22.2.2.1 Thin layer reactors

In thin layer photobioreactors, an external pump is used to circulate the liquid cultivation medium forming a very thin liquid film (6–10 mm) and ensuring very short path for light travel [37]. This configuration allows to realize very high SVR, increasing productivity and allowing a uniform light distribution [38,39]. Net aerial productivities have been reported to average between 22 and 25 g/m^2·d, with peak productivities as high as 38 g/m^2·d reported for these systems [40,41]. The gain in productivity is partly the result of the higher biomass concentrations (up to 50 g/L) that can be sustained in this configuration [41].

22.2.2.2 Biofilm reactors

Another approach to narrow the light path involves growing the cells as thin biofilms, as opposed to suspended cultures. In this case, the cells grow attached on the surface of the

510 22. Process intensification for sustainable algal fuels production

photobioreactor. The solid biofilms are either continuously wetted by circulating liquid medium over them or by periodically submerging them into the nutrient-containing liquid medium [37].

In addition to increasing SVR and improving light distribution, algal biofilms also reduce water requirements for cultivation and simplify harvesting and dewatering operations, with a significant reduction in energy requirements [42,43].

Liu et al. [44] reported biomass areal productivity reaching 15 g/m^2·d for a *Scenedesmus obliquus* algal biofilm culture. Liu et al. [44] used a combination of glass plate and filter paper as the support material for the biofilm, arranged in a multilayer configuration and with the culture media flowing through the paper, and operated their cultured at photosynthetic active radiation (PAR) irradiance of 300 μmol/m^2·s. Moll et al. [45] proposed as an alternative design using transparent foils to enclose a narrow culture volume, thus promoting the formation of biofilm and enhancing light penetration. The SVR in this configuration was around 550 m^2/m^3, which is two orders of magnitude higher than the values typically found in open and raceway ponds that have SVR below 7 m^2/m^3 [42].

A significant drawback of biofilm reactors is the formation of concentration gradients across the biofilm, which promotes heterogeneity of the cell population, affecting overall bioreactor productivity. Diffusion limitations across the biofilm result in the formation of concentration gradients (e.g., nutrients, pH, oxygen, and CO_2) that can lead to nutrient limitation or oxygen inhibitory levels in a portion of the biofilm, or even to the creation of death zones [46]. Improving the convective transport outside the biofilm, for example, using vigorous gas bubbling or a high rate cross-flow to generate intensive mixing can help to alleviate these effects [47]. Regular periodic harvesting of the biofilm is also an effective way of reducing mass transfer effects, which helps to maintain optimal biomass growth rate.

22.2.2.3 Light guides

Planar waveguides or "light guides" can distribute light throughout the culture volume [48], effectively reducing the path light needs to travel within the liquid medium. Salmean et al. [49] used an externally coupled planar waveguide to illuminate a biofilm reactor. The guides were manufactured from optically transparent cast acrylic and notches were scored at regular interval to achieve a leakiness effect [50]. It was shown that biomass productivity increased by a factor of 2.5 [49].

Sivakaminathana et al. [51] tested the use of light guides in a pilot-scale raceway pond. The guides were manufactured from polished clear polyacrylate to which a dispersion pattern was added. The addition of the guides to the raceway pond resulted in a 3.9-fold increase in areal productivity.

22.2.3 Carbon dioxide distribution

Transporting carbon dioxide from the gas phase into the aqueous medium is a potentially limiting factor for algal growth due to its low diffusion rate and the relatively low degree of mixing of the medium. In a rapidly growing culture, CO_2 uptake by algae rapidly increases the pH, leading to inhibition of further growth. It has been shown that efficient carbon utilization occurs when the supply rate of dissolved CO_2 closely matches the culture demand [52]. Consequently, CO_2 needs to be actively distributed into the cultivation system.

Handbook of Algal Biofuels

Gas mass transfer rates are affected by the geometrical design and positioning of the gas diffuser device, the two-phase flow characteristics (e.g., flowrates, bubble size, interfacial gas velocities, degree of turbulence), and the physicochemical properties (e.g., viscosity, pH, presence of surfactants or catalysts) of the liquid medium. The transport across the liquid film surrounding the gas bubbles is dominated by diffusion. The rate of mass transfer is determined by the product of the overall mass transfer coefficient (k_L), the interfacial area (a, the total surface area of the gas bubbles), and the concentration gradient across the interface. Reducing the average bubble diameter results in an increase in the interfacial area, at a constant gas flow rate, and a consequent enhancement in the $k_L a$ value.

22.2.3.1 Microbubbling and membrane diffusers

Microbubbles with diameter from 10 to 50 µm enable greater mass transfer coefficients, slower rising velocities in the liquid phase, and thus longer residence times [53,54]. Methods to generate microbubbles include applying fluidic oscillation to microporous diffusers, ultrasound waves induced acoustic cavitation, hydrodynamic cavitation induced by pressure variations in the flowing liquid, and breakthrough pressures applied to microporous membranes [53,54]. Zimmerman et al. [55] integrated a fluidic oscillator into an airlift loop bioreactor to ensure the generation of monodispersed, uniformly spaced and regularly released microbubbles. Microbubbles have also been found effective at striping the oxygen generate by photosynthesis at that can reach inhibitory levels in the reactors [56].

Membrane diffusers have been proposed to increase both CO_2 mass transfer rate and carbon utilization efficiency. Carvalho et al. [39] reported that the observed $k_L a$ values nearly doubled when using hollow fiber either hydrophobic or hydrophilic membranes. Xu et al. [57] used a flat PDMS-polysulfone membrane as liner in a raceway pond. The PDMS-polysulfone liner was used as a liquid—liquid membrane to deliver CO_2 to the bioreactor via a CO_2-loaded solvent. The pond-liner membrane provided better CO_2 delivery than hollow fibers, with a CO_2 utilization efficiency reaching up to 90% compared to 47% with the hollow fiber membrane system and 11% for just air sparging.

22.2.3.2 High alkalinity

To obviate the problems and energy costs associated with ensuring a continuous and timely supply of carbon dioxide, the direct addition of soluble carbonate or bicarbonate salts has been proposed [37,58]. This approach is, however, limited to alkali-tolerant or alkaliphilic algal strains.

Operation under high pH enables the uncoupling of CO_2 gas/liquid transport and algal growth. A scrubber is used to facilitate the transfer of CO_2 from the gas phase to the alkaline growth medium. Upon transfer to the growth medium, the dissolved CO_2 (mainly in the form of bicarbonate or carbonate ions) is circulated to the bioreactor where its concentration is depleted by the algae [37]. Ataeian et al. [59] reported that an alkaline capture and conversion system operating at high alkalinity (0.5 mol/L) and with a pH varying between 10.4 and 11.2 resulted in the areal biomass productivity reaching $15.2 \pm 1.0 \ \mathrm{g/m^2 \cdot d}$.

512   22. Process intensification for sustainable algal fuels production

22.3 Harvesting

The harvest of algal biomass from the liquid culture is reported to account for a significant fraction (up to 30%) of the total production costs [60]. Typically, biomass concentration in the liquid medium is in the range of 0.03%−0.5% of total suspended solids [61]. Increasing the biomass concentration by any of the intensification methods previously described will have a direct effect on reducing the harvesting and downstream processing costs. In addition, several intensification approaches have been proposed to directly reduce the energy requirements and cost for harvesting.

Algal biomass can be harvested through biological, mechanical, chemical, or electrical methods. Each method has advantages and shortcomings, and therefore there is no single method that can be considered ideal for all situations [62]. Currently, one of the most commonly used methods for harvesting and dewatering is centrifugation. Centrifugation is however quite energy intensive, with energy usage ranging from 1.8 to 29 MJ/m^3 of processed culture volume [63]. Consequently, although centrifugation is a commercially established technology, it is considered unsuitable in the context of biofuel production.

22.3.1 Membrane filtration

Membrane filtration is a mature technology that does not introduce foreign substances into the harvested biomass and requires relatively low energy requirements. The use of membranes for microalgal dewatering have been reviewed by Mo et al. [63] and by Zhang et al. [64]. Energy requirements for membrane filtration have been estimated to be between 0.6 and 7.2 MJ/m^3 of processed culture volume [63].

22.3.2 Flotation

Harvesting by flotation is accomplished by circulating air or gas bubbles through the culture medium and skimming the biomass at the top. It is a relatively fast, compact, and highly flexible harvesting method and it has been found to be more effective than sedimentation [65].

The addition of surfactants has been shown to increase the rate of bubble formation, flotation rate, and promotes easier biomass separation [66]. Several flotation techniques have been developed including dissolved air flotation, dispersed air flotation, microflotation, and foam flotation. Foam or froth flotation, a method that combines dispersed air flotation with foam fractionation, have been showed to be very efficient requiring only 0.05 to 0.19 MJ/m^3 of processed culture volume [67,68].

22.4 Biomass conversion to biofuel

Conversion of biomass into biofuels can be realized following multiple pathways and result in distinct products including ethanol, biodiesel, green diesel, syn-gas, biochar, and hydrogen. A naive processing approach involves breaking down the biomass to achieve

Handbook of Algal Biofuels

its complete fractionation into lipids, proteins, and carbohydrates. Each one of these components is then recovered and processed into different fuels. For example, the lipid fraction can be recovered and processed into biodiesel while the carbohydrate fraction can be converted into biogas, bioethanol or biohydrogen [69].

Fractionation and conversion processes are typically more efficient starting from dry algal biomass. However, there is a significant energy penalty if a drying step is introduced. Due to water's high latent heat of vaporization (2265 kJ/kg), drying a biomass slurry containing less than 20% total suspended solids is of limited value for biofuel applications [54]. Process intensification can be applied to significantly reduce processing costs by combining multiple processing steps, thus eliminating trade-offs, or by greatly improving the performance of individual steps.

22.4.1 Cell disruption

Conversion of algal biomass, either wet or dry, into fuels is normally enhanced by introducing a cell disruption method [70]. Cell disruption methods destabilize the cells wall and membranes enabling easier access to the cell contents. Cells disruption can be achieved via several distinct methods including mechanical (e.g., bead beating, grinding, French press, osmotic shock, homogenization), chemical (e.g., acid or alkaline hydrolysis), physical (e.g., ultrasonication, thermal shock), and biological approaches (e.g., enzyme hydrolysis). These methods, however, are costly, laborious, and energy intensive [71]. Consequently, some emerging technologies have been evaluated for the potential intensification of biomass disruption, including pulsed electric field and microwave-assisted extraction (MAE).

22.4.1.1 Pulse electric field

Pulsed electric field (PEF) has been applied to modify the properties of cell walls and membranes as the electrical field appears to interact with phospholipids in the membranes and the peptidoglycan in the cell walls [72]. PEF has been used to enhance the recovery of lipids and proteins in algal suspension, reduce the processing time, to enable the utilization of less toxic solvents, and to minimize solvent use in subsequent extraction steps [72–75]. Large-scale operation has been proven feasible with favorable energy efficiency [76,77]. PEF is, however, limited to treating nonconductive algal suspensions.

22.4.1.2 Simultaneous cell disruption and lipid extraction

MAE is an attractive method thanks to its ease of operation, speed, relative low cost, and high energy transfer efficiency [54]. In MAE, the rapid oscillation of molecules within an electric field causes friction, which generates heat [70]. MAE has been applied together with the ionic liquid ([BMIM] [H_2SO_4]) to extract lipids from wet algal biomass, resulting in an increase by one to two orders of magnitude in the extraction rate [78].

Reactive electrochemical membranes (REM) have been proposed for the simultaneous harvesting of algal biomass, cell disruption, and lipid extraction [79]. REMs are porous ceramic membranes that act as electrodes to promote anodic oxidation in addition to serving as a physical separation device [79]. The suitability of REM for large-scale operations is, however, questionable due to the catalyst cost and energy requirements [80].

Ultrasound-assisted extraction (UAE) has also been proposed as a single-step cell disruption/extraction method [70]. Adam et al. [81] demonstrated the potential of ultrasonic-assisted extraction as an ecoprocess for cell disruption and lipid extraction in a solvent-free method. Energy requirements for UAE are, however, very high [82]; although the frequency or treatment duration optimization can significantly reduce energy needs [70]. UAE has been combined with ozone pretreatment for the simultaneous recovery of lipids, proteins, carbohydrates, and phosphorus from wet slurries of *Scenesdesmus obliquus* growing in wastewater [83]. This combined pretreatment resulted in the intensification of microalgal biomass fractionation and resulted in higher overall product yields, while cutting solvent waste generation by up to 95% and overall extraction processing time by 90%.

22.4.2 Intensification of drying and oil extraction

Despite promising advances in emerging cell disruption and extraction techniques, as discussed above, significant barriers remain for the large-scale use of these emerging methods. In a life cycle basis comparison of different extraction methods, Ferreira et al. [82] concluded that hexane:isopropanol-based extraction was still the least energy-intensive method for lipid extraction from algae.

Extensive process optimization and modeling was reported by Song et al. [84] to develop and intensify the drying and oil extraction process, as shown in Fig. 22.5. In this

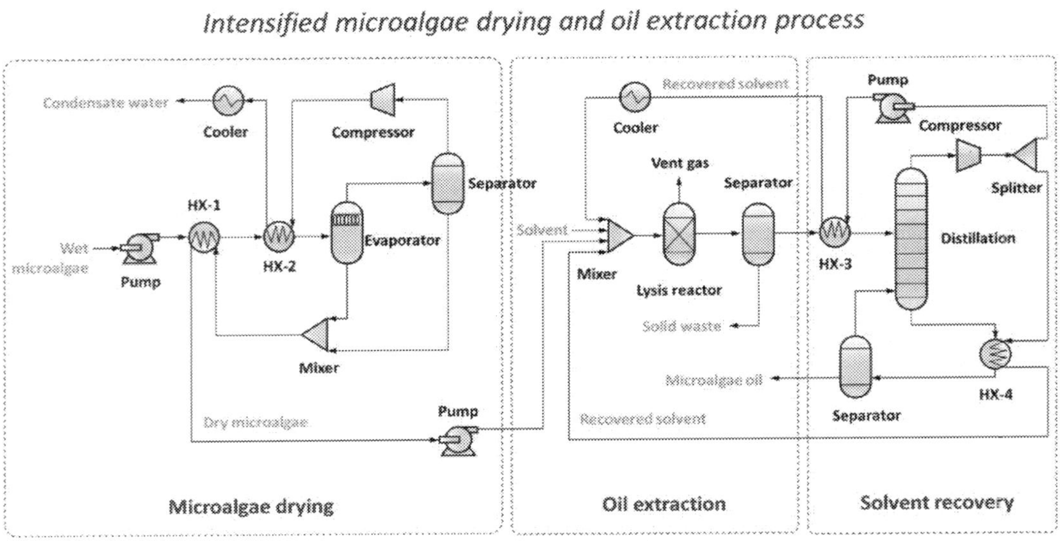

FIGURE 22.5 Process schematic of the intensified drying and oil extraction process using vapor recompression and heat integration. Source: *Reproduced with permission from C. Song, Q. Liu, N. Ji, S. Deng, J. Zhao, Y. Kitamura, Intensification of microalgae drying and oil extraction process by vapor recompression and heat integration, Bioresour. Technol., 207 (2016), 67–75.*

22.4.3 Biodiesel production

Emerging trends in biodiesel process intensification include a transition from batch processes to continuous production in continuous stirred tank reactors, the use of microchannel reactors to reduce mass transfer effects and enhance catalytic activity, cavitation reactors, and microwave reactors. Wong et al. [85] provided a critical review of the different process intensification approaches that have been applied for the conversion of lipids into biodiesel, with a particular emphasis on catalytic enhancement and novel reactor designs. Here we cover only methods for biodiesel production that are applicable to whole algal biomass, either wet or dry.

Reactive separation is a process intensification approach that combines separation and reaction steps together in a single unit operation. In situ transesterification (ISTE) is a type of reactive separation, where the oil-bearing biomass is processed in a single vessel for the extraction of lipids and their conversion to fatty acid alkyl esters [86]. ISTE can be performed on dry or wet biomass. Fig. 22.6 presents a comparison among conventional transesterification, dry ISTE, and wet ISTE.

Ghosh et al. [69] performed ISTE using dry *Chlorella* sp. biomass using homogeneous acid catalyst and methanol playing a dual role as solvent for the extraction of lipids and reactant for their transesterification. The single-step process resulted in a higher biodiesel yield than the conventional two steps transesterification process. However, other researchers have noted that ISTE requires a higher catalyst loading to achieve equivalent biodiesel yields [87].

Studies using algal biomass have shown that biodiesel yield decreases with increasing water content [88]. In wet ISTE, water acts as a retardant to catalysts, reduces the extraction efficiency, and may lead to the formation of soaps if an alkaline catalyst is used. Consequently, acid catalysts are generally preferred, and some combination methods are used to enhance cell disruption and extraction [86].

Kim et al. [86] have reviewed the different approaches followed to intensify wet ISTE of algal biomass, including the use of microwave or ultrasonic assistance, the use of cosolvents, and supercritical conditions. Similar to the ultrasonic-assisted extraction and the MAE, the use of sonication or microwaves has been found to enhance lipid solubilization in ISTE. Cheng et al. [89] found that the biodiesel production rate in the microwave-assisted process was six times faster than the conventional transesterification process.

Yadav et al. [90] introduced a process wherein carbon dioxide microbubbles were used to lyse the algal cell walls, releasing triglyceride oils, and—in a single vessel—used methanol to achieve the transesterification of the released lipids. It was found that at lower

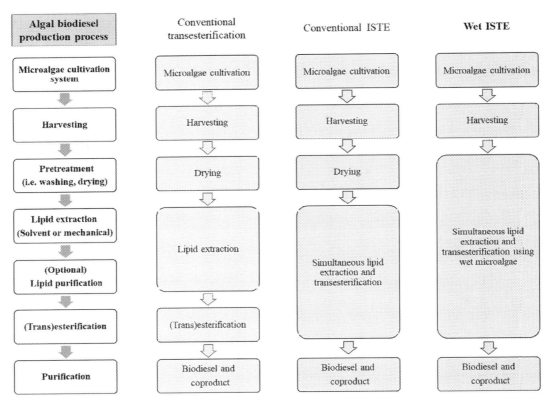

FIGURE 22.6 Comparative schematics of the conventional transesterification, conventional dry in situ transesterification (ISTE), and wet ISTE biodiesel production processes. Source: *Reproduced with permission from B. Kim, H.Y. Heo, J. Son, J. Yang, Y.-K. Chang, J.H. Lee, et al. Simplifying biodiesel production from microalgae via wet in situ transesterification: a review in current research and future prospects, Algal Res. 41 (2019) 101557.*

temperatures (353.15K–368.15K) and intermediate pressures (5–10 MPa), carbon dioxide enhances the mutual solubility of methanol and triglycerides, thus increasing the transesterification reaction rate. Process simulation tools were used to further optimize process conditions, resulting in a promising low biodiesel production cost [91].

22.4.4 Wet processing via hydrothermal liquefaction

As an alternative to the production of biodiesel via transesterification, hydrothermal liquefaction (HTL) has emerged as a promising conversion process to avoid the energy and cost-intensive thickening and drying steps. HTL can be performed directly on biomass slurries containing 15% total suspended solids or higher [54].

In HTL the algal biomass is decomposed at moderate temperatures (200°C–350°C) and at high pressures (5–20 MPa) and transformed to gas, liquid, and solid fractions. At these temperatures and high pressures, the resistance due to interphase mass transfer is mitigated, the ionic product of water (hydroxyl OH^- and hydronium H^+) increases, while its dielectric constant decreases substantially [92,93]. The target product in HTL is the liquid fraction, commonly referred to as biocrude oil, which is normally required to undergo further upgrading before being suitable as a biofuel. The objective in HTL is to maximize biocrude oil yield and quality [94].

Due to the high content of nitrogen, oxygen, and sulfur in the algal biomass, the biocrude oil obtained from the HTL process is of relatively low quality. These heteroatoms must be removed to produce a high-quality fuel. Although biocrude upgrading can be performed in a subsequent step after HTL, the use of catalysts during HTL can achieve the in-situ removal of heteroatoms [95]. Catalysts can be used to affect the distribution of HTL products and to improve overall efficiency, as the use of more selective catalyst can reduce the formation of undesirable products (e.g., char and tar). Lu et al. [96] performed the catalytic HTL of *Chlorella* biomass over mesoporous silica-based materials containing separated acidic and basic sites. Catalysts with acidic sites were more effective in denitrification, while catalysts with both acid and basic sites showed higher selectivity to hydrocarbons.

22.5 Conclusion

Process intensification of algal biofuel production processes enables significant reductions in energy requirements, land footprint, and environmental impacts. By combining multiple steps; eliminating mass, heat, and light transfer barriers; and using novel processing methods, significant improvement in process sustainability are possible. Although there have been some significant advancements in this field, further improvements are still required. Therefore there is a need for continuous research in this area.

References

[1] J. Wodley, Bioprocess intensification for the effective production of chemical products, Comp. Chem. Eng. 105 (2017) 297–307.

[2] M. Baldea, From process integration to process intensification, Comput. Chem. Eng. 81 (2015) 104–114.

[3] F.J. Keil, Process intensification, Rev. Chem. Eng. 34 (2) (2018) 135–200.

[4] J. Bielenberg, I. Palou-Rivera, The RAPID manufacturing institute −reenergizing US efforts in process intensification and modular chemical processing, Chem. Eng. Proc. Proc. Inten. 138 (2019) 49–54.

[5] T. Van Gerven, A.I. Stankiewicz, Structure, energy, synergy, time−the fundamentals of process intensification, Ind. Eng. Chem. Res. 48 (2009) 2465–2474.

[6] A. Stankiewicz, The principles and domains of process intensification, Chem. Eng. Prog. 3 (2020) 23–27.

[7] C. Posten, Design principles of photo-bioreactors for cultivation of microalgae, Eng. Life Sci. 9 (2009) 165–177.

[8] M.A. Borowitzka, Commercial production of microalgae: ponds, tanks, tubes and fermenters, J. Biotechnol 70 (1−3) (1999) 313–321.

[9] E. Molina Grima, F.G.A. Acien Fernandez, F.G. Garcia Camacho, Y. Chisti, Photobioreactors: light regime, mass transfer, and scaleup, J. Biotechnol. 70 (1−3) (1999) 231–247.

[10] A.S. Miron, F.G. Garcia Camacho, A.C. Gomez, E. Molina Grima, Y. Chisti, Bubble-column and airlift photobioreactors for algal culture, AIChE J. 46 (9) (2000) 1872–1887.

[11] D.O. Hall, F.G.A. Acien Fernandez, E.C. Guerrero, K.K. Rao, E. Molina, Grima, Outdoor helical tubular photobioreactors for microalgal production: modeling of fluid-dynamics and mass transfer and assessment of biomass productivity, Biotechnol. Bioeng. 82 (1) (2003) 62–73.

[12] E. Sierra, F.G. Acien, J.M. Acien Fernandez, J.L. Garcia, C. Gonzalez, E. Molina, Grima, Characterization of a flat plate photobioreactor for the production of microalgae, Chem. Eng. J. 138 (1–3) (2008) 136–147.

[13] A.P. Carvalho, S.O. Silva, J.M. Baptista, F.X. Malcata, Light requirements in microalgal photobioreactors: an overview of biophotonic aspects, Appl. Microbiol. Biotechnol. 89 (2011) 1275–1288.

[14] M. Castrillo, R. Díez-Montero, I. Tejero, Model-based feasibility assessment of a deep solar photobioreactor for microalgae culturing, Algal Res. 29 (2018) 304–318.

[15] M. Morweiser, O. Kruse, B. Hankamer, C. Posten, Developments and perspectives of photobioreactors for biofuel production, Appl. Microbiol. Biotechnol. 87 (2010) 1291–1301.

[16] L. Wondraczek, M. Batentschuk, M.A. Schmidt, R. Borchardt, S. Scheiner, B. Seemann, et al., Solar spectral conversion for improving the photosynthetic activity in algae reactors, Nat. Commun. 4 (2013).

[17] A. Vadiveloo, N.R. Moheimani, J.J. Cosgrove, P.A. Bahri, D. Parlevliet, Effect of different light spectra on the growth and productivity of acclimated Nannochloropsis sp. (Eustigmatophyceae), Algal Res. 8 (2015) 121–127.

[18] E.G., Nwoba, D.A. Parlevliet, D.W. Laird, K. Alameh, N.R. Moheimani, Light management technologies for increasing algal photobioreactor efficiency, Algal Res. 39 (2019) 101433.

[19] J.F. Zijffers, M. Janssen, J. Tramper, R.H. Wijffels, Design process of an area-efficient photobioreactor, Mar. Biotechnol. 10 (2008) 404–415.

[20] S. Hindersin, M. Leupold, M. Kerner, D. Hanelt, Irradiance optimization of outdoor microalgal cultures using solar tracked photobioreactors, Bioprocess. Biosyst. Eng. 36 (2013) 345–355.

[21] T. de Mooij, G. de Vries, C. Latsos, R.H. Wijffels, M. Janssen, Impact of light color on photobioreactor productivity, Algal Res. 15 (2016) 32–42.

[22] E.R. Mattos, M. Singh, M.L. Cabrera, K.C. Das, Enhancement of biomass production in Scenedesmus bijuga high-density culture using weakly absorbed green light, Biomass Bioenergy 81 (2015) 473–478.

[23] L. Wondraczek, E. Tyystjärvi, J. Méndez-Ramos, F.A. Müller, Q. Zhang, Shifting the sun: solar spectral conversion and extrinsic sensitization in natural and artificial photosynthesis, Adv. Sci. 2 (2015).

[24] M.D. Ooms, C.T. Dinh, E.H. Sargent, D. Sinton, Photon management for augmented photosynthesis, Nat. Commun. 7 (2016) 12699.

[25] C.E. Sharp, S. Urschel, X. Dong, et al., Robust, high-productivity phototrophic carbon capture at high pH and alkalinity using natural microbial communities, Biotechnol. Biofuels 10 (2017) 84.

[26] M. Strous, C. Sharp, H. De la Hoz Siegler, G. Welch, Biomass production in alkaline conditions, US Patent App. 15/428,800, 2017.

[27] J. Zorz, W.D.L. Richardson, A. Laventure, M. Haines, et al., Light manipulation using organic semiconducting materials for enhanced photosynthesis, Res. Sq. (2020). Available from: https://doi.org/10.21203/rs.3.rs-90447/v1.

[28] F. Machui, M. Hossel, N. Li, G.D. Spyropoulos, T. Ameri, R.R. Sondergaard, et al., Cost analysis of roll-to-roll fabricated ITO free single and tandem organic solar modules based on data from manufacture, Energy Environ. Sci. 7 (2014) 2792–2802.

[29] M.D. Chatzisideris, A. Laurent, G. Christoforidis, F.C. Krebs, Cost-competitiveness of organic photovoltaics for electricity self-consumption at residential buildings: a comparative study of Denmark and Greece under real market conditions, Appl. Energy 208 (2017) 471–479.

[30] Y.H. Seo, Y. Lee, D.Y. Jeon, J.-I. Han, Enhancing the light utilization efficiency of microalgae using organic dyes, Bioresour. Technol. 181 (2015) 355–359.

[31] M. Raeisossadati, N.R. Moheimani, D. Parlevliet, Luminescent solar concentrator panels for increasing the efficiency of mass microalgal production, Renew. Sustain. Energy Rev. 101 (2019) 47–59.

[32] M. Raeisossadati, N.R. Moheimani, D. Parlevliet, Red luminescent solar concentrators to enhance Scenedesmus sp. biomass productivity, Algal Res. 45 (2020) 101771.

[33] M. Janssen, J. Tramper, L.R. Mur, R.H. Wijffels, Enclosed outdoor photobioreactors: light regime, photosynthetic efficiency, scale-up, and future prospects, Biotechnol. Bioeng. 81 (2003) 193–210.

[34] M. Cuaresma, M. Janssen, C. Vílchez, R.H. Wijffels, Horizontal or vertical photobioreactors? How to improve microalgae photosynthetic efficiency, Bioresour. Technol. 102 (2011) 5129–5137.

References

519

[35] J.F. Cornet, C.G. Dussap, J.B. Gros, C. Binois, C. Lasseur, A simplified monodimensional approach for modeling coupling between radiant light transfer and growth-kinetics in photobioreactors, Chem. Eng. Sci. 50 (9) (1995) 1489–1500.

[36] A. Richmond, C.W. Zhang, Y. Zarmi, Efficient use of strong light for high photosynthetic productivity: interrelationships between the optical path, the optimal population density and cell-growth inhibition, Biomol. Eng. 20 (4–6) (2003) 229–236.

[37] K.A. Canon-Rubio, C.E. Sharp, J. Bergerson, M. Strous, H. De la Hoz Siegler, Use of highly alkaline conditions to improve cost-effectiveness of algal biotechnology, Appl. Microbiol. Biotechnol. 100 (2016) 1611–1622.

[38] K. Livansky, J. Doucha, CO_2 and O_2 gas exchange in outdoor thin layer high density microalgal cultures, J. Appl. Phycol. 8 (1995) 353–358.

[39] A.P. Carvalho, L.A. Meireles, F.X. Malcata, Microalgal reactors: a review of enclosed system designs and performances, Biotechnol. Prog. 22 (2006) 1490–1506.

[40] J. Doucha, K. Livansky, Productivity, CO_2/O_2 exchange and hydraulics in outdoor open high density microalgal (Chlorella sp.) photobioreactors operated in a Middle and Southern European climate, J. Appl. Phycol. 18 (2006) 811–826.

[41] J. Doucha, K. Livansky, High density outdoor microalgal culture, in: R. Bajpai, A. Prokop, M. Zappi (Eds.), Algal Biorefineries, Springer, New York, 2014, pp. 147–173.

[42] A. Ozkan, K. Kinney, L. Katz, H. Berberoglu, Reduction of water and energy requirement of algae cultivation using an algae biofilm photobioreactor, Bioresour. Technol. 114 (2012) 542–548.

[43] P. Schnurr, G. Espie, G. Allen, Algae biofilm growth and the potential to stimulate lipid accumulation through nutrient starvation, Bioresour. Technol. 136 (2013) 337–344.

[44] T. Liu, J. Wang, Q. Hu, P. Cheng, B. Ji, J. Liu, et al., Attached cultivation technology of microalgae for efficient biomass feedstock production, Bioresour. Technol. 127 (2013) 216–222.

[45] B. Moll, B. McCool, W. Drake, W. Purobsky, R. Adams, H. Enke, et al., Biofilm photobioreactor system and method of use, United States Patent US 8,691,538 B1, 2014.

[46] J. Wimpenny, W. Manz, U. Szewzyk, Heterogeneity in biofilms, FEMS Microbiol. Rev. 24 (2000) 661–671.

[47] C. Picard, S. Logette, J. Schrotter, P. Aimar, J. Remigy, Mass transfer in membrane aerated biofilm, Water Res. 46 (2012) 4761–4769.

[48] J.D. Sandt, Light Manipulation with Photonic Fibers and Optical Light Guides: Dynamic Structural Color and Light Distribution in Microalgae Cultures, Massachusetts Institute of Technology, 2020. https://hdl.handle.net/1721.1/127706.

[49] C. Salmean, S. Bonilla, Y. Azimi, J.S. Aitchison, D.G. Allen, Design and testing of an externally-coupled planar waveguide photobioreactor, Algal Res. 44 (2019) 101684.

[50] S.N. Genin, J.S. Aitchison, D.G. Allen, Novel waveguide reactor design for enhancing algal biofilm growth, Algal Res. 12 (2015) 529–538.

[51] S. Sivakaminathana, J. Wolf, J. Yarnold, J. Roles, I.L. Ross, E. Stephens, et al., Light guide systems enhance microalgae production efficiency in outdoor high rate ponds, Algal Res. 47 (2020) 101846.

[52] T.M. Sobczuk, F. Garcia Camacho, F. Camacho Rubio, et al., Carbon dioxide uptake efficiency by outdoor microalgal cultures in tubular airlift photobioreactors, Biotechnol. Bioeng. 67 (2000) 465–475.

[53] M.K. Lam, K.T. Lee, A.R. Mohamed, Current status and challenges on microalgae-based carbon capture, Int. J. Greenh. Gas Control. 10 (2012) 456–469.

[54] S.C. Pierobon, X. Cheng, P.J. Graham, B. Nguyen, E.G., Karakolis, D. Sinton, Emerging microalgae technology: a review, Sustain. Energy Fuels 2 (2018) 13–38.

[55] W.B. Zimmerman, B.N. Hewakandamby, V. Tesar, H.C.H. Bandulasena, O.A. Omotowa, On the design and simulation of an airlift loop bioreactor with microbubble generation by fluidic oscillation, Food Bioprod. Process. 87 (2009) 215–227.

[56] W.B. Zimmerman, M. Zandi, H.C.H. Bandulasena, V. Tesar, D.J. Gilmour, K. Ying, Design of an airlift loop bioreactor and pilot scales studies with fluidic oscillator induced microbubbles for growth of a microalgae Dunaliella salina, Appl. Energy 88 (2011) 3357–3369.

[57] X. Xu, G.J.O. Martin, S.E. Kentish, Enhanced CO_2 bio-utilization with a liquid–liquid membrane contactor in a bench-scale microalgae raceway pond, J. CO_2 Util. 34 (2019) 207–214.

[58] E.J. Lohman, R.D. Gardner, T. Pedersen, B.M. Peyton, K.E. Cooksey, R. Gerlach, Optimized inorganic carbon regime for enhanced growth and lipid accumulation in Chlorella vulgaris, Biotechnol. Biofuels 8 (2015) 82.

[59] M. Ataeian, Y. Liu, K.A. Canon-Rubio, M. Nightingale, M. Strous, A. Vadlamani, Direct capture and conversion of CO_2 from air by growing a cyanobacterial consortium at pH up to 11.2, Biotechnol. Bioeng. 116 (2019) 1604–1611.

[60] P.L. Gupta, S.M. Lee, H.J. Choi, A mini review: photobioreactors for large scale algal cultivation, World J. Microbiol. Biotechnol. 31 (2015) 1409–1417.

[61] L. Brennan, P. Owende, Biofuels from microalgae — a review of technologies for production, processing, and extractions of biofuels and co-products, Renew. Sustain. Energy Rev. 14 (2010) 557–577.

[62] A.E.M. Abdelaziz, G.B. Leite, P.C. Hallenbeck, Addressing the challenges for sustainable production of algal biofuels: II. Harvesting and conversion to biofuels, Environ. Technol. 34 (13-14) (2013) 1807–1836.

[63] W. Mo, L. Soh, J.R. Werber, M. Elimelech, J.B. Zimmerman, Application of membrane dewatering for algal biofuel, Algal Res. 11 (2015) 1–12.

[64] M. Zhang, L. Yao, E. Maleki, B.Q. Liao, H. Lin, Membrane technologies for microalgal cultivation and dewatering: recent progress and challenges, Algal Res. 44 (2019) 101686.

[65] J. Hanotu, H.C. Bandulasena, W.B. Zimmerman, Microflotation performance for algal separation, Biotechnol. Bioeng. 109 (2012) 1663–1673.

[66] M.L. Gerardo, S. Van Den Hende, H. Vervaeren, T. Coward, S.C. Skill, Harvesting of microalgae within a biorefinery approach: a review of the developments and case studies from pilot-plants, Algal Res. 11 (2015) 248–262.

[67] T. Coward, J.G.M. Lee, G.S. Caldwell, Development of a foam flotation system for harvesting microalgae biomass, Algal Res. 2 (2013) 135–144.

[68] M.A. Al-karawi, Development and intensification of a foam flotation system in harvesting microalgae for biofuel, (Ph.D. thesis), Newcastle University, 2018.

[69] S. Ghosh, S. Banerjee, D. Das, Process intensification of biodiesel production from Chlorella sp. MJ 11/11 by single step transesterification, Algal Res. 27 (2017) 12–20.

[70] J. Harris, K. Viner, P. Champagne, P.G. Jessop, Advances in microalgal lipid extraction for biofuel production: a review, Bioref. Biofuels Bioprod. 12 (2018) 1118–1135.

[71] E. Günerken, E. D'Hondt, M.H.M. Eppink, L. Garcia-Gonzalez, K. Elst, R.H. Wijffels, Cell disruption for microalgae biorefineries, Biotechnol. Adv. 33 (2) (2015) 243–260.

[72] Y.S. Lai, P. Paramsewaran, A. Li, M. Baez, B.E. Rittmann, Effects of pulsed electric field treatment on enhancing lipid recovery from the microalga Scenedesmus, Bioresour. Technol. 173 (2014) 457–461.

[73] M. Coustets, N. Al-Karablieh, C. Thomsen, J. Teissié, Flow process for electroextraction of total proteins from microalgae, J. Membr. Biol. 246 (2013) 751–760.

[74] M.D.A. Zbinden, B.S.M. Sturm, R.D. Nord, W.J. Carey, D. Moore, H. Shinogle, et al., Pulsed electric field (PEF) as an intensification pretreatment for greener solvent lipid extraction from microalgae, Biotechnol. Bioeng. 110 (6) (2013) 1605–1615.

[75] G.P. Lam, P.R. Postma, D.A. Fernandes, R.A.H. Timmermans, M.H. Vermue, M.J. Barbosa, et al., Pulsed electric field for protein release of the microalgae *Chlorella vulgaris* and *Neochloris oleoabundans*, Algal Res. 24 (2017) 181–187.

[76] N.D. Eckelberry, M.P. Green, S.A. Fraser, Systems, apparatuses, and methods for extracting non-polar lipids from an aqueous algae slurry and lipids produced therefrom, United States Patent US 2011/0095225 A1, 2011.

[77] K. de Boer, N.R. Moheimani, M.A. Borowitzka, P.A. Bahri, Extraction and conversion pathways for microalgae to biodiesel: a review focused on energy consumption, J. Appl. Phycol. 24 (2012) 1681–1698.

[78] J. Pan, T. Muppaneni, Y. Sun, H.K. Reddy, J. Fu, X. Lu, et al., Microwave-assisted extraction of lipids from microalgae using an ionic liquid solvent [BMIM] [HSO_4], Fuel 178 (2016) 49–55.

[79] L. Hua, L. Guo, M. Thakkar, D. Wei, M. Agbakpe, L. Kuang, et al., Effects of anodic oxidation of a substoichiometric titanium dioxide reactive electrochemical membrane on algal cell destabilization and lipid extraction, Bioresour. Technol. 203 (2016) 112–117.

[80] H. Sati, M. Mitra, S. Mishra, P. Baredar, Microalgal lipid extraction strategies for biodiesel production: a review, Algal Res. 38 (2019) 101413.

[81] F. Adam, M. Abert-Vian, G. Peltier, F. Chemat, "Solvent-free" ultrasound-assisted extraction of lipids from fresh microalgae cells: a green, clean and scalable process, Bioresour. Technol. 114 (2012) 457–465.

References

[82] A.F. Ferreira, A.P.S. Dias, C.M. Silva, M. Costa, Effect of low frequency ultrasound on microalgae solvent extraction: analysis of products, energy consumption and emissions, Algal Res. 14 (2016) 9–16.

[83] R.M. Gonzalez-Balderas, S.B. Velasquez-Orta, M.T. Orta, Ledesma, Biorefinery process intensification by ultrasound and ozone for phosphorus and biocompounds recovery from microalgae, Chem. Eng. Process. (2020) 107951.

[84] C. Song, Q. Liu, N. Ji, S. Deng, J. Zhao, Y. Kitamura, Intensification of microalgae drying and oil extraction process by vapor recompression and heat integration, Bioresour. Technol. 207 (2016) 67–75.

[85] K.Y. Wong, N. Jo-Han, C.T. Chong, S.S. Lam, W.T. Chong, Biodiesel process intensification through catalytic enhancement and emerging reactor designs: a critical review, Renew. Sustain. Energy Rev. 116 (2019) 109399.

[86] B. Kim, H.Y. Heo, J. Son, J. Yang, Y.-K. Chang, J.H. Lee, et al., Simplifying biodiesel production from microalgae via wet in situ transesterification: a review in current research and future prospects, Algal Res. 41 (2019) 101557.

[87] C.-L. Chen, C.-C. Huang, K.-C. Ho, P.-X. Hsiao, M.-S. Wu, J.-S. Chang, Biodiesel production from wet microalgae feedstock using sequential wet extraction/transesterification and direct transesterification processes, Bioresour. Technol. 194 (2015) 179–186.

[88] H. Im, H. Lee, M.S. Park, J.-W. Yang, J.W. Lee, Concurrent extraction and reaction for the production of biodiesel from wet microalgae, Bioresour. Technol. 152 (2014) 534–537.

[89] J. Cheng, T. Yu, T. Li, J. Zhou, K. Cen, Using wet microalgae for direct biodiesel production via microwave irradiation, Bioresour. Technol. 131 (2013) 531–535.

[90] G. Yadav, W.D. Seider, L. Soh, J. Zimmerman, Process intensification of algae oil extraction to biodiesel, Comput. Aided Chem. Eng. 44 (2018) 1699–1704.

[91] G. Yadav, L.A. Fabiano, L. Soh, J. Zimmerman, R. Sen, W.D. Seider, CO_2 process intensification of algae oil extraction to biodiesel, AICHE J. 67 (1) (2021) e16992.

[92] A.A. Peterson, F. Vogel, R.P. Lachance, M. Fröling, M.J. Antal, J.W. Tester, Thermochemical biofuel production in hydrothermal media: a review of sub- and supercritical water technologies, Energy Environ. Sci. 1 (2008) 32.

[93] R. Gautam, R. Vinu, Reaction engineering and kinetics of algae conversion to biofuels and chemicals via pyrolysis and hydrothermal liquefaction, React. Chem. Eng. 5 (8) (2020) 1320–1373.

[94] J. Yang, Hydrothermal liquefaction of biomass and process intensification, (Ph.D. thesis), Dalhousie University, 2019.

[95] O.S. Djandja, Z. Wang, L. Chen, L. Qin, F. Wang, Y. Xu, et al., Progress in hydrothermal liquefaction of algal biomass and hydrothermal upgrading of the subsequent crude bio-oil: a mini review, Energy Fuels 34 (10) (2020) 11723–11751.

[96] J. Lu, J. Wu, L. Zhang, Z. Liu, Y. Wu, M. Yang, Catalytic hydrothermal liquefaction of microalgae over mesoporous silica-based materials with site-separated acids and bases, Fuel 279 (2020) 118529.

Handbook of Algal Biofuels

CHAPTER 23

Life cycle assessment for microalgae-derived biofuels

Elham Mahmoud Ali[1,2]

[1]Department of Aquatic Environment, Suez University (SU), Suez, Egypt [2]Department of Environmental Studies, The National Authority for Remote Sensing and Space Sciences (NARSS), Cairo, Egypt

23.1 Introduction

As a universal plan, sustainability is the main subject of all developmental plans at the national and global levels which require awareness of many issues including, at the top, the ecoefficiency processes and products/services. It is also important to consider relying on nature-based energy from various materials. Therefore there is an increasing dependence on biofuels as promising alternative fuels currently and in the future. This approach will help to meet the increased developmental activities and reduce the excessive reliance on fossil fuels, thus eliminating greenhouse gases' emissions. The global energy crisis in the early 1970s created increased demand for other energy alternatives that were cost-effective and environmentally friendly. The price of crude petroleum reached its highest price of US$ 145 and 108/barrel in 2008 and 2011 [1]. These figures were the highest that crude petroleum prices had ever achieved in 30 years. The world economic crisis made some governments start to think of biofuels development and their use in petroleum fuels without needing to modify engines [2]. Biofuel production is the worldwide famous initiative that being considered to reduce the dependency on fossil fuel. In addition, the subject of sustainability became a priority issue that comes at the top of all developmental plans at both the national and global levels.

Sustainable development requires a good awareness of many environmental issues including ecoefficiency processes, the generated products/services, and the better direction to increase the reliance on natural-based energy using various materials. This global intention for sustainable development to achieve the universal sustainable development goals (SDGs) has led to the consideration of biofuel production as a promising alternative fuel currently and in the future in order to reduce the excessive reliance on fossil fuels, and hence eliminate greenhouse gas emissions. In July 2007, the Canadian government called for "a 9 investment" which devoted about CAN\$ 1.5 billion for the production biofuels in Canada (ecoENERGY for Biofuels Initiative) [3]. Recently, algal-based biofuel (ABB) (third generation) has attracted the most attention, this may be attributed to the global depletion of fossil fuel as well as the consequent climatic changes. In addition, moving to bioenergy could contribute to the millennium development goals and the SDGs through reducing poverty (by increasing energy accessibility and security) and improving life standards (by securing modern services, e.g., electricity and/or liquid oils).

Algae are fast-growing and widely distributed organisms and hence they appeared to be the most suitable alternatives for biofuel production. It is known that algae, particularly microalgae, have been naturally existing for billions of years, growing, reproducing, and storing energy within their cells as oil. Therefore it has become more crucial to encourage the researchers to study "algae" and strengthen our understanding of their numerous and unlimited benefits. Algae range from microscopic cyanobacteria ($>2\,\mu m$) to the giant kelp ($>100\,m$) and are known to reproduce and multiply quickly [4] However, although using algae as a feedstock for the production of energy has been widely considered since the mid-20th century, it has not deployed at the commercial scale until recently. Algal cultivation at regional levels for intensive production and the processing into biofuels will positively influence the populations and will create economic benefits to rural communities. As a matter of fact, there are some algal basics that led the researchers and entrepreneurs to think of biofuel production from algae, such as (1) their high productivity and biomass yields; (2) their ability to grow in wastewater, recycled water, and sea water, thus securing freshwater sources, especially in countries with limited freshwater supplies; (3) the minimal competition with arable/agriculture land as well as nutrients; and (4) their ability to recycle CO_2 from stationary gas emissions. Algae can efficiently grow in any water source, within a wide variety of climate conditions and through various methods of production, such as natural (i.e., open ponds) or artificial (i.e., photobioreactors or fermenters) habitats. Such a great ability for cultivation works as a "Job Creation Engine," opening venues for a wide variety of jobs, including researchers and engineers who contribute to the cultivation processes. As a real estimate, in the United States about 220,000 jobs were created by 2020 in this sector through "The Algal Biomass Organization" projects (https://algaebiomass.org/).

The aim of this chapter is to discuss the life cycle assessment (LCA) approach that can aid in monitoring, evaluating, and providing solutions for the challenges posed by the industrialization of algal biofuels. The present chapter focuses on evaluating the entire algal biofuel process through LCA as a systematic mechanism, compares it with other feedstocks, and provides solutions for the major challenges.

23.2 Pros and cons of algal biofuel production

A wave of increasing interest in algal cultivation has increasingly been considered due to their potential abilities to produce biofuels (e.g., ABB) which could be used for transportation, substituting the expensive and unrenewable fossil fuel. Microalgae are able to produce several types of biofuels, including methane [5], biodiesel [6,7], and biohydrogen

[7,8]. Although, relying on microalgae as a source of fuel is not a new idea [9–11], its production as well as its applications are more developed nowadays; this is mainly attributed to the petroleum price increase. The global warming phenomena and the climatic changes associated with combustion of fossil fuels are also significant reasons for the worldwide consideration of algal/microalgal biofuel [12]. Algal fuel derived from the algal energy-rich oils is an alternative to the liquid fossil fuels as well as to other common sources of biofuel, for example, sugarcane and corn. Algal-based fuels could help to reduce the national dependence on fossil fuel in some countries due to its market potential in enabling the reduction in importing fossil fuels from other countries.

Currently, several companies and government agencies are considering the conversion to bio-oils and making algae fuel production commercially viable to reduce capital and operating costs. The high productivity of algae and the year-round production, the direct utilization of combustion gas, and the potential for wastewater treatment are among the aspects that may increase the future dependence on microalgae for biofuel production. It is anticipated that a higher fraction of algal biomass could be converted to oil when compared with food crops (e.g., soybeans) with an estimate of nearly 60% compared to 2%–3%, respectively. Several studies showed that algae (some species) would generate oil yield of about 60×10^3 or 136×10^3 L for each cultivated hectare per year of wet and dry weights, respectively, giving more than 60% of their dry/wet weight [13–15]. The per unit area yield of oil from algae is estimated to be from 58,700 to 136,900 L/ha per year. These estimates mainly depend on the lipid content of the algal cell, which is 10–23 times higher than the oil palms that only produce $<6 \times 10^3$ L/ha per year [16]. In addition, algal cells can easily grow in aqueous solutions when cultivated either in open ponds or in photobioreactors with efficient access to water, CO_2, and nutrients producing biomass with a large yield. The algal oil produced could be then turned into sellable biofuel and could be utilized for transportation. In addition, algae do not need herbicides or pesticides during cultivation and hence could potentially avoid land-agriculture problems [17].

Microalgae are known for their metabolic flexibility which allows the easy modification of the biochemical pathways (e.g., synthesis of proteins, carbohydrates, or oil) and their cellular composition [18] This means that biochemical and oil contents of the cultivated algae can be changed, modified, and improved by changing the cultivation conditions. There was a report published by FAO in 2009 with all the different methodologies for algal cultivation and all kind of bio-oils and other coproducts that can be produced from algae. The report also gave detailed information on the environmental benefits of transferring to biofuels as well as the expected threats that might emerge with the production process. Microalgae are very diverse and can live within any environments where light and water are present. Hence, microalgae can grow and flourish in oceans, lakes, rivers, ice, humid soils, etc. [19]. Microalgae show a high level of biodiversity with up to several millions of species, and many thousands of species are continuously being discovered.

Algae can reproduce very rapidly and they are almost faster than any other plants. They also can be harvested daily, as the population may reach two or three times its original number within 1 day or a few hours [20]. This means that a good volume of biomass could be continuously produced with a rate greater than any other productive crops, thus increasing the potential of algae for biofuel production. In addition microalgae can potentially be used also in food, cosmetics, fertilizers, and many other industries [21]. One of the favored characteristics of algae is that they can grow within various climatic conditions and with many different cultivation and production methods; they can grow in ponds, photobioreactors, and/or fermenters. Algal cultivation can be achieved under different culturing conditions, such as photoautotrophic, mixotrophic, or heterotrophic. If they grow photoautotrophically, they utilize light to reproduce and produce new cells/biomass. However in the heterotrophic cultivation method algae only require a carbon source for energy (e.g., sugars), instead of light [22]. In mixotrophic cultivation, algae are able to use both or any of the energy sources (light or a carbon) to grow and provide biomass yield. When relying on any carbon source, algae achieve high biomass yield and this is cost-effective as it will reduce the cost of the overall process [23]. The heterotrophic and mixotrophic cultivation strategies are therefore more advantageous compared to photoautotrophic methods. It is worth mentioning that despite being diversified communities with numerous numbers of species and strains, not all species of algae are favored or suitable for biofuel production [24].

Several experimental trials show that cultivation of energy-producing crops is a promising plan for sustainable development not only due to their high consumption of conventional fuels, fertilizers, and pesticides, but also due to their less hazardous impacts on the environment and less competition for arable land when compared with the cultivation of food crops. Despite all these advantages of using microalgae for biofuel production, they are still not used worldwide at a commercial level or for industrial application; this is mainly attributed to the production cost being still too high. We need to think of options, such as provide safeguards, that help to mitigate risks and create positive effects and cobenefits [25].

Several studies have been conducted to effectively overcome most of the obstacles and challenges facing the biofuel production process and make it most efficient and ecofriendly. The LCA approach has been considered as a helpful mechanism to develop the current and future intentions toward bioenergy and can be a good tool for global economic development, particularly in low and medium developmental countries, as it enables farmers to have more income by stabilizing their production prices and can also help to secure the energy required for local communities [25]. The following section will discuss the LCA approach with regards to a biofuel production system mainly from microalgal cultures. The discussion would be focused more on bioethanol and biodiesel as the most commonly produced biofuels.

23.3 Life cycle assessment approach

LCA is a scientific and systematic approach of data analysis that assesses any potential impacts of any marketable product on the environment. It is a process or a service within the entire life cycle of use and throughout all stages. The process of LCA is mainly conducted to evaluate impacts on the environment of any product for a period of product's life, from the preparation steps to postuse stages in order to assess the whole cycle of production, increase the use efficiency, and decrease any problems/limitations. LCA is commonly referred to as a "cradle-to-grave" analysis. This together with its data-driven methodology makes the LCA approach different from other models. In our case, the LCA is used to assess the oil/lipid produced biologically from microalgae and its efficiency for replacing or substituting the fossil fuel. Biofuel production is increasingly occurring at a global level and it becomes crucial to study the environmental impact of different biofuels produced from microalgae with the integration of different biofuels and value-added products or bioremediation. Whatever the methodology applied for biofuel production, the preparation steps and/or the various feedstocks used must be assessed.

Therefore LCA helps to evaluate either the product (biofuel) or its function as it designed to perform. According to the US Environmental Protection Agency the LCA's key elements should include the definition and quantification of the raw materials used, energy source (type/quantity), emissions evolved, and wastes generated. It is also anticipated to assess the potential environmental load of these element loads (inputs/outputs) to enable providing measures or options to reduce the impacts on the environment. In contrast with the "circular economy" which is creating value for the socioeconomy as well as business through reducing, reusing, and recycling, LCA is a scientific robust tool

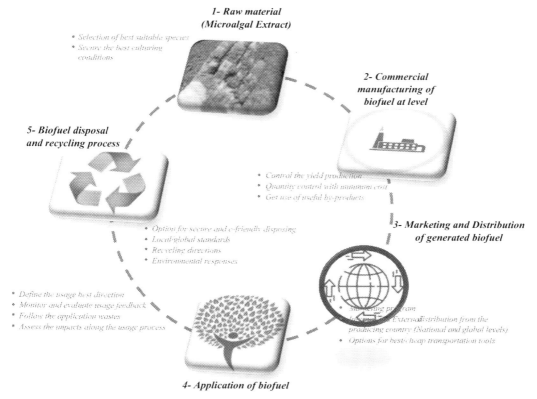

FIGURE 23.1 Diagrammatic representation of life cycle assessment, showing how it evaluates the environmental impacts of biofuel produced from microalgae from the very first to the very last. Source: *Modified after L. Golsteijn, Life cycle assessment (LCA) explained, 2020. https://pre-sustainability.com/articles/life-cycle-assessment-lca-basics/, (accessed 18.03.21).*

528   23. Life cycle assessment for microalgae-derived biofuels

working with an accountancy approach. Combining the robustness of the LCA methodology and the inspirational principles of the circular economy we can then have a more holistic and innovative approach (Fig. 23.1) [26].

LCA is currently widely used for analyzing all pathways of the biofuel production process from the very the first stage to the last; this is called in the literature from "cradle to grave." This analysis helps to improve the whole process and give directional tips for this technology in the future generation. Several studies have been devoted to such an aim (e.g., [27–30]); this allows assessment of the effects of the nitrogen supply on the cultivation of algae, the oil extraction process, as well as the biofuel energy ratio. For instance, the life cycle fossil energy ratio of biofuel generated from *Chlorella vulgaris* could be successfully improved if algae are grown under nitrogen-limited conditions. In another case, for biofuel produced from *Phaeodactylum tricornutum*, this ratio is considerably decreased when grown in a nitrogen-deficient environment [29].

23.3.1 Phases of the life cycle assessment

23.3.1.1 *Goal determination and scope definition*

In this stage, the product or service is defined, as well as the goal that determines the scope of the assessment. This includes the main objective, application of the algal biofuel, and audience.

23.3.1.2 *Inventory analysis*

Where a data compilation is performed and an inventory analysis of raw materials extractions (type of alga, species and strain—if applicable) and substances used for extraction process. Also, it defines all knowledge and data used in the LCA cycle of microalgae biofuel and all the integrated information generated out of it

23.3.1.3 *Impact assessment*

Based on the potential impacts, this step mainly classifies the various resources used and the emissions generated. Also it quantifies the relevance and importance of the predefined goal of the LCA study.

23.3.1.4 *Interpretation*

At this phase we use all information gathered to discuss the results of the LCA and determine the positive environmental impacts of the produced biofuel (e.g., contributions and relevance) and its negative effects (e.g., quality data and method limitations) on the environment to identify the potential improvement.

23.3.1.5 *Reporting*

The LCA study could be documented as a descriptive, informative, and transparent report which should be prepared according to the ISO 14044 requirements. These phases have been established as a best-practice approach over years to ensure quality LCA analyses and outcomes to any product/service from a multitude of different sectors. The whole

Handbook of Algal Biofuels

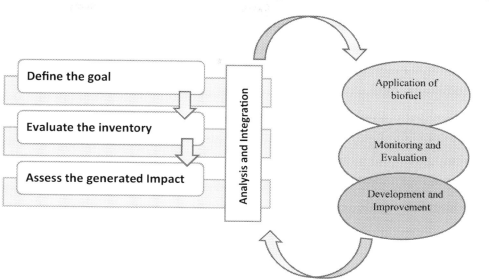

FIGURE 23.2 Stages of a life cycle assessment according to EN ISO 14040. *Source: Modified after I.V. Muralikrishna, V. Manickam, Life cycle assessment, Life Cycle Assess. (2017) 1–159, doi: 10.4324/9781315778730.*

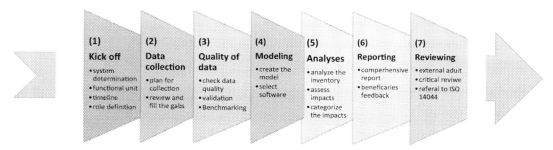

FIGURE 23.3 Diagrammatic representation of the standard steps for the life cycle assessment according to the ISO 14044.

LCA process framework and all steps could be summarized in the flow diagrams in Figs. 23.2 and 23.3 [31].

Normally, collected through data collection templates which might be made through automated ways, the data include sources of primary data (such as bills of materials/recipes, PLM software, utilization different bills, utility electric readings, any records of procurements, waste records, as well as specifications of any equipment), secondary information (e.g., database information, technical literature, journal publications, presentations in conference/workshop, and patents data), and other complementary data. Most importantly to ensure high-quality data collection, the collected data should be checked for completeness and consistency. The following table categorizes the various impacts that are potentially determined on application of the LCA approach

530   23. Life cycle assessment for microalgae-derived biofuels

TABLE 23.1 Categories, description, and evaluation methodologies of potential impacts generated with the life cycle assessment and LCIA processes.

Category	Meaning	Description
Climate change	Global warmingCarbon footprint	Determine greenhouse gas emissions (GHGs), e.g., carbon dioxide and methane, increased absorption of solar radiation on earth and the GHGs' effects on ecosystem
Eutrophication	Over fertilizationNutrients increase	Nutrient enrichment, nitrogen (N) and phosphorus (P) affect the species community structure and productivityIncreased production may lead to the formation of toxic algal bloomsHigh biomass can lead to oxygen depression in aquatic ecosystems, harming fish populations
Acidification	Acid rainLow levels of pH	Determination of acidification is a sign of increase in increased hydrogen ion (H^+) concentrationMeasures the emissions that acidify the water environment; i.e., decreasing pH (e.g., acid rain)Fish suffocation and mortality and forest destroy are among the expected impacts
Smog formation	Photochemical ozone creation	Determination of emissions contributing to the ground level of smog formation (e.g., ozone O_3)Increased levels of ground ozone can be dangerous to human, crops, and ecosystems
Particulate matter	Aerosols and dust	Measuring emissions of particulate matter (PM) with various sizes and secondary particulates (i.e., SO_2 and NO_x)Particulate matter can causes determined negative impacts to human health (e.g., respiratory illness) and increase mortality rates
Ozone depletion	Ozone hole	Measuring any air emissions contributed to ozone depletion, increasing ultraviolet rays (UVB) arriving to the Earth's surfaceUVB affects all living organisms (i.e., humans and plants)

23.3.2 Life cycle assessment approach for algal-derived biofuel

Responding to all the challenges that effectively hinder the biofuel production pathway/system and make it environmentally and economically advantageous, application of the systematic LCA approach would be a helpful mechanism to develop the current and future intension toward bioenergy conversion. Consequently, several studies used LCA for a better system/process (e.g., [32–35]. LCA can give better system enhancing criteria through several evaluation methodologies that cover many aspects (see Table 23.1). Biofuels can reduce the use of fossil fuel and emissions of greenhouse gases (GHG) compared to fossil reference [36]. GHG emissions is a main regulatory factor but it is not the only important one, water consumption is another major aspect that may be of concern [37]. McKone et al. [37] has explained some challenges that enable LCA to effectively evaluate the environmental implications and biofuel producing alternatives. There are over 200 types of feedstocks used for biofuel production to gradually replace fossil fuel, which requires a critical assessment for actual and maximum benefit [38].

23.3.2.1 Assessment of biofuel raw materials

Similar to fossil fuel, algal fuel releases carbon dioxide (CO_2) when burnt, but unlike fossil fuel it releases CO_2 that the algae have removed from the atmosphere during their

Handbook of Algal Biofuels

growth and photosynthesis process. Mostly, all biofuels used in industry are mainly oil (triglycerides) that is derived from various feedstocks or raw materials, such as sunflower or/and soybean. These triglycerides need to be transesterifed into fatty acid alkyl esters, thus changing their physicochemical properties to resemble the petroleum-based diesel and be suitable for conventional engines without any modifications [39]. The raw materials used for oil production of the second-generation biofuel are generally crops that are useful for other uses, such as food for humans and feed for animals, which creates increased demand for those materials and hence their price would increase considerably. In addition, there are other costs, such as water, fertilizers, and pesticides needed to culture conventional vegetable material and certainly have negative impacts on the environment [40].

Using oleaginous plants for fuel production can reduce pollutant emissions of GHGs by trapping the atmospheric carbon dioxide (CO_2) and use it for photosynthetic processes [41]. This means that using biofuel would reduce the net emissions of pollutants. In this regards, it is anticipated that adding 20% of biodiesel derived from soybean to petrodiesel (v/v) will typically eliminate emissions of carbon monoxide (CO) and dioxide (CO_2) [42,43]. In the third-generation biofuel, it is potentially considered that microalgae are able to absorb water excessive pollutants (N and P) [44], and also accumulate high amounts of fatty acids. Moreover, algae can provide a culture yield which might be 10 times higher than other oil-rich crops [16].

23.3.2.2 Synthesis of oil/lipid yield of microalgae

As microalgae possess a less complex cellular structure, they are much more efficient in converting solar energy into chemical energy. Compared to terrestrial plants, some selected microalgal species/strains might produce 30 times more oil per unit. Based on the optimum growth conditions, oil content produced varies from 20%−50% of the dried algal biomass if not more [45]. For instance, the lipid content of some microalgae increases when they are cultured in poor nutrient conditions, specifically nitrogen and silicon, although such conditions would block the cell division process in other species (e.g., *Euglena* and *Nannochloropsis*).

There are certain fungal (e.g., *Saccharomyces cerevisiae*) and bacterial (e.g., *Escherichia coli*) species that can be converted through genetic engineering into oily species or so-called oleaginous microbes, which are able to accumulate oil with >20% oil/dry weight. Although there are numerous species of microalgae, only a few have been explored with respect to their potential for high biomass yield and lipid content. Generally speaking, animal oils as well as plants' fatty acids are considerably different in composition to oil derived from microalgae that contain hydrocarbons, sterols, sterol esters, and wax as well as the free alcohols that cannot be saponified. The biodiesel content of fatty acid methyl esters (FAMEs) should have a value of not less than 96.5% in accordance with the EN 14103 recommendation [46].

In reality, a few microalgal species are only successful for the production of biofuels that fulfill the criteria of ester content. This is mainly attributed to the presence of some unsaponifiable constituents and limits the algal suitability for biofuel production. These limitations could be solved by further upgrading the derived bio-oil via some chemical and/or physical means. The pyrolysis process is the common method to convert algal biomass to biofuel. The direct combustion followed by the thermochemical liquefaction

532

methods are the best alternative methods for biofuel production from algae. The transesterification methodology is also a considered alternative, through which algal lipids can be converted to biofuel [47].

Algae typically store energy as oils and carbohydrates. They are estimated to produce biofuels at about two to five thousand gallons/acre/year [48]. Lipids are among the algal primary metabolites with 15%–60% of the cell dry weight and they can be classified into two types: (1) the polar (structural) type, which include phospholipids and glycolipids and (2) the nonpolar (neutral or storage) lipids, which exist in the form of triacylglyceride (TAG), storing energy, which are then transesterified producing biodiesel. The polar lipids are, however, high in polyunsaturated fatty acid, which is potentially used as food/feed. Lipids content and lipid composition in microalgae vary from one species to another, some of the species are known to have higher levels of the neutral lipid than other species [49], however, under nutrient-limited conditions, the metabolic pathway could be changed and hence the alga can store neutral lipids as TAG.

23.3.2.3 Assessment of species used for biofuel production

However, most of the microalgal species and/or strains are potentially viable for lipid production with almost 50% of the cellular dry weight according to the species growth needs and culturing conditions [16]. Therefore it is crucial to peruse a preliminary lipid analysis for the selected algal species/strain to know its lipid yield/lipid composition before selecting it for biofuel production. In some species, for example, *Botryococcus braunii* or *Schizochytrium* sp., about 80% of the cell's dry weight is lipids [19], and they can give yields more than 700 times/acre higher than other oily crops, such as sunflower and colza.

Table 23.2 shows the variations in lipid production levels from some species, based on their biomass production and the percentage of the dry weight. This knowledge allows consideration of the commercial development of high-yield biofuel [16].

23.3.2.4 Assessment of cultivation systems and reactor types

A wide range of cultivation types, including open and closed reactor systems, are available for microalgal cultivation. Selection and suitability/efficacy of the cultivation system as well as the reactor type are typically based on the species physiology, growth requirements, and optimal productivity that secure a high yield at minimum cost. To achieve an optimum yield, it is important to provide some important factors, such as (1) sufficient light, good mixing for better mass transfer, optimum temperature and oxygen, and minimum water and energy inputs [50]. Generally, growing the algae on any carbon source achieves high biomass and reduces the cost of infrastructure needed for algal growth [23] (Table 23.3).

The debate on the open versus closed system is continually undertaken in most microalgal literature (e.g., [13,51,52,53]). Cultivation of algae in open ponds is the favored system for the large-scale production and outdoor cultures, however, it is limited to specific maintainable species. One of the rarely considered issues in selecting the cultivation production system is the energy balance. It is economically important, in production, to generate

Handbook of Algal Biofuels

23.3 Life cycle assessment approach

TABLE 23.2 Assessment of the most suitable algal species biofuel production. Table includes some examples in relation to their dry weight percentage (%dw) and oil productivity.

Examples of algal species	Biomass production (g/L per day)	% Dry weight (dw)	Lipid production (mg/L per day)
Botryococcus spp.	0.02−0.25	29%−75% dw	55−112
Chlorella spp.*Chlorella protothecoidesChlorella vulgaris*	0.227−0.73	28%−29% dw15%−55% dw28.07% dw	1214.061−204.9
Ankistrodesmus spp.	0.24−034	16%−40% dw	49−56
Dunaliella tertiolecta	0.098−0.12	16.7−71 dw	698
Hantzschia		66% dw	
Nannochloris		6%−63% dw	
Nannochloropsis sp.	0.17−1.43	12%−68%dw	37.6−90
Nitzschia	ND	28%−50% dw	ND
Stichococcus:	ND	9%−59% dw	ND
Tetraselmis sp.	ND	15%−32% dw	53.1
Thalassiosira pseudonana	ND	21%−31% dw	ND
Phaeodactylum tricornutum:	ND	31% dw	ND
Scenedesmus spp.	0.07−0.16	16.7%−45% dw	0.095−26.0
Tribonema	0.17	50.23	
Cyclotella		42%	
Chlamydomonas	0.24−2.27	15.7	36.2−223

greater energy than that used in manufacture. In this regards, open culture systems have a favorable balance of energy than closed ones [54] (Table 23.4).

23.3.2.5 Assessment of harvesting and extraction processes and lipid quantification

Through the biofuel production from microalgae, there are various steps involved from the cell cultivation stage to the final production stage of the biofuel with some intermediate steps, including harvesting, dewatering, concentration, and oil extraction from microalgae. The harvesting process can be either conducted where the dewatering step is directly performed on the algae culture, or by involving algal agglomeration into macroscopic masses to facilitate the dewatering process. In the first technique while energy requirements are acceptable, there are some processes, such as flocculation, flotation, and gravity sedimentation, that are of high cost for motors and controls.

As the neutral lipids in microalgae exist as TAGs, they can be converted to FAMEs via transesterification. The accumulated lipids could be recovered when the cells rupture, thus freeing the lipids. This can be undertaken by various disruption techniques, such as autoclaving, microwave, sonication, osmotic shock, and bead beating [55]. The

Handbook of Algal Biofuels

534　23. Life cycle assessment for microalgae-derived biofuels

TABLE 23.3　Comparison features of various microalgal cultivation approaches[a].

Cultivation method	Photoautotrophic cultivation		Heterotrophic cultivation
Approach	Opened ponds	Closed photobioreactors	Closed industrial bioreactors
Advantages	Maintain temperature easily through the evaporation action cooling	Minimum loses of water1	Can use wastewater, which is rich in carbon and lignocellulosic sugars, which is inexpensive.
	Capital cost is lower than closed systems	Stable and long-term maintenance	Optimal production conditions and contamination prevention are easy to maintain
	More accessible	Support high cell density due to the high ratio of surface: volume	High biomass production
Disadvantages	Subjected to small-medium scale weather variability	Capital intensiveness	Availability and cost of the energy source (e.g., lignocellulosic sugars)
	Difficult to maintain	Less accessible	Existence of competition for sugar sources with other methods
	Require optimum light exposure	Require optimum light exposure	It is a not just a primary source for biomass
		Needs temperature maintenance	

[a]From [13,16,17,51,56,57].

microwave method is an efficient method that extracts lipids from microalgae and solvent extraction is widely used as it is cheap and simple [58]. To determine the lipid content and composition within a laboratory scale, the Bligh and Dyer method is commonly used [59,60].

Generally speaking, methods used for algal-extracted oil are typically similar to those used for seeds-derived oils. For economic viability, it is important to assess the technical practicability of the process as well as the technologies used. The dewatering stage must be applied to the microalgal culture before the oil extraction process. The selection of extraction technology for oil-derived from microalgae mainly depends on production quantity, product quality, moisture content, efficiency of extraction step, safety, and cost-effectiveness. There are also some limitations due to the presence of technical barriers, including increased cost, problems in scaling-up and optimizing the extraction efficiency [61], along with mechanical pressures and methods used for solvents extraction.

Oils derived from microalgae must be purified via separation from any other byproducts generated using hot water (50°C) [61] or organic solvents (e.g., hexane) [62,63] or by liquid–liquid separation using water–organic solvents [60,64,65]. As per the literature available, three main purification methods could be applied for purification on biofuel of the first generation [66]: (1) water washing, (2) dry washing, and (3) membrane extraction.

Handbook of Algal Biofuels

23.3 Life cycle assessment approach

TABLE 23.4 Comparative analyses for the culturing systems of production and the main characteristics of the different reactor types*.

Culture system	Description	Types	Characteristics
Open culturing system	• It is the widely applied reactor system in industry • Impossible to control contamination • Obtain low cell densities • Costs might be high due to the need to large volumes when harvesting, which increase the cost • Limitations in species that are able to tolerate extreme conditions • Difficult to maintain optimal growth conditions due to susceptibility to temperature and irradiance changes [13,51,53]	Natural Pond [56] Raceway [68]	• The harvesting is carried out in situ • Microalgae experience a minimum cost in cultivation • It is difficult to assure the product quality • Simple, round, concrete ponds or dams with about 50 m in diameter • Mixed by a rotating circular arm or by manual stirring. • The most commonly used design for commercial microalgal production • It is a recirculation channel with 15−20 cm deep built from concrete or packed earth and covered with a plastic lining • It is less expensive with low biomass productivity compared to photobioreactors • Mixed by a rotating paddlewheel with baffles placed to facilitate mixing. • CO_2 is provided by gas exchange with the surrounding air and used with less efficiently compared with photo-bioreactors
Closed cultural system	• Much expensive to build and run • Easy to control contamination and environmental parameters • Provide higher biomass concentrations • Hard to control temperature • Simply to be cleaned and sanitized • Heat is not efficient to sterilize reactors and might require chemical sterilizers • Most common designs are tubular and flat-plate PBRs	Vertical packed bed reactor (PBR) [13,52,53] Tubular reactor [66−68]	• Its size is limited by the ratio between the area and the volume • Characterized by good mixing nature but this require large mixing energy • Require less land area • Low in cost and easy to maintain and sterilize • Got high area-to-volume ratios with poor mass transfer, leading to increased O_2 and depleted CO_2 • Construction material is cheap and transparency allowing good lightening • Cells inside the tube walls decrease the light penetration and led to photoinhibition, temperature control, and fouling • There are horizontal vertical and helical tubular reactors • A conical framework is a current suggested design that improves the spatial distribution of tubes for sunlight capture

*From [13,51−53,56,66−72].

23.3.2.6 Assessment of algal residuals processing (coproducts/byproducts)

Using microalga for biofuel production is a cost-effective process, but it could be of economic added value if the process obtains other useful products together with the biofuels. Protein and oily materials are examples of these coproducts, which can be used as animal feeds. Micronutrients are another product that is used as human dietary supplements. Other examples include plastics, chemical feedstocks, lubricants, fertilizers, and even cosmetics.

The main known byproducts off biofuel production are unreacted lipids, water, alcohol, chlorophyll, metals, and glycerol [69], of which, glycerol is the most interesting. The glycerol produced can be transformed into other market valuable products such as alcohols, including propanol, isopropanol, allyl alcohol, and/or acids, including propionic acid, acrylic acid, or propylene glycol and acrolein through chemical or biological conversion (oxidation or reduction) [70,71].

Generated by-products or co-products could be digested anaerobically and produce biodiesel/biogas using hydrothermal processing that is cost-effective [72], this occur if lipid content is >40% (g lipid/g dw) [73]. When the oil extraction process completed, wastes with the remaining algal biomass could be dried and converted to "pellets" that can be used as fuel to be burnt in industrial boilers or in other power generating sources.

23.4 Benefits on application of life cycle assessment for microalgal biofuel commercial production

The LCA approach is a global analysis system that assesses the various pathways of the biofuel production process from "cradle to grave." This systematic approach monitors the global increased tendency to go with biofuel and evaluates each stage of its production—application—recycling cycle and decides the best pathway and the enhancements needed for future use. As the LCA is mainly to assess the entire process of biofuel production, a variety of technology options were used to analyze and evaluate each step of the production process. For better results of uncertainty an extensive sensitivity analysis is also performed. LCA can help assess the effects of nitrogen supply during the cultivation stage and the technology used in the oil extraction stage as well as the impacts on the life cycle fossil energy ratio of the derived biofuel, and hence provide the best culturing conditions among selected species. For example, the preference of certain conditions, for example, nitrogen-limitations for some species and not for others [29].

In a previous work conducted to present mass balances, energy, and GHGs of biofuel produced from microalgae, the mass balance shows that total nitrogen and phosphorus needed for growth could be clearly reduced by about 66% and 90%, respectively, through recycling and/or recovering processes of growth media and residual nutrients [74]. Similarly, freshwater as well as carbon requirements can also be reduced by 89% and 40%, respectively, by recirculating processes. However, it is worth mentioning that even with standard growth conditions, the performance of biofuel could not achieve the requested reductions in GHG emissions in their life cycle as anticipated from the second-generation biofuels. Moreover, when compared with the oil extracted from dried algae, extraction of lipid from wet algae achieves more than 40% improvement in fossil energy ratio of algal biofuel using subcritical cosolvents [29]. The main important outcome of the sensitivity analysis is the indication of factors and conditions influencing the LCA, including, in order: algal oil conversion rate, algae energy content, residues utilization ratio, energy demand for drying process, capacity of water mixing, and final productivity

Considering biofuels as one good alternative liquid fuel supply, many researches investigate the use of microalgae as a valuable feedstock for fuel production as a kind of national economic security. Recently, a number of studies, mostly considered making several assessments of ABB production basically to evaluate the environmental impacts, the life cycle of the GHG emissions, the energy balance, used materials, and the required infrastructures as well as the available options of process technology [68,75–77]. In addition, the LCA analyses might include the cultivation stage of the used algae within the bioreactors, comparing the energy inputs with other conventional fuels, such as gasoline and diesel, based on resources consumption (e.g., land, water, and feedstock). This work studies the economic visibility of producing biofuel from microalgae grown in open ponds and analyzes the potential environmental impacts of the production process [68,75–77]. Comparisons of GHG (greenhouse gas) emissions and costs with algal GHG emissions of $-27.6-18.2$/unit compared to 35.9 and 81.2 with canola and ULS diesel, respectively. With regards to cost, they were not as favorable as they ranged from 2.2\$ to 4.8\$ with algae, compared to 4.2\$ and 3.8\$ with canola and ULS diesel, respectively. This flags the crucial need to increase algal production rate so it can make algal biofuel economically viable and attractive. Uptake of CO_2 in agriculture biomass performs better for global warming compared to conventional diesel and biofuel produced from rapeseed and soybean. Performing the sensitivity analysis showed that cultivation of microalgae is able to fix about 25% of net GHGs emissions. Therefore on increasing the biomass yield the net energy ratio and the global warming potential would be improved. In a similar way, the LCA approach was applied to investigate the use of Jatropha (second generation) and algae (third generation) for biofuel production at a large scale in the southern regions of India and to evaluate the advantages of both source pathways in comparison to the fossil fuel diesel. The results indicated that both biosources presented vital GHG emissions with obvious reduction in energy depletion at a range of 10%–25% (for Jatropha) and 36%–40% (for algae) and the environmental performance of the biofuel produced from both species is almost comparable and clearly better when compared with fossil diesel within the geographic context of the investigation [68,75–77].

A study applied the technique of LCA in accordance to the ISO 14040 recommendations [83]. It evaluated the biofuel production from *Scenedesmus armatus* (a freshwater microalgae) and the consequent environmental implications (i.e., global warming potential) including the main four main steps of the process: cultivation, harvesting, oil extraction, and transesterification. Another extensive study [84] aimed at assessing the biofuel production process with the use of various categories of feedstocks (e.g., Jatropha, castor, cooking oil, etc.). The study is also evaluating the potential environmental problem-shift, the use of energy as well as the economic and environmental performances. Results gave positive indications with 2.3%–52.0% for net energy yields, and 74.1%–448.4% for net economic benefits, of the typical life cycle energy demands and of economic costs.

538

In general, algae are much favored in long-term production as they independently perform with a minimum demand of arable lands. However, more attention should be paid to the environmental problems associated with selecting feedstock, including competition for freshwater, opportunity for environmental toxicity, potential of photochemical oxidation, acidification problems, and aquatic eutrophication. Moreover, the sensitivity analysis helped to identify the main key processes and hence guide the plans for this technology's future improvements. Regardless, the substantial literature recommends the application of LCA approach, but it is still debatable to say that the second and third generations of biofuels are more sustainable than the first one. With some limitations to the variability and uncertainty LCA results, there is a global tendency for using it to rely on the interpretation that was previously published comparing the same impacts for fossil fuels.

23.5 Current scenario on production and application of biofuels

Biofuel is produced worldwide in several countries using various feedstocks for synthesis, ranging from edible to nonedible oils and fats. Oils derived from some microalgal species are potentially promising energy sources if provided with the efficient optimum conditions. As an example, Malaysia is the world leader in terms of the level of potential biodiesel production, followed by Indonesia, Argentina, the United States, Brazil, The Netherlands, Germany, Philippines, Belgium, and Spain. In these countries, the diverted feedstock is mostly edible oil; including soybean oil (28%), palm oil (22%), coconut oil (11%), and sunflower, rapeseed, and olive oils (5%), compared to animal fats (20%) [85]. For example, a global total biodiesel amount of about 16,000 ktons has been produced in 2009 [86], and Germany was the leading country in biofuel, mainly biodiesel production, followed by France, the United States, Brazil, Argentina, Spain, Italy, Thailand, Belgium,

TABLE 23.5 Leading countries based on biofuel production in petajoules (Statistics of 2019 [87]).

Country	Biofuel production
United States	1557.1
Brazil	992.2
Indonesia	275.5
Germany	143.4
France	113
China	111.3
Argentina	102.8
Thailand	95.6
Netherland	79.2
Spain	66.7

Poland, The Netherlands, Austria, China, Columbia, and South Korea. The amount of biofuels produced in other countries accounts for 17% of the total worldwide (http://www.biofuels-platform.ch/en/infos/production.php?id = biodiesel).

Considering bioethanol and biodiesel, the most common produced biofuels, biodiesel production of some leading countries is listed in Table 23.5. At present, biodiesel is mostly used as a transport fuel in compression ignition engines, however, it can also be utilized as a fuel for generator sets, etc. Most of the developing countries are mainly importers for the edible oils and hence cannot divert such oils for the production of biofuels; thus there must be strong consideration of other alternative feedstocks, including waste cooking oils and fats, as well as the oil derived from animal, fish, and microalgae [88].

Recently, several researches showed the great consideration of using LCA to assess the entire process of biofuel production from microalgae as a sustainable approach of evaluation that allows the global tendency of biofouling, and hence the commercialization of microalgal cultivation and production. Detailed descriptions of sustainable pathways of biofuel production from microalgae can be found in Culaba's article [89]. Some studies are vitally important to obtain an optimal system for industrial large-scale production [90] and to unify the uncertainties caused by the discrepancies of each process and to evaluate their performance with no/little uncertainties of the applied conversion systems. Other studies were successfully dedicated to using the LCA standards and categories for impacts and sensitivity analyses to upgrade the "algae—biofuel" chain performance [91]. In a study [92], the LCA analysis approach has been performed to compare two methods of transesterification; the conventional and the in situ. The results concluded there was a preference for the in situ transesterification methodology for biofuel production, which is mainly attributed to the major impact of the conventional method on the environment method through the production of the required electricity.

Applying LCA system can assess the environmental sustainability of transportation fuels; it is commonly used to address several characteristics such as climate forcing, pollutant emissions and impacts, impacts of water resources, changes in land-usage, nutrient inputs, impacts on human and ecosystems, and external costs as well as social impacts [37].

23.6 Future prospective

The continuous and constantly increasing worldwide tendency to move to biofuels, replacing conventional fuels, is encouraging researchers and stakeholders to study and evaluate the entire production process. Using the LCA analysis approach helps to make intercomparisons and to recommend the more vital options at all levels of the process stages. Selecting and cultivating the best and most suitable organism is the major step that secures the efficient quality/quantity of the remaining stages of the process. This might need high-input/output techniques, which increase the process and become economically infeasible. Relying on a genetically modified organism might be a valid option that would improve the generated strain.

Rathore et al. [38] combined the results of several LCA studies comparing the entire process and focused on the generated biofuel type based on the production purpose. This comparative study included >20 worldwide cradle-to-grave LCA studies, of which three

540   23. Life cycle assessment for microalgae-derived biofuels

used microalgae as a biofuel system feedstock and were conducted in France [76], the United States [93], and the United Kingdom [68] to mainly produce biodiesel/oil biodiesel. LCA results of the ABB system ensured a significant decrease in environmental impacts with less GHG emissions based on those three studies regardless of the special variations.

Recently, many companies have been endeavoring to achieve commercial production of algal fuel, while struggling with the high cost of the production process, since it costs >50% of fossil fuel. This is why relying on bioenergy was always a step back although biofuel research started 30–40 years ago. Fortunately, this vision has been changed and there will be a desperate demand for bioenergy as an alternative to fossil fuel. In the near-future, it will be crucial for most countries to dedicate some of its long-term projects to do biomass gasification and extract crude oil in order to be able to serve the national needs for energy. In this regard, and to minimize the fuel cost production, it is necessary to utilize wastewater, depend less on freshwater, compete less with arable lands, and use inexpensive infrastructures and bioreactors, such as disposable plastic bags and the sequestration of CO_2.

One of the key issues in implementing any LCA system is to apply a sensitivity analysis. This analyzing test can be conducted in three ways based on the IFEU (2000): (1) data uncertainty, (2) system boundaries, and (3) different comparisons. Major problems in the LCA system have been qualitatively rated and given weighting factors (1–5) that represent their severity and measure the adequacy of available solutions [94]. Reap and his coauthors concluded that special variations, local varied environments, and the availability and quality of data are the most severe problems, while defining the alternative scenarios as the better solved ones followed by environmental dynamicity

23.7 Conclusions

With the continuous increase of the price and the limited amounts of crude oil since the late 20000s, combining biofuels with fossil diesels (e.g., petrodiesel) appears a sustainable solution to reduce the dependency on oil-producing countries. The world's major crises (food and energy) have accelerated the global interest in algal culturing (algal farming) at a commercial level to generate biofuels, including biodiesel and other biocoproducts using smaller land areas that are unsuitable for agriculture. Analyzing the implications of using the first generation biofuels, several economic and environmental impacts have been determined as well as other social negative consequences. Biofuels are having various positive attempts being originally made from nontoxic raw materials that are biodegradable. In addition, they are less contributing to the global worming phenomenon reducing GHG emissions and atmospheric CO_2 and consequently they minimize air pollution.

On the other hand, to produce biofuels from oleaginous crops (such as palm and soybean) in sufficient quantity to meet with the nation's demand, a huge cultivation area would be necessary. For instance, the United States alone need to displace around 50% of its total cropland to achieve this purpose. This creates an urgent need to look for other effective, productive, and sustainable oil feedstocks that are cost-effective for the production of biofuel. In terms of biosources for energy, algae are autotrophic organisms with a wide genetic diversity and various species/strains that can be utilized to develop biofuels

Handbook of Algal Biofuels

and other valuable bioproducts [4]. Algae, like other plants, use sunlight, consume or absorb CO_2 on growing, and release O_2. Thus they are vital in recycling GHG, and reducing greenhouse gas emissions, and the biomass is utilized and converted into bioenergy. Moreover, algae can work at the same time as a purifying agent as they clean water wastes when cultivated with municipal wastewaters, sewage, animal wastes, and industrial effluents, due to their ability to thrive well in nutrient-rich waters. The production of biofuel from microalgae is becoming a sustainable solution to overcome these problems. Microalgae could reduce the emissions of CO_2 derived from coal power plants and/or wastewater pollution. It is anticipated now for further work to be undertaken to maximize the yield of the lipids-producing microalgae types (species/strain) through optimizing the cultivation conditions of the selected microalgae, thus enabling an increase in the biofuels generated.

Acknowledgment

The author is very thankful for Dr. Naglaa Zanaty, The National Authority for Remote Sensing and Space Sciences (NARSS) for her vital help in manipulating the final version of this chapter

References

[1] Ervin, M.J., Associated petroleum price data, <kentmarketingservices.com>, 2011 (accessed 23.07.11).

[2] P.S. Bindraban, E.H. Bulte, S.G. Conijn, Can large-scale biofuels production be sustainable by 2020? Agric. Syst. 101 (3) (2009) 197−199. Available from: https://doi.org/10.1016/j.agsy.2009.06.005.

[3] Natural Resources Canada, Natural Resources Canada, 2011, pp. 1−5.

[4] DOE, United States Department of Energy Office of Energy Efficiency and Renewable Energy Bioenergy Technologies Office, in: C.R. Amanda Barry, A. Wolfe, C. English, D. Lambert (Eds.), National Algal Biofuels Technology Review, June, 2016.

[5] Y. Chisti, Bioremediation—keeping the earth clean, Biotechnol. Adv. 23 (5) (2005) 371−372. Available from: https://doi.org/10.1016/j.biotechadv.2005.03.002.

[6] P.G. Roessler, L.M. Brown, T.G. Dunahay, D.A. Heacox, Jarvis, J.C. Schneider, et al., Genetic engineering approaches for enhanced production of biodiesel fuel from microalgae, Enzymatic Conversion of Biomass for Fuels Production, Vol. 566, American Chemical Society, 1994, pp. 13−255.

[7] A. Fedorov, S. Kosourov, M. Ghirardi, M. Seibert, Continuous hydrogen photoproduction by *Chlamydomonas reinhardtii* using a novel two-stage, sulfate-limited chemostat system, Appl. Biochem. Biotechnol. 121−124 (2005) 403−412. Available from: https://doi.org/10.1385/ABAB:121:1-3:0403.

[8] I. Karapinar Kapdan, F. Kargi, Bio-hydrogen production from waste materials, Enzyme Microb. Technol. 38 (2006) 569−582. Available from: https://doi.org/10.1016/j.enzmictec.

[9] N. Nagle, P. Lemke, Production of methyl ester fuel from microalgae, Appl. Biochem. Biotechnol. 24 (1) (1990) 355−361. Available from: https://doi.org/10.1007/BF02920259.

[10] S. Sawayama, S. Inoue, Y. Dote, S.Y. Yokoyama, CO_2 fixation and oil production through microalga, Energy Convers. Manage. 36 (6−9) (1995) 729−731. Available from: https://doi.org/10.1016/0196-8904(95)00108-P.

[11] Y. Chisti, An unusual hydrocarbon, J. Ramsay Soc. 27−28 (2005) 24−26. 1981.

[12] M. Gavrilescu, Y. Chisti, Chisti, Y.: Biotechnology—a sustainable alternative for chemical industryBiotechnol. Adv. 23, 471−499Biotechnol. Adv. 23 (2005) 471−499. Available from: https://doi.org/10.1016/j.biotechadv.

[13] T. Mata, A. Martins, N. Caetano, Microalgae for biodiesel production and other applications: a review, Renew. Sustain. Energy Rev. 14 (2010). Available from: https://doi.org/10.1016/j.rser.

[14] A. Jayakumar, B. Chang, B. Widner, P. Bernhardt, M. Mulholland, B. Ward, Biological nitrogen fixation in the oxygen-minimum region of the eastern tropical North Pacific ocean, ISME J. 11 (2017). Available from: https://doi.org/10.1038/ismej.2017.97.

[15] S. Gill, M. Mehmood, U. Rashid, M. Ibrahim, A. Saqib, M. Rizwan, Waste-water treatment coupled with biodiesel production using microalgae: a bio-refinery approach, Pak. J. Life Soc. Sci. 11 (2013).

[16] Y. Chisti, Biodiesel from microalgae, Biotechnol. Adv. 25 (3) (2007) 294–306. Available from: https://doi.org/10.1016/j.biotechadv. Elsevier May 01, 2007.

[17] L. Brennan, P. Owende, Biofuels from microalgae—a review of technologies for production, processing, and extractions of biofuels and co-products, Renew. Sustain. Energy Rev. 14 (2) (2010) 557–577. Available from: https://doi.org/10.1016/j.rser.2009.10.009. Pergamon.

[18] D.P. Tredici, Wild Urban Plants of the Northeast: A Field Guide Forward by Steward Pickett, Cornell University Press, 2010, p. 374.

[19] J. Deng, W. Dong, R. Socher, L.-J. Li, K. Li, F.-F. Li, ImageNet: a large-scale hierarchical image database, in: Procceedings of the IEEE Conference on Computer Vision and Pattern Recognition, 2009, pp. 248–255, doi: 10.1109/CVPR.2009.5206848.

[20] W. Khan, P. Usha, U.P. Rayirath, S. Subramanian, M.N. Jithesh, P. Rayorath, et al., Seaweed extracts as biostimulants of plant growth and development, J. Plant Growth Regul. 28 (2009) 386–399.

[21] E. Jacob-Lopes, T. Teixeira, Microalgae-based systems for carbon dioxide sequestration and industrial biorefineries, Biomass, 2010 September, doi: 10.5772/9772.

[22] M. Kamalanathan, P. Chaisutyakorn, R. Gleadow, J. Beardall, A comparison of photoautotrophic, heterotrophic, and mixotrophic growth for biomass production by the green alga *Scenedesmus* sp. (*Chlorophyceae*), Phycologia 57 (3) (2018) 309–317. Available from: https://doi.org/10.2216/17-82.1.

[23] H. Xu, X. Miao, Q. Wu, High quality biodiesel production from a microalga *Chlorella prototheecoides* by heterotrophic growth in fermenters, J. Biotechnol. 126 (4) (2006) 499–507. Available from: https://doi.org/10.1016/j.jbiotec. Dec. 2006.

[24] J.U. Grobbelaar, Physiological and technological considerations for optimising mass algal cultures, J. Appl. Phycol. 12 (3) (2000) 201–206. Available from: https://doi.org/10.1023/A:1008155125844.

[25] UNEP/GRID-Arendal, <http://www.unep.org/climatechange/mitigation/Portals/93/documents/Bioenergy/VBG_Ebook.pdf>. (accessed 30.12.12); <http://hdl.handle.net/20.500.11822/32160>, 2011.

[26] L. Golsteijn, Life cycle assessment (LCA) explained, 2020. <https://pre-sustainability.com/articles/life-cycle-assessment-lca-basics/>, (accessed 18.03.21).

[27] E. Maleche, Life cycle assessment of biofuels produced by the new integrated hydropyrolysis-hydroconversion (Ih 2) process, 2012.

[28] E. Maleche, R. Glaser, T. Marker, D. Shonnard, A preliminary life cycle assessment of biofuels produced by IH2, Environ. Prog. Sustain. Energy 33 (2014). Available from: https://doi.org/10.1002/ep.11773.

[29] H. Jian, J. Yang, Z. Peidong, Life cycle analysis on fossil energy ratio of algal biodiesel: effects of nitrogen deficiency and oil extraction technology, Sci. World J. (2015) 1–9. Available from: https://doi.org/10.1155/2015/920968.

[30] M. Matt Mihalek, J. Fan, G. Bhardwaj, R. Handler, T.N. Kalnes, D.R. Shonnard, et al., Life cycle assessments of pyrolysis-based biofuels from diverse biomass feedstocks, 2012.

[31] I.V. Muralikrishna, V. Manickam, Life cycle assessment, 2017 pp. 1–159. Available from: https://doi.org/10.4324/9781315778730.

[32] B. Muys, J. García Quijano, A new method for land use impact assessment in LCA based on ecosystem exergy concept. Internal report. Laboratory for forest, and landscape research, Leuven, Belgium, 2002. <http://www.biw.kuleuven.be/lbh/lbnl/forecoman/pdf/land%20use%20method4.pdf>.

[33] S. Kim, B.E. Dale, Regional variations in greenhouse gas emissions of biobased products in the United States—corn based ethanol and soybean oil, Int. J. Life Cycle Assess. 14 (2009) 540–546.

[34] D. Chiaramonti, L. Recchia, Is life cycle assessment (LCA) a suitable method for quantitative CO_2 saving estimations? The impact of field input on the LCA results for a pure vegetable oil chain, Biomass Bioenerg. 34 (2010) 787–797.

[35] D. Dressler, A. Loewen, M. Nelles, Life cycle assessment of the supply and use of bioenergy: impact of regional factors on biogas production, Int. J. Life Cycle Assess. 17 (2012) 1104–1115.

[36] M.F. Emmenegger, S. Pfister, A. Koehler, L.D. Giovanetti, A.P. Arena, R. Zah, Taking into account water use impacts in the LCA of biofuels: an Argentinean case study, Int. J. Life Cycle Assess. 16 (2011) 869–877.

[37] T.E. McKone, W.W. Nazaroff, P. Berck, M. Auffhammer, T. Lipman, M.S. Torn, et al., Grand challenges for life-cycle assessment of biofuels, Env. Sci. Technol. 45 (2011) 1751–1756.

References

543

[38] D. Rathore, D. Pant, A. Singh, A comparison of life cycle assessment studies of different biofuels, in: A. Singh, D. Pant, S.I. Olsen (Eds.), Green Energy and Technology, Life Cycle Assessment of Renewable Energy Sources. 2013.

[39] G. Knothe, Biodiesel and renewable diesel: a comparison, Progress Energy Combust. Sci. 36 (3) (2010) 364–373. Available from: https://doi.org/10.1016/j.pecs.2009.11.004. Pergamon.

[40] J. Smith, S.H. Schneider, M. Oppenheimer, G.W. Yohe, W. Hare, M.D. Mastrandrea, et al., Assessing dangerous climate change through an update of the Intergovernmental Panel on Climate Change (IPCC), Proc. Natl. Acad. Sci. 106 (11) (2009) 4133–4137. Available from: https://doi.org/10.1073/pnas.0812355106.

[41] I.-C. Chen, J. Hill, R. Ohlemüller, D.B. Roy, C. Thomas, Rapid range shifts of species associated with high levels of climate warming, Science 333 (2011) 1024–1026. Available from: https://doi.org/10.1126/science.1206432.

[42] D. Sheehan John, B. Terri, John, R. Paul, A look back at the United States department of energy's aquatic species, Eur. Phys. J. C 72 (6) (1998) 14 [Online]. <http://www.springerlink.com/index/10.1140/epjc/s10052-012-2043-9%5Cnhttp://arxiv.org/abs/1203.5015>.

[43] Agency, E.P., EPA guidelines for ensuring and maximizing the quality, objectivity, utility, and integrity of information disseminated by the Environmental Protection Agency, 2002.

[44] J.-P. Cadoret, O. Bernard, Lipid biofuel production with microalgae: potential and challenges, J. Soc. Biol. 202 (3) (2008) 201–211. Available from: https://doi.org/10.1051/jbio:2008022.

[45] Q. Hu, et al., Microalgal triacylglycerols as feedstocks for biofuel production: perspectives and advances, Plant J. 54 (2008) 621–639. Available from: https://doi.org/10.1111/j.1365-313X.2008.03492.x.

[46] A. Sarin, R. Arora, N.P. Singh, R. Sarin, R.K. Malhotra, K. Kundu, Effect of blends of Palm-Jatropha-Pongamia biodiesels on cloud point and pour point, Energy 34 (11) (2009) 2016–2021. Available from: https://doi.org/10.1016/j.energy.2009.08.017.

[47] C.Y. Kao, S.Y. Chiu, T.T. Huang, L. Daia, G.H. Wang, C.P. Tseng, et al., A mutant strain of microalga Chlorella sp. for the carbon dioxide capture from biogas, Biomass Bioenergy 36 (2012) 132–140. Available from: https://doi.org/10.1016/j.biombioe.2011.10.046.

[48] M.Y. Menetrez, An overview of algae biofuel production and potential environmental impact, Environ. Sci. Technol. 46 (13) (2012) 7073–7085. Available from: https://doi.org/10.1021/es300917r.

[49] J.M. Lv, L.H. Cheng, X.H. Xu, L. Zhang, H.L. Chen, Enhanced lipid production of *Chlorella vulga*ris by adjustment of cultivation conditions, Bioresour. Technol. 101 (17) (2010) 6797–6804. Available from: https://doi.org/10.1016/j.biortech.2010.03.120.

[50] A. Richmond, Microalgal biotechnology at the turn of the millennium: a personal view, J. Appl. Phycol. 12 (2000) 441–451. Available from: https://doi.org/10.1023/A:1008123131307.

[51] O. Pulz, Photobioreactors: production systems for phototrophic microorganisms, Appl. Microbiol. Biotechnol. 57 (3) (2001) 287–293. Available from: https://doi.org/10.1007/s002530100702.

[52] A. Carvalho, L. Meireles, F. Malcata, Microalgal reactors: a review of enclosed system designs and performances, Biotechnol. Prog. 22 (2006) 1490–1506. Available from: https://doi.org/10.1021/bp060065r.

[53] J. Grobbelaar, Factors governing algal growth in photobioreactors: the 'open' versus 'closed' debate, J. Appl. Phycol. 21 (2009) 489–492. Available from: https://doi.org/10.1007/s10811-008-9365-x.

[54] C. Richardson, Investigating the role of reactor design to maximise the environmental benefit of algal oil for biodiesel, University of Cape Town. <http://hdl.handle.net/11427/5409>, 2011.

[55] S.-H. Lee, J.-B. Lee, K.-W. Lee, Y.-J. Jeon, Antioxidant properties of tidal pool microalgae, *Halochlorococcum porphyrae* and *Oltamannsiellopsis unicellularis* from Jeju Island, Korea, Algae 25 (2010) 45–56. Available from: https://doi.org/10.4490/algae.2010.25.1.045.

[56] M.A. Borowitzka, Commercial production of microalgae: ponds, tanks, tubes and fermenters, J. Biotechnol. 70 (1) (1999) 313–321. Available from: https://doi.org/10.1016/S0168-1656(99)00083-8.

[57] U. Ugwu, H. Aoyagi, H. Uchiyama, Photobioreactors for mass cultivation of algae, Bioresour. Technol. 99 (2008) 4021–4028. Available from: https://doi.org/10.1016/j.biortech.2007.01.046.

[58] M. Letellier, H. Budzinski, Microwave assisted extraction of organic compounds, Analusis 27 (3) (1999) 259–270. Available from: https://doi.org/10.1051/analusis:1999116.

[59] E. Belarbi, E. Molina-Grima, Y. Chisti, A process for high yield and scaleable recovery of high purity eicosapentaenoic acid esters from microalgae and fish oil, Enzyme Microb. Technol. 26 (2000) 516–529. Available from: https://doi.org/10.1016/s0141-0229(99)00191-x.

Handbook of Algal Biofuels

[60] T. Lewis, P. Nichols, T. Mcmeekin, T. Lewis, P.D. Nichols, T.A. McMeekin, Evaluation of extraction methods for recovery of fatty acids from lipid-producing microheterotrophs, J. Microbiol. Methods 43 (2000) 107–116. Available from: https://doi.org/10.1016/S0167-7012(00)00217-7.

[61] Y. Roman-Leshkov, C. Barrett, Z. Liu, J. Dumesic, Production of dimethylfuran for liquid fuels from biomass-derived carbohydrates, Nature 447 (2007) 982–985. Available from: https://doi.org/10.1038/nature05923.

[62] R. Halim, B. Gladman, M.K. Danquah, P.A. Webley, Oil extraction from microalgae for biodiesel production, Bioresour. Technol. 102 (1) (2011) 178–185.

[63] K. Wiltshire, M. Boersma, A. Möller, H. Buhtz, Extraction of pigments and fatty acids from the green alga *Scenedesmus obliquus* (Chlorophyceae), Aquat. Ecol. 34 (2000) 119–126. Available from: https://doi.org/10.1023/A:1009911418606.

[64] R. Couto, P. Simões, A. Reis, T. da Silva, V. Martins, Y. Sánchez-Vicente, Supercritical fluid extraction of lipids from heterotrophic microalgae *Crypthecodinium cohnii*, Eng. Life Sci. 10 (2010) 158–164. Available from: https://doi.org/10.1002/elsc.200900074.

[65] C. Samorì, et al., Extraction of hydrocarbons from microalga *Botryococcus braunii* with switchable solvents, Bioresour. Technol. 101 (2010) 3274–3279. Available from: https://doi.org/10.1016/j.biortech.2009.12.068.

[66] D.Y.C. Leung, X. Wu, M.K.H. Leung, A review on biodiesel production using catalyzed transesterification, Appl. Energy 87 (4) (2010) 1083–1095.

[67] K. Miyamoto, O. Wable, J.R. Benemann, Vertical tubular reactor for microalgae cultivation, Biotechnol. Lett. 10 (10) (1988) 703–708.

[68] A.L. Stephenson, E. Kazamia, J.S. Dennis, C.J. Howe, S.A. Scott, A.G. Smith, Life-cycle assessment of potential algal biodiesel production in the United Kingdom: a comparison of raceways and air-lift tubular bioreactors, Energy Fuels 24 (7) (2010) 4062–4077. Available from: https://doi.org/10.1021/ef1003123.

[69] A. Richmond, S. Boussiba, A. Vonshak, R. Kopel, A new tubular reactor for mass production of microalgae outdoors, J. Appl. Phycol. 5 (3) (1993) 327–332. Available from: https://doi.org/10.1007/BF02186235.

[70] H. Qiang, A. Richmond, Productivity and photosynthetic efficiency of Spirulina platensis as affected by light intensity, algal density and rate of mixing in a flat plate photobioreactor, J. Appl. Phycol. 8 (2) (1996) 139–145. Available from: https://doi.org/10.1007/BF02186317.

[71] Y.-K. Lee, Microalgal mass culture systems and methods: their limitation and potential, J. Appl. Phycol. 13 (4) (2001) 307–315. Available from: https://doi.org/10.1023/A:1017560006941.

[72] A. Vonshak, Spirulina: the basic concept, Spirulina Platensis Arthrospira Physiol. Cell-Biology Biotechnol., 1997, p. 79.

[73] M. Morita, Y. Watanabe, H. Saiki, High photosynthetic productivity of green microalga *Chlorella sorokiniana*, Appl. Biochem. Biotechnol. 87 (3) (2000) 203–218. Available from: https://doi.org/10.1385/ABAB:87:3:203.

[74] M. Berrios, R.L. Skelton, Comparison of purification methods for biodiesel, Chem. Eng. J. 144 (3) (2008) 459–465. Available from: https://doi.org/10.1016/j.cej.2008.07.019.

[75] D.T. Johnson, K.A. Taconi, The glycerin glut: options for the value-added conversion of crude glycerol resulting from biodiesel production, Environ. Prog. 26 (4) (2007) 338–348.

[76] S.S. Yazdani, R. Gonzalez, Anaerobic fermentation of glycerol: a path to economic viability for the biofuels industry, Curr. Opin. Biotechnol. 18 (3) (2007) 213–219.

[77] D.R. Vardon, B.K. Sharma, G.V. Blazina, K. Rajagopalan, T.J. Strathmann, Thermochemical conversion of raw and defatted algal biomass via hydrothermal liquefaction and slow pyrolysis, Bioresour. Technol. 109 (2012) 178–187. Available from: https://doi.org/10.1016/j.biortech.2012.01.008.

[78] B. Sialve, N. Bernet, O. Bernard, Anaerobic digestion of microalgae as a necessary step to make microalgal biodiesel sustainable, Biotechnol. Adv. 27 (4) (2009) 409–416. Available from: https://doi.org/10.1016/j.biotechadv.2009.03.001.

[79] J. Yuan, A. Kendall, Y. Zhang, Mass balance and life cycle assessment of biodiesel from microalgae incorporated with nutrient recycling options and technology uncertainties, GCB Bioenergy 7 (6) (2015) 1245–1259. Available from: https://doi.org/10.1111/gcbb.12229.

[80] G. Saranya, T.V. Ramachandra, Life cycle assessment of biodiesel from estuarine microalgae, Energy Convers. Manage. X 8 (2020) 100065. Available from: https://doi.org/10.1016/j.ecmx.2020.100065.

[81] L. Lardon, A. Hélias, B. Sialve, J.P. Steyer, O. Bernard, Life-cycle assessment of biodiesel production from microalgae, Environ. Sci. Technol. 43 (17) (2009) 6475–6481. Available from: https://doi.org/10.1021/es900705j.

References

[82] R.M. Handler, D.R. Shonnard, T.N. Kalnes, F.S. Lupton, Life cycle assessment of algal biofuels: influence of feedstock cultivation systems and conversion platforms, Algal Res. 4 (1) (2014) 105–115. Available from: https://doi.org/10.1016/j.algal.2013.12.001.

[83] P. Wibula, P. Malakul, P. Pavasantc, K. Kangvansaichold, S. Papong, Life cycle assessment of biodiesel production from microalgae in thailand: energy efficiency and global warming impact reduction, Chem. Eng. Trans. 29 (2012).

[84] S. Liang, M. Xu, T. Zhang, Life cycle assessment of biodiesel production in China, Bioresour. Technol. 129 (2013) 72–77. February.

[85] Y.C. Sharma, B. Singh, Development of biodiesel: current scenario, Renew. Sustain. Energy Rev. 13 (6–7) (2009) 1646–1651.

[86] E. Santacesaria, G.M. Vicente, M. Di Serio, R. Tesser, Main technologies in biodiesel production: state of the art and future challenges, Catal. Today 195 (1) (2012) 2–13. Available from: https://doi.org/10.1016/j.cattod.2012.04.057. Elsevier.

[87] Sönnichsen. Our research and content philosophy, <https://www.statista.com/aboutus/our-research-commitment>, 2021 (accessed 18.03.21).

[88] A.O. Alabi and M. Tampier, Microalgae technologies and processes for bioenergy production in British Columbia: current technology, suitability & barriers to implementation, Final Rep. Submitt. to Br. Columbia Innov. Counc., 2009, pp. 1–88.

[89] A.B. Culaba, A.T. Ubando, P.M.L. Ching, W.-H. Chen, J.-S. Chang, Biofuel from microalgae: sustainable pathways, Sustainability 12 (19) (2020) 8009.

[90] C.-H. Sun, Q. Fu, Q. Liao, A. Xia, Y. Huang, X. Zhu, et al., Life-cycle assessment of biofuel production from microalgae via various bioenergy conversion systems, Energy, Vol. 171, Elsevier, 2019, pp. 1033–1045. C.

[91] W. Wu, Y.-C. Lei, J.-S. Chang, Life cycle assessment of upgraded microalgae-to-biofuel chains, Bioresour. Technol., 288, 2019, p. 121492.

[92] G. Uctug, D. Modi, F. Mavituna, Life cycle assessment of biodiesel production from microalgae: a mass and energy balance approach in order to compare conventional with in situ transesterification, Int. J. Chem. Eng. Appl. 8 (2017) 355–356. Available from: https://doi.org/10.18178/ijcea.2017.8.6.683.

[93] K. Sander, G.S. Murthy, Life cycle analysis of algae biodiesel, Int. J. Life Cycle Assess. 15 (2010) 704–714.

[94] J. Reap, F. Roman, S. Duncan, B. Bras, A survey of unresolved problems in life cycle assessment. Part 2: impact assessment and interpretation, Int. J. Life Cycle Assess. 13 (2008) 374–388.

CHAPTER
24

An overview of the algal biofuel technology: key challenges and future directions

Kushi Yadav, Reetu and Monika Prakash Rai

Amity Institute of Biotechnology, Amity University, Noida, India

24.1 Introduction

The fossil fuel requirement was increased upto 40% till 2014 and to overcome this requirement there is an essential need for alternative clean fuel [1]. In the last few decades, among various alternative renewable energy sources, biomass is considered a potential option as it can generate biofuel and bioproducts simultaneously [2]. Biomass-mediated biofuels include biodiesel, bioethanol, biomethanol, biohydrogen, bioethers, and bioethyltetrabutyl ether [3]. Biodiesel and bioethanol are considered vital biofuels as stated by the United States Department of energy [3]. Brazil, United States, France, Sweden, and Germany are some countries, which are leading the biofuel industry [4]. The first-generation biofuel is obtained from a feedstock of plants such as sugarcane, maize, palm, soybean, and sweet sorghum by fermenting their sugar into bioethanol with the help of yeast, additionally, biodiesel is obtained from plant oil extract due to which it is called agrofuel [5]. Agrofuel is a clean fuel but it creates competition with the food and water requirement, which became a huge challenge in biofuel production [6]. Then biofuel production shifted to second-generation fuels, where nonedible plants such as Jatropha, grass, switchgrass, silver grass, and uneatable parts of the crops are used [6]. The major challenge in this technique is the overconsumption of water and pesticides along with decreased land and water availability [2]. Algae appear as a third-generation biofuel with no competition approach but it is not economically viable, hence various improvements are required at processing methods including cultivation, harvesting, and extraction techniques [7].

Biofuel generated from microalgae is getting attention due to the fast biomass production as microalgae biofilms can be an advanced way for biomass generation, the advantage of direct energy conversion from microalgal biomass, which finally leads to biofuel processing

Handbook of Algal Biofuels
DOI: https://doi.org/10.1016/B978-0-12-823764-9.00007-8

© 2022 Elsevier Inc. All rights reserved.

548   24. algal biofuel technology

[8]. Value-added coproducts obtained during biofuel production make the process economically and environmentally viable especially while using the biorefinery approach. Various commercial benefits of microalgae are obtaining biofuel, nutrient supplements (pigments, vitamins, and proteins), CO_2 mitigation, sewage water treatment, cosmetic products, and pharmaceutical products (antimalarial, antioxidants, and antivirals) as discussed in Chapter "Byproducts recycling of algal biofuel toward bioeconomy". In the last few decades, commercial cultivation of microalgae increasing drastically to obtain biofuel and value-added products [9]. Few current research indicated the potential of microalgae after genetic modification shows an increase in biomass and lipid production along with elevated apprehend ability, but the technique requires high expenses and time [10,11].

The technical and financial challenges are debated, those need to be diminished to make algal fuel and coproducts economically and environmentally viable. Advancement in algae-based biofuel technology is discussed mainly for cultivation, harvesting, lipid extraction, and biodiesel production. Various parameters affecting different modes of algae cultivation are systematically evaluated. Conventional and advanced harvesting and lipid extraction techniques are compared and the incorporation of nanomaterials in harvesting and biodiesel synthesis are critically investigated for feasible biofuel future.

24.2 Challenges

24.2.1 Cultivation

Algae became a worldwide attraction due to its potential to generate biofuel and various value-added products, which are having high market value. As previously discussed in Chapter "Sequential algal biofuel production through whole biomass conversion", zero-waste management of algal biomass utilization is also a major point of attraction from the environmental aspect. However, there are certain challenges with algae cultivation that make the commercialization of algae expensive and complicated. In some cultivation systems, we can obtain a good yield, but energy input and maintenance costs are too high in contrast to certain cultivation systems like open cultivation system, where the process became economically viable, but we cannot get axenic algae species. Apart from biodiesel, algae can generate nutrients, antioxidants, proteins, carbohydrates, minerals as well as pigments, which decrease the biodiesel production cost and make the whole process environmentally and economically sustainable. In Fig. 24.1 the process to obtain algae-mediated biodiesel and value-added products are discussed.

24.2.1.1 Macroalgae cultivation

Macroalgae cultivation is influenced by a variety of climatic factors including light intensity, turbidity, water temperature, nutrient concentrations, pH, current, water quality, and salinity [12]. Natural seasonal variations and natural calamities such as high tides, hailstorms, cyclones, and tsunami directly affect macroalgae cultivation by changing the quality and quantity of water [13]. The risk of contamination of macroalgae during cultivation increases due to floods. Often storms and sedimentation act as a major challenge in macroalgae cultivation [14]. Obtaining a sustainable amount of biomass and value-added products become

Handbook of Algal Biofuels

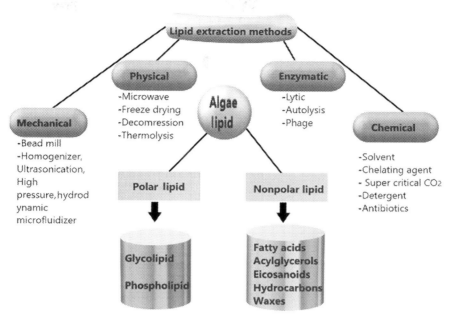

FIGURE 24.1 Algal cell disruption and lipid extraction methods.

challenging due to environmental variations. Fishing is also sea-shore oriented activity due to which there are limited areas available for large scale cultivation of macroalgae and in deep-sea area macroalgae cultivation is not economical [15]. For sustainable and economical cultivation of macroalgae under unfavorable conditions of the sea, a well-designed cultivation strategy is essential [15]. The advantages of macroalgae to be served as food and shelter to various grazers such as crustacea, fish, turtles, birds, etc., and epiphytes such as microalgae become a challenge for macroalgae cultivation [13]. Rabbitfish is a seaweed grazer specifically for *Kappaphycus alvarezii* macroalgae leads to enormous economic damage during cultivation [16]. Apart from grazers and epiphytes predators and diseases can also damage juvenile macroalgal seeds. Hence it is necessary to design an effective strategy to sterilize water and regulate environmental conditions for sustainable and economical macroalgae cultivation. In various countries, sporophyte used is cost-effective but has certain challenges such as the process is seasonal, time demanding, and so on [15]. Based on cost and time, one major challenge is site selection for macroalgae cultivation as there is a huge demand for effective macroalgae cultivation strategy for large scale biomass production.

24.2.1.2 *Microalgal cultivation*

There are different microalgae cultivation strategies among which open cultivation system and photobioreactor cultivation system is majorly used both have their benefits along with a couple of challenges. Various studies are comparing and evaluating both kinds of microalgae cultivation systems which conclude that the open cultivation system is more

550

24. algal biofuel technology

economically feasible in comparison with photobioreactors [17]. Open cultivation was found much economically suitable for the production of biologically active and dietary supplements in contrast to photobioreactors [18].

Photobioreactors are preferred by various pharmaceutical, nutraceutical, and cosmeceuticals industries because their products are consumed directly by humans. There are many toxic contaminants reported in the open cultivation system such as protozoa and bacteria which make algae-mediated products inappropriate for human consumption [17]. Under high alkaline or saline conditions, open microalgae cultivation is more viable due to less contamination probability as many contaminants cannot survive is a high alkali or saline conditions [9]. As there is a very low risk of biomass loss and contamination which makes the products suitable for human consumption [19,20]. Both open pond microalgae cultivation and photobioreactor based microalgae cultivation are financially not feasible for biodiesel production [21,22].

Another challenge with the open cultivation system is the large area requirement [9]. Microalgae cultivation is directly affected through various environmental factors such as light, pH, turbidity, salinity and so on which is a major challenge in the open cultivation system [9]. Microalgae cultivation is affected by rain as water quality and quantity change in the open cultivation system [23]. In photobioreactors, environmental factors are highly controlled in comparison to open cultivation a system which is a major challenge. Even after being economically sustainable, it is not feasible to use an open cultivation system for algae-mediated product production.

Open pond cultivation system has a challenge with pond size because increasing pond size leads to high water resistance which causes pressure to agitator mechanical part [24]. This system is not economically viable for biodiesel production due to its huge construction and agitation process cost [25]. Photobioreactors are expensive and difficult to maintain and clean which increases the chance of contamination is a major challenge. There is design complication in some photobioreactors such as bottlenecks and have limited scalability which makes it less suitable for industrial level cultivation [26]. Due to this complexity of design and maintenance issues it is difficult to use vertical column photobioreactors at a commercial scale [27]. Photobioreactors aeration design gives rise to stress damage to microalgae cells during cultivation [28].

In an open cultivation system, it is problematic to maintain monoculture as contamination risk is very high. To maintain an exotic environment for microalgae in open cultivation systems different environmental parameters are manipulated such as pH, salinity, temperature, and so on [29]. The unavailability of evaporation cooling in photobioreactors makes temperature regulation difficult [30,31]. Some other challenges with photobioreactors are penetration of light, biofilm formation on photobioreactor walls, pH gradient, dissolved oxygen, and CO_2 in photobioreactor along with hydrodynamic stresses [32−34]. There is a demand for a cost-effective and environmentally sustainable cultivation system for microalgae to obtain economical biodiesel and value-added products.

24.2.2 Harvesting

24.2.2.1 Macroalgae harvesting

Macroalgae can be harvested using both manual and mechanized methods, but the manual harvesting of the algae is the most prevalent method across the globe. The manual method can be applied both for the harvesting of naturally grown as well as artificially cultivated macroalgae. It employs the use of a fork, sickle, and net, ultimately uprooting the microalgae. Whereas mechanized methods involve the use of suction, cutters, rotating blades, and additionally requires boats and ships for the operation [30].

For instance, In France for harvesting the *Laminaria digitata*, a boat loaded with a hook-like gear known as "scoubidou" is used. It turns around itself and uproots the seaweed and then the tool is reversed in the opposite direction to unload the harvested algae onto the boat. Similarly, for harvesting the *Laminaria hyperborean*, a special type of boat is used it possesses a large hook-like device, which is dragged through the macroalgal bed and the crane attached on the rake of the boat pulls about 2 tons of macroalgae on to the boat within 2 min. Another boat termed "sablier" consists of a suction dredge mainly used for harvesting the *Lithothamnium calcareum* [31].

Vertical wet-well is also an option to be used for harvesting the seaweeds. A vertical wet-well with a depth gage is attached to a hydraulic arm moving it along the ocean bottom, it chops the macroalgae and brings it up on the boat with the help of a net. The depth gage is mainly used to avoid cutting of the plant from the very bottom as it keeps the machine 20 cm above the sea bottom. One study that tested the machine at two different shorelines concluded two very different results; one with a rate of 0.2 t/h and another with a rate of 1.125 t/h. But bringing these gigantic boats and machinery was a very tedious task and also cutting off the macroalgae required an expertise skill. Many readymade pieces of machinery like cranes are also available in the market that can be employed for the harvesting of the seaweeds. Conver, in the Netherlands, is one such brand, which manufactures such equipment that can be used to harvest the macroalgae without changes. One such machine is the Conver C430H, it is a midrange mowing boat, which comes with a T-front cutting tool. It can cut the seaweed and push it towards the net. It is a small boat, which could be handled easily and because of the small size can move near the shorelines, leading to a higher harvesting rate per hour [32]. SR and FX are the two models manufactured by the alpha boats. These boats have a cutting capacity of 1.61−1.83 m depth and 1.83−2.13 m width. They have two cutting bars, vertically at the lateral sides of the platform and horizontally beneath it. This whole platform goes inside the water and cuts the macroalgae, which is carried onto a conveyor and finally transferred to the reservoir on the boat [32].

Further, pretreatment is also done before the lipid extraction or direct conversion. It can be done using the following methods:

Removal of unwanted objects: before, further processing of the seaweeds, impurities like stones, plastics, aquatic insects, and other garbage are required to be removed. It can be done manually.

1. Grinding: It is commonly known as chopping or milling used to decrease the particle size and increase the surface area to volume ratio. Also, the small particles have a higher reaction efficiency [30].
2. Drying: It is also known as the dewatering process. It leads to stabilization of the microalgae biomass and ease in transportation. Fuel-fired boilers can also be used for drying the biomass, but it adds to the overall cost [33].

24.2.2.2 *Microalgae harvesting*

The harvesting process is more difficult, energy-consuming, and expensive in the case of microalgae due to their small size and negatively charged surface [34]. The process can be done by using both conventional techniques and advanced techniques.

24.2.2.2.1 Conventional harvesting methods

Centrifugation Centrifugation is a technique, which employs the use of centrifugal force to separate the mixed particles. It a mechanical harvesting method separating the substances based on the variable densities. Disparate varieties of centrifuges are available that can be used according to the desired efficiency. Algal biomass can be harvested using a single or two-step process. Conventionally tubular or multichamber centrifuges are used for the harvesting purpose. Efficiency wise tubular centrifuge is the most adequate for harvesting but there is no way to discharge the solid hence, could only be operated in batches making it a cumbersome process as intermittent cleaning is required for continuous use. Moreover, in terms of feasibility and cost, centrifuging a higher amount of culture is not advantageous. In concern with the algal biorefineries, disk stack centrifuges with higher gravitational forces and reduced separation time are the most widely used. But when large volumes are used, very high energy consumption takes place [35].

Flotation In this technique, air bubbles are generated, which further carry the microalgal cells to the surface of the culture from where they can be harvested easily. It depends on the strain of the microalgae to be harvested and also on the size of both the bubble and the microalgae cells. Bubbles in a diameter range of $3-10 \, \mu m$ are favorable for the process as they have a higher surface-to-volume ratio [36]. Properties like hydrophobicity and charge also determine the efficiency of air bubbles to carry the cells. The microbubble production can take place by different techniques like electrolytic flotation, dissolved air flotation, dispersed air floatation (DiAF). It involves the dissolution of pressured air in water ultimately causing nucleation of bubbles at the fall of pressure [37]. When air is passed through a porous material, microbubbles are produced, and the technique is termed as the DiAF. Although the process is energy conserving but the machinery is expensive [37]. But in general, it is more economical than the centrifugation process.

Filtration It is another conventional method used to harvest the microalgae using filters of different pore sizes. In it the culture is administered through the membrane and filter force is used for the harvesting [38]. Membrane filtration can be classified based on pore size. Filtration involving membranes in the range of $0.02-0.2 \, \mu m$ pore size is termed as ultrafiltration whereas filters with pore size $0.1-10 \, \mu m$ are referred to as microfiltration [39]. The majority of the microalgal strains can be separated using the porous membrane whereas the clogging of the filters causes a reduction in flux ultimately increasing the processing cost.

Flocculation This process induces the microalgal cells to get flocculated and settle down quickly. The technique involves the least amount of energy as no mechanical forces are required except the addition of flocculant. Based on the type of flocculant added the process can be of different types like chemical flocculation, bioflocculation, physical flocculation, etc. The chemical flocculation method involves the addition of components like metal salts, polymer, or biopolymers. As metal salts and polymers can lead to contamination of microalgal cultures hence, biopolymers are the preferred options. Various nanoparticles have also been used for flocculation purposes as the latest report by [40] depicted the use of zirconium dioxide nanoparticles for flocculating the *Chlorococcum* sp. with an efficiency of 99.59%.

Physical methods like electrocoagulation-flocculation and magnetic nanoparticles can also be used for harvesting microalgal cells. In electrocoagulation-flocculation metal ions are released from the anode and flocculation is induced whereas in magnetic nanoparticles get adsorbed at the surface microalgal cells further an external magnet is employed to harvest the microalgal flocs [41]. Then the separated nanoparticles from the cells can be recycled.

Bacteria induced flocculation is termed the bioflocculation. Moreover, if the wastewater is used as the culture medium for cultivating the microalgae, then the added bacteria can use the carbon present in the wastewater for its growth. But the demerit of this method is that bacterial cells are harvested along with the microalgal cells, which are needed to be separated ultimately increasing the downstream processing cost.

24.2.2.2.2 Advanced harvesting method

Although many conventional techniques have been discussed and explained regarding harvesting the microalgal biomass, none of the methods proved on the desired marking. Hence, the current research is focusing on the use of nanotechnology and other methods for the efficient harvesting and separation of microalgal cells. This would lead to immense advancement in biodiesel production from the algal cells and would make the whole process economically feasible [42] (Table 24.1).

Magnetic nanoparticles It is a very quick and easy process to scale up the algal cells. For instance, Fe_3O_4, a metal-oxide nanoparticle has a positive surface charge, which attracts the oppositely charged microalgal cell carrying a net negative surface charge. The concept of separation is based on the opposite charge attraction among the microalgae and nanoparticles. [48] Almomani harvested *Spirulina* sp. with the help of Fe magnetic nanoparticles and achieved an efficiency of approximately 97%. The effect of silica-coated magnetic nanoparticles on cell flocculation, lipid extraction, and linoleic acid production from *Chlorella pyrenoidosa* was studied for better harvesting and production [55].

Polymeric nanomaterials Chitosan is a positively charged biopolymer. It is a potential flocculant that is used to harvest the microalgal biomass. The basic mechanism lying behind the flocculation is charge neutralization and electrostatic bridge formation that help in the harvesting of marine algae*Nannochloropsis* sp. using ionic gelation method leading to the formation of chitosan nanopolymer [56]. Biopolymer nanocomposite of TiO_2-Chitosan was found very effective for the harvesting of *Chlorella minutissima* with an efficiency of 98%, only the problem is associated with its high cost [54] The separation took place because of the opposite charge of microalgal cells leading to precipitation [57].

Handbook of Algal Biofuels

TABLE 24.1 Effect of various nanoparticles, their size, and doses on the harvesting efficiency of microalgae.

Microalgae	Nanoparticle	Dose (mg/L)	Size (nm)	Time (min)	Harvesting efficiency (%)	Reference
Chromochloris zofngiensis	Polyethylenimine coated FE_3O_4	200	10.8 ± 4	1	95	[43]
Chlorella vulgaris	Bare iron oxide nanoparticles	100	13.1 ± 2.7	10	>95	[44]
C. vulgaris	Fe_3O_4 – quaternary ammonium salt modified plant polyphenol	5000	–	31	91	[45]
C. vulgaris	Bare Fe_3O_4	10,000	50–100	3	>90	[46]
Scenedesmus obliquus	Ferrite magnetic nanoparticles	1000	10–12	4	99	[47]
Spirulina platensis	Fe-Magnetic nanoparticles-1	30 ± 0.4	11.88 ± 1.51	10	97.4 ± 0.2	[48]
Chlorella sp. TISTR8236	Magnetic-cationic cassava starch	200–800	0.2–127 μm	3	93.77–99.15	[49]
Chlorella ellipsoidea	Polyethylenimine	20	–	2	97	[50]
Scenedesmus sp.	Fe_3O_4 NPs	140	50–100	30	95	[51]
Chlorococcum sp.	Ti NPs	15 mg/L	50	45	85.46	[43]
Nannochloropsis oculata	Organosilicone	Doped inside waveguides	–	11 days	85.95	[52]
C. vulgaris	Cationic starch nanoparticles	7.1	62	4	92	[52]
Scenedesmus ovalternus	Bare Fe_3O_4	0.5 g/g	13.1 ± 2.7	5	>95	[44]
Mixed algae	MgAC-Fe_3O_4	419–472	3.5–7.14	10	80	[53]
Chlorella minutissima	Chitosan-TiO_2	0.09 g/g	90–100	2	>98	[54]

Hybrid nanoparticles They are reported to be the most efficient source for harvesting irrespective of the ecological variations like the microalgal strain, pH, and cultivation medium. Harvesting efficiency of 99% and 80% were achieved using rod-shaped and spherical nanoparticles covered with the polydiallyl dimethyl ammonium chloride respectively [58]. The higher the positive charge, the smaller the aggregation was achieved. Similarly, in another experiment harvesting rate of 85% was attained for *Chlorella* sp. in 30 min [45]. The hybrid nanoparticle, aminoclay coated with TiO_2 MgAC-Fe_3O_4 was used to harvest a mixed microalgae culture [53].

24.3 Lipid extraction

In comparison to plants and animals, the quantity of fatty acids and lipids found more in algae, which makes it a potential source of biofuel production. There is a variety of cell disruption and lipid extraction methods are known for algal biomass, which is majorly distinguished as a physical, chemical, mechanical, and enzymatic or biological method as represented in previous studies [59,60]. On the industrial scale, the solvent-based method is widely used for lipid extraction, but the method is environmentally antagonistic due to its negative effects on human health in addition to transformation in end-product quality [61]. On comparing different solvent ratios for lipid extraction on diverse algae species 1:1 chloroform: methanol ratio (v/v) found most effective [62]. Some other conventional lipid extraction methods involve simple as well as cryogenic grinding, but it cannot be commercialized as these conventional methods are not economically viable [61].

Bead milling with bead size 0.5 mm was considered as a most efficient mechanical process for cell disruption and lipid extraction from algal biomass, but the method may not be applicable for few microalgae like *Chlorella vulgaris* [61,63,64]. At the industrial scale, the major challenge with the bead milling method for lipid extraction from algal biomass is overheating and additional processing steps [64]. In contrast to industrial scale, autoclave and homogenization techniques are suggested at laboratory scale because at large scale these techniques are not economically sustainable [64].

In the case of plant biomass, the Soxhlet method is most acceptable for oil and lipid extraction even for nutrient and chemical extraction [61]. Solvent-based lipid extraction became unfeasible due to the utilization of a high amount of toxic solvents, increasing extraction time duration along reduced lipid content [62]. The challenge of toxic solvents is reduced by switching to chemical and enzymatic methods of lipid extraction but these methods are expensive as the end product needs further purification and additional down streaming process [61]. The major challenge with the enzymatic method is a reduction in the catalytic activity of the enzymes with time and high cost of enzymes like in case of pectinases, xylanases, and cellulase [64].

Supercritical fluid extraction, pressurized liquid extraction, nanosecond pulse electric field, ultrasound-assisted extraction, and microwave-assisted extraction are some most effective and modern physical methods for cell disruption and lipid extraction from algal biomass [79–83]. In supercritical fluid extraction, CO_2 is used as a solvent under the supercritical condition to extract fatty acids from biomass, but in pressurized liquid extraction different variety of solvents are applied with high pressure to extract polar as well as nonpolar lipids [77]. The major challenges with both methods are excessive energy demand to attain high temperature and pressure that result in end product degradation, which leads to unacceptability at the industrial level [62].

During extraction of lipid from *Botryococcus braunii* through nanosecond pulse electric field method, it was reported that for lipid extraction 16.7 J/mL power is required with 50 pulses, but at large scale, this physical method is not much analyzed [78]. In ultrasound-assisted extraction, the functional frequency is between 20–100 MHz, which form bubbles and ultimately enhance their size and leads to cell wall disruption [62]. To low power demand, bath sonication is much economical than horn sonication [79]. The

556
24. algal biofuel technology

microwave-assisted extraction method is one of the most environmentally sustainable options than other methods due to various benefits such as high-speed heating leads to less input energy demand, decreased extraction time duration, fewer solvents utilization, more purified product along with better yield [61]. Lipid extraction from algal biomass with other value-added products appears more feasible with the microwave-assisted extraction method [61].

24.3.1 Biofuel production

Once the cells are disrupted and lipid extraction was done with a suitable technique, cell debris is removed for further processing of lipids. For the process of biodiesel synthesis, the extracted lipids are converted into the Fatty acid methyl esters (FAME) using the transesterification process (Reaction 1). Other than transesterification, pyrolysis, blending, microemulsions can also be used for the lipid conversion. Transesterification is the most prevalent technique and it reduces the viscosity of the Triacyl glycerides. The transesterification process involves the usage of different catalysts whether it is methanol, acid, or base catalysts, further producing glycerol as a byproduct [86,87]. It involves the triacylglycerol reacting with alcohol in the presence of acid or base catalyst leading to the formation of alkyl ester of the corresponding alcohol and glycerol [80]. Presently the most common method used commercially is the homogenous alkali-catalyzed transesterification using KOH or NaOH. But it requires an enormous amount of alcohol to increase the conversion rate. It is also stated that typical alkali catalyst-based reactions are faster than acid-based reactions. Low temperature and pressure are the prerequisites for the reaction to occur fast. Along with FAME and glycerol production, base and lipid-free fatty acids react to form in a saponification reaction, lowering the overall conversion. To overcome the hurdle of saponification, researchers have proposed acid catalyst-based transesterification reactions including the use of H_2SO_4 [81]. Pandit et al., used CaO nanocatalyst for transesterification of the lipid extracted from *Scenedesmus aromatus* [90].

Nowadays, enzyme catalysis is being considered as an alternative for performing the transesterification reaction with high specificity. Table 24.2 discusses the use of disparate catalysts in transesterification and their respective biodiesel yield. Extraction of lipid and transesterification can be combined in a single step using a direct transesterification process. Often supercritical fluids mainly supercritical methanol can be exploited for this purpose. Conventional techniques are generally impeded by the water inhibitory impact, but the use of supercritical methanol enables a solitary phase formation both with the lipid and water, ultimately reducing the deterioration effect of water. In a study by the use of supercritical ethanol to produce biodiesel was employed, where the yield peaked at 67% of total lipids at a temperature of 265°C in 20 min [91]. A ratio of 1:9 (w/v) of dry microalgae/ethanol was used to produce Fatty acid ethyl esters having parallel characteristics to FAMEs except for higher oxidative stability and calorific value. In another work, [82] used bionanocatalysts for transesterification, which is a novel technique reducing the toxic effects of nanomaterials. The studies showed an increase of 1.29 and 1.02-folds in the yield by using Bionano-$CaCO_3$ and CaO, respectively. Although transesterification is an energy-intensive process use of supercritical methanol eliminates the requirement of dry

Handbook of Algal Biofuels

24.3 Lipid extraction

TABLE 24.2   Comparison between various methods of cell disruption and lipid extraction from microalgae.

Method	Advantage	Disadvantages	Algal species	Lipid productivity (%)	References
Supercritical CO$_2$	• High-speed extraction with less toxic lipid content • Efficiency is more	• Expensive equipment and solvent required • High energy requirement	• *Shizochytrium limacinum* • *Pavlova* sp.	33.9 34	[65,66]
Soxhlet	• Easy to perform	• Low-efficiency rate • Time-consuming • Labor intensive	• *Shizochytrium limacinum* • *Nannochloropsis oculata* • *Pavlova* sp. • *Synechocystis PCC 6803*	45 40.90 45.2 48	[67–69]
Solvent extraction	• Easy and fast extraction method • Moderate efficiency rate • Product recovery by distillation or heating	• High toxicity • Inflammable • High energy and expensive solvents required	• *Synechocystis PCC 6803*	52	[67,70]
Pressurized fluid extraction	• Efficiency rate high	• Expensive due to solvents and pressurized nitrogen • High energy input	• *N. oculata*	36	[71,72]
Ultrasonic assisted extraction	• High-efficiency rate • Fast and easy	• High initial investment and maintenance cost • High energy requirement	• *Scenedesmus* sp. • *Chlorella vulgaris*	59.3 16.9	[73]
Microwave-assisted	• Very high-efficiency rate and fast • Adequate quality of an extract • Simple instrumentation	• High initial investment and maintenance cost • Extremely high energy requirement	• *Scenesdemus quadricauda*	49	[74]
Isotonic extraction	• Moderate efficiency rate	• Expensive due to "green" nonvolatile solvents • High energy input	• *Chlorella* sp.	19	[75]
Enzyme extraction	• Fast and highly specific	• High energy requirement	• *C. vulgaris*	5.27	[76]

Handbook of Algal Biofuels

558

24. algal biofuel technology

microalgae and producing a pure product, hence reducing the downstream purification costs. The transesterification technique involving supercritical fluids is quick, safe, easily controlled, and ecofriendly. But the high cost makes it economically less feasible. However, current research is targeting on several operations for producing biodiesel from microalgae (Table 24.3).

24.3.2 Economic studies

Various factors like feedstock price and the selling price of the biodiesel and byproducts produced from the microalgae directly influence the commercial feasibility of biodiesel production. Atabani et al. [90] reported the average cost of biodiesel production is $0.50/L and diesel fuel is $0.35/L [99]. The production price of biodiesel can be calculated using the following equation [91]:

Biodiesel production price

= Biodiesel production price operating costs ($/year)

−byproduct credit($/year)product year(kg/year)

TABLE 24.3 Use of different catalysts on microalgae and respective biodiesel yield.

Microalgae	Catalyst	Temperature ($^\circ$C)	Reaction time (h)	Biofuel yield (%)	References
Scenedesmus armatus	CaO nanocatalyst	75	4	90.44	[83]
Nanochloropsis oculate	Poly ethylene glycol encapsulated ZnO-Mn^{2+}	60	4	87.5	[84]
Nanochloropsis oculate	Bi$_2$O$_3$/ZrO$_2$	80	6	67.51	[85]
Nannochloropsis gaditana	16 L of water, 36.4 L hexane, 36.4 L methanol, 18.2 L 98% sulfuric acid catalyst	—	2	85.5	[86]
Chlorella sp.	Methanol, chloroform, 0.08 mol H$_2$SO$_4$	—	1.66	81.2	[87]
Chlorellap yrenoidosa	0.5 M H$_2$SO$_4$	120	3	90.9	[88]
Chlorella salina	Cellulase/lipase immobilized magnetic NPs	45	60	93.56	[53]
Nannochloropsis sp.	Mg-Zr catalyst	65	4	28	[53]
Spirulina sp.	Zn-Mg-ferrite magnetic NPs	320	1	37.1	[47]
Nannochloropsis oculata	Molecular sieve zeolite A	60	19	17	[89]

Handbook of Algal Biofuels

This cost includes the total manufacturing cost (labor, raw material, etc.), fixed charges like tax, and generalized expenses, that is, research, finance, and development [92]. After the detailed analysis of the capital cost, it was revealed that the biodiesel obtained from the vegetable oil using alkali catalyst had the lowest fixed capital cost whereas CaO-catalyzed transesterification of waste cooking oil had a lower total manufacturing cost. It is been predicted in China that the cost of biodiesel production from algal oil from 2019 to 2022 would be 2.29 USD/kg, which is marginally higher than the commercial diesel (1.08 USD/kg). After this comparative study, it would not be incorrect to state that the biodiesel production from the microalgae lacks economic feasibility [93]. Suitable microalgae strain selection with high lipid productivity could be a key step to reduce the cost and make the process economically viable [94]. Furthermore, the energy consumption cost, distillation cost associated with the separation of lipids, and expenses related to drying of biomass also affect the net biodiesel production cost from the microalgal oil. These outlays can be significantly reduced by using wet concentrated algal biomass [95]. Pretreatment is also one such remedy, which can be used to obtain biodiesel efficiently [80]. Therefore algal cell wall disruption is an essential step to enhance lipid removal requiring an additional cost of solvents. Beyond that, the parameters like the oil/alcohol molar ratio and catalyst quantity can directly have an impact on the production cost [96].

Cost-effective materials like eggshells, biochar from coconut shells, etc. can be used as an alternative to the catalysts [97]. Like in a study an alkaline heterogeneous catalyst derived from the chicken eggshell was used and attained biodiesel with a yield of 93% [98]. Harvesting is considered as one of the costliest steps among the whole process, it holds about 20%−30% of the total production cost [99]. According to an expenditure analysis done by the United States National Renewable Energy Laboratory, the total cost of algal biodiesel production was in the range of $0.53−0.85/L. Additionally, $0.42−0.97/L was the final cost of the biodiesel from algal oil as calculated by disparate government databases and sellers [99].

24.4 Future perspectives

Despite the extensive research, still there lies room for betterment. The major hurdle that exists in commercializing the biodiesel production from the algal biomass is the high cost. The diesel obtained from the algal sources is of comparable quality to the conventional diesel as per the standards. So, the only limiting factor exits are economic feasibility. The major cost associated is with the harvesting and lipid extraction, which can be reduced by using a combination of conventional and advanced techniques like the implication of nanotechnology etc. However, concerning biodiesel production from algal biomass majorly the research is focused on the decelerating toxic effects of nanomaterials. Therefore there is a scope that can be discovered by unraveling the positive impacts of nanomaterials on the process of biodiesel production from algal biomass. Similarly, using wastewater as the culture medium for cultivation can be referred to as a single measure targeting a range of problems. Parallel research is being done to improve every aspect of the pitfalls involved. Applying algae−nanohybrids in wastewater cultivation can be a feasible solution for economical biodiesel production. If the desired changes would be implemented to realize the complete energy capability of the algal biofuel systems, then it would be considered as the major step for sustainable development in the interest of mankind across the globe.

24.5 Conclusions

Algae-mediated biofuel production is still not economically viable and commercially feasible due to high energy input in cultivation, harvesting, and extraction techniques. A large number of innovative researches are taking place for the improvement of these processing steps. The utilization of wastewater as a nutrient source in algae cultivation is enormously beneficial from both environmental and economic perspectives. Recent advancement in a few of these fields like the incorporation of nanomaterials for harvesting and lipid extraction enhances the interest of the scientific community towards algae biofuel. Nanocatalysts are also taking the attention of biodiesel industries in the last few decades. Value-added coproducts obtained from algae under the biorefinery approach, making biofuel production more economical. The biorefinery is based on a zero-waste technique, which catches the attention of various industrial sectors.

Acknowledgment

The corresponding author (MPR) expresses her gratitude to the Mission innovation unit, Department of Biotechnology, New Delhi (INDIA) for financial support [file no. BT/PR31218/PBD/26/771/2019]. Authors KY, Reetu, and MPR are thankful to Amity University Uttar Pradesh, Noida for providing the required facilities.

References

[1] V.L. Mangesh, P. Tamizhdurai, P. Santhana Krishnan, S. Narayanan, S. Umasankar, S. Padmanabhan, et al., Green energy: hydroprocessing waste polypropylene to produce transport fuel, J. Clean. Prod. (2020). Available from: https://doi.org/10.1016/j.jclepro.2020.124200.

[2] A. Raheem, P. Prinsen, A.K. Vuppaladadiyam, M. Zhao, R. Luque, A review on sustainable microalgae based biofuel and bioenergy production: recent developments, J. Clean. Prod. 181 (2018) 42–59.

[3] J.C. de Carvalho, E.B. Sydney, L.F. Assú Tessari, C.R. Soccol, Culture media for mass production of microalgae, Biofuels from Algae, Elsevier, 2019, pp. 33–50. Available from: https://doi.org/10.1016/b978-0-444-64192-2.00002-0.

[4] A.E.F. Abomohra, M. Elsayed, S. Esakkimuthu, M. El-Sheekh, D. Hanelt, Potential of fat, oil and grease (FOG) for biodiesel production: a critical review on the recent progress and future perspectives, Prog. Energy Combust. Sci. 81 (2020) 100868. Available from: https://doi.org/10.1016/j.pecs.2020.100868.

[5] J. Lü, C. Sheahan, P. Fu, Metabolic engineering of algae for fourth generation biofuels production, Energy Environ. Sci. 4 (2011) 2451–2466.

[6] T.R. Brown, R.C. Brown, A review of cellulosic biofuel commercial-scale projects in the United States, Biofuels, Bioprod. Bioref. (2013). Available from: https://doi.org/10.1002/bbb.1387.

[7] E. Phillips, Algal nanotech for biofuel production, Agrica (2020). Available from: https://doi.org/10.5958/2394-448x.2020.00003.6.

[8] K. Heimann, Novel approaches to microalgal and cyanobacterial cultivation for bioenergy and biofuel production, Curr. Opin. Biotechnol. (2016). Available from: https://doi.org/10.1016/j.copbio.2016.02.024.

[9] J. Sen Tan, S.Y. Lee, K.W. Chew, M.K. Lam, J.W. Lim, S.H. Ho, et al., A review on microalgae cultivation and harvesting, and their biomass extraction processing using ionic liquids, Bioengineered (2020). Available from: https://doi.org/10.1080/21655979.2020.1711626.

References

[10] E.S. Shuba, D. Kifle, Microalgae to biofuels: 'Promising' alternative and renewable energy, review, Renew. Sustain. Energy Rev. (2018). Available from: https://doi.org/10.1016/j.rser.2017.08.042.

[11] T.A. Beacham, J.B. Sweet, M.J. Allen, Large scale cultivation of genetically modified microalgae: a new era for environmental risk assessment, Algal Res. (2017). Available from: https://doi.org/10.1016/j.algal.2017.04.028.

[12] K. Sudhakar, R. Mamat, M. Samykano, W.H. Azmi, W.F.W. Ishak, T. Yusaf, An overview of marine macroalgae as bioresource, Renew. Sustain. Energy Rev. (2018). Available from: https://doi.org/10.1016/j.rser.2018.03.100.

[13] K.N. Ingle, H. Traugott, A. Golberg, Challenges for marine macroalgal biomass production in Indian coastal waters, Bot. Mar. (2020). Available from: https://doi.org/10.1515/bot-2018-0099.

[14] M. Ganesan, K. Eswaran, C.R.K. Reddy, Farming of agarophytes in India—a long-time sustainability for the industry and preserving wild stocksin:J. Appl. Phycol. (2017). Available from: https://doi.org/10.1007/s10811-017-1128-0.

[15] J.J. Milledge, P.J. Harvey, Potential process 'hurdles' in the use of macroalgae as feedstock for biofuel production in the British Isles, J. Chem. Technol. Biotechnol. (2016). Available from: https://doi.org/10.1002/jctb.5003.

[16] K. Ingle, E. Vitkin, A. Robin, Z. Yakhini, D. Mishori, A. Golberg, Macroalgae biorefinery from *Kappaphycus alvarezii*: conversion modeling and performance prediction for India and Philippines as examples, Bioenergy Res. (2018). Available from: https://doi.org/10.1007/s12155-017-9874-z.

[17] L. Mennella, D. Tosco, F. Alberti, L. Cembalo, M. Crescimanno, T. Del Giudice, et al., Perspectives and challenges of small scale plant microalgae cultivation. Evidences from Southern Italy, Algal Res. (2020). Available from: https://doi.org/10.1016/j.algal.2019.101693.

[18] L. Amer, B. Adhikari, J. Pellegrino, Technoeconomic analysis of five microalgae-to-biofuels processes of varying complexity, Bioresour. Technol. (2011). Available from: https://doi.org/10.1016/j.biortech.2011.08.010.

[19] G.C. Zittelli, N. Biondi, L. Rodolfi, et al., Photobioreactors for mass production of microalgae, in: Handbook of Microalgal Culture: Applied Phycology and Biotechnology. Second ed., 2013. https://doi.org/10.1002/9781118567166.ch13.

[20] M.R. Tredici, L. Rodolfi, N. Biondi, N. Bassi, G. Sampietro, Techno-economic analysis of microalgal biomass production in a 1-ha Green Wall Panel (GWP®) plant, Algal Res. (2016). Available from: https://doi.org/10.1016/j.algal.2016.09.005.

[21] J.W. Richardson, M.D. Johnson, X. Zhang, P. Zemke, W. Chen, Q. Hu, A financial assessment of two alternative cultivation systems and their contributions to algae biofuel economic viability, Algal Res. (2014). Available from: https://doi.org/10.1016/j.algal.2013.12.003.

[22] W. Sawaengsak, T. Silalertruksa, A. Bangviwat, S.H. Gheewala, Life cycle cost of biodiesel production from microalgae in Thailand, Energy Sustain. Dev. (2014). Available from: https://doi.org/10.1016/j.esd.2013.12.003.

[23] M. Stark, I. O'Gara, An introduction to photosynthetic microalgae, Disrupt. Sci. Technol. (2012). Available from: https://doi.org/10.1089/dst.2012.0017.

[24] P.L. Show, M.S.Y. Tang, D. Nagarajan, T.C. Ling, C.W. Ooi, J.S. Chang, A holistic approach to managing microalgae for biofuel applications, Int. J. Mol. Sci. (2017). Available from: https://doi.org/10.3390/ijms18010215.

[25] S. Jenkins, E.G. Barrett-Lennard, Z. Rengel, Impacts of waterlogging and salinity on puccinellia (Puccinellia ciliata) and tall wheatgrass (Thinopyrum ponticum): zonation on saltland with a shallow water-table, plant growth, and Na+ and K+ concentrations in the leaves, Plant Soil 329 (2010) 91–104. Available from: https://doi.org/10.1007/s11104-009-0137-4.

[26] P.L. Gupta, S.M. Lee, H.J. Choi, A mini review: photobioreactors for large scale algal cultivation, World J. Microbiol. Biotechnol. (2015). Available from: https://doi.org/10.1007/s11274-015-1892-4.

[27] Q. Huang, F. Jiang, L. Wang, C. Yang, Design of photobioreactors for mass cultivation of photosynthetic organisms, Engineering (2017). Available from: https://doi.org/10.1016/J.ENG.2017.03.020.

[28] L. Wolf, T. Cummings, K. Müller, M. Reppke, M. Volkmar, D. Weuster-Botz, Production of β-carotene with *Dunaliella salina* CCAP19/18 at physically simulated outdoor conditions, Eng. Life Sci. (2020). Available from: https://doi.org/10.1002/elsc.202000044.

[29] G.S. Murthy, Overview and assessment of algal biofuels production technologies, in: Biofuels, 2011. https://doi.org/10.1016/B978-0-12-385099-7.00019-X.

562   24. algal biofuel technology

[30] G. Roesijadi, S.B. Jones, L.J. Snowden-Swan, Y. Zhu, Macroalgae as A Biomass Feedstock: A Preliminary Analysis, Pacific Northwest National Lab.(PNNL), Richland, WA, 2010.

[31] L. Mesnildrey, M. Lesueur, C. Jacob, K. Frangoudes, Seaweed industry in France, Report. Interreg Progr. NETALGAE, Les Publ. Du Pôle Halieut. AGROCAMPUS OUEST, 2012.

[32] M. Ghadiryanfar, K.A. Rosentrater, A. Keyhani, M. Omid, A review of macroalgae production, with potential applications in biofuels and bioenergy, Renew. Sustain. Energy Rev. 54 (2016) 473−481.

[33] T. Bruton, H. Lyons, Y. Lerat, M. Stanley, M.B. Rasmussen, A review of the potential of marine algae as a source of biofuel in Ireland, Sustainable Energy, 2009.

[34] A.E.-F.A.E.F. Abomohra, W. Jin, V. Sagar, G.A.G.A. Ismail, Optimization of chemical flocculation of Scenedesmus obliquus grown on municipal wastewater for improved biodiesel recovery, Renew. Energy 115 (2018) 880−886. Available from: https://doi.org/10.1016/j.renene.2017.09.019.

[35] A.J. Dassey, S.G. Hall, C.S. Theegala, An analysis of energy consumption for algal biodiesel production: comparing the literature with current estimates, Algal Res. (2014). Available from: https://doi.org/10.1016/j.algal.2013.12.006.

[36] J. Kim, G. Yoo, H. Lee, J. Lim, K. Kim, C.W. Kim, et al., Methods of downstream processing for the production of biodiesel from microalgae, Biotechnol. Adv. 31 (2013) 862−876. Available from: https://doi.org/10.1016/j.biotechadv.2013.04.006.

[37] A.I. Barros, A.L. Gonçalves, M. Simões, J.C.M. Pires, Harvesting techniques applied to microalgae: a review, Renew. Sustain. Energy Rev. 41 (2015) 1489−1500. Available from: https://doi.org/10.1016/j.rser.2014.09.037.

[38] T. Mathimani, N. Mallick, A comprehensive review on harvesting of microalgae for biodiesel—key challenges and future directions, Renew. Sustain. Energy Rev. 91 (2018) 1103−1120. Available from: https://doi.org/10.1016/j.rser.2018.04.083.

[39] J.J. Milledge, S. Heaven, A review of the harvesting of micro-algae for biofuel production, Rev. Environ. Sci. Bio/Technol. 12 (2013) 165−178.

[40] A. Khanra, S. Vasistha, M. Prakash Rai, ZrO_2 nanoparticles mediated flocculation and increased lipid extraction in chlorococcum sp. for biodiesel production: a cost effective approach, Materials Today: Proceedings, Elsevier Ltd., 2020, pp. 1847−1852. Available from: https://doi.org/10.1016/j.matpr.2020.05.290.

[41] D. Vandamme, I. Foubert, K. Muylaert, Flocculation as a low-cost method for harvesting microalgae for bulk biomass production, Trends Biotechnol. 31 (2013) 233−239. Available from: https://doi.org/10.1016/j.tibtech.2012.12.005.

[42] Y.-C. Lee, H.U. Lee, K. Lee, B. Kim, S.Y. Lee, M.-H. Choi, et al., Aminoclay-conjugated TiO_2 synthesis for simultaneous harvesting and wet-disruption of oleaginous Chlorella sp, Chem. Eng. J. 245 (2014) 143−149. Available from: https://doi.org/10.1016/j.cej.2014.02.009.

[43] K. Gerulová, A. Bartošová, L. Blinová, K. Bártová, M. Dománková, Z. Garaiová, et al., Magnetic Fe_3O_4-polyethyleneimine nanocomposites for efficient harvesting of Chlorella zofingiensis, Chlorella vulgaris, Chlorella sorokiniana, Chlorella ellipsoidea and Botryococcus braunii, Algal Res. 33 (2018) 165−172. Available from: https://doi.org/10.1016/j.algal.2018.05.003.

[44] P. Fraga-García, P. Kubbutat, M. Brammen, S. Schwaminger, S. Berensmeier, Bare iron oxide nanoparticles for magnetic harvesting of microalgae: from interaction behavior to process realization, Nanomaterials 8 (2018). Available from: https://doi.org/10.3390/nano8050292.

[45] Y. Zhao, Q. Fan, X. Wang, X. Jiang, L. Jiao, W. Liang, Application of Fe_3O_4 coated with modified plant polyphenol to harvest oleaginous microalgae, Algal Res. 38 (2019) 101417. Available from: https://doi.org/10.1016/j.algal.2019.101417.

[46] L.D. Zhu, E. Hiltunen, Z. Li, Using magnetic materials to harvest microalgal biomass: evaluation of harvesting and detachment efficiency, Environ. Technol. (United Kingdom) 40 (2019) 1006−1012. Available from: https://doi.org/10.1080/09593330.2017.1415379.

[47] D. Egesa, C.J. Chuck, P. Plucinski, Multifunctional role of magnetic nanoparticles in efficient microalgae separation and catalytic hydrothermal liquefaction, ACS Sustain. Chem. Eng. 6 (2018) 991−999. Available from: https://doi.org/10.1021/acssuschemeng.7b03328.

[48] F. Almomani, Algal cells harvesting using cost-effective magnetic nano-particles, Sci. Total Environ. 720 (2020) 137621. Available from: https://doi.org/10.1016/j.scitotenv.2020.137621.

[49] K. Jangyubol, K. Kasemwong, T. Charoenrat, S. Chittapun, Magnetic−cationic cassava starch composite for harvesting Chlorella sp. TISTR8236, Algal Res. 35 (2018) 561−568. Available from: https://doi.org/10.1016/j.algal.2018.09.027.

Handbook of Algal Biofuels

References

[50] Y. Yang, J. Hou, P. Wang, C. Wang, L. Miao, Y. Ao, et al., Interpretation of the disparity in harvesting efficiency of different types of *Microcystis aeruginosa* using polyethylenimine (PEI)-coated magnetic nanoparticles, Algal Res. 29 (2018) 257–265. Available from: https://doi.org/10.1016/j.algal.2017.10.020.

[51] A. Abo Markeb, J. Llimós-Turet, I. Ferrer, P. Blánquez, A. Alonso, A. Sánchez, et al., The use of magnetic iron oxide based nanoparticles to improve microalgae harvesting in real wastewater, Water Res. 159 (2019) 490–500. Available from: https://doi.org/10.1016/j.watres.2019.05.023.

[52] Y. Sun, Y. Huang, Q. Liao, A. Xia, Q. Fu, X. Zhu, et al., Boosting nannochloropsis oculata growth and lipid accumulation in a lab-scale open raceway pond characterized by improved light distributions employing built-in planar waveguide modules, Bioresour. Technol. 249 (2018) 880–889. Available from: https://doi.org/10.1016/j.biortech.2017.11.013.

[53] B. Kim, V.K. Bui, W. Farooq, S.G. Jeon, Y.-K. Oh, Y.-C. Lee, Magnesium aminoclay-Fe_3O_4 (MgAC-Fe_3O_4) hybrid composites for harvesting of mixed microalgae, Energies 11 (2018). Available from: https://doi.org/10.3390/en11061359.

[54] R. Dineshkumar, A. Paul, M. Gangopadhyay, N.D.P. Singh, R. Sen, Smart and reusable biopolymer nanocomposite for simultaneous microalgal biomass harvesting and disruption: integrated downstream processing for a sustainable biorefinery, ACS Sustain. Chem. Eng. 5 (2017) 852–861. Available from: https://doi.org/10.1021/acssuschemeng.6b02189.

[55] S. Vasistha, A. Khanra, M.P. Rai, Progress and challenges in biodiesel production from microalgae feedstock, in: Microalgae Biotechnology for Development of Biofuel and Wastewater Treatment., 2019. Available from: https://doi.org/10.1007/978-981-13-2264-8_14.

[56] M.S. Farid, A. Shariati, A. Badakhshan, B. Anvaripour, Using nano-chitosan for harvesting microalga Nannochloropsis sp, Bioresour. Technol. 131 (2013) 555–559. Available from: https://doi.org/10.1016/j.biortech.2013.01.058.

[57] Y.-C. Lee, K. Lee, Y.-K. Oh, Recent nanoparticle engineering advances in microalgal cultivation and harvesting processes of biodiesel production: a review, Bioresour. Technol. 184 (2015) 63–72. Available from: https://doi.org/10.1016/j.biortech.2014.10.145.

[58] J.K. Lim, D.C.J. Chieh, S.A. Jalak, P.Y. Toh, N.H.M. Yasin, B.W. Ng, et al., Rapid magnetophoretic separation of microalgae, Small 8 (2012) 1683–1692. Available from: https://doi.org/10.1002/smll.201102400.

[59] Y.S. Shin, H. Il Choi, J.W. Choi, J.S. Lee, Y.J. Sung, S.J. Sim, Multilateral approach on enhancing economic viability of lipid production from microalgae: a review, Bioresour. Technol. 258 (2018) 335–344. Available from: https://doi.org/10.1016/j.biortech.2018.03.002.

[60] H. Rajhi, D. Puyol, M.C. Martínez, E.E. Díaz, J.L. Sanz, Vacuum promotes metabolic shifts and increases biogenic hydrogen production in dark fermentation systems, Front. Environ. Sci. Eng. (2016). Available from: https://doi.org/10.1007/s11783-015-0777-y.

[61] R. Kapoore, T. Butler, J. Pandhal, S. Vaidyanathan, Microwave-assisted extraction for microalgae: from biofuels to biorefinery, Biol. (Basel) (2018). Available from: https://doi.org/10.3390/biology7010018.

[62] R.R. Kumar, P.H. Rao, M. Arumugam, Lipid extraction methods from microalgae: a comprehensive review, Front. Energy Res. (2015). Available from: https://doi.org/10.3389/fenrg.2014.00061.

[63] R.V. Kapoore, S. Vaidyanathan, Towards quantitative mass spectrometry-based metabolomics in microbial and mammalian systems, Philos. Trans. R. Soc. A Math. Phys. Eng. Sci. (2016). Available from: https://doi.org/10.1098/rsta.2015.0363.

[64] M. Gong, A. Bassi, Carotenoids from microalgae: a review of recent developments, Biotechnol. Adv. (2016). Available from: https://doi.org/10.1016/j.biotechadv.2016.10.005.

[65] C.H. Cheng, T.B. Du, H.C. Pi, S.M. Jang, Y.H. Lin, H.T. Lee, Comparative study of lipid extraction from microalgae by organic solvent and supercritical CO_2, Bioresour. Technol. (2011). Available from: https://doi.org/10.1016/j.biortech.2011.08.064.

[66] A. Patel, L. Matsakas, K. Sartaj, R. Chandra, Extraction of lipids from algae using supercritical carbon dioxide, in: Green Sustainable Process for Chemical and Environmental Engineering and Science, 2020. Available from: https://doi.org/10.1016/b978-0-12-817388-6.00002-7.

[67] J. Sheng, R. Vannela, B.E. Rittmann, Evaluation of methods to extract and quantify lipids from Synechocystis PCC 6803, Bioresour. Technol. (2011). Available from: https://doi.org/10.1016/j.biortech.2010.08.007.

[68] S. Tang, C. Qin, H. Wang, S. Li, S. Tian, Study on supercritical extraction of lipids and enrichment of DHA from oil-rich microalgae, J. Supercrit. Fluids (2011). Available from: https://doi.org/10.1016/j.supflu.2011.01.010.

[69] M.A. López-Bascón-Bascon, M.D. Luque de Castro, Soxhlet extraction, in: Liquid-Phase Extraction, 2019. Available from: https://doi.org/10.1016/B978-0-12-816911-7.00011-6.

[70] M.J. Jiménez Callejón, A. Robles Medina, M.D. Macías Sánchez, L. Esteban Cerdán, P.A. González Moreno, E. Navarro López, et al., Obtaining highly pure EPA-rich lipids from dry and wet Nannochloropsis gaditana microalgal biomass using ethanol, hexane and acetone, Algal Res. (2020). Available from: https://doi.org/10.1016/j.algal.2019.101729.

[71] S. Pieber, S. Schober, M. Mittelbach, Pressurized fluid extraction of polyunsaturated fatty acids from the microalga Nannochloropsis oculata, Biomass Bioenergy (2012). Available from: https://doi.org/10.1016/j.biombioe.2012.10.019.

[72] D.A. Esquivel-Hernández, I.P. Ibarra-Garza, J. Rodríguez-Rodríguez, S.P. Cuéllar-Bermúdez, M. de, J. Rostro-Alanis, et al., Green extraction technologies for high-value metabolites from algae: a review, Biofuels, Bioprod. Bioref. (2017). Available from: https://doi.org/10.1002/bbb.1735.

[73] L. Vernès, M. Vian, F. Chemat, Ultrasound and microwave as green tools for solid-liquid extraction, in: Liquid-Phase Extraction, 2019. Available from: https://doi.org/10.1016/B978-0-12-816911-7.00012-8.

[74] C. Onumaegbu, A. Alaswad, C. Rodriguez, A. Olabi, Modelling and optimization of wet microalgae Scenedesmus quadricauda lipid extraction using microwave pre-treatment method and response surface methodology, Renew. Energy (2019). Available from: https://doi.org/10.1016/j.renene.2018.09.008.

[75] M. Vanthoor-Koopmans, R.H. Wijffels, M.J. Barbosa, M.H.M. Eppink, Biorefinery of microalgae for food and fuel, Bioresour. Technol. (2013). Available from: https://doi.org/10.1016/j.biortech.2012.10.135.

[76] A. Zuorro, G. Maffei, R. Lavecchia, Optimization of enzyme-assisted lipid extraction from Nannochloropsis microalgae, J. Taiwan. Inst. Chem. Eng. (2016). Available from: https://doi.org/10.1016/j.jtice.2016.08.016.

[77] S.P. Jeevan Kumar, G. Vijay Kumar, A. Dash, P. Scholz, R. Banerjee, Sustainable green solvents and techniques for lipid extraction from microalgae: a review, Algal Res. 21 (2017) 138−147. Available from: https://doi.org/10.1016/j.algal.2016.11.014.

[78] B. Hosseini, A. Guionet, H. Akiyama, H. Hosano, Oil extraction from microalgae by pulsed power as a renewable source of energy, IEEE Trans. Plasma Sci. (2018). Available from: https://doi.org/10.1109/TPS.2018.2845902.

[79] M. Al hattab, A. Ghaly, Microalgae oil extraction pre-treatment methods: critical review and comparative analysis fundamentals of renewable energy and applications, J. Fundam. Renew. Energy Appl. (2015). Available from: https://doi.org/10.4172/20904541.1000172.

[80] K.A. Salam, S.B. Velasquez-Orta, A.P. Harvey, A sustainable integrated in situ transesterification of microalgae for biodiesel production and associated co-product-a review, Renew. Sustain. Energy Rev. 65 (2016) 1179−1198. Available from: https://doi.org/10.1016/j.rser.2016.07.068.

[81] E.A. Ehimen, Z.F. Sun, C.G. Carrington, Variables affecting the in situ transesterification of microalgae lipids, Fuel 89 (2010) 677−684. Available from: https://doi.org/10.1016/j.fuel.2009.10.011.

[82] K. Rathinasamy, K. Rani, A. kumar Balasubramaniem, I. Moorthy, A. Dhakshinamoorthy, P. Varalakshmi, Green energy from Coelastrella sp. M-60: bio-nanoparticles mediated whole biomass transesterification for biodiesel production, Fuel 279 (2020) 118490. Available from: https://doi.org/10.1016/j.fuel.2020.118490.

[83] P.R. Pandit, M.H. Fulekar, Biodiesel production fromScenedesmus armatususing egg shell waste as nanocatalyst, Mater. Today Proc. 10 (2019) 75−86. Available from: https://doi.org/10.1016/j.matpr.2019.02.191.

[84] J. Raj, B. Bharathiraja, B. Vijayakumar, S. Arokiyaraj, J. Iyyappan, R. Praveen Kumar, Biodiesel production from microalgae Nannochloropsis oculata using heterogeneous poly ethylene glycol (PEG) encapsulated ZnOMn^{2+} nanocatalyst, Bioresour. Technol. 282 (2019) 348−352.

[85] N. Rahman, A. Ramli, K. Jumbri, Y. Uemura, Biodiesel production from N. oculata microalgae lipid in the presence of Bi_2O_3/ZrO_2 Catalysts, Waste Biomass Valoriz. (2019). Available from: https://doi.org/10.1007/s12649-019-00619-8.

[86] S. Torres, G. Acien, F. García-Cuadra, R. Navia, Direct transesterification of microalgae biomass and biodiesel refining with vacuum distillation, Algal Res. 28 (2017) 30. Available from: https://doi.org/10.1016/j.algal.2017.10.001.

[87] M. Karimi, Exergy-based optimization of direct conversion of microalgae biomass to biodiesel, J. Clean. Prod. 141 (2017) 50−55. Available from: https://doi.org/10.1016/j.jclepro.2016.09.032.

[88] C. Hechun, Z. Zhang, X. Wu, X. Miao, Direct biodiesel production from wet microalgae biomass of Chlorella pyrenoidosa through in situ transesterification, Biomed. Res. Int. 2013 (2013) 930686. Available from: https://doi.org/10.1155/2013/930686.

References

[89] Y. Li, S. Lian, D. Tong, R. Song, W. Yang, Y. Fan, et al., One-step production of biodiesel from Nannochloropsis sp. on solid base Mg–Zr catalyst, Appl. Energy 88 (2011) 3313–3317. Available from: https://doi.org/10.1016/j.apenergy.2010.12.057.

[90] A.E. Atabani, A.S. Silitonga, I.A. Badruddin, T.M.I.I. Mahlia, H.H. Masjuki, S. Mekhilef, A comprehensive review on biodiesel as an alternative energy resource and its characteristics, Renew. Sustain. Energy Rev. 16 (2012) 2070–2093. Available from: https://doi.org/10.1016/j.rser.2012.01.003.

[91] S. Rezania, B. Oryani, J. Park, B. Hashemi, K.K. Yadav, E.E. Kwon, et al., Review on transesterification of nonedible sources for biodiesel production with a focus on economic aspects, fuel properties and by-product applications, Energy Convers. Manage. 201 (2019) 112155.

[92] V. Rahimi, M. Shafiei, Techno-economic assessment of a biorefinery based on low-impact energy crops: a step towards commercial production of biodiesel, biogas, and heat, Energy Convers. Manage. 183 (2019) 698–707. Available from: https://doi.org/10.1016/j.enconman.2019.01.020.

[93] J. Sun, X. Xiong, M. Wang, H. Du, J. Li, D. Zhou, J. Zuo, Microalgae biodiesel production in China: a preliminary economic analysis, Renew. Sustain. Energy Rev. (2019). Available from: https://doi.org/10.1016/j.rser.2019.01.021.

[94] M.L. Menegazzo, G.G. Fonseca, Biomass recovery and lipid extraction processes for microalgae biofuels production: a review, Renew. Sustain. Energy Rev. 107 (2019) 87–107. Available from: https://doi.org/10.1016/j.rser.2019.01.064.

[95] F. Naghdi, L. González, W. Chan, P. Schenk, Progress on lipid extraction from wet algal biomass for biodiesel production, Microb. Biotechnol. 9 (2016). Available from: https://doi.org/10.1111/1751-7915.12360.

[96] M.N. Hussain, T. Al Samad, I. Janajreh, Economic feasibility of biodiesel production from waste cooking oil in the UAE, Sustain. Cities Soc. 26 (2016) 217–226. Available from: https://doi.org/10.1016/j.scs.2016.06.010.

[97] Z.-E. Tang, S. Lim, Y.-L. Pang, H.-C. Ong, K.-T. Lee, Synthesis of biomass as heterogeneous catalyst for application in biodiesel production: state of the art and fundamental review, Renew. Sustain. Energy Rev. 92 (2018) 235–253. Available from: https://doi.org/10.1016/j.rser.2018.04.056.

[98] J. Goli, O. Sahu, Development of heterogeneous alkali catalyst from waste chicken eggshell for biodiesel production, Renew. Energy 128 (2018) 142–154. Available from: https://doi.org/10.1016/j.renene.2018.05.048.

[99] S. Nagarajan, S.K. Chou, S. Cao, C. Wu, Z. Zhou, An updated comprehensive techno-economic analysis of algae biodiesel, Bioresour. Technol. 145 (2013) 150–156. Available from: https://doi.org/10.1016/j.biortech.2012.11.108.

Handbook of Algal Biofuels

CHAPTER 25

History and recent advances of algal biofuel commercialization

Ali Noor[1] and Fouzia Naseer[2]

[1]Department of Biological Sciences, Karakoram International University, Gilgit-Baltistan, Pakistan [2]Department of Botany, University of Karachi, Karachi, Pakistan

25.1 Introduction and history of biofuel production

During geologic processes the remnant of organic matter, that is, fossils containing carbon and hydrogen, are formed, usually in the Earth's crust. These have the ability to be combusted to produce fuel [1]. For a long time fossil fuels have been consumed as the foremost energy source owing to its high energy worth and abundant presence. On the other hand, it is a nonrenewable source of energy and the world now currently prefers many renewable sources for generating energy, for example, water power, solar, geothermal, biomass, and wind energy. Fossil fuel attained from the remnants of plants or animals, namely, natural gas, petroleum, and coal, as shown in Fig. 25.1A−D[2], needs millions of years to naturally replenish, while other sources of renewable sources usually take less time to replenish. That is the reason that fossil fuels are now becoming more expensive and are not an economically justified source [3,4].

Several forms of solid, liquid, and gaseous biofuels, for example, biodiesel, bioethanol, biogas, and biohydrogen, are now becoming the best economical energy sources [5]. Actually, biofuel is a liquid or gaseous form of the latest and sustainable renewable energy sources due to its continuously replenishable quality (ethanol and biodiesel are the most common forms of the recent era). Biofuel is derived from biomass. Gaseous biofuel is usually used to generate heat and power while liquid biofuels are utilized in the transportation sector [6−9].

Biofuel as renewable source of energy is the best solution to be adopted to satisfy the human populations' increased requirements, not only related to their basics also directed toward better economic development and output. But still information related to the development of long-term, renewable energy is limited [10].

FIGURE 25.1 Millions of years require to convert fossils into fuel need, three different types of fossil fuels i.e. (A) dead skeleton bone in desert area source national geographic source: shutterstock.com (B) natural gas source: national geographic (C) Petroleum source: ZME Science and (D) Coal source: Coal Power Impacts.

25.2 Recent advancement in large-scale biofuel production

For a long time petroleum and coal have been used as primary sources of energy because of their presence in large quantities and their low cost [11]. Although fossil fuel contributes almost 80% of energy source, this fuel produces damaging environmental effects as well because it releases unsafe, unhealthy, and risky gases that usually damage biodiversity and the climate. Nonetheless biofuel has also some serious threats for the forestry, soil form, food production, and large amounts of water reservoir for, that reason along with terrestrial source of crop energy, it is dire need to find out some other types of better biomass producing sources to overcome the negative impact of biofuel [7].

Although the production of biofuel is growing day by day, but the success of this biofuel production will depends on several aspects like market price compete with fossil fuel rate , affordable foodstock material for the production of biofuel, advanced techniques , environment friendly and best fitted as substitute source of fossil fuel in worldwide. It has been reported that Brazil, the United States, and the EU (European Union) generate about 90% of its total production but need to progress in production of biofuel in Malaysia, China, and other countries due to its raw material and feedstocks requirement in future, for the extraction of vegetable oil, corn and sugar production Fig. 25.2 [12].

Researchers of many fields are focusing their attention on the identification of the best algal species and their cultivation for the production of fuel energy and bio-based products [13]. However, the production of algal biodiesel requires further consideration from

25.3 Pilot-scale and large-scale trials of algal biofuel production

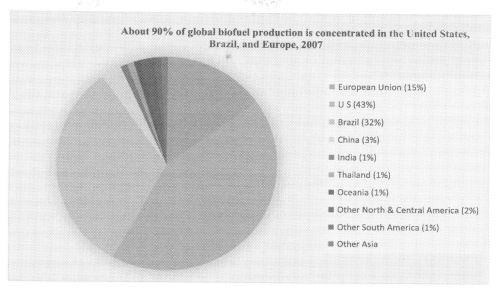

FIGURE 25.2 *Fig. 25.2 Percentage of biofuel productionSource: A Global Perspective. Economic research service, United States Department of Agriculture. (https://www.ers.usda.gov/amber-waves/2007/november/the-future-of-biofuels-a-global-perspective/.*

researchers. In the process of the conversion of algal oil into a biodiesel form, the identification of appropriate species, along with its farming and harvesting are the initial steps, and then it might be possible to extract algal oil and develop standardization [14].

Studies have revealed that algal species with faster growth and the presence of appropriate chemical constituents are preferred for aquaculture [15]. The photosynthetic algae produce biomass with metabolites, for example, sugars, oils, and lipids. These primary metabolites are diagnosed as very beneficial for various food products, in the pharmaceutical field, in industries, and in fuel generation for vehicles. Algae produce secondary metabolites in extreme environmental stress conditions, which is helpful for the production of biofuel with advantages in other areas [16]. The generation of lipids and oils from algae is 30% better than any other terrestrial crop [17]. Initially several fast growing algal species with oil (may possibly convert into biodiesel) are selected, which are able to be converted into biodiesel [14].

The proteins are the main components which are usually found in all the microalgae, while the presence of carbohydrates and lipids are also quite high in various algal species [15]. Microalgae are nowadays all over the world and have become of interest for use in pharmaceutical industries and biofuel production, being beneficial for our environment and the economy [18].

25.3 Pilot-scale and large-scale trials of algal biofuel production

Biofuel production is necessary to overcome the energy requirements; algal biofuel production bridges the gap in the energy demand. The production of different algal products, particularly biofuel production from algae, is a tremendous biotechnological innovation. However, the production of microalgal biofuel is not viable for commercial uses yet at a large scale, due to the low storage amount of lipid mass in microalgal cultivation. Thus the selection of lipid-rich algae for cultivation has become the key research topic for microalgal biodiesel production [19]. For large-scale production of biofuel from algae different trials have been started by different institutions and industries.

The mass production of algae has been extensively explored for applications in the food industry, aquaculture, bioenergy, and particularly biodiesel production [20]. Algae are auspicious alternative sources of lipid for the production of biofuel. The selection and cultivation of algal species with high lipid content are important for the production of biofuel. Different industries and researchers have made efforts to find methods to cultivate the high lipid containing algal species to promote the environment-friendly algal biofuel production at the large scale, to meet the increasing demands for energy. In the literature on the growth rate of microalgal and the productivities of lipid 55 species of microalgae have been used, which include 17 Chlorophyta, 11 Bacillariophyta, and five Cyanobacteria as well as other taxa [21].

The advantages of algal production compared with superior plants have been documented [22,23]. Many algal species have been manufactured and compounds accumulated with high economic value, for example, lipids, carbohydrates, proteins, and different pigments. Different biomass production systems have also been tailored together with different operation levels [24], which can be incorporated fully by mechanically controlled facilities for maximum production. Algal cultivation has gained a lot of consideration in current decades [25]. The algal cultivation at the large scale is practiced in shallow open pond tanks, raceways, or in circular forms.

The Energy Department of the United States supports research and development to manage the challenges in algal biofuel production in their Multi-Year Program Plan (MYPP). The main and long-term aims of MYPP for algal biofuel research are the production of algal biofuel capacity upto 5 million gallons per year at the domestic level by 2030 and the establishment of technologies to produce algal biofuel intermediates at a cost of $3/gallon of gasoline equivalent by 2020. The main distinguishing characteristics of this project were the use of wastewater and wastewater nutrients.

California Polytechnic State University, San Louis Obispo (Cal Poly) sought to develop the production capability of biofuel from intermediates from microalgae grown on municipal wastewater at a 20-acre algae-based wastewater processing facility in Delhi, California, which consists of 7 acres of raceway ponds. It is connecting algal biofuel production with wastewater treatment, exploiting the ample waste nutrients and carbon present in wastewater and minimizing the use of clean water in the algal biorefinery system [26].

25.4 Top companies of biofuel production from different feedstocks

Biofuels are emerging sources of energy, many companies around globe have engaged the researchers, engineers and modern technologies for production of biofuels such as ethanol, biodiesel, biogasoline, green diesel, biogas, etc. are produced from a variety of feedstock including agriculture crops, wood pellets, residues of forests, wood fuel, charcoal, wastes of industries, municipal waste, manures and organic waste and others.

According to the staff writer of NS Energy on 15 November 2019 World Gas Renewables Reviews, published by Italian oil and gas major Eni, in 2018 about 2616 thousand barrels/day biofuel were produced around the globe. The United States and Brazilian biofuel production companies were remained dominated in production. Biogas production in 2018 around the world 1890 thousand barrels/day was recorded, while the total biodiesel 702 barrels/day in the year was produced [27].

25.4.1 Eni Gela biorefinery in Europe

Staff writer of NS Energy on 27 September, 2019 has mentioned that multinational oil and gas company of Italy Eni Spa has started its 750,000 tones Gela biorefinery in Europe, which will be able to increase the utilization quantities of raw material for example, used vegetables, fate of animals, algae and other by products to produce of renewable biodiesel or hydrotreated vegetable oil (HVO) (Fig. 25.3A).

The refinery can take up 100% loads of raw material of second -generations, from waste advanced material and used vegetable oil [28].

25.4.2 Australian Renewable Fuels Limited

Australian Renewable Fuel Limited (ARF) is larger company was established in 2001 by Amadeus Energy Limited. According to the ARF production of world fuel production of biodiesel is roundabout 5 billion liters per year. The company has completed the constriction of biofuel plant in 2006 which produces currently 44.5 million liters of biofuel per year. The Adelaide and Picton facilities utilize different feedstocks like vegetable oils, tallow and cooking oils, which are considered as cheap sources(Fig. 25.3B), [29].

25.4.3 Blue fire renewables

Blu Fire Renewables was established in 2006 its head quarter is found in Irvine, California, United States. The company utilized the Arkenol patented process for producing ethanol from wood wastes, wheat straws, rice and other resides of agriculture. The farm is only viable throughout the world cellulose to ethanol production. It has ranges from producing alcoholic barrages and fuel -ethanol to produce citric acid and xantham gum as food uses. However, the low cost of competing petroleum-based biofuel and the high cost of sugar have held the production of chemicals mostly confined to producing ethanol from corn sugar till-date.

The farm has improved a significant to develop a well-known conversion technology known as Concentrated Acid Hydrolysis, and the process is completed and ready for implementation on commercial basis. The technology is unique in that, it enables for the 1st time, wildly available cellulosic substance or common biomass to be converted into sugar in sustainable manner, which are providing a low-cost raw material for fermentation/ chemical conversion into any several kind useful chemicals (Fig. 25.3C).

FIGURE 25.3 (A) Eni has inaugurated its biorefinery in Gela. (B) Australian Renewable Fuels Limited. (C) Production plant of Blue Fire Renewables. (D) Cosan's Costa Pinto sugar cane mill and ethanol distillery plant at Piracicaba, São Paulo, Brazil. Source: *(A) Photo: courtesy of LEEROY Agency/Pixabay. (B) https://www.bordermail.com.au/story/3714413/buyers-fuel-biodiesel-plant-hope/. (C) http://www.bfreinc.com. (D) https://en.wikipedia.org/wiki/Cosan#/media/File.*

The usage ability of low-cost feedstock to produce products that sell in high value markets provide good business, can be placed any geographical urban and rural regions. Due to the usage of reasonable fuel energy, the production of minimal waste streams has significant benefits for the environment and minimal permitting requirements. [30].

25.4.4 Cosan Limited

Cosan Limited company was established in 1936 situated in Sao Paulo, Brazil. Initially the company engaged in the supply and commercialization of bioethanol, sugar fuels for energy. With its subsidiary Cosan SA Industria e Comercio the farm is active in five sectors. In Raizan Fuel sector, the farm engaged in marketing and distribution of fuel via a franchised network of gasoline station with the brand names Shell and Esso. The Raizan Energy segment engaged in sale and production of many products manufactured by sugarcane, including high polarity sugar and ethanol, as well as the energy cogeneration from sugarcane material and research for production of new technologies. According to CEO Arnold Klann the production process of ethnol from different feedstocks, such as wood wastes, rice, wheat straws, and many other agriculture residues, is patented and only viable throughout the world for manufacturing ethanol from cellulose. The company is the fifth largest in the world in terms of ethanol production (Fig. 25.3D), [31].

25.5 Top companies of algal products commercialization

Algae are photosynthetic and oxygen producing unicellular and multicellular organisms with diverse characteristics like the storage of essential nutrients, vitamins, minerals, microorganisms, and consumption of CO_2. The use of healthy food and supplementary diets and change in human lifestyle have changed the perspectives of the algal industry. The algal product producing companies have been starting to produce different edible items, healthcare remedies, cosmetic products, and jewelry items. The algal production market at the global level is estimated to grow a considerable CAGR of 4.2% from 2018 to 2025. [32]. Due to the increasing demand of natural products and efforts of marketing agents to formulate the consumable products of algae to satisfy quality, taste, texture, and nutritional demand of users. Previously production of biofuel was not considered as production cost was high, however, the present advancement in technology and feasibility algal biofuel is possible to use as alternative of fossil fuel.

Drastic increase in human population and demand of food and health care remedies, the manufacturers of algal products boost up the algal based products to overcome the demand of people and addition for livestock, consumption of algae-based health care drugs etc. Many reputed companies are involved and provide the algal based products, some are as fallows,

25.5.1 Earthrise Nutritionals

The world leading company Earthrise Nutritional was found in 1976 situated in California, United States by a group of visionary partners to develop the blue–green algae

"Spirulina" as a food resource for the world. The group members have started the cultivation of the Spirulina in the hot desert area southeastern part of California later changed the name to Earthrise Nutritionals.

Later, in 1981 Earthrise have made a partnership with a Japanese company "Dainippon" Ink and Chemicals, a diversified company with agreement to promote microalgae for food. This unique corporation between California entrepreneur and Japanese corporative businessman has flourished a tremendous growth of spirulina production and as well as market expansion world-wide. In 2005 Earthrise Nutritional was fully owned by Daippon Ink and Chemicals.

Spirulina is one of the most saturated nutrient packed superfoods on land. This interesting environment friendly algae has been used as a nutritious food supplement for thousands of years (http://www.earthrise.com).This important food supplement of algae has been produced and supplied by the Earthrise Nutritionals to different countries of the world [33].

25.5.2 Yunnan Green A Biological Project Co. Ltd

Yunnan Green A Biological Project Co. Ltd. was found in 1997 headquarter is situated in Kunming, China. The company is engaged in developing, cultivation, breading, production, marketing and research of algal products. The Yunnan Green A Biological Project Co., Ltd acts as subsidiary of Yunnan Spirin Biotechnology Co. Ltd. The annual microalgal production products of company (*Spirulina*, *Chlorella*, and *Haematococcus pluvialis*) is 3000 tons. The company supply spirulina in powder and tablet form and phycocyanin extract. The company also uses spirulina and *H. pluvialis* in different fields like products of healthcare, medicine, cosmetics and foods.

The research and production development base is situated near the Chenghai Lake, Yunnan province, which is one of the three biggest alkaline naturally spirulina producing lakes of the world [34].

25.5.3 Inner Mongolia Rejuve Biotech Co. Ltd

Inner Mongolia Rejuve Biotech Co. Ltd. Was founded in 2006 headquarter is situated at Wukan, China. The company is engaged cultivation, harvesting, drying and production of spirulina. The company encompasses 630,000 m^2 of algae breeding area. The annual production of spirulina powder is 1100 ton and tablet are 100 ton. The company supplies product to different countries of the world [34].

25.5.4 Fuqing King Dnarmsa Spirulina Co. Ltd

Fuqing King Dnarmsa Spirulina Co. Ltd. company was found in 1995 situated in Fuqing, China. The Firm has seven cultivation farms, one research institute of algae and one deep product processing workshop. Per year product of the company is 1600-ton spirulina and 400 ton of chlorella [34].

25.5.5 Far East Microalgae Industries Co. Ltd

The company was established in 1976 found in Taipei, Taiwan. Far East Microalgae Industries Co. Ltd. (FEMICO) working on development and marketing of algal products. The company serves three nutritional products of microalgae in form of tablets, powder, and extracts, such as red algae, spirulina, and chlorella. The covered area of company is 140,00 m^2 and the capacity production is 1000 tons of chlorella and 200 tons of spirulina per year [34].

25.5.6 Cyanotech Corporation

Cyanotech Corporation was established in 1984 at Kailua Kona, United State, the farm engaged in production, cultivation and processing of microalgal products for human health and food supplements. The farm serves microalgae, like Bioastin and spirulina. It has 90-acre microalgae cultivation ponds. [34].

Day by day increase growth in human population in all over the world directly affect necessities like industrialization, vehicles and other energy requirements, that's how as natural resource fossil fuel has no capability to fulfill increased demands rapidly, because it needs long time to create, therefore search for substitute energy resource become essential in the scenario when energy resource shortage is occurred [14]. High yielding potential and lower competition chances for food and resources algal biofuel have most beneficial quality in the upcoming era which can face crisis situation due to the climatic change and global energy shortage [35,36]. Although alga is suitable material for the biofuel generation due to many reasons, but it is considered an expensive process than other kind of fossil fuel. Selection and harvesting of appropriate algae for the production of biofuel is the first task before any process, for minimizing high cost for developing in fuel form many workers concentrate to convert harvesting biomass directly [37].

Some advantageous benefits of algal biomass for the production of biofuel such as high oil yielding property, availability throughout the year under stream climatic conditions, less water consumption in cultivation, not difficult method of cultivation that is, no need of herbicide and pesticide, high abortion potential of carbon dioxide [38–40].

25.6 Top companies of biofuel production from algae

In recognition of decreasing of fossil fuel reserves, drastic increase in price and negative impact in environment, the algae biofuel production is gained intensive attention. Algae fuel is also termed algal fuel, oilgae, algal oleum, and third generation of biofuel. High fuel price, competing demands of biofuel sources, foods and world food crisis have ignited attention in algae cultivation for producing biodiesel, bioethanol, bio gasoline, bio butanol, bioethanol, vegetable oil and other biofuels [41]. The microalgal biofuel has been proposed as source of energy with high renewable potential. The algal biofuel production is a high concerned in renewable bioenergy research in the new millennium, especially in recent decade, between 2007 and 2017.

576

Microalgae are contained a huge amount of carbohydrates, proteins, biolipids and other valuable substances, which have potential for the application as feedstocks in many industrial fields, for making biofuels, healthcare products, fertilizers etc. [42–44].

In current scenario, the algae biofuel production is an expanding sector, several different companies have engaged in production of biofuel from algae based materials, some of which are as under,.

25.6.1 Sapphire Energy Limited

The company was established in 2007 situated in San Diego, California, United States. The farm is engaged in turns algae into green crude material for fuel. The Green Crude Farm is the first world's commercial demonstration algae into energy facility and also known as Integrated Algal Bio Refinery started work in 2018 now functioning in Luna Country, near Columbus, New Mexico. The Green Crude Farm of Sapphire integrates the entire value chain of fuel algal-based material through cultivation and conservation of prepared green crude material, with converging biotechnology, agriculture and energy. The company announced in 2008 that it could turned algae into oil, producing a green-colored crude yielding ultra-clean variety of diesel and gasoline without reduction of biofuel production. According to Chief Executive Jason Pyle of Sapphire company green raw material could be processed in existing oil refineries that would be powerful and decrease the pollution of existing verity of fuel [45].

The announcement of the company official is the latest development for the companies and researchers to fiend the methods for cut harmful emissions of fuel [46].

25.6.2 Solix Biofuels company

The company was established in 2006 situated in Fort Collins Colorado, United States. The company is engaged in cultivation of algae for producing biofuel, different chemicals and other important products.

Most recently company working as Solix Biosystem, the establishment of company was algal biofuel and successfully developed the deployed robust Algal Growth System (AGS) based on its patented, extended-surface area closed photobioreactor panels. AGS is a prone system, which is applicable for the production of algal biomass and is applicable to a brad verity of algal species [47].

The Solix photobioreactors for algae manufacture based upon twenty-years or research work of Aquatic Species Program initiating at the National Renewable Energy Laboratory and are enormously scalable according to the company objectives. Cultivation of algae in plastic bags, drastically minimize the chance of infestation. The low novel energy temperature control system keeps the algae with in a range of temperature, maximize the growth of algae (Fig. 25.4A, B), [48].

The basic object of company to focus on the understanding the algae and to apply the expertise for commercial purposes.

Officials Colorado State and Solix are collaborating with New Belgium Brewing Co. to use excess CO_2 from the plant of brewery to test the biodiesel of algal substances. Algae

FIGURE 25.4 A and B. Cultivation and collection process of algae biomass (https://news.algaeworld.org/2012/08/solix-biosystems-gets-31-million-for-commercial-algae-oil-plant/) C. Algenol algae producing field. (https://www.google.com/AlgenolBiofuelspictures).

cells are obtained from the fluid with a centrifuge. Once obtained, the oil will be extracted and the obtained oil can then be refined into fuel as biodiesel through the same transesterification method which is currently used to refine other vegetative oil sources into biodiesel.

Solix officials estimate that the widespread construction of its photobioreactor process could fulfill the requirement for the US biodiesel consumption about 4 million barrels per day through algae growth of less than 0.5% land area of United State.

According to the chief executive officer of Solix Doug Henston, biofuel from algae are still being developing, yet a strong case can be for worldwide demonstration of algae as an energy crop.

25.6.3 Algenol Biofuels

The Algenol Biofuels company was founded in 2006 situated in Florida. The main objective of the company is to cultivate from which ethanol can be directly obtained, without need to kill the organisms. Such type of technic minimize the overall required energy to harvest the end product and helps to balance the ratio of carbon produced from burning of ethanol with amount of carbon utilized in producing it (Fig. 25.4C).

Important point of the company is efficiency measures of biofuel production technology, the quantity of produced fuel per acre. In order to avoid threatening the supply of food, while still meeting the energy demand of the world. In current requirement of energy, the production of biofuel amount from one-acre land needs more than 10,000 US gallons. Algenol technology has reached it to 6000 to 8000 gallons per acre, with predictions of 10,000 gallons being produced in the near future.

The technology that Algenol uses is termed as Direct to Ethanol and relatively based on simple process, the algae which are grown in saline-water, are providing CO_2 and sugar, then they used waste material for the production of energy and by product as ethanol.

The company maintains facilities in both Europe and the United States. Algenol also keeps a subsidiary unit in Germany known as Algenol Biofuels Germany, the location is specific in the field of algae, a subsidiary unit is also found in Switzerland [49].

25.6.4 Solazyme Inc

Solazyme company was established in 2003 situated in South San Francisco, California United States. The main objective of the company is to utilize the microalgae to produce a renewable sources of energy transportation of biofuel. It is a bioproduct producer from algae, has enough biofuel trading with focusing of algae based oil production on food personal care industries.

Solazyme currently renamed as Terra Via, the company lists the present low prices, changing the residues around the benefits of biofuels [50].

25.7 Biofuel production and its impact on environment

Nowadays most of the world's human population is affected by various types of pollution. The US Environmental Protection Agency has provided a list of six pollutant compounds, that is, carbon monoxide (CO), nitrogen oxides (NO and NO_2), sulfur dioxide (SO_2) and ozone (O_3), particulate matter and lead (Pb), that produce generally adverse effects on the environment, economy, and on human health (Fig. 25.5) [51].

The environment is continuously affected by the emission of toxic gases and other compounds from fossil fuels used by millions of vehicles, industrial harmful gases, and waste.

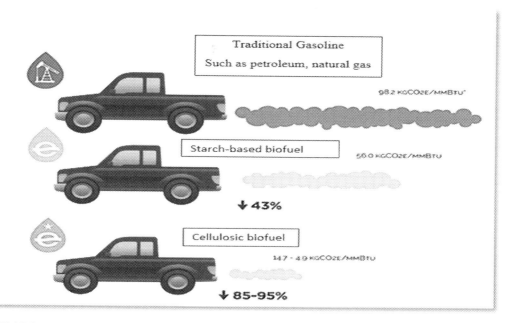

FIGURE 25.5 Illustration of traditional and biofuel difference in transport by British thermal units, Emissions of sulfur dioxide and nitrogen dioxide. Source: *Encyclopædia britannica.com*.

Ozone layer depletion is affected by pollution which also causes various types of diseases in humans, plants, and animal. To overcome these problems biofuel is a better option to adopt which could minimize our losses due to the emergence of toxic compounds. The other benefit of biofuel is its renewable ability. Fig. 1.8 showed some damaging effect of pollutant emerging from fossil fuel, while Fig. 25.6 illustrates transport smoke from using different types of biofuel, showing the difference in using fossil fuel and biofuel in vehicles [52].

The world has become a global village, and the daily consumption of energy has become high all over the world. Consequently the generation and discoveries of new resources to fulfill this requirement for energy and associated facilities is essential in order to attain modern necessities. For this purpose, the utilization of renewable energy sources is a tremendous idea that also supports the mitigation of climate change [14,37].

Renewable fuels grow faster and more cheaply than nonrenewable fossil fuel, while also being better for human health and a clean environment [37]. Fuel generated from algal biomass is considered best renewable source in terms of climate-friendliness, with the ability to satisfy the global fuel demand on a long-term basis [53].

With the increasing awareness of environmental issues and fossil fuel depletion, biofuel and biodiesel in modern age are becoming more attractive sources of energy [12]. Organic waste has become one of the important causes of environmental pollution. The utilization of waste valorization techniques has generated several beneficial products [54]. Atmospheric carbon dioxide can be transformed into some other valuable products by microalgae (Fig. 25.7) [55].

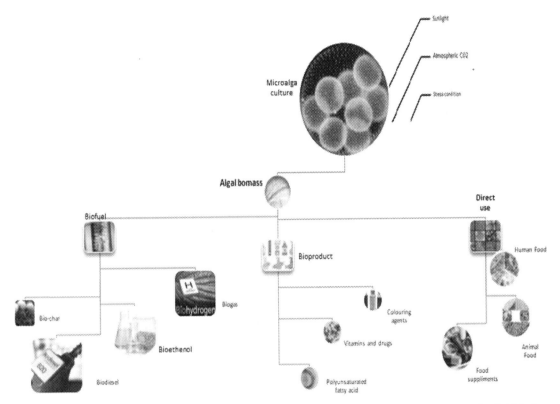

FIGURE 25.6 Illustration of the differences between traditional fuel and biofuel in transport by British thermal units.

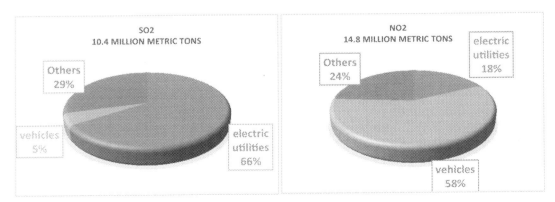

FIGURE 25.7 *Percentage of biofuel productionSource: A Global Perspective. Economic research service, United States Department of Agriculture.(https://www.ers.usda.gov/amber-waves/2007/november/the-future-of-biofuels-a-global-perspective/)*

25.8 Challenges of biofuel commercialization from algae

Despite the need and importance of biofuel as energy source, many technical challenges and disadvantages are being faced in order to provide sufficient biofuel to meet the worldwide demand of energy. The current increasing requirement for fuel cannot be compensated by fossil fuel, due to the decreasing fossil fuel reserves [56]. Increasing global warming, pollution, and increasing oil prices have led to the demand for renewable energy sources. Biofuel is an environment-friendly alternative to fossil fuel, but there are many problems and drawbacks in the production and commercialization of biofuel from algae.

The production of biofuel from algae at a large scale requires lot of equipment, tools, energy, water, and nutrients for each step. The cultivation of algae at a large scale and agriculture could use the same ratio of nutrients [57].

Replacing only 5% algal biofuel with gasoline or diesel could require 123 billion to 143 trillion liters of water. Wastewater use not only prevents competition by crops and food, but also provides some amounts of phosphorus and nitrogen. However, the main drawback of wastewater is that it contain algal pathogens, predators, heavy metals, and other pollutants. The supply of wastewater to treatment plants needs intensive energy and other requirements, making it expensive, and the water cannot be recycled completely [57]. To produce each gallon of biodiesel via algae cultivation requires about $14-21$ kg of carbon dioxide [57], $6-15 \times 10^9$ kg of nitrogen, and $1-2 \times 10^9$ kg of phosphorus [58]. Besides these, various industrial wastes of carbon dioxide contain heavy metals such as arsenic, lead, mercury, cadmium, etc. [56].

Current political targets for the use of biofuel are driven by environmental concerns about the emissions of carbons, security concerns, especially in the United States, and the cost of imported fuels. In the meantime, the economic case for biofuels has been increased by the unstoppable increase in crude oil price to US$ 100 per barrel [59]. Many environmental groups, NGOs in developing, and some scientists have started questioning the wisdom of the recent rush to biofuels. It has been pointed out that the industrial production of biofuels in economies like the EU may generate more carbon than it captures [60].

Besides these, some other challenges are also creating hurdles for the production of biofuel from algae, like the choice of algal strain, isolation, and refinement. Site selection and identification of natural microalgal assemblies is usually a tedious and time-consuming process. For the identification of species, analyses of both morphological and genetic characteristics are required [61].

The extraction cost of energy from algae is ten times more than the extraction from soybean oil. The cost was reported in 2012 to be US$ 27/gallon [62]. The high cost of pretreatment is one of the major problems for the production of biofuel at a large scale, from the available resources, in addition to high capital and operation costs [63].

25.9 Future prospective of biofuel

Biofuel is a flammable solid, liquid, or gaseous substance, directly obtained from recently produced materials (plant origin), and is used for power generation. The most current biofuel is derived from plants. In recent years the production of biofuel has rapidly increased at a global level, due to the increase in fuel oil and mutual government policy initiatives. This spectacle has been mostly obvious in the high energy consuming regions such as the United States, European Union, and regions of primary producers as Malaysia, Brazil, Indonesia, etc. [64].

Amigo energy, the US fossil fuel generation will remain dominate in the world, it is surprising fact that when people come to utilize biofuel, the United States established natural gas to rule over the world which having high energy power than coal [2]. The United States nowadays generates energy from about 34% natural gas, 30% coal, 20% nuclear, 15% different renewables, and 1% oil; these figures show that renewable generation is not acceptable. The National Academies of Sciences also mentioned in their report that fossil energy utilization in the United States is almost 81% of the overall energy [65]. Due to the minimal available fossil resources and the environmental pollution, the increasing demand for biofuel is enhancing its value [66]. But compatible biofuel for existing transport that can be derived from sustainable environment resources is needed. For this purpose various species of bacteria are considered to be best to produce fuel with their unique and significant properties. [67].

Containing organic and nutrient-rich substances, waste food material is a valuable potential resource for the production of biofuel [68,69]. Thus pilot plants waste can be converted for the production of fuel around the world. For this purpose an initiative has been taken by the UK chemical company Ineos to try to utilize leftovers into biofuels commercially. The aim of this work is to consume raw residual waste material, and to utilize material which can be recycled [70,71].

In this connection according to BBI International, another study has been made by the Worcester Polytechnic Institute with the help of a US Department of Energy grant [72].

Generally organic waste has various nutrients, for example, carbohydrate, lipid, phosphate, vitamin, and fatty acids. Most of the nutrients can be converted into bioethanol, biodiesel, and bio-oil [73]. Consuming food wastes is considered best to resolve waste disposal, energy security, and energy scarcity in upcoming years [69]. They also contain significant quantities of sugar compounds which are considered the best raw material for bioethanol production [71].

Due to this reason the collection of kitchen food waste is the initial step. It is then dried by various techniques, and then via gas-chromatography-mass spectrometry lipid extraction is employed to obtain fatty acid [74].

The use of wastewater for the feedstock cultivation in the production of biofuel is also being considered by researchers to maintain the safety of freshwater resources [75]. Due to its significant qualities, especially regarding environmental issues, BTL technology could be used to develop biofuel on industrial scale [76].

The investigations with regard to biofuel production have suggestions that responsible attitudes of researchers and industry holders towards the best utilization of algal biomass in the production of biofuel and other useful products especially in the coastal areas is the today's serious demand of the people to protect the world from pollution. Most of the third world countries with a coastline should utilize algal biomass to generate biofuel as their topmost priority. Algal biomass is a renewable source that could be the best economic revolution in developing countries because of its significant importance, environment-friendly potential, and the production quality.

Polluters could use their organic wastes for the production of liquid fuels and in this way both the issues can be resolved simultaneously. Biofuels are considered as atmospherically clean energy. Over time pollution is becoming one of the biggest issues in the world. This issue can be resolved by using organic wastes and algal biomass for fuel production. Biofuel is the best solution to create a healthy and clean atmosphere because each day fossil fuel contributes a high amount of toxic gases and poisonous chemicals to the atmosphere, which damages economic infrastructure including causing public health issues, soil fertility problems, water issues, land waste, and agricultural development threats.

Biofuels are the best renewable source of energy. For this purpose, governments and local bodies should make priority-based short term agendas to take advantage of this biomass. The serious attention of researchers and funding agencies is necessary in this connection when discussing the detailed advantages of biofuel production, especially algal biofuel. Worldwide funding is also required for the quick achievement of developing this fuel.

Currently Pakistan imports almost 8.1 million tons (US$ 9.4 billion per annum) of fossil fuel. Because fossil fuel resources are depleting, the cost of fossil fuel is likely to become too high to afford; specialists in the field predict fossil fuels will be unaffordable for most countries by 2050. Thus fossil fuel resources will become unable to fulfill all necessities worldwide [77]. Biofuel development using different biomass sources will be the main route to achieving our target not only in terms of economics but also in terms of controlling environmental pollution.

References

[1] Turgeon and E. Morse, National geographic. Biomass energy. Encyclopaedic Entry, <https://www.nationalgeographic.org/encyclopedia/biomass-energy/> 2012 (accessed 25.12.20).

[2] (A). skeleton in desert with single bones around <Skeleton Desert Single Bones Around Stock Photo (Edit Now) 165473642 (shutterstock.com)> (accessed 25-12-2020) [2](B).National Geographic, Encyclopedic entry. Natural gas <https://www.nationalgeographic.org/encyclopedia/natural-gas/#burgan-field>. (accessed 25-12-2020). [2] (C). ZME Science, Green petroleum from McClintock Well 1, the oldest oil well still in production. Image credits Drake Well Museum <https://www.zmescience.com/science/what-is-petroleum/>. (accessed 25-12-2020). (D) Coal Power Impacts, <https://www.ucsusa.org/resources/coal-power-impacts>, 2017 (accessed 25-12-2020).

[3] T. Stevens, Amigo energy renewable energy versus fossil fuels: 5 essential facts, from <https://amigoenergy.com/blog/renewable-energy-versus-fossil-fuels/> (accessed 25.12.20).

[4] M. Enzler, Lenntech.com. Fossil fuels. <https://www.lenntech.com/greenhouse-effect/fossil-fuels.htm> (accessed 25.12.20).

[5] S. Behera, R. Singh, R. Arora, N.K. Sharma, M. Shukla, S. Kumar, Scope of algae as third generation biofuels, Front. Bioeng. Biotechnol. 2 (90) (2015).

[6] Energy.gv. Biofuels basics | Department of Energy. Office of energy efficiency & renewable energy. <https://www.energy.gov/eere/bioenergy/biofuels-basics> (accessed 25.12.20).

[7] Y. Gudkova, Some recent advances in biofuel research. Permaculture research institute. Permaculture news. <https://www.permaculturenews.org/2017/08/17/recent-advances-biofuel-research/>, 2017 (accessed 28.12.20).

[8] Europeanbusinessreview, Biofuel-and-their-uses/. The European Business Review, Empowering communications globally, <https://www.europeanbusinessreview.com/the-three-different-types-of-biofuel-and-their-uses/>, 2020 (accessed 25.12.20).

[9] C. Nunez, Biofuels, explained (Promising but sometimes controversial, alternative fuels offer a path away from their fossil-based counterparts). National geographic. <https://www.nationalgeographic.com/environment/global-warming/biofuel/#>, 2019 (accessed 25.12.20).

[10] Student Energy, Biofuel 101. <https://www.youtube.com/watch?v = ZGmwtDffc74&feature = youtu.be>, 2015, 17 May, (accessed 25.12.20).

[11] O.C. Kopp, Coal (fossil fuel). Encyclopædia Britannica. <https://www.britannica.com/science/coal-fossil-fuel>, 2020 (accessed 26.12.20).

[12] W.T. Coyle, The future of biofuels: a global perspective. Economic research service, United States Department of Agriculture. <https://www.ers.usda.gov/amber-waves/2007/november/the-future-of-biofuels-a-global-perspective/>, 2007 (assessed 16.12.20).

[13] M. Hannon, J. Gimpel, M. Tran, B. Rasala, S. Mayfield, Biofuels from algae: challenges and potential, Biofuels 1 (2010) 763−784.

[14] A. Choudhary, K. Rachan, K. Krishnendu, D. Vinod, "Algal" biodiesel: future prospects and problems 68 (2011) 44−51.

[15] G.S. Costard, R.R. Machado, E. Barbarino, R.C. Martino, S.O. Lourenço, Chemical composition of five marine microalgae that occur on the Brazilian coast, Int. J. Fish. Aquac. Acad. J. 4 (9) (2012) 191−201.

[16] S.M. Shanab, E.A. Shalaby, Algal Chemical Compounds, Lap Lambert Academic publishing, 2016.

[17] J. Sheehan, T. Dunahay, J.R. Benemann, P. Roessler The renewable energy laboratory, Golden, CO. Prepared for: Office of Fuels Development, United States Department of Energy. A look back at the United States Department of Energy's aquatic species program−Biodiesel from algae, 1998.

[18] M.I. Khan, J.H. Shin, J.D. Kim, The promising future of microalgae: current status, challenges, and optimization of a sustainable and renewable industry for biofuels, feed, and other products, Microb. Cell Fact. 17 (36) (2018).

[19] X. Wen, K. Du, Z. Wang, X. Peng, L. Luo, H. Tou, et al., Effective cultivation of microalgae for biofuel production: a pilot-scale evaluation of a novel oleaginous microalga Graesiella sp. WBG-1, Biotechnol. Biofuel 9 (1) (2016) 1−12.

[20] Q. Hu, Industrial production of microalgal cell mass and secondary products-major industrial species, in: A. Richmond (Ed.), Handbook of Microalgal Culture: Biotechnology and Applied Phycology, Wiley, Oxford, 2004, pp. 264−272.

[21] M.J. Griffiths, S.T.L. Harrison, Lipid productivity as a key characteristic for choosing algal species for biodiesel production, J. Appl. Phycol. 21 (2009) 493−507.

[22] L. Barsanti, P. Coltelli, P.V. Evangelista, A.M. Frassanito, N. Vesenti, P. Gualtieri, The world of algal, in: V. Evangelista, L. Barsanti, A.M. Frassanito, V. Passareli, P. Gualtieri (Eds.), Algal Toxins Nature Occurrence, Effect and Detection, 1st (ed.), Springer Science + Business Media B.V, Netherlands, 2008, pp. 1−16.

[23] A. Vonshak, Microalgal biotechnology: is it an economic success? in: E.J. Da. Silva, A.C. Ratlege, Sasson (Eds.), Biotechnology: Economic and Social Aspects, 1st (ed.), Cambridge University Press, Great Britain, 2009, pp. 70−81.

[24] F.K. El-Baz, H. Hanaa Abd El Baky, Pilot scale of microalgal production using photobioreactor, photosynthesis - from its evolution to future improvements in photosynthetic efficiency using nanomaterials, Juan Cristóbal García Cañedo and Gema Lorena López Lizárraga, Intech Open (2018). Available from https://doi.org/10.5772/intechopen.78780. <https://www.intechopen.com/books/photosynthesis-from-its-evolution-to-future-improvements-in-photosynthetic-efficiency-using-nanomaterials/pilot-scale-of-microalgal-production-using-photobioreactor>.

[25] J. Degen, A. Uebele, A. Retze, U. Schmid-Staiger, W. Trosch, A novel airlift photobioreactor with baffles for improved light utilization through the flashing light effect, J. Biotechnol. 92 (2) (2001) 89−94.

[26] Ananyms, Final report scale-up of algal biofuel production using waste nutrients, California Polytechnic State University San Luis Obispo, California Advanced Algae Systems Program Bioenergy Technologies Office United States Department of Energy (2018). <https://www.google.com/search?source = univ&tbm = isch&q = Australian + Renewable + Fuels + Limited + image>. Retrieved on 25th December 2020.

[27] Staff writer of NS Energy on 15 November 2019 world gas renewables reviews, published by Italian oil and gas major Eni. <https://www.nsenergybusiness.com/features/top-biofuel-production-countries> (accessed 05.01.21).

[28] Staff writer of NS Energy on 27 September, 2019, world gas renewables reviews, published by Italian oil and gas major Eni. <https://www.eni.com/en-IT/global-energy-scenarios/world-gas-renewables-review-second-volume.html>. (https://www.nsenergybusiness.com/news/eni-gela-bio-refinery-europe/) (accessed 05.01−21.

[29] <https://www.bordermail.com.au/story/3714413/buyers-fuel-biodiesel-plant-hope> (accessed 20.01.21).

[30] <http://www.bfreinc.com> (accessed 29.12.20).

References

585

[31] \<https://en.wikipedia.org/wiki/Cosan#/media/File:Panorama_Usina_Costa_Pinto_Piracicaba_SA\> (accessed 31.12.20).

[32] \<https://www.alliedmarketresearch.com/algae-products-market\> (acceesed 26.12.20).

[33] \<https://www.cmtc.com/made-in-california-profile/earthrise-nutritionals\> (accessed 06.01.21).

[34] \<https://meticulousblog.org/top-10-companies-in-spirulina-market\> (accessed 27.12.20).

[35] M.Y. Menetrez, An overview of Algae biofuel production and potential environmental impact, Environ. Sci. Technol. 46 (13) (2012) 7073−7085.

[36] M. Daroch, S. Geng, G. Wang, Recent advances in liquid biofuel production from algal feedstocks, Appl. Energy 102 (2013) 1371−1381.

[37] M. Rajvanshi, R. Sayre, Recent advances in algal biomass production. (2020) Biomass.

[38] P. Spolaore, C. Joannis-Cassan, E. Duran, A. Isambert, Commercial applications of microalgae, J. Biosci. Bioeng. 101 (2006) 87−96.

[39] G.C. Dismukes, D. Carrieri, N. Bennette, G.M. Ananyev, M.C. Posewitz, Aquatic phototrophs: efficient alternatives to land-based crops for biofuels, Curr. Opin. Biotechnol. 19 (2008) 235−240.

[40] G. Dragone, B. Fernandes, A.A. Vicente, J.A. Teixeira, Third generation biofuels from microalgae, in: A. Mendez-Vilas (Ed.), Current research, Technology and Education Topics in Applied Microbiology and Microbial Biotechnology, Formatex, Madrid, 2010, pp. 315−1366.

[41] S.C. Bhatia (Ed.), Advanced Renewable System, Woodhead Publishing India PVT. LTD, 2014.

[42] E.W. Becker, Micro-algae as a source of protein, Biotechnol. Adv. 25 (2007) 207−210. Available from: https://doi.org/10.1016/j.biotechadv.2006.11.002.

[43] H. Chen, T. Qiu, J.F. Rong, C.L. He, Q. Wang, Microalgal biofuel revisited: an informatics-based analysis of developments to date and future prospects, Appl. Energy 155 (2015) 585−598. Available from: https://doi.org/10.1016/j.apenergy.2015.06.055.

[44] M. Giordano, Q. Wang, Microalgae for industrial purposes, in: S. Vaz Jr (Ed.), Biomass and Green Chemistry: Building a Renewable Pathway, Springer, Cham, 2018, pp. 133−167. Available from: https://doi.org/10.1007/978-3-319-66736-2_6.

[45] \<http://articles.latimes.com/2008/may/29/business/fi-greencrude29\> (accessed 02.01.21).

[46] \<https://www.nbcnews.com/id/wbna39898804#.UyfU4WT5mXc\> (accessed 02.01.21).

[47] \<https://news.algaeworld.org/2012/08/solix-biosystems-gets-31-million-for-commercial-algae-oil-plant/\> (\<https://www.oilmonster.com/company/solix-biofuels/42327\>) (accessed 02.01.21).

[48] \<https://www.google.com/Solix + Biofuels + company + pictures&tbm = isch&source = iu&ictx = 1&fir\> (accessed 02.01.21).

[49] \<http://biofuel.org.uk/Algenol.html\> (accessed 31.12.20).

[50] Biofuel International. Published by Woodcote Media Ltd, Marshal House, 124 Middleton Road Morden, Surrey. SM4 6 RW. \<https://biofuels-news.com/news/solazyme-abandons-algal-biofuels-refocuses-on-food/\>, 2016

[51] J.A. Nathanson, Water pollution, Encyclopædia Britannica, \<https://www.britannica.com/science/water-pollution\>, 2020 (accessed 06.01.21).

[52] Vital the essential perspective, get grounded in the facts: environmental and health benefits of biofuels. \<https://vitalbypoet.com/stories/environmental-and-health-benefits-of-biofuels\>, 2018 (accessed 26.12.20).

[53] V. Stanislav, V. Christina, Composition, properties and challenges of algae biomass for biofuel application: an overview, Fuel. 181 (2016) 1−33.

[54] H.I. Abdel-Shafya, M.S.M. Mansou, H.I. Abdel-Shafy, S.M. Mona, Mansour, solid waste issue: sources, composition, disposal, recycling, and valorization, Egypt J. Pet. 27 (4) (2018) 1275−1290 (\<http://www.sciencedirect.com/science/article/pii/S1110062118301375\>) ISSN 1110-0621.

[55] A. Richmond, Q. Hu (Eds.) Handbook of microalgal culture: applied phycology and biotechnology, 2nd Edition (2013) (ISBN: 978-0-470-67389-8).

[56] M.G. Saad, D.S. Noura, Z.S. Mohamed, S.M. Hesham, Algal biofuels: current status and key challenges, Energies Rev, 2019. 12-01920.

[57] Office of Energy Efficiency and Renewable Energy. National Algal Biofuels Technology Roadmap; United States Department of Energy: Washington, DC, USA, 2010.

[58] D. Pimentel, T.W. Patzek, Ethanol production using corn, switchgrass, and wood; Biodiesel production using soybean and sunflower, Nat. Resour. Res. (14)(2005) 65−76.

Handbook of Algal Biofuels

[59] In 2003, crude oil prices on the New York Mercantile Exchange(NYMEX) were below \$25/barrel but by late 2007 they had increased 3.sixfold to over \$96/barrel.

[60] D.J. Murphy, Future prospects for oil palm in the 21st century: biological and related challenges, Eur. J. Lipid Sci. Technol 109 (2007) 296–306.

[61] T. Mutanda, D. Ramesh, S. Karthikeyan, S. Kumari, A. Anandraj, F. Bux, Bioprospecting for hyper-lipid producing microalgal strains for sustainable biofuel production, Bioresour. Technol 102 (2011) 57–70.

[62] P. Savage, Algae under pressure and in hot water, Science 338 (2012) 1039.

[63] W.Y. Cheah, R. Sankaran, P.L. Show, N.B.T.I.T.N.B. Tg, K.W. Chew, A. Culaba, et al., Pretreatment methods for lignocellulosic biofuels production: current advances, challenges and future prospects, Biofuel Res. J. 25 (2020) 1115–1127. Available from: https://doi.org/10.18331/BRJ2020.7.1.4.

[64] D.J. Murphy, Future prospects for biofuels, <https://www.researchgate.net/publication/228389985_Future_prospects_for_biofuels> (accessed 03.01. 21).

[65] The National Academies of Science, Engenieering and medicines Fossil fuel, <http://needtoknow.nas.edu/energy/energy-sources/fossil-fuels/>, 2020 (accessed 25.12.20).

[66] M.R. Riazi, D. Chiaramonti, Biofuels Production and Processing Technology, CRC Press, 2018, pp. 1–710.

[67] L.S. Gronenberg, R.J. Marcheschi, J. C Liao, Next generation biofuel engineering in prokaryotes, Curr. Opin. Chem. Biol. 17 (3) (2013) 462–471.

[68] S. Li, X. Yang, 20 - Biofuel production from food wastes, in: R. Luque, C. Sze Ki Lin, K. Wilson, J. Clark (Eds.), Handbook of Biofuels Production, (Second Edition), Woodhead Publishing, 2016, pp. 617–653.

[69] S.K. Karmee, C.S.K. Lin, Valorisation of food waste to biofuel: current trends and technological challenges, Sustain Chem. Process 2 (22) (2014).

[70] E. Davis, The biofuel future? royal society of chemistry. <https://www.chemistryworld.com/features/the-biofuel-future/3004815.article>, 2009 (accessed 04.01.20).

[71] D. Matsakas, M.Loizidou Kekos, P. Christakopoulos, Utilization of household food waste for the production of ethanol at high dry material content, Biotechnol. Biofuels 7 (4) (2014).

[72] BBI international, biomass- researchers improve method to convert food waste into biofuels.<http://biomassmagazine.com/articles/15171/researchers-improve-method-to-convert-food-waste-into-biofuels> (accessed 25.12.20).

[73] A.K. Karmee, Liquid biofuels from food waste: current trends, prospect and limitation, Renew. Sustain. Energy Rev 53 (2016) 945–953.

[74] S. Barik, K.K. Paul, D. Priyadarshi. Utilization of kitchen food waste for biodiesel production. Published under licence by IOP Publishing Ltd. IOP Conference Series: Earth and Environmental Science, 167, in 8th International Conference on Environment Science and Engineering (ICESE 2018) 11–13 March 2018, Barcelona, Spain, Conf. Ser. Earth Environ. Sci. 167 012036, 2018.

[75] M. Arshad, M. Abbas, Future biofuel production and water usage, in: M. Arshad (Ed.), Perspectives on Water Usage for Biofuels Production, Springer, Cham, 2018.

[76] IFP Energies Nouvelles. What future for biofuels? <https://www.ifpenergiesnouvelles.com/issues-and-foresight/decoding-keys/renewable-energies/what-future-biofuels> (accessed 03.01.21).

[77] M. Ahmad, H.A. Jan, S. Sultana, M. Zafar, M.A. Ashraf, K. Ullah. Prospects for the production of biodiesel in Pakistan, biofuels - status and perspective, Krzysztof Biernat, Intech Open, DOI: 10.5772/59318. <https://www.intechopen.com/books/biofuels-status-and-perspective/prospects-for-the-production-of-biodiesel-in-pakistan>, 2015 (accessed 06.01.21).

CHAPTER

26

Biointelligent quotient house as an algae-based green building

Anas Tallou[1], Khalid Aziz[2], Mounir El Achaby[3], Sbihi Karim[4] and Faissal Aziz[5,6]

[1]Polydisciplinary Laboratory of Research and Development, Faculty of Sciences and Techniques, Sultan Moulay Slimane University of Beni Mellal, Beni-Mellal, Morocco [2]Materials, Catalysis and Valorization of Natural Resources, Faculty of Sciences, University Ibn Zohr, Agadir, Morocco [3]Materials Science and Nano-Engineering Department, Mohammed VI Polytechnic University, Benguerir, Morocco [4]Laboratory of Biotechnology, Materials and Environment, Natural Substances and Environment Unit, Faculty Polydisciplinary of Taroudant, University Ibn Zohr, Taroudant, Morocco [5]Laboratory of Water, Biodiversity and Climate Changes, Faculty of Sciences Semlalia, Cadi Ayyad University, Marrakech, Morocco [6]National Centre for Research and Study on Water and Energy (CNEREE), Cadi Ayyad University, Marrakech, Morocco

26.1 Introduction

Nowadays, humanity faces substantial environmental challenges and climate change due to our daily activities (transport, industry, and economy). The world is looking and seeking new renewable and environmentally ecofriendly options [1]. Also, there is high demand to reach a circular economy system that boosts economic and industrial growth, ensures the quality of human life, and protects the environment [2] according to the sustainable development goals (SDGs). Three major groups of these SDGs show the demand for new human resources, the sustainability of these resources, and the redistribution of benefits equally [3].

Despite the global shift toward renewable energy production, electricity production from renewable energies is only 22.8%, while solar energy accounts for 0.9%, wind energy for 3.1%, and 1.8% of energy comes from biomass (bioenergy). In addition, the

Handbook of Algal Biofuels
DOI: https://doi.org/10.1016/B978-0-12-823764-9.00009-1

© 2022 Elsevier Inc. All rights reserved.

demand for biomass energy (heat, electricity, and transport) is growing steadily and despite the important role of biomass resources in economic development, the biomass contribution is small [4]. The building sector's total energy consumption is about 40%, reaching 30% of the total annual greenhouse gas emissions. Besides, in the next 20 years, the greenhouse gas emissions from buildings will double. For this reason, to reduce the greenhouse gas emissions that come from buildings, new environmental technologies should be implemented as one of the strategies to mitigate climate change in this sector. In order to boost and improve the energy performance of buildings, a global new approach should be adopted, for example, approaching all dimensions of efficient building design to improve overall efficiency [4].

On the other hand, the use of algae as a biomass resource to produce energy in buildings is still in its first stage and needs to be implemented widely [5]. The vertical façades can be used as smart photosynthetic surfaces that produce bioenergy and mitigate the current situation of climate change impacts [5]. Compared to higher plants, algae are considered to be at least 10 times more efficient in terms of the photosynthesis process [5,6]. Algae's performance can be increased and boosted when integrated in building façades, which can result in healthy, livable, productive, and smart buildings [5]. This approach promises that the buildings will be converted into bioenergy producers [4].

The United States Department of Energy (USDE) declared that buildings devoured about 40% of energy during 2015. Sustainable and smart buildings have become increasingly important, especially since there is a high demand to minimize building energy use and its harmful effects on the environment. It is known that the façades have a part in reducing loads for cooling and minimizing the thermal performance of the building. According to ASHRAE fundamentals (SI) 2009, the index of temperature conductance of the walls is between 0.2 and 1.1 W/m^2K and the index of heat transfer of windows is from 0.6 to 0.8 W/m^2K. Therefore the heat transfer of the windows is superior to the heat transfer of walls. In this regard, with the effort done in order to improve the performances of envelopes, authors and researchers studied the algae façades technology to be implemented in buildings. The first implementation of algae in buildings was in the BIQ (biointelligent quotient) system or building in Hamburg, Germany in 2013. When the façades are oriented to the southwest and the southeast, they can produce energy for the buildings. The algae façade can insulate the building and can provide shade even if there is sunlight. Even though this project targets the presentation of instructions and standards to adopt the algae façades systems, other applications have not been implemented and the feasibility of the system is still under study. In this regard, some researchers are exploring how algae façades are implemented in buildings by understanding the building systems' needs and requirements by analyzing the energy and waste mainstream, including the algae façades. Therefore such a system's objective is to evaluate and investigate the feasibility of the algae façades system in the building [7].

Because of the sustainable technology requirements and regulations, microalgae systems were studied in order to be used in energy production and wastewater treatment. Algae are considered the organisms that produce energy through photosynthesis, which is more sufficient in terms of energy than biofuel derived from a flower, corn, soybean, etc. Algae generate biomass through photosynthesis, which is considered a source of biogas produced in anaerobic conditions. In terms of CO_2 elimination, waste management capacity, organic fertilizers, ecosystem raw materials, and biological derivatives, algae have been identified as a renewable energy source used in these contexts. The algal approach has been investigated and studied for application in closed, open, and mixed systems. Closed algae systems are divided into tubular design, flat plate design, and other designs. Those algae systems have been studied in order to be applied in smart buildings with additional efforts for performance improvement of building envelopes. According to closed-loop technology, due to the process's outputs, inputs are activated. It is expected that the algae façade will transform the building into becoming a protector of the environment. The first time that algae façades were implemented in a building was in Hamburg, Germany, in 2013. This innovative idea of algae façade application in the building is very efficient for energy production, environment protection, and greenhouse gas reduction. However, it is still in its conceptual stage and needs to be applied widely. This chapter reports an overview on the need for green and sustainable buildings, through highlighting the benefit of this ecofriendly technology on climate change mitigation via also its combination with renewable energy, and then presents two successful case studies of green building implementation, one in Hamburg, Germany and the other in Sydney, Australia.

26.2 Green buildings

A rapidly growing number of papers published during the past decade have demonstrated extensive research on green and smart buildings. Also, a developing degree of public conscience and consciousness of green buildings was noted. Nonetheless, there was a large debate on what green buildings are and what they should cover. Perceiving the significance of green building practices on "going green" and "building sustainability" has been presented for a long time. In any case, construction is still a sector that consumes many energies depending on published works. This could be because of the uninvolved construction experts embracing sustainable solutions. Confronting the rising energy expenses and developing natural concerns, the interest in green building practices with insignificant ecological effects has been raised recently [8]. The United Nations (UN) Environment Program for Sustainable Buildings and Construction announced that the building field alone uses about 40% of the world's energy, consumes 12% of potable water, and causes about 30% of greenhouse gas emissions [9].

Green building has many significant sustainable and socioeconomic benefits on societies. The positive impacts of green building reside in minimizing the energy used in construction, satisfying human well-being requirements, providing new job opportunities, directly and indirectly protecting our natural resources, reducing greenhouse gas emissions, and using and reusing clean energy sources [9]. Green building advantages in addition to energy preservation and environmental protection are also well recognized; for example, avoiding noise, water pollution, dust, and solid and liquid waste during the construction period are the most important. In contrast, normal buildings or unsustainable buildings consume about 40% of total energy, as reported by the World Business Council for Sustainable Development and produce large quantities of the greenhouse gas emissions responsible for the current global warming of the Earth's surface [10]. In 2035 carbon emissions worldwide will reach about 42.4 billion tons with a 43% increase from 2007.

Moreover, green buildings' modeling and innovation will comprise the consumption of typical energy and assets, creating noise, pollutants, and greenhouse gas emissions. When building structures arrive at their end, the removal of structures is likewise connected with energy utilization and solid waste generation. For example, in Austria the waste produced from the building sector accounts for 16.6 million tons. This represented 38% of the total waste, 43% of which was dumped into open landfills. The expanding requirement for landfills presents another challenge to all countries with restricted lands [10]. The International Energy Agency forecasts that the institutional and business buildings will increase multiple times by 2050. The sustainable plan in design and facilities is generally supported through green building certification systems, Green Mark, Energy and Environmental Design Leadership, the Building Research Establishment Environmental Assessment Method, and Green Star. Until this point, there have been in excess of 200 distinctive green building system plans around the planet, with a gauge of 1,000,000 affirmed projects. These accreditation plans require normal updates to adapt the continuous interest to improve buildings [10,11]. Also, it is vital to intently screen, keep up, and improve the guaranteed green building exhibition through postcertificate activities to ensure the expected performance and ecological advantages. Dr. De Wilde defined the green building or the sustainable building from three different views: a design perspective on the building as an article, a cycle perspective on working as a development activity, and an art focus where execution includes the ideas of green. Technically, building execution alludes to how well a green building plays out the undertakings or satisfies its capacities. When a building becomes operational, it requests that this building reach the performances indicated initially. In such a manner, the affirmation plans can maintain a solid situation in surveying and assessing existing buildings functioning on a continuous premise and show that the guaranteed buildings are proceeding as planned [9,12]. Significantly, an exhibition hole is regularly seen in these buildings, principally alluding to the overutilization of energy as an extent of the planned energy request assessment. Also, the European buildings must accept the Energy Performance Certificate (EPC), which is viewed as a significant advance to assess a building's performance. It also shows that there are frequently enormous differences between the natural metered energy utilization in buildings and energy utilized in demonstrating for consistency purposes. The EPC also suggests a disparity seen in Energy Performance Certified structures regarding determined energy use and CO_2 discharge levels [13].

26.3 Renewable energy applications in green buildings

Generally, it was perceived to develop innovative and new methodologies to reduce carbon dioxide (CO_2) discharges due to energy use in buildings and construction activity. Taking into account that the energy use in buildings affects the sustainable development of the environment, the consultation with regard to energy is closely associated with feasible development. Hence, environment-friendly energy sources, such as the sun, winds, waves, algae, anaerobic digestion, and composting, are profoundly persuasive in the upgrade of sustainability. The investigations show that the green buildings are interwoven with successful energy plans and have progressed by incorporating advancements to cut energy utilization in warming, cooling, power, and by the use of on-site ecofriendly energy sources. With a view to the supportable energy execution of green structures, the proper use of sustainable power batteries in the building is crucial. In this sense, sun-based manageability contexts have consistently been vital in the advancement of green buildings [9].

In 2004 a study showed an effective method for developing sustainable buildings and improving their particular measurements by creating a consistent model [14]. Furthermore, this study [14] developed and analyzed a solar cooking system's performance using vacuum-tube collectors with heating pipes that comprise a refrigerant as fluid. The aptitude to cook was attained by pointing out that the appropriate cooking time varies depending on a refrigerant's collective selection and climatic parameters. Ultimately, the huge proficiency of the created cooking framework contrasted with the regular concentrators, and also box cookers were demonstrated for preheating the framework. Sunlight-based connectors were used as a fundamental segment of green buildings. Some researchers' separate experiments examined impact using different refrigerants on the warm change of a two-stage heat pump sunlight-based authority. Refrigerant-charged homegrown high-temperature water frameworks can be just produced by including general level plate sunlight-based collectors while providing the energy needs. Hence, it is prescribed to use the different outcomes in additional plans and advancement of sun-oriented homegrown high temp water frameworks. Cooling and warming frameworks are considered as a significant central purpose of recent investigations into green buildings. The investigation by Ensen [15] analyzed the energetic adequacy of the ground-coupled heat pump framework for warming applications. This examination demonstrates that huge enhancements of exergy productivity can be acquired while zeroing in on adjustments of the source temperature. The development has assumed significant commitments within the field of green structure escalation. Appropriately, the examination by Ozgen [16] developed a strategy for opening using an adequate plate made from alluminium.

The collector proficiency can be considerably upgraded by increasing the liquid speed and improving the warmth move coefficient between the safeguard plate and the air. This way bolsters the probability of performing different tests to find the impact of adaptable factors on the adequacy of sun-powered air heaters. Green buildings are equipped to work in various stages. The investigation by Balbay [17] likewise proposes an elective procedure for the removal of snow from extensions and asphalt. They use a ground source heat pump in Turkey. A separate framework is equipped for keeping different mishaps from occurring. Therefore the utilization of sustainable energy technologies can be considered for the improvement of future ecourban communities. Sustainable energy innovations could be significantly useful in terms of social, ecological, and economic concerns (Table 26.1). Evaluating the new insightful endeavors, the primary fixation is on the use of wind power age and the utilization of solar-based energy, that is, sun-based photovoltaics and sun-based nuclear energy frameworks. Be that as it may, it is still urgently required to investigate the conditions for the growth of new sustainable power. It is accepted that in the 21st century, cities and urban regions should be greener; thus advancing practical urban communities has become a central point of contention for some

26. Biointelligent quotient house as an algae-based green building

TABLE 26.1 Sustainable development targets and green buildings.

Objectives	Sustainable development goals
Bioresources and energy efficiency	Goal 7: Affordable and clean energy
Pollution reducing	GOAL 13: Climate action
Integration and harmonization with the environment	GOAL 12: Responsible consumption and production
Green and sustainable buildings	GOAL 11: Sustainable cities and communities

agricultural nations. The idea of sustainability or green building is an expansive global issue, including different interrelated sectors such as the environment, economy, and society (Fig. 26.1) [18]. The meaning of practical urban areas could be clarified by recognizing the role of manageability. Indeed, this manageability speaks to another methodology that recognizes the idea of "green building" in light of a reevaluating cycle designed to interface the full implementation of current urban communities to the climate, innovation, economy, society, and human well-being. It is at last presumed that supportability includes the three major constituents of environmental, social, and economic sustainability, while the individually referenced parts are significantly bound up with the conditions of the upgrade of prosperity for the occupants.

A complete analysis [19] has highlighted the significant negative environmental impacts of building in the United States, similar to other developed countries. Therefore the most significant negative consequences that require innovative solutions are energy consumption and gas emissions. The sustainability of buildings is essentially considered to be a systematic approach to adapting and developing architecture regarding environmental, economic, and sociocultural concerns. Sustainability envelops financial seriousness explicitly while also looking at hybrid energy frameworks. As of late, the thought of regular assets and energy conservation is now a worldwide issue and challenge because of a climate change consequences and the exhaustion of energy assets. Different studies have been completed corresponding to green buildings, zeroing in on energy support ideas, collected energy, and environment-friendly power assets.

A study by Levin [19] contends that there is a need for green assembled climate advancement, with 40% of the total energy use around the planet linked to building. This study shows the significance of natural manageability inside green building conditions to find harmony between energy requirements and energy assets. Concerning the huge part of green building on future urban communities' manageability, it is essential to consider supportable construction. Along these lines, the economic development models should be introduced to the draftsmen and engineers for utilization in the whole development cycle of green and sustainable building. Kibert presents significant rules of sustainable development: reduction in asset utilization, reuse of assets, reuse of materials, preservation of the indigenous habitat, elimination of harmful thinking about economic productivity. It is also necessary to instruct clients toward a feasible plan for lessening the energy utilization level of structures and their destructive effect on their habitats. Eventually, the economic systems should be created depending on the principal focuses of feasible improvements for a low-carbon future. Since the US

Handbook of Algal Biofuels

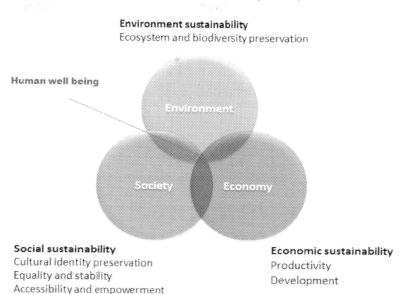

FIGURE 26.1 Sustainable development concept.

Congress of 1992, an evaluation of energy efficiency in buildings and attempts has been made to educate professionals and ordinary people about the importance of green buildings energy, energy use, and the environment's impact. The study expresses that this degree of utilization and preservation is exceptionally connected with mechanical advancements, innovation receptions, the client's way of life, economic development, and so on. Henceforth, it is reasonable to communicate the severe need to examine the energy utilization level of structures and develop new answers for accomplishing supportability in constructed conditions [18].

26.4 Biointelligent quotient in Hamburg, Germany

The BIQ building project is considered to be the first algal smart building in the world. This project was supervised and implemented by Arup in partnership with the German consultancy SSC (Strategic Science Consult) during the International Building Exhibition in 2013 in Hamburg, Germany. This project was the prize winner in the Land of Ideas competition in 2013/2014. The project's design implements flat-plate PBRs for the first time to produce heat and biogas under anaerobic conditions [20] in order to create a sustainable building where the building's energy needs will be provided using the microalgae cultivation. "SolarLeaf," which is the brand of the photobioreactors façade, is considered the first façade system that aims to cultivate microalgae to produce biomass and bioenergy as renewable energy resources. To reach an adequate building system in terms of energy efficiency, the SolarLeaf façades use the process of photosynthesis [4]. The flat glass panels are first filled with the different algae species, which capture heat and sunlight and consume carbon dioxide and nutritional elements to produce bioenergy and biomass. The BIQ is a cubic residential building and the microalgae photobioreactor façades are oriented to the southwest and the southeast for better production [4]. That building concept is composed of the CO_2 and the nutrients supply, biomass filtering process, the temperature control, culture liquid circulation, storage, and distribution systems. Algal biomass is transported to the biogas plant to be digested and transformed to methane and biogas and then converted into electricity. Also, this building system can start automatically, which reduces the maintenance cost [4].

The BIQ house, in Hamburg, Germany (Fig. 26.2) has a surface of 200 m^2 of closed PBR and 120 façades of algae and heat used to produce energy, and this is why the system is considered a low-energy building. The algae façade system creates a microclimate around the building because the temperature inside is controlled, it reduces noise from the outside, and provides local shading. Microalgae were developed in planetarium interfaces in solar-glass bioreactors, and the resulting thermal energy frame was used as a closed-loop frame for the power center at the basement. In addition, the separation of biomass can be done by floating, while heat can be separated from the water generated by algae using a simple heat exchanger. It is worth noting that the Hamburg biointelligent quotient building's energy was far less than that caused by traditional solar-panel systems. The heat generated by these panels is 38% efficient. In contrast, the conventional solar thermal source has an efficiency between 60% and 65%. The biomass is only 10% in terms of efficiency, whereas a conventional PV is 12%–15% efficient in comparison [22].

26.5 University of technology Sydney green building case study

The flat photobioreactors are considered efficient for algal cultivation and need only minimal maintenance. SolarLeaf's bioreactors are composed of four glass layers. The two

FIGURE 26.2 Bioreactive façade-algae (biointelligent quotient house, Hamburg, opened April 2013) [21].

internal parts have a 24-L bore for intermediary circulation. On both sides of these parts, the weathered parts of argon's insulation help to minimize heat loss. The front glass panel consists of white antireflective glass, while the mirror at the rear element can incorporate the treatments of different glass properties [23]. Down each photobioreactor, the air is compressed and then delivered at different intervals. After that, the gas rises to the top as big air bubbles, producing a water flow at the top of the basin that stimulate algae to take sunlight and carbon dioxide. During this time, the air inside photobioreactors helps in mixing algae to avoid sedimentation and to ensure the maximum exposition to the light [23].

Another case study of these building systems was undertaken at the University of Technology Sydney. The building is an innovative and elegant design used as a landmark in Sydney, Australia. It is located on one of the main roads in Sydney. The tower is elevated with a vertical footprint of 1500 m^2 divided into 10 floors, and it is exposed to the sun to receive maximum radiation throughout the day. Each floor has a height of 5 m and a width of 30 m. This building is also equipped with a biodigester to produce biogas from anaerobic digestion of algae [22,24]. In the illustration below (Fig. 26.3), the building in the center of Sydney is presented in three images to show the UTC building first, then the impression of the vertical area of 150 m^2 chosen for solar panels, and closed PBRs deployment. In Fig. 26.3, the vertical area of 150 m^2 that is suited for using the closed photobioreactor system is presented. In the closed PBR system, the 150 m^2 space is divided on the tower's vertical surface into five equal parts of the 300 m^2 site. Every unit has two floors of vertical surface area (150 m$^2 \times 2$) containing photobioreactor tubes that are linked to the reactor system which controls the flow of water and oxygen between two floors. The PBR system is equipped with tubes of 0.1-m diameter, with a vertical distance between these tubes of 0.05 m. Every reactor has 10 loops, with a tube length that reaches 80 m. The system has a total height of 5 m and no covering is proposed because it influences the productivity and effectiveness of the process (reducing light). The biomass delivered by this system will be moved to an anaerobic digester to generate biogas, which is also utilized as a bioenergy source. The wastewater of this case study building will likewise be transferred to this digester to be purified. The cost of the PBR system is presented in Table 26.2.

26.6 Conclusion

With regard to new building concepts, reducing the energy consumption and covering the energy needs for each building system are in high demand worldwide, especially the use of renewable energies such as energy from algal biomass. The first example of these buildings was introduced in Hamburg in Germany. Many other smart buildings have started to enter the market with new technologies and new designs and methods of cultivation. But there is still an enormous need for new solutions and performances with the low-cost installation necessary for deployment on the market. Microalgae are known to be simple cells and they can live and reproduce in all conditions and different types of water. The algae require for their growth only water, light, and CO_2, and can produce important yields of energy and biomass. There are many cultivation methods, but they are still insufficient in terms of cost and yield production rates. Therefore more work and new techniques should be developed and implemented to improve the existing technology. For instance, a more advanced system of

FIGURE 26.3 The vertical area (150 m^2) selected to deploy solar panels and closed photobioreactors.

TABLE 26.2 Costs of assembling the photobioreactor system.

System	Cost
Photobioreactor (material + implementation)	8857 EUR
Anaerobic digester or bioreactor [20]	253,432 EUR
Total cost	262,289 EUR
Operational cost + maintenance cost per annum	8414.58 EUR
Insurance per annum	994.72 EUR
Variable cost of digester	26,886.60 EUR
Total variable cost	36,295.9 EUR

cultivation should be developed to enhance the productivity of microalgae. The BIQ building is a promising innovative idea that become more important day after day and its ideas need to be implemented widely to aid in the reduction in use of the depleting fossil fuels.

References

[1] Y. Li, et al., Effects of digestion time in anaerobic digestion on subsequent digestate composting Submitted to Bioresource Technology, Bioresour. Technol. (2018). Available from: https://doi.org/10.1016/j.biortech.2018.04.098.

[2] K. Winans, A. Kendall, H. Deng, The history and current applications of the circular economy concept, Renew. Sustain. Energy Rev. 68 (2017) 825–833. Available from: https://doi.org/10.1016/j.rser.2016.09.123. no. September 2016.

[3] Y. Si, et al., Science of the total environment revealing the water-energy-food nexus in the upper yellow river basin through multi-objective optimization for reservoir system, Sci. Total Environ. 682 (2019) 1–18. Available from: https://doi.org/10.1016/j.scitotenv.2019.04.427.

[4] G.M. Elrayies, Microalgae: prospects for greener future buildings, Renew. Sustain. Energy Rev. 81, no. September 2017. (2018) 1175–1191. Available from: https://doi.org/10.1016/j.rser.2017.08.032.

[5] R. Remigio, S. Martin, Typologie des espaces-frontières à l'heure de la globalisation, Belgeo (2013) 0–16. Available from: http://belgeo.revues.org/10546.

[6] E.M. Charalampidou, S.A. Hall, S. Stanchits, H. Lewis, G. Viggiani, Characterization of shear and compaction bands in a porous sandstone deformed under triaxial compression, Tectonophysics 503 (1–2) (2011) 8–17. Available from: https://doi.org/10.1016/j.tecto.2010.09.032.

[7] S. Chang, D. Castro-lacouture, F. Dutt, P.P. Yang, Framework for evaluating and optimizing algae façades, Energy Proc. (2017). Available from: https://doi.org/10.1016/j.egypro.2017.12.677.

[8] D.T. Doan, A. Ghaffarianhoseini, N. Naismith, T. Zhang, A. Ghaffarianhoseini, J. Tookey, A critical comparison of green building rating systems, Build. Environ. 123 (2017) 243–260. Available from: https://doi.org/10.1016/j.buildenv.2017.07.007.

[9] Y. Geng, W. Ji, Z. Wang, B. Lin, Y. Zhu, A review of operating performance in green buildings: energy use, indoor environmental quality and occupant satisfaction, Energy Build. 183 (2019) 500–514. Available from: https://doi.org/10.1016/j.enbuild.2018.11.017.

[10] M. Shan, B. gang Hwang, Green building rating systems: global reviews of practices and research efforts, Sustain. Cities Soc. 39 (2018) 172–180. Available from: https://doi.org/10.1016/j.scs.2018.02.034.

[11] A. Samanci, A. Berber, Experimental investigation of single-phase and twophase closed thermosyphon solar water heater systems, Sci. Res. Essays 6 (4) (2011) 688–693. Available from: https://doi.org/10.5897/SRE09.072.

[12] D.X. Zhao, B.J. He, C. Johnson, B. Mou, Social problems of green buildings: from the humanistic needs to social acceptance, Renew. Sustain. Energy Rev. 51 (2015) 1594–1609. Available from: https://doi.org/10.1016/j.rser.2015.07.072.

[13] Z. Afroz, H. Burak Gunay, W. O'Brien, A review of data collection and analysis requirements for certified green buildings, Energy Build. 226 (2020) 110367. Available from: https://doi.org/10.1016/j.enbuild.2020.110367.

[14] M. Esen, Thermal performance of a solar cooker integrated vacuum-tube collector with heat pipes containing different refrigerants, Sol. Energy 76 (6) (2004) 751–757. Available from: https://doi.org/10.1016/j.solener.2003.12.009.

[15] H. Esen, M. Inalli, M. Esen, K. Pihtili, Energy and exergy analysis of a ground-coupled heat pump system with two horizontal ground heat exchangers, Build. Environ. 42 (10) (2007) 3606–3615. Available from: https://doi.org/10.1016/j.buildenv.2006.10.014.

[16] F. Ozgen, M. Esen, H. Esen, Experimental investigation of thermal performance of a double-flow solar air heater having aluminium cans, Renew. Energy 34 (11) (2009) 2391–2398. Available from: https://doi.org/10.1016/j.renene.2009.03.029.

[17] A. Balbay, M. Esen, Experimental investigation of using ground source heat pump system for snow melting on pavements and bridge decks, Sci. Res. Essays 5 (24) (2010) 3955–3966.

Handbook of Algal Biofuels

[18] A. Ghaffarianhoseini, N.D. Dahlan, U. Berardi, A. Ghaffarianhoseini, N. Makaremi, M. Ghaffarianhoseini, Sustainable energy performances of green buildings: a review of current theories, implementations and challenges, Renew. Sustain. Energy Rev. 25 (2013) 1–17. Available from: https://doi.org/10.1016/j.rser.2013.01.010.

[19] H. Levin, Systematic evaluation and assessment of building environmental performance (SEABEP), in: Proceedings of the 2nd International Conference on Building, no. June [Online]. 1997, pp. 3–10.<http://buildingecology.net/index_files/publications/SystemicEvaluationandAssessmentSEABEP.pdf>.

[20] A. Tallou, F.P. Salcedo, A. Haouas, M.Y. Jamali, K. Atif, F. Aziz, et al., Assessment of biogas and biofertilizer produced from anaerobic co-digestion of olive mill wastewater with municipal wastewater and cow dung, Environ. Technol. Innov. 20 (2020) 101152. Available from: https://doi.org/10.1016/j.eti.2020.101152.

[21] A.K.C. Chan, Tackling global grand challenges in our cities, Engineering 2 (1) (2016) 10–15. Available from: https://doi.org/10.1016/J.ENG.2016.01.003.

[22] N. Biloria, Y. Thakkar, Integrating algae building technology in the built environment: a cost and benefit perspective, Front. Archit. Res. (xxxx)(2020). Available from: https://doi.org/10.1016/j.foar.2019.12.004.

[23] A. Farhat, Role of advanced building façade system considering sustanaibility & renewabale energy practices: analyzing the need of sustainibility with technology, Int. J. Adv. Innov. Res., no. March 2016, 2018.

[24] A. Tallou, A. Haouas, M.Y. Jamali, K. Atif, F. Aziz, S. Amir, Review on cow manure as renewable energy, in Smart Village Technology, Springer, Cham, 2020.

CHAPTER
27

National Renewable Energy Laboratory

Sbihi Karim[1,2], Aziz Faissal[2,3] and El Baraka Noureddine[4]

[1]Laboratory of Biotechnology, Materials and Environment, Natural Substances and Environment Unit, Faculty Polydisciplinary of Taroudant, University Ibn Zohr, Taroudant, Morocco [2]National Center for Research and Study on Water and Energy (CNEREE), University Cadi Ayyad, Marrakech, Morocco [3]Laboratory of Water, Biodiversity and Climate Changes, Semlalia Faculty of Sciences, Marrakech, Morocco [4]Laboratory of Biotechnology, Materials and Environment, Physicochemistry of Natural Environments, Materials and Environment Team, Polydisciplinaire Faculty of Taroudant, Taroudant, Morocco

27.1 Introduction

The United States' energy change was founded while relying on the gas pump. When the country's energy security appeared unsure, the leaders started questioning their dependence on imported petroleum. Was it reasonable for the United States to produce national, safe, and economical energy using renewable sources? Keep novel renewable energy technologies in the country and boost up the economy by creating companies, products, and jobs? Energy transformation has been the task of the National Renewable Energy Laboratory (NREL) since 1977 [1].

Moreover, NERL is among the 17 US research labs administered by the US Department of Energy (DOE) and the only one assigned to guarding the United States' energy security within new unlimited natural energy resources and power production. Founded in Colorado, this lab transforms how we produce, use, save, and circulate energy. Renewable energy sources, such as solar, wind, water, and geothermal, are among the primary research domains [2–6]. NREL is aiming to develop energy productivity with excellent production, development technologies, and federal energy management. NREL is developing biofuels, hydrogen infrastructure, and fuel cells to increase renewable transportation options. Inevitably, by combining the power of data and computing, and concentrating on combined solutions, including grid modernization and security, NREL is reinventing the

Handbook of Algal Biofuels
DOI: https://doi.org/10.1016/B978-0-12-823764-9.00006-6

© 2022 Elsevier Inc. All rights reserved.

Earth's energy systems. NREL's equipment and verifying the models and suggesting innovative ideas and products [1].

Furthermore, NREL contributes crucial technical support to communities big and small, assisting them with their most critical energy difficulties. The energy innovations produced increase the economy and reinforce energy security. Although the oil embargo of the 1970s is in the past, energy problems will continue to challenge us in the years ahead. As the population continues to grow, emerging technology and appliance are introduced to the system. NREL needs to evaluate the impacts, strengthen security, and heighten resilience inside crucial systems [1,7]. Moreover, as environmental menaces increase and population needs in cities grow, we require our energy resources to be adaptive, independent, and responsive. The world's most important sources, that is, water and food, are under threat. We require efficient and sustainable processes of producing energy. By research, innovation, and analytic collaborations, NREL continues to predict these problems, and provide solutions. NREL constructs the foundations of the potential fields of the future and inspires the nation's economic growth [1].

In this context, this chapter aims to present the NREL and one of their algal biofuels projects (microalgal isolation, characteristics, and limitation of industrial application) to show the great promise of microalgae as an essential part of renewable energy.

27.2 National Renewable Energy Laboratory

NREL is under public ownership, and is a contractor-operated facility concentrating on a specific area in the research and development of renewable energy and sustainable energy. This agreement permits a private entity to operate the laboratory on behalf of the federal government. NREL receives congressional financing to use for research and development projects.

27.2.1 National Renewable Energy Laboratory history

Founded in 1974, NREL started functioning as a Research Institute for Solar Energy in 1977. Under the Jimmy Carter presidency, its works continued beyond solar energy research and development as it sought to popularize the awareness of existing technologies, such as passive solar. During the Ronald Reagan presidency, the budget of the institute was cut by almost 90%, the workforce was reduced, and the work of the laboratory decreased [8].

Renewed interest in the energy issue strengthened the institute's position in later years, but funding varied. Expected Congressional budget shortfalls in 2011 resulted in a voluntary buyout scheme with the loss of 100—150 employees [9]. The fiscal 2016 budget was US\$ 427.4 million, down from a high of US\$ 536.5 million 5 years earlier [10]. Budget adjustments have sometimes forced NREL to slash staffing [11]. After its founding, the Solar Energy Research Institute has been administered under contract by MRIGlobal [12]. In September 1991, the solar energy research institute was named a national laboratory of the United States Department of Energy by President George H.W. Bush and changed its title to NREL.

Actually, Battelle and MRIGlobal created Alliance for Sustainability, LLC to acquire the NREL management and operating contract. The management and operating contractor, Alliance, is entirely responsible for the DOE, Office of Energy Efficiency and Renewable Energy for ensuring NREL's production [13].

In November 2015, Dr Martin Keller became NREL's ninth director and now serves as both the laboratory director and Alliance's president. He replaced Dan Arvizu, who parted in September 2015 after ten years in this role [14].

27.2.2 Mission and programs

The NREL develops clean energy and energy efficiency technologies and uses advanced science and engineering, and provides knowledge and innovations to integrate energy systems at all scales.

NREL's mission (Fig. 27.1) is to advance the US DOE and the US' energy goals to obtain new renewable alternatives to supply all houses, industries, and vehicles.

Through its Laboratory Directed Research and Development (LDRD) program, NREL actively continues algal biofuel projects, including national and international collaborations and several privately financed projects. Among these projects, NREL addresses critical challenges to completing an economical production process for algal biodiesel.

This research supports NREL's goal of turning the energy issues of today into solutions for tomorrow.

27.2.3 Bioenergy

NREL's bioenergy science and technology group conducts a wide range of biomass research at the molecular level by optimizing the biorefinery method to carry biofuels and bioproducts to the market.

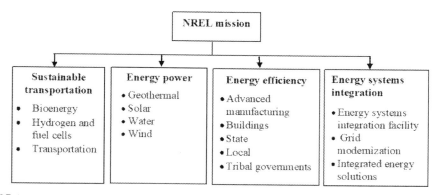

FIGURE 27.1 National Renewable Energy Laboratory's mission [15].

Bioenergy fuels, biobased materials, and power from plants and waste represent a crucial and nearly untapped supply of a potentially enormous natural renewable resource. Many applications, such as transport, power generation, or as a feedstock for chemicals, fiber, and other goods, can replace or displace imported fossil fuels.

Biorefineries that generate a concept still in its infancy with multiple products but with tremendous potential could ultimately contribute to developing a robust, domestic biobased industry. The broader use of biomass would produce new revenues for farmers and the rural economy.

NREL's biofuel research focuses on conversion technologies for two main biofuels: bioethanol and biodiesel. NREL's alternative fuel research and development objective are developing cost-effective, environmentally sustainable technologies to turn biomass into substitute transportation fuels and fuel additives.

Different research bioenergy areas developed in NREL include, for instance, analysis and characterization, biochemical processes and bioenergetics (Fig. 27.2).

27.2.4 Algal biofuels

Microalgae are microscopic photosynthetic algae able to transform biomass and oil to atmospheric carbon dioxide (CO_2). They offer enormous promise to create an essential contribution to renewable energy and fulfill the need to adequately decrease our reliance on the Earth's diminishing natural energy resources. The latest NREL algal biodiesel program has developed from the knowledge and background obtained from the Aquatic Species Program's (ASP) research into microalgal biofuels that was founded in 1978 with financing from the United States DOE.

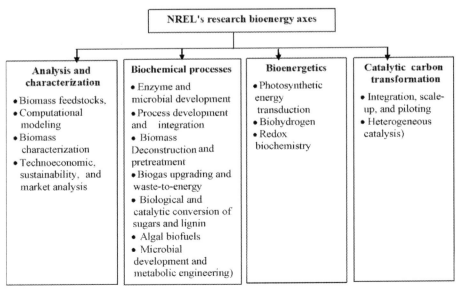

FIGURE 27.2 Different research bioenergy areas developed in National Renewable Energy Laboratory [15].

NREL is evolving technologies and helping to train a new generation of workers to commercialize biofuels from algae.

NREL focuses on understanding the current cost of generating algal biofuels and using that knowledge to identify cost control measures and improve them. NREL work is spread across the entire value chain, from the production algae to biofuel and bioproducts upgrading.

Coal, petroleum and natural gas have been the primary sources of energy for centuries. However, tomorrow's fuels will be produced from natural sources such as algae.

NREL is an expert in the development of new technology. Researchers transform biomass into biofuels. Algal cells consist of a mixture of biopolymers, including cellulose, hemicellulose, and lignin. These polymers are a source of cellulose sugars that can be converted to diesel-like ethanol. Cellulose ethanol produced from vegetal residues is very promising for reducing our dependence on imported oil and guarantees our energy autonomy.

Via a multiyear private-sector development program, NREL has undertaken significant pretreatment, enzymatic hydrolysis, and fermentation to aid the price objective for biochemistry transformation, demonstrating that biofuel can be competitively priced with other transport fuels. Researchers have additionally commenced projects at NREL's combined biorefinery research facility, which has facilitated the industry to advance applications to commercialize biofuel. Also, the laboratory is progressing in perfecting specific innovations with its collaborators, leading to the new era of biofuels.

The second main objective of NREL is thermochemical conversion (employing temperature and chemistry) to accelerate biomass transformation into liquid transport fuels through inorganic catalysts. NREL brings its experience in thermochemical technology to innovative maxima by approaching the goal of creating renewable aircraft fuels for an industry (every year, almost 1.7 billion barrels of petroleum jet fuel are consumed).

The third important biofuel production platform of NREL is microalgae. This technology utilizes a large range of methods, mixing algae cultivation with thermochemical and biochemical processes to recover intermediates and the production of fuels, including finished products such as chemicals, nutraceuticals, and even food.

Algae can be cultivated indoors, but to scale it up for biofuels, the work needs to move outside, where strains are heavily impacted by varying sunlight, carbon dioxide, and nutrients in the water. Researchers use NREL's computer simulation capability to consider dynamic algae cultivation variables to ameliorate algae farms' positions and conditions.

Collaborating with NREL, researchers can use all the laboratory resources to develop the technologies required to determine industrial difficulties.

27.3 History of National Renewable Energy Laboratory algal biofuels projects

Many projects concerning algal biofuels are sponsored by NREL's LDRD program. These projects discuss the problems that encompass the process of the production of algal biofuels. Some examples of the project are given below [16]:

27.3.1 Establishment of a 400+ bioenergy-focused microalgae strain collection using rapid, high-throughput methodologies

NREL, Colorado School of Mines, and the National Research Council of Canada have identified various bioenergy-focused strains. High-throughput technologies are applied to find different microalgal species for biodiesel utilization. About 400 isolated microalgal species have been produced using novel techniques and are currently being tested for high-growth and lipid-accumulation isolates. Different microalgae collections (marine, freshwater, and hypersaline environments) have been isolated from diverse US locations.

27.3.2 Molecular foundations of algal biofuel production: proteomics and transcriptomics of algal oil production

NREL couples two approaches (proteomics and transcriptomics) to ultimately determine algal biofuels production's molecular foundation, including different environmental parameters for oil accumulation in green algae. The first study was seeking to solve critical questions about algal oil production to develop cost-effective and algal-based jet fuel.

NREL incorporates two approaches to improve the excellent knowledge of the variations in the transition to rapid growth and high accumulation of lipids, which gives NREL genetic explanations for biofuels synthesis and guides species development strategies. NREL has approved experimental promoter-gene sequences isolated from a green algae species and uses chloroplast genome sequencing information to increase lipid production, and to establish further an algal genetic toolbox.

27.3.3 Evaluation of regulated enzymatic disruption of algal cell walls as an oil extraction technology

NREL characterizes cell-wall-degrading enzymes adequately to enable internal oil bodies to escape and be easily collected for their activity in disrupting algal cell walls. This technology could be a cost-effective and straightforward alternative for harvesting lipid algal biomass.

27.3.4 Development of novel microalgal production and downstream processing technologies for alternative biofuels application

In this project, NREL is studying the usability of algal biofuels as petroleum-based fuel substitutes. The essential step incorporates studying the biochemical-based conversion routes of carbohydrates and lipids to fuels. NREL also analyzes the optimization of cost-effective thermochemical conversion and the extraction of oil biomass. Additionally, practical research of advanced thermochemical conversion technologies is aiming to produce biofuel algal with intermediate products.

27.3.5 Efficient use of algal biomass residues for anaerobic digestion biopower production coupled with nutrient recycle

NREL and Washington State University are researching biogas production optimization through anaerobic digestion from algal residues. This project aims to explain the criteria for wastewater treatment by anaerobic digestion and the bioavailability and recycling of nutrients such as nitrogen and phosphorus after treatment to promote algae growth.

27.3.6 Development of robust and high-throughput characterization technologies for algal biomass

NREL scientists are creating a novel approach for studying process-relevant biofuel constituents in algal biomass to evaluate algal biofuels' production process's performance using technoeconomic analyses, making NERL at the forefront of compositional analysis of novel feedstocks and quantifying essential constituents in a circle. Further, NREL produces a high-throughput spectroscopic prediction system to characterize algal biomass in a few minutes [16].

27.4 Principal project

27.4.1 Project presentation

The necessity to produce sustainable and renewable energy sources is more critical than ever, as concerns about US energy security and global warming increase. Microalgae as feedstocks that transform solar radiation and CO_2 into chemical intermediates such as oils, hydrocarbons, and carbohydrates that can be extracted and readily converted into biofuels is one exciting renewable energy scenario. Screening of large numbers of diverse strains is vital to recognize microalgae strains exhibiting superior production of biofuels. About 35,000 algal organisms have been reported to have been described globally. However, in nature, as many as 200,000 to 1 million species may occur. Consequently, only a small amount of this vast biological and genetic diversity has been studied. Given the tremendous amount of potentially important biodiversity and remains undiscovered, a microalgal bioprospecting project used fluorescence enabled cell sorting (FACS) to quickly isolate and classify 500 distinct microalgal strains in the southwestern part of the United States from a variety of aquatic environments. Screening with confocal fluorescence microscopy of these isolates and a neutral lipid-specific lipophilic dye identified many microalgal strains with significant oil production capabilities [17].

This project is a collaboration of the C2B2, NREL, Colorado State University, University of Colorado, and the Colorado School of Mines. This is part of the House Bill's 06−1322 broader state-wide initiative to build a Colorado collaboration, which envisages the four organizations working together as part of its energy strategy. The first project of many envisaged in this overall initiative is with the Colorado School of Mines. The project focuses on developing high-throughput methods to rapidly isolate and purify novel

606

27. National Renewable Energy Laboratory

strains of microalgae (specifically green algae and diatoms) from water samples collected from specific aquatic environments [17].

27.4.2 Project description

The NERL is developing this project in two-phases. The Phase I goal is to isolate and characterize 500 microalgal strains utilizing rapid, high-throughput methodologies through four key points: sample unique aquatic environments; sort microalgal cells using FACS; characterize algae based on morphology and lipid accumulations; and cryopreserve microalgae isolates for long-term storage. There is also a second phase, involving lipid production analysis of most promising strains with key five points: select the most promising strains based on preliminary growth rates and lipid content; optimize growth media to enhance lipid productivity; perform GC-FID and GC-Ms to obtain detailed fatty acid methyl ester (FAME) profiles; use novel techniques utilizing near infrared spectrophotometry developed at NREL to rapidly profile lipid content. 18S rRNA gene sequences will be determined for precise taxonomic classifications.

Locations of sample collection sites:
Targeted sites from diverse places in the southwest.
Site determination:

1. Locally: based on the presence of visible algal growth
2. Regionally: documentation of productivity and visible growth.

71 Samples collected:

1. 32 from Colorado
2. 9 from Utah
3. 8 from New Mexico
4. 1 from Sea of Cortez
5. 6 from Arizona
6. 5 from California
7. 10 from Nevada

27.5 Microalgae isolation and characteristics during the project

27.5.1 Water sample collection and analysis

During two sampling expeditions, 71 water samples (250 mL of each sample) from separate sites across the Southwest United States were obtained in spring and summer of 2008 (Samples 1−47) and 2009 (samples 48−71). Ten micrometer pore-sized plankton nets were kept suspended in each water site for 1 min and then allowed to reduce to concentrate the captured algae. Biofilm samples were obtained after collecting the site water by collecting all biomats, including scraping different rocks in each water site. The site water sample container was specifically applied to these biofilms and sediment materials, and the mixture was closed and carefully mixed by strong shaking for 1 min. This suspension,

Handbook of Algal Biofuels

like algal grazers and other debris, was then filtered to eliminate larger particles. Filtration was obtained by employing a stacked sieve (two coupled 10.2 cm diameter with immobilized 100 and 30 mesh in the upper and lower sieve, respectively) [18].

The lower sieve filtrate was then assembled and used multiparameter environmental water quality measurement to test specific conductivity and pH on-site. Each filtered sample was split into two aliquots of 30 mL and put into conical vials of 50 mL. With one 0.6 cm hole for aeration, the lids were perforated. For a transport time of upto 2 weeks, samples were then put into an illuminated cooler and stored vertically to avoid spillage. Between each application, the collection equipment was thoroughly disinfected by pulverizing with 90% v/v isopropyl alcohol accompanied by a quick washing with sterile Milli-Q deionized water. Thirty-two white LEDs ensured the 12:12 light/dark cycle to generate 40 microns m^{-2}s^{-1} PAR (photosynthetically active radiation measured using an LI-250A light meter). The cooler's bottom had 7.6 cm of crushed ice replaced every day, and the temperature of the cooler was kept between 4.4°C and 10°C. Samples were suspended 8 cm above the ice layer in test tube racks. Using a Global Positioning System (GPS) receiver, registered sample site latitude and longitude coordinates and site positions were traced employing Garmin Base Camp v2.0.3 software [18].

27.5.2 Laboratory conditions for growth algae

Each sample and enrichment were stored in 10-mL test tubes in a bright growing room for 1 month of incubation. Fluorescent 5000K T8 lights were positioned 15 cm under the glass shelving and faced upward to illuminate samples from below through the shelves. The conditions of chamber culture were fixed in approximately 65 μmol/m^2 per second of PAR, 16:18 light-dark cycle and the temperature and humidity were maintained at 22°C and 90%, respectively. FACS-generated microtiter plate cultures were cultivated for 1 month in a humidified plant culture room.

Shake flask cultures, including 150 mL of algal cells, were cultivated in a bright growing chamber in 500 mL of unbleached shake bottles closed by silicone sponge covers to avoid contamination but allowing for gas exchange. With 5000K T8 fluorescent lights positioned 30 cm over the shaker base, the agitation was fixed at 120 rotations per minute with the same culture conditions mentioned above [18].

27.5.3 Laboratory water sampling preparation and enrichment

Enrichment by vortexing field samples was carried out on samples, and then 100 μL of the sample was transferred into 5 mL of sterilized culture medium to preserve sterility and provide the transfer of gas. Furthermore, the sample was incubated for several weeks up to 1 year before FACS in the cultivation room. To preserve algal cultures, an inoculation of 5 mL of culture medium was provided with 100 μL enrichment of the sample every 6−10 weeks. Microalgae medium ingredients were modified to nonlimiting conditions by standardization of vitamins, trace metals, nitrate, and silica for mBBM, mASP2 [19], mDF/2 [20] mPLM, mWLM, and mMLM [21], and mS2 (c10, c25, c40, c55, and c70) [22]. Before FACS, Samples 1−11 were enriched with mBBM algal cultures medium, but before

FACS, samples 12–47 were not enriched. Samples 48–71 were enriched with all of the eleven various culture media, and repeated serial transfers maintained these 264 enrichments in the growth chamber before the FACS was carried out [18].

27.5.4 Fluorescent activated cell sorting for isolation of single cells

A custom BD FACS AriaTM operating BD FACSDivaTM v6.0 (BD Biosciences) program was applied to perform FACS to examine fresh and enriched water samples and various species algal. Three types of laser were included in the FACSAriaTM hardware configuration for particle interrogation, namely, a consistent sapphire solid-state laser, an JDS Uniphase HeNe Air-Cooled laser, and a point-source solid-state laser: 13 mW 488 nm, 11 mW 633 nm, and 10 mW 407 nm, respectively. In addition to forwarding dispersion (FSC) and side scatter, the FACSAriaTM could detect up to 10 separate particle fluorescence channels (SSC). An ND1.5 (neutral density) light attenuation filter was placed before the FSC detector. A 407 nm trigon photomultiplier tube (PMT) channel with a custom 650 nm long-pass optical filter was used to detect chlorophyll autofluorescence for these experiments (Chroma). The fluorescence of samples (1–47) was observed on a 488 nm octagonal PMT channel equipped with a 530/30 nm optical bandpass filter using a lipophilic dye Bodipy 493/503. PMT voltage parameters were set to 10 3 arbitrary units (AUs) before running algal samples, by Sphero Rainbow Calibration Particles (BD Biosciences) of 3.0–3.4 μm in size. Sphero Rainbow beads were applied to set the FSC and SSC voltages to 100 AUs [18].

The drop time to sort using 6-μm Accudrop fluorescent beads (BD Biosciences) was determined in the two-tube sorting layout following the manufacturer's instructions. It should be noted that the FSC threshold parameter was already set to 2000 and the window extension to 4.0. The sorting of individual microalgal cells was performed by arranging the 96-well plates. Two different FACS procedures were applied to separate algal cells, one for each sample. For samples 1–47, FACS has been employed to replicate individual algal cells within 96-well microtiter culture plates including 240 μL of culture media containing mBBM or mASP2 for freshwater or saline samples. Before FACS analysis of the sample, it was filtered and the filtrate was collected in BD Falcon flow cytometer test tubes. It was then colored with Bodipy 493/503 at a terminal assay concentration of 10 μg/mL and allowed to incubate at room temperature before sorting was required, which was often within 2 h of staining. The samples and dilutions were then analyzed on the BD FACSAria to obtain an event rate of between 200 and 800 events per second at a sample flow rate of 1–2 based on a relative flow scale of 1–11 on the BD FACSDivaTM v6.0 control program. Based on positive chlorophyll and bodipy fluorescence, event samples were selected for filtering. After filtration, the 96-well plates for clonal isolation were covered with radiation sterilized film for cell culture applications to reduce water evaporation from the medium, which was important without film and reduced cross-contamination. The enclosed plates were then incubated in the growth chamber for upto 4 weeks, during which visible algal growth was observed [18].

Fluorescence enabled cell sorting was used to clone separate algal cells in 96-well microtiter culture plates comprising 240 μL of culture media used in the enrichment per well for the samples (48–71). Each sample enrichment had one row (12 wells) of a 96-well plate designated for the sorting of a single event. Just the 115 durable enrichments of the 264 enrichments were considered for FACS. The others were eliminated. To produce an event rate between 200 and 800 events per second at a relative flow rate of 1–2, samples were diluted before FACS and filtered with a 35-μm cell strainer beforehand without staining with Bodipy. The classification of event samples for sorting (into each of the 24 wells of a 96-well plate) is dependent only on chlorophyll autofluorescence. Moreover, to increase gas exchange, the plates were closed with their microtiter covers and incubated for upto 4 weeks in a Sanyo humidified plant chamber. The growth room conditions used were around $65\,\mu$moles/m^2 per second PAR, 16:18 light-dark cycle, and temperature and humidity maintained at 22°C and 90%, respectively. Diascopic light microscopy for morphological characterization was used to examine microalgae based on morphology and classify specific species candidates to include them in the culture collection [18].

27.5.5 Culture maintenance

The FACS process appeared in 360 clonal algal species being isolated, continuously maintained in a liquid medium by serial passage, and continued in the bright cultivation chamber. Before eventual cryopreservation and revivification of the samples, the stock cultures underwent continuous maintenance every 6–10 weeks with subcultures and dilution (1:500) for strain preservation. After 6 months or 1 year of freezing, the viability of cryopreserved microalgae was tested.

27.5.6 Summary of National Renewable Energy Laboratory project results

A promising renewable energy scenario includes using photosynthetic microalgae to produce biofuels. In sunlight, algae convert energy into different storage materials, including triacylglycerols, which can easily be converted into surrogates for diesel fuel. It is essential to grow economically viable algal biofuels to optimize the targeted bioenergy carriers' accumulation via selected feedstock strains. To identify promising production isolates, the NREL has created, analyzed, and improved modern high-throughput cell-sorting techniques to identify promising production isolates from diverse ecosystems in the Southwest United States.

This project succeeded in the isolation of 360 distinct microalgae strains. NREL conducted initial culture collection isolation to classify major biodiesel phenotypes, including neutral lipid accumulation and rapid growth rates. Also, NREL identified appropriate cultivation media and examined cryopreservation techniques necessary for long-term microalgae storage. This precious source of biodiversity is a crucial resource that can be used to select bioenergy feedstock strains and provide fundamental advances in NREL's understanding of photoautotrophic diversity, biology, and metabolism. In the consequent work, NREL established a small-scale photobioreactor process to allow the growth of several strains in light under regulated conditions (including air and CO_2 inputs) to evaluate these growth and lipid productivity strains. One hundred promising strains were selected for examination, and 74 produced sufficient biomass for the analysis of dry weight and lipid

610   27. National Renewable Energy Laboratory

content (by FAME concentration) after 15 days of cultivation. Cells at this stage in the growth process were presumed to be rich in nutrients and therefore required to have a high lipid content [17].

27.6 Limitation of industrial application

It is well agreed that microalgae's industrial production must be optimized to be economically competitive with higher plants because large quantities of medium growth, water, and energy are required. The natural selection of traits that enhance the fitness of organisms is due to biological diversity in nature. Algae have developed the capacity to withstand continuing nutrient exposure, instability, high temperatures, diverse salinity and pH ranges, varying light levels, other species niche competition, predatory grazing, contaminants, and day-to-day evolutions. This degree of environmental adaptability highlights complex metabolic strategies that can prevent and promote algae as a feedstock in response to environmental stressors. Now and in the future, the industry will continue to face challenges. In particular, the algal activity can range in size from individual bioreactor arrays generating speciality chemicals and nutraceuticals to the widespread processing of food products and biofuels on a farm scale. To fund development and conduct ecological life cycle research, accurate evaluation of their potential economic and environmental footprint would be crucial. There is no harmonized descriptive language to explain the various technologies being proposed for scaled algal farms, nor were measurement methodologies specifically developed. The absence of an appropriate popular language and process has created confusion in expressing attributes and represents a barrier to industry expansion. However, since not every species is oleaginous, lipid productivity is essential in selecting algae to produce an economically viable microalgal biofuels industry [23]. As a result, several study efforts aim to genetically engineer well-characterized microalgae to maximize triacylglycerol storage lipids production, which requires considerable time and resource expenditure.

27.7 Conclusion of the project

The growth of microalgae as a feedstock for biofuel production is an encouraging scenario for renewable energy that many believe can provide a considerable portion of our energy demands for transport. Nevertheless, this technology's viable commercial production will eventually involve the use of algal strains that generate chemical intermediates much more consistently than several species mentioned to date. Numerous works are currently concentrated on genetically optimizing metabolic pathways to improve lipid efficiency in laborious and time-consuming tasks in well-studied species. Though the challenge could be alleviated using natural selection and identifying native species of microalgae that previously possess the metabolic characteristics required for commercial feedstock production. To date, considering the potential utility that these species have, the diversity of microalgae has been primarily unexplored, and targeted bioprospecting activities will be able to discover practical metabolic strategies that can be leveraged to produce commercial biofuels. The aim of the project defined in this analysis has been to:

1. Create a selection of microalgae from different water chemistries by establishing protocols based on recent techniques.
2. Determine culture media that give excellent algal growth with low-cost chemical compositions.
3. Apply high-throughput methods for selecting microalgae, including abundant intracellular lipid reserves with rapid growth rates.
4. Find the long-term maintenance of cultures using good cryopreservation technique.

The first phase of this project successfully established a collection of bioenergetic microalgae cultures, which had robust growth in low-cost growth media, and realized a successful cryopreservation technique for these species' long-term maintenance. Also, there was the creation of a database to catalog the metadata acquired for each strain. In general, the extraordinary work of bioprospecting and isolation has resulted in establishment of a collection of microalgae for further biofuel research. The collection of various microalgae from different aquatic habitats (freshwater, hypersaline, various water pH, and carbonate levels), contribute to assemble information about the different species for a possible future retrieval and interpretation. The scientific data management system was developed to associate all the experimental metadata found. Also, to estimate these metadata, the environmental conditions and water chemistry in which oil-producing algae can grow rapidly were determined and will allow the discovery new microalgae that can be productive as biofuel feedstocks. The consequent study determination quantifies the capacity for bioenergy production of microalgae in the collection and tries to establish relationships between water chemistries and phenotypes. This work's results remain a starting point for further research for the characterization of promising strains from a bioenergy perspective.

The second phase of this work concerned the precise characterization, using high-throughput methodologies, of numbers of microalgae in the culture collection. A high-throughput spectrophotometric microscale growth assay was used in the first attempt to distinguish the robust growth strains in this collection. To evaluate the microscale assay's capacity to identify the microalgae growth production at industry scale, 30 representative strains were scaled upto shake flask cultures. Gravimetric approaches to larger-scale harvested biomass can be more practical for measuring microalgae growth and producing more robust results.

A high-throughput photobioreactor's development represents the second effort to screen the collection and isolate promising microalgae to produce large biomass for parallel gravimetric analysis for upto 50 species. Ultimately, the HTPBR was created to control experimental variables that immediately influence the increasing biomass, such as changes in PAR levels, gas sparing and mixing ratio flow rates, atmosphere gas (CO_2, O_2), growth medium compositions, and temperature. Under these conditions, those variables considered most relevant to the production of biofuels could be isolated and changed, and the results of these variations on growth rates and lipid productivity of up to 50 microalgae could be tested in parallel.

The HTPBR was seeded with 100 separate microalgal species, using N-replete media without vitamins, twice for 2 weeks per run. The biomass was then harvested for dry

612
27. National Renewable Energy Laboratory

weight and nonpolar lipid processing gravimetric analysis. Several strains have been described as having rapid growth and accumulation of lipids. For a third, N-deplete run, the top 50 strains were selected but were unfortunately not completed.

In conclusion, this project's results led to the successful establishment of a biofuel microalgal collection. It led to the vigorous research and the creation of high-throughput characterization tools to evaluate 100 strains for biomass and lipid productivity in parallel. Furthermore, this project has produced many promising strains for more characterization research. Prospective works should concentrate on carrying out more comprehensive biochemical, genetic, and further industrial growth studies to determine capacity under numerous commercially applicable parameters.

References

[1] NREL, Welcome to NREL, <https://www.nrel.gov/news/video/welcome-to-nrel-text.html>, 2019 (accessed 10.09.20).

[2] A. Zakutayev, Design of nitride semiconductors for solar energy conversion, J. Mater. Chem A 4 (2016) 6742–6754. Available from: https://doi.org/10.1039/c5ta09446a.

[3] J. Helsen, Y. Guo, J. Keller, P. Guillaume, Experimental investigation of bearing slip in a wind turbine gearbox during a transient grid loss event, Wind Energy 19 (2016) 2255–2269. Available from: https://doi.org/10.1002/we.1979.

[4] D. Inman, E. Warner, D. Stright, J. Macknick, C. Peck, Estimating biofuel feedstock water footprints using system dynamics, J. Soil. Water Conserv. 71 (2016) 343–355. Available from: https://doi.org/10.2489/jswc.71.4.343.

[5] D.J. McTigue, D. Wendt, K. Kitz, J. Gunderson, Nh Kincaid, G. Zhu, Assessing geothermal/solar hybridization - integrating a solar thermal topping cycle into a geothermal bottoming cycle with energy storage, Appl. Therm. Eng. 171 (2020) 12.

[6] O.J. Guerra, J. Zhang, J. Eichman, P. Denholm, J. Kurtz, B.M. Hodge, The value of seasonal energy storage technologies for the integration of wind and solar power, Energy Environ. Sci. 13 (2020) 1754–5692. Available from: https://doi.org/10.1039/d0ee00771d.

[7] M. Lieve, L. Laurens, A. New Algae Technical, Standards focus group: summarizing and guiding the algae state of the art, Algal Res. 53 (2021) 4.

[8] National Renewable Energy Laborsatory –NREL 25 Years of Research Excellence 1977–2002, 2002 July (archive).

[9] National Renewable Energy Lab in Golden to cut 100–150 jobs through buyouts. <https://www.denverpost.com/2011/10/03/national-renewable-energy-lab-in-golden-to-cut-100-150-jobs-through-buyouts/>, 2011 (accessed 10.09.2020).

[10] Funding History, National Renewable Energy Laboratory. <https://www.nrel.gov/about/funding-history.html>, 2017 (accessed 10.09.2020).

[11] NREL cutting four percent of workforce, lays off solar researchers. <https://www.denverpost.com/2015/10/07/nrel-cutting-four-percent-of-workforce-lays-off-solar-researchers/>, 2015 (accessed 10.09.2020)

[12] Alliance for Sustainable Energy, MRIGlobal. Archived from the original on July 28, 2014. (accessed 22.07.14).

[13] Alliance for Sustainable Energy, LLC Documents. Alliance for Sustainable Energy, LLC Documents. 2014 (accessed 23.07.14).

[14] NREL director Dan Arvizu to retire in September, BizWest. <https://bizwest.com/2015/03/20/nrel-director-dan-arvizu-to-retire-in-september/>, 2015 (accessed 10.09.2020).

[15] Mission and Programs, <https://www.nrel.gov/about/mission-programs.html>, 2020 (accessed 10.09.2020).

[16] Algal Biofuels R&D at NREL, <https://www.nrel.gov/docs/fy12osti/56309.pdf>, 2012 (accessed 10.09.2020).

[17] CRADA Report, Cooperative Research and Development Final Report, 2013 November.

[18] L.G. Elliott, C. Feehan, L.M.L. Laurens, P.T. Pienkos, A. Darzins, M.C. Posewitz, Establishment of a bioenergy-focused microalgal culture collection, Algal Res. 1 (2012) 102–113. Available from: http://doi.org/10.1016/j.algal.2012.05.002.

[19] R.A. Andersen, J.A. Berges, P.J. Harrison, M.M. Watanabe, Recipes for freshwater and seawater media, in: R. A. Andersen (Ed.), Algal Culturing Techniques, 2005, pp. 429–532.

[20] H. Glover, Effects of Iron deficiency on *Isochrysis galbana* (Chrysophyceae) and *Phaeodactylum tricornutum* (Bacillariophyceae), J. Phycol. 13 (1977) 208–212.

[21] W.H. Thomas, D.L.R. Seibert, M. Alden, P. Eldridge, Cultural requirements, yields and light utilization efficiencies of some desert saline microalgae, Nova Hedwigia 83 (1986) 60–69.

[22] J. Sheehan, T. Dunahay, J. Benemann, P.G. Roessler, A look back at the United States department of energy's acquatic species program - biodiesel from algae, Close Out Re-port TP (1998) 580–24190.

[23] M. Griffiths, S. Harrison, Lipid productivity as a key characteristic for choosing algal species for biodiesel production, J. Appl. Phycol. 21 (2009) 493–507.

CHAPTER

28

Aquatic species program

Faissal Aziz[1,2], Anas Tallou[3], Karim Sbihi[4], Khalid Aziz[5] and Nawal Hichami[6]

[1]Laboratory of Water, Biodiversity, and Climate Change, Faculty of Sciences Semlalia, Cadi Ayyad University, Marrakech, Morocco [2]National Center for Research and Studies on Water and Energy (CNEREE), Cadi Ayyad University, Marrakech, Morocco [3]Polydisciplinary Laboratory of Research and Development, Faculty of Sciences and Techniques, Sultan Moulay Slimane University of Beni Mellal, Beni Mellal, Morocco [4]Laboratory of Biotechnology, Materials and Environment, Natural Substances and Environment Unit, Faculty Polydisciplinary of Taroudant, University Ibn Zohr, Taroudant, Morocco [5]Materials, Catalysis and Valorization of Natural Resources, Faculty of Sciences, University Ibn Zohr, Agadir, Morocco [6]Laboratory of Biotechnology and Sustainable Development of Natural Resources, Sultan Moulay Slimane University, Beni Mellal, Morocco

28.1 Introduction

The recent concerns of oil resources and reserves experts have raised many longer-term supply concerns. Society's dependence on fossil fuels, mainly in the transport sector, with regard to resource depletion and global warming has resulted in significant advances in renewable energy research. This research aims at the substitution or replacement of fossil fuels by various renewable energies [1]. However, biofuels such as bioethanol produced from corn and biodiesel from crops like soybeans have led to several issues. Indeed, the rise of these biofuels has led to increased food prices with many impacts on the population, increased land use promoting deforestation worldwide, and increased pollution from fertilizers and pesticides. Therefore in connection with these findings and the energy crisis, other efforts have made in searching for alternative solutions, such as biodiesel production from microalgae oils [1,2].

These research efforts were initially directed toward producing microalgae with autotrophic metabolism, that is to say, using solar radiation (energy source) and CO_2 as a source of carbon. Impressive results were obtained while allowing production on infertile

Handbook of Algal Biofuels
DOI: https://doi.org/10.1016/B978-0-12-823764-9.00022-4

© 2022 Elsevier Inc. All rights reserved.

or neglected land. Some companies and research centers have been working for several years to produce biodiesel from microalgae, mostly from autotrophic metabolism. Autotrophic production depends on sunlight and optimal temperatures that have been found primarily at latitudes near the equator [3].

From 1978 until 1996, the Department of Energy's (DOE) Office of Fuels Development financed the new energy production program. The Algae Species Program (ASP) plan's primary focus was based on biofuel production from lipids present in algae cultivated in the ponds, using resulted CO_2 from coal-fired power plants. Throughout 20 years of ASP, great progress was made in manipulating the metabolic rate of algae and the engineering of microalgae cultivation methods and techniques. The ASP analyzed factors affecting the ability to produce oil. Experts were not merely concerned about discovering algae that produce large quantities of oil, but they also studied algae grown under different temperatures, salinity, and pH conditions. No information was available on algae in terms of those restrictions before the start of the program. In the beginning, experts attempted to create such a collection. Algae were accumulated from locations in the west, northwest, and the southeast United States. They specified over 3000 varieties of algae organisms. The range was ultimately narrowed down to 300 species, generally eco-friendly algae and diatoms, after testing, isolation, and characterization attempts [3,4].

This chapter provides an overview of the ASP's history and its project on microalgal isolation and characteristics, including how the ASP is related to the National Renewable Energy Laboratory (NREL). This chapter discusses the challenges faced by the industrial application of the developed technologies by this program. The ASP project conclusions are stated at the end of this document.

28.2 Introduction to US department of energy

The DOE is one of the United States' wealthiest departments with a most diverse history. While only in operation since 1977, the DOE traces its history back to the Manhattan Project's attempt to create the nuclear bomb in the Second World War and to the numerous energy-related projects that had historically been spread through various federal agencies.

28.2.1 History of the department of energy

The 1977 DOE Act established one of the federal government's most exciting and dynamic agencies. Started on October 1, 1977, for the first time under a single organization, the 12th cabinet-level department pulled together two coexisting programmatic practices under federal control. Defense priorities covered the Manhattan Nuclear Bomb scheduling, installation, and testing of the project and a combination of energy activities spread across the federal government. At the beginning of the Second World War in 1939 Albert Einstein informed President Franklin D. Roosevelt about the recent studies demonstrating that the nuclear chain reaction could make the development of "potent bombs." After that, Roosevelt established a federal program of research. The US Army Engineers founded the Manhattan Department of engineering in 1942 to develop and invent the world's first nuclear bomb. The Manhattan Project was established and is considered the start of distinguished scientists collaborating with industrial, military, and social sectors to transform initial technological innovations into a brand-new type of weapon.

When the existence of this national secret project was disclosed to Americans after the bombings of Hiroshima and Nagasaki, the majority of people were shocked when they knew that such a large, government-run, top-secret project existed. The project recruited about 130,000 employees and had received investment of about $2.2 billion by the end of the war. After the war, Congress was interested in a lengthy discussion on atomic energy's civilian and military control. Under the Atomic Energy Act of 1946, the Atomic Energy Commission (AEC) was established to oversee the Manhattan Project's massive research findings and industrial complex. At the beginning of the cold war, AEC concentrated on constructing and developing nuclear weapons and reactors. In 1954 the Atomic Energy Act terminated the government's exclusive use of atoms and began developing industrial nuclear power, allowing the AEC to exert leverage over the new industrial development.

Atomic Energy Commission announced several projects aimed at establishing peaceful applications of nuclear technology in 1950s. This technology gave rise to the International AEC and other commitments and the growing domestic energy industry, which the Eisenhower administration believed was linked to nuclear energy production in Europe and other regions. The distribution of radioisotopes from Oak Ridge's X-10 Graphite Reactor was another application program. Such radioactive isotopes have been used in chemistry and physics, as well as in industry and agriculture. In the biomedical field, radioisotopes have been used to treat cancer and as radioactive substances to study biological processes. The AEC was responsible for the construction of the world's first large-scale nuclear power plant at Shippingport, Pennsylvania, in 1954 in yet another demonstration of nuclear power for peaceful purposes. By the mid-1960s, 75 nuclear power stations had been placed into operation.

Prior to the 1970s, when electricity was relatively cheap and abundant, the government played a minor role in national energy policy development. The private sector met the majority of the country's energy needs. Americans have traditionally expected private enterprises to finance their manufacturing, distribution, marketing, and pricing strategies—the federal government controlled energy prices in the absence of free-market conditions. Two significant changes radically altered the federal government's position in the energy sector during the 1970s. Firstly, during the 1970s, an energy shortage accelerated various government reorganizations as the executive and legislative constitutions struggled to integrate governmental energy policies and services better. Second, the activities planned by AEC for the increase and commercialization of nuclear technology portrayed the government's effort and concerns over the need to determine nuclear authorization. The formation of the Administration for Energy Research and Development and the Nuclear Regulatory Commission in 1974 accelerated the AEC's dissociation.

The establishment of the DOE in 1977 centralized the bulk of federal energy operations and laid the groundwork for an comprehensive and long-term national energy policy. Long-term, high-risk energy infrastructure research and development, marketing of federal power, energy efficiency, different nuclear weapons programs, regulatory programs of energy, and a national energy data collection and analysis program were all been transferred to the DOE. The DOE has changed the focus of its direction and attention over its

40-year existence as the nation's priorities have evolved. The department promoted energy production and control during the late 1970s. Research, development, and generation of nuclear weapons became a focus in the 1980s. At the end of the Cold War, the department concentrated on environmental cleanup, nuclear nonproliferation, and supply security. The department's preference in the 2000s was to focus on science and technology alternatives for energy, environmental, and nuclear challenges in order to ensure the nation's safety and prosperity. The DOE has started to look into how to transform the country's energy system and maintain leadership in renewable energy technologies, as well as how to advance world-class science and engineering as a foundation for economic success and how to improve nuclear security through defense, nonproliferation, and environmental actions [5].

28.2.2 Department of energy mission

The DOE strives to ensure the security and success of America by focusing on revolutionary technology and science solutions for its energy, environment, and nuclear challenges. This operation is completed by performing multiple functions and achieving multiple objectives (Fig. 28.1) [6].

28.2.3 Organization

The DOE is managed and supervised by the Secretary of Energy of the United States, a political appointee of the US President (Fig. 28.2). A US Deputy Secretary of Energy, also

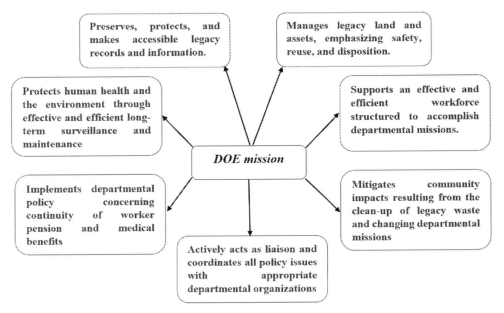

FIGURE 28.1 Department of energy mission.

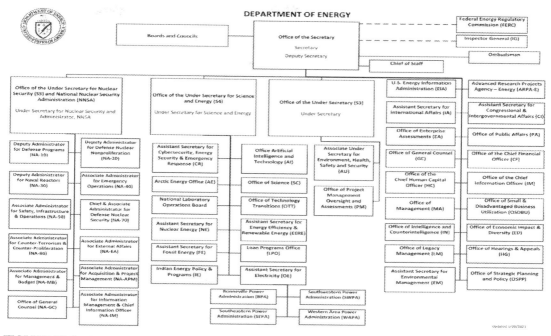

FIGURE 28.2 Department of energy organization chart [7].

named by the President, who performs the secretary's responsibilities in his absence, supports the Energy Secretary in managing the department.

There are three undersecretaries in the department, each appointed by the President, who supervise the major areas of the department's work. The President similarly designates seven officials with the grade of Assistant Secretary of Energy, responsible for the department's major organizational elements. The Energy Secretary defines their functions and responsibilities.

28.3 History of the algae species program

The first laboratories focusing on developing solar technology and energy were created by the Solar Energy Research Institute (SERI) in the United States in 1978. The power crises in the early and mid-1970s prompted the creation of this laboratory. During the Carter administration, all national electricity tasks were consolidated under the newly established DOE's auspices [2,4]. The DOE launched a study on the use of vegetation as a fuel production source, as part of its many research projects developing all forms of solar energy. Today, the fuels development department's work under the Assistant Secretary for Energy Efficiency and Renewable Energy is funded by DOE to manage the biofuels program. The program has worked on an extensive number of renewable fuels (ethanol and methanol,

620   28. Aquatic species program

biogas and biodiesel). ASP analyzed elements in the biofuels program to create alternative types of all-natural oil for the creation of biodiesel from organic matter conversion [4].

Researchers had first to figure out how to collect and characterize algae that complied with the technology's minimal requirements. From 1980 to 1987, data were collected and tested over 7 years. When a large amount of information about the various types of oil generated by algae and their capabilities became available, this system starts to shift its focus to understand the biochemistry and physiological behaviors of oil production in algae. The next logical step is to apply what we've learned so far in order to figure out how to genetically modify algae's metabolic functions to boost oil production [2]. Researchers first figured out how to collect and identify algae that met the technology's minimal requirements. From 1980 to 1987, data was collected and tested. When a large amount of information about the different types of oil-producing algae and their capabilities became available, this system began to shift its focus to understanding the biochemistry and physiology of oil production in algae. The next logical step was to use this knowledge to find ways to genetically manipulate algae's metabolic process to improve oil production [2]. Organisms isolated from shallow habitats were also expected to be more resistant to salinity and temperature.

Meanwhile, subcontractors and researchers had been collecting organisms through the southeast of the United States [4]. The program's scientists had developed their methodologies for collecting and assessing algal species by 1984. These included saline media mixtures and an adapted analysis that replicated typical freshwater conditions in the southwest. A rapid assessment test to distinguish high algae oil-producing algae from low algae oil-producing algae was available in 1985 [4].

The top priority of the collection work shifted to discovering algae that were temperature tolerant in the project's final years. The collection contained over 3000 different species of algae in 1987. When the collection efforts dwindled, it became evident that the current species would likely meet most of the technological requirements [3,4]. The ASP program started studies on the biochemical and physiological process of oil generation from algae about halfway through the collection efforts to know how to boost current organisms' effectiveness. Several ASP subcontractors were unable to pinpoint the "lipid trigger." These experiments confirmed previous findings that a small amount of nitrogen could raise the known level of oil in various algae species. The discovery of cell structure also supported the theory that a trigger caused a rapid buildup of oil droplets in the cells [2,4]. However, the studies conducted simultaneously by scientists of the NREL and the program subcontractors concluded that a simple little trigger concerning lipid manufacturing exists. As an alternative, we found that environmental stresses, such as nitrogen depletion, trigger cell division, with no direct slowing of oil production. This demonstrated that there was no simple way to increase oil production without sacrificing overall productivity due to reduced cellular development. Nutrient depletion as a means of inducing oil production may have some merit. Most experiments conducted by NREL proposed that cellular growth kinetics and lipid accumulation are straightforward. Having a better comprehension of such kinetics may allow the increase of oil productivity by carefully managing the timing of nutrient depletion and cell harvesting [3].

By 1986 the NREL's scientists revealed Si depletion to be a real solution to augment oil levels inside diatoms (microalgae) [4]. They discovered that anytime Si is consumed, cell division decreased because Si is a part of diatoms cellular wall surfaces. The rate of oils

Handbook of Algal Biofuels

manufacture remained constant as soon as Si depletion occurred, although development rates regarding the cells fell in diatom *Cyclotella cryptica* (5–25 μm). Other research studies identified a couple of aspects within species:

1. Si-depleted cells use more of their absorbed carbon for lipids synthesis and less to make carbohydrates.
2. Si-depleted cells transform nonlipid cell compounds to lipids progressively over time.

For diatoms, it is easy to store carbon in lipid form rather than carbohydrate form [3]. The outcome of the experiments conducted is to improve the path of cells employed for storage (Fig. 28.3):

Through the photosynthesis process, the algae cells can formulate carbon. Many pathways are metabolic, through which the carbon dioxide may go, resulting in the synthesis of various forms of compounds used by the algae cell. These pathways contain different enzymes, each one of which catalyzes a series of biochemical reactions. Two feasible outcomes for carbon are revealed in Fig. 28.3. They express the two forms that carbon usually takes [3,4]. Researchers at NREL began to identify essential nutrients within the synthesis path of a lipid. These are enzymes whose degree of activity is influenced by the cell from which natural oils were developed. These nutrients can be considered to be regulators or spigots that control the flow of CO_2 along the pathway. The higher activity is an enzyme that helps to raise oil production prices. Closing the spigot allows better carbon movement to oil synthesis when algal tissue increases the task of useful enzymes. Detecting these enzymes is crucial for understanding the mechanisms that control oil production [4,8].

By 1988, according to researchers, increases in the known quantities of the enzyme acetyl CoA Carboxylase (ACCase) were correlated with lipid buildup during Si exhaustion. They also discovered that the high grade is linked to an additional phrase for gene encryption because of this enzyme. These findings prompted researchers to concentrate their efforts on isolating this enzyme and cloning the gene that controls its expression.

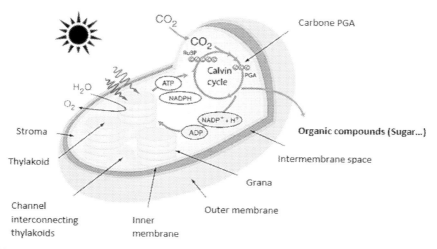

FIGURE 28.3 Algae cell photosynthesis.

Finally, scientists succeeded in cloning the ACCase gene and managed to develop genetic instruments [4,8]. By the 1990s, genetic engineering had become the emphasis and the priority of the ASP program. While there was success in overexpressing ACCase in diatoms, more methods and techniques were also created for gene expression in eco-friendly algae and diatoms. Another area of research targeted distinguishing essential enzymes involved in the synthesis of space carbohydrates. Instead of overexpressing the specific enzymes, experts hoped to inactivate all of them. Returning to the "spigot" analogy, this method was actually like shutting off the movement of carbon dioxide to carbohydrates, hoping it would force carbon to run on the lipid synthesis path (Fig. 28.3). This important work resulted in developing a particular enzyme that is multifunctional for the carbohydrates synthesis pathway. NREL patented this chemical and its particular gene in 1996 [4,8].

Prior to the ASP, the DOE completed the task on algae technology for energy production. In 1976 the vitality Studies and Development Administration (that merged with DOE) funded a project at the University of California Berkeley's Richmond area station to assess a microalgae-based wastewater treatment plant in order to produce fuel. Over several years, the Richmond Field facility demonstrated algae collection methods and exercised control over the variety of algae that developed in open ponds [4]. The ASP focused on investigating microalgae, while the focus had already switched from wastewater treatment to alga farm-specific procedures. From 1980 to 1987, this system financed two actions to develop a large-scale programs for microalgae. Additional work was undertaken at the University of Hawaii, where the "Algae Raceway Production System" was patented. Berkeley's design was successfully implemented in California's wastewater treatment plant in 1983. Ponds of 100 m^2 were used to study various fundamental functional issues, like the effects of liquid stream activities, light-intensity, demolished oxygen levels, pH, and algae cultivation methods. With competitive bidding, this system involved processes in choosing a concept for scale-up of algae mass cultivation after studies carried out in California and Hawaii. The High-Rate Pond system was chosen for scale-up after being evaluated at UC Berkeley. The "Outdoor Test Establishment" (OTF) was designed and built in a wastewater treatment plant in Roswell, New Mexico. At Roswell, 1000 m^2 ponds were effectively controlled from 1988 to 1990. This research demonstrated how to construct high-rate ponds with extremely efficient CO_2 utilization ($>90\%$). Native algae species that naturally took up more space inside the ponds showed ideal results. Furthermore, when using both silica and nitrogen depletion techniques, the OTF discovered high levels of oils in algae. While the daily production rate met the program's objective of 50 g/m^2 per day, total production was lower (around 10 g/m^2 per day) due to the number of cold temperature periods experienced in this location. However, the project developed a demonstration for a large-scale open pond system. The establishment was turned off in 1990 and has not been operated since. Various other outside studies were also financed during the program, including a 3-year algal biodiesel production in the Middle East. In addition, a Georgia Institute of Technology study occurred in the 1980s [4,9]. This work consisted of a combination of software modeling and experimental work. The project contributed to the development of the Algal Pond Model (APM), a computer software for modeling to predict a system's performance [10]. Two investigation methods were carried out in the ASP program: resource assessments and engineering design. Engineering models render some insight into the relevant question of oil production impacts of algae on petroleum use from limited resources. These designs inform us of things concerning these innovative systems [4,8,9].

The planned system began investigating the issue of reference availability for new algae technology in 1982. The requirements and methods that should be used in the assessment were outlined in preliminary reports. In 1985 Argonne National Lab (ANL) conducted research that resulted in maps of suitable algae production areas in the United States based on climate, land, and water supply. In 1990 for the first time, the available CO_2 reserves were estimated. These estimations proposed enough CO_2 to be available in the reports where climate circumstances had been appropriate to produce from 2 to 7 quads of fuel production every year. The cost of CO_2 is estimated to be approximately between 7 and 74 euros per ton of CO_2 [4,9]. This study did not consider any regional information and instead made conclusions from general CO_2 availability data collected across a large area. A research study was also funded in 1990 to assess land and water availability for algae innovation in New Mexico. The difficulties encountered in these types of scientific studies were addressed in this study. The resource availability results can either provide a comprehensive but general perspective on available data or be more detailed in strategy but only partially in the review of most resources. ASP program funding since its creation is shown in Fig. 28.4. The maximum funding was in 1985 with € 2,228,580, and the minimum funding was € 165 000 in 1991 [4,9].

28.4 Microalgal isolation and characteristics

Microalgae are microscopic organisms that reproduce by photosynthesis. Microalgae grow at a faster rate than macroalgae and terrestrial plants, making them far more

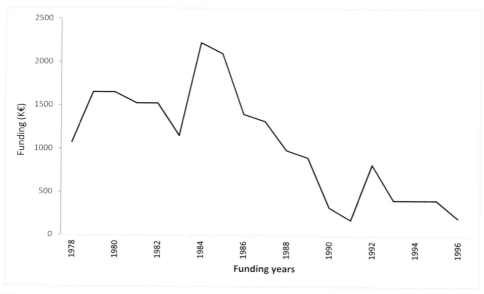

FIGURE 28.4   Funding history of the algae species program.

624

successful. Asexual plant cell division is the primary mode of reproduction for microalgae, though sexual replication can occur in a variety of forms under certain conditions. The pigment composition, biochemical content, life cycle, and ultrastructure of many microalgae types differ [1,2]. For the ASP, five groups of microalgae were studied:

1. Cyanobacteria (Class Cyanophyceae): Cyanophyceae or blue–green algae are widespread prokaryotic algae with no nucleus, of which about 7500 species are known to date. They are usually small in size, 10 μm or less. They can be found in virtually all habitats (freshwater, saltwater, brackish water, and soil) due to their ability to withstand extreme temperatures and resistance to desiccation. The main restriction of Cyanophyceae is that they require light for their growth and are mainly mandatory autotrophic organisms. However, some species may be able to oxidize certain sugars in the absence of light. Cyanophyceae primarily store their energy in the form of starch (amyloidosis and amylopectin) or oils [1,3].

2. Green algae (Class Chlorophyceae): Chlorophyceae are found in all habitat types and starch and oil are their principal energy reserves. Among this group, several microalgae have photoautotrophic, photoheterotrophic, and chemoheterotrophic metabolism. The most studied species with chemoheterotrophic metabolism is *Chlorella vulgaris*. This species is very interesting for biodiesel production because of its exciting percentage of fatty acids. The primary sources of assimilable carbon are in the form of sugars and acetic acid. Other species in this class have exciting potential for the production of algofuels, including *Chlorella prototheoides*. Approximately 8000 species of green algae are estimated to be in existence. However, these algae cannot accumulate lipids and thus were not valuable for ASP [1,3].

3. Golden-brown algae (Class Chrysophyceae): Chrysophyceae are generally found in freshwater. One hundred and fifty types and about 800 species represent this class. These algae's nutrition modes are very varied. Still, some have a heterotrophic way of nutrition, such as *Ochromonas danica*, which includes 27.4% fatty acids, and their source of carbon is glucose [1,3].

4. Eustigmatophytes: this community presents a significant element of the "picoplankton," tiny cells (2–4 μm). *Nannochloropsis* is one of the few marine forms and is expected to be found in oceans worldwide. Chlorophyll-a may be the only type present in the cells, although a few xanthophylls act as accessory photosynthetic pigments [1,3].

5. Prymnesiophytes (Class Prymnesiophyceae): this community of algae, also called the haptophytes, has roughly 500 known species. They are generally aquatic organisms and it considers one of the dominant species in the tropical ocean. Lipids and chrysolaminarin are the significant storage formats. The coccolithophorids are recognized by calcareous machines surrounding the cell wall structure [1,3].

6. Diatoms (Class Bacillariophyceae): diatoms are often the dominant group of microalgae in phytoplankton populations and are extremely widespread in all habitat types. More than 100,000 species are known. They are unicellular and measure from 2 μm to 1 mm. Diatoms store their reserves in the form of chrysolaminaran, a polysaccharide, as well as oils. They are known for their fatty acid content, and for several years, scientists believed that lipids were their only reserve compound. The major components of these lipids are triglycerides [1,3]. Microalgae such as *Nitschia* sp. and *Navicula pelliculosa* are interesting for biodiesel

production in heterotrophic mode given the importance of the percentage of fatty acids. Among diatom species, many have either mandatory or optional heterotrophic metabolism (mainly pennal diatoms), while some have photoheterotrophic metabolism (mainly centric diatoms). Diatoms with heterotrophic or mixotrophic metabolism are very diverse in terms of the rate of assimilation of carbohydrates as well as the type of carbohydrate used. Organic substrates supporting their growth vary, while the presence of vitamins appears to be essential for most diatoms [1,3].

Sampling and isolation of microalgae from typical habitats are techniques that are well-established. According to their habitats, microalgae differ in their method of cultivation under lab conditions. A step that is important in microalgal isolation is to control those conditions. Microalgae are sensitive to biological problems, such as heat, pH, and salinity. Some varieties of microalgae need specific nutrients, for example, diatoms require silica-supplemented media.

28.4.1 Sampling and collection

The collection of microalgal species is the essential step in microalgal isolation. Microalgae can live and can be found in different environmental conditions and thus various sampling methods are used (syringe sampling, brushing, scraping, etc.). Microalgae single-cell isolation occurs when a single cell is extracted from a sample using a micropipette under microscopic observation. To distinguish between microalgal cells and other particles in aquatic samples, ultrapure droplets are required [4,5].

28.4.2 Serial dilution

Serial dilution is the most popular method used to isolate microalgae from their environment. This dilution factor is often constant and causes the concentration to move logarithmically geometrically. A group of sterilized test tubes containing the necessary nutrients can be used for dilution. Depending on the algae's habitats and specific parameters, the medium can be supplied to pressure certain algal culture areas selectively. The precision of a quantity of cell culture measured during the transfer from one medium to another determines the technique's efficiency. This technology is useful for algal monoculture production [4,5].

28.4.3 Streak plate method

Streaking is a systemic method in the presence of different organisms for microalgal isolation. This particular technology calls for the preparation of aseptic agar nutrient plates by the introduction of a sterile inoculation loop to the liquid sample or the tweaking with a sterile pin instrument of a morphologically distinct colony from a later membrane surface. For most algal species, particularly diatoms, small cyanobacteria, filamentous algae, and soil microalgae, this technique is considered successful. Streak plate is the ideal method of isolation of axenic culture [4,5].

28.4.4 Density centrifugation

The centrifugation technique uses gravity in order to separate organisms and microorganisms into different sizes. Most of the time, this method is used to separate larger organisms from microalgal cells. It categorizes the various microalgae species into various groups. This method is used to concentrate the total number of cells in a microalgal strain. Combining this with other isolation methods can result in concentrated cell isolation. The rate and duration of centrifugation vary depending on the microalgal species being studied. This technique, on the other hand, can damage smooth cells due to shear stress [4,5].

28.4.5 Enrichment media

The nutritional needs of microalgae vary depending on their original living environment and cellular physiology. Single microalgae species isolation could be achieved using improved microalgal cultures with selection pressure. Specific nutrient sources, soil extracts, and other nutrients such as nitrate, phosphorus, and trace elements are all examples of commonly used enrichment compounds. pH control is a common strategy for achieving bacteria-free cultures. The organic compounds like fungus extract and casein from various fruits and vegetables could also be added to the medium. One or more nutritional elements required for microalgae development and growth may be lacking in the original habitat. Essential ingredients are recycled or supplied by other microorganisms' physiological activity in nature. Sampling reduces the recycling of critical elements and can result in the death of microalgae [4,5].

28.4.6 Micromanipulation

Micromanipulation is a recent effective technique for microalgal isolation that enables single-celled action and culture. Mainstream microalgal isolation methods do not report agglutination, and therefore colony development arises from multiple cells instead of a single cell. Typically, this micromanipulation technique uses capillary tubes to identify cells and transfer these to aseptic water. This is a manual process that is difficult and demands precision and perfection [4,5].

28.4.7 Automated techniques

Fluorescence-activated cell sorting (FACS) coupled with flow cytometry is a fast technique to isolate microalgal species and purification from different organisms. FACS depends on the light spread and fluorescence generated from a thin fluid flow transfer through one or more laser supports. The laser beam is absorbed by cells, scattered, and sends out fluorescence that provides data on cell size, shape, and photosynthetic characteristics. This technique has gained acceptance and celebrity due to the performance and efficiency, whereby it can sort and characterize 5000−10,000 cells/second [8].

Handbook of Algal Biofuels

28.5 Relationship between National Renewable Energy Laboratory and algae species program

In 1978 in Golden, Colorado the Carter administration established the SERI. This institute was the first federal laboratory intended to focus on solar energy production. The SERI was renamed the NREL in September 1991, by President George H.W. Bush. The lab was established in response to energy shortages in the early and mid-1970s. Around the same time, the Carter administration brought together all federal energy operations under the newly formed DOE. Among the numerous projects to improve solar energy sources, today's biofuels program is endorsed and conducted under the DOE's Assistant Secretary for Energy Efficiency and Renewable Energy by the Office of Fuels Development in the Transportation Technology Office. Over time, the program has concentrated on a wide variety of alternative fuels, including ethanol and methanol, biogas and biodiesel. The ASP was a part of the biofuels program's research to develop alternative natural oil sources for biodiesel production.

SERI's in-house researchers collected, screened, and characterized microalgal strains, which was a major undertaking for ASP researchers in the 1980s. In addition, more than 3000 algal species were obtained from other culture collections or collected from sites across the continental United States and Hawaii [11–24]. The implementation of the SERI Culture Collection as a genetic source was a significant achievement for the ASP. However, due to a funding deficiency, a significant portion of the collection was lost. Prior to work done by ASP researchers at NREL, very little work was done on improving microalgae strains, especially in terms of creating a commercial organism. However, significant progress has been made in order to identify the environmental and also the genetic factors that lead to the accumulation of lipid in microalgae, as well as in the ability to manipulate these factors to produce strains with desirable traits.

The ASP made significant progress in understanding the biochemical and biological synthesis of lipids in algae, as well as having success in the genetic engineering of algae. The development of a transformation system for an oleaginous microalgal species was a major goal of the ASP, and a significant amount of effort was put into this project during the early 1990s at NERL [25–32]. Over 18 years, DOE invested approximately $25 million to research various aquatic organisms to generate renewable energy [33]. ASP has been productive in demonstrating the viability of algal culture as an oil source, resulting in significant technological advancements. These advancements have made within the isolation and characterization of algal strains [33], algal physiology, also biochemistry studies [34,35], genetic engineering [36,37], engineering and process growth, and outdoor demonstration-scale culture of algal mass [38].

Although ASP has made considerable advances during its 18-year history, the program has been discontinued to reduce federal spending and the projected future cost of algal oil production in the range of € 33–50 per container relative to € 16 per container for crude oil in 1995. This year, DOE announced a complicated decision to reduce funding for algae research within the biofuels program. The DOE, under pressure to reduce its budget, opted to concentrate its precious resources more specifically on one or two key areas, the most important being bioethanol production. The objective of this work is to bring closure to the biofuels program's algae research. This chapter describes and evaluates all the work performed in the last 16 years of the ASP. This incorporates NREL researchers' work in other laboratories in Golden and the subcontracted research and development efforts of private companies and universities across the country. The program emphasized the importance of understanding and optimizing the biological mechanisms of algal lipid accumulation, as well as creating innovative, cost-effective culture and process engineering solutions to isolate lipids from very diluted biomass suspensions. In 1998 a detailed analysis of the project was completed [33]. Funding for research into algal biofuels was very limited in the years that followed the ASP, and progress was slow. Algal research, on the other hand, has exploded significantly in recent years. While government agencies started to show signs of increased funding, several new groups in academic, industrial, and national laboratories, such as NREL, have begun to work in this field which was largely funded by private investors and industrial sources.

28.6 Limitations of industrial applications

This project has shown laboratory and large-scale successes of the developed technology and the better transfer of technology and experience between partners, which is already confirmed by implementing various real installations in several countries. In addition, the project faced enough challenges that limited the industrialization of the technologies developed and the recommendations raised. In this part, these industrial application limits will be pointed out and discussed.

28.6.1 High-cost

The aspects that most significantly affect the cost are biological and certainly not engineering-related. Such analysis points to extremely productive organisms' requirements ready to near-theoretical amounts of total absorbed sunlight transformation to biomass. Even with intense presumptions regarding biological productiveness, the cost of biodiesel tends to be two times greater than existing petroleum diesel fuel costs. This cost comparison with familiar energy sources has changed over time. According to Laurens [39], in 1996 it was concluded that it was not financially practical to generate biodiesel from microalgae. Even if the perfect scenarios of photosynthesis process productivity are used, the cost could be twice as much as the same amount of petroleum diesel. As of 1996 petroleum diesel's price has increased largely, and biofuels made from microalgae must now be economically viable. However, this assumes that other factors have remained relatively constant, which unfortunately they have not; for example, fertilizer prices have increased significantly over this period. According to the ANL; NREL; PNNL report [40], the US DOE's biomass program has started a project to find regular quantitative measurements for algal biofuel manufacturing to build an "integrated baseline," through harmonizing and combining the program's national resource assessment, technoeconomic analysis (TEA), and life cycle analysis models. This report shows that the harmonized TEA model's main cost drivers are for ponds and pond liners. Therefore these are crucial things to enhance to recognize financial stability.

Nevertheless, this might be a complicated attempt as the ponds were previously tiny by design [41].

Consequently, a complete 50% decrease in prices, as presumed for cropping and extraction, could not be as conveniently attainable for pond expenses. Alternatively, a 30% expense reduction is supposed to be doable by relaxing the land grading and digging demands or through a necessary redesign of the pond process. One last strategy is the patented pond and liner design manufactured by Phyco Biosciences, which uses a simple trench-type pond with an inexpensive mechanical setup process for liners. The company claims that this method reaches a "30 + % decrease in investment costs" relative to conventional raceway ponds [42].

28.6.2 Resource availability

Biodiesel production and CO_2 reduction could be assisted by land, water, and CO_2 resources. The ASP evaluated the search queries of available resources for biodiesel from microalgae on a regular basis. This is by no means a minor undertaking. These resource assessments necessitate a thorough examination of the ideal environment, as well as land and resource accessibility. According to these studies, many potential land, water, and CO_2 resources exist to keep this innovation going. Algal biodiesel could easily provide many "quads" of biodiesel—significantly more than current oilseed crops could. Microalgae systems use a fraction of the water that traditional oilseed crops do. One quad of gas could be generated by 200,000 hectares. Despite the fact that the technology must overcome numerous research and development challenges before becoming viable, it is clear that resource constraints are not an issue.

28.6.3 Cultivation challenges

Microalgae can be grown in multiple ways, including heterotrophically, autotrophically, and mixotrophically. The cultivation of autotrophic microalgae has been developed for the marketing of selected microalgae species, such as *Spirulina* and *Chlorella*. ASP has shown that autotrophic cultivation in an open pond system could be a promising technique for the cultivation of microalgae for biofuel production [43]. However, existing methods have faced numerous challenges, including cell density, water evaporation, pathologic bacteria, and nutrient deficiency. Heterotrophic cultivation does not directly use solar energy or CO_2, which significantly increased the cost of a large-scale system. Heterotrophic cultivation should not be used for large-scale microalgae cultivation for biofuel production; however, it can be used for the rapid generation of seed cultures. Microalgae were grown autotrophically for CO_2 consumption and then metabolized heterotrophically for oil generation, according to a new culturing technique for *C. protothecoides* [44]. The advanced risk of bacterial contamination in the heterotrophic culture after scale-up is a real barrier to using this technique [45]. When manufacturing biofuels from microalgae becomes more industrialized, glycerol generation as a biodiesel source may become a problem.

28.6.4 The disconnect between the lab and the field

The obstacle occurs when choosing, improving, and maintaining the algal strains essential for large-scale, low-cost microalgae production [46]. The ASP implemented immense efforts in this field, with the isolation, screening, conserving, lab-scale work, big-scale cultivation, and genetic improvement of microalgal species. In general, laboratory outcomes were not sufficient to anticipate the large-scale efficiency. The strains more well protected inside the open-air occupy the ponds, often for long times. One inference from the outdoor culturing field was that strains maintained in laboratory cultures are commonly not too efficient in open ponds [43].

We can deduce from the outdoor assessment of the algae production system that preserving laboratory organisms outside is very challenging. Algal species that demonstrated extreme attention after assay in the laboratory were not robust under the field's circumstances. The excellent strategy for the effective production of consistent algae strains was to allow a autochthon contaminant into the ponds.

28.6.5 Dust issue

The Salton Sea is an area example that demonstrates the dust issue, due to the evaporation of the Salton Sea, a large portion of the coast is exposed, causing problems with air pollution from dirt, heavy metals, and other toxins in the surrounding area. By covering vast coastal regions, large-scale implementation of open ponds or photobioreactors, such as those manufactured by Solix, may help to mitigate this issue.

28.6.6 Commercialization potential

The technological issues mentioned describe significant barriers toward the improvement of financially viable biofuels from algae. Until a large-scale process has been created and shown, numerous problems remain, such as pond building components, combining, and ideal pond scale. With the National Algal Biofuels System Roadmap Workshop in 2008 the Biomass Program began scoping strategies to discover the critical technical obstacles to low-cost and scalable algae-based biofuels. The generated roadmap identified technological challenges to overcome in order to accelerate the development of a scalable, low-cost, and environment-friendly national algal biofuels industry. These issues are divided into three categories: algae feedstock production and distribution, algae feedstock operation and transformation, and promotion of infrastructure for siting and circulation.

28.7 Conclusions of the project

Significant effort were made between the 1980s and 1990s by ASP researchers and partners to select or promote microalgal strains that showed characteristics showing cost-effective biomass and lipid generation. High efficiency, rich-lipid, competitiveness in the ecosystem, and resistance to different ranges of temperature and salinity fluctuations are all desirable properties in these strains. While a range of algal strains was explored for growth and lipid generation characteristics, the ideal prospects were revealed in two microalgae classes, Chlorophyceae (green algae) and Bacillariophyceae (diatoms). In each group, organisms were collected and identified that demonstrated high productivity, the ability to develop in large-scale culture, and lipid buildup in response to nutrient stress. Nonetheless, diatoms could be a better candidate organisms for biofuel production for a variety of reasons. Diatoms had the highest lipid content (40%−60% of the AFDW). Given the increasing number of potential species identified, no single species was found to have the best features. Perhaps the most important point to remember is that the conditions that encourage high productivity and rapid growth (nutrient sufficiency) and the conditions that increase lipid accumulation (nutrient limitation), as discussed in this chapter, are unique. To overcome this barrier, more research in the area of genetic control of algal strains to increase photosynthetically or augment constitutive quantities of lipid deduction in algal strains is needed.

In terms of total energy inputs, nutrient use (e.g., CO_2), water needs, growing technologies, and basic system models, the ASP Microalgal Mass Culture Work concluded that there appears to be no underlying knowhow or financial problem that might limit the technical feasibility of microalgae culture. Furthermore, in terms of full biomass and algal lipids (oils) currently achieved through the ASP, productivities are significantly higher than those stated and even expected prior to the ASP, but still far below theoretical expectations and the demands for cost-effective stability. To correctly use the derivable advantages from microalgae, a good comprehension of the parameters affecting the development rate, biomass, and lipid aggregation is crucial for improving productivity. The selection and enhancement of extreme yield varieties, design and building of cultivation techniques with reduced cost and minimal energy requirements, practical strategies, green extraction techniques, and wet transformation operations can all enhance the microalgae industry's durability. Specific issues still impede large-scale generation of microalgae for value-added products such as biofuels at some phases along the production process cycle. Besides these problems, the potential for enhanced productiveness at reduced cost is high.

To sum up, any potential research and development strategy for microalgae CO_2 capture and biofuels generation should begin using the development of the microalgae "biocatalysts." The objective can be toward developing strains via genetic engineering or different strain enhancement techniques that accomplish substantial solar transformation benefits and deliver maximum lipid (oil) microalgal biomass, as needed by economic studies. The primary advice for a future research and development strategy is to consider biocatalyst development efforts, building on the skills established by the ASP. This is a crucial moment for a feasible approach to promote and proclaim microalgae biodiesel generation during the wastewater treatment systems; research and development in this field are also recommended. Finally, global warming's international nature enables the thoughtfulness of such technologies' universal outcome.

Acknowledgments

This work was done in the frame of the bilateral Morocco−Tunisia R&D cooperation project (2021−2023) entitled: "Development of an innovative hybrid process for the treatment and recovery of industrial wastewater."

References

[1] J. Sheehan, D. Terri, B. John, R. Paul, A. Look, Back at the United States department of energy's aquatic species, Eur. Phys. J. C. 72 (2012) 14. Available from: http://www.springerlink.com/index/10.1140/epjc/s10052-012-20439%5Cn; http://arxiv.org/abs/1203.5015.

[2] A. Singh, D. Pant, S.I. Olsen, P.S. Nigam, Key issues to consider in microalgae based biodiesel production, Energy Educ. Sci. Technol. A Energy Sci. Res. 29 (2012) 687−700.

[3] A.W.D. Larkum, S.E. Douglas, J.A. Raven, Photosynthesis in algae, J. Phycol. 40 (2004) 1178−1180. Available from: https://doi.org/10.1111/j.1529-8817.2004.40601.x.

[4] Y. Sasaki, A. Kozaki, M. Hatano, Link between light and fatty acid synthesis: thioredoxin-linked reductive activation of plastidic acetyl-CoA carboxylase, Proc. Natl. Acad. Sci. United States A. 94 (1997) 11096−11101. Available from: https://doi.org/10.1073/pnas.94.20.11096.

[5] A Brief History of the Department of Energy, Office of Legacy Management, <https://www.energy.gov/lm/doe-history/brief-history-department-energy>, 2021 (accessed 08.02.21).

[6] Mission, <https://www.energy.gov/lm/mission>, 2021, (accessed 20.01.21).

[7] Organization Chart, <https://www.energy.gov/leadership/organization-chart>, 2021 (accessed 20.01.21).

[8] J.M. Biddy, C. Scarlata, C.r Kinchin, Chemicals from biomass: a market assessment of bioproducts with near-term potential, NREL/TP-5100−65509, (nrel.gov), 2016.

[9] R. Revelle, H.E. Suess, Carbon dioxide exchange between atmosphere and ocean and the question of an increase of atmospheric CO, during the past decades, Tellus 9 (1957).

[10] H.O. Buhr, S.B. Miller, A dynamic model of the high-rate algal-bacterial wastewater treatment pond, Water Res 17 (1983) 29−37. Available from: https://doi.org/10.1016/0043-1354(83)90283-X.

[11] W.R. Barclay, Microalgal technology and research at SERI: species collection and characterization, in: Aquatic Species Program Review: Proceedings of the April 1984 Principal Investigators' Meeting, Solar Energy Research Institute, Golden, Colorado, SERI/CP-231−2341, 1984, pp. 152−159.

[12] W. Barclay, J. Johansen, P. Chelf, N. Nagle, R. Roessler, P. Lemke, Microalgae culture collection 1986−1987, Solar Energy Research Institute, Golden, Colorado, SERI/SP232−3079, 1986, p. 147.

[13] B. Barclay, N. Nagle, K. Terry, Screening microalgae for biomass production potential: Protocol modification and evaluation, FY 1986 Aquatic Species Program Annual Report, Solar Energy Research Institute, Golden, Colorado, SERI/CP-231−3071, 1987, pp. 23−40.

[14] B. Barclay, N. Nagle, K. Terry, P. Roessler, Collecting and screening microalgae from shallow, inland saline habitats, in: Aquatic Species Program Review: Proceedings of the March 1985 Principal Investigators' Meeting, Solar Energy Research Institute, Golden, Colorado, SERI/CP-231−2700, 1985, pp. 52−68.

[15] W.R. Barclay, N.J. Nagle, K.L. Terry, S.B. Ellingson, M.R. Sommerfeld, Characterization of saline groundwater resource quality for aquatic biomass production: a statistically-based approach, Water Res. (1988) 373−379.

[16] M.R. Sommerfeld, S.B. Ellingson, Collection of high energy yielding strains of saline microalgae from southwestern states, FY 1986 Aquatic Species Program Annual Report, Solar Energy Research Institute, Golden, Colorado, SERI/CP-231−307, 1987, pp. 53−66.

[17] B. Barclay, N. Nagle, K. Terry, Screening microalgae for biomass production potential: protocol modification and evaluation, FY 1986 Aquatic Species Program Annual Report, Solar Energy Research Institute, Golden, Colorado, SERI/SP-231−3071, 1986, pp. 22−40.

[18] W.R. Barclay, N.J. Nagle, K.L. Terry, J.C. Weissman, R.P. Goebel, Potential of new strains of marine and inland saline-adapted microalgae for aquaculture, J. World Aquac. Soc. 18 (1987) 218−228.

[19] J. Johansen, P. Lemke, W. Barclay, N. Nagle, Collection, screening, and characterization of lipid producing microalgae: progress during fiscal year 1987, FY 1987 Aquatic Species Program Annual Report, Solar Energy Research Institute, Golden, Colorado, SERI/SP-231−3206, 1987, pp. 27−42.

[20] D. Berglund, B. Cooksey, K.E. Cooksey, L.R. Priscu, Collection and screening of microalgae for lipid production: possible use of a flow cytometer for lipid analysis, FY 1986 Aquatic Species Program Annual Report, Solar Energy Research Institute, Golden, Colorado, SERI/SP-231−3071, 1987, pp. 41−52.

[21] M. Sommerfeld, S. Ellingson, P. Tyler, Screening microalgae isolated from the southwest for growth potential and lipid yield, FY 1987 Aquatic Species Program Annual Report, Solar Energy Research Institute, Golden, Colorado, SERI/SP-231−3206, 1987, pp. 43−57.

References

[22] A. Ben-Amotz, T.G. Tornabene, Chemical profile of algae with emphasis on lipids of microalgae, in: Aquatic Species Program Review: Proceedings of the March 1983 Principal Investigators' Meeting, Solar Energy Research Institute, Golden, Colorado, SERI/CP-231−1946, 1983, pp. 123−134.

[23] T.G. Tornabene, Chemical profile of microalgae with emphasis on lipids, in: Aquatic Species Program Review: Proceedings of the April 1984 Principal Investigators' Meeting, Solar Energy Research Institute, Golden, Colorado, SERI/CP-231−2341, 1984, pp. 64−78.

[24] T.G. Tornabene, J.R. Benemann, Chemical profiles on microalgae with emphasis on lipids, in: Aquatic Species Program Review: Proceedings of the March 1985 Principal Investigators' Meeting, Solar Energy Research Institute, Golden, Colorado, SERI/CP-231−2700, 1985, pp. 83−99.

[25] P.W. Bergeron, R.E. Corder, A.M. Hill, H. Lindsey, M.Z. Lowenstein, S.E.R.I. Biomass Program annual technical report: 1982, Solar Energy Research Institute, Golden, Colorado, SERI/TR-231−1918, 1983, p. 83.

[26] S. Lien, Studies on the production and accumulation of oil and lipids by microalgae, in: Proceedings of the SERI Biomass Program Principal Investigators' Review Meeting, Aquatic Species Program Reports, June 23−25, 1982, Solar Energy Research Institute, Golden, Colorado, SERI/CP-231−1808, 1982, pp. 45−54.

[27] J. Ohlrogge, J. Jaworski, D. Post-Beittenmiller, G. Roughan, P. Roessler, K. Nakahira, Regulation of flux through the fatty acid biosynthesis pathway, in: N. Murata, C.R. Somerville (Eds.), In Biochemistry and Molecular Biology of Membrane and Storage Lipids of Plants, American Society of Plant Physiologists, Rockville, MD, 1993, pp. 102−112.

[28] P.G. Roessler Purification and characterization of acetyl-CoA carboxylase from the diatom *Cyclotella cryptica*, in: W.S. Bollmeier, S. Sprague (Eds.), Aquatic Species Program Annual Report, Solar Energy Research Institute, SERI/SP-231−3579, 1989, pp. 121−129.

[29] J.C. Schneider, P.G. Roessler, A novel acyltransferase activity in an oleaginous alga, in: L.C. Kader, P. Mazliak (Eds.), Plant Lipid Metabolism, Kluwer Academic Publishers, the Netherlands, 1995, pp. 105−107.

[30] R.G. Walsh, Microalgal technology research at SERI: the biochemistry of lipid trigger mechanisms in microalgae—alteration of lipid metabolism during transition from rapid growth to rapid lipid accumulation, in: Aquatic Species Program Review: Proceedings of the April 1984 Principal Investigators Meeting, Solar Energy Research Institute, Golden, Colorado, SERI/CP231−2341, 1984, pp. 170−183.

[31] L.M. Brown, T.G. Dunahay, E.E. Jarvis, Applications of genetics to microalgae production, Dev. Ind. Microbiol. 31 (1990) 271−274.

[32] T.G. Dunahay, S.A. Adler, J.W. Jarvik, Transformation of microalgae using silicon carbide whiskers, in: R. Tuan (Ed.), Methods in Molecular Biology: Recombinant Gene Expression Protocols, 62, Humana Press, Totowa, NJ, 1997, pp. 503−509.

[33] P.G. Roessler, L.M. Brown, T.G. Dunahay, D.A. Heacox, E.E. Jarvis, J.C. Schneider, et al., Genetic engineering approaches for enhanced production of biodiesel fuel from microalgae, in: M.E. Himmel, J. Baker, R.P. Overend (Eds.), Enzymatic Conversion of Biomass for Fuels Production, American Chemical Society, 1994, pp. 256−270.

[34] J.T. Sheehan, J. Dunahay, J.R. Benemann, P.G. Roessler, A look back at the United States department of energy's aquatic species program—biodiesel from algae. <http://govdocs.aquake.org/cgi/reprint/2004/915/9150010.pdf>, 1998.

[35] P.G. Roessler, UDP-glucose pyrophosphorylase activity in the diatom *Cyclotella cryptica*: pathway of chrysolaminarin biosynthesis, J. Phycol. 23 (1987) 494−498.

[36] P.G. Roessler, Changes in the activities of various lipid and carbohydrate biosynthetic enzymes in the diatom *Cyclotella cryptica* in response to silicon deficiency, Arch. Biochem. Biophys. 267 (1988) 521−528.

[37] P.G. Roessler, Purification and characterization of acetyl-CoA carboxylase from the diatom *Cyclotella cryptica*, Plant. Physiol. 92 (1990) 73−78.

[38] P.G. Roessler, J.B. Ohlrogge, Cloning and characterization of the gene that encodes acetyl-coenzyme A carboxylase in the alga *Cyclotella cryptica*, J. Biol. Chem. 268 (1993) 19254−19259.

[39] L.M.L. Laurens, R.J. Shields, R.W. Lovitt, K.J. Flynn, H.C. Greenwell, Placing microalgae on the biofuels priority list: a review of the technological challenges, 7 (2010) 703−726. doi: 10.1098/rsif.2009.0322.

[40] ANL, NREL, PNNL, Renewable diesel from algal lipids: an integrated baseline for cost, emissions, and resource potential from a harmonized model, in: ANL/ESD/12−4; NREL/TP-5100−55431; PNNL-21437, Argonne National Laboratory, Argonne, IL; National Renewable Energy Laboratory, Golden, CO; Pacific Northwest National Laboratory, Richland, WA, 2012.

634
28. Aquatic species program

[41] T.J. Lundquist, I.C. Woertz, A realistic technology and engineering assessment of algae biofuel production, Energy Biosciences Institute, University of California, 2011.

[42] Phyco Biosciences, Phyco Biosciences Super Trough System, <http://www.phyco.net/Technology>, 2012 (accessed 15.03.12).

[43] J. Sheehan, T. Dunahay, J. Benemann, P. Roessler, A look back at the United States department of energy's aquatic species program-biodiesel from algae, Report NREL/ TP-580—24190, National Renewable Energy Laboratory, Golden, CO, 1998.

[44] T. Lopes da Silva, P. Moniz, C. Silva, A. Reis, The dark side of microalgae biotechnology: a heterotrophic biorefinery platform directed to ω-3 rich lipid production, Microorganisms 7 (2019) 670. Available from: https://doi.org/10.3390/microorganisms7120670.PMID:31835511;PMCID:PMC6956277.

[45] M. Dębowski, M. Zieliński, J. Kazimierowicz, N. Kujawska, S. Talbierz, Microalgae cultivation technologies as an opportunity for bioenergetic system development-advantages and limitations, Sustainability 12 (2020) 9980. Available from: https://doi.org/10.3390/su12239980.

[46] M. Musa, G.A. Ayoko, A. Ward, C. Rösch, R.J. Brown, T.J. Rainey, Factors affecting microalgae production for biofuels and the potentials of chemometric methods in assessing and optimizing productivity, Cells 8 (2019) 851. Available from: https://doi.org/10.3390/cells8080851.

CHAPTER
29

Algal fuel production by industry: process simulation and economic assessment

Sayeda M. Abdo[1], Entesar Ahmed[2], Sanaa Abo El-Enin[3], Guzine El Diwan[3], K.M. El-Khatib[3], Gamila H. Ali[1] and Rawheya A. Salah El Din[2]

[1]Water Pollution Research Department, National Research Centre, Giza, Egypt [2]Botany and Microbiology Department, Faculty of Science, Al-Azhar University (Girls Branch), Cairo, Egypt [3]Chemical Engineering and Pilot Plant Department, National Research Centre, Cairo, Egypt

29.1 Introduction

Greenhouse gas emissions (mainly carbon dioxide), which affect climate changes, are one of the problems facing the world in the modern age. The most important reasons are the high dependence on fossil fuels for many aspects of life, such as power generation, transportation, industry, and deforestation. The presence of renewable biofuel in comparison with total fuel demand was found to be unsatisfactory to replace fossil fuels [1,2]. Recently it was reported that about 87% of global CO_2 emitted from human activities is due to the use of fossil resources, with oil, natural gas, and coal providing 36%, 20%, and 43%, respectively. Therefore it became imperative to decrease carbon emissions by utilizing sustainable energy resources globally but integrating them at the local level. During the past decade, biomass has attracted interest as a source of bioenergy as well as bioproducts. There have been four generations, including edible and nonedible feedstocks, of bioenergy development [2]. The importance of using microalgae for biodiesel production is to substitute and reduce fossil fuel consumption. It is interesting to highlight that cultivation of some microalgae species allows over 50% of the oil that can be extracted and transferred into fuel compared to classical bioenergy crops and seaweed. Macroalgae's

Handbook of Algal Biofuels
DOI: https://doi.org/10.1016/B978-0-12-823764-9.00029-7

635

© 2022 Elsevier Inc. All rights reserved.

636
29. Algal fuel production by industry: process simulation and economic assessment

disadvantage as that they tend to produce more carbohydrates than lipids and their oil content is not in excess of 5% dry weight, plus they are expensive for biodiesel production. Meanwhile, microalgae can produce several types of fuel such as methane, kerosene (aircraft fuel), biogas, green diesel, biobutanol, biodiesel, and bioethanol. Also, in terms of life cycle assessment (LCA) there is an increase in the useful coproducts obtained from nonlipid biomass after the process of biofuels [3–5].

Algae are considered the most promising sustainable sources of oils, food, fuel, feed, and other coproducts. Algae are very attractive due to the vast range of benefits related with how and where they grow [6]. There are many international businesses and projects related to the production of biofuel. They include Sapphire Energy Inc., Joule Unlimited Inc., Cellana Inc., Synthetic Genomics Inc, Algenol Biotech LLC, and TerraVia Inc. (formerly Solazyme). Both Cellana Inc. and Sapphire Energy Inc. are commercially producing a plant and already have a pilot plant. Both entities produce biofuel and after extraction the algal biomass is defatted into coproducts including omega-3 supplements and animal and aquaculture feed. Now TerraVia Inc. is employing its heterotrophic strains to produce an omega-9 supplement [7].

29.2 Life cycle assessment toward microalgae industrialization

LCA (Life Cycle Assessment) is intended to reflect all of the impacts of the procedure, from "cradle to grave," that is, it studies the whole process starting from all inputs and ending with the destination of all items and waste. LCAs are associated with energy [such as energy return on energy investment (EROI)] and sustainability studies. Currently, the main concern is with regard to greenhouse gas (GHG) emissions; expressed as CO_2 equivalent including nitrous oxides, methane, and other secondary greenhouse gases. LCA methods have been broadly developed [8], and LCA studies are currently a main prerequisite for all renewable energy processes. This process allows finding the transmission of pollutants from one step to another or one type of environmental impact to another. The European Renewable Energy Directive [9] defined in a 2009 LCA as an suitable indication method for assessing the environmental impacts of biofuels for reducing greenhouse gases by 50% in 2017. Third-generation biofuels use feedstocks that do not compete with crop plants. This third category includes microalgae which have a large photosynthetic yield and are able to store fats and carbohydrate, and also have potential to be developed in a controlled environment.

Biofuels from microalgae do not compete with food crops and have limited environmental impacts. The use for microalgal growth of carbon dioxide that is emitted directly from industry is a promising feature of flue gas mitigation [10,11]. However, this is considered a challenging technology. Microalgal industrial production was developed to produce high-value macromolecules (such as betacarotene) or nutritional supplements (many species of microalgae such as *Spirulina* or *Chlorella* are found as tablets in health stores); so the environmental impact has never been a concern. Moreover, the volume of the microalgal industry for bioactive compounds production is much smaller than the volume required to produce the fuel. Afterward, there is a requirement to evaluate the probable environmental performance of potential production systems, to prioritize the improvements

Handbook of Algal Biofuels

needed for each process. LCA is currently fundamental to planning an improved, energy-efficient, and environment-friendly biofuel production system. Recently, biofuels produced from microalgae have attracted the international scientific and economic communities, and numerous environmental, active, or economic assessments have been published, with different energy vectors and different assumptions [12].

Several publications of LCA for biofuel production and its environmental impacts were used as references. Since there is no actual industrial data with regard to producing energy from microalgae, the models used and extrapolations to define the production systems are associated with the functional units (FU) of LCAs for microalgae production and perimeters. The notion of the FU is the main characteristic of LCA that permits significant as well as reasonable comparisons between studies or between different technological opportunities [13]. Some examples of the studies are briefly described in the following:

1. Kadam [14]: using LCA for comparison between power generation from coal alone or coal and microalgal biomass.
2. Lardon et al. [15]: using LCA for biodiesel production at nitrogen stress conditions in open raceway ponds. In addition to a comparison between using wet or dry algal biomass for lipid extraction.
3. Baliga and Powers [16]: using LCA for biodiesel production in photobioreactors (PBR) in a cold climate.
4. Batan et al. [17]: based on the GREET model (greenhouse gases, transportation energy, and emissions) they used LCA to produce biodiesel in PBR flat plate and tubular PBR.
5. Clarens et al. [18]: the comparative LCA of used energy for microalgae and crops as a biofuel feedstock. Microalgae were grown in raceways with chemical fertilizers.
6. Jorquera et al. [19]: using LCA for different cultivation technology of microalgae including open raceways,
7. Sander and Murthy [20]: LCA of microalgae biodiesel production. The microalgae cultivation was procced in two steps; PBR as the first one and the second step in open raceways.
8. Stephenson et al. [21]: LCA comparing the production of biodiesel in PBR and open raceways, defatted algal biomass was subjected to an anaerobic digestion process to use as fertilizers.
9. Brentner et al. [22]: combinatorial LCA for industrial biodiesel production from microalgae. Starting from extraction and ending with using biomass as fertilizers.
10. Campbell et al. [23]: LCA and economic analysis for the production of biodiesel in the open ponds. The process of nitrogen fertilizer synthesis produces pure CO_2 which can be used as a carbon source.
11. Clarens et al. [24]: LCA algal biodiesel and bioelectricity can be used for transportation.

The four bioenergies produced and compared are: (1) anaerobic digestion of microalgal biomass for bioelectricity production; (2) using oilcake after using anaerobic digestion and biodiesel production to generate bioelectricity; (3) producing biodiesel in addition to the combustion of oilcake to produce bioelectricity; (4) combustion of biomass for bioelectricity production. Also, there are multiple ways to provide nutrients and we can compare four

638　29. Algal fuel production by industry: process simulation and economic assessment

of them as follows: (1) pure CO_2; (2) CO_2 captured from a local coal power plant; (3) CO_2 in flue gas; (4) CO_2 in flue gas and nutrients in the wastewater.

1. Collet et al. [25]: LCA for microalgae biomass to produce biogas using anaerobic digestion.
2. Hou et al. [26]: LCA of biodiesel production from algae compared with soybean and jatropha.
3. Khoo et al. [27]: LCA of biodiesel from microalgae; cultivation is carried out in two phases, first in PBR, then in an open raceway.
4. Yang et al. [28]: LCA of biodiesel production limited to water and nutrients consumption.

As described before regarding the 15 selected researches, two categories were taken into consideration: production of either biomass or bioenergy. The majority of the studies are aimed at producing microalgal biodiesel as the main power output. The quantities of produced biodiesel are reported with different units: volume [16], mass [21], or energy content [15]. Unfortunately, there is no comparison between the energy consumed and the mass density of algal oil and algal methyl ester. Finally, six studies are seen as the most promising among all earlier studies dedicated to biodiesel production, whereas the use of the fuel is not involved in the perimeter [16,17,20,22,27,28]. On the other hand, five are well considered, where the utilization of the fuel is included [15,21,23,24,26]. According to the LCA method, many parameters should be defined, such as functional unit and study perimeter Also the system should be defined, along with in each process the technical inputs and outputs, the energy and resource consumption, and the emissions. With the lack of real industrial data on the cultivation or transformation of microalgae, the selected studies often depend on either extrapolation of results at the laboratory scale, or on adaptation from similar processes at different conditions or with different feedstocks, or modeling.

It is well-known that nutrient requirements depend not only on the species but also on the stimulation of fat or carbohydrate storage using stress-induced processes. Also, phosphorus and nitrogen ratios can vary greatly during stress conditions [29]. Assumptions about required fertilizers vary greatly between species, and between different publications on the same species. Nitrogen requirements vary from 10.9 [12] to 20.32 g/kg per DM [21], and in limited conditions, from 9.41 [14] to 77.6 g/kg per DM without stress [24]. Phosphorus requirements differ from 2.4 to 2.58 g/kg per DM in limited circumstances [15,27], respectively, and from 0.02 to 71 g/kg per DM when lacking stress [14,28] respectively. All authors agree on the high consumption of nutrients in the cultivation of microalgae but there are disagreements on the methods for providing it. Some authors such as Sander and Murthy [20] and Clarens et al. [18] state that nitrogen and phosphorus requirements may be fully or partly achieved through adding wastewater to the medium.

Microalgae culture is achieved in two types of systems: raceway open pond (ROP) or PBR. RORPs are shallow ponds (10–50 cm deep) with a paddle wheel. These open ponds can be constructed from concrete or cut from the ground [23] and can be recovered by a plastic liner polyethylene [25] or polyvinyl chloride. They are generally open but can be found under a greenhouse. This system is commonly used to produce microalgae for the

Handbook of Algal Biofuels

manufacture of foodstuffs [30,31]. PBRs are closed systems that allow for culture condensation. Diverse designs of PBRs can be used, such as tubular, flat sheet [19], or rustic made from simple polyethylene bags soaked in a thermostatic water bath [17]. Algal growth media can be selected independently of the planting system. According to the selected algal species, algae can be grown in freshwater, brackish water, or seawater. The use of wastewater has been suggested by several authors [18], providing a dual advantage of an untreated source of water and nutrients. Referring to the microalgae that grow on wastewater, they cannot be used as feedstock for fish or livestock. One of the greatest challenges in the production of microalgae for bioenergy is the great water consumption which is recorded as a major environmental concern. Thus some authors suggest growing algae in seawater, in order to obtain an infinite resource [27]. Brackish water is also utilized from groundwater in a few systems [24]. It is worth mentioning that freshwater is still required in these systems to stabilize salinity. The biofuel company Solix has grown microalgae in flat, immersed plastic bags. Solix produces 5000—8000 gallons of algal oil per acre per year according to its website [32]. A large amount of oxygen can affect algal growth negatively, so a supplementary tank is usually added to separate excess oxygen [33]. At the same time, pollution is eradicated [34]. The main difficulty with these controlled facilities is their cost [35]. Other challenges include high temperature, difficulty of expansion, biocontamination, and cell damage over time [3]. Rashid et al. [36] showed that using open pond systems for biofuel production is better than PBR. Unfortunately, open pond systems are more susceptible to pollution but the use of saltwater types may mitigate this risk.

29.3 Operating conditions

Environmental conditions are well-known to affect growth rates, as can mineral composition C/N ratio, availability of nutrients, or the occurrence of stress. For example, nitrogen starvation affects fat storage in certain microalgae species. However, in spite of high algal cell division the lipid content is increased and thus mass productivity is lower. Thus it must be emphasized that all parameters are interrelated and will not be defined based on independent assumptions or sources. A significant variation in yield, fat percentage, or nutritional requirements was observed between the different studies. In some publications, the authors propose imposing nitrogen deprivation on algae. The problem of low growth rate under nutrient stress can be overcome through microalgae cultivation in two steps. First, the microalgal biomass is grown with repleted nitrogen for a high growth rate. Then the microalgae undergo nitrogen depletion in order to increase their lipid content [27].

The growth rate of microalgal species is strongly influenced by conditions of the environment (temperature and light) [37]. Microalgae growth rate can be reduced by stress protocols used to induce lipid accumulation due to nutrient deprivation [38]. The assumptions made in the LCA studies showcase this broad spectrum. In ORW, the growth rates vary from 25 to $40.6 \, g/m^2$ d [18,25], respectively. In PBR, the yield is much higher and ranges from 270 to 1536 g/3.d [22]. The growth rate of microalgae was increased when cultivated in PBR [19]. The concentration of microalgae ranges from 0.5 to 1.67 g/L [21] in ORW, and from 1.02 to 8.3 g/L [21] in PBR. The lipid contents varied widely from 17.5% [15] to 50% [14] without nitrogen deprivation, and from 25% [27] to 50% with deficiency of nitrogen.

To reach a high algal biomass yield, carbon dioxide must be supplied to the growth medium. The pH of the media is one of the most affecting parameters and it must be regulated, also microalgae can be very tolerant of the source of CO_2 [39]. Meanwhile, the ability of the microalgae to consume CO_2 according to its dissolution efficiency is highly dependent on the farming system. Carbon dioxide is either supplied from the purified compressed gas or from the flue gas to the local power plant, after capture or directly. The amounts of carbon dioxide in the flue gas vary from 5% to 15% [23]. It is common to report a lack of knowledge regarding the long-term consequences for algae and a culture facility for flue gas use. However, Yoo et al. [40] showed that *Botryococcus braunii* and *Scenedesmus* sp. can be grown using flue gas as a carbon source. The active costs of the injection and the head losses have to be taken into consideration. The compression of large amounts of gas is needed from the flue gas into the growth medium to reduce the efficiency of the gas injection system. Hence there is a clear tradeoff in terms of energy consumption between the previous purification and gas injection. Some authors include purification and transportation costs in their study [24].

Carbon dioxide emissions certainly occur in ORW due to the low efficiency of the injection system and release of natural gas from the medium. Four publications take into consideration these losses, with their CO_2 emissions, and were found to be equal to 0.07%, 30%, 10%, and 10% [14,21,23,25]. Some studies only look at emissions of other gases. They consider 0.11% of nitrogen volatilize without specific forms of emissions [23]. Hou et al. [26] stated that 0.5% is volatile like NH_3. Also, Batan et al. [17] report the volatilization of NH_3 without quantification.

It is greatly recognized that the concentration step is one of the most important bottlenecks in bioenergy production from microalgae. Selected studies are evaluating diverse techniques to gain concentration, dehydration, and occasionally dry algal biomass. At the end, the conversion process before biofuel production depends on the dry matter (DM) content. For example, the anaerobic digestion of bulk microalgae requires a low dry matter content, from 5% to 14% [25]. Abdo et al. [41] conducted a survey of the percentage of oils present in many green algae in addition to some blue−green algae isolated from the Egyptian environment. The results showed that *Microcystis aeruginosa* had the highest oil content of 30%, followed by the first speaker at 21% oil. This is the first and most important step when choosing a specific species to use as a source of biofuels.

Only a few data are available to produce a stock of microalgae oil extraction. Characterization of lipid is based on the solvent used for extraction and the techniques used to lyophilize algae. Oil extraction and esterification are mostly based on methods for vegetable (such as rapeseed or soybeans) oil production and transesterification. Some studies illustrate a pretreatment stage, depending on the mixture. Disruption of the cell walls increases the extraction efficiency and digestibility of the extraction residue [21]. Extraction of the triglycerides is achieved using an organic solvent, followed by separation of the hexane, fat, and water phases, and finally the oil/hexane mixture is purified by distillation. Most of the hexane is recovered during distillation, and thus only a small amount is lost by volatilization.

Esterification of triglyceride is obtained using methanol and alkaline stimulation. Methyl esterification requires high temperature, mixing, and a base, mostly potassium hydroxide. Also, a simultaneous saponification reaction is needed using water to reduce the reaction yield. Thus there is a tradeoff between the energy that is invested in dewatering and drying of the biomass, and the energy required to extract the fraction of the fats and to treat them, with the reaction output greatly affected by the water content. Different methods were suggested, such as supercritical CO extraction of fats or on-site esterification. The two methods can have very high water content. Another study dealt with in situ esterification and supercritical methanol proposed a way to overcome this problem. This last option was identified in the proposed LCA-based optimization [22].

The conversion of microalgal biomass into biogas is obtained by anaerobic digestion of the oil. The potential of methane varies greatly, depending on species composition and degradability [42]. In studies by Collet et al. [25] and Brentner et al. [22], the methane conversion ranged from 0.262 to 0.800 m^3 CH_4.kg per DM, respectively. It must be noted that *Scenedesmus'* maximum theoretical value is higher than this previous last value [42]. In anaerobic digestion energy consumption is neglected most of the time. Unfortunately, the long hydraulic retention periods required for digestion of low biodegradable materials (10—40 days) represent large active mixing and heating efforts. The electricity consumption is evaluated to be 0.47 and 0.39 MJ/kg per DM and heat consumption is estimated to be 2.45 MJ/kg per DM [22,25].

One of the expressions of the LCA methodology is relating each economic and environmental flow to the functional unit reference flow. However, many processes involved in the production of a functional unit can lead to the production of many products. There are two possible ways to deal with the multiple functions of the system: assignment or substitution. A task approach is to disperse the natural burden of the starting pathway among all the common products of the multioutput handle. This dispersion ought to be based on the foremost delicate model, for illustration, mass, financial value, or dynamic substance of items. The parameter extension (or substitution) choice comprises the coproduct to the functional unit [43].

The oil extraction process produces an extraction residue (oil cake); some authors prefer to use an energy-dependent assignment at this level. Other authors have chosen to use anaerobic digestion directly. In aquaculture or livestock food, carbohydrates could be replaced by oil cakes for the production of bioethanol [44,45].

Methyl ester and glycerin are produced during oil esterification; their economic and energy allocation is often used. Glycerin is used mainly as a heat source. Anaerobic digestion produces biogas, solid, and liquid digestion. The byproducts can be considered either waste or a soil conditioner as well as fertilizer. Meanwhile, the digestible liquid can be recycled into the culture device and replaced with the mineral fertilizer fraction needed for the microalgae. The heat is produced from conversion of biogas, which is used on-site to heat the digesters and/or convert it into electricity. Electricity is consumed at the site, and the surplus is injected into the grid [43].

29.4 Algal biodiesel

Algal biodiesel has not gained commercial value until now because of its high investment costs. Feedstock expense represents 80% of its total production expenses, which is the obstruction for biodiesel production. Other publication revealed that biodiesel produced by microalgae is being used in coincidence with conventional diesel at a level of more than 15%, thus its efficiency needs to be improved to displace the conventional diesel [44,46]. The presence of certain fatty acids in the microalgae lipid composition as palmitic acid (C16:0), stearic (C18:0), oleic (C18:1), linoleic (C18:2), and linolenic acid (C18:3), is a good signal to produce qualified biodiesel. The biosynthesis lipid of microalgae alters according to the limiting conditions to increase the percentage of neutral lipids (20%—50%) in the procedure of triacylglycerol (TAG) which is store in cytosolic bodies of lipid [46].

The immaturity of the technology proved challenging in determining the economic viability of biofuel production from microalgae. The economic viability analysis of biofuels from microalgae feedstocks was completed by several Technical and Economic Assessments (TEA) [47]. Norsker et al. [48] reported that the cost of $ 4520/tonne^{-1}, with improvements it could decrease to $ 740 per ton.

The majority of TEAs have assumed the use of an open pond production system. An alternative open growth system for algae production is the Algae Grass Purifier. Relatively, the ATS systems have a simple design for easily harvesting the produced biomass using farm equipment [49]. The ATS uses an inclined substrate that allows contaminated water to flow along with the algae, which in turn take in inorganic compounds. In the open systems large-scale the costs of microalgae feedstocks production should be estimated. Development the model of the systems engineering process and they're integrated with the economic modeling to assess the cost of biomass production in ATS and open racecourse growth systems. The direct comparison between the production and harvesting cost of biomass during the growth is one of the objectives. The prototype building module facilitated the integration of downstream processing through hydrothermal liquefaction for an economic evaluation of fuel production.

Comparing the financial viability of producing biofuels relative to fossil fuels, conventional biofuels revealed higher production costs and thus uncompetitive retail prices [50,51]. For example, Hill et al. [51] proved that biodiesel produced from soybeans in 2005, was 20% higher than the wholesale price of diesel, while ethanol production was 5% higher than the wholesale price of gasoline. After, 2005 there was an increase by upto 30% in fossil fuel prices allowed these biofuels to be increasingly competitive [52]. Prices mostly favored fossil fuels so far. Costs in the market do not don't reflect the potential nonmarket benefits related with the generation and utilization of biofuels [51]. This data in an inefficient resource and a shortfall in the biofuels supply, assuming net positive externalities [43] can lead to an economically efficient amount and price of biofuels. In theory, subsidies related to biofuels represent the external benefits of the lower net environmental impact for fossil fuels and the profits from increased fuel access and national/regional energy independence known in economic terms as positive externalities [45].

First and second-generation biofuels, which are for the most part incapable to compete within the absence of subsidies, microalgal biofuels are not as of now in competition with fossil fuels. However, it may be a potential fuel source due to its combined energy characteristics [53] and has been of interest in airline pilot metrics. Moreover, research has indicated potential improvements to both agriculture [54] and remediation [55], with the recent focus on reducing capital costs through low-cost specially designed to treat microalgae. Cost reductions can also be achieved if sources of carbon dioxide, nutrients, and water were lower costs or recycled within production [56]. In any case, with current production restricted to smaller R&D ventures, the possibility of these thoughts in commercial generation has not been explored.

There are numerous byproducts related to microalgae production which have commercial value. Fats (which can be changed over into biofuels) adds up as it were around 30% of the collected biomass, the biomass after extraction can be utilized as animal feed or other energy-related items such as ethanol [43], gas, or indeed hydrogen which can be utilized for fuel? Generation of high-value-added products from remaining biomass has the commercial advantage of microalgae over ordinary biofuel feedstocks. Long run commercial viability of microalgae as a biofuel may moreover depend on the suitable commercial utilization of these coproducts.

Microalgae biofuel businesses influence social and financial benefits which will contribute to a socially economical result. Social sustainability incorporates, among other angles, the plausibility of a more evenhanded dispersion of financial benefits over society, counting regional and urban communities [20], and improved quality of life. This is often in differentiation to businesses on fossil powers that depend on restricted assets and conventional biofuels generation from microalgae which can to give specialties for related work development over ability levels, comparable to those related with ordinary biofuels [43,57].

Microalgae based businesses present an opportunity for financial development in nonurban and territorial regions. public and private investment in bioenergy ventures is frequently centered on work and pay openings for companies and nearby communities, especially in territorial locales [58]. It has been proposed that there are critical openings for feasible development of agroindustries and livelihoods through conventional biofuels [59]. In expansion to supplementing regular industry incomes, synergies from biostabilization of profluent build-ups, and the generation of usable joint items (such as feed and fertilizers) [60].

To understand the size and importance of the microalgae industry to economic growth below are some examples of larger projects:

1. ExxonMobil, United States, will invest in developing synthetic genomics with $600 million, a microalgae biofuels process using genetically modified algae that excrete hydrocarbons.
2. Sapphire Energy, Inc., United States, >$300 million in building 120-hectare to produce biofuels from microalgae using open ponds.
3. Algenol Biofuels, Inc., FL, United States, *$70million private, $25 million government funds, for bioethanol production from microalgae grown in the photobioreactor.
4. AuroraAlgae, Inc., United States and Australia, *$70 million private funding to produce nutritional products; uses seawater systems and open ponds.
5. Eni, Italy, operates a *0.5-ha pilot plant project for microalgae for biodiesel/oil using mainly open ponds, upto 2000 m^2 (budget not demonstrated).
6. The European Commission FP7 Program supports three 10-ha algae transportation biofuel projects, *$7 million in European Commission (EC) funding each, one in Spain.
7. Solix BioSystems, Inc., CO, United States, with about $70 million in investments, has built a *0.5-ha PBR system to produce algal oils for biofuels and nutritional products.
8. Cellana, Inc., HI, United States (started as a joint venture with Shell Oil, since withdrawn) built a*0.5-ha algal oil production pilot plant with PBRs and ponds (*$40 million Est).
9. Joule Unlimited, Inc., MA, United States, a recent start-up (*$30 million venture capital), claims extraordinary productivities with cyanobacteria excreting hydrocarbons in PBRs.

29.5 Process simulation

1. The study of Abdo et al. [61] used Aspen HYSYS V7.0 software developed by AspenTech Inc (2004) to evaluate the economic assessment.
2. To produce a process simulation analysis the following steps should be used:
   a. Most ingredients must be known, such as methanol, sulfuric acid, sodium hydroxide.
   b. The operation costs and investments will be established.
3. Finally, for evaluation the process the ROI % and breakeven point will be calculated.

The industrial production of biodiesel in Egypt is in its early stages. Factories have been built with small or medium capacities. The study passed 10,000 tons/year biodiesel.

29.6 Process description

Microalgae biodiesel production process is divided into steps: harvesting, extraction, and transesterification. (Figs. 29.1, 29.2, 29.3).

A simulation of Aspen HYSYS V8.4 software was used to construct a model of the process. A process model's setup first step is the definition of chemical components. The

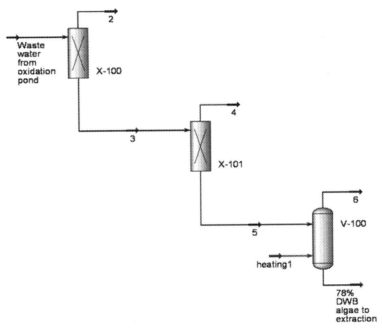

FIGURE 29.1 Harvesting (X-100), centrifuges (X-101), and dryers (V-100). [61,62].

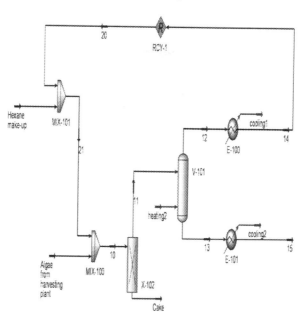

FIGURE 29.2 Extraction, microalgae/solvent mixer (MIX-100), recycled/makeup solvent mixer (MIX-101), settling tank (X-102), hexane evaporator (V-101), recycling cooler (E-100), and oil cooler (E-101). [61,62].

microalgae oil obtained will be considered to be Triolein ($C_{57}H_{104}O_6$) and the product of biodiesel was methyl oleate ($C_{19}H_{36}O_2$). The Hypo Manager tool of HYSYS was used to define these components which were not shown in the library. The model of NRTL thermodynamics was adopted as a result of the presence of methanol as a very high polar component.

Product purification:

The main units for the process are reactors, separators, distillation columns, pumps, and heat exchangers. The cultivation process of microalgae uses an open pond system which requires the removal of 20% of the algae regularly to preserve a steady growth rate. Then filtering the removed water is executed in two stages (centrifuge then decanter centrifuge) and it dries before the step extracting the oil. Although the accessibility to full knowledge of the process kinetics was not possible, the transesterification reaction was showcased by a simple reactor model for oil conversion 97% to FAME. Multistage distillation was used to recover methanol. A simple unit of flash could not simulate the biodiesel purity because of the big difference between the boiling point of methanol (65°C) and that of FAME (320°C) at 1 atmosphere. The produced biodiesel purity was 99.65wt.% and glycerin was 86% pure (Fig. 29.4).

Equipment sizing: the size of the process equipment was determined according to the principles described in the literature [63].

Distillation column: the size of the distillation column diameter was determined using the Souders−Brown equation.

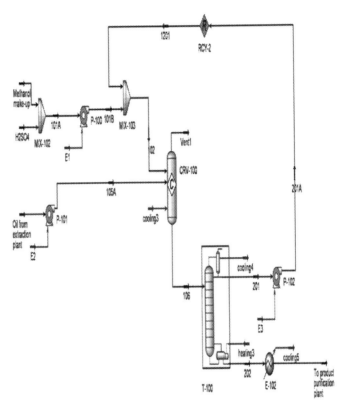

FIGURE 29.3 Transesterification (MIX-102), recycled alcohol/catalyst and fresh alcohol mixer (MIX-103), alcohol and catalyst pump (P-100), oil pump (P-101), recycled alcohol pump (P-102), transesterification reactor (CRV-100), methanol recovery distillation tower (T-100), and cooler (E-102). [61,62].

29.7 Economic assessment

There are numerous components for project assessment such as financial, environmental, and social variables. Financial execution is an imperative figure in surveying practicality. The financial performance of a biodiesel plant (e.g., fixed capital cost, total production cost, the breakeven cost for biodiesel) can be decided by essentially deciding certain variables, such as plant capacity, used innovation, raw material cost, and chemical costs. All specifications and prices are shown in Table 29.1 according to the Egyptian local market prices.

According to Abdo et al. [60] the definition of the cost of capital estimates the economics and is classified in this review as an "estimated study." It is based on the following:

1. Developing a process flow diagram.
2. Approximate sizing of major process equipment.
3. The estimate of this study had a predictable precision between +30% to −20%. Thus the results of such a preliminary evaluation may not accurately reflect the final

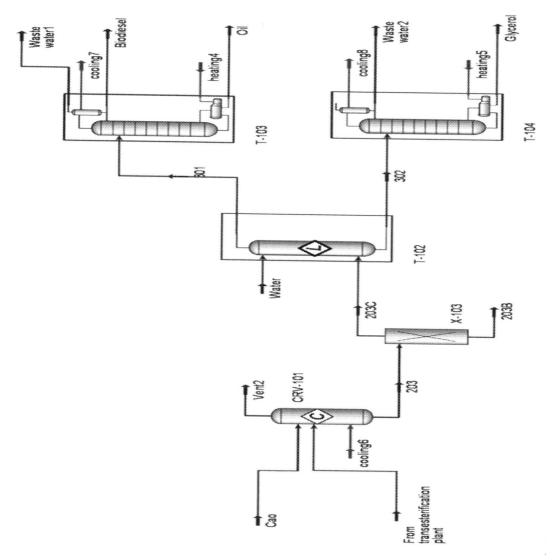

FIGURE 29.4 Purification for product: neutralization tank (CRV-101), settling tank (X-103), water washing tower (T-102), biodiesel purification distillation tower (T-103), and glycerol purification distillation tower (T-104). [61,62].

648   29. Algal fuel production by industry: process simulation and economic assessment

TABLE 29.1 Cost of raw materials, products, utilities, and basics used in the process for *Microcystis aeruginosa* and *Spirulina maxima* [61,62].

Item	Microcystis aeruginosa	Spirulina maxima
Raw materials		
Hexane ($/ton)	791	791
Methanol ($/ton)	440	440
H_2SO_4 ($/ton)	2398	2398
CaO ($/ton)	150	150
Water ($/ton)	0.007	0.007
Products		
Biodiesel ($/ton)	1235	1235
Cake ($/ton)	300	300
Glycerol ($/ton)	1200	1200
Utilities		
LPS ($/ton)	6.8	6.8
HPS ($/ton)	10	10
Cooling water ($/m^3)	0.007	0.007
Electricity ($/kw.hr)	0.04	0.04
% oil in algae	19.3%	7%
Type of growing	Freshwater	Wastewater
Algae conc. (mg/L)	100	1700

profitability of a chemical plant but can be used as a tool for comparison between multiple process alternatives.

4. Total capital investment is divided into two parts:
   a. Fixed capital investment is determined as the investment required to get the plant ready for startup and includes costs of all equipment, construction, and ancillary buildings. In addition to contractor fees and contingencies.
   b. Working capital investment, it is described as the investment required to operate the plant for a period of 3 months, in which all variables are adjusted in these 3 months until the plant is ready for real production.

Table 29.2 reveals the total capital investment along with the amount of the main equipment costs used with *M. aeruginosa* and *Spirulina maxima*. The results showed that the total cost of capital investment for *S. maxima* is lower. This was due to the addition of the cost of creating an open pond in the case of *M. aeruginosa*.

It is necessary to determine the price of production to be able to calculate the profit and set the unit price. Table 29.3 demonstrates the direct and indirect costs of production.

Handbook of Algal Biofuels

29.7 Economic assessment

649

TABLE 29.2 Equipment cost, fixed capital cost, and total capital investment [61,62].

Item	Microcystis aeruginosa	Spirulina maxima
Equipment Cost $		
Cultivation system	313,014,628	7,008,030
Centrifuge	1,117,013	6,535,714
Decanter centrifuge	3,921,429	—
Solar dryer	191,926	4,08,486
Reactors	496,361	496,361
Distillation towers	609,869	609,869
Evaporator	3,372,363	1,098,876
L-L extractor	167,324	167,324
Pumps	45,327	45,327
Heat exchangers	130,281	179,294
Mixers	505,862	71,911
Gravity settlers	436,418	593,254
Other	523,037	1,395,223
Purchased cost of equipment (PCE)	324,531,837	18,609,669
Contingency = 0.1 X PCE	32,453,184	1,860,967
Auxiliary facility = 0.25 X PCE	8,113,296	465,242
Fixed capital investment (FCI)	365,098,317	20,935,878
Working capital investment = 0.15 X FCI	54,764,748	3,140,382
Total capital investment	419,863,064	24,07,6260

TABLE 29.3 Cost analysis of biodiesel produced from Microcystis aeruginosa and Spirulina maxima [61,62].

Item	Microcystis aeruginosa	Spirulina maxima
Products		
Biodiesel ($/ton)	12,350,000	1235
Cake ($/ton)	25,586,407	1200
Glycerol ($/ton)	1,155,840	300
total selling Cost ($)	39,092,247	67,255,863
Rate of return (ROR%)	− 56.579	86.92
Payback period (years)	− 1.915	1.100

650
29. Algal fuel production by industry: process simulation and economic assessment

Direct manufacturing cost (DMC) includes the cost of raw materials and labor cost. It was estimated to be $ 493,458 in the case of *S. maxima*, while it was $ 507,097 in the case of *M. aeruginosa*, calculated assuming that the operator worked 48 weeks per year and there were three 8-h shifts per day for the continuous plant duration. This increase was due to adding the cost of nutrients in the case of *M. aeruginosa* where it was grown in the laboratory while *Spirula maxima* was grown in wastewater without adding nutrients.

Indirect manufacturing cost includes sales, distribution, overheads, and research and development. It equals 25% of DMC.

Return on investment (ROI) and breakeven point:

To earn the ROI, the net profit must be evaluated. This economic evaluation was based on the following assumptions:

1. The biodiesel plant was assumed to be 8000 h/year.
2. In the simulation, the pump efficiency was assumed to be 70%.
3. Low-pressure steam was used as a heating medium. Water was the medium of cooling.

Its specifications and prices are shown in (Table 29.3) (all costs shown are in US dollars). Equipment prices were updated from 2001 or 2007 to 2013 available values using the chemical engineering plant index, where I2013 = 567.3, I2001 = 397, and I2007 = 525 [63]. All chemical costs, including raw materials, catalysts and products, are shown in Table 29.3 according to the Egyptian local market prices.

References

[1] A. Goli, A. Shamiri, A. Talaiekhozani, N. Eshtiaghi, N. Aghamohammadi, M.K. Aroua, An overview of biological processes and their potential for CO_2 capture, J. Environ. Manage. 183 (2016) 41−58.

[2] A. Raheem, P. Prinsen, A.K. Vuppaladadiyam, M. Zhao, R. Luque, A review on sustainable microalgaebased biofuel and bioenergy production: recent developments, J. Clean. Prod. 181 (2018) 42−59.

[3] O.M. Adeniyi, U. Azimov, A. Burluka, Algae biofuel: current status and future applications, Renew. Sustain. Energy Rev. 90 (2018) 316−335.

[4] M. Bošnjaković, N. Sinaga, The perspective of large-scale production of algae biodiesel, Appl. Sci. 10 (2020) 8181. Available from: https://doi.org/10.3390/app10228181.

[5] A.G. Silva, R. Carter, F.L.M. Merss, D.O. Correa, J.V.C. Vargas, A.B. Mariano, et al., Life cycle assessment of biomass production in microalgae compact photobioreactors, GCB Bioenergy 7 (2015) 184−194.

[6] B.O. Abo, E.A. Odey, M. Bakayoko, L. Kalakodio, Microalgae to biofuels production: a review on cultivation, application and renewable energy, Rev. Environ. Health 34 (1) (2019) 91−99.

[7] S. Dickinson, M. Mientus, D. Frey, A. Amini-Hajibashi, S. Ozturk, F. Shaikh, et al., A review of biodiesel production from microalgae, Clean. Technol. Environ. Policy 19 (2017) 637−668.

[8] ARB. Proposed regulation to implement the low carbon fuel standard, Volume I Staff Report: Initial Statement of Reasons, California Environmental Protection Agency and Air Resources Board, 2009.

[9] European Union, Directive 2009/28/EC of the European Parliament and of the Council of 23 April 2009 on the promotion of the use of energy from renewable sources and amending and subsequently repealing Directives 2001/77/EC and 2003/30/EC, Off. J. Eur. Union. 5 (2009).

[10] Y. Chisti, Biodiesel from microalgae, Biotechnol. Adv. 25 (2007) 294−306.

[11] M.E. Huntley, D.G. Redalje, CO_2 mitigation and renewable oil from photosynthetic microbes: a new appraisal, Mitig. Adapt. Strateg. Glob. Change 12 (2007) 573−608.

[12] P. Collet, D. Spinelli, L. Lardon, A. Hélias, J.P. Steyer, O. Bernard, Life-cycle assessment of microalgal-based biofuels, in: A. Pandey, D.J. Lee, Y. Chisti, C.R. Soccol (Eds.), Biofuels from Algae, Elsevier, USA, 2013, pp. 287−312.

[13] H.A. Udo de Haes, S.A. Wegener, R. Heijungs, Similarities, differences and synergisms between HERA and LCA- an analysis at three levels, Hum. Ecol. Risk Assess. 12 (3) (2006) 431−449.

Handbook of Algal Biofuels

References

651

[14] K.L. Kadam, Environmental implications of power generation via coal-microalgae cofiring, Energy 27 (10) (2002) 905−922.

[15] L. Lardon, A. Hélias, B. Sialve, J.P. Steyer, O. Bernard, Life-cycle assessment of biodiesel production from microalgae, Environ. Sci. Technol. 43 (17) (2009) 6475−6481.

[16] R. Baliga, S.E. Powers, Sustainable algae biodiesel production in cold climates, Int. J. Chem. Eng. (2010). Volume Article ID 102179, 13 pages. Available from: https://doi.org/10.1155/2010/102179.

[17] L. Batan, J. Quinn, B. Willson, T. Bradley, Net energy and greenhouse gas emission evaluation of biodiesel derived from microalgae, Environ. Sci. Technol. 44 (20) (2010) 7975−7980.

[18] A.F. Clarens, E.P. Resurreccion, M.A. White, L.M. Colosi, Environmental life cycle comparison of algae to other bioenergy feedstocks2010 Environ. Sci. Technol. 44 (5) (2010) 1813−1819.

[19] O. Jorquera, A. Kiperstok, E.A. Sales, M. Embiruçu, M.L. Ghirardi, Comparative energy life-cycle analyses of microalgal biomass production in open ponds and photobioreactors, Bioresour. Technol. 101 (2010) 1406−1413.

[20] K. Sander, G.S. Murthy, Life cycle analysis of algae biodiesel, Int. J. Life Cycle Assess. 15 (2010) 704−714.

[21] A.L. Stephenson, E. Kazamia, J.S. Dennis, C.J. Howe, S.A. Scott, A.G. Smith, Life-cycle assessment of potential algal biodiesel production in the United Kingdom: a comparison of raceways and air-lift tubular bioreactors, Energy Fuels 24 (7) (2010) 4062−4077.

[22] L.B. Brentner, M.J. Eckelman, J.B. Zimmerman, Combinatorial life cycle assessment to inform process design of industrial production of algal biodiesel, Environ. Sci. Technol. 45 (16) (2011) 7060−7067.

[23] P.K. Campbell, T. Beer, D. Batten, Life cycle assessment of biodiesel production from microalgae in ponds, Bioresour. Technol. 102 (1) (2011) 50−56.

[24] A.F. Clarens, H. Nassau, E.P. Resurreccion, M.A. White, L.M. Colosi, Environmental impacts of algae-derived biodiesel and bioelectricity for transportation, Environ. Sci. Technol. 45 (17) (2011) 7554−7560.

[25] P. Collet, A. Hélias, L. Lardon, M. Ras, R.-A. Goy, J.-P. Steyer, Life-cycle assessment of microalgae culture coupled to biogas production, Bioresour. Technol. 102 (1) (2011) 207−214.

[26] J. Hou, P. Zhang, X. Yuan, Y. Zheng, Life cycle assessment of biodiesel from soybean, jatropha and microalgae in China conditions, Renew. Sustain. Energy Rev. 15 (9) (2011) 5081−5091.

[27] H.H. Khoo, P.N. Sharratt, P. Das, R.K. Balasubramanian, P.K. Naraharisetti, S. Shaik, Life cycle energy and CO_2 analysis of microalgae-to-biodiesel: preliminary results and comparisons, Bioresour. Technol. 102 (10) (2011) 5800−5807.

[28] J. Yang, M. Xu, X. Zhang, Q. Hu, M. Sommerfeld, Y. Chen, Life-cycle analysis on biodiesel production from microalgae: water footprint and nutrients balance, Bioresour. Technol. 102 (1) (2011) 159−165.

[29] R. Geider, J. La Roche, Redfield revisited: variability of C: N: P in marine microalgae and its biochemical basis, Eur. J. Phycol. 37 (1) (2002) 1−17.

[30] H. Shimamatsu, Mass production of *Spirulina*, an edible microalga, Hydrobiologia 512 (2004) 39−44.

[31] J.A. Del Campo, M. García-González, M.G. Guerrero, Outdoor cultivation of microalgae for carotenoid production: current state and perspectives, Appl. Microbiol. Biotechnol. 74 (6) (2007) 1163−1174.

[32] P. Schlagermann, G. Göttlicher, R. Dillschneider, R. Rosello-Sastre, C. Posten, Composition of algal oil and its potential as biofuel, J. Combust. (2012). Volume Article ID 285185, 14 pages. Available from: https://doi.org/10.1155/2012/285185.

[33] I. Rawat, R.R. Kumar, T. Mutanda, F. Bux, Biodiesel from microalgae: a critical evaluation from laboratory to large scale production, Appl. Energy 103 (2013) 444−467.

[34] G. Huang, F. Chen, D. Wei, X. Zhang, G. Chen, G. Biodiesel production by microalgal biotechnology. Appl. Energy 87 (1) (2010) 38−46.

[35] Q. Liao, J.S. Chang, C. Herrmann, A. Xia (Eds.), Bioreactors for Microbial Biomass and Energy Conversion, Springer, 2018, p. 468.

[36] N. Rashid, Y.F. Cui, M.S.U. Rehman, J.I. Han, Enhanced electricity generation by using algae biomass and activated sludge in microbial fuel cell, Sci. Total Environ. 456−459 (2013) 91−94.

[37] P.G. Falkowski, J.A. Raven, Aquatic Photosynthesis, second ed., Princeton University Press, 2013, p. 488.

[38] T. Lacour, A. Sciandra, A. Talec, P. Mayzaud, O. Bernard, Neutral lipid and carbohydrate productivities as a response to nitrogen status in *Isochrysis* sp. (T-ISO; Haptophyceae): starvation versus limitation, J. Phycol. 48 (3) (2012) 647−656.

[39] J. Doucha, F. Straka, K. Lívanský, Utilization of flue gas for cultivation of microalgae (*Chlorella* sp.) in an outdoor open thin-layer photobioreactor, J. Appl. Phycol. 17 (2005) 403−412.

[40] C. Yoo, S.Y. Jun, J.Y. Lee, C.Y. Ahn, H.M. Oh, Selection of microalgae for lipid production under high levels carbon dioxide, Bioresour. Technol. 101 (1) (2010) S71−S74.

Handbook of Algal Biofuels

[41] S.M. Abdo, E. Ahmed, S.A. El-Enin, R.A. Salah El Din, G. El-Diwani, G. Ali, Growth rate and fatty acids profile of 19 microalgal strains isolated from river Nile for biodiesel production, J. Algal Biomass Util 4 (4) (2013) 51–59.

[42] B. Sialve, N. Bernet, O. Bernard, Anaerobic digestion of microalgae as a necessary step to make microalgal biodiesel sustainable, Biotechnol. Adv. 27 (4) (2009) 409–416.

[43] A. Doshi, Economic analyses of microalgae biofuels and policy implications in Australia, (Ph.D. thesis), Queensland University of Technology, Australia, 2017, p. 294.

[44] M.U. Akhtar, A. Ali Khan, W. Jahangir Khan, T. Furqan, Microalgae as sources of biofuel production through waste water treatment, Nov. Res. Microbiol. J. 3 (5) (2019) 464–470.

[45] S.A. El-Mekkawi, S.M. Abdo, F.A. Samhan, G.H. Ali, Optimization of some fermentation conditions for bioethanol production from microalgae using response surface method, Bull. Natl. Res. Cent. 43 (2019) 164–171.

[46] F.K. El-Baz, M.S. Gad, S.M. Abdo, K.A. Abed, I.A. Matter, Performance and exhaust emissions of a diesel engine burning algal biodiesel blends, Int. J. Mech. Mechatron. Eng. 16 (3) (2016) 151–158.

[47] J. Hoffman, Techno-economic assessment of micro-algae production systems, (MSc. thesis), Utah State University, USA, 2016, p. 36.

[48] N.-H. Norsker, M.J. Barbosa, M.H. Vermuë, R.H. Wijffels, Microalgal production—a close look at the economics, Biotechnol. Adv. 29 (1) (2011) 24–27.

[49] C. Pizarro, W. Mulbry, D. Blersch, P. Kangas, An economic assessment of algal turf scrubber technology for treatment of dairy manure effluent, Ecol. Eng. 26 (4) (2006) 321–327.

[50] A. Demirbaş, Biofuels sources, biofuel policy, biofuel economy and global biofuel projections, Energy Convers. Manage. 49 (8) (2008) 2106–2116.

[51] J. Hill, E. Nelson, D. Tilman, S. Polasky, D. Tiffany, Environmental, economic, and energetic costs and benefits of biodiesel and ethanol biofuels, Proc. Natl. Acad. Sci. 103 (300) (2006) 11206–11210.

[52] W. Coyle, The future of biofuels: a global perspective, Amber Waves 5 (5) (2007) 24–29.

[53] M. Vanthoor-Koopmans, R.H. Wijffels, M.J. Barbosa, M.H.M. Eppink, Biorefinery of microalgae for food and fuel, Bioresour. Technol. 135 (2013) 142–149.

[54] R. Davis, A. Aden, P.T. Pienkos, Techno-economic analysis of autotrophic microalgae for fuel production, Appl. Energy 88 (10) (2011) 3524–3531.

[55] G. Pokoo-Aikins, A. Nadim, M.M. El-Halwagi, V. Mahalec, Design and analysis of biodiesel production from algae grown through carbon sequestration, Clean. Technol. Environ. Policy 12 (3) (2010) 239–254.

[56] R.C. Pate, G.T. Klise, B. Wu, Resource demand implications for United States algae biofuels production scale-up, Appl. Energy 88 (10) (2011) 3377–3388.

[57] M. Khanna, G. Hochman, D. Rajagopal, S. Sexton, D. Zilberman, Sustainability of food, energy and environment with biofuels. CAB Rev. 4 (28) (2009) 1–10. Available from: https://doi.org/10.1079/PAVSNNR20094028.

[58] J. Domac, K. Richards, S. Risovic, Socio-economic drivers in implementing bioenergy projects, Biomass Bioenergy 28 (2005) 97–106.

[59] M.R. Anuar, A.Z. Abdullah, Challenges in biodiesel industry with regards to feedstock, environmental, social and sustainability issues: a critical review, Renew. Sustain. Energy Rev. 58 (2016) 208–223.

[60] F. Alam, A. Date, R. Rasjidin, S. Mobin, H. Moria, A. Baqui, Biofuel from algae—is it a viable alternative? Proc. Eng. 49 (2012) 221–227.

[61] S.M. Abdo, S.A.A. Abo El-Enin, K.M. El-Khatib, M.I. El-Galad, S.Z. Wahba, G. El-Diwani, et al., Preliminary economic assessment of biofuel production from microalgae, Renew. Sustain. Energy Rev. 55 (2016) 1147–1153.

[62] S.M. Abdo, Biofuel production from Egyptian freshwater algae, 2014, Ph.D. Thesis, Al-Azhar University (Girls Branch), Cairo, Egypt.

[63] R. Turton, R.C. Bailie, W.B. Whiting, J.A. Shaeiwitz, D. Bhattacharyya, Analysis, Synthesis, and Design of Chemical Processes, fourth ed., Prentice Hall, 2012, p. 1007.

Index

Note: Page numbers followed by "*f*" and "*t*" refer to figures and tables, respectively.

A

ABB. *See* Algal-based biofuel (ABB)
ABE fermentation. *See* Acetone–butanol–ethanol fermentation (ABE fermentation)
ABS. *See* Algal bacterial symbiosis (ABS)
ACC. *See* Average cellular chlorophyll (ACC)
Acetabularia, 68–70, 133
Acetogenesis, 229
Acetogenic bacteria, 342
Acetogens, 229
Acetone (C_3H_6O), 265
Acetone–butanol–ethanol fermentation (ABE fermentation), 265, 271–272, 335, 391–392
Acetyl CoA, 59
    Acetyl CoA-dependent pathway, 59
    pool, 197–198
Acetyl-CoA carboxylase (ACCase), 98–99, 197–198, 621–622
Acetyltransferase, 98–99
Achnanthes elongata, 18
Achnanthidium
    A. dolomiticum, 21–22
    A. lineare, 21–22
    A. pfisteri, 21–22
    A. pyrenaicum, 21–22
    A. trinode, 21–22
Acid number, 215
Acid-catalyzed protein NH_3, 289–290
Acidic treatment, 239–240, 240*t*
Acidogenesis, 229, 392–393
Acidominococcus (AsCpf1), 109
Acidophilic algae, 30
Acidophilic organisms, 30
Acinetobacter caloaceticus, 491
Acrochaetium sp., 480
Activated sludge systems, lipase enzymes role in, 493
Active solar tracking systems, 435–436
Active systems of solar water heater, 430
Acutodesmus, 15
    A. obliquus, 71
Acyl carrier protein (ACP), 98–99, 197–198
Acyl chain elongation, 98–99
Acyl CoA-independent pathway, 59

Acylglycerol phosphate acyltransferase (AGPAT), 99
AD. *See* Anaerobic digestion (AD)
Adaptive laboratory evolution (ALE), 468
Adenosine triphosphate/nicotinamide adenine dinucleotide phosphate synthesis (ATP/NADPH synthesis), 80
Administration for Energy Research and Development, 617
ADP-glucose pyrophosphorylase (AGPases), 100
Advanced harvesting method, 553–554. *See also* Conventional harvesting method
    hybrid nanoparticles, 554
    magnetic nanoparticles, 553
    polymeric nanomaterials, 553
    effect of various nanoparticles, 554*t*
AEC. *See* Atomic Energy Commission (AEC)
Aerial algae, 14–16, 16*f*
Aeroterrestrial algae, 17
AFC. *See* Algal fuel cells (AFC)
Agar, 45, 137
    red algae, 43
Agaropectin, 45
Agarophytes, 45
Agricultural crops, 126–127
Agrivoltaic system, 426
Agrofuel, 253–254, 547
Agrophotovoltaic (APV), 426
AGS. *See* Algal Growth System (AGS)
Aheterocysteae, 4–5
Aiolocolax pulchella, 25–26
Air humidity, 14
Airborne algae, 15
Airborne cyanobacteria, 15
Aircraft-range, 354–355
Akashiwo sanguine, 26
Akontae, 12–14
Alcohol-to-jet (ATJ), 333. *See also* Oil-to-jet (OTJ)
    conversion pathways, 333–336
        assessment of life cycle, 336
        economic perspective, 335–336
        process description, 334–335
Alcohols, 332, 335
    oligomerization, 334

Alcohols (*Continued*)
production, 335–336
LCA for, 336
solubility, 210
Algae, 1–2, 50, 123, 196, 251–252, 284–285, 333, 385–386, 421–422, 503, 636
aerial, 14–16, 16*f*
aeroterrestrial, 17
algae by-products applications, 409–411, 409*f*
algae-based biofuel technology, 548
algae-based municipal wastewater treatment process, 153–154
HRAP, 154
waste stabilization pond systems, 154
algae-based nonenergy field, 131–135
anticancer activity of algal substances, 134
antiviral activity of algal compounds, 133
medicinal uses of algae drugs, 132–133
pharmaceuticals, 131–132
pigments and carotenoids, 134–135
products of species of algae and application, 132*t*
algae-based wastewater treatment plants, 153–154
effluent from industrial wastewater treatment plants and microalgae, 154
algae-mediated biodiesel, 548
algae-mediated products, 550
algae–biofuel, 539
applications of, 253*f*
biojet fuel, 345–346
biomass, 79–80
carbohydrates in, 66–67
synthesis, 67–70
carotenoids, 410
fuel, 575
global distribution and habitats of, 14–31
aerial algae, 14–16
algae living in extreme environments, 26–31
aquatic algae, 18–22
parasitic algae, 25–26
symbiotic algae, 22–25
terrestrial algae, 17–18
medicinal uses of algae drugs, 132–133
metabolic process, 620
as monitor of water quality, 141
polysaccharides, 406–407
production system, 630
as source of antioxidant properties, 133–134
taxonomic characteristics of, 2, 3*t*
thermal decomposition, 284–285
Algae façades technology, 588
"Algae for green fuel" concept, 386
Algae Grass Purifier, 642
Algae living in extreme environments, 26–31, 28*f*

"Algae Raceway Production System", 622–623
Algae Species Program (ASP), 616, 619–623
algae cell photosynthesis, 621*f*
funding history of algae species program, 623*f*
"AlgaeBase", 42
Algal
assessment of Algal residuals process, 536
to biofuels pathways, 504–505
biorefinery system, 570
carbohydrates, 406
cells, 386, 433, 603
conversion process, 503
cultivation process, 433, 503, 525
environmental conditions effect on algal lipids, 205–208
genetic toolbox, 604
growth media, 638–639
hydrogen production, 388
monoculture production, 625
oil production, 604
physiology and cultivation, 79–80
artificial growth of seaweeds, 86–87
cost analysis of algae cultivation, 87–88
factors influencing growth and biochemical composition, 81–83, 82*t*
integrated cultivation system, 88–89
microalgae cultivation system, 83–85
photosynthetic efficiency of algae, 80–81
taxonomy, 11–14
Algal bacterial symbiosis (ABS), 154
system for wastewater treatment, 154–157
microalgae–bacteria symbiosis mechanism, 155
microalgae–bacteria system, 155
microalgal–bacteria relation and production of biofuel, 155–157
Algal biodiesel, 391, 568–569, 641–643
and other types of physicochemical properties, 186
Algal biofuel, 253–254, 602–603
challenges, 548–554
cultivation, 548–550
harvesting, 550–554
commercialization
biofuel production and impact on environment, 578–580
challenges of biofuel commercialization from algae, 581
future prospective of biofuel, 581–583
history of biofuel production, 567
pilot-scale and large-scale trials, 569–570
recent advancement in large-scale biofuel production, 568–569
top companies of algal products commercialization, 573–575

top companies of biofuel production from algae, 575–578

top companies of biofuel production from different feedstocks, 570–573

in desalination process, role of, 175–176

future perspectives, 559–560

lipid extraction, 555–559

production, 376–377, 504–505, 569–570

integrated microalgae biorefinery concept with dark fermentation, 399f

mass balance of *Chlamydomonas mexicana* biomass, 398f

processes of, 387–396, 387f

recent trends in, 396–400

projects, 600

history of NREL, 603–605

research, 570

Algal biomass, 131, 133–134, 183–186, 254, 421–422, 429, 432, 438–439, 504–505, 512, 555, 582

ABS system for wastewater treatment, 154–157

advantage for biodiesel and bioethanol production, 130

algae-based wastewater treatment plants, 153–154

biomass utilization, 150f

bioremediation, 152–153

biosorption and bioaccumulation, 157–162

carbon dioxide biosequestration using microalgae, 151–152

challenges facing algal biomass for biofuel production, 131

compound, 285–287

lipid in, 197–208

biosynthesis in, 197–199

content and fatty acid profiles of, 199–204

strategies of carbon dioxide sequestration, 150–151

"Algal Biomass Organization", 524

Algal fuel cells (AFC), 387, 394–396

configuration of algal fuel cells, 395f

Algal fuel production by industry

algal biodiesel, 641–643

economic assessment, 646–650

life cycle assessment toward microalgae industrialization, 636–639

operating conditions, 639–641

process description, 644–645

extraction, 645f

harvesting, 644f

transesterification, 646f

process simulation, 644

Algal Growth System (AGS), 576

Algal industries, 124–125

commercial production, 124–125

companies depend on algae in world, 137–138

cosmetics, 135–136

energy production, 126–131

algae biomass for biofuel production, challenges facing, 131

advantage of algal biomass for biodiesel and bioethanol production, 130

bio-oil, 129–130

biobutanol production, 129

biofuels, 126–129

biohydrogen, 130

industrial application of algae and, 126f

food ingredients and polymers, 136–137

agar, 137

alginate, 136

aquaculture feed, 137

carragenans, 136–137

main chemical compounds and bioactive compounds in, 125–126

wastewater treatment by marine algae, 138–141

Algal microbial fuel cells (AMFC), 364. *See also* Photosynthetic microalgal microbial fuel cells (PMMFC)

Algal Pond Model (APM), 622–623

Algal-based biofuel (ABB), 523–524

Algal-based jet fuel, 604

Algal-derived biofuel, LCA approach for, 530–536

algal residuals processing, assessment of, 536

biofuel raw materials, assessment of, 530–531

categories, description, and evaluation methodologies, 530t

cultivation systems and reactor types, assessment of, 532–533

harvesting and extraction processes and lipid quantification, assessment of, 533–535

species assessment used for biofuel production, 532

synthesis of oil/lipid yield of microalgae, 531–532

Algenol Biofuels, 578

Algenol Biotech LLC, 636

Alginate, 44, 125–126, 136

Alginic acid, 44, 125, 228

Algomics, 110–111

Aliinostoc, 7

Aliterellaceae, 9–11

Alkali treatment, 239–240, 240t

Alkaline

habitats, 30–31

marsh, 31

stimulation, 641

Alkaliphilic algae, 30

Allophycocyanin, 44

Alphaα-amylase (AMY3), 100

Alteromonas sp., 479–480

Alumina (Al_2O_3), 334

656 Index

American Society for Testing and Materials (ASTM), 186, 211, 346–347
AMFC. *See* Algal microbial fuel cells (AMFC)
Amigo energy, 582
Amino acids, 58, 291–292
Amoebophrya, 26
Amphidinium operculatum, 155
Amphora sp., 170
   A. coffeaeformis, 170
AMY3. *See* Alphaα-amylase (AMY3)
Anabaena, 8, 17
   A. sphaerica, 133
Anabaeniolum, 24
Anabaenopsis, 30–31
   A. arnoldii, 169
Anaerobic digestion (AD), 225–226, 228–233, 256–257, 387, 443, 641
   microbiology of, 228–232
      acetogenesis, 229
      acidogenesis, 229
      hydrolysis, 229
      main stages for biogas production, 231f
      methanogenesis, 229–232
   optimization of, 232
   process parameters, 232–233
      codigestion, 233
      temperature and digester configuration, 232–233
   of seaweed, 232
Anaerobic reactor, 490
Anaerobic sludge blanket reactor (ASB), 491
Analysis of cumulative effect of enzymatic prehydrolysis, 491–492
   application of lipase obtained from microorganisms in lipid and oil bioremediation, 492t
Ancylonema nordenskioeldii, 27–29
3,6-Anhydro-D-galactose, 45
3,6-Anhydro-L-galactose, 45
ANL. *See* Argonne National Lab (ANL)
Anomeric carbon atom, 67
Anomoeoneis sphaerophora, 29–31
Anopheles, 18
Anticancer activity of algal substances, 134
Antioxidants, 133–134
   algae as source of antioxidant properties, 133–134
Antiviral activity of algal compounds, 133
Aphanothece, 15
   A. halophytica, 29–31
APM. *See* Algal Pond Model (APM)
Aquaculture, 426–427
   feed, 137
   land-based cultivation systems, 269–270
      ponds, 270
      seaweed cultivation in sea, 270

species-specific cultivation method, 270
      tanks, 269–270
   production, 124
   seaweed cultivation, 269–270
Aquaculture Raceway Integrated Design (ARID), 110–111
Aquatic algae, 18–22
Aquatic ecosystem, 388–389
Aquatic species program, 602, 619–623
   limitations of industrial applications, 628–630
      commercialization potential, 630
      cultivation challenges, 629
      disconnect between lab and field, 630
      dust issue, 630
      high-cost, 628–629
      resource availability, 629
   microalgal isolation and characteristics, 623–626
   relationship between NREL and algae species program, 627–628
   US department of energy, 616–619
Aquavoltaics, 426–427
Aqueous phase reforming (APR), 344
Aquifex aeolicus, 312
Arbitrary units (AU), 608
ARF. *See* Australian Renewable Fuel Limited (ARF)
Argonne National Lab (ANL), 623
ARID. *See* Aquaculture Raceway Integrated Design (ARID)
Arisarum vulgarum, 25
Arkenol patented process, 571
Arthrospira, 151–152, 256, 409–410, 441
   A. fusiformis, 30–31, 72–73
   A. platensis, 65–66, 72–73, 169
ARTP. *See* Atmospheric room temperature plasma (ARTP)
Arum italicum, 25
ASB. *See* Anaerobic sludge blanket reactor (ASB)
Ascophyllum, 25–26, 46
   A. nodosum, 44, 50, 128, 225, 267–269
Ascophylum nodosum. *See* Marine brown algae (*Ascophylum nodosum*)
Asexual plant cell division, 623–625
ASP. *See* Algae Species Program (ASP)
Aspen HYSYS V8.4 software, 644–645
Aspen Plus processes, 356
Aspergillus sp., 493
   A. awamori, 490
Assistant Secretary for Energy Efficiency and Renewable Energy, 619–620
Assistant Secretary of Energy, 619
Astaxanthin, 103
Asterocapsa, 8

ASTM. *See* American Society for Testing and Materials (ASTM)
ATJ. *See* Alcohol-to-jet (ATJ)
ATJ-SPK (2018), 346–347
Atmospheric room temperature plasma (ARTP), 465–467, 466*f*
Atomic Energy Commission (AEC), 617
ATP/NADPH synthesis. *See* Adenosine triphosphate/nicotinamide adenine dinucleotide phosphate synthesis (ATP/NADPH synthesis)
AU. *See* Arbitrary units (AU)
Aulosira, 8
Australian Renewable Fuel Limited (ARF), 571
Autotrophic metabolism, 615–616
Auxenochlorella sp., 487
  A. prothecoides, 457–458
  A. protothecoides, 64, 156
Average cellular chlorophyll (ACC), 371
Azotobacter, 479–480

B

B-phycoerythrin, 44
Bacillariophyceae, 9, 12–14, 407, 479–480
Bacillariophyta, 12–14, 30–31
Bacillus sp., 485–486, 491
  B. licheniformis, 157
  B. subtilis, 490
Bacteria, 478, 550
  bacteria-free cultures, 626
Bacteroidetes, 483–485
Bare soft substrates, 19
Barriers, 47–49
Batrachospermum, 18, 22
Battery-powered electric vehicles, 304–305
BBM. *See* Bold's Basal Medium (BBM)
BD FACS AriaTM, 608
BD FACSDivaTM v6.0, 608
BECCS. *See* Bioenergy with carbon capture and sequestration (BECCS)
Benzothiophenes (C_8H_6S), 347–348
Beta position, 67
β-carotene, 377, 410, 443
β-carotene oxygenase (BKT), 103
β-D-galactose to glucose-1-phosphate, 68
BG 11 medium, 172
BI. *See* Biodegradability index (BI)
Bio intelligent quotient system (BIQ system), 588
Bio-oil, 129–130, 182, 293–294, 332
  by-products from Bio-oil fuel production, 412
  production, 338–339, 451–452
Bioaccumulation, 159–162
  biocoagulation, 161
  Biodegradation/bioconversion, 161–162

metal bioaccumulation induction to lipid production, 160–161
Bioactive compounds in algae, 125–126
  conversion, 264–265
Bioalcohol, 182, 254–256
  bioethanol, 255
  C_4H_9OH, 255–256
conversion, 264–265
Bioastin, 575
Biobased jet fuel, 349, 351, 353–354
  fuel fluidity, 332
  production, 332
Biobutanol (C_4H_9OH), 255–256, 391–394, 393*f*
  from microalgae, 265–267, 266*t*, 271–273
  production, 129
  various fuels resulting from microalgae, 256*f*
Biocatalyst-mediated esterification, 489
Biochemical compounds of algae, 57–58
  carbohydrates, 66–73
  lipids and fatty acids in algae, 58–66
    factors affecting triacylglycerol synthesis, 62–66
    lipid groups in algal cells, 58–59
    TAG synthesis, 59–62
  proteins, 73
Biochemical conversion pathways, 333, 343
Biochemical measuring methods (BMP), 233–236
Biocoagulation, 161
Bioconversion process, 161–162
Biocrude oil, 422–423, 439
Biodegradability index (BI), 236
Biodegradation, 161–162
Biodiesel, 128, 156, 182–183, 186, 209–210, 282, 386, 390–391, 413, 422–423, 515–516, 526, 538–539, 615
  algae for, 195–196
    biodiesel characteristics, 211–216
    different methods of transesterification, 208–211
    economic feasibility, 217
    lipids in algal biomass, 197–208
  algal biomass for, 130
  biodiesel-derived microalgae, 265
  biodiesel–bioethanol production, 467
  comparative schematics of conventional transesterification, 516*f*
  environmental impact of, 413–414
  characteristics, 211–216
    boiling point, flash point, and calorific value, 215
    cetane number, acid number, iodine number, and sulfur content, 215
    cloud point, cold filter plugging point, and pour point, 215–216
    kinematic viscosity and density, 216
    OS, 216
    water and sediment content, 216

658   Index

Bioeconomy, 547–548
  in waste management, 408–409
Bioelectricity production, 376–377
Bioenergetic microalgae cultures, 611
Bioenergy, 601–602
  bioenergy-focused microalgae strain collection, 604
  bioenergy-focused strains, 604
  crops, 635–636
  different research bioenergy areas, 602f
  products enhancement, 98–99
  seaweed, 635–636
  system, 442
Bioenergy with carbon capture and sequestration (BECCS), 457–458
Bioengineered ferredoxin–hydrogenase fusion, 319
Bioethanol, 157, 255–256, 391–394, 422–423, 526, 615
  biosynthesis, 406
  from microalgae, 265, 270–271
  production, 127–128
    algal biomass for, 130
  simplified metabolic pathways of clostridial acetone–butanol–ethanol fermentation, 392f
Biofilm, 50
  reactors, 509–510
  samples, 606–607
Biofuels, 57–58, 79–80, 124, 126–129, 174–175, 181–182, 226, 251, 281–282, 413–414, 451–452, 567
  bioalcohols conversion, 264–265
  biodiesel, 128
  bioethanol production, 127–128
  biomass production, 259–263
  biomass recovery/harvesting microalgal biomass, 263–264
    bulk harvesting, 263
    filtration, 264
    flocculation, 264
    flotation, 264
  biomethane production, 128–129
  challenges of biofuel commercialization from algae, 581
  comparison between various methods of cell disruption and lipid extraction from microalgae, 557t
  current status, 253–254
    methodologies for developing better biofuel technology, 254t
  dewatering process and biomass extraction, 264
  feedstocks, 385
  freezing-point, 351
  generation by MBC, 486–487, 486t
  impact on environment, 578–580
    atmospheric carbon dioxide, 580f

    differences between traditional fuel and biofuel, 580f
    emissions of sulfur dioxide and nitrogen dioxide, 579f
  microalgal–bacteria relation and production of, 155–157
  production, 225–226, 259–265, 385–386, 523–524, 556–558, 569–570, 578–580, 636
    algae biomass for, 131
    history of, 567, 568f
  raw materials assessment, 530–531
    algal biodiesel and other types of physicochemical properties, 186
    algal biomass, 183–186
    combustion and emission parameters, 186–187
    mechanisms of, 183, 184t
  use of different catalysts on microalgae and respective biodiesel yield, 558t
Biogas, 157, 182, 422–423
Biohydrogen, 130, 156, 387–388, 422–423
  biosystem and semiartificial system for, 320–325
    complex of photosystem I and NiFe-hydrogenases via PsaE, 320–321
    fabrication of PsaD-hoxYH complex, 325
    hydrogenase-ferredoxin fusion, 320
    photosystem I-hydrogenase complex via nanowire from phylloquinone, 324–325
    wiring photosystem I through nanoconstruction, 321–323
  production from algae
    and applications, 304
    benefits of biohydrogen and future prospects, 306
    biohydrogen as efficient future fuels, 305–306
    development of HydESS, 305
    direct cellular biohydrogen production, 306–314
    enhancing hydrogen production in microalgae by gene technology, 316–319
    hydrogen as new vehicle energy source, 304–305
    photosynthetic hydrogen production— cyanobacteria, 314–315
Biointelligent quotient house (BIQ), 593–594, 594f
  algae-based green building, 589–590
    BIQ in Hamburg, Germany, 593–594
    renewable energy applications, 590–593
    university of technology Sydney green building case study, 594–595
Biojet conversion method, 345–346
Biojet fuels production from algae, 332–333
  algae biojet fuel, 345–346
  biojet fuel performance characteristics, 346–350
    combustion characteristics, 348–350
    stability of thermal oxidation, 347–348

Index

659

fuel compatibility with current fueling system of aircraft, 350–355
process simulation, 355–356, 356f
Biological pretreatment, 241–242
Biological sequestration, 151
Biological wastes, 405–406
Biomass, 155–156, 174–175, 236, 238–239, 269, 281–282, 340–341, 405–406, 429, 438–439, 587–588
to electricity conversion using solar radiation, 429
conversion, 282, 405–406, 548
pathways, 333
conversion to biofuel, 512–517
biodiesel production, 515–516
cell disruption, 513–514
intensification of drying and oil extraction, 514–515
wet processing via hydrothermal liquefaction, 516–517
jet fuel conversion pathways, 333–345
alcohol-to-jet conversion pathways, 333–336
oil-to-jet conversion pathways, 336–340
process of gas-to-jet fuel, 340–343
process of sugar to jet fuel, 343–345
production, 154, 259–263, 260t, 547–548
CPPU, 262
heterotrophic biomass production, 263
hybrid two-stage production unit, 263
mixotrophic biomass production, 263
OPPU, 262
program, 630
recovery, 263–264
revenue production, 262
Biomass to liquid (BTL), 340–343, 582
Biomethane, 388–390
production, 128–129
theoretical biomethane yield from lipids, proteins, and carbohydrates, 389f
Biorefineries, 385–386, 560, 602
Bioremediation, 152–153
Biosorption, 157–159
Biotinyl carbon dioxide (B-CO_2), 59–62
BIQ. See Biointelligent quotient house (BIQ)
BIQ system. See Bio intelligent quotient system (BIQ system)
BKT. See Beta3-carotene oxygenase (BKT)
Black band disease, 26
Blastodinium, 26
Blastophysa rhizopus, 24
Bligh and Dyer method, 533–534
Blue Fire Renewables, 571–573, 572f
Blue photosynthetic pigment, 125
Blue snow, 27–29

Blue–green algae (Cyanophyceae), 2–5, 256, 503
BMP. See Biochemical measuring methods (BMP)
Boiling point (BP), 215
Bolbocoleon piliferum, 24
Bold and Wynne's classification of algae, 12–14
Bold's Basal Medium (BBM), 172
Botrydium, 17
Botryococcus, 58, 252
B. branuii, 58, 63, 71, 151–152, 252, 531–532, 555–556, 640
BP. See Boiling point (BP)
Brachysira calcicola, 21–22
Brachytrichia, 4–5
Brackish water, 429, 638–639
Brake thermal efficiency (BTE), 186
Branched alkanes (farnesan), 352–353
Branchipus, 18
Brasilonema, 8
British Ministry of Defense (MOD), 346
Brown algae (Saccharina), 19, 125, 271–272
Brown macroalgae, 127, 271–272
Bryophyta, 131
BTL. See Biomass to liquid (BTL)
Building Research Establishment Environmental Assessment Method, 590
Bulk harvesting, 263
Burkholderia sp., 211, 491
B. arboris, 491
Butanediol, 342
Butanol production, 129
By-products recycling of algal biofuel
algal-based technology, 406f
applications, 409–411
from bio-oil fuel production, 412
biorefinery products from microalgae, 408f
economic feasibility of microalgae biodiesel, 414–415
environmental impact
of biodiesel, 413–414
of by-products, 413–414
from ethanol production, 411–412
future research focus and perspectives, 415–416
generation of biofuel and valuable by-products from microalgae, 407f
glycerol by-products of biodiesel productions, 412
microalgal by-products of biomasses conversion processes, 411
microalgal-based protein by-products, 413
pharmacological effects of algal carbohydrates, 406t

C

C-phycoerythrin, 44
CA. See Carbonic anhydrase (CA)
Calcareous algae, 48

660   Index

Calcium, 63, 207
Calorific value (CV), 215
Calothrix, 8, 17, 24
Calotrichaceae, 9—11
Calvin—Benson cycle, 318—319, 465
Candida rugosa, 492—493
Canthaxanthin, 103, 377
CAPP treatment. *See* Cold atmospheric pressure
    plasma treatment (CAPP treatment)
Carbohydrates, 44—45, 66—73, 149—150, 182, 239, 265,
    406
  agar, 45
  in algae, 66—67
    structure of main sugars, 67*f*
  alginates, 44
  Carrageenan, 45
  enhancement of carbohydrate accumulation in
    microalgae, 100
  factors affecting carbohydrate synthesis, 70—73
    carbon sources, 71
    light sources and light intensity, 72
    low-dose gamma radiation, 72—73
    nutrients and trace metals, 70—71
    pH, 72
    salinity, 72
    temperature, 71—72
  fucans, 44
  laminarin, 44
  synthesis in algae, 67—70
    Leloir pathway, 68*f*
    total carbohydrate contents of microalgae and
     macroalgae, 69*t*
  ulvan, 44
Carbon, 64, 71, 340—341
  carbon-rich fuels, 303
  MBC role in mitigation, 481—485, 482*f*
Carbon dioxide (CO_2), 169, 181, 206—207, 225—226,
    255, 348, 392—393, 429, 456—458, 530—531,
    590—591, 602, 640
  bioenergy with carbon capture and sequestration
    strategy, 458*f*
  biosequestration using microalgae, 151—152
  distribution, 510—511
    high alkalinity, 511
    microbubbling and membrane diffusers, 511
  life cycle emissions, 336
  microbubbles, 515—516
  reduction in CO_2 emission and electricity generation,
    487—488
  strategies of carbon dioxide sequestration, 150—151
    biological sequestration, 151
    nonbiological methods, 151
    phytosequestration, 151

Carbon monoxide (CO), 186, 348, 531, 578
Carbon nanotubes (CNT), 425
Carbonate salts, 48
Carbonic anhydrase (CA), 105, 457—458, 461
Carotenoids, 100—103, 134—135
Carpoblepharis, 18
Carrageenan, 45, 136—137
  red algae, 43
Carteria, 27—29
Catalytic hydrothermolysis (CH), 333, 338
Catalytic process, 332, 405—406, 489
Caulerpa, 46
Caulerpa cylindracea, 491
Caulerpa peltata, 211, 215
Caulerpales cellulose, 43
CCS. *See* CO_2 sequestering and storage (CCS)
Cell
  cell-based immobilization, 489
  disruption, 389, 513—514
    pulse electric field, 513
    simultaneous cell disruption and lipid extraction,
     513—514
  structures, 2
  walls, 42—43
Cellana Inc., 636
Cellular biohydrogen production, 306—314
  hydrogenases, 312—314
  photosynthetic electron transport chain in natural
    system, 307—308
  photosystem I, 311—312
  photosystem II, 309—311
Cellular ion ratios, 168—169
Cellular metabolic pathway engineering, 107—109
Cellular signaling process, 453—454
Cellulose, 43, 385
  compound, 285—287
  ethanol, 603
Centrifugation, 411—412, 443, 552
Cephaleuros, 25
  C. biolophus, 25
  C. minimus, 25
  C. parasiticus, 25
  C. pilosa, 25
Ceramiales, 25—26
Cetane number, 215
CFD. *See* Computational fluid dynamics (CFD)
CFPP. *See* Cold filter plugging point (CFPP)
CH. *See* Catalytic hydrothermolysis (CH)
Chaeloceros sp., 63—64, 136, 256, 407
  C. brevis, 71
  C. gracilis, 414—415
  C. mualleri, 452—453, 461—463
Chaetomorpha, 40, 46

Index

661

Chaetonema, 18
Chaetophora elegans, 103–104
Chaetophorales, 18
Chamaesiphon, 19–20, 22
Chamaesiphonales, 4–5
Chapman's classification of algae, 12–14
Chara, 18, 22, 39
Characiopsis, 18
Characium, 18
Charophyceae, 9
Charophycophyta, 12–14
Charophyta, 12–14
Chasmolithic algae, 18
Chelonicola sp., 18
Chemical flocculation method, 552–553
Chemical oxygen demand (COD), 364, 398–399, 479
Chemical vapor deposition (CVD), 425–426
Chitosan, 553
Chlainomonas kolii, 27–29
Chlamydomonas sp., 17, 19–20, 27–29, 68–70, 104–105, 127, 141, 156, 211, 316, 319–320, 393–394
  C. mexicana, 207, 397, 407–408
  C. nivalis, 27–29, 170
  C. perigranulata, 257–259
  C. pyrenoidosa, 131
  C. reinhardtii, 63, 71, 81, 97–98, 100–105, 107–111, 155, 157, 197, 263, 318–319, 387–388, 480
    C. reinhardtii nac2–26, 318
    C. reinhardtii sta1, 100
  C. yellowstonensis, 27–29
Chlamydomonas cre-miR1174. 2 miRNA, 105
Chlorella sp., 15, 17, 25, 58, 63, 68–72, 83, 88–89, 128, 132, 135–137, 141, 151–152, 173–174, 206–208, 215, 256, 262, 284–285, 290–291, 318, 393–394, 407, 409–410, 441, 443, 461–463, 479–480, 515, 554, 629, 636–637
  C. ellipsoidea, 72
  C. emersonii, 64–65, 414–415
  C. fusca, 464–465
  C. kessleri, 64–65, 71, 464–465
  C. minutissima, 409, 553
  C. protothecoides, 128, 414–415, 624
  C. pyrenodosa, 64, 103, 128, 457–458
  C. pyrenoidosa, 463–467
  C. saccharophila, 409
  C. sorokiniana, 64–66, 71, 103, 151–152, 463
  C. vanabilis, 407
  C. viscose, 64–65
  C. vulgaris, 25, 63–65, 71–73, 83, 103, 128, 131, 135, 156, 160–161, 169, 173, 207–208, 211, 217, 364, 371, 409–410, 413–415, 438, 452–453, 458–459, 461, 466–467, 485–487, 528, 555
  C. vulgaris LEB-104, 151–152

C. zofingiensis, 81, 100–103, 205–207
Chlorochytrium, 25
Chlorococcum, 15, 17, 89, 136, 173, 257–259, 552–553
  C. infusionum, 391
  C. littorale, 128, 151–152, 468
  C. nivale, 452–453
Chloroflexus, 169
Chlorogloeopses spp., 135
Chlorohydra viridissima, 25
Chloromonadineae, 12–14
Chloromonadophyta, 12–14
Chloromonas, 27–29
  C. nivalis, 27–29
  C. polyptera, 27–29
  C. rostafinski, 27–29
  C. rubra, 27–29
  C. rubroleosa, 27–29
Chlorophyceae, 9, 11–14, 46, 407, 624
Chlorophycean, 43
Chlorophycophyta, 12–14
Chlorophyll, 43, 377, 431
  chlorophyll-a, 431
Chlorophyta (green algae), 11–14, 17, 30–31, 41–43, 100, 267
Chlorophytes, 316
Chlorophyts, 131
Chlorosarcina antarctica, 27–29
Chlorosarcinopsis sp., 88–89
Chlorospermae (green algae), 12–14
Chondrus crispus, 136
Chromophycota, 9
Chroococcaceae, 2–4
Chroococcales, 4–6, 9
Chroococcidiopsidales, 9–11
Chroococcus, 15, 30–31, 169
  C. turgidus, 133
Chrysophyceae (golden-brown or golden algae), 12–14, 407, 624
Chrysophycophyta, 12–14
Chrysophyta, 12–14, 17, 27–30
Chu-10 medium, 172
CI engines. *See* Compression-ignition engines (CI engines)
Circular economy, 527–528, 587
Cladophora, 18, 22, 40, 87
  C. fasciculis, 139–140
  C. glomerata, 22
Cladophorales, 18
Clamydonomas reinhardtii, 71–72
Closed bioreactor systems, 481–483
Closed loop tracking systems, 435
Closed photobioreactor production unit (CPPU), 262
Closed photobioreactors (PBRs), 87–88, 432

662

Index

Closed-loop technology, 588–589
Clostridium, 129, 271–272
    C. acetobutylicum, 322, 324, 391–392
    C. pasteurianum, 313, 321
    C. saccharoperbutylacetonicum, 394
Cloud point (CP), 215–216
CNT. *See* Carbon nanotubes (CNT)
CO_2 sequestering and storage (CCS), 149–150
Coal-fired power plants, 616
Coastal rocky shores, 48
Coccobotrys, 24
Coccogoneae, 2–4
Coccomyxa subellipsoidea C-169, 105
COD. *See* Chemical oxygen demand (COD)
Codigestion, 233
Codium, 46
    C. tomentosum, 128, 267–269
Coelastrum, 264
Cold atmospheric pressure plasma treatment (CAPP treatment), 62, 66, 466–467
Cold filter plugging point (CFPP), 215–216
Coleochaete, 18
Colorado collaboration, 605–606
Combustion, 440–441
    characteristics of biobased jet fuels, 348–350
        DCN, 349–350
        gaseous emissions, 349
        PM emissions, 349
        smoke point, 348–349
    conversion of distinct species of algae by appropriate thermochemical process, 442*t*
    developments, 186–187
Commercialization
    challenges of biofuel commercialization from algae, 581
    potential of aquatic species program, 630
Composition based on classification, seaweeds, 42–46
Compression-ignition engines (CI engines), 183, 186
Computational fluid dynamics (CFD), 355–356
Concentrated Acid Hydrolysis, 571
Concentrated solar cells (CSC), 424–425
Continuously stirred tank bioreactor (CSTR), 487
Conventional fuel, 183, 186, 421–422
Conventional harvesting method, 552–553. *See also* Advanced harvesting method
    centrifugation, 552
    filtration, 552
    flocculation, 552–553
    flotation, 552
Conventional transesterification process, 515
Conventional wastewater treatments, 377
Copper, 207
Corallina strain, 133

Corallinales, 25–26
Coralline macroalgae, 48
Corticated, 40
Cosan Limited company, 573
Cosmarium leave, 133
Cosmetics, algal in, 135–136
    hair care, 136
    sunscreen, 135
    whitening, 135
Cost analysis of algae cultivation, 87–88
Cost-effective torrefaction techniques, 294
Coupling biological waste treatments, 414
CP. *See* Cloud point (CP)
CPF1 gene (cryptochrome), 104–105
"Cradle to grave" analysis, 526–527, 536, 636
CRISPR/Cas9, 98, 107–109
Crustose calcareous algae, 48–49
Crustose thallus, 40
Cryophilic algae, 27
Cryopreservation techniques, 609–610
Crypthecodinium cohnii, 398
Cryptogamia, 11
Cryptonemia, 137
Cryptophyceae, 12–14
Cryptophycin, 134
Cryptophycophyta, 12–14
Cryptophyta, 12–14, 30–31
Cryptophytic algae, 17
CSC. *See* Concentrated solar cells (CSC)
CSTR. *See* Continuously stirred tank bioreactor (CSTR)
Ctenocladus circinatus, 29–31
Cultivation, 414, 548–550
    algal cell disruption and lipid extraction methods, 549*f*
    challenges in Aquatic species program, 629
    macroalgae cultivation, 548–549
    microalgal cultivation, 549–550
    system, 409, 504–505
        assessment of, 532–533
Culture maintenance, 609
Cumbersome process, 552
CV. *See* Calorific value (CV)
CVD. *See* Chemical vapor deposition (CVD)
Cyanidiophytina, 11
Cyanidioschyzon merolae, 316
Cyanidium caldarium, 170
Cyanobacteria, 6, 17–18, 70, 130, 134, 314–315, 624
    modern classification of, 7–9
Cyanobacterial cell cultures, 366
Cyanobacterial taxonomy, 9
    history of, 2–5
    modern age of, 5–6
    present status and future of, 9–11
        phylogenetic positioning, 10*f*

Index

663

Cyanobacterium, 263
Cyanochloronta, 12–14
Cyanocyta korschikoffiana, 24
Cyanomargaritaceae, 9–11
Cyanophora paradoxa, 24
Cyanophyceae, 8–9, 407, 624
Cyanophycota, 12–14
Cyanophyta, 12–14, 17
Cyanoprokaryotes, 1–11
  cyanobacterial taxonomy
    history of, 2–5
    modern age of, 5–6
    present status and future of, 9–11
  development of polyphasic approach, 6–7
  global distribution and habitats of, 14–31
    aerial algae, 14–16
    algae living in extreme environments, 26–31
    aquatic algae, 18–22
    parasitic algae, 25–26
    symbiotic algae, 22–25
    terrestrial algae, 17–18
  history and present-day algal taxonomy, 11–14
  modern classification of cyanobacteria, 7–9
  taxonomic characteristics of, 2, 3*t*
Cyanosarcina, 8
  C. chroococcoides, 15
Cyanospira, 30–31
Cyanotech Corporation, 575
Cyanothece, 8
Cyclotella cryptica, 620–621
Cylindrospermum, 8, 17
Cymbella diminuta, 21–22
Cymbopleura austriaca, 21–22
Cystoseira, 48–49
Cytochrome b6f (Cyt-b6f), 308
Cytochrome c6 (Cyt-c6), 308, 311

D

D-galactose, 45, 125–126
D-mannuronic acid, 44
D-sugars, 66
D-xylose, 125–126
Dactylococcopsis hungarica, 27–29
Dapisostemonaceae, 9–11
DAT. Double axis tracking systems (DAT);. *See* Dual
    axis solar tracking systems (DAT)
Data mining process, 374–376
Date-based tracking systems, 436
DC. *See* Direct current (DC)
dCas9 protein, 107–109
DCN. *See* Derived cetane number (DCN)
Dead Sea, 171
Dehydrase (DH), 98–99
Delicatophycus delicatulus, 21–22

Delicatophycus minutus, 21–22
Denaturing gradient gel electrophoresis (DGGE), 374
Density
  of biodiesel, 216
  centrifugation of microalgal isolation, 626
Denticula elegans, 21–22
Denticula tenuis, 21–22
Deoxyribonucleic acid (DNA), 6, 41, 411–412
  sequencing of microbial genomes, 373–376
    phylogenetic analysis, 375*f*
    time expenditures for assessing levels of human
      fecal contamination, 376*f*
Derived cetane number (DCN), 348–350
Dermocarpales, 4–5
Desalination, 169
Desertifilaceae, 9–11
Desikacharya, 7
Desmodesmus, 15, 19–20, 130, 173, 289–290, 466–467
  D. intemedius, 64
Desmosiphon, 8
Dewatering process, 533
  and biomass extraction, 264
DGAT. *See* Diacylglycerol acyltransferase (DGAT)
DGGE. *See* Denaturing gradient gel electrophoresis
    (DGGE)
DHA. *See* Docosahexaenoic acid (DHA)
Diacylglycerol acyltransferase (DGAT), 99, 198–199
Diacylglycerols (DAGs), 59, 197
DiAF. *See* Dispersed air floatation (DiAF)
Diafuel (biofuel from diatoms), 377–378
Diascopic light microscopy, 608
Diatomacea (diatoms), 12–14
Diatoms (Bacillariophyceae), 256, 624–625
Dichothrix, 21–22
Dictyota, 40, 46
Diglycerides, 58–59
Dihydroxyacetone phosphate, 59
Diketopiperazine (DKP), 287–289
Dilabifilum, 24
2,5-Dimethylfuran (DMF), 344–345
Dinoflagellates, 12–14, 19
Dinophyceae, 12–14
Dinophyta, 12–14, 100
Direct current (DC), 423
Direct manufacturing cost (DMC), 650
Disaccharides, 67
Dispersed air floatation (DiAF), 552
Dissolved oxygen (DO), 370
Distillation process, 411–412, 429
Distribution of seaweeds, 46–49
  barriers, 47–49
  seaweeds production, 49
  temperature, 46–47
Distributional patterns, 31

664   Index

Diurnal cycles, 436–437
DKP. *See* Diketopiperazine (DKP)
DM. *See* Dry matter (DM)
DMC. *See* Direct manufacturing cost (DMC)
DNA. *See* Deoxyribonucleic acid (DNA)
Docosahexaenoic acid (DHA), 455
DOE. *See* US Department of Energy (DOE)
Downstream process, 411, 504
DPH1 gene (endogenous phytochrome), 104–105
Drinking water scarcity, 429
Dry matter (DM), 640
Dry weight (DW), 252
Drying process, 412, 428–429, 551
Dual axis solar tracking systems (DAT), 433–435
Dunaliella
    D. salina, 169
    D. tertiolecta, 110–111, 174–175, 205–206
Dunaliella salina, 168–169
Dunaliella sp., 29–31, 68–70, 129–130, 134–135, 155,
    175–176, 256, 262, 393–394, 407, 410, 441, 443
    D. bardawil, 72, 155
    D. parva, 29
    D. salina, 29, 104–105, 410, 443
    D. tertiolecta, 66, 71–72, 173, 205–206
        D. tertiolecta SAD-13.86LEB-52, 151–152
    D. viridis, 29
Dunaniella salina, 170, 174
Dye-sensitized solar cells (DSSC), 377–378, 424–425
Dyes removal process, 139–140

E
Earthrise nutritionals, 573–574
EC. *See* European Commission (EC)
Ecklonia radiate, 409
Eco-friendly biogas production from algal biomass
    anaerobic digestion, 228–233
    challenges of biogas production from algae, 243
    pretreatments, 233–243
    structural and chemical composition of seaweeds,
        227–228
Economic assessment of algal fuel production, 646–650
Economic studies of lipid extraction, 558–559
Edaphophytic algae, 17
Eichler's classification of algae, 12–14
Eicosapentaenoic acid (EPA), 45, 455
Electric vehicles (EV), 363–364
Electrocoagulation-flocculation, 553
Electrokinetic disintegration, 238
Electron paramagnetic resonance (EPR), 312–313
Electrooptical sensors, 436
Elysia chlorotica, 25
Emission parameters, 186–187
Encyonopsis cesatii, 21–22

Encyonopsis microcephala, 21–22
Endedaphic algae, 17
Endolithic algae, 18, 24–25
Endothermic biomass (CHaOb), 291
Endozoic algae, 18, 25
Energy
    conversion process, 477–478
    energy-related projects, 616
    transfer process, 453–454
    transformation, 599
Energy and Environmental Design Leadership, 590
Energy Independence and Security Act, 255
Energy Performance Certificate (EPC), 590
Energy return on energy investment (EROI), 270–271,
    636
Energy Secretary, 619
Engler classification of algae, 12–14
Enhancing biomass production, 306
Eni Gela biorefinery in Europe, 571
Enoyl reductase (ER), 98–99
Enrichment media, 626
Enteromorpha, 22, 68–70, 133, 267
    E. intestinalis, 267–269
Environment Committee of European Parliament, 414
Environmental conditions effect on algal lipids,
    205–208
    fatty acid profile, 201t
    nutrients effect, 205–207
    oil content of some algal species, 199t
    other environmental conditions, 207–208
Environmental pollution, 181, 477–478
Enzymes, 210
EPA. *See* Eicosapentaenoic acid (EPA)
EPC. *See* Energy Performance Certificate (EPC)
Epidaphic algae, 17
Epilithic habitats, 19–20
Epimetallous algae, 15
Epipelic habitats, 19–20
Epiphytic algae, 15, 18
Epipsammic habitats, 19–20
Epixylous algae, 15
Epizoic algae, 15–16, 18
EPR. *See* Electron paramagnetic resonance (EPR)
EPS. *See* Extracellular Polymeric Substances (EPS)
ER. *See* Enoyl reductase (ER)
EROI. *See* Energy return on energy investment (EROI)
Escherichia coli, 100, 153, 156, 364, 374–376
Esterification of triglyceride, 641
Estuaries, 19
Ethanol, 342
Ethanol production, by-products from, 411–412
Euaerial terms, 14–15
Eucheuma, 86–87, 136, 267–269

Euglena, 141
  E. gracilis, 160–161
  E. sanguine, 19–20
Euglenophyceae (euglenoids), 12–14
Euglenophycophyta, 12–14
Euglenophycota, 9
Euglenophyta, 12–14, 17, 30–31
Eukaryota, 9, 12–14
Eukaryotic algae, 11, 17
Eunotia borealpina, 21–22
Eunotia exigua, 21–22
Eunotia minor, 21–22
Eunotia tenella, 21–22
Euphycophyta, 12–14
European Commission (EC), 643
European Committee for Standardization, 211
European Renewable Energy Directive, 636
Eustigmatophyceae, 9
Eustigmatophytes, 624
Euterrestrial algae, 17
EV. *See* Electric vehicles (EV)
Exiguobacterium sp., 485–486
Exococcus, 8
Exoelectrogenic bacteria, 365
Extracellular Polymeric Substances (EPS), 486–487

F

F/2 medium, 172–173
FACS. *See* Fluorescence-activated cell sorting (FACS)
Farm photovoltaic systems, 428
Farming system, 640
Fast pyrolysis, 284, 440
Fat, oil, and grease (FOG), 389–390
Fatty acid methyl ester (FAME), 186, 531, 556, 606
Fatty acid synthase (FAS), 98–99
Fatty acids (FAs), 58, 197–200
  in algae, 58–66
FC. *See* Fuel cells (FC)
Fe-hydrogenases, 313–314
Feedback control system, 435
Feedstock, 641–642
  cultivation, 582
FeFe-hydrogenases, 312–313
FEMICO, 575
Fermentation process, 332, 406, 451–452
Ferredoxin (Fd), 307–308, 311, 315
Ferredoxin-NADP-reductase (FNR), 320
Ferrochelatase genes (hemH genes), 318–319
FFAs. *See* Free fatty acids (FFAs)
Filtration, 264, 411–412, 552
First-generation feedstocks, 195–196
First-generation fuels, 57–58

First-generation solar cells, 424
Fischer–Tropsch synthesis (FT synthesis), 332
Fisher Tropsch biomass to liquid (FT-BTL), 340–341
  process of converting raw biomass, 341*f*
  process of converting syngas, 342*f*
"Fit for all sizes" mode, 7
Fit-for-purpose properties, 346–347
Fixed capital investment, 648
Flagellatae, 12–14
Flash point (FP), 215, 354
Floatovoltaic (FV), 426–427
Flocculant, 552–553
Flocculation, 264, 552–553
Flotation, 264, 512, 552
Flow cytometry, 626
Flue gas, 640
Fluorescence-activated cell sorting (FACS), 605, 626
  for isolation of single cells, 608–609
FNR. *See* Ferredoxin-NADP-reductase (FNR)
FOG. *See* Fat, oil, and grease (FOG)
Foliose, 40
Food ingredients and polymers, 136–137
  agar, 137
  alginate, 136
  aquaculture feed, 137
  carrageenans, 136–137
Fortieaceae, 9–11
Fossil energy sources, 126
Fossil fuel, 57–58, 181, 195, 251, 255, 281, 303, 477–478,
    567, 635–636
  algal biodiesel and types of physicochemical
      properties, 186
  algal biomass, 183–186
  combustion and emission parameters, 186–187
    evaluation of numerous properties, 187*t*
  mechanisms of, 183, 184*t*
Fossil fuels, 567
Fourth-generation solar cells, 425–426
FP. *See* Flash point (FP)
Fractionation process, 513
Francisella novicida (FnCpf1), 109
Free fatty acids (FFAs), 197, 336–337
Freezing point, 351–353, 352*t*
Fritch's classification of algae, 12–14
Fritschiella, 17
Frustulia crassinervia, 21–22
FSJF. *See* Fully synthesized jet fuel (FSJF)
FT synthesis. *See* Fischer–Tropsch synthesis (FT
    synthesis)
FT-BTL. *See* Fisher Tropsch biomass to liquid (FT-BTL)
Fucans, 44
Fucoidane acid, 125–126

Index

Fucucus Spiralis, 267—269
Fucus sp., 40, 46, 225
  F. vesiculosus, 44, 236—238
Fuel
  compatibility with current fueling system of aircraft, 350—355
    distillation property, 354
    fuel density, 355
    fuel-metering and aircraft-range, 354—355
    low-temperature-fluidit, 351—354
    lubricity, 351
    volume-swells of seal material, 350
  energy, 568—569
  physicochemical properties, 182—183
  volatility, 353—354
Fuel cells (FC), 304—305
Fully synthesized jet fuel (FSJF), 347—348
Functional units (FU), 637
Fuqing King Dnarmsa Spirulina Co. Ltd, 574
Furanose ring, 67
FV. *See* Floatovoltaic (FV)

G

G3P. *See* Glycerol-3-phosphate (G3P)
GA. *See* Glycolaldehyde (GA)
Galdieria sulphuraria, 480—481
Gamma radiation, 463
Gamma rays (γ rays), 66
γ-aminobutyric acid, 64
GAPDH gene, 104—105
Gas exchange chambers, 152
Gas fermentation, process of, 342
Gas to jet (GTJ), 333
Gas-chromatography-mass spectrometry lipid extraction, 582
Gas-to-jet fuel, 340—343. *See also* Sugar to jet fuel
  assessment of life cycle, 343
  economic perspective, 342—343
  process illustration, 340—342
    process of FT-BTL, 340—341
    process of gas fermentation, 342
Gaseous biofuel, 567
Gaseous emissions produced by jet fuels, 349
Gasification process, 332, 412, 440
  thermochemical conversion, 291—292, 292f
Gasoline equivalent (GGE), 335
Geitleriaceae, 9—11
Geitlerinema sulphureum, 27
Gelidiella, 46
Gelidium, 68—70, 133, 137
  G. sesquipedale, 139—140
  Gelidium J. V. Lamouroux, 267—269
Gene silencing, 104—105, 107—109

Genetic manipulation of microalgae, 97—98
  future perspectives of genetic engineering in microalgae, 103—105
    completely sequenced algal genome, 112t
    some databases and websites used in algal biotechnology, 113t
  for generation of energy and value-added metabolites, 98—103
    enhancement of bioenergy products, 98—99
    enhancement of carbohydrate accumulation in microalgae, 100
    other value-added compound production, 100—103
Genome editing, advance methods in, 103—110
  CRISPR/Cas9, 107—109
  RNAi, 103—105
  TALENs, 109—110
  ZFNs for targeted genome editing, 105—107
Genomic approaches implementation for wastewater treatment, 487—488
Geotrichum candidum, 490—491
Gervasia, 8
GGE. *See* Gasoline equivalent (GGE)
GHG emissions. *See* Global greenhouse gas emissions (GHG emissions); Greenhouse gases emissions (GHG emissions)
Gigartina stellata, 136
Gigartinales, 25—26
Glaucophyta, 9, 100
Global energy crisis, 386, 523
Global energy demands, 181
Global greenhouse gas emissions (GHG emissions), 332—333
Global Positioning System (GPS), 607
Global seaweeds diversity
  basis of seaweeds classification, 40—41
  diverse groups of seaweeds, 41—42
  global distribution of seaweeds, 46—49
    barriers, 47—49
    seaweeds production, 49
    temperature, 46—47
  seaweeds composition based on classification, 42—46
  symbiotic relation of seaweeds with other marine organisms, 49—50
Global warming, 421, 524—525, 589—590
Gloeobacterales, 9
Gloeobacteriophycidae, 8
Gloeocapsa, 15, 17
Gloeocapsopsis, 21—22
Gloeothece, 21—22
Glucan-water dikinases (GWD), 100
Glucose, 64, 127—128
Glycerin, 641

Glycerol by-products of biodiesel productions, 412
Glycerol phosphate acyltransferase (GPAT), 99
Glycerol-3-phosphate (G3P), 59, 198–199
Glycoengineering, 110–111
Glycolaldehyde (GA), 318
Glycolipids, 58–59
Glycolysis, 67–68
Glycose, 287–289
Gold nanoparticles (AuNPs), 81
Golden algae (Chrysophyceae), 256
Golden-brown algae, 377–378, 624
Gomphonema, 22
    G. elegantissimum, 21–22
    G. lateripunctatum, 21–22
Gongrosira incrustans, 19–20
GPAT. *See* Glycerol phosphate acyltransferase (GPAT)
GPS. *See* Global Positioning System (GPS)
Gracilaria, 40, 46, 68–70, 87, 137, 267
    G. changii, 87
    G. edulis, 87
    G. lemaneiformis, 71
    G. vermiculophylla, 236
Gracilariales, 25–26
Graphene (G), 425
Grateloupia lanceolata, 24
Great Salt Lake, 171
Green algae, 17, 40, 68, 173, 256, 263, 624
Green building certification systems, 590
Green Crude Farm, 576
Green Mark, 590
Green microalgae, 487
Green snow, 27–29
Green Star, 590
Greenhouse gases emissions (GHG emissions), 343, 421, 479–480, 523, 530, 587–588, 635–636
Grinding process, 551
GTJ. *See* Gas to jet (GTJ)
Guluronic acids (G acids), 125
Gunnera, 24
GWD. *See* Glucan-water dikinases (GWD)

H

H_2-ase uptake (HupLS), 315
Haematococcus, 441
    H. pluvialis, 100–103, 134–135, 443, 508, 574
Hair care, algal in, 136
Haloleptolyngbya, 30–31
Halomonas sp., 155
Halophilic algae, 29, 169–170
    for water desalination, 169, 170f
        algal biofuel role in desalination process, 175–176
        economic feasibility, 174–175
        efficiency of microalgae and seaweeds for water desalination, 173

        isolation of halophilic microalgae, 171–173
    marine environment, 169–171
Halophilic microalgae, 169, 386
    isolation of, 171–173
Halospirulina tapeticola, 29
Halotia, 8
Halymeniales, 25–26
Hapalosiphon, 15
Haplonema, 8
Haptophyta, 12–14
Haptophytes. *See* Prymnesiophytes
Harnessing solar radiation for algal cultivation, 432, 433t
Harvesting, 512, 533, 550–554
    flotation, 512
    macroalgae, 550–551
    membrane filtration, 512
    microalgae, 552–554
    microalgal biomass, 263–264
Harvesting and extraction processes, assessment of, 533–535
Harvey's classification of algae, 12–14
Hassalia, 15
Hazardous waste sequestering by MBC, 485–486
HDCJ. *See* Hydrotreated depolymerized cellulosic jet (HDCJ)
Heat release rate (HRR), 186
Heavy metals removal
    by algae, 139–140
    by MBC, 485–486
HEFA. *See* Hydrogenated esters and fatty acids (HEFA)
hemH genes. *See* Ferrochelatase genes (hemH genes)
Hemicellulose, 239, 285–287, 385
Heptoses, 66
Heribaudiella fluviatilis, 19–20
Heterococcus, 24
Heterocysteae, 4–5
Heterogeneous catalysis, 334–335
Heterohormogonium schizodichotomum, 27
Heterokontae, 12–14
Heterokontophyta, 9
Heteroscytonemataceae, 9–11
Heterotrophic bacteria, 374, 481
Heterotrophic biomass production, 263
Hexoses, 66
HHV. *See* Higher heating values (HHV)
High rate algal pond systems (HRAP), 154
High rate algal ponds (HRAP), 154
High temperature Fischer–Tropsch (HTFT), 341
High-throughput photobioreactor, 611
High-throughput technologies, 604
Higher heating values (HHV), 227, 290–291
Homocysteae, 4–5

668 Index

Horizontal with tilted modules single axis solar tracking systems (HTSAT), 436
Hormogonales, 4–5
Hormogoneae, 2–4
HRAP. *See* High rate algal pond systems (HRAP); High rate algal ponds (HRAP)
HRJ. *See* Hydroprocessing of renewable jet (HRJ)
HRR. *See* Heat release rate (HRR)
HRT. *See* Hydrolytic retention time (HRT)
HTFT. *See* High temperature Fischer–Tropsch (HTFT)
HTL. *See* Hydrothermal liquefaction (HTL)
HTSAT. *See* Horizontal with tilted modules single axis solar tracking systems (HTSAT)
Humidophila
  H. contenta, 21–22
  H. perpusilla, 21–22
HupLS. *See* H$_2$-ase uptake (HupLS)
HVO. *See* Hydrotreated vegetable oil (HVO)
Hybrid Chemistry approach (HyChem approach), 356
Hybrid two-stage production unit, 263
HyChem approach. *See* Hybrid Chemistry approach (HyChem approach)
Hyd ESS. *See* Hydrogen-dependent energy storage systems (Hyd ESS)
Hyd-Hyd conjugates, 322–323
Hydra viridissima, 25
Hydrocarbon(s), 58–59, 343
  catalytic upgrading, sugars to, 344
  chain, 285–287
  fuels, 344
Hydrochar, 439–440
Hydrochloric acid (HCl), 239
Hydrocolloids, 58
Hydrodeoxygenation, 338–339
Hydrodictyon, 19–20
Hydrogen (H$_2$), 318, 392–393
  as new vehicle energy source, 304–305
  production
    and applications, 304
    capacity, 303
    elimination of competing pathways, 319
    in microalgae by gene technology, 316–319
    overcoming O$_2$ sensitivity of hydrogenase, 316–319
Hydrogen sulfide (H$_2$S), 141
Hydrogen-dependent energy storage systems (Hyd ESS), 305
Hydrogenases, 312–314
  Fe-hydrogenases, 313–314
  FeFe-hydrogenases, 313
  hydrogenase-ferredoxin fusion, 320
  improving hydrogenases activity, 314
  NiFe-hydrogenases, 312–313

Hydrogenated esters and fatty acids (HEFA), 333, 337–338
  HEFA-SPK, 345–346
Hydrolysis, 229
Hydrolytic retention time (HRT), 389
Hydrolytic starch degradation, 100
Hydroprocessing of renewable jet (HRJ), 336–337, 339
Hydroprocessing technologies, 332
Hydroterrestrial algae, 17
Hydrothermal carbonization, 439–440
Hydrothermal conversion process, 439–440
Hydrothermal gasification, 291, 439–440
Hydrothermal liquefaction (HTL), 282, 284, 287–291, 336–338, 439, 516
  concept of, 287*f*
  potential algal biomass reaction pathways, 288*f*
  of various macroalgae species, 290*t*
  of various microalgae species, 289*t*
Hydrothermal process, 338–339
Hydro–thermolysis processes, 332
Hydrotreated depolymerized cellulosic jet (HDCJ), 336–339
Hydrotreated vegetable oil (HVO), 571
Hydrotreatment process, 339
5-Hydroxymethylfurfural (HMF), 344–345
Hydrurus foetidus, 21–22
Hyella, 8
Hypersaline environments, 169–170
Hypnea, 40, 46, 267–269
Hypneacervicorn, 137
Hypo Manager tool, 644–645
Hypolithic algae, 17

I

I/S ratios. *See* Inoculum/substrate ratios (I/S ratios)
ICE. *See* Internal combustion engine (ICE)
Ignition delay (ID), 186
Immobilization method, 127
Immobilized enzyme in lipid bioremediation, 493
Immune system modulation, 406–407
Impact assessment, 528
In situ hydrothermal liquefaction, 391
In situ transesterification (ISTE), 515
Indirect land use transition drives (ILUC drives), 225
Indole acetic acid, 480
Induction period (IP), 348
Industrial sustainable bioenergy system, 385–386
Industrial wastewater treatment plants, effluent from, 154
Inner Mongolia Rejuve Biotech Co. Ltd, 574
Inoculum/substrate ratios (I/S ratios), 233–236
"Inorganics-in-organics" cells. *See* Fourth-generation solar cells

Integrated Algal Bio Refinery, 576
Integrated cultivation system, 88−89, 89*t*
Intensification of drying and oil extraction, 514−515, 514*f*
Internal combustion engine (ICE), 304−305
International Air Transport Association Guidance Material, 346
International Energy Agency, 363−364
International Organization for Standardization (ISO), 636
International Space research station, 374−376
Interpretation (life cycle assessment phase), 528
Inventory analysis, 528
Iodine number, 215
Ion beam radiation, 463, 464*t*
Ionic gelation method, 553
Ionic stress, 168−169
IP. *See* Induction period (IP)
Ipomoea, 18
Iron, 71, 207
ISO. *See* International Organization for Standardization (ISO)
Isochrysis, 256, 407
  I. galbana, 63, 70, 455, 461
Isochrysis galbana
Isokontae, 12−14
ISTE. *See* In situ transesterification (ISTE)

J

Jania rubens, 65
Jatropha, 253−254, 390−391
Jet fuel, 350
  oxidation stability, 348
Jet Fuel Thermal-Oxidation-Stability Test (JFTOT), 347−348
JFTOT. *See* Jet Fuel Thermal-Oxidation-Stability Test (JFTOT)
Johnson medium, 172
Joule Unlimited Inc., 636

K

Kappaphycus, 86−87, 267
  K. alvarezii, 136, 267−269, 548−549
KAS. *See* Ketoacyl-ACP synthase (KAS)
Kelps, 46
Kerosene, 344
Ketoacyl reductase (KR), 98−99
Ketoacyl synthase (KS), 98−99
Ketoacyl-ACP synthase (KAS), 197−198
Kinematic viscosity (KV), 351
  of algae biojet fuel, 353
  of biodiesel, 216
Kirchneriella, 409
Klebsormidium, 15
Koliella, 27−29

Kosakonia, 490
KR. *See* Ketoacyl reductase (KR)
KS. *See* Ketoacyl synthase (KS)
KV. *See* Kinematic viscosity (KV)

L

L-guluronic acid, 44
L-sugars, 66
La Mala valley, 171
Laboratory Directed Research and Development program (LDRD program), 601
Laboratory-grown strains, 6−7
Lachnospiraceae (LbCpf1), 109
Lagerheimiella, 8
Lambda-carrageenan, 45
Laminaria, 46, 68−70, 225, 236−238, 364−365
  L. digitata, 86, 128, 232−233, 551
  L. hyperborean, 551
  L. hyperobea, 271−272
  L. japonica, 185−186, 267, 388
  L. saccharina, 130
Laminaria digitata, 236
Laminarin, 44
Land-based cultivation systems, 269−270
  ponds, 270
  seaweed cultivation in sea, 270
  species-specific cultivation method, 270
  tanks, 269−270
lba genes. *See* Leghemoglobin genes (lba genes)
LCA. *See* Life cycle assessment (LCA)
ldhA gene, 319
LDRD program. *See* Laboratory Directed Research and Development program (LDRD program)
Lead (Pb), 578
LED. *See* Light emitting diodes (LED)
Lee's classification of algae, 9
Leghemoglobin genes (lba genes), 318−319
Leifsonia sp., 156
Lemanea, 22
Lentic water bodies, 19−20
Leptolyngbya, 8
Leptolyngbyaceae, 6
LHC. *See* Light-harvesting complex (LHC)
Life cycle assessment (LCA), 255, 333, 524, 635−636
  approach, 526−536, 527*f*
    for algal-derived biofuel, 530−536
  benefits on LCA application for microalgal biofuel commercial production, 536−538
  current scenario on production and application of biofuels, 538−539
    leading countries based on biofuel production, 538*t*
  cycle of microalgae biofuel, 528
  future prospective, 539−540

670   Index

Life cycle assessment (LCA) (*Continued*)
LCA-based optimization, 641
methods, 636
phases of, 528—529
goal determination and scope definition, 528
impact assessment, 528
interpretation, 528
inventory analysis, 528
reporting, 528—529
pros and cons of algal biofuel production, 524—526
stages, 529*f*
standard steps for, 529*f*
studies, 636
toward microalgae industrialization, 636—639
Light, 14, 459—461
absorption mechanism in algae, 431—432
attenuation, 509—510
biofilm reactors, 509—510
guides, 510
thin layer reactors, 509
dilution and saturation, 507—508
scheme of luminescent solar concentrator, 509*f*
solar tracking systems, 507
spectral filtration and shifting, 507—508
fluctuation and mixing, 506
innovations on adjusting the wavelength of irradiance, 462*t*
intensity, 72, 208, 507
light-mediated stress, 460*t*
path length reduction, 509—510
sources, 72
Light effect on PMMFC performance, 370—373
Light emitting diodes (LED), 62, 370—371
and PMMFC, 370—373
average cellular chlorophyll content, 372*f*
electric current productions, 373*f*
Light-harvesting complex (LHC), 81, 466—467
Lignin, 285—287, 344
Lignocellulosic biomass, 338—339, 393—394
Lignocellulosic wastes, 385
Lingulodinium polyedrum, 155
1,4-Linked 3, 6-anhydro-α-L-galactopyranose, 45
1,3-Linked β-Dgalactopyranose, 45
Linoleic acid, 83—84
Lipase, 478
catalyzed hydrolysis, 491
enzymes role in activated sludge systems, 493
pretreatment, 493
role in wastewater treatment, 488—493
Lipid contaminated wastewater conversion into value-added products, 493
Lipid(s), 45—46, 149—150, 182, 604, 639
in algae, 58—66

algal biomass, 604
in algal biomass, 197—208
biosynthesis in algal biomass, 197—199
content and fatty acid profiles of algal biomass, 199—204
environmental conditions effect on algal lipids, 205—208
simplified de novo fatty acid and TAG pathways, 198*f*
assessment of lipid quantification, 533—535
biotechnology, 478
extraction, 555—559
biofuel production, 556—558
economic studies, 558—559
simultaneous, 513—514
lipid-free algal biomass, 397
neutral, 197
nonpolar, 197
polar, 197
production
metal bioaccumulation induction to, 160—161
pathway, 98—99
structural, 197
Lipomyces starkey, 493
Liquefaction process, 412
Liquefied petroleum gas (LNG), 339
Lithomyxa, 8
Lithophyllum yessoense, 46
Lithophytic algae, 15
Littoral zone, 20
Long-chain PUFAs, 58—59
Loriella, 8
Lotic water bodies, 20—21
Low-dose gamma radiation and carbohydrate synthesis, 72—73
Low-temperature Fischer—Tropsch (LTFT), 341
Low-temperature-fluidity of biobased jet fuel, 351—354
flash point, 354
freezing point, 351—353
fuel volatility, 353—354
kinematic viscosity at—20°C, 353
LTFT. *See* Low-temperature Fischer—Tropsch (LTFT)
Lubricity of jet fuels, 351
Luedeking—Piret model, 483—485
Luminescent solar concentrators, 508
Lutein, 377
Lycopene, 100—103
Lyngbya, 15
L. limnetica, 15
L. majuscule, 407
Lyophilize algae, 640

Index

M

Macroalgae, 196, 226, 251–252, 252*t*, 271–272, 284–285. *See also* Microalgae
  cultivation, 548–549
  harvesting, 550–551
Macronutrients, 452–453
MAE method. *See* Microwave-assisted extraction method (MAE method)
Magnesium, 63, 207
Magnetic nanoparticles, 553
Malonyl CoA (MCoA), 59, 197–198
  synthesis, 98–99
Manganese, 207
Manhattan Project, 616–617
Mannuronic acids (M acids), 125
Marine algae, 123
  wastewater treatment by, 138–141
Marine brown algae (*Ascophylum nodosum*), 125
Marine environment, 169–171
Marine macroalgae, 125
Marl lakes, 31
Mastigocladus, 4–5
  M. laminosus, 27
MBC. *See* Microalgal–bacterial consortia (MBC)
MCoA. *See* Malonyl CoA (MCoA)
Mediterranean Sea, 46
Melanospermae (brown algae), 12–14
Membrane
  diffusers, 511
  filtration, 512, 552
Meridion circulare, 21–22
Mesorhizobium sp., 155
Mesotaenium, 17, 27–29
  M. berggrenii, 27–29
Metabolic by-products, 42–43
Metal bioaccumulation induction to lipid production, 160–161
MetE. *See* Methionine synthase (MetE)
Methane (CH$_4$), 128, 157, 182
Methanogenesis, 229–232
Methanol, 210, 641
Methionine synthase (MetE), 155
Methyl ester, 641
Methyl esterification, 641
Methyl oleate (C$_{19}$H$_{36}$O$_2$), 644–645
Methyl tertiary butyl ether (MTBE), 255
Methylotrophic methanogens, 231–232
MFC. *See* Microbial fuel cell (MFC)
MG3P. *See* Monoacylglycerol-3-phosphate (MG3P)
Micro green algae, 168–169
Microalgae
  enhancement of carbohydrate accumulation in, 100

Microalgae, 20, 68–71, 79–82, 87, 97–99, 123, 153, 168–169, 174–175, 183–186, 196, 208–209, 226, 251–252, 252*t*, 256–259, 267–270, 292–293, 356, 407, 451–452, 478, 525, 602, 623–625, 635–636. *See also* Macroalgae
  biobutanol from, 265–267, 271–273
    brown and red alga fermentation for, 272*t*
  biodiesel
    economic feasibility of, 414–415
    production process, 644
  bioethanol from, 265, 270–271
    different species of different groups, 271*t*
  biofilms, 547–548
  biofuel businesses, 643
  biomass, 81
  carbohydrate content of macroalgae, 266*t*
  cells, 83
  cultivation process, 645
  cultivation system, 83–85, 84*f*
    comparison of different types, 85*t*
  culture, 638–639
  diverse types, 257*t*
  efficiency for water desalination, 173
  effluent from, 154
  enhancing hydrogen production in, 316–319
  harvesting, 552–554
    advanced harvesting method, 553–554
    conventional harvesting methods, 552–553
  isolates, 125
  isolation, 606–610
    culture maintenance, 609
    fluorescent activated cell sorting for isolation of single cells, 608–609
    laboratory conditions for growth algae, 607
    laboratory water sampling preparation and enrichment, 607–608
    summary of NREL project results, 609–610
    water sample collection and analysis, 606–607
  oil, 615
    extraction, 640
  strains, 125
  systems, 588–589
  total carbohydrates of some representative microalgal strains, 258*t*
Microalgae–bacteria
  relation and production of biofuel, 155–157
    biodiesel, 156
    biogas and bioethanol, 157
    biohydrogen, 156
  symbiosis mechanism, 155
  system impact on production of algal biomass and associated compounds, 155

672 Index

Microalgal
  biofuel, 569—570
    production, 467
  biomass, 547—548
  bioprospecting project, 605
  by-products of biomasses conversion processes, 411
  cells, 552
    monomer degradation, 291—292
  cultivation, 549—550, 569—570
  industrial production, 636—637
  isolation, 623—626
    automated techniques, 626
    density centrifugation, 626
    enrichment media, 626
    micromanipulation, 626
    sampling and collection, 625
    serial dilution, 625
    streak plate method, 625
  lipid, 407—408
  metabolism, 451—452
  microalgal-based biofuel production, 452
  microalgal-based biofuel technology, 281—282
  microalgal-based protein by-products, 413
  microalgal—bacterial interaction, 479
  starch, 127
  systems, 415—416
Microalgal—bacterial consortia (MBC), 478
  biofuel and bioproducts generation by MBC, 486—487
  distribution and role in wastewater treatment, 481—486
    heavy metal removal and sequestering of hazardous waste by MBC, 485—486
    MBC role in mitigation of carbon and nutrients, 481—485, 482f
  interchange of substrates, intercellular communication, and horizontal gene, 479—481
  lipase role in wastewater treatment, 488—493
  reduction in CO_2 emission and electricity generation, 487—488
Microbial fuel cell (MFC), 364, 394—395
Microbial lipase-mediated biocatalysis, 489—491
Microbial remediation, 152—153
Microbiology of anaerobic digestion, 228—232
Microbiota, 49
Microbubbling, 511
Microcoleus, 17
  M. chthonoplastes, 29
Microcystis, 8, 479—480
  M. aeruginosa, 640
Microfibrils, 227
Micromanipulation, 626
Micromonas pusilla, 316

Micronutrients, 452—453, 536
Microorganism-based techniques, 479
Microprocessors, 436
Microthrix parvicella, 493
Microwave pretreatment, 236—238
Microwave-assisted extraction method (MAE method), 513, 555—556
Mineral and hot springs, 27
MinION, 374
Mixotrophic biomass
  growth mode, 263
  production, 263
MOD. See British Ministry of Defense (MOD)
Mojavia, 8
Monoacylglycerol-3-phosphate (MG3P), 59
Monoacylglycerols, 197
Monocrystal solar cell, 424
Monodus sp., 136
Monoglycerides, 58—59
Monoraphidium sp., 64, 173, 207
Monosaccharides, 66
Monostroma, 86, 267
Monounsaturated fatty acids (MUFAs), 64, 197
Mrakia blollopsis, 491
MTBE. See Methyl tertiary butyl ether (MTBE)
Mucor, 490
Muddy shores, 19
MUFAs. See Monounsaturated fatty acids (MUFAs)
Multi-Year Program Plan (MYPP), 570
Multiwall carbon nanotubes (MWCNT), 425
Murine sarcoma virus (MuSV-124), 133
Mus musculus, 374—376
MuSV-124. See Murine sarcoma virus (MuSV-124)
Mutations, 465
MWCNT. See Multiwall carbon nanotubes (MWCNT)
MYPP. See Multi-Year Program Plan (MYPP)
Myxophyceae, 12—14
Myxophycophyta, 12—14
Myxosarcina, 8
  M. chroococcoides, 15

N

N-acyl-homoserine lactones, 480
NADPH-dehydrogenase (NDH-1), 319
Nannochloropsis, 58, 64, 72, 97—98, 105, 137, 211, 289—290, 371, 409—410, 414—415, 455, 458—459, 463, 553, 624
  N. gaditana, 72, 157, 211, 413
  N. oceanica, 66, 81, 407—408, 486—487
  N. oculata, 64, 83, 173—175, 206—207, 464—465
  N. oculate, 151—152, 455
  N. salina, 183
Nanocrystal-based solar cells, 424—425

Nanomaterials, 556–558
  organic-based nanomaterials, 425
  polymeric, 553
Nanoparticles (NPs), 242–243
  treatment, 242–243
Nanopore DNA technology, 365, 374–376
Nanosecond pulse electric field method, 555–556
Nanotechnology, 553
National Renewable Energy Laboratory (NREL), 576, 599–603, 616
  algal biofuels, 602–603
  bioenergy, 601–602
  history, 600–601, 601f
  limitation of industrial application, 610
  microalgae isolation and characteristics, 606–610
  mission and programs, 601
  NREL algal biofuels projects, history of, 603–605
    bioenergy-focused microalgae strain collection, 604
    development of novel microalgal production, 604
    development of robust and high-throughput characterization technologies, 605
    efficient use of algal biomass residues for anaerobic digestion, 605
    evaluation of regulated enzymatic disruption, 604
    proteomics and transcriptomics of algal oil production, 604
  principal project, 605–606
    project description, 606
    project presentation, 605–606
  relationship between NREL and algae species program, 627–628
National Research Council of Canada, 604
Natrun Valley, 171
Natural calamities, 548–549
Navicula, 141
  N. pelliculosa, 624–625
  N. subinflatoides, 29–31
Navicula exilis, 21–22
NCI. *See* Net cash income (NCI)
NDH-1. *See* NADPH-dehydrogenase (NDH-1)
Neochloris sp., 58, 89, 479–480
  N. oleoabundans, 63–65, 71–72, 205–206
Net cash income (NCI), 87–88
Neutral lipids, 197
NHEJ. *See* Nonhomologous end-joining (NHEJ)
Nickel–iron-hydrogenases (NiFe-hydrogenases), 312–313
  NiFe-H2ase HoxYH, 325
  NiFe-hydrogenases via PsaE, 320–321
  NiFeSe-hydrogenases, 313
  structures of functional active site, 313f
NiFe-hydrogenases. *See* Nickel–iron-hydrogenases (NiFe-hydrogenases)

Nitella, 18
Nitrogen, 63, 70, 73, 205–206
Nitrogen oxides (NO), 186
Nitschia sp., 624–625
Nizimuddinia zanardini, 236
Nizschia, 141
  N. fonticola, 21–22
  N. frustulum, 29–31
Nodularia, 17
  N. spumigena, 29–31
Nonbiological methods, 151
Noncoralline algae, 48
Nonhomologous end-joining (NHEJ), 109–110
Nonpolar lipids, 58–59, 197
Nostoc, 8, 15, 17, 24, 27–29, 135
  N. cicadae, 24
Nostoc-like genera, 7
Nostocaceae, 5
Nostocales, 4–5, 9–11, 314–315
Nostochophycidae, 8
Nostochopsaceae, 5
NPs. *See* Nanoparticles (NPs)
NREL. *See* National Renewable Energy Laboratory (NREL)
Nuclear Regulatory Commission, 617
Nuclear technology, 617
Nutrient(s)
  depletion, 620
  effect on algal lipids, 205–207
  MBC role in mitigation, 481–485, 482f
  nutrient-mediated stress, 453–454
    advantages, challenges, and prospective solutions of, 454f
  removal, 140–141, 153
  stress, 452–454
Nutrients, 70–71

O

O-ring seals, 350
OC. *See* Organic carbon (OC)
Ochromonas, 131
  O. danica, 624
  O. itoi, 27–29
  O. smithii, 27–29
Ochrophyta, 41, 46
Ochrophytes, 40, 44
Oculatellaceae, 9–11
Odontidium hyemale, 21–22
Odontidium mesodon, 21–22
OEC. *See* Oxygen evolution complex (OEC)
Oedocladium, 17
Oedogoniales, 18
Oil extraction process, 528, 534, 641

674

Index

Oil-bearing biomass, 515
Oil-to-jet (OTJ), 333. *See also* Alcohol-to-jet (ATJ)
   conversion pathways, 336–340
      assessment of life cycle, 339–340
      economic perspective, 339
      HDCJ, 338–339
      HEFA, 337–338
      process of CH, 338
oil/Lipid yield of microalgae, synthesis of,
   531–532
Oilseed crops, 128
Oleaginous plants, 531
Oleic acid, 59, 83–84
Omics tools, 110–111
Oocardium stratum, 21–22
Open bioreactor systems, 481–483
Open loop tracking systems, 435
Open pond cultivation system, 550
Open pond production unit (OPPU), 262
Open pond systems (OPS), 262, 638–639
Open ponds bioreactors, 507
Open raceway ponds (OPRs), 87–88
Ophrydium, 25
OPPU. *See* Open pond production unit (OPPU)
OPRs. *See* Open raceway ponds (OPRs)
OPS. *See* Open pond systems (OPS)
Organic carbon (OC), 263
Organic solar cells, 508
OS. *See* Oxidation stability (OS)
Oscillatoria, 8, 141
   O. acuminate, 15
   O. curviceps BDU92191, 161–162
   O. geminate, 27
   O. limnetica, 133
   O. pilicola, 15–16
   O. terebriformis, 27
Oscillatoriaceae, 5
Oscillatoriales, 9–11
Oscillatoriophycidae, 8
Osmotic stress, 168–169
Ostreococcus lucimarinus, 316
Ostreococcus tauri, 197, 316
OTJ. *See* Oil-to-jet (OTJ)
Outdoor Test Establishment (OTF), 622–623
Oxidation stability (OS), 216, 348
Oxidation–reduction process, 378
Oxidative pentose phosphate pathway, 67–68
Oxidative phosphorylation, 67–68
Oxidative treatment, 241
Oxygen (O_2), 315, 366
   atoms, 309
   sensitivity of hydrogenase, 316–319
      genetic manipulation, 317t

      increased O_2 consumption/sequestration,
       318–319
      partial PSII inactivation, 318
Oxygen evolution complex (OEC), 309–311
Oxygenates, 344
Oxygenic photosynthesis, 506
Oxynema acuminatum, 15
Ozonation treatment, 241
Ozone (O_3), 241, 578
   layer depletion, 578–579

P

PA. *See* Phosphatidic acid (PA)
Padina, 40, 43
   P. pavonica, 65
Palmaria palmate, 87
Palmariales, 25–26
Palmitic acid, 59, 83–84
Palmitoleic acid, 83–84
PAMFC. *See* Photosynthetic algae-mediated microbial
   fuel cells (PAMFC)
PAP. *See* Phosphatidic acid phosphatase (PAP)
Papenfuss's classification of algae, 12–14
PAR. *See* Photosynthetic active radiation (PAR);
      Photosynthetically active radiation (PAR)
Parachlorella beijerinckii, 64–65
Parasitic algae, 18, 25–26
Parker's classification of algae, 12–14
Particulate matter (PM), 413
   PM emissions, 348–349
Pascher's classification of algae, 12–14
Passive solar tracking systems, 435–436
Paulschulzia pseudovolvox, 394–395
Pavlova, 137
   P. lutheri, 63, 455, 458–459
   P. pinguis, 409
   P. tricornutum, 63
PBR. *See* Photobioreactor (PBR)
PBRs. *See* Closed photobioreactors (PBRs)
Pc. *See* Plastocyanin (Pc)
PDS. *See* Phytoene desaturase (PDS)
Pediastrum, 19–20
PEF. *See* Pulse electric field (PEF)
Pelagibaca bermudensis, 156
Pelagic zone, 20
Pelvetia canaliculated, 267–269
Penicillium
   P. chrysogenum, 490
   P. restrictum, 492
Pentoses, 66
PEP. *See* Phosphoenolpyruvate (PEP)
Peroxide treatment, 241
Petalonema, 8

Index

675

Petroleum
  jet fuel, 347–348, 350
  petroleum-based biofuel, 571
  petroleum-based diesel, 530–531
  petroleum-derived jet fuel, 332
PHA. *See* Polyhydroxyalkanoate (PHA)
Phaeodactylum, 137
  P. tricornutum, 58–59, 63, 66, 97–98, 100–105,
      109–111, 205–206, 316, 414–415, 455, 528
Phaeophyceae, 9, 12–14, 41–43
Phaeophycean algae cell, 43
Phaeophycean species, 46
Phaeophycophyta, 12–14
Phaeophyta, 12–14, 41, 267
Pharao desertorum, 17
Pharmaceutical industries, 124
Pharmaceuticals, algae in, 131–132
PhE. *See* Photosynthetic efficiency (PhE)
Phenolic compounds, 228
Pheridia tenuis, 169
Phormidium, 15, 17
  P. autumnale, 409
  P. corallyticum, 26
  P. incrustatum, 21–22
  P. tenue, 27
Phosphatidic acid (PA), 59, 198–199
Phosphatidic acid phosphatase (PAP), 99
Phosphoenolpyruvate (PEP), 99
Phosphoglucan water dikinases (PWD), 100
Phospholipids, 58–59
Phosphorescent materials, 508
Phosphorus, 63, 70, 206, 453–454
Photobioreactor (PBR), 84–85, 169, 183–185, 263,
      370–371, 422–423, 468, 481–483, 507, 577,
      609–610, 637
  cultivation system, 549–550
  intensification of, 506–511
    carbon dioxide distribution, 510–511
    light attenuation and light path length reduction,
      506, 509–510
    light saturation and light dilution, 506–508
Photocurrent
  biosystem for, 320–325
  semiartificial system for, 320–325
Photoheterotrophic metabolism, 624–625
Photomultiplier tube (PMT), 608
Photosynthesis process, 183, 422–423, 451–452, 588, 621
Photosynthetic active radiation (PAR), 153, 510
Photosynthetic algae, 569
Photosynthetic algae-mediated microbial fuel cells
      (PAMFC), 487
Photosynthetic efficiency (PhE), 183, 432
  of algae, 80–81
Photosynthetic electron transport chain in natural
      system, 307–308

Photosynthetic hydrogen production, 314–315
Photosynthetic microalgae, 366, 410–411
Photosynthetic microalgal microbial fuel cells
      (PMMFC), 365–369
  different algae used in microbial fuel cell as algal
      microbial fuel cell, 369t
  DNA sequencing of microbial genomes, 373–376
  future of PMMFC using diatoms, 377–378
  integrated approaches of, 376–377
    bioelectricity production, 376–377
    production of value-added chemicals, 377
    recycling wastewater, 377
  effect of light on the performance of, 370–373
    light-emitting diodes and, 370–373
  schematic illustration of microbial fuel cell with
      microalgae configurations, 366f
  schematic representation of electric vehicles being
      driven by electricity, 379f
  schematic representation of electron shuttling, 365f
  seven different configurations of photosynthetic
      microbial fuel cells, 368f
Photosynthetic organisms synthesize carbohydrates, 66
Photosynthetic storage materials, 42–43
Photosynthetic unit (PSU), 80
Photosynthetically active radiation (PAR), 430
Photosystem I, 311–312, 320–321
  through nanoconstruction, 321–323
    models of hydrogenase-PsaE-PS complex, 322f
    schematic shows nanocomplexes, 323f
  photosystem I-hydrogenase complex via nanowire
      from phylloquinone, 324–325
    proposed scheme for hydrogen production via PS
      I-NQ, 324f
Photosystem II (PSII), 309–311, 318
Photovoltaic power generation, 426
Photovoltaic solar panels, 423
Photovoltaic technology, 424
Phyco Biosciences, 629
Phycobiliproteins, 44
Phycocyanin, 44
Phycoerythrins, 44
Phycoremediation, 152–153
Phyllosiphon, 25
Physical stress, 454–467
  challenges and future directions of, 467–468
  for enhanced biofuel production
    atmospheric room temperature plasma, 465–467
    carbon dioxide, 456–458
    light, 459–461
    magnetic field, 464–465
    pH, 458–459
    physical stress influence on biofuel potential of
      microalga, 455f
    radiation, 461–463
    temperature, 455–456

Index

Physical stress (*Continued*)
  nutrient stress, 452–454
Phyta, 12–14
Phytoene desaturase (PDS), 100–103
Phytoene synthase (PSY), 100–103
Phytosequestration, 151
"Picoplankton", 624
Pigments, 42–44, 58–59, 134–135
  pigmentation of three major phyla of seaweeds, 43t
Plackett-Burman experimental design, 127
Placoma, 8
Planktolyngbya limnetica, 15
Planothidium frequentissimum, 21–22
Planothidium lanceolatum, 21–22
Plant hormones, 64
Planting system, 638–639
Plastocyanin (Pc), 308, 311
Plastoquinone (PQ), 307
Plectonema boryanum, 315
Pleurocapsales, 4–5, 9
Plocamiales, 25–26
Plyextremophiles, 31
PM. *See* Particulate matter (PM)
PMMFC. *See* Photosynthetic microalgal microbial fuel
  cells (PMMFC)
PMT. *See* Photomultiplier tube (PMT)
Polar aligned single axis solar tracking systems
  (PSAT), 436
Polar lipids, 58–59, 197
Polyaniline (polyA), 376–377
Polycrystal solar cell, 424
Polyhydroxyalkanoate (PHA), 486–487
Polymer-based solar cells, 424–425
Polymeric nanomaterials, 553
Polymerization process, 412
Polymers, 136–137
Polyphasic approach, development of, 6–7, 11
Polyphenols, 58
Polyphosphoric acid, 334–335
Polypyrrole (poly P), 376–377
Polysaccharides, 67, 125, 227–228, 236–238
Polysiphonia, 44
  P. caespitosa, 25–26
  P. lanosa, 25–26
  P. virgate, 18
Polystyrene (PS), 81
Polyunsaturated FAs (PUFAs)
Polyunsaturated fatty acid (PUFA), 45, 58–59, 197, 455,
  486–487
Ponds, aquaculture seaweed cultivation in, 270
Porphyra, 40, 43–44, 46, 135, 267
Porphyridium, 44
Porphyridium, 44, 71, 133, 406–407
  P. purpureum, 70, 155
Porphyrosiphon, 15

Potamogeton pectinatus, 18
Potential value-added products, 281–282
Poulinea lepidochelicola, 18
Pour point (PP), 215–216
PoX. *See* Pyruvate oxidase (PoX)
PQ. *See* Plastoquinone (PQ)
Prantl's classification of algae, 12–14
Prasinophyceae, 9, 479–480
Prasinophyta, 12–14
Prescott's classification of algae, 12–14
Pretreatment, 233–243, 559
  biological pretreatment, 241–242
  chemical treatment, 239–241
    alkali or acidic treatment, 239–240
    oxidative, 241
    ozonation, 241
    peroxide treatment, 241
  inhibitor removal, 243
  NPs treatment, 242–243
  physical treatment, 233–239
    electrokinetic disintegration, 238
    extrusion, 238–239
    mechanical treatment, 233–236, 234t
    microwave pretreatment, 236–238
    size reduction, 236
    thermal treatment, 236
    ultrasonic treatment, 238
Process intensification
  algal to biofuels pathways, 504–505
  biomass conversion to biofuel, 512–517
  harvesting, 512
  intensification of photobioreactors, 506–511
  principles and domains of, 503–504
Process simulation, 644
Prochlorophycota, 12–14
Prokaryota, 9
Proteins, 73, 182, 413
Proteobacteria, 483–485
Proteomics and transcriptomics of algal oil production,
  604
Protosiphon, 17
Prototheca wickerhamii, 26
Prototheca zopfii, 26
Protozoa, 550
Prymnesiophyta, 9
Prymnesiophytes, 624
Prymnesium parvum, 131
PS. *See* Polystyrene (PS)
PsaD-hoxYH complex, fabrication of, 325
PSAT. *See* Polar aligned single axis solar tracking
  systems (PSAT)
Pseudanabaena, 8
Pseudanabaenaceae, 6
Pseudoaerial cyanobacteria, 14–15
Pseudoalteromonas sp, 480

Pseudochlorella sp., 407–408
Pseudomonadaceae, 486–487
Pseudomonas sp., 156, 479–480, 485–487, 490–491
   P. aeruginosa, 480, 490–492
   P. chrysogenum, 491
   P. tricornutum, 458–459
PSII. *See* Photosystem II (PSII)
PSU. *See* Photosynthetic unit (PSU)
Pterocladia, 46
PUFA. *See* Polyunsaturated fatty acid (PUFA)
Pulse electric field (PEF), 513
Purification process, 412
Purple–brown snow, 27–29
PV systems. *See* Solar photovoltaic systems (PV systems)
PWD. *See* Phosphoglucan water dikinases (PWD)
Py-GC/Ms. *See* Pyrolyzer combined with gas chromatography/mass spectrometry (Py-GC/Ms)
Pyranose ring, 67
Pyrolysis, 283–287, 338–339, 412, 440, 441*f*, 531
  experiment setup for algae biomass, 284*f*
  of various macroalgae species, 286*t*
  of various microalgae species, 285*t*
Pyrolyzer combined with gas chromatography/mass spectrometry (Py-GC/Ms), 284
Pyrrhophycophyta, 12–14
Pyrrophyta, 12–14
Pyruvate oxidase (PoX), 319

R

R-phycoerythrin, 44
Rabbitfish, 548–549
Raceway open pond (ROP), 638–639
Raceway pond, 262, 441
Radiation, 461–463
  gamma, 463
  ion beam, 463
  ultraviolet, 461–463
Radioactive isotopes, 617
Raizan Energy segment, 573
Ralstonia eutropha, 320–321
Raoultella sp., 491
Raphidonema, 27–29
Raphidonema brevirostre, 27–29
Raphidophyceae, 9
Rapid Thermal Processing, 338–339
Reactive electrochemical membranes (REM), 513–514
Reactive oxygen species (ROS), 160–161, 456
Reactor types, 532–533
ReadiDiesel, 338
ReadiJet, 338
Recycling
  economic feasibility of, 408–409
  wastewater, 377

RED. *See* Renewable Energy Directive (RED)
Red algae, 19, 25–26, 40, 125–126, 269
Red macroalgae, 271–272
"Red rust", 25
Red snow, 27–29
REM. *See* Reactive electrochemical membranes (REM)
Renewable aircraft fuels, 603
Renewable biofuels, 183, 405–406
  Energy Independence and Protection Act for, 270–271
Renewable energies, 303
  applications in green buildings, 590–593
    sustainable development concept, 593*f*
    sustainable development targets and green buildings, 592*t*
Renewable energy, 57–58, 126, 305, 405–406, 547, 567, 599
Renewable Energy Directive (RED), 225–226
Renewable fuels, 579, 619–620
Research Institute for Solar Energy, 600
Resource availability, 629
Response surface methodology (RSM), 355–356
Return on investment (ROI), 442, 650
Rhizobium sp., 479–480
Rhizopus oryzae, 493
Rhodochorton, 44
Rhodococcus sp., 156
Rhodophyceae, 12–14, 43–44
Rhodophycean species, 46
Rhodophycophyta, 12–14
Rhodophycota, 12–14
Rhodophyta, 11–14, 17, 30, 42–43, 45–46, 100
Rhodophyta (red algae), 41, 267
Rhodophyta japonica, 267
Rhodophytina, 11
Rhodopseudomonas palustris, 364
Rhodosorus, 44
Rhodospermae (red algae), 12–14
Rhodymenia pseudopalmata, 72
Rhodymeniales, 25–26
Ribulose bisphosphate carboxylase, 461
Ribulose-1, 5-bisphosphate carboxylaseoxygenase (RuBisCo), 457–458
River habitats, 22
Rivularia, 18–20
Rivulariaceae, 5
RNA interference (RNAi), 103–105
  biochemical production by transgenic microalgae, 104*t*
ROI. *See* Return on investment (ROI)
ROP. *See* Raceway open pond (ROP)
ROS. *See* Reactive oxygen species (ROS)
Rotbunte Tiefenbiocönose, 20–21
Round's classification of algae, 12–14
RSM. *See* Response surface methodology (RSM)

678

Index

S

16S rRNA genes, 6–8
"Sablier", 551
Saccharide, 66
Saccharina latissima, 86–87, 228, 267–269
Saccharomyces cerevisiae, 127, 157, 270–271, 391
Saccorhiza polyschides, 267–269
Salinity, 65, 168–170, 638–639
  affecting carbohydrate synthesis, 72
  stress, 207
Salt wastewater, 171
Saltmarshes, 19
Salton Sea, 630
Sand flats, 19
Sanguina nivaloides, 27–29
Saphophytic algae, 17
Sapphire Energy Limited, 576
Sargassum, 46, 48–49, 133
  S. latifolium, 139–140
  S. muticum, 227, 267–269
  S. piluliferum, 45–46
SAT. *See* Single axis solar tracking (SAT)
Saturated fatty acids (SFAs), 197
Scenedesmus sp., 15, 19–20, 63, 68–70, 88–89, 151–152,
    160–161, 169, 173, 175–176, 208, 393–394, 440,
    453–454, 640
  S. almeriensis, 413
  S. aromatus, 556
  S. destricola, 452–453
  S. dimorphus, 151–152
  S. obliquus, 63–65, 71–72, 99, 130, 141, 151–152, 173,
    175–176, 205–207, 217, 292–293, 371, 387–388,
    390–391, 396, 410, 413, 438, 463
    S. obliquus SJTU 3, 457–458
    S. obliquus XJ002, 81
    S. obliquus, 458–459, 514
    S. platensis, 371
Schizochytrium sp., 531–532
  S. limacinum, 65
Schizophyta (blue–green algae), 12–14
Schizothrix, 8, 21–22
  S. braunii, 15
Scientific data management system, 611
Scotiella nivalis, 27–29
"Scoubidou", 551
SCWG. *See* Supercritical Water Gasification (SCWG)
Scytonema, 7, 15, 17, 21–22
Scytonemataceae, 5
SDG. *See* Sustainable development goals (SDG)
Seawater, 50, 86–87
Seaweed, 39, 47, 68–70, 79–80, 125, 157–159, 196, 225,
    269–270
  artificial growth of, 86–87, 86f, 87f
  classification, 40–41
  cultivation, 86

  in sea, 270
  distribution of seaweeds, 46–49
    barriers, 47–49
    seaweeds production, 49
    temperature, 46–47
  efficiency for water desalination, 173
  grazer, 548–549
  production, 49
  seaweed-associated bacteria, 49
  seaweeds composition based on classification,
    42–46
    carbohydrates, 44–45
    cell wall, 43
    lipids, 45–46
    pigments, 43–44
  structural and chemical composition of, 227–228
    chemical composition variability, 228
    moisture and salt content, 227
    polysaccharides, 227–228
    structure composition, 227
Seaweed, anaerobic digestion of, 232
Second-generation biofuels, 57–58
Second-generation solar cells, 424
Semiartificial system for photocurrent and
    biohydrogen productions, 320–325
Senedesmus sp. LX1, 206
Separate particle fluorescence channels (SSC), 608
Sequential algal biofuel production
  different processes of, 387–396
    algal fuel cells, 394–396
    biodiesel, 390–391
    bioethanol/biobutanol (bioalcohols), 391–394
    biohydrogen, 387–388
    biomethane, 388–390
  recent trends in, 396–400
SERI. *See* Solar Energy Research Institute (SERI)
Serial dilution, 625
Serratia marsescens, 491
SFAs. *See* Saturated fatty acids (SFAs)
Shewanella oneidensis, 364
Short-chain polyunsaturated FAs, 58–59
Silica–alumina catalysts, 334–335
Silicon wafer, 424
Single axis solar tracking (SAT), 433–436
Single-wall carbon nanotubes (SWCNT), 425
Siphonaceous algae, 43
Situ transesterification, 209–210
Skeletonema, 137
SMF. *See* Static magnetic field (SMF)
Smith's classification of algae, 12–14
Smithora, 44
Smithsonimonas abbotii, 27–29
Smoke point, 348–349
Soda lakes, 30–31
Sodium chloride (NaCl), 169–170

Solar cells, 423–426, 425f
  different generations of, 424f
  first-generation, 424
  fourth-generation, 425–426
  second-generation, 424
  third-generation, 424–425
Solar cooker, 429–430
Solar distillation technology, 429
Solar dryers, 428–429
Solar Energy Research Institute (SERI), 619–620
Solar Lake, 171
Solar panel, 426, 427f
Solar photovoltaic systems (PV systems), 428
Solar radiation, biomass to electricity conversion using, 429
Solar radiation for potential algal biomass production
  applications of, 426–430
    agrophotovoltaic, 426
    aquavoltaic, 426–427
    biomass to electricity conversion using solar radiation, 429
    solar cooker, 429–430
    solar distillation, 429
    solar dryers, 428–429
    solar photovoltaic (PV) systems, 428
    solar tractors, 427–428
    solar water heater, 430
    solar water pumping, 428
  conversion of solar radiation to algal biomass, 430–433
  different modes of operation of solar tracker coupled with photobioreactor, 437–438
    schematic diagram of functioning of microalgae cultivation pilot plant, 438f
  solar cells, 423–426
  solar panel, 426
  solar to heat for thermochemical conversion of algal biomass, 438–441
    classification of thermochemical processes, 439–441
  technoeconomic considerations for different routes of conversion of algae, 441–443
Solar tracking system, 433–438, 507
  classification based on tracking strategy, 436
  efficacy of solar tracker in harnessing solar energy for algal cultivation, 436–437
  scheme of cone-shape light guides, 508f
  solar trackers classification, 435–436
    classification based on degree of freedom, 436
    classification based on driving systems, 435–436
    classification based on their control, 435
Solar tractors, 427–428
Solar water heater, 430
Solar water pumping, 428

"SolarLeaf", 593–594
  bioreactors, 594–595
Solazyme Inc, 578
Solix Biofuels company, 576–578, 638–639
Solventogenesis, 392–393
Sorghum, 126–127
Souders–Brown equation, 645
Soybean oil, 339–340
Species assessment used for biofuel production, 532, 533t
Species-specific cultivation method, 270
Spectral filtration and shifting, 507–508
Spirogyra spp., 19–20, 257–259, 443
Spirulina platensis, 263
Spirulina sp., 68–70, 130, 135, 137, 169, 262–264, 284–285, 289–291, 312, 407, 409–410, 465, 553, 629, 636–637
  S. maxima, 70, 211
  S. occulta, 289–290
  S. platensis, 30–31, 71, 133, 135, 151–152, 215, 289–290, 366, 464–467
"Spirulina" (blue–green algae), 573–575
Spirulinales, 9
SPK. *See* Synthetic paraffinic kerosene (SPK)
Spring habitats, 21–22
SSC. *See* Separate particle fluorescence channels (SSC)
Staphylococcus aureus, 491
Stappia sp., 156
Starch synthesis, 100
Static magnetic field (SMF), 464–465
Stearic acid, 83–84
Stenotrophomonas sp., 479–480
Stentor, 25
Stephanokontae, 12–14
Stichococcus, 24, 409
Stigeoclonium, 18, 22
Stigonema, 15, 17
Stigonemataceae, 5
Stigonematales, 4–5
Stomatochroon, 25
Streak plate method, 625
Stream habitats, 22
Streptococcus pyogenes-derived Cas9 nuclease, 109
Streptomyces sp., 485–486
Streptophyta, 11
Structural lipids, 197
Sugar, 332, 344–345
Sugar to jet (STJ), 333
Sugar to jet fuel, 343–345. *See also* Gas-to-jet fuel
  assessment of life cycle, 345
  economic prospects, 344–345
  sugar to jet processes, 343–344
    direct sugar to hydrocarbons, 344
    process of sugars to hydrocarbons catalytic upgrading, 344

Suhria vittata, 18
Sulfitobacter species, 480
Sulfonic acid, 125–126
Sulfur dioxide (SO_2), 578
Sulfur t, 63, 71, 215
Sunscreen, algal in, 135
Supercritical Water Gasification (SCWG), 291
Surface to volume ratio (SVR), 429, 509
Sustainable development goals (SDG), 523–524, 587
SVR. *See* Surface to volume ratio (SVR)
SWCNT. *See* Single-wall carbon nanotubes (SWCNT)
Symbiotic algae, 22–25
Symbiotic relation of seaweeds, 49–50
Synecho cystis, 376–377
Synechococcales, 6, 9–11
Synechococcophycidae, 8
Synechococcus sp., 8, 30–31, 169, 319
   Synechococcus elongatus, 27
   Synechococcus PCC 7942, 312
Synechocystis sp., 30–31, 320–321
   Synechocystis PCC 6803, 319–321, 324–325
Syngas, 332, 340–341, 439–440
Synthetic Genomics Inc, 636
Synthetic paraffinic kerosene (SPK), 337–338
Synurophyceae, 9

T

T-front cutting tool, 551
Tabellaria flocculosa, 21–22
TAG. *See* Triacylglycerol (TAG)
TAGs. *See* Triacylglycerides (TAGs)
TALENs. *See* Transcription activator-like effector
     nucleases (TALENs)
TALEs. *See* Transcription activator-like effectors
     (TALEs)
Tank-to-wake analysis (TTW analysis), 333
Tanks, aquaculture seaweed cultivation in, 269–270
Tapinothrix crustacean, 21–22
Tapinothrix varians, 21–22
Taxonomy, 6, 40
   cyanobacterial, 2–5
TCL. *See* Thermochemical liquefaction (TCL)
TE. *See* Thioesterase (TE)
Technical and Economic Assessments (TEA), 642
Techno economic analysis (TEA), 628–629
Temperature
   affecting carbohydrate synthesis, 71–72
   configuration in AD, 232–233
   in global distribution of seaweeds, 46–47
   physical stress for enhanced biofuel production,
     455–456
   of TAG synthesis, 64
TerraVia Inc, 636

Terrestrial algae, 17–18
Tertiolecta, 290–291
Tetracystis, 17
Tetraselmis, 72, 137, 409, 461
   T. suecica, 64, 66, 71–72, 414–415, 485–486
Tetraspora, 18
Tetroses, 66
TGA. *See* Thermogravimetry analysis (TGA)
Thalassiosira pseudonana, 316
Thalassiosira sp., 136
Thalassiosiraae, 137
Thalassohaline environments, 169–170
Thermal treatment of seaweed, 236, 237t
Thermochemical conversion of algal biomass, 281–283,
   283f
   direct combustion, 293
   economic feasibility, 293–294
   gasification, 291–292
     typical gasification process, 292f
   HTL, 287–291
   pyrolysis, 283–287
   torrefaction, 292–293
Thermochemical conversion process, 439, 603–604
Thermochemical liquefaction (TCL), 282–283
Thermochemical process, 438–441, 439f
Thermogravimetry analysis (TGA), 284
Thermophilic algae, 27
Thermosynechococcus elongates, 309, 321
Thermosynechococcus vulcanus, 309
Thermosyphon, 430
Thin layer reactors, 509
Thin-film solar cells. *See* Second-generation solar cells
Thioesterase (TE), 98–99
Third-generation solar cells, 424–425
Thirdgeneration biofuels, 123, 281–282
Thraustochytrium sp., 64
Thylakoid membrane (TM), 308
Tilden's classification of algae, 12–14
Tilted single axis solar tracking systems (TSAT), 436
Time-based tracking systems, 436
Titanium dioxide (TiO_2), 425
Tolypothrix distorta, 19–20
Top companies
   of algal products commercialization, 573–575
     Cyanotech Corporation, 575
     Earthrise Nutritionals, 573–574
     FEMICO, 575
     Fuqing King Dnarmsa Spirulina Co. Ltd, 574
     Inner Mongolia Rejuve Biotech Co. Ltd, 574
     Yunnan Green A Biological Project Co. Ltd, 574
   of biofuel production from algae, 575–578
     Algenol Biofuels, 578
     Sapphire Energy Limited, 576

Solazyme Inc, 578
Solix Biofuels company, 576–578
of biofuel production from different feedstocks, 570–573
 Australian Renewable Fuels Limited, 571
 Blue Fire Renewables, 571–573
 Cosan Limited, 573
 Eni Gela biorefinery in Europe, 571
Torrefaction, 285–287, 440
 thermochemical conversion, 292–293
Total capital investment, 648
Total dissolved solids, 173
Toxin bioremediation, 413–414
Trace metals, 63, 70–71
Transcription activator-like effector nucleases (TALENs), 109–110, 316
Transcription activator-like effectors (TALEs), 109–110
Transesterification, 208–211, 390, 451–452, 531–532, 556
Transmembrane subunits, 311
Trebouxia, 24
Trentepohlia, 15
Trepacantha barbata, 58–59
Triacylglycerides (TAGs), 287–289
Triacylglycerol (TAG), 58–59, 98–99, 156, 197, 641–642
 minimum and maximum values of total lipid and fatty acid constituents, 60t
 simplified scheme of triacylglycerol synthesis in algae, 62f
 synthesis, 59–62
  carbon sources, 64
  factors affecting, 62–66
  light intensity and sources, 64–65
  low-dose cold atmospheric pressure plasma, 66
  low-dose rate of ionizing radiation, 66
  nutrients and trace metals, 63
  pH, 65
  salinity, 65
  temperature, 64
Tribonema minus, 456–457
Tricarboxylic acid cycle, 67–68
Trichocoleaceae, 9–11
Trichormus, 8, 17
 T. azollae, 25
Triglycerides, 58–59, 338
 esterification of, 641
Triolein ($C_{57}H_{104}O_6$), 644–645
TSAT. *See* Tilted single axis solar tracking systems (TSAT)
TTW analysis. *See* Tank-to-wake analysis (TTW analysis)
Turbinaria, 46, 48–49
Typha, 18

U

UFAs. *See* Unsaturated fatty acids (UFAs)
Ulothrix, 22
Ulothrix sp., 485–486
Ulotrichales, 18
Ultrasonic treatment, 238
Ultrasonication, 364–365
Ultrasound, 238
Ultrasound-assisted extraction (UAE), 514
Ultraviolet radiation, 461–463
Ulva sp., 40, 46, 68–70, 129
 U. fasciata, 71
 U. intestinalis, 183, 397
 U. lactuca, 127–129, 140–141, 225, 271–272, 364
 U. linza, 65
 U. pertusa, 45–46, 87
 U. prolifera, 72
 U. rigida, 64, 267–269
 U. rotundata, 86
Ulvan, 44
Ulvella leptochaete, 24
Unburned hydrocarbon (UHC), 186, 349
Undaria, 40, 46, 86
 U. pinnatifida, 267
United Nations Environment Program, 589
United States Department of Energy (USDE), 588
United States National Renewable Energy Laboratory, 559
University of technology Sydney green building case study, 594–595, 596f, 596t
Unsaturated fatty acids (UFAs), 197
Upstream process, 504
Urbanization, 456–457
Uronic acid, 125–126
US Army Engineers, 616–617
US Department of Energy (DOE), 124–125, 599–600, 616–619
 department of energy mission, 618
 history of, 616–618
 organization, 618–619
US Deputy Secretary of Energy, 618–619
US Environmental Protection Agency, 578
UV-mediated stress induction, 461–463

V

Vallisneria, 18
Value-added chemicals, production of, 377
Value-added compound production, 100–103, 101t
Value-added products, 79–80
Vaucheria, 17, 22, 24
 V. litorea, 25
Violaxanthin, 103
Viscosity of biodiesel, 216
Volatile fatty acids (VFA), 229

682   Index

Volume-swells of seal material, 350
Volvox, 19–20

W

Washington State University, 605
Waste glycerol (WG), 64, 397
Waste stabilization pond systems (WSPs), 154
Wastewater (WW), 79–80, 88–89, 152–153, 398
  bioremediation, 154, 478
  systems, 485–486
Wastewater treatment, 124, 153, 373, 479, 605
  algae-based wastewater treatment plants, 153–154
    algae-based municipal wastewater treatment process, 153–154
    effluent from industrial wastewater treatment plants and microalgae, 154
  algal bacterial symbiosis system for, 154–157
  implementation of genomic approaches for, 487–488, 488t
  lipase role in, 488–493
    in activated sludge systems, 493
    analysis of cumulative effect of enzymatic prehydrolysis, 491–492
    conversion of lipid contaminated wastewater into value-added products, 493
    immobilized enzyme and whole-cell biocatalysts in lipid bioremediation, 493
    lipase mediated esterification of free fatty acids to alkyl esters, 489f
    microbial lipase-mediated biocatalysis, 489–491
    potential microbial strains used for processing of complex oil and lipid, 491
  by marine algae, 138–141
    algae as monitor of water quality, 141
    removal of heavy metals and dyes, 139–140
    removal of nutrients, 140–141
Water
  bodies, 79–80, 426–427
    lentic, 19–20
    lotic, 20–21
  desalination
    efficiency of microalgae and seaweeds for, 173
    halophilic algae for, 167–169, 170f
  molecules, 309
  sample collection and analysis, 606–607
  scarcity, 167
"Weizmann organism", 391–392

Well-to-tank analysis (WTT analysis), 333
Well-to-wake analysis (WTW analysis), 333
West's classification of algae, 12–14
Westiellopsis, 17
Westiellopsis prolifica, 15
Whitening, algal in, 135
Whole-cell biocatalysts in lipid bioremediation, 493
Wild stock production, 124
Working capital investment, 648
World Business Council for Sustainable Development, 589–590
World Gas Renewables Reviews, 571
World Meteorological Organization, 487
WSPs. *See* Waste stabilization pond systems (WSPs)

X

X-10 Graphite Reactor, 617
Xanthophyceae (yellow-green algae), 12–14
Xanthophyta, 12–14
Xenobiotic compounds, 488

Y

Yeast, 127
Yellow snow, 27–29
Yunnan Green A Biological Project Co. Ltd, 574

Z

Zeaxanthin, 103, 377
Zero-waste technique, 386, 548, 560
Zerovalent Fe NPs (Nzvi), 242–243
ζ-carotene desaturase (ZDS), 100–103
Zinc, 207
Zinc figure nucleases (ZFNs), 105–107, 316
  metabolic pathway of microalgal, 106f
  pictorial presentation of genome editing, 107f
  showing multichannel pathway, 106f
  for targeted genome editing, 105–107
Zinc oxide, 340–341
Zirconium dioxide, 552–553
Zoochlorellae, 25
Zoogloea sp., 485–486
Zooparasitic algae, 26
Zooxanthellae, 25
ZSM-5 zeolite catalyst, 334–335
Zygnema, 19–20
Zygogonium ericetorum, 17–18

Printed in the United States
by Baker & Taylor Publisher Services